Philosophy of Mathematics

Handbook of the Philosophy of Science

General Editors

Dov Gabbay
Paul Thagard
John Woods

AMSTERDAM • BOSTON • HEIDELBERG • LONDON • NEW YORK • OXFORD
PARIS • SAN DIEGO • SAN FRANCISCO • SINGAPORE • SYDNEY • TOKYO
North Holland is an imprint of Elsevier

Philosophy of Mathematics

Edited by

Andrew D. Irvine
University of British Columbia,
Vancouver, Canada

AMSTERDAM • BOSTON • HEIDELBERG • LONDON • NEW YORK • OXFORD
PARIS • SAN DIEGO • SAN FRANCISCO • SINGAPORE • SYDNEY • TOKYO
North Holland is an imprint of Elsevier

North Holland is an imprint of Elsevier
30 Corporate Drive, Suite 400, Burlington, MA 01803, USA
Linacre House, Jordan Hill, Oxford OX2 8DP, UK
Radarweg 29, PO Box 211, 1000 AE Amsterdam, The Netherlands

First edition 2009

Copyright © 2009 Elsevier B.V. All rights reserved

No part of this publication may be reproduced, stored in a retrieval system
or transmitted in any form or by any means electronic, mechanical, photocopying,
recording or otherwise without the prior written permission of the publisher

Permissions may be sought directly from Elsevier's Science & Technology Rights
Department in Oxford, UK: phone (+44) (0) 1865 843830; fax (+44) (0) 1865 853333;
email: permissions@elsevier.com. Alternatively you can submit your request online by
visiting the Elsevier web site at http://elsevier.com/locate/permissions, and selecting
Obtaining permission to use Elsevier material

Notice
No responsibility is assumed by the publisher for any injury and/or damage to persons
or property as a matter of products liability, negligence or otherwise, or from any use
or operation of any methods, products, instructions or ideas contained in the material
herein. Because of rapid advances in the medical sciences, in particular, independent
verification of diagnoses and drug dosages should be made

British Library Cataloguing in Publication Data
A catalogue record for this book is available from the British Library

Library of Congress Cataloging-in-Publication Data
A catalog record for this book is available from the Library of Congress

ISBN: 978-0-444-51555-1

For information on all North Holland publications
visit our web site at books.elsevier.com

Printed and bound in Hungary

09 10 11   11 10 9 8 7 6 5 4 3 2 1

*Cover Art: University of British Columbia Library, Rare Books and Special Collections,
from Oliver Byrne, The First Six Books of* The Elements of Euclid *in which Coloured
Diagrams and Symbols are used instead of Letters for the Greater Ease of Learners
(London: William Pickering, 1847)*

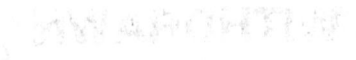

GENERAL PREFACE

Dov Gabbay, Paul Thagard and John Woods

Whenever science operates at the cutting edge of what is known, it invariably runs into philosophical issues about the nature of knowledge and reality. Scientific controversies raise such questions as the relation of theory and experiment, the nature of explanation, and the extent to which science can approximate to the truth. Within particular sciences, special concerns arise about what exists and how it can be known, for example in physics about the nature of space and time, and in psychology about the nature of consciousness. Hence the philosophy of science is an essential part of the scientific investigation of the world.

In recent decades, philosophy of science has become an increasingly central part of philosophy in general. Although there are still philosophers who think that theories of knowledge and reality can be developed by pure reflection, much current philosophical work finds it necessary and valuable to take into account relevant scientific findings. For example, the philosophy of mind is now closely tied to empirical psychology, and political theory often intersects with economics. Thus philosophy of science provides a valuable bridge between philosophical and scientific inquiry.

More and more, the philosophy of science concerns itself not just with general issues about the nature and validity of science, but especially with particular issues that arise in specific sciences. Accordingly, we have organized this Handbook into many volumes reflecting the full range of current research in the philosophy of science. We invited volume editors who are fully involved in the specific sciences, and are delighted that they have solicited contributions by scientifically-informed philosophers and (in a few cases) philosophically-informed scientists. The result is the most comprehensive review ever provided of the philosophy of science.

Here are the volumes in the Handbook:

Philosophy of Science: Focal Issues, edited by Theo Kuipers.

Philosophy of Physics, edited by John Earman and Jeremy Butterfield.

Philosophy of Biology, edited by Mohan Matthen and Christopher Stephens.

Philosophy of Mathematics, edited by Andrew D. Irvine.

Philosophy of Logic, edited by Dale Jacquette.

Philosophy of Chemistry and Pharmacology, edited by Andrea Woody, Robin Hendry and Paul Needham.

Philosophy of Statistics, edited by Prasanta S. Bandyopadhyay and Malcolm Forster.

Philosophy of Information, edited by Pieter Adriaans and Johan van Benthem.

Philosophy of Technological Sciences, edited by Anthonie Meijers.

Philosophy of Complex Systems, edited by Cliff Hooker.

Philosophy of Ecology, edited by Bryson Brown, Kent Peacock and Kevin de Laplante.

Philosophy of Psychology and Cognitive Science, edited by Pau Thagard.

Philosophy of Economics, edited by Uskali Mki.

Philosophy of Linguistics, edited by Ruth Kempson, Tim Fernando and Nicholas Asher.

Philosophy of Anthropology and Sociology, edited by Stephen Turner and Mark Risjord.

Philosophy of Medicine, edited by Fred Gifford.

Details about the contents and publishing schedule of the volumes can be found at http://www.johnwoods.ca/HPS/.

As general editors, we are extremely grateful to the volume editors for arranging such a distinguished array of contributors and for managing their contributions. Production of these volumes has been a huge enterprise, and our warmest thanks go to Jane Spurr and Carol Woods for putting them together. Thanks also to Andy Deelen and Arjen Sevenster at Elsevier for their support and direction.

CONTENTS

General Preface **Dov Gabbay, Paul Thagard and John Woods**	v
Preface **Andrew D. Irvine**	ix
List of Contributors	xv
Les Liaisons Dangereuses **W. D. Hart**	1
Realism and Anti-Realism in Mathematics **Mark Balaguer**	35
Aristotelian Realism **James Franklin**	103
Empiricism in the Philosophy of Mathematics **David Bostock**	157
A Kantian Perspective on the Philosophy of Mathematics **Mary Tiles**	231
Logicism **Jaakko Hintikka**	271
Formalism **Peter Simons**	291
Constructivism in Mathematics **Charles McCarty**	311
Fictionalism **Daniel Bonevac**	345
Set Theory from Cantor to Cohen **Akihiro Kanamori**	395

Alternative Set Theories **Peter Apostoli, Roland Hinnion, Akira Kanda and Thierry Libert**	461
Philosophies of Probability **Jon Williamson**	493
On Computability **Wilfried Sieg**	535
Inconsistent Mathematics **Chris Mortensen**	631
Mathematics and the World **Mark Colyvan**	651
Index	703

PREFACE

One of the most striking features of mathematics is the fact that we are much more certain about what mathematical knowledge we have than about what mathematical knowledge is knowledge *of*. Mathematical knowledge is generally accepted to be more certain than any other branch of knowledge; but unlike other scientific disciplines, the subject matter of mathematics remains controversial.

In the sciences we may not be sure our theories are correct, but at least we know what it is we are studying. Physics is the study of matter and its motion within space and time. Biology is the study of living organisms and how they react and interact with their environment. Chemistry is the study of the structure of, and interactions between, the elements. When man first began speculating about the nature of the Sun and the Moon, he may not have been sure his theories were correct, but at least he could point with confidence to the objects about which he was theorizing. In all of these cases and others we know that the objects under investigation — physical matter, living organisms, the known elements, the Sun and the Moon — exist and that they are objects within the (physical) world.

In mathematics we face a different situation. Although we are all quite certain that the Pythagorean Theorem, the Prime Number Theorem, Cantor's Theorem and innumerable other theorems are true, we are much less confident about what it is to which these theorems refer. Are triangles, numbers, sets, functions and groups physical entities of some kind? Are they objectively existing objects in some non-physical, mathematical realm? Are they ideas that are present only in the mind? Or do mathematical truths not involve referents of any kind? It is these kinds of questions that force philosophers and mathematicians alike to focus their attention on issues in the philosophy of mathematics.

Over the centuries a number of reasonably well-defined positions have been developed and it is these positions, following a thorough and helpful overview by W. D. Hart,[1] that are analyzed in the current volume. The realist holds that mathematical entities exist independently of the human mind or, as Mark Balaguer tells us, realism is "the view that our mathematical theories are true descriptions of some real part of the world."[2] The anti-realist claims the opposite, namely that mathematical entities, if they exist at all, are a product of human invention. Hence the long-standing debate about whether mathematical truths are discovered or invented. Platonic realism (or Platonism) adds to realism the further provision that mathematical entities exist independently of the natural (or physical) world.

---

[1] W. D. Hart, "Les Liaisons Dangereuses", this volume, pp. 1–33.
[2] Mark Balaguer, "Realism and Anti-realism in Mathematics," this volume, pp. 35–101.

Aristotelian realism (or Aristotelianism) adds the contrary provision, namely that mathematical entities are somehow a part of the natural (or physical) world or, as James Franklin puts it, that "mathematics is a science of the real world, just as much as biology or sociology are."[3] Platonic realists such as G.H. Hardy, Kurt Gödel and Paul Erdös are thus regularly forced to postulate some form of nonphysical mathematical perception, distinct from but analogous to sense perception. In contrast, as David Bostock reminds us, Aristotelian realists such as John Stuart Mill typically argue that empiricism – the theory that all knowledge, including mathematical knowledge, is ultimately derivable from sense experience – "is perhaps most naturally combined with Aristotelian realism."[4]

The main difficulty associated with Platonism is that, if it is correct, mathematical perception will appear no longer to be compatible with a purely natural understanding of the world. The main difficulty associated with Aristotelianism is that, if it is correct, a great deal of mathematics (especially those parts of mathematics that are not purely finitary) will appear to outrun our (purely finite) observations and experiences. Both the Kantian (who holds that mathematical knowledge is synthetic and *a priori*) and the logicist (who holds that mathematics is reducible to logic, and hence that mathematical knowledge is analytic) attempt to resolve these challenges by arguing that mathematical truths are discoverable by reason alone, and hence not tied to any particular subject matter. As Mary Tiles tells us, Kant's claim that mathematical knowledge is synthetic *a priori* has two separate components. The first is that mathematics claims to provide *a priori* knowledge of certain objects because "it is the science of the forms of intuition"; the second is that "the way in which mathematical knowledge is gained is through the synthesis (construction) of objects corresponding to its concepts, not by the analysis of concepts."[5] Similarly, initial accounts of logicism aimed to show that, like logical truths, mathematical truths are "truths in every possible structure" and it is for this reason that they can be discovered *a priori*, simply because "they do not exclude any possibilities."[6] Exactly how much, if any, of such programs can be salvaged in the face of contemporary meta-theoretical results remains a matter of debate. Constructivism, the view that mathematics studies only entities that (at least in principle) can be explicitly constructed, attempts to resolve the problem by focusing mathematical theories solely on activities of the human mind. In Charles McCarty's helpful phrase, constructivism in mathematics ultimately boils down to a commitment to the "business of practice rather than of principle."[7] Critics claim that all three positions — Kantianism, logicism and constructivism — ignore large portions of mathematics' central subject matter. (Constructivism in particular, because of the emphasis it places upon verifiability, is regularly accused of failing to account for the impersonal, mind-independent

---

[3] James Franklin, "Aristotelian Realism," this volume, pp. 103–155.
[4] David Bostock, "Empiricism in the Philosophy of Mathematics," this volume, pp. 157–229.
[5] Mary Tiles, "A Kantian Perspective on the Philosophy of Mathematics," this volume, pp. 231–270.
[6] Jaakko Hintikka, "Logicism," this volume, pp. 271–290.
[7] Charles McCarty, "Constructivism in Mathematics," this volume, pp. 311–343.

parts of mathematics.)

Formalism, the view that mathematics is simply the "formal manipulations of essentially meaningless symbols according to strictly prescribed rules,"[8] goes a step further, arguing that mathematics need not be considered to be about numbers or shapes or sets or probabilities at all since, technically speaking, mathematics need not be *about* anything. But if so, an explanation of how we obtain our non-formal, intuitive mathematical intuitions, and of how mathematics integrates so effectively with the natural sciences, seems to be wanting. Fictionalism, the view that mathematics is in an important sense dispensable since it is merely a conservative extension of non-mathematical physics (that is, that every physical fact provable in mathematical physics is already provable in non-mathematical physics without the use of mathematics), can be attractive in this context. But again, it is a theory that fails to coincide with the intuitions many people — including many working mathematicians — have about the need for a realist-based semantics. As Daniel Bonevac tells us, even if fictionalist discourse in mathematics is largely successful, we are still entitled to ask why "that discourse, as opposed to other possible competitors, succeeds"; and as he reminds us in response to such a question, any citation of a fact threatens to collapse the fictionalist project into either a reductive or modal one, something not easily compatible with the fictionalist's original aims.[9]

The moral appears to be that mathematics sits uncomfortably half way between logic and science. On the one hand, many are drawn to the view that mathematics is an axiomatic, *a priori* discipline, a discipline whose knowledge claims are in some way independent of the study of the contingent, physical world. On the other hand, others are struck by how mathematics integrates so seamlessly with the natural sciences and how it is the world — and not language or reason or anything else — that continually serves as the main intuition pump for advances even in pure mathematics.

In fact, in spite of its abstract nature, the origins of almost all branches of mathematics turn out to be intimately related to our innumerable observations of, and interactions with, the ordinary physical world. Counting, measuring, grouping, gambling and the many other activities and experiences that bring us into contact with ordinary physical objects and events all play a fundamental role in generating new mathematical intuitions. This is so despite the sometimes-made claim that mathematical progress has often occurred independently of real-world applications. Standardly cited advances such as early Greek discoveries concerning the parabola, the ellipse and the hyperbola, the advent of Riemannian geometries and other non-Euclidean geometries well in advance of their application in contemporary relativistic physics, and the initial development of group theory as long ago as the early 1800s themselves all serve as telling counterexamples to such claims. Group theory, it turns out, was developed as a result of attempts to solve simple polynomial equations, equations that of course have immediate application in

---

[8] Peter Simons, "Formalism," this volume, pp. 291–310.
[9] Daniel Bonevac, "Fictionalism," this volume, pp. 345–393.

numerous areas. Non-Euclidean geometries arose in response to logical problems intimately associated with traditional Euclidean geometry, a geometry that, at the time, was understood to involve the study of real space. Early Greek work studying curves resulted from applied work on sundials. Mathematics, it seems, has always been linked to our interactions with the world around us and to the careful, systematic, scientific investigation of nature.

It is in this same context of real-world applications that fundamental questions in the philosophy of mathematics have also arisen. Paradigmatic over the past century have been questions associated with issues in set theory, probability theory, computability theory, and theories of inconsistent mathematics, all now fundamentally important branches of mathematics that have grown as much from a dissatisfaction with traditional answers to philosophical questions as from any other source. In the case of set theory, dissatisfaction with our understanding of the relationship between a predicate's intension and its extension has led to the development of a remarkably simple but rich theory. As Akihiro Kanamori reminds us, set theory has evolved "from a web of intensions to a theory of extension *par excellence*."[10] At the same time, striking new developments continue to be made, as we see in work done by Peter Apostoli, Roland Hinnion, Akira Kanda and Thierry Libert.[11] In the case of probability theory, the frustrating issue of how best to interpret the basic concepts of the theory has long been recognized. But as Jon Williamson suggests, Bayesianism, the view that understands probabilities as "rational degrees of belief", may help us bridge the gap between objective chance and subjective belief.[12] Wilfried Sieg[13] and Chris Mortensen[14] give us similarly exciting characterizations of developments in computability theory and in the theory of inconsistent mathematics respectively.

Over the centuries the philosophy of mathematics has traditionally centered upon two types of problem. The first has been problems associated with discovering and accounting for the nature of mathematical knowledge. For example, what kind of explanation should be given of mathematical knowledge? Is all mathematical knowledge justified deductively? Is it all *a priori*? Is it known independently of application? The second type of problem has been associated with discovering whether there exists a mathematical reality and, if so, what about its nature can be discovered? For example, what is a number? How are numbers, sets and other mathematical entities related? Are mathematical entities needed to account for mathematical truth? If they exist, are mathematical entities such as numbers and functions transcendent and non-material? Or are they in some way a part of, or reducible to, the natural world? During much of the twentieth century it was the first of these two types of problem that was assumed to be fundamental. Logicism, formalism and intuitionism all took as their starting point the presupposition that

---

[10] Akihiro Kanamori, "Set Theory from Cantor to Cohen," this volume, pp. 395–459.

[11] Peter Apostoli, Roland Hinnion, Akira Kanda and Thierry Libert, "Alternative Set Theories," this volume, pp. 461–491.

[12] Jon Williamson, "Philosophies of Probability," this volume, pp. 493–533.

[13] Wilfried Sieg, "Computability," this volume, pp. 535–630.

[14] Chris Mortensen, "Inconsistent Mathematics," this volume, pp. 631–649.

it was necessary to account for the absolute certainty that was assumed to be present in all genuine mathematical knowledge. As a result, all three schools emphasized that they could account for the resolution of antinomies, such as Russell's paradox, in a satisfactory way. All three hoped that such a crisis in the foundations of mathematics could be guaranteed never to happen again. Their disagreements were over matters of strategy, not over ultimate goals. Only in the latter parts of the century was there a shift away from attempting to account for the certainty of mathematical knowledge towards other areas in the philosophy of mathematics. This leaves us, as Mark Colyvan says, "with one of the most intriguing features of mathematics,"[15] its applicability to empirical science, and it on this topic that the current volume ends.

For their help in preparing this volume, my thanks goes to Jane Spurr and Carol Woods as well as to the series editors, Dov Gabbay, Paul Thagard and John Woods, but most especially to the contributors for their hard work, generosity of spirit, and especially their redoubtable expertise in such a broad range of fascinating and important topics.

Andrew D. Irvine
University of British Columbia

---

[15] Mark Colyvan, "Mathematics and the World," this volume, pp. 651–702.

CONTRIBUTORS

Peter Apostoli
University of Pretoria, RSA.
peter_cornerstone@yahoo.ca

Mark Balaguer
California State University, Los Angeles, USA.
mbalagu@calstatela.edu

Daniel Bonevac
University of Texas, Austin, USA.
bonevac@mail.utexas.edu

David Bostock
Merton College, Oxford, UK.

Mark Colyvan
University of Sydney, Australia.
mcolyvan@usyd.edu.au

James Franklin
University of New South Wales, Australia.
j.franklin@unsw.edu.au

W. D. Hart
University of Illinois at Chicago, USA.
hart@uic.edu

Roland Hinnion
Université libre de Bruxelles, Belgium.
rhinnion@ulb.ac.be

Jaakko Hintikka
Boston University, USA.
hintikka@bu.edu

Andrew D. Irvine
University of British Columbia, Canada.
a.irvine@ubc.ca

Akira Kanda
Omega Mathematical Institute
kanda@cs.toronto.edu

Akihiro Kanamori
Boston University, USA.
aki@math.bu.edu

Thierry Libert
Université Libre de Bruxelles, Belgium.
tlibert@ulb.ac.be

Charles McCarty
Indiana University, USA.
dmccarty@indiana.edu

Chris Mortensen
Adelaide University, Australia.
chris.mortensen@adelaide.edu.au

Wilfried Sieg
Carnegie Mellon University, USA.
sieg@cmu.edu

Peter Simons
Trinity College, Dublin, Ireland.
psimons@tcd.ie

Mary Tiles
University of Hawaii at Manoa, USA.
mtiles@hawaii.edu

Jon Williamson
University of Kent at Canterbury, UK.
j.williamson@kent.ac.uk

LES LIAISONS DANGEREUSES

W. D. Hart

Mathematics and philosophy are roughly coeval in our historical imagination. Plato's dialogues form the oldest surviving extended body of work in the canon of western philosophy. Euclid's *Elements* is the oldest surviving intact monument in the evolution of our mathematics. Plato taught Aristotle, who died in 322 B.C., and Euclid's floruit is around 300 B.C., so the gap from Plato to Euclid is like that from grandparent to grandchild, and from nearly two and a half millennia later, that gap looks small.

There was of course philosophy before Plato. We have fragments from the presocratics, and Plato made his teacher Socrates the star of most of his dialogues. There was mathematics before Euclid. He seems to have been as much an editor as a mathematician, and probably not the first. The Greeks had invented or discovered proof centuries before; just think of the Pythagorean theorem or the proof of the irrationality of the square root of two. In Plato's day, Theatetus seems to have proved that there are exactly five regular solids, a gorgeous result that impressed Plato enough to give Theatetus a leading role in a major dialogue. As proofs proliferate, patterns start to emerge, and aficionados want to organize the profusion of arguments into a coherent whole developed logically from a minimal stock of assumptions. Doubtless there were such editions of geometry before Euclid, but his *Elements* is the work whose authority lasted through the centuries.

Aristotle seems to have been less impressed by mathematics than Plato. But Aristotle did begin the systematic study of logic. His account of syllogisms is now usually assimilated to our monadic quantification theory (and truth functional logic is usually credited to the later Stoic philosophers).[1] An interest in logic could have arisen from the effort to piece disparate proofs together into a unified and coherent system, though syllogistic is a pretty thin description of the reasoning deployed in ancient geometry. Still, we should not be impatient, since it was not until the nineteenth century that people like de Morgan[2] and Peirce began to work out a systematic understanding of relations, which was crucial in the logicist regimentation of mathematics.

But besides starting systematic logic, Aristotle also articulated a version, or a vision, of the axiomatic method. In the *Posterior Analytics* he describes a real body of knowledge as deduced by infallible logic from axioms. The axioms

---

[1] William and Martha Kneale, *The Development of Logic* (Oxford: Clarendon Press, 1962).

[2] All horses are animals, from which it follows that all heads of horses are heads of animals. De Morgan observed that Aristotle's syllogistic does not suffice to certify the validity of this inference, which turns on relations and polyadic quantification.

should have an immediate appeal, and the logic should transmit this appeal to the theorems. As we said, Aristotle wrote before Euclid. But he might have been trying to articulate an ideal he saw struggling to emerge from editions of geometry older than Euclid. And one wonders whether Euclid might have been struggling to realize an ideal earlier articulated in Aristotle.

This is our theme, the dangerous liaisons between mathematics and philosophy. They are not just coeval, like strangers or distant acquaintances who happen to have been born in the same town within a short span of time. They have also been bedfellows, sometimes strange even if not made so by politics. We will sketch some of their offspring. Some does not mean all; ours will not be a family tree, but a selection of hybrids. And because I am a philosopher who admires mathematics but does not claim to be a mathematician, most of these compounds will have more philosophical elements than mathematical.

In some ways, the axiomatic method can seem like proof writ large. To be sure, a proof aims to establish a single theorem, while in an axiomatic system we prove a sequence of theorems. In the heat of live mathematics, one does not practice axiomatically. One does not copy one's premises out of a constitution written down and approved by the founding mathematical fathers and mothers. One starts instead from what is clear, and clarity here probably means what one's peers will accept without complaint. So one needs to be sensitive to one's peers, and for pretty much all of us this requires being admitted to the community of peers through an education. But once the community has approved a body of proofs, some of the peers may set out to regiment it. This process includes collecting the clear starting premises that passed muster, selecting from them some from which the rest can be derived, and so on until we have axioms from which a sequence of theorems follow, where of course some later theorems are deduced from earlier. Once such a system is established, incorporability of a new argument in it can become a standard for being a proof. Euclid set such a standard in geometry for centuries, and set theory (usually in Zermelo-Frankel form) did so for mathematics generally in the twentieth century.

This is a rather sociological description of axiomatization. Philosophers and mathematicians share a taste for long and abstract chains of reasoning, but they often differ in how they get started. Mathematicians seem to like their premises to be shared, perhaps throughout their community, or as close to that as possible. That way the community can be expected to follow their reasoning. There are philosophers, like Aristotle and Kant, who seem not to want to frighten the horses, but they may be trying to calm things down after earlier philosophers like Plato and Hume have stirred them up by going where the reasoning led from premises for which they may have claimed more popularity than was generally recognized. At any rate, philosophy looks more contentious than mathematics. But however much they disagreed elsewhere, Plato and Aristotle seem to have agreed that a version of the axiomatic method describes an ideal for knowledge.

Even if it is not perfectly clear whether this ideal starts life in mathematics or in philosophy, the axiomatic method is a mode of exposition that has become a tried and true device in the mathematical repertoire. It was exaggerated by philosophers into the ideal of foundations of knowledge, or this or that department of knowledge. A vivid example is Spinoza writing his *Ethics* in *more geometrico*. It is perhaps ironic to note that the Latin word "mos" from which "more" declines means custom or usage, which seems more sociological than Spinoza probably had in mind. (Anyone trying to formalize Spinoza's system by modern lights is in for a bad time.) The basic philosophical idea seems to be that there is a right way to organize for justification truths, beliefs, or knowledge. This idea has gripped philosophical imaginations for centuries.

The ideal can be articulated in different ways. Sometimes the right order is the right order in which to justify our beliefs or knowledge. In Descartes's urban renewal of knowledge we are to rebuild from clear and distinct ideas of indubitable certainty like the *cogito*. In more empiricist philosophers like Locke, Berkeley, and Hume we are to begin from sense experience, and increasingly their problem is whether we can get beyond our impressions without losing the certainty that made perception an appealing foundation.

It was this empirical spectre of skepticism that startled the horses and woke Kant from his dogmatic slumbers. To trace out firm foundations for knowledge, he looked to the surest systematic body of knowledge going, and from the Greeks on, mathematics had always been the best-developed system of the most absolute truth known with the greatest certainty. In Kant's day and before, mathematics meant first and foremost geometry, and geometry meant Euclid's system not just of planes but also of the space in which we live and move and have our being. The idea of other spaces is later and quite unkantian, and the mathematics of number (beyond elementary number theory like the infinity of the primes) achieves independence only in the nineteenth century. Kant does of course give sensibility a basic role in contributing to knowledge. But it is his conception of the character of geometrical knowledge that not only gets his critical philosophy going, but also sets an agenda for many later and rather unkantian philosophers.

To exposit this conception we need some distinctions.[3] Assume the anachronistically labeled traditional analysis of knowledge as justified true belief. Epistemology is much more about justification than knowledge. Kant calls knowledge *a posteriori* when it is justified, even in part, by appeal to sense experience. Knowledge is *a priori* when it is knowledge but not *a posteriori*, that is, not justified even in part by experience. Kant thought that mathematics, that is, geometry, and logic are systematic bodies of *a priori* knowledge. We will consider an argument for this thought in a moment.

Consider next judgments. This is Kant's usual term for mental states like beliefs (such as that grass is green or seven is prime) and thoughts. Around the turn of

---

[3] Kant draws his distinctions in the introduction to *The Critique of Pure Reason*, trans. Norman Kemp Smith (London: MacMillan, 1963). Moore discussed propositions in chapter 3 of *Some Main Problems of Philosophy* (New York: Collier Books, 1962). The basic Tarski piece is "The Concept of Truth in Formalized Languages," in *Logic, Semantics, and Metamathematics*, trans. J. H. Woodger (Oxford: Clarendon Press, 1956). For Austin, see his "Truth," in *Philosophical Papers*, ed. J. O. Urmson and G. J. Warnock (Oxford: Clarendon Press, 1961).

the twentieth century, G. E. Moore and Russell will replace judgments by propositions, which are platonic abstracta like numbers rather than mental. Frege's thoughts are more like Russell's propositions than Kant's judgments. During the twentieth century, philosopher-logicians like Tarski will replace both judgments and propositions with sentences. Sentences are linguistic items where judgments and propositions were supposed to be independent of language. (Around 1950 J. L. Austin will try to replace sentences with statements thought of as actions performed using sentences.) Kant divides judgments into analytic and synthetic. Analytic judgments are reminiscent of Locke's trifling propositions (not to be confused with russellian propositions) and Hume's relations of ideas.

One way to move in on analyticity is through examples. An example of Moore's is the claim that all bachelors are unmarried. The Social Science Research Council would be ill advised to fund a door-to-door survey in which bachelors are asked whether they are married, the results are tallied, and finally the bold hypothesis that all of them are unmarried is advanced. This would be a waste because, so the story goes, being unmarried is part of what it means to be a bachelor.

It seems clear that there is some sort of difference between the claim that bachelors are unmarried and the claim that bachelors are more flush financially than husbands. Controversy sets in when we try to articulate the difference. Kant gave two accounts of analyticity. On one, the predicate of an analytic judgment is contained in its subject. Note three points about this account. First, it seems to presuppose that all judgments are of subject-predicate form. Whatever grammarians may say, Russell was excited by the revelation in the logic reforming around him of other forms, especially quantificational, of judgment. We follow Russell, so Kant's account may seem too narrow to us. Second, his account presupposes that judgments have subjects and predicates. That is, Kant seems to be reading sentence structure back into judgments. One role in which Kant's judgments or Russell's propositions or Tarski's sentences are cast is as bearers of the truth values; these are the things that are true or false. Whether it is true that Socrates was snub-nosed depends in part on the man Socrates and what his nose was like. That is, the truth bearers (or vehicles, as Austin called them) need to be articulated into bits smaller than whole truth vehicles. Sentences wear such an articulation into smaller bits, words, on their inscribed faces. It seems all but irresistible to read this articulation back into the judgments or propositions expressed by sentences. But then Tarski's choice of sentences as truth vehicles seems more up front than Kant's judgments or Russell's propositions. Third, Kant's trope of the predicate of a judgment being contained in its subject is clearly a metaphor, and this leaves us without a literal account of analyticity. On Kant's other account of analyticity, the denial of an analytic judgment cannot be thought without contradiction. Never mind that thought here seems to assume the analyticity of logic without argument. What might be worth noting here is the relativity of this account to which premises we are allowed. If, for example, it is one of our premises that bachelors are richer than husbands, we will not be able to think the denial of the judgment that bachelors are richer than husbands without contradiction. That would make the judgment analytic contrary to the motivation for the notion.

For much of the twentieth century, the quick gloss on analyticity was that a sentence (proposition, judgment) is analytic if it is true by virtue of the meanings of the words in the sentence (used to express the proposition or judgment). This gloss seems confused. Analyticity is at best a mode of justification, not of truth. It is an ancient and honorable view that truth is correspondence to fact; for a sentence to be true, the world should be as the sentence says it is. Moreover, truth is univocal. That is why the conjunction of two truths from areas however disparate is nonetheless true. The truth that bachelors are unmarried (or, for that matter, bachelors) is as much about bachelors as the claim that bachelors are richer than husbands, and if true, they are so because bachelors are unmarried, and richer than husbands. What was distinctive about the claims attracting the label analytic was epistemic, a matter of justification rather than the nature of a kind of truth. The judgment that bachelors are unmarried would then be analytic if knowledge of the meanings of the words used to express the judgment sufficed without experience of its subject matter (bachelors) to justify belief in the judgment.

This provisional story is no better than our grasp of what knowledge of the meanings of words comes to, and that grasp is at best pretty shaky. How much it makes analytic is unclear. Kant said it is synthetic that all bodies are heavy. (The synthetic judgments are those that are not analytic. Analyticity wears the trousers in its distinction, as being *a posteriori* does in its; they get a positive account, and their opposites are defined by negation.) This example is plausible if we are reluctant to build gravitational attraction into the meaning of the word "body." But it would be embarrassing to have to give a justification for this reluctance. Kant said it is analytic that all bodies are extended. This is rather an odd example for Kant to give. Like any eighteenth-century intellectual, Kant admired Newton. Indeed, part of Kant's objective was to secure certainty for much of Newton's physics, and Kant was not innocent of that science. Anyone familiar with Newton will remember how much he makes of mass-points. Does Kant mean to exile the mass-points from the bodies by definition? Is he defining mass-points out of existence? Analyticity is often a cloak for arbitrary legislation.

With two binary distinctions, we get four compounds. Kant ruled out the analytic *a posteriori*. A survey could amass evidence that all bachelors are unmarried, but such cases do not seem worth fretting over. The synthetic *a posteriori* would include most of the natural science, the physics, astronomy, and chemistry coming to be in Kant's day. Since *a posteriori* knowledge is justified by experience, we have at least the beginnings of a story about how such science is known. But since *a priori* knowledge is defined negatively as not *a posteriori,* there is a question how it could be justified. Kant thought logic is known *a priori*, but is analytic, and so justified from the meanings of logical words like "if" and "all" and "is." Kant wrote during the low-water mark of the history of logic. The achievements of the Schoolmen had been largely rejected during the enthusiasm of the Renaissance, and what was left was Aristotle's syllogistic and some Stoic truth function theory. It is hardly a blunder to think Barbara (If all cats are vertebrates and all vertebrates are animals, then all cats are animals) can be certified by an elaboration of the meanings of "all" and "are," but logic did not stick at its kantian low-water mark.

The moneybox was *a priori* knowledge of synthetic truths. Such truths are not known from experience, nor are they justified from the meanings of the words used to express them. So how are they known? How, Kant asks, is synthetic *a priori* knowledge possible? This question is the pretext for the critical philosophy, and Kant's answer is transcendental idealism. But Kant's question has purchase only if there is synthetic *a priori* knowledge. To appreciate one example whose consideration goes back to the nineteenth century, we should make a third distinction, this time between necessary and contingent truths. People don't usually read for long standing up, so you are probably not standing as you read this. If so, it is true that you are not standing. But you could have been, so that truth is contingent. You are also identical with yourself, and that is not something you could fail to be, so that truth is necessary. The contingent truths could be otherwise, but the necessary ones could not. At B3 in the first *Critique*, Kant says that experience teaches us that a thing is so and so, but not that it cannot be otherwise. In the ways we can see color and shape, or feel shape and texture, we have no experience of necessity or (nonactual) possibility, only of actuality. So, Kant thought, knowledge of necessity is *a priori*. Kripke[4] later observed that we sometimes can infer a necessary truth from two premisses, one known *a posteriori*, so if *a priori* knowledge rules out all justification by experience, some necessary truths are known *a posteriori*. Fair enough; these examples Kant missed out, and even in them the necessity in the conclusion comes from the premiss known *a priori*, which vindicates Kant somewhat. The nineteenth-century example is that nothing could be red all over and green all over at the same time and place. That these colors exclude each other does seem to be a necessary truth; in general, determinants (for example, being six feet tall and being seven feet tall) of a determinable (for example, height) exclude one another necessarily. We cannot imagine an object having both, and the imagination is the royal road to knowledge of possibility and necessity (even if it does not always accord with, say, materialist prejudices). Granted that it is necessary that nothing is at once red all over and green all over, knowledge of it would be *a priori*. But, the story continues, the colors red and green are simple and basic enough that no definitions of them in more basic terms are available, so there are no definitions of them from which to show that this necessity is analytic. It is then a synthetic necessity known *a priori*. This example has been much discussed and no plausible definitions of the colors that would show it analytic have been generally accepted.[5] On the other hand, it bucks the trend

---

[4] Saul A. Kripke, *Naming and Necessity* (Cambridge: Harvard University Press, 1972).

[5] See, for example, Arthur Pap, "Logical Nonsense," *Philosophy and Phenomenological Research* 9 (1948), 269-83; "Are All Necessary Propositions Analytic?" *Philosophical Review* 50 (1949), 299-320; *Elements of Analytic Philosophy* (New York: MacMillan, 1949), chap. 16b; Hilary Putnam, "Reds, Greens, and Logical Analysis," *Philosophical Review* 65 (1956), 206-17; Pap, "Once More: Colors and the Synthetic a Priori," *Philosophical Review* 66 (1957), 94-99; Putnam, "Red and Green All Over Again: A Rejoinder to Arthur Pap," *Philosophical Review*

in most twentieth-century analytic philosophy that necessity is our creature; there is no necessity out there in nature independent of us.

That determinants of the same determinable exclude one another necessarily yields a relatively scattered fund of examples, so if that were the only synthetic *a priori* knowledge, it would have been less front and center on the postkantian philosophical agenda. But Kant thought that mathematics is synthetic *a priori*. That it is *a priori* might seem evident, since mathematicians do not perform experiments on prime numbers, nor do they make expeditions to examine exotic ones; they just sit around and think, and what could be more *a priori*? His argument for the synthetic is more dubious. His example is that $7 + 5 = 12$. He says that the concept of the sum of 7 and 5 contains nothing save the union of the two numbers into one, and in this no thought is being taken as to what that single number may be which combines both. He says that the concept of 12 is by no means already thought in merely thinking this union of 7 and 5, and analyze our concept of such a sum as we please, still we will never find the 12 in it. Instead, we must go outside the concepts of 7, 5, and addition, and call in examples (like intuitions of fingers or points) to see the number 12 come into being. It is probably anachronistic to be too fussy about whether Kant is here discussing psychology (ideas of 7, 5, sum, and 12) or semantics (the meanings of numerals and function signs). But in the way a dictionary definition for "bachelor" seems forthcoming and uncontroversial, definitions for the numerals and function signs are more problematic. We can, to be sure, label some claims in which they are used definitions, and from these deduce some conventional arithmetic wisdom logically, but even if all this is necessarily true, where is the semantics independent of the philosophy at issue to settle whether these claims are analytic or synthetic? No one has stated such a semantics that convinces many others.

But maybe Hume can help Kant out here. In part IX of his *Dialogues Concerning Natural Religion*,[6] Hume turns to the *a priori* arguments for the existence of God, like Anselm's ontological argument. Usually Hume is as patient as all get out at criticism; he likes to give his opponent all the rope he wants with which to hang himself. But here Hume is brisk. We might try to articulate what is eating Hume by saying that no existence proposition is analytic; you cannot make things exist however you define terms purporting to denote them. Of course Hume did not use Kant's term of art "analytic." In Hume's vocabulary we could say existence is always a matter of fact, never a relation of ideas. But the thesis that no existence proposition is analytic seems to be one of the few constants in philosophical consciences. For almost any philosophical view, one can find a stretch in Russell's life, for example, where he believed that view; nevertheless, not even Wittgenstein could con Russell into analytic existence.

Let us offer Kant Hume's thesis in kantian terms: no existence judgment is analytic. Now note that there are many existence claims in mathematics, witness the infinity of primes, the five regular solids, and undecidable propositions of *Principia*

---

66 (1957), 100-03.

[6]David Hume, *Dialogues Concerning Natural Religion* (New York: Hofner Press, 1948).

Mathematica and related systems. It would follow that some mathematical truths are synthetic, and so granting that they are known *a priori*, mathematics provides a fund of synthetic *a priori* knowledge. How extensive a fund would remain to be seen, but if it requires making meaning out, the prospects are dim.

Frege agreed with Kant that geometry is synthetic *a priori*. Note here Euclid's fifth and most famous postulate. In the form made familiar by the eighteenth-century editor Playfair,[7] this postulate says that given a line L in a plane and a point P in that plane not on L, there is one and only one line in the plane through P parallel to L. This is clearly an existence claim, and so by our earlier argument not analytic but synthetic. But that argument would not have persuaded Frege. Frege[8] said that the distinctions between the *a priori* and *a posteriori* and between the analytic and synthetic concern not the content of the judgment but the justification for making the judgment. (In a footnote to this remark, Frege adds that he does not mean to assign new senses to these terms, but only to state accurately what earlier writers, Kant in particular, meant by them. The reader may wish to be careful about whether Kant would have agreed with Frege on this point.) Frege says that when a proposition is called *a posteriori* or analytic in his sense, this is not a judgment about how we might form the content of the proposition in our consciousness or about how one might come to believe it, but about the ultimate ground upon which rests the justification for holding it to be true. This is an in-your-face remark because it assumes flat out that there is a unique and ultimate ground on which the justification for a proposition rests. Frege perhaps takes the figure of foundations of knowledge more seriously than any other philosopher. Frege initiated the analytic style of philosophy, and an important trend in its development is a loss of confidence in the idea that knowledge has foundations; in this process we diminished our exaggeration of the mathematician's axiomatic method.

Frege says that to settle whether a truth is analytic or synthetic, *a priori* or *a posteriori*, we must find the proof of the proposition and follow it right back to the primitive truths. So a truth has a single, unique proof that begins from a unique stock of primitive truths (*Urwahrheiten*). The presuppositions are so pronounced that one wonders whether they are as metaphysical as epistemic, whether the right order of truths is their logical order in being as much as the order in which we may justify them. If the primitive truths from which the proof of the proposition proceeds are nothing but general logical laws and definitions, then the truth is analytic. (Here he reminds us that when a singular term, for example, is defined by a definite description, the definition is admissible only if the predicate in the definite description is true of at least and at most one thing, and we will need to find the primitive truths on which rest these conditions for the admissibility of the definition.) If the primitive truths belong to the sphere of a special science,

---

[7]Heath says Playfair's axiom is stated by Proclus. See Euclid, *The Thirteen Books of the Elements*, trans. and intro. Sir Thoms Heath, vol. 1, 2nd ed. (New York: Dover), p. 220.

[8]Gottlob Frege, *The Foundations of Arithmetic*, trans. J. L. Austin (Oxford: Blackwell, 1959), sec. 3.

then the truth is synthetic. This account makes analyticity turn on generality rather than meaning. But the generality cannot be that all the quantifiers in the primitive truths are universal, so the primitive truths would be true even if there were nothing at all, and so cannot prove any existence claims. For Frege is out to show that the truths of elementary number theory are analytic, and these require the existence of infinitely many natural numbers. For a truth to be *a posteriori* it must be impossible to prove it without appeal to facts, that is, truths that cannot be proved and are not general because they contain assertions about particular objects. Note the absence here of any mention of justification by sense experience; the focus is instead on singular truths about particular objects that in some absolute sense cannot be proved, so it is almost as metaphysical as epistemic. For a truth to be *a priori*, its proof should proceed exclusively from general laws that neither need nor admit of proof. (Here Frege adds in a footnote an argument for general primitive truths.) So on Frege's view, the analytic and the *a priori* both descend from utter generality, and we expect a substantial overlap between them. He does say that the general law underlying the analytic should be logical, while those underlying the *a priori* should neither need nor admit of proof. He may have meant there are general primitive truths besides those of logic underlying, say, geometry. The intriguing notion here is of general laws that neither need nor admit of proof. Need here might mean that we can know, and so be justified in believing, these general laws without proof, and then the interesting question is how we know them. To say that they do not admit of proof raises the question how we would know of a general truth that it cannot be proved, and there is no standard way to prove such an absolute claim.

Frege's account makes it analytic that logic is analytic. His strategy was to take over ad hominem from Kant the premiss that logic is analytic. Then he would reduce the mathematics of elementary number theory to logic. While Euclid had axiomatized plane geometry millennia before, Dedekind had only recently axiomatized the natural numbers.[9] Frege had to define Dedekind's primitive terms (zero, the successor function, and being a number) in purely logical terms, and then deduce Dedekind's axioms from logic and these definitions. This reduction may be called logicism in a narrow sense. Given the reduction and the analyticity of logic, the analyticity of elementary number theory follows, and that may be called logicism in a broader sense. Logicism in a broader sense would deprive Kant's critical philosophy of some of its presuppositions. From time to time, a philosopher claims that thus-and-suches (material objects, say) are reducible to so-and-so (like sense experience). What was new with Frege, and still impressive to this day, is that instead of blustering, he gets on with it. Thereby he began a constructional tradition in analytic philosophy that includes Russell, Carnap, Tarski, Goodman, and David Lewis.[10]

---

[9]Richard Dedekind, "The Nature and Meaning of Numbers," in *Essays on the Theory of Numbers*, trans. W. W. Beman (New York: Dover, 1963).

[10]Besides *Principia Mathematica*, see also Russell's *Our Knowledge of the External World* (New York: Mentor, 1960); Rudolf Carnap, *The Logical Structure of the World*, trans. Rolf

Frege viewed his reduction as an extension of the arithmetization of analysis, which was a central focus of nineteenth-century mathematics. Analysis is the part of mathematics with the calculus (differentiation and integration) at its core. Newton and Leibniz had invented or discovered the calculus in the seventeenth century, and Newton put it to work in his physics. That physics was perhaps the single most notable event in the emergence of natural science, which after the French Revolution replaced theology as the most prestigious part of knowledge. But in *Principia*, Newton did not express himself in terms of the new calculus. Instead his arguments are mostly novel but recognizable extensions of Euclid, whose authority was unabated.[11] The public face of the calculus was geometrical. The only numbers the Greeks were comfortable recognizing as such were positive whole numbers (and even one was maybe insufficiently plural). They talked about proportions but not fractions or rationals, and they did not have decimals for irrationals. Where we now talk about functions, like the square of a number, mathematicians from the Greeks until Kant's day might talk about curves, by which they meant literal curves like the parabola. Differentiation was about tangents to a curve, and integration was about area under a curve, and area was less a number than a patch of the plane.

Euclid's geometry is mostly finitary. Angles get bisected and polygons have finitely many sides. Archimedes had given some gorgeous limiting arguments, but their infinitary aspect was much of what made them so striking. But as the calculus developed during the eighteenth century, its infinitary aspect became inescapable. In our expositions, for example, one of the earliest concepts distinctive of the calculus is the notion of, say, a point being the limit of an infinite sequence of points. The problem is that while in finitary Euclidean geometry, intuition (visual imagination) by and large did not lead people recognizably astray, intuition had begun to founder on paradox among infinitary processes by the turn of the nineteenth century.[12] Here is an example that Henri Lebesque says was current among schoolboys near the turn of the twentieth. In triangle ABC, let D be the midpoint of side AB, E of side BC, and F of side AC. Join D and E, and join E and F.

A. George (Berkeley: University of California Press, 1969); Tarski on truth cited above for the notion of material adequacy conditions; Nelson Goodman, *The Structure of Appearance* (Indianapolis: Bobbs-Merrill, 1966); David Lewis, *Counterfactuals* (Cambridge: Harvard University Press, 1973).

[11] Francois De Gandt, *Force and Geometry in Newton's Principia*, trans. Curtis Wilson (Princeton: Princeton University Press, 1995).

[12] Henri Lebesgue, *Measure and the Integral*, ed. Kenneth O. May (San Francisco, London, Amsterdam: Holden-Day, 1966).

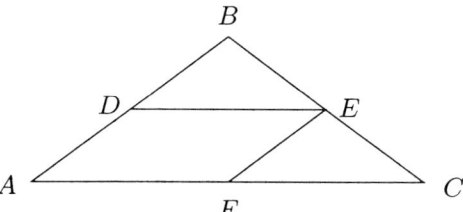

It is familiar from Euclid that quadrilateral $ADEF$ is a parallelogram, and AF is as long as DE, and EF is as long as DA. Hence, the broken line $CFEDB$ is as long as the two sides CA and AB together. Now repeat in the two little triangles FEC and DBE the argument just carried our in ABC to get a four tooth jagged line equal to CA and AB together. Carry this repetition out ad infinitum, and the length never changes from that of CA and AB together. But the limit of these jagged lines certainly looks to be side BC. So the sum of two sides of a triangle is not greater than the third, and a straight segment is not the shortest distance between two points. One wonders how the author of the antinomies of pure reason would have reacted to this infinitary paradox. There may be some relief to be had in noting that the angles of the teeth never flatten but are always equal to the angle at A. So the limit of the infinite sequence of jagged lines, assuming we can figure out what this limit is, is not the straight segment BC, but rather a line that all too often has no single unique direction. Analysis came to dote on such examples during the nineteenth century.

A line each of whose points is arbitrarily close to BC but mostly has no unique direction (say from B to C or vice versa) is not easy to visualize. To understand such things, mathematicians turned from intuition to understanding. The calculus was transposed from geometry to number. With hindsight we can see Descartes pointing a way to do this. Analytic geometry is usually credited to him. We are taught to begin cartesian coordinates with the number line. This hybrid is rooted in a one-to-one correspondence between the points on an infinite Euclidean straight line and (what we now call) the real numbers that preserves order, that is, such that point p is left of point q if and only if the number assigned to p is less than that assigned to q. It is not patent how much of our conception of the reals Descartes shared, so it is probably unfair to ask how he knew there is such a one-to-one correspondence. Where we are taught analytic geometry as a way to use numbers to answer geometrical questions, Descartes used it as a way to answer numerical questions with geometry. But once we lose confidence in the capacity of geometrical intuition to answer infinitary questions, one strategy would be to reverse direction and take real numbers more seriously.

To do so, we want a non-geometrical account of the reals, and that is where the arithmetization of analysis comes from. Let us briefly rehearse the familiar saga. Start with the natural numbers, the whole numbers 0, 1, and so on. We get negative numbers by arranging to subtract 5 from 3. So let us see how to handle the ordered pair $\langle n, k \rangle$ of natural numbers as if it were the difference $n - k$ no matter whether n is bigger than k or not. A little algebra shows that $\langle n, k \rangle$ and

$\langle p, q \rangle$ have the same difference when $n + q$ is $p + k$, addition being familiar on the natural numbers. So we specify a relation R to hold between a pair $\langle n, k \rangle$ and a pair $\langle p, q \rangle$ just in case $n + q = p + k$. Then R is reflexive (any pair is R to itself), symmetric (if $\langle n, k \rangle$ is R to $\langle p, q \rangle$, $\langle p, q \rangle$ is R to $\langle n, k \rangle$), and transitive (if $\langle n, k \rangle$ is R to $\langle p, q \rangle$ and $\langle p, q \rangle$ is R to $\langle r, s \rangle$, then $\langle n, k \rangle$ is R to $\langle r, s \rangle$). Such a relation is called an equivalence relation. Being as tall as is an equivalence relation, and it partitions people into exclusive and exhaustive groups of people equally tall. The philosophy starts to creep in if we think of these groups as the heights, like six foot six. Our R partitions the ordered pairs of natural numbers into groups, or equivalence classes as they are called, some think of as the integers. Write $[\langle n, k \rangle]$ for the equivalence class of the pair $\langle n, k \rangle$. We should raise addition from the natural numbers to the integers, and the evident way to do so is to set the sum of $[\langle n, k \rangle]$ and $[\langle p, q \rangle]$ equal to $[\langle n + p, k + q \rangle]$. Then for any $[\langle p, q \rangle]$, the sum of $[\langle p, q \rangle]$ and $[\langle n, n \rangle]$ is $[\langle p, q \rangle]$, so $[\langle n, n \rangle]$ is the zero of the integers. For any $[\langle p, q \rangle]$, the sum of $[\langle p, q \rangle]$ and $[\langle q, p \rangle]$ is the zero of the integers, so $[\langle q, p \rangle]$ is the negative of $[\langle p, q \rangle]$. To subtract an integer $[\langle p, q \rangle]$ from $[\langle n, k \rangle]$, add $[\langle q, p \rangle]$ to $[\langle n, k \rangle]$. The integers $[\langle n, o \rangle]$ are a copy of the naturals, and the integers $[\langle o, n \rangle]$ are their negatives. In a flush of enthusiasm, some call this constructing the integers from the naturals.

Constructing the rationals is similar except that this time we start wanting to be able to divide any integer i, even 3, by any non-zero integer j, even 5. We take ordered pairs $\langle i, j \rangle$ of integers, and we say $\langle i, j \rangle$ is Q to $\langle p, q \rangle$ if and only if the product iq equals the product jp, which should hold when i is to j as p is to q. Q is an equivalence relation, and its equivalence classes are the rational numbers. The equivalence relation is as old as Eudoxus's theory of proportions, but we probably did not step all the way to its equivalence classes, the rational numbers, until the nineteenth century. The next step is to the real numbers. One way to see the need is through the square root of two. By Pythagoras's theorem, this should be the length of the diagonal of a square whose side is of length one, but one can find in Euclid a proof that no ratio of, in effect, rational numbers can have 2 as its square. There is a story that the Pythagoreans knew this proof long before Euclid but hushed it up because it put the root of 2 beyond their reach. Here are two ways to flesh out the rationals with irrationals. The first is Dedekind cuts.[13] A Dedekind cut is a pair $\langle L, R \rangle$ of sets of rationals such that every rational is in L or R, no rational is in both, and every member of L is less than every member of R. Picture L as the left part, and R the right, produced by cutting the rational number line in two somewhere. If L is the rationals less than or equal to one but R is those greater, this cut corresponds to the rational real one. Note that L has a greatest member, namely one. Now let L be the rationals whose squares are less than or equal to 2, while R is the rationals whose squares are greater than 2. This time L has no greatest member, and R, no least. This cut corresponds to the irrational real, root 2.

But square roots are an unrepresentative ground for irrationals. Decimals give a

---

[13] Dedekind, "Continuity and Irrational Numbers," in *Essays on the Theory of Numbers*, trans. W. W. Beman (New York: Dover, 1963).

better picture. It is not difficult to show that the decimal for any rational number is either finite (like .25 for 1/4) or repeating (like .333... for 1/3), and conversely. So the decimal for the root of 2 is neither finite nor repeating. But we can compute its decimal as far as we like. Its first n digits are those in the Arabic numeral for the least natural number k such that $(k+1)^2 > 2(10^{2n})$. So it starts off 1.414213.... We can break this decimal up into the sequence $1, 1.4, 1.41, 1.414, 1.4142, \ldots$. All the members of this sequence are finite decimals and so represent rationals. These rationals are all less than, say, 1.5, but they never decrease. So they get squeezed closer and closer together, which makes us expect them to be pushing up against a limit, a least number greater than or equal to all of them. But there is no such rational number, so we need an irrational limit. A more general version of this issue is that a non-decreasing sequence bounded above should have a least upper bound. This property is called completeness, and the rationals do not have it. But now suppose $L_1, L_2, \ldots$ are the left halves of an infinite sequence of Dedekind cuts such that for each n there is a member L_{n+1} greater than every member of L_n (so the sequence is increasing), but there is a rational greater than every member of every L_n (so the sequence is bounded above). Then the union of all the $L_1, L_2, \ldots$ fixes the left half of the cut that is the desired least upper bound, or limit, of the sequence of cuts. The reals are complete, and their completeness under limits is part of their centrality in analysis.

We can also approach the reals and completeness by thinking about squeezing.[14] Let S be a non-empty set we will call a space. A metric on S is a binary function d that assigns to any two points x and y in S a number intended to represent the distance between x and y. Usually these numbers are real, but since we are constructing the reals, let us start off with rational distances. It is required that the values of d be non-negative, that the distance between x and y be zero if and only if x is y, that the distance between x and y be the distance between y and x, and that the sum of the distance from x to y and that from y to z be less than or equal to that from y to z. This last is called the triangle inequality; the sum of two sides of a triangle is greater than the third. For any number x, its absolute value $|x|$ is x if x is positive, but $-x$ if x is negative. (Intuitively the absolute value of x is its distance from 0.) If for any rationals x and y we set $d(x,y) = |x-y|$, then d is a metric on the space of rationals. In our sequence $1, 1.4, 1.41, 1.414, \ldots$ of rationals, the distance between successive members gets smaller. In fact for any n however big there comes a stage in the sequence after which any two members of the sequence are less than $1/n$ apart. Such a sequence is called a Cauchy sequence. Cauchy sequences get squeezed. Let $p_1, p_2, \ldots$ be a sequence of points in a metric space. A point p is a limit of the sequence if for every n however big there is a k such that for m greater than or equal to k, the distance between p_m and p is less than $1/n$. Another version of completeness requires that every Cauchy sequence have (or converge to) a limit. Our sequence $1, 1.4, 1.41, \ldots$ is Cauchy but does not converge. But now let $a_1, a_2, \ldots$ and $b_1, b_2, \ldots$ be two sequences. Say that these

---

[14]See, for example, Patrick Suppes, *Axiomatic Set Theory* (Princeton: D. Van Nostrand, 1960), chap. 6.

sequences are C to each other if for every n however big there is a k such that for m greater than or equal to k, the distance from a_m to b_m is less than $1/n$. When sequences are C they come together. C is an equivalence relation, and we can construct the reals as the C equivalence classes of the Cauchy sequences. Then our sequence has a limit, namely, its C equivalence class.

So we can construct integers from natural numbers, rationals from integers, and reals from rationals. Kronecker said God made the natural numbers; all the rest is the work of man.[15] But construction and work are metaphors here. We construed integers, for example, as equivalence classes of ordered pairs of natural numbers. Nowadays we take equivalence classes and ordered pairs as sets. This is also true of sequences. We do not build sets; there are too many of them, and they are too abstract, for that. Instead, we assume them. Besides prying the various sorts of numbers out of geometry, we also unify them as applications of set theory to the natural numbers.

If we go on to qualify, we might say both Frege and Cantor complete the process by reducing natural numbers to sets. The first qualification is that Frege works not with sets but with functions.[16] He recognizes two truth values, truth and falsity, and he defines a concept as a function whose value is always a truth value. So the concept of humanity is the function whose value is truth for the argument Socrates but falsity for the number seven as argument. The value range of a function is roughly its graph, so the value range of humanity is the curve passing through truth over people but falsity over everything else. Though Frege would not like it, we can recover a set as the part of the domain of a concept on which it takes the value truth. All Frege's functions have the same domain, the universe of absolutely everything, so our device attributes to Frege what Gödel describes as the conception of sets as all ways to divide the universe in two.[17] Frege describes numbers as objects belonging to concepts, and if we replace his concepts by sets, we could say both Frege and Cantor think of cardinal numbers as answers to the question how many members does a set have. For both the central notion is of a function f that maps a set A one-to-one onto a set B. If f assigns different values in B to different arguments in A, it is one-to-one, and since there is then no collapsing of different arguments into a single value, B is at least as big as A. If every member of B is also a value of f for some argument in A, the f maps A *onto* B, and since everything in B is hit by f at least once, A is at least as big as B. So if f is both one-to-one and onto, B is at least as big as A, and A is at least as big as A, and they are of the same size. We can use one-to-one

---

[15] Herman Weyl, *Philosophy of Mathematics and Natural Science* (New York: Atheneum, 1963), p. 33, cites Kronecker as saying God created the integers, but Kronecker does not give us enough credit. Constructing the integers is so like constructing the rationals that we should extend Kronecker's trope.

[16] Gottlob Frege, "Function and Concept," in *Translations from the Philosophical Writings of Gottlob Frege,* trans. Peter Geach and Max Black (Oxford: Blackwell, 1960).

[17] Kurt Gödel, "What is Cantor's Continuum Problem?" in *Collected Works*, vol. II, ed. Solomon Federman, John Dawson, Stephen Kleene, Gregory Moore, Robert Solovay, and Jean van Heijenoort (New York: Oxford University Press, 1990), p. 180.

and onto functions to explain having the same number of members without a prior account of number. You can exhibit such a function by putting your finger tips together, thereby showing you have as many digits on one hand as the other without counting either.

The as-many-as relation is an equivalence relation between sets, and we could try the equivalence class of A under it as the (cardinal) number of (members of) A. That was roughly Frege's approach, but Cantor construes the number of members of A as something we abstract from the sets the same size as A.[18] Abstraction in this traditional sense is metaphysically and epistemically less sophisticated than the equivalence class construction. (There is an interesting critique of abstraction in Peter Geach's *Mental Acts*.)[19]

Frege and Cantor go in different directions from the number of members of a set. Cantor was after infinite numbers. Euclid took it as an axiom that a whole is always greater than any of its (proper) parts. When Galileo noted that doubling maps the natural numbers one-to-one onto the even numbers, so there are as many even numbers as natural numbers even though the even numbers do not exhaust the natural numbers, Euclid's authority sufficed for Galileo to infer that there is no completed totality comprised of either the even numbers or the naturals, and Leibniz further concluded that there are no infinite numbers either.[20] Such views are of a piece with Aristotle's doctrine that the infinite can only ever be potential (though a possibility that cannot be actual seems contradictory). Dedekind inverted this conventional wisdom. He defined an infinite set as one the same size as one of its proper subsets.[21] (One set is a subset of another if all members of the first are members of the second, and it is proper if there are members of the second not in the first.) Dedekind also gave a truly dodgy proof that there is an infinite set: we have, he said, an idea of each of our ideas, but we also have an idea of ourselves, who are not ideas, so the set of our ideas is infinite.[22] Russell agonized over this argument[23] instead of just denying that we have an idea of each of our ideas. Nowadays we usually just assume the existence of an infinite set.

A number is infinite when it is the number of members of an infinite set. It is a marked advantage of constructing numbers from sets that it makes sense of infinite numbers. Cantor is perhaps most famous for proving that there are different infinite numbers. He proved, for example, that there are more real numbers than natural numbers. A set A is smaller than or equal in size to a set B if there is a

---

[18] Georg Cantor, *Contributions to the Founding of the Theory of Transfinite Numbers*, trans. Philip Jourdain (New York: Dover, n.d.; originally published 1915), p. 86.

[19] Peter Geach, *Mental Acts* (London: Routledge and Kegan Paul, 1957).

[20] Herman Weyl, *The Philosophy of Mathematics and Natural Science* (New York: Atheneum, 1963), pp. 47–48.

[21] Dedekind, "The Nature and Meaning of Numbers," in *Essays on the Theory of Numbers*, trans. W. W. Beman (New York: Dover, 1963).

[22] Dedekind, in *Essays on the Theory of Numbers*, article 66.

[23] Bertrand Russell, *Introduction to Mathematical Philosophy* (London: George Allen and Unwin, 1919), pp. 138-40.

one-to-one function that assigns to each member of A a member of B. B is bigger than A if A is smaller than or equal in size to B but B is neither smaller than nor equal in size to A. A number n is less than or equal to a number k if there are sets A and B such that n is the number of members of A, k is the number of members of B, and A is smaller than or equal in size to B, and k is larger than n if n is less than or equal to k but k is neither less than nor equal to n. So Cantor had at least two infinite numbers.

But he had more. He defined the power set of a set as the set of all its subsets. It is called the power set because if it is finite and has, say, n members, then in forming an arbitrary subset of it, there are only two things to be done with any member of the set — either put it in the subset or leave it out. So it has 2^n subsets. An induction shows that 2^n is always bigger than n, so the power set of a finite set is always larger than the set. Cantor showed that this holds for infinite sets too. He proved this by what is now called a diagonal argument, a mode of reasoning Cantor discovered, though there seem to be many more diagonal arguments in recursion theory than in set theory. Diagonal arguments remind some people of the liar paradox and can be controversial. But Cantor's theorem that the power set of a given set is always strictly larger than the given set is now as received as any other theorem. It gives us a wealth of infinite sizes. Let A_0 be the set of natural numbers and for each n, let A_{n+1} be the power set of A_n. Then $A_0, A_1, \ldots$ give us as many infinite numbers as natural numbers. Next let A be the union of all of $A_0, A_1, \ldots$. If A were smaller than or equal in size to some A_n, it would be smaller than A_{n+1}, but since A_{n+1} is a subset of A, A_{n+1} is smaller than or equal in size to A, and thus A cannot be smaller than A_{n+1}. Hence, A is larger than all of $A_0, A_1, \ldots$. Now we iterate power set from A as we did from A_0. Indeed we go on to iterating power set and union into an indefinite distance. For each infinite number n, there are more than n infinite numbers (just as for each natural number k, there are more than k natural numbers). David Hilbert called this wealth of infinite numbers, and its mathematics, Cantor's paradise.[24]

Suppose we took Frege's truth values, truth and falsity, to be the numbers 0 and 1. The characteristic function of a set A of natural numbers is the function that assigns 0 (truth) to members and 1 (falsity) to non-members, so the characteristic function of A is reminiscent of Frege's concept of A-ness. We are used to decimals written with the ten Arabic numerals, but we could as well write them with the binary numerals 0 and 1 favored by computers. Then the binary decimal for a non-negative real less than 1 is a list of the values of the characteristic function for a set of natural numbers, and conversely. So there are as many reals in that interval as there are members of the power set of natural numbers. Sending each non-negative real x to $x/(x+1)$ and each negative x to $-(x/(x-1))$ shows there are as many reals as there are reals between -1 and 1, and sending point p to $f(p)$ as in

---

[24]David Hilbert, "On the Infinite," trans. Stefan Bauer-Mengelberg, in Jean van Heijenoort, *From Frege to Gödel: A Sourcebook in Mathematical Logic, 1879-1931* (Cambridge: Harvard University Press, 1967), p. 376.

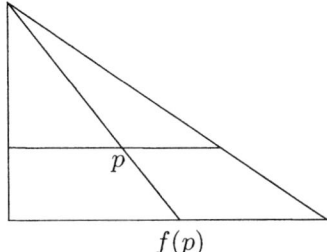

shows that any two bounded intervals contain the same number of points. So Cantor's theorem generalizes his result that there are more real than natural numbers. He conjectured that there is no infinite number between the size of the naturals and the size of the reals. This is called the Continuum Hypothesis, and Hilbert put it first on the agenda for mathematics in the twentieth century. In the 1930s Gödel showed that the continuum hypothesis is consistent with set theory (if set theory is consistent), and in the 1960s Paul Cohen showed that its negation is consistent with set theory (if set theory is consistent). We will not settle the continuum hypothesis without agreeing on new axioms.

Where Cantor was after the infinite, Frege was after the finite. To reduce the natural numbers, Frege had to define Dedekind's primitive notions. Zero is the number of things not identical with themselves, and (for present purposes) the successor of n is the number of numbers less than or equal to n. The infinite numbers show that the natural numbers do not exhaust the cardinal numbers, so Frege needed to separate the natural numbers from the cardinals. He defines them as the members of all sets of which zero is a member and the successor of a member is always a member. This definition makes the analyticity of mathematical induction go down all too smoothly.

Frege's and Cantor's projects both founder in paradox.[25] To reduce the mathematics of natural numbers to logic, Frege had to beef up logic. Axiom V of his logic says that all and only the objects falling under one concept fall under another just in case the value ranges of those concepts are identical. This requires that every concept have a value range, or in more familiar terms, that every predicate have an extension, the set of all and only the things of which the predicate is true. This last is often called comprehension, as if a predicate comprehends, or collects, an extension. Traditional logic was shot through with talk of extensions of concepts, so Frege's V could easily seem at home in logic. But thinking about Cantor's diagonal argument, Russell wondered about the extension of the non-self-membership predicate. The set of all lions is not a lion, but the set of all sets is a set. Russell's set R collects all sets like the first, that is, all those that are not members of themselves. But then R is a member of R if and only if R is not a member of R. This reasoning is known as Russell's paradox. Russell sent it to Frege in a letter in 1902. Frege lived until 1925, but to my mind nothing he did after 1902

---

[25] All three original presentations of the paradoxes of set theory occur in van Heijenoort's *Sourcebook*, cited in the previous note.

measures up to what he did before. In Cantor's case, consider the universe, the set U of absolutely anything. The power set of U is, by Cantor's theorem, larger than U, but it is also a subset of U, and so smaller than or equal in size to U. This reasoning is known as Cantor's paradox, and Cantor sent it to Dedekind in a letter in 1895. There is also a third paradox of set theory about ordinal numbers and known as the Burali-Forth paradox after the man who published it in 1897.

To those who take the figure of foundations of mathematics seriously, the paradoxes of set theory are a crisis. But any crisis was philosophical. Frege's axiom V is not analytic. It is not necessarily true, and it is not known *a priori*. It is none of these things because it is just plain false. Russell's paradox is a proof by reductio ad absurdum that it is false.

If someone says more or less out of the blue that the prime minister of Estonia is at this very moment seated rather than standing, it seems sensible to withhold judgment at least for a bit. Evidence may be forthcoming, but it hasn't yet. But there are claims where, as it were, saying is believing. Playfair's version of the axiom of parallels says that given a line L in a plane and a point P in the plane not on L, there is one and only one line in the plane through P parallel to L. People largely innocent of geometry typically agree to Playfair's axiom without argument. The same acquiescence typically greets Euclid's axiom that the whole is larger than any of its (proper) parts. Back when set theory was only naïve, the same acquiescence was also accorded assuming that for every predicate there is a set of all and only those things of which the predicate is true. We might try to rehabilitate an abused notion by taking intuitions as beliefs for which we perhaps should have justification but do not and yet nevertheless hold.[26] On this account, intuition is not some faculty whose exercise justifies belief and yields knowledge in mysterious ways. On the contrary, intuitions are beliefs held without justification but held anyway.

Suppose a person holds a belief without justification. Suppose he is asked not whether this belief is true but why he believes it. Often such a person responds by digging in his heels and repeating his intuition with increasing assurance. Suppose now that everyone else (or almost everyone else) who considers the issue agrees with him, and no one (or almost no one) disagrees or withholds judgment. What is emerging may be the sociology of analyticity and *a priori* knowledge. If so, these notions are table pounding in fancy dress. Necessary truth may differ somewhat because claims that a truth is necessary can be tested against fertile imaginations. With claims that a truth is analytic one may get nothing but question-begging semantic claims, and with claims that a truth is known *a priori* one usually gets nothing at all.

The case is worse when we are expected to believe that a claim is true because it is analytic, necessary, or known *a priori*. Axioms lack proof, at least in the systems of which they are axioms. If the axiom is not independent of the others but deducible from them, then the question just shifts over to the axioms from

---

[26] Compare W. V. Quine, *Word and Object* (Cambridge: Technology Press of the Massachusetts Institute of Technology, and London and New York: John Wiley and Sons, 1960), p. 36 fn.

which it is deduced. Lacking proof means lacking proof, not having some super but ineffable proof. So even if, like everyone else, I believe it, I still am better off if I acknowledge that I just do not have a proof of it. Kant thought Playfair's axiom was synthetic *a priori*, Frege thought comprehension was analytic, and Galileo appealed to Euclid's axiom that the whole is greater than the part to deny the actual infinite. Our best bet nowadays is that all three axioms are false. A shrewd philosopher is wary when an interesting claim is called analytic, necessary, or *a priori*, since often the really interesting stuff is to be broached by working out how the claim could be false.

When set theory was naïve, it assumed only comprehension (every predicate has an extension) and extensionality (sets with the same members are identical).[27] Once comprehension had been refuted, set theory needed reform. It is striking that Russell, who refuted comprehension, never quite gave it up. Instead he denied meaning to the claim that the set of all lions is not a member of itself. He converted the two-way split between truth and falsity into a three-way split between these two and meaninglessness. For any predicate counted as meaningful by his theory of types, he allowed himself to form a singular term denoting its extension, but his allegiance to comprehension was so implicit that he never stated that all meaningful predicates have extensions. Russell's focus on meaning, and especially his denial of meaning to claims we have no trouble understanding, is in no small way responsible for the coming of philosophy of language. Comparing the claim that the set of all lions is not a member of itself with Lewis Carroll's line that 'twas brillig and the slithy toves did gyre and gimbel in the wabe, we might want to ask an analogue of Prichard's question[28] whether moral philosophy rests on a mistake.

When Russell was working out *Principia Mathematica* with Whitehead, his main opponent in the journals was Henri Poincaré. Poincaré's well-deserved prestige in France coupled with his hostility to the new logic made that logic unpopular in France for decades, which was a real loss to logic. But however much Russell and Poincaré differed, they agreed in assimilating what we now in hindsight distinguish as the paradoxes of set theory to the semantic paradoxes. These last are illustrated by a sentence which says of itself that it is false, for if it is false, it has the property it ascribes to itself and so is true, while if it is true, then because it says it is false, it must be false. The analogy was between self-membership and self-reference in Russell's paradox and the liar paradox. We owe the distinction between the two sorts of paradox to Ramsey.[29] He said logic need not address the semantic paradoxes, and can solve the other by arranging things in types with non-sets on the bottom, and sets of things of type n in type $n + 1$. It is striking that only a few years later Tarski[30] proposed to solve the semantic paradoxes with

---

[27] See Paul R. Halmos, *Naive Set theory* (Princeton: D. Van Nostrand, 1960).

[28] H. A. Prichard, "Does Moral Philosophy Rest on a Mistake?" in *Moral Obligation* (Oxford: Clarendon Press, 1949).

[29] Frank Ramsey, "The Foundations of Mathematics," in *The Foundations of Mathematics*, ed. R. B. Braithewaite (Paterson, N.J.: Littlefield, Adams, 1960).

[30] Alfred Tarski, "The Concept of Truth in Formalized Languages" cited in note 3 above.

levels of language in which talk not about language is on the bottom, and talk about language of level n is conducted in language of level $n + 1$. From a sufficiently abstract point of view, the similarity between Ramsey's simplified types and Tarski's levels of language makes one wonder whether Russell and Poincaré were on to a good thing.

It is a basic tenet of the theory of types that no proposition may say anything about itself. Nor could there be a claim, say C, about absolutely everything, since then C would have to be about C too, and that would violate the first tenet. Let K be the claim that no claim is about absolutely everything. It follows from K that K is not about absolutely everything, so it certainly looks as though the theory of types violates itself. Such self-destruction is surprisingly frequent among philosophical theories, and it is certainly a weapon that we want to remain in the philosopher's critical armory. We want, for example, to be able to fault a theory T of theories that gives a good account of all theories except T, and any line like the theory of types that purports to resist such self-applications should for that reason be resisted.

In *Principia* Russell had to reconcile two conflicting objectives. On the one hand he had to weaken logic enough to prevent the derivation of paradox, but on the other hand he had to keep logic strong enough for the mathematics of number to be reducible to it and to keep open the gates to Cantor's paradise. To do so, he needed four assumptions that bothered him, or should have. He recognized that he needed axioms of choice, infinity and reducibility, and he used his version of comprehension even if he did not quite recognize it explicitly. For the philosopher it matters right now less what these assumptions say than that they are assumptions of existence. Russell was not comfortable with the line that existence propositions can be analytic. Indeed he looks to toy with inverting Frege's syllogism: since mathematics is synthetic and mathematics is reducible to logic, logic is synthetic. In the preface to *Principia* he said that the chief reason in favor of a theory on the principles of mathematics must always be inductive.[31] By this he meant that instead of the conventional mathematical wisdom being justified by deduction from the so-called foundations of mathematics, the foundation is justified only if it suffices for the deduction of the conventional mathematical wisdom. This inversion is an important moment in the critique of the philosopher's exaggeration of the mathematician's axiomatic method, even if it was not properly appreciated for a while. But eventually Quine[32] will say that no statement is any more intrinsically a postulate than is a point in Ohio intrinsically a starting point. From Playfair's axiom we can show that if two angles of one triangle equal two of another, the triangles are similar. But if we replace Playfair's axiom with this theorem, then

---

[31] Bertrand Russell and Alfred North Whitehead, *Principia Mathematica*, 2nd ed. (Cambridge: Cambridge University Press, 1925), vol. I, p. v.

[32] W. V. Quine, "Two Dogmas of Empiricism," in *From a Logical Point of View*, 2nd ed. (Cambridge: Harvard University Press, 1961), p. 35. Frege and Quine express extreme views about foundations of knowledge. There are probably pairs of claims on which there would be a consensus as to which is epistemically more basic. But we do not have a settled body of principles of epistemic priority.

we can deduce Playfair's axiom.[33]

While Russell's theory of types was the reform of set theory best known among philosophers, it was not much favored among mathematicians. It is hideous and probably immune to thorough understanding. Hilbert's student Zermelo inaugurated in 1908 what has become pretty much received set theory.[34] He had to restrict comprehension, so he assumed instead that for every definite property P and any set x, there is a set of all those members of x having P. This is called separation since it separates out the members of x with P. There is no settled mathematical consensus on properties, and about a decade later Frankel reformulated separation in terms of the predicates of a first-order formal language.[35] Separation is weak enough that it needs supplementing by existence assumptions. The more interesting of these include infinity (there is an infinite set), union (given a set, there is a union of all its members), and the big one, power set (every set has a power set). Skolem added replacement, which says that if the domain of a function is a set, so is its range.[36] Kripke pictures sets as corrals, and we might similarly picture them as lassos, except that extensionality would forbid two lassos roping the same set. Functions can be pictured as collections of arrows from the domain to the range. If we have lassoed the domain, slide the lasso along the arrows to lasso the range.

Zermelo-Frankel set theory, or ZF, had emerged by the early 1920s, though in justice it should be called ZFS set theory to give credit to Skolem. In 1929 von Neumann worked with a structure in which there are no infinite sequences x_1, x_2, ... of sets such that x_{n+1} is a member of x_n, that is, there are no infinite descending membership chains.[37] By around 1960 this assumption had been incorporated into ZF. Indeed in those days some set theorists leaned on this assumption, called foundation, and some remarks of Gödel's in the 1940s,[38] to disparage some of Quine's work on comparative set theory.[39] Consensus creates bullies. Foundation does make set theoretic life easier, but if we explore its denial we can model self-referential propositions.[40] If the proposition that Socrates is bald is the ordered pair whose first member is Socrates and whose second member is the set of bald men, we can say the proposition is true if its first member is

---

[33] George David Birkhoff and Ralph Beatley, *Basic Geometry*, 3rd ed. (New York: Chelsea Publishing Company, 1959).

[34] Ernst Zermelo, "Investigations in the Foundations of Set Theory I," in van Heijenoort's *Sourcebook,* cited in note 24. See note 36 also.

[35] Abraham A. Fraenkel, "The Notion 'Definite' and the Independence of the Axiom of Choice," in van Heijenoort's *Sourcebook,* cited in note 24.

[36] Thoralf Skolem, "Some Remarks on Axiomatized Set Theory," in van Heijenoort's *Sourcebook,* cited in note 24. This address is a major contribution.

[37] John von Neumann, "Ueber eine Widerspruchsfreiheitsfrage in der axiomatischen Mengenlehre," *Journal für reine und angewandte Mathematik* 160 (1929), 227-41.

[38] See note 17.

[39] W. V. Quine, *Set Theory and Its Logic*, rev. ed. (Cambridge, Mass.: Belknap Press, 1969).

[40] See Peter Aczel, *Non-Well-Founded Sets* (Stanford, Cal.: Center for the Study of Language and Information, 1988,) Lecture Notes No. 14, and Jon Barwise and John Etchemendy, *The Liar* (Oxford: Oxford University Press, 1987).

a member of its second. A proposition that says of itself that it is a proposition could then be an ordered pair p whose first member is p and whose second is the set of propositions. Foundation is neither analytic nor known *a priori*.

Zermelo's axiom of separation turns Russell's paradox into a proof that the universe does not exist. For if there were a set of which absolutely everything, or even just every set, were a member, the predicate for non-self-membership would separate Russell's paradoxical set from it. So "is self-identical" and "is a set" are predicates that do not have extensions. This does not mean that they have empty extensions, since then they would have extensions, albeit empty ones. It means that there is no set of all those things identical with themselves, and there is no set of all sets. Nor does any set have a complement whose members are the things not in the set, for otherwise the union of a set with its complement would give us back the universe. Where Russell and Poincaré blamed the paradoxes on self-membership and its ilk like self-reference, ZF shies away from collections it deems too big to exist. One might call the loss of the universe the revenge of the potential infinite. By replacement, no collection as big as the universe can exist, and this rules out a set of all (cardinal) numbers. Suppose there were a set N of all cardinal numbers. For every cardinal k, the set $N(k)$ of ordinals with fewer than k predecessors is a set of cardinality k, so we have a function that maps the cardinals one-to-one into the sets. On the other hand, for every set A there is a cardinal $n(A)$ which is the number of members of A. (In ZF we use the axiom of choice to show this, but set theory would be in a sorry state without it, so often we assume ZFC, which is ZF with the axiom of choice. Choice says that for any set of non-empty sets there is a function whose value at each of these sets is a member of it. Russell observed that while from each of an infinity of pairs of shoes we can pick the right shoe, we need choice to pick from an infinity of pairs of socks.[41] Since Zermelo first articulated choice in 1904, there has been controversy about it. In the 1920s, for example, Tarski and Banach used choice to show that any sphere the size of a pea can be cut into no more than four pieces that, without the inflation of topology, can be reassembled into a sphere the size of the sun.)[42] Given that every set has a unique number of members, we have a function mapping the sets one-to-one into the cardinals. It follows (by the Schroder-Bernstein theorem) that there are exactly as many sets as cardinals, so "is a cardinal number" has no extension either. Note also that if N existed, it should have a number n of members. Since $N(k)$ is always a subset of N, each k is less than or equal to n. But by Cantor's theorem, the power set of N would have a number of members bigger than n. ZF dodges (the usual proofs of) Cantor's paradox by not asserting the existence of the universe or N.

Not accepting the universe has consequences for logic. Frege and Peirce liberated logic by isolating the quantifiers, all and some, from the applied quantifiers, all

---

[41] Bertrand Russell, *Introduction to Mathematical Philosophy* (London: George Allen and Unwin, 1919), p. 126. Frankel's proof of the independence of the axiom of choice in the paper mentioned in note 35 above is a formalization of Russell's observation.

[42] Stan Wagon, *The Banch-Tarski Paradox* (Cambridge: Cambridge University Press, 1985).

*A*s and some *A*s, of Aristotle's syllogistic. This made quantification into polyadic predicates possible (everything is R to something), and that is the life's blood of the new logic. One should expect some fraternization between the universe and the universal quantifier. For example, in the (admittedly peculiar) set theory in "New Foundations," Quine accepts the universe and explains universal quantification by saying that everything is F if and only if the set of Fs is the universe.[43] Without the universe, universal quantification will be limited. We mostly follow Tarski in using models (or structures) to interpret languages.[44] A model is a set theoretic object with a domain required to be a (non-empty) set. The predicates of the language are interpreted by assigning them extensions (of suitable polyadicity) in the domain, the constants, by assigning them members of the domain, and mutatis mutandis for the function signs. The universal quantifier is interpreted as all members of the domain. (Some go so far as to infer from this dependence of quantification on the choice of domain that the quantifiers are not logical constants.) Suppose we want to say that everything is self-identical. Not accepting the universe means that we are restricted to saying that every member of this or that set taken as domain of a model is self-identical. An axiom of ZF assures us that everything is a member of a set, namely, the set whose only member is that thing. This is called its unit set or singleton. So for everything there is a model whose domain has that thing as a member, and we can use that model to say that thing is self-identical. Tarski offers us a notion of validity under which a claim is valid if it comes out true under all interpretations in all models, so by saying the claim that everything is self-identical is valid we might seem to have an indirect way of saying what we originally wanted to say. But then we wonder how to interpret the universal quantifiers in "Everything is a member of a set" or "Such and such a claim comes out true under all interpretations in all models." We meant absolutely everything by "everything," and we are not allowed to. There are as many models as sets, so we are not allowed to quantify over them all at once. Neither Russell's theory of types nor ZF accepts the universe, and this restraint trips logic up enough to make us look for ways to dodge the paradoxes of set theory while hanging on to the universe.[45]

Frege took zero as the number of things not identical with themselves, and on our set theoretic construal of his reduction, this turns out to be the unit set of the empty set, the set with no members. Then one, the successor of zero, is the number of numbers less than or equal to zero, so since only zero is less than or equal to zero, one is the set of all unit sets. But since everything has a unit set, the set of unit sets is the same size as the universe. So the number one does not exist, and neither do any other numbers except zero. It would be a distinct embarrassment for a reduction of mathematics to logic if it turned out

---

[43] W. V. Quine, "New Foundations for Mathematical Logic," in *From a Logical Point of View*, 2nd ed. (Cambridge: Harvard University Press, 1961), p. 94.

[44] Alfred Tarski, "On the Concept of Logical Consequence," in *Logic, Semantics, and Metamathematics*, trans. J. H. Woodger (Oxford: Clarendon Press, 1956).

[45] The late Raúl Orayen often stressed this difficulty.

that zero is the only cardinal number. Remember Russell's insight that instead of the conventional mathematical wisdom being justified by deduction from the so-called foundations of mathematics, the foundation is justified only if it suffices for the deduction of the conventional mathematical wisdom. So how are we to save the numbers? John von Neumann did so by selecting from each of Frege's equivalence classes a standard representative.[46] Since Frege's zero is the unit set of the empty set, von Neumann's zero is the empty set. Then von Neumann takes the successor of n to be the union of n with its unit set. Thus each n has n members. So where Frege took a set A's having n members as A's being a member of n, von Neumann takes it as A's being the same size as n. Zermelo also took zero as the empty set, but he took the successor of n as its unit set. This works for the natural numbers too, and it has been argued that because there is no unique right way to reduce numbers to sets, numbers are not sets after all. Well, maybe. But von Neumann's finite cardinals generalize beautifully into the infinite while Zermelo's do not. Enthusiasm for von Neumann's construction tempts one to think he discovered which sets the cardinal numbers really are. It is as if Cantor's nineteenth-century theory of transfinite cardinals reveals which sets the prime numbers Euclid proved to be infinite in number really are (though it has not yet helped to figure out the distribution of the primes).

Logicism like Frege's is a kind of platonism. Platonism is the metaphysical view that there are non-mental, non-physical abstract objects like numbers that do not depend on us or anything mental or physical for their existence. Plato's theory of forms was the first recorded metaphysical platonism. Very abstract objects like numbers or platonic forms are utterly inert, so since, as Grice[47] argues, perception is by its nature causal, there is an epistemological problem for empiricists about how we might justify belief in and about abstract objects. Plato struggles with this problem in the *Symposium*, the *Meno*, and the *Republic*. Set theory can be viewed as a contemporary version of the theory of forms. Both forms and sets answer to predicates, and comparing Russell's paradox with the third man argument shows that as set theory has a problem with self-membership, the theory of forms has a problem with self-participation. One difference may be that set theory has an explicit commitment to extensionality, an issue Plato does not seem to have addressed. We should be clear that even sets of concrete objects are not to be identified with the physical aggregates of the members. An example of Quine's points out that while the set of states in the US and the set of counties in the US have the same physical aggregates, they have no members in common and so are different sets.[48]

It may seem curious that while logicism is a form of platonism, Frege held that logic and mathematics are analytic. The curiosity is most acute for those content

---

[46] J. Donald Monk, *Introduction to Set Theory* (New York: McGraw-Hill, 1969), has a good exposition of von Neumann's theory.

[47] H. P. Grice, "The Causal Theory of Perception," reprinted in *Perceiving, Sensing and Knowing*, ed. Robert J. Swartz (New York: Doubleday, 1965).

[48] W. V. Quine, *Mathematical Logic*, rev. ed. (New York: Harper Torchbooks, 1962), p. 120.

with the quick gloss of analyticity as truth by virtue of meaning. If one takes truth by meaning as an alternative to truth by correspondence to fact, the analyticity of mathematics looks like a way to deny that the theorem that there are infinitely many primes is true only if there are infinitely many primes, and so numbers, and so to avoid platonism. For Ayer in *Language, Truth and Logic,* a virtue of the analyticity of the *a priori* is the exit it offers to platonism.[49]

For Frege, of course, analyticity is not truth by meaning but derivability from logic, and his logic is pretty platonic. But platonism makes some people nervous; they may be letting their ontological (what are the basic sorts of things?) view be governed tacitly by the prevailing naturalist cosmology (how are the basic bits unified into the world?) put by Hume as the idea that causation is the cement of the universe.[50] For if abstract objects like numbers are utterly inert, how could causation cement them to the mental and physical stuff in the system of the world? Maybe membership is another cement of the world, not only binding sets in a system, but also binding the concrete to the abstract as in sets of concrete things. Then reducing mathematical objects to sets would have a unifying cosmological advantage.[51]

But suppose one favors naturalism. Medieval nominalists like William of Occam rejected platonic forms and universals in favor of linguistic items like predicates, and in the philosophy of mathematics those who favor signs like numerals over abstracta like numbers are sometimes called formalists. (After the paradoxes, Hilbert inaugurated a program to prove the consistency of reformed mathematical systems. This was also called formalism since it took these systems as systems of notation. But Hilbert thought consistency proofs would help justify belief in abstracta. He was not a metaphysical formalist.) If numerals are to have the epistemic appeal of concreta, they should be actual physical inscriptions and utterances, and as Frege observed long ago,[52] there will never be enough inscriptions and utterances to do duty for the infinity of natural numbers. One move often made at this point is to replace actual inscriptions with possible ones, or with set theoretically defined sequences of actual inscriptions, and then the epistemic appeal of numerals begins to fade.[53]

But truth is the crux. Formalists often take truth in mathematics to be not correspondence to fact but provability. Proof looks to be a linguistic activity and the favored way mathematicians settle mathematical questions. Of course, if proof is deduction from axioms, there will arise a question about how axioms are settled,

---

[49] Alfred Jules Ayer, *Language, Truth and Logic*, 2nd ed. (New York: Dover, 1946), chap. IV.

[50] The phrase comes from David Hume's abstract of *A Treatise of Human Nature*, ed. L. A. Selby-Bigge, 2nd ed., ed. P. H. Nidditch (Oxford: Oxford University Press, 1978), p. 662, as John L. Mackie made plain in *The Cement of the Universe: A Study of Causation* (Oxford: Oxford University Press, 1974).

[51] Thus Quine sees set theory as unifying rather than founding mathematics.

[52] Gottlob Frege, "Frege Against the Formalists," in *Translations from the Philosophical Writings of Gottlob Frege*, trans. Peter Geach and Max Black (Oxford: Blackwell, 1960), p. 222.

[53] See W. V. Quine, "Ontological Relativity," in *Ontological Relativity and Other Essays* (New York: Columbia University Press, 1969), pp. 41-42.

but let us bracket that question. For there is a juicier confrontation between the formalist conception of mathematical truth as provability and Gödel's Incompleteness Theorem (1931). Let us sketch what the theorem says informally. Consider any system of proof with three properties. First, it should be consistent. (Actually, Gödel used a stronger property in 1931, but in 1936 Rosser showed that consistency suffices.) Given that contradictions are false, any system in which proof suffices for truth should be consistent. Second, it should encode calculations on natural numbers, like addition and multiplication, for which there are algorithms. A due respect for the conventional mathematical wisdom urges encoding, for example, $7 + 5 = 12$. Third, there should be an algorithm for whether a patch of discourse in the language of the system is or is not a proof. Church argues that the point of proof is to settle mathematical questions decisively and it cannot do so unless there is such an algorithm.[54] Then, for any such system we can write down a sentence in the language of the system such that neither it nor its negation is provable in the system. Since, by the law of the excluded middle, one of these is true, there is an unprovable truth of the system.

Philosophy is probably supple enough for any of its interesting theses to dodge knock-down refutation by a precise scientific result. Gödel's first incompleteness theorem may knock down formalism, but perhaps the formalist can get up again. One maneuver might be to stretch the requirement for an algorithm for what counts as a proof by allowing some infinitely large proofs. If we allow inferences of a universally quantified sentence from all its instances, then any truth with enough instances will be provable but proofs will in general be infinitely large and our epistemic access to them will be as mathematical as our grasp of other infinities. A palliative here might be to allow ourselves only infinite patches of proofs that can be grasped by finite algorithms or acceptable generalizations of such. Jon Barwise wrote an excellent exposition of some ways of doing this in *Admissible Sets and Structures*,[55] but as yet there is no received systematic philosophical critique of such expanded proof in print. Going second order is another stratagem. In a first-order system one quantifies only into subject position or other positions in a sentence occupied by a singular term, while in a second-order system one may also quantify into predicate position. Dedekind wrote down a single second-order sentence from which every truth of elementary number theory follows, so a formalist might try saying that being a truth of elementary number theory is following from Dedekind's sentence. The rub is in the logic. For first order systems, Gödel (and others) showed that we can write down rules of inference such that for any premisses, anything that follows from those premisses may actually be deduced from them by the rules, and conversely. In a first-order system any sentence that follows from a set of premisses can be deduced from them; this is what the completeness of first-order logic says. But completeness fails second-order. So even

---

[54] Alonzo Church, *Introduction to Mathematical Logic*, vol. I (Princeton: Princeton University Press, 1956), pp. 50 ff.

[55] Jon Barwise, *Admissible Sets and Structures* (Berlin, Heidelberg, New York: Springer-Verlag, 1975).

though a truth of elementary number theory follows from Dedekind's sentence, showing that it does is just as much a mathematical problem as any other. We are not guaranteed proofs to which we have epistemic access that would obviate platonism.

Gödel himself suggested another maneuver.[56] His theorem is about formalized deductive theories. Maybe his theorem really shows that genuine proof in medias mathematical res eludes thoroughgoing formalization. Consider our favorite such formal system S and add to the language of S a unary sentence operator D (with grammar like negation) such that for any sentence p in the language of S, Dp is intended to mean that it is provable (in the full-blooded mathematical way) that p. We should have enough confidence in S to add to it a rule allowing us to infer from any theorem p of S to Dp, and we should have enough confidence in proof to add to S an axiom saying that if Dp, then p. If we write the dash for negation and the ampersand for conjunction, it is then not difficult to show that

$$D\neg D(p \& \neg p)$$

is a theorem of S. This says that it is absolutely provable that a contradiction is not absolutely provable, that is, absolute provability proves its own consistency. But Gödel's second incompleteness theorem shows that no system of proof satisfying the three conditions stated above proves its own consistency. From this one might infer that the absolute provability the formalist wants is not proof in a formalized system. On the other hand one might also wonder just what such absolute provability looks like, and how to be sure one has recognized it correctly. Such uncertainty may drain some of the epistemic appeal from formalism. Gödel's second incompleteness theorem also looks to impede Hilbert's program for proving the consistency of reformed mathematical systems, since especially for strong systems like set theory, the second incompleteness theorem seems to say that proving the consistency of the system requires methods stronger than those of the system, thus depriving the consistency proof of its epistemic purpose. For now, formalism is still down, though for all we know it will rise again. The history of philosophy is not of solving problems but of extending the dialectic they generate.

Gödel's first incompleteness theorem shows that mathematical truth outstrips provability. But by how much? The third condition on systems to which it applies requires that the relation that holds when and only when a stretch of discourse is a proof of a statement be decidable, that is, that there is an algorithm for whether some discourse is or is not a proof of a sentence. The property of theoremhood arises from this relation by existential quantification since a sentence is a theorem if and only if there is a proof of it. Suppose the system S in question is a typical first-order formalization of elementary number theory, and assume S consistent. We can show S is not decidable, that is, there is no algorithm for theoremhood in S. So provability is one quantifier away from decidability. Gödel proved incompleteness

---

[56] Kurt Gödel, "An Interpretation of the Intuitionistic Propositional Calculus," in *Collected Works*, vol. I., ed. Solomon Federman, John Dawson, Stephen Kleene, Gregory Moore, Robert Solovay, and Jean van Heijenoort (New York: Oxford University Press, 1986), pp. 301-03.

by arithmetizing (or gödelizing) syntax. This involves a function g with three properties. First, g maps expressions in the language of S (like sentences and proofs) one-to-one into the natural numbers. Next, g is computable, that it, there is an algorithm that given an expression e, yields the number g assigns to e. Last, g is effective backwards. This requires two algorithms, one that given a natural number n settles whether g assigns an expression to n, and a second that, if g assigns an expression to n, finds it. There are infinitely many different ways to gödelize theories like S. The number g assigns e is called its gödel number. Since the set of theorems of S is not decidable, neither is the set of gödel numbers of theorems. So an undecidable set of numbers is one quantifier from decidability.

Let A and B be sets of natural numbers. Turing showed how to reconstrue mathematically the subjunctive that if there were an algorithm for membership in A, then there would be one for membership in B. In that case we say A is (Turing) reducible to B. If A is reducible to B and B is reducible to A, then their membership problems are of equal difficulty, and they are (Turing) equivalent. This is an equivalence relation, and it partitions the sets of natural numbers into the degrees of (Turing) unsolvabilty. One degree is higher than another if a problem in the second is reducible to one in the first but not vice versa. Let d_0 be the degree of gödel numbers of theorems of S. We can show that there is a sequence $d_0, d_1 \ldots$ of degrees such that for all n, d_{n+1} is higher than d_n and there is in d_n a relation among natural numbers that is the extension of a formula from the language of S in the usual model of S and has n quantifiers in its prenex. That is, each additional quantifier we allow increases the degree of unsolvability of relations we can express with formulae in the language of S. We can also show that the set of gödel numbers of sentences of S true in the usual model is undecidable. Let d be its degree. Then d is higher than all of $d_0, d_1 \ldots$. So truth in number theory is infinitely many degrees, or quantifiers, higher than provability in that theory. Truth outstrips provability by a good bit.[57]

A first-order deductive theory is axiomatic if there is an algorithm for whether a formula in the language of the theory is or is not an axiom of it. At least prima facie, only the axiomatic theories are of epistemic interest, for only there are we sure of algorithms for whether a patch of discourse is or is not a proof of a sentence, and that, as Church argued, is necessary for proof to satisfy its raison d'être of settling questions decisively. By 1944, many axiomatic theories had been shown undecidable, and Post noted that any two of them were of the same degree of unsolvability (the one we called d_0). He asked whether any two undecidable axiomatic theories are of the same degree.[58] This is called Post's Problem, and it gave the part of logic called recursion theory a life of its own. In the mid-1950s, a young American, Friedberg, and a young Russian, Muchnik, discovered indepen-

---

[57] Compare Hartley Rogers Jr., *Theory of Recursive Functions and Effective Computability* (New York: McGraw-Hill, 1967), sec. 14.7.

[58] Emil Post, "Recursively Enumerable Sets of Positive Integers and their Decision Problems," reprinted in *The Undecidable*, ed. Martin Davis (Hewlitt, N.Y.: Raven Press, 1965), pp. 304-37. This address is another major contribution.

dently a method called the priority method that suffices to solve Post's Problem. There are undecidable axiomatic theories of different degrees of unsolvability. So the structure of these degrees becomes of at least mathematical interest. Sacks, for example, showed that they are dense, that is, there are sequences of such degrees such that between any two there is a third.[59]

But such degrees could be of more than mathematical interest only. A mineralogist can learn some things just by looking at rocks, but to figure out chemical formulae, for example, he will probably use more indirect methods; how does the rock interact with this or that acid, for example. When a cosmologist or astrophysicist is looking into the curvature of spacetime, he does not pick up a handful of spacetime and measure how bent it is. He looks instead to discriminating observable effects of such curvature. In psychology, introspection is not much trusted as a way to examine reason. So one strategy would be to look to discriminating observable products of reason. Since the Greeks invented or discovered reason, deductive theories have been among its most salient products. A critique of reason might seek structures occupied by such theories. A critique of pure reason might look to structures of all possible such theories. It would be fun if recursion theory could provide a denotation for Kant's title.

On the mathematical side, logicism developed out of the arithmetization of analysis, and analytic geometry was a crucial link in arithmetizing the curves of analysis. It was basic to analytic geometry that there be a one-to-one correspondence between the real numbers and the points of a line that preserves order (so less-to-greater goes to left-to-right). But even if the space around us is flat, how do we know such a correspondence exists? Could the lines in space actually need more numbers as coordinates?

Think back to learning what a derivative is. You are looking at the parabola that is the curve for the square function, x^2, and you want to know the slope of this curve above the abscissa x. Slope is rise over run, so you imagine increasing x by tiny amount dx. The new square is $x^2 + 2x dx + (dx)^2$, so the rise is $2x dx + (dx)^2$ when the run is dx. Over means division so the slope is $2x + dx$, and to get the slope at x itself you want this increment dx to be so small it can be ignored, which would put the slope at $2x$, the passing answer on your calculus test.

This argument feels dodgy. To divide by dx it should not be zero, but to ignore it in $2x + dx$ is to treat it as indistinguishable from zero. So dx is one of the notorious infinitesimals of the calculus invented or discovered by Newton and Leibniz independently. Berkeley made fun of infinitesimals in *The Analyst* as ghosts of departed quantities. For anything we have said so far, infinitesimals do seem like an embarrassment, and yet the calculus is crucial in making Newton's physics the first real natural science. Part of what people like Cauchy and Weierstrass did during the arithmetization of analysis was to replace infinitesimals with the ε - δ methods notorious in elementary calculus classes. Then a number a is the derivative of a function f at an argument x if for any positive ε however small there is a

---

[59]See Robert I. Soare, *Recursively Enumerable Sets and Degrees* (New York: Springer-Verlag, 1987).

positive δ small enough that for arguments y within distance δ of x, $f(y)$ is within distance ε of $f(x)$.

So we can avoid infinitesimals. But should we? In the 1960s Abraham Robinson discovered that we need not.[60] Let M be a structure, or model, whose domain is the set R of the (familiar) real numbers. The distinguished individuals of M are all the members of R, the relations, all those of R, and mutatis mutandis for functions. A language of which M is an interpretation is pretty big, but it can still be first order. One unary predicate $N(x)$ of such a language would have as extension the set of all natural numbers, and a binary predicate $x = y$ would have the identity relation as extension. Let T' be the set of all sentences of this language true in M. Add to the language a single new constant a, and form T by adding to T all of

$$N(a), a \neq \underline{0}, a \neq \underline{1}, \ldots$$

where $\underline{0}, \underline{1}, \ldots$ are constants of the language denoting the natural numbers in R. For any finite subset of T', let n be the largest natural number such that the sentence $a \neq \underline{n}$ is a member of this set. Then letting a denote $n+1$ makes all of this subset true in M. So by the compactness of first-order languages, T' has a model M' whose domain R' is the same size as R. Let b be the denotation of a in R'. Then b has to obey all the laws of natural numbers in T, so b cannot be less than zero, nor can it lie between n and $n+1$, nor can it be any of $0, 1, 2, \ldots$. But it must be comparable in size to all of $0, 1, 2, \ldots$, so it can only be greater than all of them.

Nor is it alone out there since each natural number other than zero has a predecessor, and each natural number has a successor. But none of $b - n$ can be a natural k since then $b = k + n$, contrary to a sentence in T' true in M'. So b is surrounded by a clump $\ldots, b-2, b-1, b, b+1, b+2, \ldots$ all of whose members lie above $0, 1, 2, \ldots$. Nor is there a least clump above $0, 1, \ldots$. For b is odd or even. Suppose it is even; a similar argument works if it is odd. Then $b = 2k$ for some k. As before if $k = b - n$ for some n, $k + n + b = 2k$, so $k = n$ and then $b = 2n$, contrary to a sentence in T' true in M'; nor is k among $0, 1, \ldots$. Moreover, given a clump around c and another around d, assume $c + d$ is even, and the clump around $c + d/2$ lies wholly between the two given clumps. So the natural numbers of M' start with a copy of the usual, or standard, natural numbers. These are followed by the clumps, any one of which looks like the integers (negative, zero, positive) in usual order, and the clumps are ordered like the rationals or reals (a dense linear order without top or bottom). The standard negative numbers look like the standard positive integers in reverse order, so the integers of M' look like a dense linear order without endpoints, but of clumps.

Since b is not zero, it has a reciprocal $1/b$, and since b is greater than all of $1, 2, 3, \ldots$, its reciprocal $1/b$ is less than all of $1, 1/2, 1/3, \ldots$. But since b is

---

[60] See Robert Goldblatt, *Lectures on the Hyperreals* (New York: Springer-Verlag, 1998). His exposition is more in the mathematical mainstream than ours, and both differ from Abraham Robinson, *Non-standard Analysis* (Princeton: Princeton University Press, 1996).

positive, so is $1/b$, so it is greater than 0. Hence $1/b$ is greater than 0 but less than all of $1, 1/2, 1/3, \ldots$. So $1/b$ is infinitesimal. We found a ghost of departed fractions.

In secondary school most of us thought of real numbers in terms of decimals. Let a digit be one of the numbers 0, 1, 2, 3, 4, 5, 6, 7, 8, and 9. Picture the set I of integers shrinking in size to the left and increasing to the right. A sequence of digits is a function f whose domain is I, whose values are digits, and such that for some integer n, $f(m) = 0$ for all $m < n$, and such that for no integer n is $f(m) = 9$ for all $m > n$ (this last so we do not have two decimals, like .250... and .249..., for any real, like $\frac{1}{4}$). Then for any real r there is a unique sequence f of digits such that r is the sum of $f(n)/10^n$ for all integers n. This can be formalized as a sentence of T true in M.[61] So it is also true in M', but in M' there are more integers, and sequences of digits are longer. The members of R' are called the hyperreals. Let r be a hyperreal and let f be its sequence of digits. If $f(n)$ is positive for some non-standard negative integer n, then r is infinite. If for all negative n and all standard non-negative n, $f(n)$ is zero, then r is infinitesimal. For a standard real r in R, let $\underline{r}$ be its name in the language M interprets, and let r^* be the hyperreal in R' denoted by $\underline{r}$ in M'. Then we think of r as the standard part of r^*. Two hyperreals r and s are close if their difference is infinitesimal. Being close is an equivalence relation, so it partitions the hyperreals into equivalence classes. The equivalence class of a given hyperreal is called its halo; it is a set of hyperreals close to the given hyperreal.

In familiar expositions of the calculus we say that a standard real a is the limit of a sequence a_n of standard reals if for any positive ε however small there is a k such that for n greater than or equal to k, the distance between a_n and a is less than ε. This is equivalent to saying that for infinite n, a_n is within the halo of a, or a_n is close to a. (Here we slip back and forth between a and a^*.) To say that a is the derivative of f at x is equivalent to saying that when y and x are close but differ, then the rise from $f(x)$ to $f(y)$ divided by the run from x to y is in the halo of a. When hyperreals are close but differ, they have the same standard part, and the standard part of their difference is zero, but their difference is not zero, and so is an acceptable divisor. That is how the hyperreals avoid the embarrassment of older arguments with infinitesimals.

For any standard reals r and s such that r is less than s, the halo of r^* is bounded above by s^* but has no least upper bound among the hyperreals. So no one-to-one correspondence between the standard reals and the hyperreals preserves order. Hence, the standard reals and the hyperreals cannot both be cartesian coordinates for the points on a line. What would space be like if the hyperreals were the right coordinates?

Pick a standard measure. This could be a meter or a yard stick, but it could also be an angstrom, or a lightyear stick. If we lay off our stick a first time to the

---

[61] In the language of M we can name the unique binary function that assigns to each real r and each integer n the digit $f(n)$, as we called it, in the decimal for r. So we need not say that for each r there is such a sequence f of digits, which would be second order.

left from here, then a second, and so on through all and only the standard natural numbers, then do the same to the right, in front, in back, up and down, then we will have laid off axes that suffice to locate any point in space as we supposed it back in school solid geometry before we heard of curvature or hyperreals. But if the hyperreals are the right coordinates, we will only have measured out a clump of hyperreals along each axis, and this blob will hardly be all of space.

So if the hyperreals are the right coordinates, space partitions into infinitely large blobs each with three clumps of hyperreals as its coordinate axes. A gold nugget that exactly fills out one such blob would use up all of our familiar space in school solid geometry. But if the hyperreals are the right coordinates, there may for all we know be such a nugget some way off (in fact an infinite number of miles) to the west. Could we tell? On Newton's theory of gravity, bodies attract one another instantaneously, so if two bodies popped into being simultaneously and far apart, there would be no lag time before each attracted the other gravitationally. (If gravity acted by sending a pull to the attracted body, then on Newton's theory the pull would have to be in more than one place at once, so if nothing can be in more than one place at a time, Newton's gravity has to act at a distance, that is, without transmission.) On Newton's theory, then, we would have been pulled hard to the west long ago by that nugget in a super gold rush. But on Einstein's theory, gravitational attraction travels at the speed of light, which is finite. Let c be the speed of light in miles per hour. If that nugget lies d miles to the west, but has existed for fewer than d/c hours, then the nugget's gravitational pull has not yet had time to seize us. Since d is infinite and the age of the universe is finite, we are safe for a while.

Infinitesimal, and infinite hyperreals, are probably too small in the first case and too big in the second for spatiotemporal and physical phenomena of those sizes to be obvious to middle-sized wet goods like us. But it would be too parochial and positivist to infer from that alone that the standard reals are the right coordinates for space. Besides, ingenious experiments and observations may detect what is not obvious. Whether the standard reals or the hyperreals (or some other numbers) are the right coordinates for space is settled by the nature of space. Space, not people, picks which numbers work as its coordinates, and space is a matter of fact, not a relation of ideas. It is nether analytic nor known *a priori* how to arithmetize space.

Both the standard reals and the hyperreals exist, or at least a clear-headed Platonist should be happy to grant that both exist. A mathematician should be free to study either, as he is free to study complex numbers or quaternions.[62] But consider how our study of the standard reals emerged historically from our study of geometry. One wonders whether settling whether the standard reals or the

---

[62] There is an interesting sense in which the reals, the complex numbers and the quaternions exhaust an important subject. See I. N. Herstein, *Topics in Algebra* (New York: Blaisdell Publishing Company, 1964), chap. 7, sec. 3. But the reals are applied all over the place, and the complex numbers are applied in many places, while the quaternions are pretty much just a curiosity and comparatively neglected.

hyperreals are the right coordinates for space might not channel the development of mathematical analysis, and in at least that sense settle which are the real reals.

The topics on which we have touched by no means exhaust the interactions between philosophy and mathematics. The Löwenheim-Skolem Theorem, for example, has occasioned philosophy, and intuitionism, predicativism, and finitism are philosophical positions with mathematical aspects. But the samples on which we have touched unify more or less under the issues of analyticity, *a priori* knowledge, and foundations, and they illustrate what can come of dalliance between mathematics and philosophy.

ACKNOWLEDGEMENTS

I am grateful to John Baldwin, Mihai Ganea, Jon Jarrett, Mitzi Lee, Constance Meinwald, and Matthew Moore for advice. I am grateful to Faith Hart and Charlotte Jackson for help in preparing the text. I am grateful to the philosophy of mathematics discussion group at the University of Illinois at Chicago and to the audience at the Universidad Nacional Mayor de San Marcos in Lima, Peru, during May 2007 for helpful discussion of earlier drafts of this essay.

REALISM AND ANTI-REALISM IN MATHEMATICS

Mark Balaguer

The purpose of this essay is (a) to survey and critically assess the various metaphysical views — i.e., the various versions of realism and anti-realism — that people have held (or that one might hold) about mathematics; and (b) to argue for a particular view of the metaphysics of mathematics. Section 1 will provide a survey of the various versions of realism and anti-realism. In section 2, I will critically assess the various views, coming to the conclusion that there is exactly one version of realism that survives all objections (namely, a view that I have elsewhere called *full-blooded platonism*, or for short, FBP) and that there is exactly one version of anti-realism that survives all objections (namely, *fictionalism*). The arguments of section 2 will also motivate the thesis that we do not have any good reason for favoring either of these views (i.e., fictionalism or FBP) over the other and, hence, that we do not have any good reason for believing or disbelieving in abstract (i.e., non-spatiotemporal) mathematical objects; I will call this the weak epistemic conclusion. Finally, in section 3, I will argue for two further claims, namely, (i) that we could *never* have any good reason for favoring either fictionalism or FBP over the other and, hence, could never have any good reason for believing or disbelieving in abstract mathematical objects; and (ii) that there is no fact of the matter as to whether fictionalism or FBP is correct and, more generally, no fact of the matter as to whether there exist any such things as abstract objects; I will call these two theses the strong epistemic conclusion and the metaphysical conclusion, respectively.

(I just said that in section 2, I will argue that FBP and fictionalism survive all objections; but if I'm right that there is no fact of the matter as to whether FBP or fictionalism is correct, then it can't be that these two views survive *all* objections, for surely my no-fact-of-the-matter argument constitutes an objection of some sort to both FBP and fictionalism. This, I think, is correct, but for the sake of simplicity, I will ignore this point until section 3. During sections 1 and 2, I will defend FBP and fictionalism against the various traditional objections to realism and anti-realism — e.g., the Benacerrafian objections to platonism and the Quine-Putnam objection to anti-realism — and in doing this, I will write as if I think FBP and fictionalism are completely defensible views; but my section-3 argument for the claim that there is no fact of the matter as to which of these two views is correct does undermine the two views.)

Large portions of this paper are reprinted, with a few editorial changes, from my book, *Platonism and Anti-Platonism in Mathematics* (Oxford University Press,

1998)[1] — though I should say that there are also several new sections here. Now, of course, because of space restrictions, many of the points and arguments in the book have not been included here, but the overall plan of this essay mirrors that of the book. One important difference, however, is this: while the book is dedicated more to developing my own views and arguments than to surveying and critiquing the views of others, because this is a survey essay, the reverse is true here. Thus, in general, the sections of the book that develop my own views have been pared down far more than the sections that survey and critique the views of others. Indeed, in connection with my own views, all I really do in this essay is briefly sketch the main ideas and arguments and then refer the reader to the sections of the book that fill these arguments in. Indeed, I refer the reader to my book so many times here that, I fear, it might get annoying after a while; but given the space restrictions for the present essay, I couldn't see any other way to preserve the overall structure of the book — i.e., to preserve the defenses of FBP and fictionalism and the argument for the thesis that there is no fact of the matter as to which of these two views is correct — than to omit many of the points made in the book and simply refer the reader to the relevant passages.

1 A SURVEY OF POSITIONS

Mathematical realism (as I will use the term here) is the view that our mathematical theories are true descriptions of some real part of the world. *Mathematical anti-realism*, on the other hand, is just the view that mathematical realism is false; there are lots of different versions of anti-realism (e.g., formalism, if-thenism, and fictionalism) but what they all have in common is the view that mathematics does not have an ontology (i.e., that there are no objects that our mathematical theories are about) and, hence, that these theories do not provide true descriptions of some part of the world. In this section, I will provide a survey of the various versions of realism and anti-realism that have been endorsed, or that one might endorse, about mathematics. Section 1.1 will cover the various versions of realism and section 1.2 will cover the various versions of anti-realism.

1.1 *Mathematical Realism*

Within the realist camp, we can distinguish *mathematical platonism* (the view that there exist abstract mathematical objects, i.e., non-spatiotemporal mathematical objects, and that our mathematical theories provide true descriptions of such objects) from *anti-platonistic realism* (the view that our mathematical theories are true descriptions of concrete, i.e., spatiotemporal, objects). Furthermore, within anti-platonistic realism, we can distinguish between *psychologism* (the view that our mathematical theories are true descriptions of mental objects) and *mathematical physicalism* (the view that our mathematical theories are true descriptions

---

[1] I would like to thank Oxford University Press for allowing the material to be reprinted.

of some non-mental part of physical reality). Thus, the three kinds of realism are platonism, psychologism, and physicalism. (One might think there is a fourth realistic view here, namely, Meinongianism. I will discuss this view below, but for now, let me just say that I do not think there is fourth version of realism here; I think that Meinongianism either isn't a realistic view or else is equivalent to platonism.)

I should note here that philosophers of mathematics sometimes use the term 'realism' interchangeably with 'platonism'. This, I think, is not because they deny that the logical space of possible views includes anti-platonistic realism, but rather, because it is widely thought that platonism is the only really tenable version of realism. I think that this is more or less correct, but since I am trying to provide a comprehensive survey, I will cover anti-platonistic realism as well as platonistic realism. Nontheless, since I think the latter is much more important, I will have far more to say about it. Before I go into platonism, however, I will say a few words about the two different kinds of anti-platonistic realism — i.e., physicalism and psychologism.

1.1.1 Anti-platonistic realism (physicalism and psychologism)

The main advocate of mathematical physicalism is John Stuart Mill [1843, book II, chapters 5 and 6]. The idea here is that mathematics is about ordinary physical objects and, hence, that it is an empirical science, or a natural science, albeit a very general one. Thus, just as botany gives us laws about plants, mathematics, according to Mill's view, gives us laws about all objects. For instance, the sentence '2 + 1 = 3' tells us that whenever we add one object to a pile of two objects, we will end up with three objects. It does not tell us anything about any abstract objects, like the numbers 1, 2, and 3, because, on this view, there are simply no such things as abstract objects. (There is something a bit arbitrary and potentially confusing about calling this view 'physicalism', because Penelope Maddy [1990b] has used the term 'physicalistic platonism' to denote her view that set theory is about sets that exist in spacetime — e.g., sets of biscuits and eggs. We will see below that her view is different from Mill's and, indeed, not entirely physicalistic — it is platonistic in at least some sense of the term. One might also call Mill's view 'empiricism', but that would be misleading too, because one can combine empiricism with non-physicalistic views (e.g., Resnik and Quine have endorsed empiricist platonist views[2]); moreover, the view I am calling 'physicalism' here is an *ontological* view, and in general, empiricism is an epistemological view. Finally, one might just call the view here 'Millianism'; I would have no objection to that, but it is not as descriptive as 'physicalism'.)

Recently, Philip Kitcher [1984] has advocated a view that is similar in certain ways to Millian physicalism. According to Kitcher, our mathematical theories are about the activities of an ideal agent; for instance, in the case of arithmetic, the activities involve the ideal agent pushing blocks around, i.e., making piles of

---
[2]The view is developed in detail by Resnik [1997], but see also Quine (1951, section 6).

blocks, adding blocks to piles, taking them away, and so on. I will argue in section 2.2.3, however, that Kitcher's view is actually better thought of as a version of anti-realism.

Let's move on now to the second version of anti-platonistic realism — that is, to psychologism. This is the view that mathematics is about mental objects, in particular, ideas in our heads; thus, for instance, on this view, '3 is prime' is about a certain mental object, namely, the idea of 3.

One might want to distinguish two different versions of psychologism; we can call these views *actualist psychologism* and *possibilist psychologism* and define them in the following way:

> *Actualist Psychologism* is the view that mathematical statements are about, and true of, actual mental objects (or mental constructions) in actual human heads.[3] Thus, for instance, the sentence '3 is prime' says that the mentally constructed object 3 has the property of primeness.
>
> *Possibilist Psychologism* is the view that mathematical statements are about what mental objects it's possible to construct. E.g., the sentence 'There is a prime number between 10,000,000 and (10,000,000! + 2)' says that it's possible to construct such a number, even if no one has ever constructed one.

But (according to the usage that I'm employing here) possibilist psychologism is not a genuinely psychologistic view at all, because it doesn't involve the adoption of a psychologistic *ontology* for mathematics. It seems to me that possibilist psychologism collapses into either a platonistic view (i.e., a view that takes mathematics to be about abstract objects) or an anti-realist view (i.e., a view that takes mathematics not to be about anything — i.e., a view like deductivism, formalism, or fictionalism that takes mathematics not to have an ontology). If one takes possible objects (in particular, possible mental constructions) to be real things, then presumably (unless one is a Lewisian about the metaphysical nature of possibilia) one is going to take them to be abstract objects of some sort, and hence, one's possibilist psychologism is going to be just a semantically weird version of platonism. (On this view, mathematics is about abstract objects, it is objective, and so on; the only difference between this view and standard platonism is that it involves an odd, non-face-value view of *which* abstract objects the sentences of mathematics are about.) If, on the other hand, one rejects the existence of possible objects, then one will wind up with a version of possibilist psychologism that is essentially anti-realistic: on this view, mathematics will not have an ontology. Thus, in this essay, I am going to use 'psychologism' to denote actualist psychologism.

By the way, one might claim that actualist psychologism is better thought of as a version of anti-realism than a version of realism; for one might think that

---

[3] Obviously, there's a question here about *whose* heads we're talking about. Any human head? Any decently trained human head? Advocates of psychologism need to address this issue, but I won't pursue this here.

mathematical realism is most naturally defined as the view that our mathematical theories provide true descriptions of some part of the world *that exists independently of us human beings*. I don't think anything important hangs on whether we take psychologism to be a version of realism or anti-realism, but for whatever it's worth, I find it more natural to think of psychologism as a version of realism, for the simple reason that (in agreement with other realist views and disagreement with anti-realist views) it provides an ontology for mathematics — i.e., it says that mathematics is *about objects*, albeit mental objects. Thus, I am going to stick with the definition of mathematical realism that makes actualist psychologism come out as a version of realism. However, we will see below (section 2.2.3) that it is indeed true that actualist psychologism bears certain important similarities to certain versions of anti-realism.

Psychologistic views seem to have been somewhat popular around the end of the nineteenth century, but very few people have advocated such views since then, largely, I think, because of the criticisms that Frege leveled against the psychologistic views that were around back then — e.g., the views of Erdmann and the early Husserl.[4] Probably the most famous psychologistic views are those of the intuitionists, most notably Brouwer and Heyting. Heyting for instance said, "We do not attribute an existence independent of our thought ... to ... mathematical objects," and Brouwer made several similar remarks.[5] However, I do not think we should interpret either of these philosophers as straightforward advocates of actualist psychologism. I think the best interpretation of their view takes it to be an odd sort of hybrid of an actualist psychologistic view of mathematical assertions and a possibilist psychologistic view of mathematical negations. I hope to argue this point in more detail in the future, but the basic idea is as follows. Brouwer-Heyting intuitionism is generated by endorsing the following two principles:

(A) A mathematical assertion of the form 'Fa' means 'We are actually in possession of a proof (or an effective procedure for producing a proof) that the mentally constructed mathematical object a is F'.

(B) A mathematical sentence of the form '$\sim P$' means "There is a derivation of a contradiction from 'P' ".

Principle (A) commits them pretty straightforwardly to an actualist psychologistic view of assertions. But (B) seems to commit them to a possibilist psychologistic view of negations, for on this view, in order to assert '$\sim Fa$', we need something that entails that we *couldn't* construct the object a such that it was F (not merely that we *haven't* performed such a construction) — namely, a derivation of a contradiction from 'Fa'. I think this view is hopelessly confused, but I also think

---

[4]See, for instance, Husserl [1891] and Frege [1894] and [1893–1903, 12–15]. Husserl's and Erdmann's works have not been translated into English, and so I am not entirely certain that either explicitly accepted what I am calling psychologism here. Resnik [1980, chapter 1] makes a similar remark; all he commits to is that Erdmann and Husserl — and also Locke [1689] — came close to endorsing psychologism.

[5]Heyting [1931, 53]; and see, e.g., Brouwer [1948, 90].

it is the most coherent view that is consistent with what Brouwer and Heyting actually say — though I cannot argue this point here. (By the way, none of this is relevant to Dummett's [1973] view; his version of intuitionism is not psychologistic at all.)[6,7]

1.1.2 Mathematical platonism

As I said above, platonism is the view that (a) there exist abstract mathematical objects — objects that are non-spatiotemporal and wholly non-physical and non-mental — and (b) our mathematical theories are true descriptions of such objects. This view has been endorsed by Plato, Frege, Gödel, and in some of his writings, Quine.[8] (One might think that it's not entirely clear what thesis (a) — that there exist abstract objects — really amounts to. I think this is correct, and in section 3.2, I will argue that because of this, there is no fact of the matter as to whether platonism or anti-platonism is true. For now, though, I would like to assume that the platonist thesis is entirely clear.)

There are a couple of distinctions that need to be drawn between different kinds of platonism. The most important distinction, in my view, is between the traditional platonist view endorsed by Plato, Frege, and Gödel (we might call this *sparse platonism*, or *non-plenitudinous platonism*) and a view that I have developed elsewhere [1992; 1995; 1998] and called *plenitudinous platonism*, or *full-blooded platonism*, or for short, FBP. FBP differs from traditional platonism in several ways, but all of the differences arise out of one bottom-level difference concerning the question of *how many* mathematical objects there are. FBP can be expressed very intuitively, but perhaps a bit sloppily, as the view that the mathematical realm is plenitudinous; in other words, the idea here is that all the mathematical objects that (logically possibly) *could* exist actually *do* exist, i.e., that there actually exist mathematical objects of all logically possible kinds. (More needs to be said about what exactly is meant by 'logically possible'; I address this in my [1998, chapter 3, section 5].) In my book, I said a bit more about how to define FBP, but Greg Restall [2003] has recently argued that still more work is

---

[6]Intuitionism itself (which can be defined in terms of principles (A) and (B) in the text) is not a psychologistic view. It is often assumed that it goes together naturally with psychologism, but in work currently in progress, I argue that intuitionism is independent of psychologism. More specifically, I argue that (i) intuitionists can just as plausibly endorse platonism or anti-realism as psychologism, and (ii) advocates of psychologism can (and indeed should) avoid intuitionism and hang onto classical logic. Intuitionism, then, isn't a view of the metaphysics of mathematics at all. It is a thesis about the semantics of mathematical discourse that is consistent with both realism and anti-realism. Now, my own view on this topic is that intuitionism is a wildly implausible view, but I will not pursue this here because it is not a version of realism or anti-realism. (And by the way, a similar point can be made about *logicism*: it is not a version of realism or anti-realism (it is consistent with both of these views) and so I will not discuss it here.)

[7]Recently, a couple of non-philosophers — namely, Hersh [1997] and Dehaene [1997] — have endorsed views that sound somewhat psychologistic. But I do not think these views should be interpreted as versions of the view that I'm calling psychologism (and I should note here that Hersh at least is careful to distance himself from this view).

[8]See, e.g., Plato's *Meno* and *Phaedo*; Frege [1893–1903]; Gödel [1964]; and Quine [1948; 1951].

required on this front; I will say more about this below, in section 2.1.3.

I should note here that the non-plenitudinousness of traditional platonism is, I think, more or less unreflective. That is, the question of whether the mathematical realm is plenitudinous was almost completely ignored in the literature until very recently; but despite this, the question is extremely important, for as I have argued — and I'll sketch the argument for this here (section 2.1) — platonists can defend their view if and only if they endorse FBP. That is, I have argued (and will argue here) that (a) FBP is a defensible view, and (b) non-plenitudinous versions of platonism are not defensible.

I don't mean to suggest, however, that I am the only philosopher who has ever defended a view like FBP. Zalta and Linsky [1995] have defended a similar view: they claim that "there are as many abstract objects of a certain sort as there possibly could be." But their conception of abstract objects is rather unorthodox, and for this reason, their view is quite different, in several respects, from FBP.[9] Moreover, they have not used FBP in the way that I have, arguing that platonists can solve the traditional problems with their view if and only if they endorse FBP. (I do not know of anyone else who has claimed that the mathematical realm is plenitudinous in the manner of FBP. In my book [1998, 7-8], I quote passages from Hilbert, Poincaré, and Resnik that bring the FBP-ist picture to mind, but I argue there that none of these philosophers really endorses FBP. Hilbert and Poincaré don't even endorse platonism, let alone FBP; Resnik does endorse (a structuralist version of) platonism, but it's unlikely that he would endorse an FBP-ist version of structuralistic platonism. It *may* be that Shapiro would endorse such a view, but he has never said this in print. In any event, whatever we end up saying about whether these philosophers endorse views like FBP, the main point is that they do not give FBP a prominent role, as I do. On my view, as we have seen, plenitudinousness is the key prong in the platonist view, and FBP is the only defensible version of platonism.)

A second divide in the platonist camp is between *object-platonism* and *structuralism*. I have presented platonism as the view that there exist abstract mathematical *objects* (and that our mathematical theories describe such objects). But this is not exactly correct. The real core of the view is the belief in the abstract, i.e., the belief that there is something real and objective that exists outside of spacetime and that our mathematical theories characterize. The claim that this abstract something is a collection of *objects* can be jettisoned without abandoning platonism. Thus, we can say that, strictly speaking, mathematical platonism is the view that our mathematical theories are descriptions of an abstract *mathematical realm*, i.e., a non-physical, non-mental, non-spatiotemporal aspect of reality.

Now, the most traditional version of platonism — the one defended by, e.g., Frege and Gödel — is a version of object-platonism. Object-platonism is the view that the mathematical realm is a system of abstract mathematical *objects*, such as numbers and sets, and that our mathematical theories, e.g., number theory and set theory, describe these objects. Thus, on this view, the sentence '3 is prime'

---
[9]See also Zalta [1983; 1988].

says that the abstract object that is the number 3 has the property of primeness. But there is a very popular alternative to object-platonism, *viz.*, structuralism. According to this view, our mathematical theories are not descriptions of particular systems of abstract objects; they are descriptions of abstract *structures*, where a structure is something like a *pattern*, or an "objectless template" — i.e., a system of *positions* that can be "filled" by any system of objects that exhibit the given structure. One of the central motivations for structuralism is that the "internal properties" of mathematical objects seem to be mathematically unimportant. What is mathematically important is structure — i.e., the relations that hold between mathematical objects. To take the example of arithmetic, the claim is that any sequence of objects with the right structure (i.e., any ω-*sequence*) would suit the needs of arithmetic as well as any other. What structuralists maintain is that arithmetic is concerned not with some particular one of these ω-sequences, but rather, with the structure or pattern that they all have in common. Thus, according to structuralists, there is no *object* that is the number 3; there is only the fourth position in the natural-number pattern.

Some people read Dedekind [1888] as having held a view of this general sort, though I think that this is a somewhat controversial interpretation. The first person to explicitly endorse the structuralist thesis as I have presented it here — i.e., the thesis that mathematics is about structure and that different systems of objects can "play the role" of, e.g., the natural numbers — was Benacerraf [1965]. But Benacerraf's version of the view was anti-platonistic; he sketched the view very quickly, but later, Hellman [1989] developed an anti-platonistic structuralism in detail. The main pioneers of platonistic structuralism — the view that holds that mathematics is about structures and positions in structures and that these structures and positions are real, objective, and abstract — are Resnik [1981; 1997] and Shapiro [1989; 1997], although Steiner [1975] was also an early advocate.

In my book, I argued that the dispute between object-platonists and structuralists is less important than structuralists think and, indeed, that platonists don't need to take a stand on the matter. Resnik and Shapiro think that by adopting structuralism, platonists improve their standing with respect to both of the great objections to platonism, i.e., the epistemological objection and the non-uniqueness objection, both of which will be discussed in section 2.1. But I have argued (and will sketch the argument here) that this is false. The first thing I have argued here is that structuralism doesn't do any work in connection with these problems after all (in connection with the epistemological problem, I argue this point in my [1998, chapter 2, section 6.5] and provide a brief sketch of the reasoning below, in section 2.1.1.4.3; and in connection with the non-uniqueness problem, I argue the point in my [1998, chapter 4, section 3] and provide a sketch of the reasoning below, in section 2.1.2.3). But the more important thing I've done is to provide FBP-ist solutions to these two problems that work for both structuralism and object-platonism [1998, chapters 3 and 4]; below (section 2.1), I will quickly sketch my account of how FBP-ists can solve the two problems; I will not take the space to argue that FBP is consistent with structuralism as well as with object-platonism, but the point is entirely obvious.[10]

The last paragraph suggests that there is no reason to favor structuralism over object-platonism. But the problem here is even deeper: it is not clear that structuralism is even *distinct* from object-platonism in an important way, for as I argue in my book (chapter 1, section 2.1), positions in structures — and, indeed, structures themselves — seem to be just special kinds of mathematical *objects*. Now, in light of this point, one might suggest that the structuralists' "objects-versus-positions" rhetoric is just a distraction and that structuralism should be defined in some other way. One suggestion along these lines, advanced by Charles Parsons,[11] is that structuralism should be defined as the view that mathematical objects have no internal properties, i.e., that there is no more to them than the relations that they bear to other mathematical objects. But (a) it seems that mathematical objects do have non-structural properties, e.g., being non-spatiotemporal and being non-red; and (b) the property of having only structural properties is *itself* a non-structural property (or so it would seem), and so the above definition of structuralism is simply incoherent. A second suggestion here is that structuralism should be defined as the view that the internal properties of mathematical objects are not mathematically *important*, i.e., that structure is what is important in mathematics. But whereas the last definition was too strong, this one is too weak. For as we'll see in section 2.1.2, traditional object-platonism is perfectly consistent with the idea that the internal properties of mathematical objects are not mathematically important; indeed, it seems to me that just about everyone who claims to be an object-platonist would *endorse* this idea. Therefore, this cannot be what separates structuralism from traditional object-platonism. Finally, structuralists might simply define their view as the thesis that mathematical objects are positions in structures that can be "filled" by other objects. But if I'm right that this thesis doesn't do any work in helping platonists solve the problems with their view, then it's not clear what the motivation for this thesis could be, or indeed, why it is philosophically important.[12]

I think it is often convenient for platonists to speak of mathematical theories as describing structures, and in what follows, I will sometimes speak this way. But as I see it, structures are mathematical objects, and what's more, they are made up of objects. We can think of the elements of mathematical structures as "positions" if we want to, but (a) they are still mathematical *objects*, and (b) as

---

[10] I have formulated FBP (and my solutions to the problems with platonism) in object-platonist terms, but it is obvious that this material could simply be reworded in structuralistic FBP-ist terms (or in a way that was neutral between structuralism and object-platonism).

[11] See the first sentence of Parsons [1990].

[12] Resnik has suggested to me that the difference between structuralists and object-platonists is that the latter often see facts of the matter where the former do not. One might put this in terms of property possession again; that is, one might say that according to structuralism, there are some cases where there is no fact of the matter as to whether some mathematical object a possesses some mathematical property P. But we will see below (sections 2.1.2–2.1.3) that object-platonists are not committed to all of the fact-of-the-matter claims (or property-possession claims) normally associated with their view. It will become clearer at that point, I think, that there is no important difference between structuralism and object-platonism.

we'll see below, there is no good *reason* for thinking of them as "positions".

1.2 Mathematical Anti-Realism

Anti-realism, recall, is the view that mathematics does not have an ontology, i.e., that our mathematical theories do not provide true descriptions of some part of the world. There are lots of different versions of anti-realism. One such view is *conventionalism*, which holds that mathematical sentences are analytically true. On this view, '$2 + 1 = 3$' is like 'All bachelors are unmarried': it is true solely in virtue of the meanings of the words appearing in it. Views of this sort have been endorsed by Ayer [1946, chapter IV], Hempel [1945], and Carnap [1934; 1952; 1956].

A second view here is *formalism*, which comes in a few different varieties. One version, known as *game formalism*, holds that mathematics is a game of symbol manipulation; on this view, '$2 + 1 = 3$' would be one of the "legal results" of the "game" specified by the axioms of PA (i.e., Peano Arithmetic). The only advocates of this view that I know of are those, e.g., Thomae, whom Frege criticized in his *Grundgesetze* (sections 88–131). A second version of formalism — *metamathematical formalism*, endorsed by Curry [1951] — holds that mathematics gives us truths about what holds in various formal systems; for instance, on this view, one truth of mathematics is that the sentence '$2 + 1 = 3$' is a theorem of the formal system PA. One might very well doubt, however, that metamathematical formalism is a genuinely anti-realistic view; for since this view says that mathematics is *about theorems and formal systems*, it seems to entail that mathematics has an ontology, in particular, one consisting of sentences. As a version of realism, however — that is, as the view that mathematics is about actually existing sentences — the view has nothing whatsoever to recommend it.[13] Finally, Hilbert sometimes seems to accept a version of formalism, but again, it's not clear that he really had an anti-realistic view of the metaphysics of mathematics (and if he did, it's not clear what the view was supposed to be). I think that Hilbert was by far the most brilliant of the formalists and that his views on the philosophy of mathematics were the most important, insightful, and original. But I also think that the metaphysical component of his view — i.e., where he stood on the question of realism — was probably the least interesting part of his view. His finitism and his earlier view that axiom systems provide definitions are far more important; I will touch on the axiom-systems-are-definitions thesis later on, but I will not discuss this

---

[13] One might endorse an anti-platonistic version of this view (maintaining that mathematics is about sentence *tokens*) or a platonistic version (maintaining that mathematics is about sentence *types*). But (a) the anti-platonistic version of this view is untenable, because there aren't *enough* tokens lying around the physical world to account for all of mathematical truth (indeed, to account even for finitistic mathematical truth). And (b) the platonistic version of this view has no advantage over traditional platonism, and it has a serious disadvantage, because it provides a non-standard, non-face-value semantics for mathematical discourse that flies in the face of actual mathematical practice (I will say more about this problem below, in section 2.2.2).

view (or Hilbert's finitism) in the present section, because neither of these views is a version of anti-realism, and neither entails anti-realism. As for the question of Hilbert's metaphysics, in the latter portion of his career he seemed to endorse the view that finitistic arithmetical claims can be taken to be about sequences of strokes — e.g., '2 + 1 = 3' can be taken as saying something to the effect that if we concatenate '||' with '|', we get '|||' — and that mathematical claims that go beyond finitary arithmetic can be treated instrumentally, along the lines of game formalism. So the later Hilbert was an anti-realist about infinitary mathematics, but I think he is best interpreted as a platonist about finitary arithmetic, because it is most natural to take him as saying that finitary arithmetic is about stroke *types*, which are abstract objects.[14,15]

Another version of anti-realism — a view that, I think, can be characterized as a descendent of formalism — is *deductivism*, or *if-thenism*. This view holds that mathematics gives us truths of the form 'if A then T' (or 'it is necessary that if A then T') where A is an axiom, or a conjunction of several axioms, and T is a theorem that is provable from these axioms. Thus, for instance, deductivists claim that '2 + 1 = 3' can be taken as shorthand for the sentence '(it is necessary that) if the axioms of arithmetic are true, then $2 + 1 = 3$'. Thus, on this view, mathematical sentences come out true, but they are not *about* anything. Putnam originally introduced this view, and Hellman later developed a structuralist version of it. But the early Hilbert also hinted at the view.[16]

Another anti-realistic view worth mentioning is Wittgenstein's (see, e.g., his [1956]). His view is related in certain ways to game formalism and conventionalism, but it is distinct from both. I do not want to try to give a quick formulation of this view, however, because I do not think it is possible to do this; to capture the central ideas behind Wittgenstein's philosophy of mathematics would take quite a bit more space. (I should point out here that Wittgenstein's view can be interpreted in a number of different ways, but I think it's safe to say that however we end up interpreting the view, it is going to be a version of anti-realism.)

Another version of anti-realism that I don't want to try to explain in full is due to Chihara [1990]. Chihara's project is to reinterpret all of mathematics, and it would take a bit of space to adequately describe how he does this, but the basic anti-realist idea is very simple: Chihara's goal is to replace sentences involving ontologically loaded existential quantification over mathematical objects (e.g., 'there is a set x such that...') with assertions about what open-sentence tokens it is possible to construct (e.g., 'it is possible to construct an open sentence

---

[14]See Hilbert [1925] for a formulation of the formalism/finitism that he endorsed later in his career. For his earlier view, including the idea that axioms are definitions, see his [1899] and his letters to Frege in [Frege, 1980].

[15]The idea that mathematics is about symbols — e.g., strokes — is a view that has been called *term formalism*. This view is deeply related to metamathematical formalism, and in particular, it runs into a problem that is exactly analogous to the problem with metamathematical formalism described above (note 13).

[16]See Putnam [1967a; 1967b], Hellman [1989], and Hilbert [1899] and his letters to Frege in [Frege, 1980].

x such that...'). Chihara thinks that (a) his reinterpreted version of mathematics does everything we need mathematics to do, and (b) his reinterpreted version of mathematics comes out true, even though it has no ontology (i.e., is not about some part of the world) because it merely makes claims about what is *possible*. In this respect, his view is similar to certain versions of deductivism; Hellman, for instance, holds that the axioms of our mathematical theories can be read as making claims about what is possible, while the theorems can be read as telling us what would follow if the axioms were true.

Another version of anti-realism — and I will argue in section 2.2 that this is the best version of anti-realism — is *fictionalism*. This view differs from other versions of anti-realistic anti-platonism in that it takes mathematical sentences and theories at *face value*, in the way that platonism does. Fictionalists agree with platonists that the sentence '3 is prime' is about the number 3^{17} — in particular, they think it says that this number has the property of primeness — and they also agree that if there is any such thing as 3, then it is an abstract object. But they disagree with platonists in that they do not think that there is any such thing as the number 3 and, hence, do not think that sentences like '3 is prime' are true. According to fictionalists, mathematical sentences and theories are fictions; they are comparable to sentences like 'Santa Claus lives at the North Pole.' This sentence is not true, because 'Santa Claus' is a vacuous term, that is, it fails to refer. Likewise, '3 is prime' is not true, because '3' is a vacuous term — because just as there is no such person as Santa Claus, so there is no such thing as the number 3. Fictionalism was first introduced by Hartry Field [1980; 1989]; as we'll see, he saw the view as being wedded to the thesis that empirical science can be nominalized, i.e., restated so that it does not contain any reference to, or quantification over, mathematical objects. But in my [1996a] and [1998], I defend a version of fictionalism that is divorced from the nominalization program, and similar versions of fictionalism have been endorsed by Rosen [2001] and Yablo [2002].

One obvious question that arises for fictionalists is this: "Given that '$2+1=3$' is false, what is the difference between this sentence and, say, '$2+1=4$'?" The difference, according to fictionalism, is analogous to the difference between 'Santa Claus lives at the North Pole' and 'Santa Claus lives in Tel Aviv'. In other words, the difference is that '$2+1=3$' is part of a certain well-known mathematical story, whereas '$2+1=4$' is not. We might express this idea by saying that while neither '$2+1=3$' nor '$2+1=4$' is true *simpliciter*, there is another truth predicate (or pseudo-truth predicate, as the case may be) — *viz.*, 'is true in the story of mathematics' — that applies to '$2+1=3$' but not to '$2+1=4$'. This seems to be the view that Field endorses, but there is a bit more that needs to be said on

---

[17] I am using 'about' here in a *thin* sense. I say more about this in my book (see, e.g., chapter 2, section 6.2), but for present purposes, all that matters is that in this sense of 'about', 'S is about b' does not entail that there is any such thing as b. For instance, we can say that the novel *Oliver Twist* is about an orphan named 'Oliver' without committing to the existence of such an orphan. Of course, one might also use 'about' in a *thicker* way; in this sense of the term, a story (or a belief state, or a sentence, or whatever) can be about an object only if the object exists and the author (or believer or speaker or whatever) is "connected" to it in some appropriate way.

this topic. In particular, it is important to realize that the above remarks do not lend any metaphysical or ontological distinction to sentences like '2 + 1 = 3'. For according to fictionalism, there are *alternative* mathematical "stories" consisting of sentences that are not part of standard mathematics. Thus, the real difference between sentences like '2 + 1 = 3' and sentences like '2 + 1 = 4' is that the former are part of *our* story of mathematics, whereas the latter are not. Now, of course, fictionalists will need to explain why we use, or "accept", this particular mathematical story, as opposed to some alternative story, but this is not hard to do. The reasons are that this story is pragmatically useful, that it's aesthetically pleasing, and most important, that it dovetails with our conception of the natural numbers.

On the version of fictionalism that I defend, sentences like '3 is prime' are simply false. But it should be noted that this is not essential to the view. What is essential to mathematical fictionalism is that (a) there are no such things as mathematical objects, and hence, (b) mathematical singular terms are vacuous. Whether this means that sentences like '3 is prime' are false, or that they lack truth value, or something else, depends upon our theory of vacuity. I will adopt the view that such sentences are false, but nothing important will turn on this.[18]

It is also important to note here that the comparison between mathematical and fictional discourse is actually not central to the fictionalistic view of mathematics. The fictionalist view that we're discussing here is a view about mathematics only; it includes theses like (a) and (b) in the preceding paragraph, but it doesn't say anything at all about fictional discourse. In short, mathematical fictionalism — or at any rate, the version of fictionalism that I have defended, and I think that Field would agree with me on this — is entirely neutral regarding the analysis of fictional discourse. My own view (though in the present context this doesn't really matter) is that there are important differences between mathematical sentences and sentences involving fictional names. Consider, e.g., the following two sentence tokens:

(1) Dickens's original token of some sentence of the form 'Oliver was F' from *Oliver Twist*;

(2) A young child's utterance of 'Santa Claus lives at the North Pole'.

Both of these tokens, it seems, are untrue. But it seems to me that they are very different from one another and from ordinary mathematical utterances (fictionalistically understood). (1) is a bit of pretense: Dickens knew it wasn't true when he uttered it; he was engaged in a kind of pretending, or literary art, or some such thing. (2), on the other hand, is just a straightforward expression of a false belief. Mathematical fictionalists needn't claim that mathematical utterances are analogous to either of these utterances: they needn't claim that when

---

[18] It should be noted here that fictionalists allow that *some* mathematical sentences are true, albeit vacuously so. For instance, they think that sentences like 'All natural numbers are integers' — or, for that matter, 'All natural numbers are zebras' — are vacuously true for the simple reason that there are no such things as numbers. But we needn't worry about this complication here.

we use mathematical singular terms, we're engaged in a bit of make-believe (along the lines of (1)) or that we're straightforwardly mistaken (along the lines of (2)). There are a number of different things fictionalists can say here; for instance, one line they could take is that there is a bit of imprecision in what might be called our communal intentions regarding sentences like '3 is prime', so that these sentences are somewhere between (1) and (2). More specifically, one might say that while sentences like '3 is prime' are best read as being "about" abstract objects — i.e., thinly about abstract objects (see note 17) — there is nothing built into our usage or intentions about whether there really do exist abstract objects, and so it's not true that we're explicitly involved in make-believe, and it's not true that we clearly intend to be talking about an actually existing platonic realm. But again, this is just one line that fictionalists could take. (See my [2009] for more on this and, in particular, how fictionalists can respond to the objection raised by Burgess [2004].)

One might think that '3 is prime' is less analogous to (1) or (2) than it is to, say, a sentence about Oliver uttered by an informed adult who intends to be saying something true about Dickens's novel, e.g.,

(3) Oliver Twist lived in London, not Paris.

But we have to be careful here, because (a) one might think (indeed, I do think) that (3) is best thought of as being about Dickens's *novel*, and not Oliver, and (b) fictionalists do *not* claim that sentences like '3 is prime' are about the story of mathematics (they think this sentence is about 3 and is true-in-the-story-of-mathematics, but not true *simpliciter*). But some people — e.g., van Inwagen [1977], Zalta [1983; 1988], Salmon [1998], and Thomasson [1999] — think that sentences like (3) are best interpreted as being about Oliver Twist, the actual literary character, which on this view is an abstract object; a fictionalist who accepted this platonistic semantics of (3) could maintain that '3 is prime' is analogous to (3).

Finally, I end by discussing Meinongianism. There are two different versions of this view; the first, I think, is just a terminological variant of platonism; the second is a version of anti-realism. The first version of Meinongianism is more well known, and it is the view that is commonly ascribed to Meinong, though I think this interpretation of Meinong is controversial. In any event, the view is that our mathematical theories provide true descriptions of objects that have some sort of being (that *subsist*, or that *are*, in some sense) but do not have full-blown existence. This sort of Meinongianism has been almost universally rejected. The standard argument against it (see, e.g., [Quine 1948]) is that it is not genuinely distinct from platonism; Meinongians have merely created the illusion of a different view by altering the meaning of the term 'exist'. On the standard meaning of 'exist', any object that *is* — that has any being at all — exists. Therefore, according to standard usage, Meinongianism entails that mathematical objects exist (of course, Meinongians wouldn't assent to the sentence 'Mathematical objects exist', but this, it seems, is simply because they don't know what 'exist' means);

but Meinongianism clearly doesn't take mathematical objects to exist in space-time, and so on this view, mathematical objects are abstract objects. Therefore, Meinongianism is not distinct from platonism.[19]

The second version of Meinongianism, defended by Routley [1980] and later by Priest [2003], holds that (a) things like numbers and universals don't exist at all (i.e., they have no sort of being whatsoever), but (b) we can still say true things about them — e.g., we can say (truly) that 3 is prime, even though there is no such thing as 3. Moreover, while Azzouni [1994] would not use the term 'Meinongianism', he has a view that is very similar to the Routley-Priest view. For he seems to want to say that (a) as platonists and fictionalists assert, mathematical sentences — e.g., '3 is prime' and 'There are infinitely many transfinite cardinals' — should be read at face value, i.e., as being about mathematical objects (in at least some thin sense); (b) as platonists assert, such sentences are true; and (c) as fictionalists assert, there are really no such things as mathematical objects that exist independently of us and our mathematical theorizing. I think that this view is flawed in a way that is similar to the way in which the first version of Meinongianism is flawed, except that here, the problem is with the word 'true', rather than 'exists'. The second version of Meinongianism entails that a mathematical sentence of the form 'Fa' can be true, even if there is no such thing as the object a (Azzouni calls this a sort of truth by convention, for on his view, it applies by stipulation; but the view here is different from the Ayer-Hempel-Carnap conventionalist view described above). But the problem is that it seems to be built into the standard meaning of 'true' that if there is no such thing as the object a, then sentences of the form 'Fa' cannot be literally true. Or equivalently, it is a widely accepted criterion of ontological commitment that if you think that the sentence 'a is F' is literally true, then you are committed to the existence of the object a. One might also put the point here as follows: just as the first version of Meinongianism isn't genuinely distinct from platonism and only creates the illusion of a difference by misusing 'exists', so too the second version of Meinongianism isn't genuinely distinct from fictionalism and only creates the illusion of a difference by misusing

---

[19]Priest [2003] argues that (a) Meinongianism is different from traditional platonism, because the latter is non-plenitudinous; and (b) Meinongianism is different from FBP, because the former admits as legitimate the objects of *inconsistent* mathematical theories as well as consistent ones; and (c) if platonists go for a plenitudinous view that also embraces the inconsistent (i.e., if they endorse what Beall [1999] has called *really full-blooded platonism*), then the view looks more like Meinongianism than platonism. But I think this last claim is just false; unless Meinongians can give some appropriate content to the claim that, e.g., 3 is but doesn't exist, it seems that the view should be thought of as a version of platonism. (I should note here that in making the above argument, Priest was very likely thinking of the *second* version of Meinongianism, which I will discuss presently, and so my argument here should not be thought of as a refutation of Priest's argument; it is rather a refutation of the idea that Priest's argument can be used to save first-version Meinongianism from the traditional argument against it. Moreover, as we'll see, I do not think the second version of Meinongianism is equivalent to platonism, and so Priest's argument will be irrelevant there.) Finally, I might also add here that just as there are different versions of platonism that correspond to points (a)–(c) above, so too we can define analogous versions of Meinongianism. So I don't think there's any difference between the two views on this front either.

'true'; in short, what they call truth isn't *real* truth, because on the standard meaning of 'true' — that is, the meaning of 'true' in *English* — if a sentence has the form '*Fa*', and if there is no such thing as the object *a*, then '*Fa*' isn't true. To simply stipulate that such a sentence *is* true is just to alter the meaning of 'true'.

2 CRITIQUE OF THE VARIOUS VIEWS

I will take a somewhat roundabout critical path through the views surveyed above. In section 2.1, I will discuss the main criticisms that have been leveled against platonism; in section 2.2, I will critically assess the various versions of anti-platonism, including the various anti-platonistic versions of realism (i.e., physicalism and psychologism); finally, in section 2.3, I will discuss a lingering worry about platonism. I follow this seemingly circuitous path for the simple reason that it seems to me to generate a logically pleasing progression through the issues to be discussed — even if it doesn't provide a clean path through realism first and anti-realism second.

2.1 *Critique of Platonism*

In this section, I will consider the two main objections to platonism. In section 2.1.1, I will consider the epistemological objection, and in section 2.1.2, I will consider the non-uniqueness (or multiple-reductions) objection. (There are a few other problems with platonism as well, e.g., problems having to do with mathematical reference, the applications of mathematics, and Ockham's razor. I will address these below.) As we will see, I do not think that any of these objections succeeds in refuting platonism, because I think there are good FBP-ist responses to all of them, though we will also see that these objections (especially the epistemological one) do succeed in refuting non-full-blooded versions of platonism.

2.1.1 *The Epistemological Argument Against Platonism*

In section 2.1.1.1, I will formulate the epistemological argument; in sections 2.1.1.2–2.1.1.4, I will attack a number of platonist strategies for responding to the argument; and in section 2.1.1.5, I will explain what I think is the correct way for platonists to respond.

2.1.1.1 Formulating the Argument While this argument goes all the way back to Plato, the *locus classicus* in contemporary philosophy is Benacerraf's [1973]. But Benacerraf's version of the argument rests on a causal theory of knowledge that has proved vulnerable. A better formulation of the argument is as follows:

(1) Human beings exist entirely within spacetime.

(2) If there exist any abstract mathematical objects, then they exist outside of spacetime.

Therefore, it seems very plausible that

(3) If there exist any abstract mathematical objects, then human beings could not attain knowledge of them.

Therefore,

(4) If mathematical platonism is correct, then human beings could not attain mathematical knowledge.

(5) Human beings have mathematical knowledge.

Therefore,

(6) Mathematical platonism is not correct.

The argument for (3) is everything here. If it can be established, then so can (6), because (3) trivially entails (4), (5) is beyond doubt, and (4) and (5) trivially entail (6). Now, (1) and (2) do not deductively entail (3), and so even if we accept (1) and (2), there is room here for platonists to maneuver — and as we'll see, this is precisely how most platonists have responded. However, it is important to notice that (1) and (2) provide a strong *prima facie* motivation for (3), because they suggest that mathematical objects (if there are such things) are totally inaccessible to us, i.e., that information cannot pass from mathematical objects to human beings. But this gives rise to a *prima facie* worry (which may or may not be answerable) about whether human beings could acquire knowledge of abstract mathematical objects (i.e., it gives rise to a *prima facie* reason to think that (3) is true). Thus, we should think of the epistemological argument not as *refuting* platonism, but rather as issuing a challenge to platonists. In particular, since this argument generates a *prima facie* reason to doubt that human beings could acquire knowledge of abstract mathematical objects, and since platonists are committed to the thesis that human beings can acquire such knowledge, the challenge to platonists is simply to explain *how* human beings could acquire such knowledge.

There are three ways that platonists can respond to this argument. First, they can argue that (1) is false and that the human mind is capable of, somehow, forging contact with the mathematical realm and thereby acquiring information about that realm; this is Gödel's strategy, at least on some interpretations of his work. Second, we can argue that (2) is false and that human beings can acquire information about mathematical objects via normal perceptual means; this strategy was pursued by the early Maddy. And third, we can accept (1) and (2) and try to explain how (3) could be false anyway. This third strategy is very different from the first two, because it involves the construction of what might be called a *no-contact* epistemology; for the idea here is to accept the thesis that human beings cannot come into any sort of information-transferring contact with mathematical objects — this is the result of accepting (1) and (2) — and to try to explain how humans could nonetheless acquire knowledge of abstract objects. This third strategy has been the most popular among contemporary philosophers. Its advocates include Quine, Steiner, Parsons, Hale, Wright, Resnik, Shapiro, Lewis, Katz, and myself.

In sections 2.1.1.2–2.1.1.4, I will describe (and criticize) the strategy of rejecting (1), the strategy of rejecting (2), and all of the various no-contact strategies in the literature, except for my own. Then in section 2.1.1.5, I will describe and defend my own no-contact strategy, i.e., the FBP-based epistemology defended in my [1995] and [1998].

2.1.1.2 Contact with the Mathematical Realm: The Gödelian Strategy of Rejecting (1)

On Gödel's [1964] view, we acquire knowledge of abstract mathematical objects in much the same way that we acquire knowledge of concrete physical objects: just as we acquire information about physical objects via the faculty of sense perception, so we acquire information about mathematical objects by means of a faculty of *mathematical intuition*. Now, other philosophers have endorsed the idea that we possess a faculty of mathematical intuition, but Gödel's version of this view involves the idea that the mind is non-physical in some sense and that we are capable of forging contact with, and acquiring information from, non-physical mathematical objects. (Others who endorse the idea that we possess a faculty of mathematical intuition have a *no-contact* theory of intuition that is consistent with a materialist philosophy of mind. Now, some people might argue that Gödel had such a view as well. I have argued elsewhere [1998, chapter 2, section 4.2] that Gödel is better interpreted as endorsing an immaterialist, contact-based theory of mathematical intuition. But the question of what view Gödel actually held is irrelevant here.)

This reject-(1) strategy of responding to the epistemological argument can be quickly dispensed with. One problem is that rejecting (1) doesn't seem to help solve the lack-of-access problem. For even if minds are immaterial, it is not as if that puts them into informational contact with mathematical objects. Indeed, the idea that an immaterial mind could have some sort of information-transferring contact with abstract objects seems just as incoherent as the idea that a physical brain could. Abstract objects, after all, are causally inert; they cannot generate information-carrying signals at all; in short, information can't pass from an abstract object to *anything*, material or immaterial. A second problem with the reject-(1) strategy is that (1) is, in fact, true. Now, of course, I cannot argue for this here, because it would be entirely inappropriate to break out into an argument against Cartesian dualism in the middle of an essay on the philosophy of mathematics, but it is worth noting that what is required here is a very strong and implausible version of dualism. One cannot motivate a rejection of (1) by merely arguing that there are real mental states, like beliefs and pains, or by arguing that our mentalistic idioms cannot be reduced to physicalistic idioms. One has to argue for the thesis that there actually exists immaterial human mind-stuff.

2.1.1.3 Contact in the Physical World: The Maddian Strategy of Rejecting (2) I now move on to the idea that platonists can respond to the epistemological argument by rejecting (2). The view here is still that human beings are capable of acquiring knowledge of mathematical objects by coming into contact with them, i.e., receiving information from them, but the strategy now is not to bring human beings up to platonic heaven, but rather, to bring the inhabitants of platonic heaven down to earth. Less metaphorically, the idea is to adopt a naturalistic conception of mathematical objects and argue that human beings can acquire knowledge of these objects via *sense perception*. The most important advocate of this view is Penelope Maddy (or rather, the *early* Maddy, for she has since abandoned the view).[20] Maddy is concerned mainly with set theory. Her two central claims are (a) that sets are spatiotemporally located — a set of eggs, for instance, is located right where the eggs are — and (b) that we can acquire knowledge of sets by perceiving them, i.e., by seeing, hearing, smelling, feeling, and tasting them in the usual ways. Let's call this view *naturalized platonism*.

I have argued against naturalized platonism elsewhere [1994; 1998, chapter 2, section 5]. I will just briefly sketch one of my arguments here.

The first point that needs to be made in this connection is that despite the fact that Maddy takes sets to exist in spacetime, her view still counts as a version of *platonism* (albeit a non-standard version). Indeed, the view *has* to be a version of platonism if it is going to be (a) relevant to the present discussion and (b) tenable. Point (a) should be entirely obvious, for since we are right now looking for a solution to the epistemological problem with *platonism*, we are concerned only with platonistic views that reject (2), and not anti-platonistic views. As for point (b), if Maddy were to endorse a thoroughgoing anti-platonism, then her view would presumably be a version of physicalism, since she claims that there do exist sets and that they exist in spacetime, right where their members do; in other words, her view would presumably be that sets are purely physical objects. But this sort of physicalism is untenable. One problem here (there are actually many problems with this view; see section 2.2.3 below) is that corresponding to every physical object there are infinitely many sets. Corresponding to an egg, for instance, there is the set containing the egg, the set containing that set, the set containing *that* set, and so on; and there is the set containing the egg and the set containing the egg, and so on and on and on. But all of these sets have the same physical base; that is, they are made of the exact same matter and have the exact same spatiotemporal location. Thus, in order to maintain that these sets are different things, Maddy has to claim that they differ from one another in *non-physical* ways and, hence, that sets are at least partially non-physical objects. Now, I suppose one might adopt a psychologistic view here according to which sets are *mental* objects (e.g., one might claim that only physical objects exist "out there

---

[20]See Maddy [1980; 1990]. She abandons the view in her (1997) for reasons completely different from the ones I present here. Of course, Maddy isn't the first philosopher to bring abstract mathematical objects into spacetime. Aside from Aristotle, Armstrong [1978, chapter 18, section V] attempts this as well, though he doesn't develop the idea as thoroughly as Maddy does.

in the world" and that we then come along and somehow construct all the various different sets in our minds); but as Maddy is well aware, such views are untenable (see section 2.2.3 below). Thus, the only initially plausible option for Maddy (or indeed for anyone who rejects (2)) is to maintain that there is something non-physical and non-mental about sets. Thus, she has to claim that sets are abstract, in some appropriate sense of the term, although, of course, she rejects the idea that they are abstract in the traditional sense of being non-spatiotemporal.

Maddy, I think, would admit to all of this, and in my book (chapter 2, section 5.1) I say what I think the relevant sense of abstractness is. I will not pursue this here, however, because it is not relevant to the argument that I will mount against Maddy's view. All that matters to my argument is that according to Maddy's view, sets are abstract, or non-physical, in at least some non-trivial sense.

What I want to argue here is that human beings cannot receive any relevant perceptual data from naturalized-platonist sets (i.e., sets that exist in spacetime but are nonetheless non-physical, or abstract, in some non-traditional sense) — and hence that platonists cannot solve the epistemological problem with their view by rejecting (2). Now, it's pretty obvious that I can acquire perceptual knowledge of physical objects and aggregates of physical matter; but again, there is more to a naturalized-platonist set than the physical stuff with which it shares its location — there is something *abstract* about the set, over and above the physical aggregate, that distinguishes it from the aggregate (and from the infinitely many other sets that share the same matter and location). Can I perceive this abstract component of the set? It seems that I cannot. For since the set and the aggregate are made of the same matter, both lead to the same retinal stimulation. Maddy herself admits this [1990, 65]. But if I receive only one retinal stimulation, then the perceptual data that I receive about the set are identical to the perceptual data that I receive about the aggregate. More generally, when I perceive an aggregate, I do not receive *any* data about *any* of the infinitely many corresponding naturalized-platonist sets that go beyond the data that I receive about the aggregate. This means that naturalized platonists are no better off here than traditional platonists, because we receive no more perceptual information about naturalized-platonist sets than we do about traditional non-spatiotemporal sets. Thus, the Benacerrafian worry still remains: there is still an unexplained epistemic gap between the information we receive in sense perception and the relevant facts about sets. (It should be noted that there are a couple of ways that Maddy could respond to this argument. However, I argued in my book (chapter 2, section 5.2) that these responses do not succeed.)

2.1.1.4 Knowledge Without Contact We have seen that mathematical platonists cannot solve the epistemological problem by claiming that human beings are capable of coming into some sort of contact with (i.e., receiving information from) mathematical objects. Thus, if platonists are to solve the problem, they must explain how human beings could acquire knowledge of mathematical objects without the aid of any contact with them. Now, a few different no-contact platonists (most notably, Parsons [1980; 1994], Steiner [1975], and Katz [1981; 1998]) have started out their arguments here by claiming that human beings possess a (no-contact) faculty of mathematical intuition. But as almost all of these philosophers would admit, the epistemological problem cannot be solved with a mere appeal to a no-contact faculty of intuition; one must also explain how this faculty of intuition could be reliable — and in particular, how it could lead to *knowledge* — given that it's a *no-contact* faculty. But to explain how the faculty that generates our mathematical intuitions and beliefs could lead to knowledge, despite the fact that it's a no-contact faculty, is not significantly different from explaining how we could acquire knowledge of mathematical objects, despite the fact that we do not have any contact with such objects. Thus, no progress has been made here toward solving the epistemological problem with platonism.[21] (For a longer discussion of this, see my [1998, chapter 2, section 6.2].)

In sections 2.1.1.4.1–2.1.1.4.3, I will discuss and criticize three different attempts to explain how human beings could acquire knowledge of abstract objects without the aid of any information-transferring contact with such objects. Aside from my own explanation, which I will defend in section 2.1.1.5, these three explanations are (as far as I know) the only ones that have been suggested. (It should be noted, however, that two no-contact platonists — namely, Wright [1983, section xi] and Hale [1987, chapters 4 and 6] — have tried to solve the epistemological problem *without* providing an explanation of how we could acquire knowledge of non-spatiotemporal objects. I do not have the space to pursue this here, but in my book (chapter 2, section 6.1) I argue that this cannot be done.)

2.1.1.4.1 Holism and Empirical Confirmation: Quine, Steiner, and Resnik

One explanation of how we can acquire knowledge of mathematical objects despite our lack of contact with them is hinted at by Quine [1951, section 6] and developed by Steiner [1975, chapter 4] and Resnik [1997, chapter 7]. The claim here is that we have good reason to believe that our mathematical theories are true, because (a) these theories are central to our overall worldview, and (b) this worldview has been repeatedly confirmed by empirical evidence. In other words, we don't need contact with mathematical objects in order to know that our theories of these objects are true, because *confirmation is holistic*, and so these theories are confirmed every day, along with the rest of our overall worldview.

One problem with this view is that confirmation holism is, in fact, false. Confirmation may be holistic with respect to the *nominalistic* parts of our empirical theories (actually, I doubt even this), but the mathematical parts of our empir-

---

[21] Again, most platonists who appeal to a no-contact faculty of intuition would acknowledge my point here, and indeed, most of them go on to offer explanations of how no-contact intuitions could be reliable (or what comes to the same thing, how we could acquire knowledge of abstract mathematical objects without the aid of any contact with such objects). The exception to this is Parsons; he never addresses the worry about how a no-contact faculty of intuition could generate knowledge of non-spatiotemporal objects. This is extremely puzzling, for it's totally unclear how an appeal to a no-contact faculty of intuition can help solve the epistemological problem with platonism if it's not conjoined with an explanation of reliability.

ical theories are *not* confirmed by empirical findings. Indeed, empirical findings provide no reason whatsoever for supposing that the mathematical parts of our empirical theories are true. I will sketch the argument for this claim below, in section 2.2.4, by arguing that the nominalistic contents of our empirical theories could be true even if their platonistic contents are fictional (the full argument can be found in my [1998, chapter 7]).

A second problem with the Quine-Steiner-Resnik view is that it leaves unexplained the fact that mathematicians are capable of acquiring mathematical knowledge without waiting to see if their theories get applied and confirmed in empirical science. The fact of the matter is that mathematicians acquire mathematical knowledge *by doing mathematics*, and then empirical scientists come along and use our mathematical theories, which we already know are true. Platonists need to explain how human beings could acquire this pre-applications mathematical knowledge. And, of course, what's needed here is precisely what we needed to begin with, namely, an explanation of how human beings could acquire knowledge of abstract mathematical objects despite their lack of contact with such objects. Thus, the Quinean appeal to applications hasn't helped at all — platonists are right back where they started.

2.1.1.4.2 Necessity: Katz and Lewis A second version of the no-contact strategy, developed by Katz [1981; 1998] and Lewis [1986, section 2.4], is to argue that we can know that our mathematical theories are true, without any sort of information-transferring contact with mathematical objects, because these theories are *necessarily* true. The reason we need information-transferring contact with ordinary physical objects in order to know what they're like is that these objects could have been different. For instance, we have to look at fire engines in order to know that they're red, because they could have been blue. But on the Katz–Lewis view, we don't need any contact with the number 4 in order to know that it's the sum of two primes, because it is necessarily the sum of two primes.

This view has been criticized by Field [1989, 233–38] and myself [1998, chapter 2, section 6.4]. In what follows, I will briefly sketch what I think is the main problem.

The first point to note here is that even if mathematical truths are necessarily true, Katz and Lewis still need to explain how we know that they're true. The mathematical realm might have the particular nature that it has of necessity, but that doesn't mean that we could know what its nature is. How could human beings know that the mathematical realm is composed of structures of the sort we study in mathematics — i.e., the natural number series, the set-theoretic hierarchy, and so on — rather than structures of some radically different kind? It is true that *if* the mathematical realm is composed of structures of the familiar sort, then it follows of necessity that 4 is the sum of two primes. But again, how could we know that the mathematical realm is composed of structures of the familiar kind?

It is important that this response not be misunderstood. I am not demanding here an account of how human beings could know that there exist any mathematical objects at all. That, I think, would be an illegitimate skeptical demand; as is argued in Katz's [1981, chapter VI] and my [1998, chapter 3], all we can legitimately demand from platonists is an account of how human beings could know the *nature* of mathematical objects, *given* that such objects exist. But in demanding that Katz and Lewis provide an account of how humans could know that there are objects answering to our mathematical theories, I mean to be making a demand of this latter sort. An anti-platonist might put the point here as follows: "Even if we assume that there exist mathematical objects — indeed, even if we assume that the mathematical objects that exist do so of necessity — we cannot assume that *any* theory we come up with will pick out a system of actually existing objects. Platonists have to explain how we could know *which* mathematical theories are true and which aren't. That is, they have to explain how we could know which kinds of mathematical objects exist."

The anti-platonist who makes this last remark has overlooked a move that platonists can make: they can say that, in fact, we *can* assume that any purely mathematical theory we come up with will pick out a system of actually existing objects (or, more precisely, that any such theory that's *internally consistent* will pick out a system of objects). Platonists can motivate this claim by adopting FBP. For if all the mathematical objects that possibly *could* exist actually *do* exist, as FBP dictates, then every (consistent) purely mathematical theory picks out a system of actually existing mathematical objects. It is important to note, however, that we should not think of this appeal to FBP as showing that the Katz-Lewis necessity-based epistemology can be made to work. It would be more accurate to say that what's going on here is that we are *replacing* the necessity-based epistemology with an FBP-based epistemology. More precisely, the point is that once platonists appeal to FBP, there is no more reason to appeal to necessity at all. (This point is already implicit in the above remarks, but it is made very clear by my own epistemology (see section 2.1.1.5 below, and my 1998, chapter 3), for I have shown how to develop an FBP-based epistemology that doesn't depend upon any claims about the necessity of mathematical truths.) The upshot of this is that the appeal to necessity isn't doing any epistemological work at all; FBP is doing all the work. Moreover, for the reasons already given, the necessity-based epistemology cannot be made to work without falling back on the appeal to FBP. Thus, the appeal to necessity seems to be utterly unhelpful in connection with the epistemological problem with platonism.

But this is not all. The appeal to necessity is not just epistemologically unhelpful; it is also *harmful*. The reason is that the thesis that our mathematical sentences and theories are necessary is dubious at best. Consider, for instance, the null set axiom, which says that there exists a set with no members. Why should we think that this sentence is necessarily true? It seems pretty obvious that it isn't logically or conceptually necessary, for it is an existence claim, and such claims aren't logically or conceptually true.[22] Now, one might claim that

---

[22]I should note, however, that in opposition to this, Hale and Wright [1992] have argued that the existence of mathematical objects is conceptually necessary. But Field [1993] has argued

our mathematical theories are *metaphysically* necessary, but it's hard to see what this could really amount to. One might claim that sentences like '2 + 2 = 4' and '7 > 5' are metaphysically necessary for the same reason that, e.g., 'Cicero is Tully' is metaphysically necessary — because they are true in all worlds in which their singular terms denote, or something along these lines — but this doesn't help at all in connection with existence claims like the null set axiom. We can't claim that the null set axiom is metaphysically necessary for anything like the reason that 'Cicero is Tully' is metaphysically necessary. If we tried to do this, we would end up saying that 'There exists an empty set' is metaphysically necessary because it is true in all worlds in which there exists an empty set. But of course, this is completely unacceptable, because it suggests that *all* existence claims — e.g., 'There exists a purple hula hoop' — are metaphysically necessary. In the end, it doesn't seem to me that there is any interesting sense in which 'There exists an empty set' is necessary but 'There exists a purple hula hoop' is not.

2.1.1.4.3 Structuralism: Resnik and Shapiro Resnik [1997, chapter 11, section 3] and Shapiro [1997, chapter 4, section 7] both claim that human beings can acquire knowledge of abstract mathematical structures, without coming into any sort of information-transferring contact with such structures, by simply constructing mathematical axiom systems; for they argue that axiom systems provide *implicit definitions* of structures. I want to respond to this in the same way that I responded to the Katz–Lewis appeal to necessity. The problem is that the Resnik-Shapiro view does not explain how we could know *which* of the various axiom systems that we might formulate actually pick out structures that exist in the mathematical realm. Now, as was the case with Katz and Lewis, if Resnik and Shapiro adopt FBP, or rather, a structuralist version of FBP, then this problem can be solved; for it follows from (structuralist versions of) FBP that any consistent purely mathematical axiom system that we formulate will pick out a structure in the mathematical realm. But as was the case with the Katz-Lewis epistemology, what's going on here is not that the Resnik-Shapiro epistemology is being salvaged, but rather that it's being replaced by an FBP-based epistemology.

It is important to note in this connection that FBP is not built into structuralism; one could endorse a non-plenitudinous or non-full-blooded version of structuralism, and so it is FBP and not structuralism that delivers the result that Resnik and Shapiro need. In fact, structuralism is entirely irrelevant to the implicit-definition strategy of responding to the epistemological problem, because one can claim that axiom systems provide implicit definitions of collections of mathematical objects as easily as one can claim that they provide implicit definitions of structures. What one needs, in order to make this strategy work, is FBP, not structuralism. (Indeed, I argue in my book (chapter 2, section 6.5) that similar remarks apply to everything Resnik and Shapiro say about the epistemology of mathematics: despite their rhetoric, structuralism doesn't play an essential role in their arguments, and so it is epistemologically irrelevant.)

---

convincingly that their argument is flawed.

Finally, I should note here, in defense of not just Resnik and Shapiro, but Katz and Lewis as well, that it may be that the views of these four philosophers are best interpreted as involving (in some sense) FBP. But the problem is that these philosophers don't *acknowledge* that they need to rely upon FBP, and so obviously — and more importantly — they don't *defend* the reliance upon FBP. In short, all four of these philosophers could have given FBP-based epistemologies without radically altering their metaphysical views, but none of them actually did.

(This is just a sketch of one problem with the Resnik-Shapiro view; for a more thorough critique, see my [1998, chapter 2, section 6.5].)

2.1.1.5 An FBP-Based Epistemology Elsewhere [1992; 1995; 1998], I argue that if platonists endorse FBP, then they can solve the epistemological problem with their view without positing any sort of information-transferring contact between human beings and abstract objects. The strategy can be summarized as follows. Since FBP says that all the mathematical objects that possibly could exist actually do exist, it follows that if FBP is correct, then all consistent purely mathematical theories truly describe some collection of abstract mathematical objects. Thus, to acquire knowledge of mathematical objects, all we need to do is acquire knowledge that some purely mathematical theory is *consistent*. (It doesn't matter how we come up with the theory; some creative mathematician might simply "dream it up".) But knowledge of the consistency of a mathematical theory — or any other kind of theory, for that matter — does not require any sort of contact with, or access to, the objects that the theory is about. Thus, the Benacerrafian lack-of-access problem has been solved: we can acquire knowledge of abstract mathematical objects without the aid of any sort of information-transferring contact with such objects.

Now, there are a number of objections that might occur to the reader at this point. Here, for instance, are four different objections that one might raise:

1. Your account of how we could acquire knowledge of mathematical objects seems to assume that we are capable of *thinking about* mathematical objects, or *dreaming up stories about* such objects, or *formulating theories about* them. But it is simply not clear how we could do these things. After all, platonists need to explain not just how we could acquire *knowledge* of mathematical objects, but also how we could do things like have *beliefs* about mathematical objects and *refer* to mathematical objects.

2. The above sketch of your epistemology seems to assume that it will be easy for FBP-ists to account for how human beings could acquire knowledge of the consistency of purely mathematical theories without the aid of any contact with mathematical objects; but it's not entirely clear how FBP-ists could do this.

3. You may be right that if FBP is true, then all consistent purely mathematical theories truly describe *some* collection of mathematical objects, or some part

of the mathematical realm. But *which* part? How do we know that it will be true of the part of the mathematical realm that its authors intended to characterize? Indeed, it seems mistaken to think that such theories will characterize *unique* parts of the mathematical realm at all.

4. All your theory can explain is how it is that human beings could *stumble onto* theories that truly describe the mathematical realm. On the picture you've given us, the mathematical community accepts a mathematical theory T for a list of reasons, one of which being that T is consistent (or, more precisely, that mathematicians believe that T is consistent). Then, since FBP is true, it turns out that T truly describes part of the mathematical realm. But since mathematicians have no conception of FBP, they do not know *why* T truly describes part of the mathematical realm, and so the fact that it does is, in some sense, *lucky*. Thus, let's suppose that T is a purely mathematical theory that we know (or reliably believe) is consistent. Then the objection to your epistemology is that you have only an FBP-ist account of

(M1)   our ability to know that *if* FBP is true, *then* T truly describes part of the mathematical realm.[23]

You do not have an FBP-ist account of

(M2)   our ability to know that T truly describes part of the mathematical realm,

because you have said nothing to account for

(M3)   our ability to know that FBP is true.

In my book (chapters 3 and 4), I respond to all four of the above worries, and I argue that FBP-ists can adequately respond to the epistemological objection to platonism by using the strategy sketched above. I do not have the space to develop these arguments here, although I should note that some of what I say below (section 2.1.2) will be relevant to one of the above objections, namely, objection number 3.

In addition to the above objections concerning my FBP-ist epistemology, there are also a number of objections that one might raise against FBP itself. For instance, one might think that FBP is inconsistent with the *objectivity* of mathematics, because one might think that FBP entails that, e.g., the continuum hypothesis (CH) has no determinate truth value, because FBP entails that both CH and ∼CH truly describe parts of the mathematical realm. Or, indeed, one might think that because of this, FBP leads to *contradiction*. In my book (chapters 3 and 4), and my [2001] and [2009], I respond to both of these worries — i.e., the worries about objectivity and contradiction — as well as several other worries about FBP. Indeed, I argue not just that FBP is the best version of platonism there is, but that

---

[23]The FBP-ist account of (M1) is simple: we can learn what FBP says and recognize that if FBP is true, then *any* theory like T (i.e., any consistent purely mathematical theory) truly describes part of the mathematical realm.

it is entirely defensible — i.e., that it can be defended against all objections (or at any rate, all the objections that I could think of at the time, except for the objection inherent in my argument for the claim that there is no fact of the matter as to whether FBP or fictionalism is true (see section 3 below)). I do not have anywhere near the space to develop all of these arguments here, though, and instead of trying to summarize all of this material, I simply refer the reader to my earlier writings. However, I should say that responses (or at least partial responses) to the two worries mentioned at the start of this paragraph — i.e., the worries about objectivity and contradiction — will emerge below, in sections 2.1.2–2.1.3, and I will also address there some objections that have been raised to FBP since my book appeared. (I don't want to respond to these objections just yet, because my responses will make more sense in the wake of my discussion of the non-uniqueness problem, which I turn to now.)

2.1.2 The Non-Uniqueness Objection to Platonism

2.1.2.1 Formulating the Argument Aside from the epistemological argument, the most important argument against platonism is the non-uniqueness argument, or as it's also called, the multiple-reductions argument. Like the epistemological argument, this argument also traces to a paper of Benacerraf's [1965], but again, my formulation will diverge from Benacerraf's. In a nutshell, the non-uniqueness problem is this: platonism suggests that our mathematical theories describe *unique* collections of abstract objects, but in point of fact, this does not seem to be the case. Spelling the reasoning out in a bit more detail, and couching the point in terms of arithmetic, as is usually done, the argument proceeds as follows.

(1) If there are any sequences of abstract objects that satisfy the axioms of Peano Arithmetic (PA), then there are infinitely many such sequences.

(2) There is nothing "metaphysically special" about any of these sequences that makes it stand out from the others as *the* sequence of natural numbers.

Therefore,

(3) There is no unique sequence of abstract objects that is the natural numbers.

But

(4) Platonism entails that there *is* a unique sequence of abstract objects that is the natural numbers.

Therefore,

(5) Platonism is false.

The only vulnerable parts of the non-uniqueness argument are (2) and (4). The two inferences — from (1) and (2) to (3) and from (3) and (4) to (5) — are both fairly trivial. Moreover, as we will see, (1) is virtually undeniable. (And note that we cannot make (1) any less trivial by taking PA to be a second-order theory and, hence, categorical. This will only guarantee that all the models of PA are isomorphic to one another. It will not deliver the desired result of there being only one model of PA.) So it seems that platonists have to attack either (2) or (4). That is, they have to choose between trying to *salvage* the idea that our mathematical theories are about unique collections of objects (rejecting (2)) and *abandoning* uniqueness and endorsing a version of platonism that embraces the idea that our mathematical theories are not (or at least, might not be) about unique collections of objects (rejecting (4)). In section 2.1.2.4, I will argue that platonists can successfully solve the problem by using the latter strategy, but before going into this, I want to say a few words about why I think they can't solve the problem using the former strategy, i.e., the strategy of rejecting (2).

2.1.2.2 Trying to Salvage the Numbers I begin by sketching Benacerraf's argument in *favor* of (2). He proceeds here in two stages: first, he argues that no sequence of *sets* stands out as *the* sequence of natural numbers, and second, he extends the argument so that it covers sequences of other sorts of objects as well. The first claim, i.e., the claim about sequences of sets, is motivated by reflecting on the numerous set-theoretic reductions of the natural numbers. Benacerraf concentrates, in particular, on the reductions given by Zermelo and von Neumann. Both of these reductions begin by identifying 0 with the null set, but Zermelo identifies $n+1$ with the singleton $\{n\}$, whereas von Neumann identifies $n+1$ with the union $n \cup \{n\}$. Thus, the two progressions proceed like so:

$$\emptyset, \{\emptyset\}, \{\{\emptyset\}\}, \{\{\{\emptyset\}\}\}, \ldots$$

and

$$\emptyset, \{\emptyset\}, \{\emptyset, \{\emptyset\}\}, \{\emptyset, \{\emptyset\}, \{\emptyset, \{\emptyset\}\}\}, \ldots$$

Benacerraf argues very convincingly that there is no non-arbitrary reason for identifying the natural numbers with one of these sequences rather than the other or, indeed, with any of the many other set-theoretic sequences that would seem just as good here, e.g., the sequence that Frege suggests in his reduction.

Having thus argued that no sequence of sets stands out as *the* sequence of natural numbers, Benacerraf extends the point to sequences of other sorts of objects. His argument here proceeds as follows. From an arithmetical point of view, the only properties of a given sequence that *matter* to the question of whether it is the sequence of natural numbers are *structural* properties. In other words, nothing about the individual objects in the sequence matters — all that matters is the structure that the objects jointly possess. Therefore, any sequence with the right structure will be as good a candidate for being the natural numbers as any other sequence with the right structure. In other words, any *ω-sequence* will be as good a candidate as any other. Thus, we can conclude that no one sequence of objects stands out as *the* sequence of natural numbers.

It seems to me that if Benacerraf's argument for (2) can be blocked at all, it will have to be at this second stage, for I think it is more or less beyond doubt that no sequence of *sets* stands out as *the* sequence of natural numbers. So how can we attack the second stage of the argument? Well, one strategy that some have followed is to argue that all Benacerraf has shown is that numbers cannot be *reduced* to objects of any other kind; e.g., Resnik argues [1980, 231] that while Benacerraf has shown that numbers aren't sets or functions or chairs, he hasn't shown that numbers aren't objects, because he hasn't shown that numbers aren't *numbers*. But this response misses an important point, namely, that while the first stage of Benacerraf's argument is couched in terms of reductions, the second stage is not — it is based on a premise about the arithmetical irrelevance of non-structural properties. But one might think that we can preserve the spirit of Resnik's idea while responding more directly to the argument that Benacerraf actually used. In particular, one might try to do this in something like the following way.

"There is some initial plausibility to Benacerraf's claim that only structural facts are relevant to the question of whether a given sequence of objects is the sequence of natural numbers. For (a) only structural facts are relevant to the question of whether a given sequence is *arithmetically adequate*, i.e., whether it satisfies PA; and (b) since PA is our best theory of the natural numbers, it would seem that it captures *everything we know* about those numbers. But a moment's reflection reveals that this is confused, that PA does *not* capture everything we know about the natural numbers. There is nothing in PA that tells us that the number 17 is not the inventor of Cocoa Puffs, but nonetheless, we know (pre-theoretically) that it isn't. And there is nothing in PA that tells us that numbers aren't sets, but again, we know that they aren't. Likewise, we know that numbers aren't functions or properties or chairs. Now, it's true that these facts about the natural numbers aren't *mathematically important* — that's why none of them is included in PA — but in the present context, that is irrelevant. What matters is this: while Benacerraf is right that if there are any sequences of abstract objects that satisfy PA, then there are many, the same cannot be said about our *full conception of the natural numbers* (FCNN). We know, for instance, that no sequence of sets or functions or chairs satisfies FCNN, because it is built into our conception of the natural numbers that they do not have members, that they cannot be sat on, and so forth. Indeed, we seem to know that no sequence of things that aren't natural numbers satisfies FCNN, because part of our conception of the natural numbers is that they are natural numbers. Thus, it seems that we know of only *one* sequence that satisfies FCNN, *viz.*, the sequence of natural numbers. But, of course, this means that (2) is false, that one of the sequences that satisfies PA stands out as *the* sequence of natural numbers."

Before saying what I think is wrong with this response to the non-uniqueness argument, I want to say a few words about FCNN, for I think this is an important notion, independently of the present response to the non-uniqueness argument. I say more about this in my [1998] and my [2009], but in a nutshell, FCNN is just the collection of everything that we, as a community, believe about the natural numbers. It is not a formal theory, and so it is not first-order or second-order, and it does not have any axioms in anything like the normal sense. Moreover, it is likely that there is no clear fact of the matter as to precisely which sentences are contained in FCNN (although for *most* sentences, there *is* a clear fact of the matter — e.g., '3 is prime' and '3 is not red' are clearly contained in FCNN, whereas '3 is not prime' and '3 is red' are clearly not). Now, I suppose that one might think it is somehow illegitimate for platonists to appeal to FCNN, or alternatively, one might doubt the claim that it is built into FCNN that numbers aren't, e.g., sets or properties. I cannot go into this here, but in my book [1998, chapter 4], I argue that there is, in fact, nothing illegitimate about the appeal to FCNN, and I point out that in the end, my own response to Benacerraf doesn't depend on the claim that it is built into FCNN that numbers aren't sets or properties.

What, then, is wrong with the above response to the non-uniqueness argument? In a nutshell, the problem is that this response begs the question against Benacerraf, because it simply helps itself to "the natural numbers". We can take the point of Benacerraf's argument to be that if all the ω-sequences were, so to speak, "laid out before us", we could have no good reason for singling one of them out as *the* sequence of natural numbers. Now, the above response does show that the situation here is not as grim as Benacerraf made it seem, because it shows that *some* ω-sequences can be ruled out as definitely *not* the natural numbers. In particular, any ω-sequence that contains an object that we recognize as a non-number — e.g., a function or a chair or (it seems to me, though again, I don't need this claim here) a set — can be ruled out in this way. In short, any ω-sequence that doesn't satisfy FCNN can be so ruled out. But platonists are not in any position to claim that all ω-sequences but one can be ruled out in this way; for since they think that abstract objects exist *independently of us*, they must admit that there are very likely numerous kinds of abstract objects that we've never thought about and, hence, that there are very likely numerous ω-sequences that satisfy FCNN and differ from one another only in ways that no human being has ever imagined. I don't see any way for platonists to escape this possibility, and so it seems to me very likely that (2) is true and, hence, that (3) is also true.

(I say a bit more on this topic, responding to objections and so on, in my book (chapter 4, section 2); but the above remarks are good enough for our purposes here.)

2.1.2.3 Structuralism Probably the most well-known platonist response to the non-uniqueness argument — developed by Resnik [1981; 1997] and Shapiro [1989; 1997] — is that platonists can solve the non-uniqueness problem by merely adopting a platonistic version of Benacerraf's own view, i.e., a platonistic version of *structuralism*. Now, given the way I formulated the non-uniqueness argument above, structuralists would reject (4), because on their view, arithmetic is not about some particular sequence of *objects*. Thus, it might seem that the non-uniqueness problem just doesn't arise at all for structuralists.

This, however, is confused. The non-uniqueness problem *does* arise for structuralists. To appreciate this, all we have to do is reformulate the argument in (1)–(5) so that it is about *parts of the mathematical realm* instead of objects. I did this in my book (chapter 4, section 3). On this alternate formulation, the two crucial premises — i.e., (2) and (4) — are rewritten as follows:

(2′) There is nothing "metaphysically special" about any part of the mathematical realm that makes it stand out from all the other parts as *the* sequence of natural numbers (or natural-number positions or whatever).

(4′) Platonism entails that there *is* a unique part of the mathematical realm that is the sequence of natural numbers (or natural-number positions or whatever).

Seen in this light, the move to structuralism hasn't helped the platonist cause at all. Whether they endorse structuralism or not, they have to choose between trying to salvage uniqueness (attacking (2′)) and abandoning uniqueness, i.e., constructing a platonistic view that embraces non-uniqueness (attacking (4′)). Moreover, just as standard versions of object-platonism seem to involve uniqueness (i.e., they seem to accept (4) and reject (2)), so too the standard structuralist view seems to involve uniqueness (i.e., it seems to accept (4′) and reject (2′)). For the standard structuralist view seems to involve the claim that arithmetic is about *the* structure that all ω-sequences have in common — that is, *the* natural-number structure, or pattern.[24] Finally, to finish driving home the point that structuralists have the same problem here that object-platonists have, we need merely note that the argument I used above (section 2.1.2.2) to show that platonists cannot plausibly reject (2) also shows that they cannot plausibly reject (2′). In short, the point here is that since structures exist independently of us in an abstract mathematical realm, it seems very likely that there are numerous things in the mathematical realm that count as structures, that satisfy FCNN, and that differ from one another only in ways that no human being has ever imagined.

In my book (chapter 4) I discuss a few responses that structuralists might make here, but I argue that none of these responses works and, hence, that (2′) is every bit as plausible as (2). A corollary of these arguments is that contrary to what is commonly believed, structuralism is wholly irrelevant to the non-uniqueness objection to platonism, and so we can (for the sake of rhetorical simplicity) forget about the version of the non-uniqueness argument couched in terms of parts of the mathematical realm, and go back to the original version couched in terms of mathematical objects — i.e., the version in (1)–(5). In the next section, I will

---

[24]Actually, I should say that this is how *I interpret* the standard structuralist view, for to the best of my knowledge, no structuralist has ever explicitly discussed this point. This is a bit puzzling, since one of the standard arguments for structuralism is supposed to be that it provides a way of avoiding the non-uniqueness problem. I suppose that structuralists just haven't noticed that there are general versions of the non-uniqueness argument that apply to their view as well as to object-platonism. They seem to think that the non-uniqueness problem just disappears as soon as we adopt structuralism.

sketch an argument for thinking that platonists can successfully respond to the non-uniqueness argument by rejecting (4), i.e., by embracing non-uniqueness; and as I pointed out in my book, structuralists can mount an exactly parallel argument for rejecting (4'). So again, the issue of structuralism is simply irrelevant here.

(Before leaving the topic of (2) entirely, I should note that I do not think platonists should *commit* to the truth of (2). My claim is that platonists should say that (2) is very likely true, and that we humans could never know that it was false, but that it simply doesn't matter to the platonist view whether (2) is true or not (or more generally, whether any of our mathematical theories picks out a unique collection of objects). This is what I mean when I say that platonists should reject (4): they should reject the claim that their view is committed to uniqueness.)

2.1.2.4 The Solution: Embracing Non-Uniqueness The only remaining platonist strategy for responding to the non-uniqueness argument is to reject (4). Platonists have to give up on uniqueness, and they have to do this in connection not just with arithmetical theories like PA and FCNN, but with all of our mathematical theories. They have to claim that while such theories truly describe collections of abstract mathematical objects, they do not pick out *unique* collections of such objects (or more precisely, that if any of our mathematical theories does describe a unique collection of abstract objects, it is only by blind luck that it does).

Now, this stance certainly represents a departure from traditional versions of platonism, but it cannot be seriously maintained that in making this move, we *abandon* platonism. For since the core of platonism is the belief in abstract objects — and since the core of mathematical platonism is the belief that our mathematical theories truly describe such objects — it follows that the above view is a version of platonism. Thus, the only question is whether there is some reason for thinking that platonists cannot make this move, i.e., for thinking that platonists are *committed* to the thesis that our mathematical theories describe unique collections of mathematical objects. In other words, the question is whether there is any *argument* for (4) — or for a generalized version of (4) that holds not just for arithmetic but for all of our mathematical theories.

It seems to me — and this is the central claim of my response to the non-uniqueness objection — that there *isn't* such an argument. First of all, Benacerraf didn't give any argument at all for (4).[25] Moreover, to the best of my knowledge, no one else has ever argued for it either. But the really important point here is that, *prima facie*, it seems that there couldn't *be* a cogent argument for (4) — or for a generalized version of (4) — because, on the face of it, (4) and its generalization are both highly implausible. The generalized version of (4) says that

---

[25] Actually, Benacerraf's [1965] paper doesn't even assert that (4) is true. It is arguable that (4) is implicit in that paper, but this is controversial. One might also maintain that there is an argument for (4) implicit in Benacerraf's 1973 argument for the claim that we ought to use the same semantics for mathematese that we use for ordinary English. I will respond to this below.

(P) Our mathematical theories truly describe collections of abstract mathematical objects

entails

(U) Our mathematical theories truly describe *unique* collections of abstract mathematical objects.

This is a *really* strong claim. And as far as I can tell, there is absolutely no reason to believe it. Thus, it seems to me that platonists can simply accept (P) and reject (U). Indeed, they can endorse (P) together with the *contrary* of (U); that is, they can claim that while our mathematical theories do describe collections of abstract objects, none of them describes a unique collection of such objects. In short, platonists can avoid the so-called non-uniqueness "problem" by simply *embracing* non-uniqueness, i.e., by adopting *non-uniqueness platonism* (NUP).

In my book (chapter 4, section 4) — and see also my [2001] and [2009] in this connection — I discuss NUP at length. I will say just a few words about it here. According to NUP, when we do mathematics, we have objects of a certain kind in mind, namely, the objects that correspond to our *full conception* for the given branch of mathematics. For instance, in arithmetic, we have in mind objects of the kind picked out by FCNN; and in set theory, we have in mind objects of the kind picked out by our full conception of the universe of sets (FCUS); and so on. These are the objects that our mathematical theories are about; in other words, they are the *intended* objects of our mathematical theories. This much, I think, is consistent with traditional platonism: NUP-ists claim that while our mathematical theories might be satisfied by all sorts of different collections of mathematical objects, or parts of the mathematical realm, they are only really about the *intended* parts of the mathematical realm, or the *standard* parts, where what is intended or standard is determined, very naturally, by our intentions, i.e., by our full conception of the objects under discussion. (Sometimes, we don't have any substantive pretheoretic conception of the relevant objects, and so the intended structures are just the structures that satisfy the relevant axioms.) But NUP-ists differ from traditional platonists in maintaining that in any given branch of mathematics, it may very well be that there are multiple intended parts of the mathematical realm — i.e., multiple parts that dovetail with all of our intentions for the given branch of mathematics, i.e., with the FC for the given branch of mathematics.

Now, according to NUP, when we do mathematics, we often don't worry about the fact that there might be multiple parts of the mathematical realm that count as intended for the given branch of mathematics. Indeed, we often ignore this possibility altogether and proceed as if there is just one intended part of the mathematical realm. For instance, in arithmetic, we proceed as if there is a unique sequence of objects that is the natural numbers. According to NUP-ists, proceeding in this way is very convenient and completely harmless. The reason it's convenient is that it's just intuitively pleasing (for us, anyway) to do arithmetic in this way, assuming that we're talking about a unique structure and thinking about that structure in the normal way. And the reason it's harmless is that we simply aren't interested in the differences between the various ω-sequences that satisfy FCNN. In other words, because all of these sequences are structurally equivalent, they are indistinguishable with respect to the sorts of facts and properties that we are trying to characterize in doing arithmetic, and so no harm can come from proceeding as if there is only one sequence here.

One might wonder what NUP-ists take the truth conditions of mathematical sentences to be. Their view is that a purely mathematical sentence is *true simpliciter* (as opposed to true in some specific model or part of the mathematical realm) iff it is true in all of the intended parts of the mathematical realm for the given branch of mathematics (and there is at least one such part of the mathematical realm). (This is similar to what traditional (U)-platonists say; the only difference is that NUP-ists allow that for any given branch of mathematics, there may be *numerous* intended parts of the mathematical realm.) Now, NUP-ists go on to say that a mathematical sentence is *false simpliciter* iff it's false in all intended parts of the mathematical realm. Thus, NUP allows for failures of bivalence (and I argue in my [2009] that this does not lead to any problems; in particular, it doesn't require us to stop using classical logic in mathematical proofs). Now, some failures of bivalence will be mathematically uninteresting — e.g., if we have two intended structures that are isomorphic to one another, then any sentence that's true in one of these structures and false in the other will be mathematically uninteresting (and note that within the language of mathematics, there won't even be such a sentence). But suppose that we develop a theory of Fs, for some mathematical kind F, and suppose that our concept of an F is not perfectly precise, so that there are multiple structures that all fit perfectly with our concept of an F, and our intentions regarding the word 'F', but that aren't structurally equivalent to one another. Then, presumably, there will be some mathematically interesting sentences that are true in some intended structures but false in others, and so we will have some mathematically interesting failures of bivalence. We will have to say that there is no fact of the matter as to whether such sentences are true or false, or that they lack truth value, or some such thing. This *may* be the case right now with respect to the continuum hypothesis (CH). It may be that our full conception of set is compatible with both ZF+CH hierarchies and ZF+∼CH hierarchies. If so, then hierarchies of both sorts count as intended structures, and hence, CH is true in some intended structures and false in others, and so we will have to say that CH has no determinate truth value, or that there is no fact of the matter as to whether it is true or false, or some such thing. On the other hand, it may be that there is a fact of the matter here. Whether there is a fact of the matter depends upon whether CH or ∼CH follows from axioms that are true in all intended hierarchies, i.e., axioms that are built into our conception of set. Thus, on this view, the question of whether there is a fact of the matter about CH is a *mathematical* question, not a philosophical question. Elsewhere [2001; 2009], I have argued at length that (a) this is the best view to adopt in connection with CH, and (b) NUP (or rather, FBP-NUP) is the only version of realism that yields this view of CH.[26]

This last sentence suggests that platonists have independent reasons for favoring NUP over traditional (U)-platonism — i.e., that it is not the case that the only reason for favoring NUP is that it provides a solution to the non-uniqueness objection. There is also a second independent reason here, which can be put in the following way: (a) as I point out in my book (chapter 4, section 4), FBP leads very naturally into NUP — i.e., it fits much better with NUP than with (U)-platonism — and (b) as we have seen here (and again, this point is argued in much more detail in my book (chapters 2 and 3)), FBP is the best version of platonism there is; indeed, we've seen that FBP is the only tenable version of platonism, because non-full-blooded (i.e., non-plenitudinous) versions of platonism are refuted by the epistemological argument.

But the obvious question that needs to be answered here is whether there are any good arguments for the opposite conclusion, i.e., for thinking that traditional (U)-platonism is superior to NUP, or to FBP-NUP. Well, there are many arguments that one might attempt here. That is, there are many objections that one might raise to FBP-NUP. In my book, I responded to all the objections that I could think of (see chapter 3 for a defense of the FBP part of the view and chapter 4 for a defense of the NUP part of the view). Some of these objections were discussed above; I cannot go through all of them here, but in section 2.1.3, I will respond to a few objections that have been raised against FBP-NUP since my book appeared, and in so doing, I will also touch on some of the objections mentioned above.

In brief, then, my response to the non-uniqueness objection to platonism is this: the fact that our mathematical theories fail to pick out unique collections of mathematical objects (or *probably* fail to do this) is simply not a problem for platonists, because they can endorse NUP, or FBP-NUP.

I have now argued that platonists can adequately respond to both of the Benacerrafian objections to platonism. These two objections are widely considered to be the only objections that really challenge mathematical platonism, but there are some other objections that platonists need to address — objections not to FBP-NUP in particular, but to platonism in general. For instance, there is a worry about how platonists can account for the applicability of mathematics; there are worries about whether platonism is consistent with our abilities to refer to, and have beliefs about, mathematical objects; and there is a worry based on Ockham's razor. I responded to these objections in my book (chapters 3, 4, and 7); I cannot discuss all of them here, but below (section 2.3) I will say a few words about the

---

[26]These remarks are relevant to the problem of accounting for the objectivity of mathematics, which I mentioned in section 2.1.3.5. It is important to note that FBP-ists can account for lots of objectivity in mathematics. On this view, sentences like '3 is prime' are objectively true, and indeed, sentences that are undecidable in currently accepted mathematical theories can be objectively true. E.g., I think it's pretty clear that the Gödel sentence for Peano Arithmetic and the axiom of choice are both true in all intended parts of the mathematical realm. But unlike traditional platonism, FBP also allows us to account for how it *could* be that *some* undecidable sentences do not have objective truth values, and as I argue in my [2001] and [2009], this is a strength of the view.

Ockham's-razor-based objection.

2.1.3 Responses to Some Recent Objections to FBP-NUP

2.1.3.1 Background to Restall's Objections Greg Restall [2003] has raised some objections to FBP-NUP. Most of his criticisms concern the question of how FBP is to be *formulated*. In my book [1998, section 2.1], I offered a few different formulations of FBP, although I wasn't entirely happy with any of them. I wrote:

> The idea behind FBP is that the ordinary, actually existing mathematical objects exhaust all of the logical possibilities for such objects; that is, that there actually exist mathematical objects of all logically possible kinds; that is, that all the mathematical objects that logically possibly *could* exist actually *do* exist; that is, that the mathematical realm is plenitudinous. Now, I do not think that any of the four formulations of FBP given in the previous sentence avoids all ... difficulties ..., but it seems to me that, between them, they make tolerably clear what FBP says.

I'm now no longer sure that these definitions are unacceptable — this depends on what we say about *logical possibilities*, and *kinds*, and how clear we take 'plenitudinous' to be. Moreover, to these four formulations, I might add a fifth, suggested to me by a remark of Zalta and Linsky: There are as many mathematical objects as there logically possibly could be.[27] In any event, I want to stand by what I said in my book: together, these formulations of FBP make it clear enough what the view is.

Restall doesn't object to any of these definitions of FBP; rather, he objects to two other definitions — definitions that, in my book, I explicitly distanced myself from. One of these definitions is a statement of second-order modal logic. After making the above informal remarks about FBP, I said that I do not think "that there is any really adequate way to formalize FBP", that "it is a mistake to think of FBP as a formal theory", and that "FBP is, first and foremost, an informal philosophy of mathematics" (p. 6). But having said this, I added that one might try to come close to formalizing FBP with this:

(1) $(\forall Y)(\Diamond(\exists x)(Mx \& Yx) \supset (\exists x)(Mx \& Yx))$ — where 'Y' is a second-order variable and 'Mx' means 'x is a mathematical object'.

The second definition of FBP that Restall attacks can be put like this:

(0) Every logically consistent purely mathematical theory truly describes a part of the mathematical realm. (Note that to say that T truly describes a part P of the mathematical realm is not just to say that P is a *model* of T, for

---

[27] This isn't an exact quote, but see their [1995, 533] for a similar remark.

theories can have very unnatural models;[28] rather, the idea here is that if T truly describes P, then T is intuitively and straightforwardly *about P* — that is, P is a part of the mathematical realm that is, so to speak, *lifted straight off* of the theory, and not some convoluted, unnatural model.)

Now, as we saw above, it is true that thesis (0) *follows from* FBP and, indeed, that (0) is an important feature of my FBP-ist epistemology; but I never intended to use (0) as a *definition* of FBP (I make this point in my book (chapter 1, endnote 13)). One reason for this is as follows: if (0) is true, then it requires explanation, and as far as I can see, the explanation could only be that the mathematical realm is plenitudinous.[29] Thus, by defining FBP as the view that the mathematical realm is plenitudinous, I am simply zeroing in on something that is, in some sense, prior to (0); so again, on this approach, (0) doesn't define FBP — it *follows* from FBP. Moreover, this way of proceeding dovetails with the fact that FBP is, at bottom, an *ontological* thesis, i.e., a thesis about which mathematical objects exist. The thesis that the mathematical realm is plenitudinous (which is what I take FBP to be) is an ontological thesis of this sort, but intuitively, (0) is not; intuitively, (0) is a thesis about mathematical theories, not mathematical objects.

Nonetheless, Restall's objections are directed toward (1) and (0), taken as definitions. Now, since I don't endorse (1) or (0) as definitions, these objections are irrelevant. Nonetheless, I want to discuss Restall's objections to show that they don't raise any problems for the definitions I do use (or any other part of my view). So let us turn to his objections now.

2.1.3.2 Restall's Objection Regarding Formalization Restall begins by pointing out that if FBP-ists are going to use a definition along the lines of (1), they need to insist that the second-order predicate Y be a mathematical predicate. I agree with this; as I made clear in the book, FBP is supposed to be restricted to *purely* mathematical theories, and so, obviously, I should have insisted that Y be purely mathematical. Thus, letting 'Math (Y)' mean 'Y is a purely mathematical property', we can replace (1) with

(3) $(\forall Y)[(\text{Math}(Y) \,\&\, \Diamond(\exists x)(Mx \,\&\, Yx)) \supset (\exists x)(Mx \,\&\, Yx)]$.

Restall then goes on to argue that (3) is unacceptable because it is contradictory; for, Restall argues, since CH and ∼CH are both logically possible, it follows from (3) that CH and ∼CH are both true.

As I pointed out above (section 2.1.1.5), this worry arises not just for definitions like (3), but for FBP in general. In particular, one might worry that because FBP

---

[28] Moreover, T could truly describe a part of the mathematical realm that isn't a *model* at all; e.g., one might say of a given set theory that it truly describes the part of the mathematical realm that consists of all pure sets. But there is no model that corresponds to this part of the mathematical realm, because the domain of such a model would be the set of all sets, and there is no such thing.

[29] Alternatively, one might try to explain (0) by appealing to Henkin's theorem that all syntactically consistent first-order theories have models, but this won't work; see my book (chapter 3, note 10) for more on this.

entails that all consistent purely mathematical theories truly describe collections of abstract objects, and because ZF+CH and ZF+∼CH are both consistent purely mathematical theories, FBP entails that CH and ∼CH are both true. I responded to this objection in my book (chapter 3, section 4); I won't repeat here everything I said there, but I would like to briefly explain how I think FBP-ists can respond to this worry. (And after doing this, I will also say a few words about the status of (3) in this connection.)

The main point that needs to be made here is that FBP does not lead to contradiction, because it does not entail that either CH or ∼CH is true. It entails that they both truly describe parts of the mathematical realm, but it does not entail that they are *true*, for as we saw above, on the FBP-NUP-ist view, a mathematical statement is *true simpliciter* iff it is true in all intended parts of the mathematical realm (and there is at least one such part); so truly describing a part of the mathematical realm is not sufficient for truth. A second point to be made here is that while FBP entails that both ZF+CH and ZF+∼CH truly describe parts of the mathematical realm, there is nothing wrong with this, because on this view, they describe *different* parts of that realm. That is, they describe different hierarchies. (Again, this is just a sketch of my response to the worry about contradiction; for my full response, see my book (chapter 3, section 4).)

What do these considerations tell us about formalizations like (3)? Well, it reveals another problem with them (which we can add to the problems I mentioned in my book), namely, that such formalizations fail to capture the true spirit of FBP because they don't distinguish between *truly describing a part of the mathematical realm* and *being true*. To solve this problem, we would have to replace the occurrences of 'Yx' in (3) with "'Yx' truly describes x", or something to this effect. But of course, if we did this, we would no longer have a formalization of the sort I was considering.

2.1.3.3 Restall's Objection Regarding FCNN

Next, Restall argues against the following potential definitions of FBP:

(5) Every consistent mathematical theory has a model; and

(7) Every consistent mathematical theory truly describes some part of the mathematical realm.

I wouldn't use either of these definitions, however; if I were going to use a definition of this general sort, I would use (0) rather than (5) or (7). Again, I don't think of (0) as definitional, but if I were going to fall back to a definition of this general kind, it would be to (0) and not to (5) or (7). I disapprove of (5) because it uses 'has a model' instead of 'truly describes part of the mathematical realm', and as I pointed out above, these are not equivalent; and I disapprove of (7) because it isn't restricted to *purely* mathematical theories. Because of this, Restall's objections to (5) and (7) are irrelevant.

At this point, however, Restall claims that even if we restrict our attention to purely mathematical theories — and hence, presumably, move to a definition like

(0) — two problems still remain. I will address one of these problems here and the other in the next section. The first alleged problem can be put like this: (a) if FBP applies only to purely mathematical theories, then it won't apply to FCNN; but (b) if FBP doesn't apply to FCNN, "then we need some *other* reason to conclude that FCNN truly describes some mathematical structure" (Restall, 2003, p. 908).

My response to this is simple: I never claimed (and don't need the claim) that FCNN truly describes part of the mathematical realm. The purpose of the FBP-NUP-ist's appeal to FCNN is to limit the set of structures that count as *intended* structures of arithmetic; the claim, put somewhat roughly, is that a structure counts as an intended structure of arithmetic just in case FCNN truly describes it.[30] But it is not part of FBP-NUP that FCNN *does* truly describe part of the mathematical realm. If it doesn't truly describe any part of the mathematical realm (even on the assumption that FBP is true), then that's a problem for *arithmetic*, not for the FBP-NUP-ist philosophy of arithmetic — it means that there is something wrong with our conception of the natural numbers, because it means that (even if FBP is true) there are no structures that correspond to our number-theoretic intentions and, hence, that our arithmetical theories aren't true. Now, for whatever it's worth, I think it's pretty obvious that there *isn't* anything wrong with our conception of the natural numbers, and so I think that if FBP is true, then FCNN does truly describe part of the mathematical realm. For (a) it seems pretty obvious that FCNN is consistent, and given this, FBP entails that the purely mathematical part of FCNN (i.e., the part consisting of sentences like the axioms and theorems of PA, and sentences like 'Numbers aren't sets') truly describes part of the mathematical realm; and (b) I think it's pretty obvious that the mixed part of FCNN (i.e., the part containing sentences like 'Numbers aren't chairs') is more or less trivial and, in particular, that it doesn't rule out all of the parts of the mathematical realm that are truly described by the purely mathematical part of FCNN; it is just very implausible to suppose that there are mixed sentences built into the way that we *conceive* of the natural numbers that rule out *all* of the "candidate structures" (from the vast, plenitudinous mathematical realm) that are truly described by the purely mathematical part of FCNN. Of course, this is *conceivable* — it *could* be (in some sense) that it's built into FCNN that 2 is such that snow is purple. But this just seems very unlikely. (Of course, it is also very unlikely that it's built into FCNN that 2 is such that snow is white; our conception of 2 is pretty obviously neutral regarding the color of snow, although I think it does follow from our conception of 2 that it isn't *made* of snow.) In any event, if the above remarks are correct, and if FBP is true, then it is *very likely* that FCNN truly describes part of the mathematical realm. But again, the FBP-NUP-ist doesn't need this result.

---

[30] I say this is "somewhat rough" because it is a bit simplified; in particular, it assumes that FCNN is consistent. I say a few words about how to avoid this assumption in my [2001], especially in endnotes 5, 18, and 20 (and the corresponding text).

2.1.3.4 Restall's Objection Regarding Non-Uniqueness The second alleged problem that still remains after we restrict FBP to purely mathematical theories (and the last problem that Restall raises) is that definitions of FBP along the lines of (0) are inconsistent with NUP. Restall claims that if NUP is true, and if we have a standard semantics, so that only one thing can be identical to the number 3, then mathematical theories don't truly describe their objects in the manner of (0).

First of all, it strikes me as an utter contortion of issues to take this as an objection to (0)-type definitions of FBP. Restall's objection can be put in the following way: "If you embrace (0)-type FBP and NUP, then you'll have to endorse the thesis that

(M) The numeral '3' doesn't have a unique reference; i.e., there are multiple things that are referents of '3'.

But (M) is absurd, for if '3' refers to two different objects x and y, then we'll have $x = 3$ and $y = 3$ and $x \neq y$, which is a contradiction. Therefore, we have to give up on (0)-type FBP or on NUP." It seems to me, however, that it is clearly NUP, and not FBP, that is the culprit in giving rise to (M); for (a) any version of NUP, whether it is FBP-ist or not, will run into (M)-type problems, but (b) this is not true of FBP — if it is not combined with NUP, it will not run into any such problem. Conclusion: this argument isn't an argument against FBP, or (0)-type definitions of FBP, at all; rather, it is an argument against NUP.

Nonetheless, as an argument against NUP, it is worth considering. Now, the first point I want to make in this connection is that the overall problem here is one that I addressed in my book. I pointed out myself that FBP-NUP entails (M), and I spent several pages (84–90) arguing that it is *acceptable* for platonists to endorse (M) and responding to several arguments for the contrary claim that it is *not* acceptable for platonists to endorse (M). Restall has a different argument for thinking (M) unacceptable, however, and so I want to address his argument.

Restall's argument against (M) is that it leads to contradiction, because if '3' refers to two different objects x and y, then we'll have $x = 3$ and $y = 3$ and $x \neq y$. But in fact, my FBP-NUP-ist view doesn't lead to this contradiction. Of course, there are some theories that endorse (M) that *do* lead to this contradiction. Consider, for instance, a theory that (a) talks about two different structures — e.g., $0^*, 1^*, 2^*, 3^*, \ldots$; and $0', 1', 2', 3' \ldots$ — that both satisfy FCNN and, hence, are both candidates for being the natural numbers, and (b) says that '$3 = 3^*$', '$3 = 3'$', and '$3^* \neq 3'$' are all true. This theory is obviously contradictory. But this isn't my FBP-NUP-ist view; in particular, FBP-NUP doesn't lead to the result that sentences like '$3 = 3^*$' and '$3 = 3'$' are true. Why? Because neither of these sentences is true in all intended parts of the mathematical realm — which, recall, is what is required, according to FBP-NUP, for a mathematical sentence to be true, or true *simpliciter*. Sentences like '$3 = 3^*$' and '$3 = 3'$' are true in some intended structures, but they are not true in *all* intended structures.

(Of course, according to FBP-NUP, sentences like this aren't *false simpliciter* either, and so we have here a failure of bivalence, though of course, not a mathematically interesting or important failure of bivalence. See section 2.1.2.4 above.)

2.1.3.5 Colyvan and Zalta: Non-Uniqueness vs. Incompleteness

It is worth noting that if they wanted to, FBP-ists could avoid committing to NUP and (M). To see how, notice first that FBP-ists can say that among all the abstract mathematical objects that exist in the plenitudinous mathematical realm, some are *incomplete objects*. (Some thought would need to be put into defining 'incomplete', but here's a quick definition off the top of my head that might need to be altered: an object o is *incomplete with respect to the property P* iff there is no fact of the matter as to whether o possesses P.) Given this, and on the assumption that FCNN does truly describe part of the mathematical realm, FBP-ists could claim that FCNN picks out a *unique* part of the mathematical realm, namely, the part that (a) satisfies FCNN and (b) has no features that FCNN doesn't entail that it has. Call this view *incompleteness-FBP*. Zalta [1983] endorses a version of platonism that's similar to this in a couple of ways (but also different in a few important ways — e.g., on his view, FCNN doesn't play any role at all), and in a review of my book, he and co-author Mark Colyvan [1999] point out that no argument is given in my book for thinking that NUP-FBP is superior to incompleteness-FBP.

Colyvan and Zalta are right that I didn't address this in my book, so let me say a few words about why I think FBP-ists should favor NUP-FBP over incompleteness-FBP. It seems to me that incompleteness-FBP would be acceptable only if it were built into our intentions, in ordinary mathematical discourse, that we are speaking of objects that don't have any properties that aren't built into our intentions. Now, of course, it is an empirical question whether this *is* built into our intentions, but it seems to me implausible to claim that it is. If I am right about this, then in fact, our arithmetical intentions just don't zero in on unique objects. Now, I suppose one might object that regardless of whether the above kind of incompleteness is built into our intentions, *uniqueness* is built into our intentions, so that if FCNN doesn't pick out a unique part of the mathematical realm, then it doesn't count as being true. But I think this is just false. If God informed us that there are two different structures that satisfy FCNN and that differ from one another only in ways that no human being has ever imagined (and presumably these differences would be non-structural and, hence, mathematically uninteresting), I do not think the mathematical community (or common sense opinion) would treat this information as falsifying our arithmetical theories. Indeed, I think we wouldn't care that there were two such structures and wouldn't feel that we needed to choose between them in order to make sure that our future arithmetical claims were true. And this is evidence that a demand for uniqueness is not built into FCNN. In other words, it suggests that NUP doesn't fly in the face of our mathematical intentions and that it is perfectly acceptable to say, as NUP-FBP-ists do, that in mathematics, truth *simpliciter* can be defined in terms of truth in all intended parts of the mathematical realm.

2.2 Critique of Anti-Platonism

2.2.1 Introduction: The Fregean Argument Against Anti-Platonism

There are, I suppose, numerous arguments against mathematical anti-platonism (or, what comes to the same thing, in favor of mathematical platonism), but it seems to me that there is only one such argument with a serious claim to cogency. The argument I have in mind is due to Frege [1884; 1893–1903], though I will present it somewhat differently than he did. The argument is best understood as a pair of embedded inferences to the best explanation. In particular, it can be put in the following way:

(i) The only way to account for the truth of our mathematical theories is to adopt platonism.

(ii) The only way to account for the fact that our mathematical theories are applicable and/or indispensable to empirical science is to admit that these theories are true.

Therefore,

(iii) Platonism is true and anti-platonism is false.

Now, *prima facie*, it might seem that (i) is sufficient to establish platonism by itself. But (ii) is needed to block a certain response to (i). Anti-platonists might claim that the alleged fact to be explained in (i) — that our mathematical theories are true — is really no fact at all. More specifically, they might respond to (i) by denying that our mathematical theories are true and endorsing *fictionalism* — which, recall, is the view that (a) mathematical sentences like '2 + 1 = 3' do purport to be about abstract objects, but (b) there are no such things as abstract objects, and so (c) these sentences are not true. The purpose of (ii) is to argue that this sort of fictionalist response to (i) is unacceptable; the idea here is that our mathematical theories have to be true, because if they were fictions, then they would be no more useful to empirical scientists than, say, the novel *Oliver Twist* is. (This argument — i.e., the one contained in (ii) — is known as the *Quine-Putnam indispensability argument*, but it does trace to Frege.[31])

I think that the best — and, in the end, the only tenable — anti-platonist response to the Fregean argument in (i)–(iii) is the fictionalist response. Thus, what I want to do here is (a) defend fictionalism (I will do this in section 2.2.4, as well as the present section), and (b) attack the various non-fictionalistic versions of anti-platonism (I will argue against non-fictionalistic versions of anti-realistic anti-platonism in section 2.2.2, and I will argue against the two realistic versions of anti-platonism, i.e., physicalism and psychologism, in section 2.2.3). Now, in connection

---

[31]Frege appealed only to *applicability* here; see his [1893-1903, section 91]. The appeal to *indispensability* came with Quine (see, e.g., his [1948] and [1951]) and Putnam [1971; 1975].

with task (a) — i.e., the defense of fictionalism — the most important objection that needs to be addressed is just the Quine-Putnam objection mentioned in the last paragraph. I will explain how fictionalists can respond to this objection in section 2.2.4. It is worth noting, however, that there are a few other "minor" objections that fictionalists need to address. Here, for instance, are a few worries that one might have about fictionalism, aside from the Quine-Putnam worry:

1. One might worry that fictionalism is not genuinely anti-platonistic, i.e., that any plausible formulation of the view will involve a commitment to abstract objects. E.g., one might think that (a) fictionalists need to appeal to modal notions like *necessity* and *possibility* (or perhaps, *consistency*) and (b) the only plausible ways of interpreting these notions involve appeals to abstract objects, e.g., possible worlds. Or alternatively, one might claim that when fictionalists endorse sentences like "'3 is prime' is true-in-the-story-of-mathematics," they commit to abstract objects, e.g., sentence types and stories. (One might also worry that Field's nominalization program commits fictionalists to spacetime points and the use of second-order logic, and so one might think that, for these reasons, the view is not genuinely anti-platonistic; but we needn't worry here about objections to Field's nominalization program, because I am going to argue below that fictionalists don't need to — and, indeed, *shouldn't* — rely upon that program.)

2. One might worry that fictionalists cannot account for the objectivity of mathematics; e.g., one might think that fictionalists can't account for how there could be a correct answer to the question of whether the continuum hypothesis (CH) is true or false.

3. One might worry that fictionalism flies in the face of mathematical and scientific practice, i.e., that the thesis that mathematics consists of a body of truths is inherent in mathematical and scientific practice.

In my book (chapter 1, section 2.2, chapter 5, section 3, and the various passages cited in those two sections), I respond to these "minor" objections to fictionalism — i.e., objections other than the Quine-Putnam objection. I will not take the space to respond to all of these worries here, but I want to say just a few words about worry 2, i.e., about the problem of objectivity.

The reader might recall from section 2.1.1.5 that an almost identical problem of objectivity arises for FBP. (The same problem arises for both FBP and fictionalism because both views entail that from a purely *metaphysical* point of view, ZF+CH and ZF+∼CH are equally "good" theories; FBP says that both of these theories truly describe parts of the mathematical realm, and fictionalism says that both of these theories are fictional.) Now, in section 2.1.2.4, I hinted at how FBP-ists can respond to this worry, and it is worth pointing out here that fictionalists can say essentially the same thing. FBP-ists should say that whether ZF+CH or ZF+∼CH is correct comes down to the question of which of these theories (if either) is true in all of the intended parts of the mathematical realm, and that this in turn comes down to whether CH or ∼CH is inherent in *our notion of set*. Likewise, fictionalists should say that the question of whether CH is "correct" is determined by whether it's part of the story of set theory, and that this is determined by whether CH would have been true (in all intended parts of the mathematical realm) if there had existed sets, and that this in turn is determined by whether CH is inherent in our notion of set. So even though CH is undecidable in current set theories like ZF, the question of the correctness of CH could still have an objectively correct answer, according to fictionalism, in the same way that the question of whether 3 is prime has an objectively correct answer on the fictionalist view. But fictionalists should also allow, in agreement with FBP-ists, that it *may* be that neither CH nor ∼CH is inherent in our notion of set and, hence, may be that there is no objectively correct answer to the CH question. (I say a bit more about this below, but for a full defense of the FBP-ist/fictionalist view of CH, see my [2001] and [2009], as well as the relevant discussions in my book (chapter 3, section 4, and chapter 5, section 3).)

Assuming, then, that the various "minor" objections to fictionalism can be answered, the only objection to that view that remains is the Quine-Putnam indispensability objection. In section 2.2.4, I will defend fictionalism against this objection. (Field tried to respond to the Quine–Putnam objection by arguing that mathematics is not indispensable to empirical science. In contrast, I have argued, and will argue here, that fictionalists can (a) admit (for the sake of argument) that there *are* indispensable applications of mathematics to empirical science and (b) account for these indispensable applications from a fictionalist point of view, i.e., without admitting that our mathematical theories are true.) Before I discuss this, however, I will argue against the various non-fictionalistic versions of anti-platonism (sections 2.2.2–2.2.3).

2.2.2 Critique of Non-Fictionalistic Versions of Anti-Realistic Anti-Platonism

In the next two sections, I will critique the various non-fictionalistic versions of anti-platonism. I will discuss non-fictionalistic versions of anti-realistic anti-platonism in the present section, and I will discuss realistic anti-platonism (i.e., physicalism and psychologism) in the next section, i.e., section 2.2.3.

Given the result that the Quine-Putnam worry is the only important worry about fictionalism, it is easy to show that no version of anti-realistic anti-platonism possesses any advantage over fictionalism. For it seems to me that all versions of anti-realism encounter the same worry about applicability and indispensability that fictionalism encounters. Consider, for example, deductivism. Unlike fictionalists, deductivists try to salvage mathematical truth. But the truths they salvage cannot be lifted straight off of our mathematical theories. That is, if we take the theorems of our various mathematical theories at *face value*, then according to deductivists, they are not true. What deductivists claim is that the theorems of our mathematical theories "suggest" or "represent" certain closely related mathematical assertions that *are* true. For instance, if T is a theorem of Peano

Arithmetic (PA), then according to deductivists, it represents, or stands for, the truth '$AX \supset T$', or '$\Box(AX \supset T)$', where AX is the conjunction of all of the axioms of PA used in the proof of T. Now, it should be clear that deductivists encounter the same problem of applicability and indispensability that fictionalists encounter. For while sentences like '$AX \supset T$' are true, according to deductivists, AX and T and PA are *not* true, and so it is still mysterious how mathematics could be applicable (or, indeed, indispensable) to empirical science.

Now, one might object here that the problem of applicability and indispensability that deductivists face is *not the same* as the problem that fictionalists face, because deductivists have their "surrogate mathematical truths", i.e., their conditionals, and they might be able to solve the problem of applicability by appealing to these truths. But this objection is confused. If these "surrogate mathematical truths" are really *anti-platonistic* truths — and they have to be if they are going to be available to deductivists — then fictionalists can endorse them as easily as deductivists can, and moreover, they can appeal to them in trying to solve the problem of applicability. The only difference between fictionalists and deductivists in this connection is that the former do not try to use any "surrogate mathematical truths" to *interpret mathematical theory*. But they can still *endorse* these truths and appeal to them in accounting for applicability and/or indispensability. More generally, the point is that deductivism doesn't provide anti-platonists with *any* truths that aren't available to fictionalists. Thus, deductivists do not have any advantage over fictionalists in connection with the problem of applicability and indispensability.[32]

In my book (chapter 5, section 4), I argue that analogous points can be made about *all* non-fictionalist versions of anti-realistic anti-platonism — e.g., conventionalism, formalism, and so on. In particular, I argue that (a) all of these views give rise to *prima facie* worries about applicability and indispensability, because they all make the sentences and theories of mathematics factually empty in the sense that they're not "about the world", because they all maintain that our mathematical singular terms are *vacuous*, i.e., fail to refer; and (b) none of these views has any advantage over fictionalism in connection with the attempt to solve the problem of applications, because insofar as these views deny the existence of mathematical objects, their proponents do not have available to them any means of solving the problem that aren't also available to fictionalists.

These remarks suggest that, for our purposes, we could lump all versions of anti-realistic anti-platonism together and treat them as a single view. Indeed, I argued in my book (chapter 5, section 4) that if I replaced the word 'fictionalism' with the expression 'anti-realistic anti-platonism' throughout the book, all the same points could have been made; I would have had to make a few stylistic changes, but

---

[32]Thus, for instance, fictionalists are free to endorse Hellman's [1989, chapter 3] account of applicability. For whatever it's worth, I do not think that Hellman's account of applicability is a good one, because I think that the various problems with the conditional interpretation of mathematics carry over to the conditional interpretation of empirical theory. I will say a few words about these problems below.

nothing substantive would have needed to be changed, because all the important features of fictionalism that are relevant to the arguments I mounted in my book are shared by all versions of anti-realistic anti-platonism.

But I did not proceed in that way in the book; instead, I took fictionalism as a representative of anti-realistic anti-platonism and concentrated on it. The reason, very simply, is that I think there are good reasons for thinking that fictionalism is the *best* version of anti-realistic anti-platonism. One argument (not the only one) can be put in the following way.

The various versions of anti-realistic anti-platonism do not differ from fictionalism (or from one another) in any metaphysical or ontological way, because they all deny the existence of mathematical objects. (This, by the way, is precisely why they don't differ in any way that is relevant to the arguments concerning fictionalism that I develop in my book.) With a couple of exceptions, which I'll discuss in a moment, the various versions of anti-realism differ from fictionalism (and from one another) only in the interpretations that they provide for mathematical theory. But as soon as we appreciate this point, the beauty of fictionalism and its superiority over other versions of anti-realism begin to emerge. For whereas fictionalism interprets our mathematical theories in a very standard, straightforward, face-value way, other versions of anti-realism — e.g., deductivism, formalism, and Chihara's view — advocate controversial, non-standard, non-face-value interpretations of mathematics that seem to fly in the face of actual mathematical practice. Now, in my book (chapter 5, section 4), I say a bit about *why* these non-standard interpretations of mathematical theory are implausible; but since I don't really need this result — since I could lump all the versions of anti-realism together — I will not pursue this here. (It is worth noting, however, that in each case, the point is rather obvious — or so it seems to me. If we see the various non-standard interpretations of mathematics as claims about the semantics of actual mathematical discourse, they just don't seem plausible. E.g., it doesn't seem plausible to suppose, with deductivists, that ordinary utterances of '3 is prime' really mean '(Necessarily) if there are natural numbers, then 3 is prime'. If we're just doing empirical semantics (that is, if we're just trying to discover the actual semantic facts about actual mathematical discourse), then it seems very plausible to suppose that '3 is prime' means that 3 is prime — which, of course, is just what fictionalists say.[33])

There are two versions of non-fictionalistic anti-realism, however, that *don't*

---

[33] At least one advocate of reinterpretation anti-realism — namely, Chihara — would admit my point here; he does not claim that his theory provides a good interpretation of actual mathematical discourse. But given this, what possible reason could there be to adopt Chihara's view? If (a) the fictionalistic/platonistic semantics of mathematical discourse is the correct one, and (b) there's no reason to favor Chihara's anti-realism over fictionalism — after all, it encounters the indispensability problem, provides no advantage in solving that problem, and so on — then isn't fictionalism the superior view? It seems to me that if point (a) above is correct, and if (as fictionalists and Chihara agree) there are no such things as abstract objects, then fictionalism is the correct view of actual mathematics. Chihara's view might show that we could have done mathematics differently, in a way that would have made our mathematical assertions come out true, but I don't see why this provides any motivation for Chihara's view.

offer non-standard interpretations of mathematical discourse. But the problems with these views are just as obvious. One view here is the second version of Meinongianism discussed in section 1.2 above; advocates of this view agree with the platonist/fictionalist semantics of mathematese; the only point on which they differ from fictionalists is in their claim that the sentences of mathematics are true; but as we saw in section 1.2, second-version Meinongians obtain this result only by using 'true' in a non-standard way, maintaining that a sentence of the form 'Fa' can be true even if its singular term (i.e., 'a') doesn't refer to anything. The second view here is conventionalism, which holds that mathematical sentences like '3 is prime' are analytically true. Now, advocates of this view *might* fall back on a non-standard-interpretation strategy, maintaining that the *reason* '3 is prime' is analytic is that it really means, say, 'If there are numbers, then 3 is prime' — or whatever. But if conventionalists don't fall back on a reinterpretation strategy, then their thesis is just implausible, and for much the same reason that second-version Meinongianism is implausible: if we read '3 is prime' (or better, 'There is a prime number between 2 and 4') at face value, then it's clearly not analytic, because (a) in order for this sentence to be true, there has to exist such a thing as 3, and (b) sentences with existential commitments are not analytic, because they cannot be conceptually true, or true in virtue of meaning, or anything else along these lines.

One might object to the argument that I have given here — i.e., the argument for the supremacy of fictionalism over other versions of anti-realism — on the grounds that fictionalism *also* runs counter to mathematical practice. In other words, one might think that it is built into mathematical and/or scientific practice that mathematical sentences like '3 is prime' are *true*. But in my book (chapter 5, section 3), I argue that this is not the case.

(This is just a sketch of my argument for taking fictionalism to be the best version of anti-realism; for more detail, see my book (chapter 5, section 4) and for a different argument for th supremacy of fictionalism over other versions of anti-realism, see my [2008].)

2.2.3 Critique of Realistic Anti-Platonism (i.e., Physicalism and Psychologism)

In this section, I will argue against the two realistic versions of anti-platonism, thus completing my argument for the claim that fictionalism is the only tenable version of anti-platonism. I will discuss psychologism first and then move on to physicalism.

I pointed out in section 1.1.1 that psychologism is a sort of watered-down version of realism; for while it provides an ontology for mathematics, the objects that it takes mathematical theories to be about do not exist independently of us and our theorizing (for this reason, one might even deny that it is a version of realism, but this doesn't matter here). Because of this, psychologism is similar in certain ways to fictionalism. For one thing, psychologism and fictionalism both involve the idea that mathematics comes entirely from us, as opposed to something independent of us. Now, of course, fictionalists and psychologists put the idea here in different ways: fictionalists hold that our mathematical theories are fictional stories and, hence, not true, whereas advocates of psychologism allow that these theories are true, because the "characters" of the fictionalist's stories exist in the mind; but this is a rather *empty* sort of truth, and so psychologism does not take mathematics to be *factual* in a very deep way. More importantly, psychologism encounters the same worry about applicability and indispensability that fictionalism encounters; for it is no less mysterious how a story about ideas in our heads could be applicable to physical science than how a fictional story could be so applicable.

What, then, does the distinction between psychologism and fictionalism really come to? Well, the difference certainly *doesn't* lie in the assertion of the *existence* of the mental entities in question. Fictionalists admit that human beings do have ideas in their heads that correspond to mathematical singular terms. They admit, for instance, that I have an idea of the number 3. Moreover, they admit that we can make claims about these mental entities that correspond to our mathematical claims; corresponding to the sentence '3 is prime', for instance, is the sentence 'My idea of 3 is an idea of a prime number'. The only difference between fictionalism and psychologism is that the latter, unlike the former, involves the claim that our mathematical theories are *about* these ideas in our heads. In other words, advocates of psychologism maintain that the sentences '3 is prime' and 'My idea of 3 is an idea of a prime number' say essentially the *same thing*, whereas fictionalists deny this. Therefore, it seems to me that the relationship between fictionalism and psychologism is essentially equivalent to the relationship between fictionalism and the versions of anti-realistic anti-platonism that I discussed in section 2.2.2. In short, psychologism interprets mathematical theory in an empty, non-standard way in an effort to salvage mathematical truth, but it still leads to the Quine-Putnam indispensability problem in the same way that fictionalism does, and moreover, it doesn't provide anti-platonists with any means of solving this problem that aren't also available to fictionalists, because it doesn't provide anti-platonists with any entities or truths that aren't available to fictionalists.

It follows from all of this that psychologism can be handled in the same way that I handled the various versions of non-fictionalistic anti-realism and, hence, that I do not really need to refute the view. But as is the case with the various versions of non-fictionalistic anti-realism, it is easy to see that fictionalism is superior to psychologism, because the psychologistic interpretation of mathematical theory and practice is implausible. The arguments here have been well-known since Frege destroyed this view of mathematics in 1884. First of all, psychologism seems incapable of accounting for any talk about the class of *all* real numbers, since human beings could never construct them all. Second, psychologism seems to entail that assertions about very large numbers (in particular, numbers that no one has ever thought about) are all untrue; for if none of us has ever constructed some very large number, then any proposition about that number will, according to psychologism, be vacuous. Third, psychologism seems incapable of accounting for mathematical *error*: if George claims that 4 is prime, we cannot argue with him, because he is presumably saying that *his* 4 is prime, and for all we know, this could very well be *true*.³⁴ And finally, psychologism turns mathematics into a branch of psychology, and it makes mathematical truths contingent upon psychological truths, so that, for instance, if we all died, '2 + 2 = 4' would suddenly become untrue. As Frege says, "Weird and wonderful ... are the results of taking seriously the suggestion that number is an idea."³⁵

Let me turn now to Millian physicalism. The idea here, recall, is that mathematics is simply a very general natural science and, hence, that it is about ordinary physical objects. Thus, just as astronomy gives us laws concerning all astronomical bodies, so arithmetic and set theory give us laws concerning all objects and piles of objects. The sentence '2 + 1 = 3', for instance, says that whenever we add one object to a pile of two objects, we end up with a pile of three objects.

Let me begin my critique of physicalism by reminding the reader that in section 2.1.1.3, I argued that because (a) there are infinitely many numerically distinct sets corresponding to every physical object and (b) all of these sets share the same physical base (i.e., are made of the same matter and have the same spatiotemporal location), it follows that (c) there must be something non-physical about these sets, over and above the physical base, and so it could not be true that sets are purely physical objects. A second problem with physicalism is that there simply isn't enough physical stuff in the universe to satisfy our mathematical theories. ZF, for instance, tells us that there are infinitely many transfinite cardinals. It is not plausible to suppose that this is a true claim about the physical world. A third problem with physicalism is that (a) it seems to entail that mathematics is an empirical science, contingent on physical facts and susceptible to empirical falsification, but (b) it seems that mathematics is not empirical and that its truths cannot be empirically falsified. (These arguments are all very quick; for a more thorough argument against the Millian view, see my book (chapter 5, section 5).)

Some of the problems with Millian physicalism are avoided by Kitcher's view [1984, chapter 6]. But as I argue in my book (chapter 5, section 5), Kitcher avoids these problems only by collapsing back into an *anti-realistic* version of anti-platonism, i.e., a view that takes mathematical theory to be *vacuous*. In particular, on Kitcher's view — and he readily admits this [1984, 117] — mathematical theories make claims about non-existent objects, namely, ideal agents. Thus, since Kitcher's view is a version of anti-realism, it can be handled in the same way that I handled all of the other versions of non-fictionalistic anti-realism: (a) I do not have to provide a refutation of Kitcher's view, because it would be acceptable to lump it together with fictionalism; and (b) while Kitcher's view has no advantage

---

³⁴One might reply that the notion of error can be analyzed in terms of non-standardness, but I suspect that this could be cashed out only in terms of *types*. That is, the claim would have to be that a person's theory of arithmetic could be erroneous, or bad, if her concepts of 1, 2, 3, etc. were not of the culturally accepted types. But to talk of types of 1's, 2's, 3's, etc. is to collapse back into platonism.

³⁵See Frege [1884, section 27]. Just about all of the arguments mentioned in this paragraph trace to Frege. His arguments against psychologism can be found in his [1884, introduction and section 27; 1893-1903, introduction; 1894 and 1919].

over fictionalism (it still encounters the indispensability problem, delivers no way of solving that problem that's not also available to fictionalists, and so on), we do have reason to favor fictionalism over Kitcher's view, because the latter involves a non-standard, non-face-value interpretation of mathematical discourse that flies in the face of actual mathematical practice. (Once again, this is just a sketch of my argument for the claim that fictionalism is superior to Kitcher's view; for more detail, see my book (chapter 5, section 5).)

2.2.4 Indispensability

I have now criticized all of the non-fictionalistic versions of anti-platonism, but I still need to show that fictionalists can respond to the Quine–Putnam indispensability argument (other objections to fictionalism were discussed in section 2.2.1). The Quine-Putnam argument is based on the premises that (a) there are indispensable applications of mathematics to empirical science and (b) fictionalists cannot account for these applications. There are two strategies that fictionalists can pursue in trying to respond to this argument. The first strategy, developed by Field [1980], is to argue that

(NI) Mathematics is *not indispensable* to empirical science; and

(AA) The mere fact that mathematics is applicable to empirical science — i.e., applicable in a dispensable way — can be accounted for without abandoning fictionalism.

Most critics have been willing to grant thesis (AA) to Field,[36] but (NI) is extremely controversial. To motivate this premise, one has to argue that all of our empirical theories can be *nominalized*, i.e., reformulated in a way that avoids reference to, and quantification over, abstract objects. Field tries to do this by simply showing how to carry out the nominalization for one empirical theory, namely, Newtonian Gravitation Theory. Field's argument for (NI) has been subjected to a number of objections,[37] and the consensus opinion among philosophers of mathematics seems to be that his nominalization program cannot be made to work. I am not convinced that Field's program cannot be carried out — the most important objection, in my opinion, is Malament's [1982] objection that it is not clear how Field's program can be extended to cover quantum mechanics, but in my [1996b], and in my book (chapter 6), I explain how Field's program can be so extended — but I will not pursue this here, because in the end, I do not think fictionalists should respond to the Quine-Putnam objection via Field's nominalization strategy. I think they should pursue another strategy.

The strategy I have in mind here is (a) to grant (for the sake of argument) that there *are* indispensable applications of mathematics to empirical science — i.e.,

---

[36] But see Shapiro [1983] for one objection to Field's argument for (AA), and see Field [1989, essay 4] for a response.

[37] Malament [1982] discusses almost all of these objections, but see also Resnik [1985] and Chihara [1990, chapter 8, section 5].

that mathematics is hopelessly and inextricably woven into some of our empirical theories — and (b) to simply *account* for these indispensable applications from a fictionalist point of view. I developed this strategy in my book (chapter 7), as well as my [1996a] and [1998b]; the idea has also been pursued by Rosen [2001] and Yablo [2002], and a rather different version of the view was developed by Azzouni [1994] in conjunction with his non-fictionalistic version of nominalism. I cannot even come close here to giving the entire argument for the claim that fictionalists can successfully block the Quine–Putnam argument using this strategy, but I would like to rehearse the most salient points.

The central idea behind this view is that because abstract objects are causally inert, and because our empirical theories don't assign any causal role to them, it follows that the truth of empirical science depends upon two sets of facts that are entirely independent of one another, i.e., that hold or don't hold independently of one another. One of these sets of facts is purely platonistic and mathematical, and the other is purely physical (or more precisely, purely nominalistic). Consider, for instance, the sentence

(A) The physical system S is forty degrees Celsius.

This is a *mixed* sentence, because it makes reference to physical and abstract objects (in particular, it says that the physical system S stands in the Celsius relation to the number 40). But, trivially, (A) does not assign any causal role to the number 40; it is not saying that the number 40 is *responsible* in some way for the fact that S has the temperature it has. Thus, if (A) is true, it is true in virtue of facts about S and 40 that are entirely independent of one another, i.e., that hold or don't hold independently of one another. And again, the same point seems to hold for all of empirical science: since no abstract objects are causally relevant to the physical world, it follows that if empirical science is true, then its truth depends upon two entirely independent sets of facts, *viz.*, a set of purely nominalistic facts and a set of purely platonistic facts.

But since these two sets of facts are *independent* of one another — that is, hold or don't hold independently of one another — it could very easily be that (a) there does obtain a set of purely physical facts of the sort required here, i.e., the sort needed to make empirical science true, but (b) there are no such things as abstract objects, and so there *doesn't* obtain a set of purely platonistic facts of the sort required for the truth of empirical science. In other words, it could be that the *nominalistic content* of empirical science is correct, even if its platonistic content is fictional. But it follows from this that mathematical fictionalism is perfectly consistent with the claim that empirical science paints an essentially accurate picture of the physical world. In other words, fictionalists can endorse what I have called *nominalistic scientific realism* [1996a; 1998, chapter 7; 1998b]. The view here, in a nutshell, is that there do obtain purely physical facts of the sort needed to make empirical science true (regardless of whether there obtain mathematical facts of the sort needed to make empirical science true); in other words, the view is that the physical world holds up *its end* of the "empirical-science bargain".

Nominalistic scientific realism is different from standard scientific realism. The latter entails that our empirical theories are strictly true, and fictionalists cannot make this claim, because that would commit them to the existence of mathematical objects. Nonetheless, nominalistic scientific realism is a genuinely *realistic* view; for if it is correct — i.e., if there does obtain a set of purely physical facts of the sort needed to make empirical science true — then even if there are no such things as mathematical objects and, hence, our empirical theories are (strictly speaking) not true, the physical world is nevertheless *just the way empirical science makes it out to be*. So this is, indeed, a kind of scientific realism.

What all of this shows is that fictionalism is consistent with the actual role that mathematics plays in empirical science, whether that role is indispensable or not. It simply doesn't matter (in the present context) whether mathematics is indispensable to empirical science, because even if it is, the picture that empirical science paints of the physical world could still be essentially accurate, even if there are no such things as mathematical objects.

Now, one might wonder what mathematics is doing in empirical science, if it doesn't need to be true in order for empirical science to be essentially accurate. The answer, I argue, is that mathematics appears in empirical science as a *descriptive aid*; that is, it provides us with an easy way of saying what we want to say about the physical world. In my book, I argue that (a) this is indeed the role that mathematics plays in empirical science, and (b) it follows from this that mathematics doesn't need to be true in order to do what it's supposed to do in empirical science.

(Again, this is just a quick summary; for the full argument that fictionalism can be defended against the Quine–Putnam argument along these lines, see my book (chapter 7), as well as my [1996a] and [1998b].)

(Given that I think that Field's response to the Quine–Putnam argument may be defensible, why do I favor my own response, i.e., the response just described in the last few paragraphs? Well, one reason is that my response is simply less controversial — i.e., it's not open to all the objections that Field's response is open to. A second reason is that my response fits better with mathematical and scientific practice (I argue this point in my book (chapter 7, section 3)). A third reason is that whereas Field's strategy can yield only a piecemeal response to the problem of the applications of mathematics, I account for all applications of mathematics at the same time and in the same way (again, I argue for this in my book (chapter 7, section 3)). And a fourth reason is that unlike Field's view, my view can be generalized so that it accounts not just for the use made of mathematics in empirical science, but also for the use made there of *non-mathematical*-abstract-object talk — e.g., the use made in belief psychology of 'that'-clauses that purportedly refer to propositions (the argument for this fourth reason is given in my [1998b]).)

2.3 Critique of Platonism Revisited: Ockham's Razor

I responded above to the two Benacerrafian objections to platonism, i.e., the epistemological objection and the non-uniqueness objection. These are widely regarded as the two most important objections to platonism, but there are other objections that platonists need to address. For one thing, as I pointed out above, there are a number of objections that one might raise against FBP-NUP in particular; I discussed these above (section 2.1) and in more detail in my book (chapters 3 and 4). But there are also some remaining objections to platonism in general; e.g., there is a worry about how platonists can account for the applicability of mathematics, and there are worries about whether platonism is consistent with our abilities to refer to, and have beliefs about, mathematical objects. In my book, I responded to these remaining objections (e.g., I argued that FBP-NUP-ists can account for the applicability of mathematics in much the same way that fictionalists can, and I argued that they can solve the problems of belief and reference in much the same way that they solve the epistemological problem). In this section, I would like to say just a few words about one of the remaining objections to platonism, in particular, an objection based on Ockham's razor (for my full response to this objection, see my book (chapter 7, section 4.2)).

I am trying to argue for the claim that fictionalism and FBP are both defensible and that they are equally well motivated. But one might think that such a stance cannot be maintained, because one might think that if both of these views are really defensible, then by Ockham's razor, fictionalism is superior to FBP, because it is more parsimonious, i.e., it doesn't commit to the existence of mathematical objects. To give a bit more detail here, one might think that Ockham's razor dictates that if *any* version of anti-platonism is defensible, then it is superior to platonism, regardless of whether the latter view is defensible or not. That is, one might think that in order to motivate platonism, one needs to refute every different version of anti-platonism.

This, I think, is confused. If *realistic* anti-platonists (e.g., Millians) could make their view work, then they could probably employ Ockham's razor against platonism. But we've already seen (section 2.2.3) that realistic anti-platonism is untenable. The only tenable version of anti-platonism is *anti-realistic* anti-platonism. But advocates of this view, e.g., fictionalists, cannot employ Ockham's razor against platonism, because they simply throw away the facts that platonists claim to be explaining. Let me develop this point in some detail.

One might formulate Ockham's razor in a number of different ways, but the basic idea behind the principle is the following: if

(1) theory A explains everything that theory B explains, and

(2) A is more ontologically parsimonious than B, and

(3) A is just as simple as B in all non-ontological respects,

then A is superior to B. Now, it is clear that fictionalism is more parsimonious than FBP, so condition (2) is satisfied here. But despite this, we cannot use Ockham's razor to argue that fictionalism is superior to FBP, because neither of the other two conditions is satisfied here.

With regard to condition (1), FBP-ists will be quick to point out that fictionalism does not account for everything that FBP accounts for. In particular, it doesn't account for facts such as that 3 is prime, that $2 + 2 = 4$, and that our mathematical theories are true in a face-value, non-factually-empty way. Now, of course, fictionalists will deny that these so-called "facts" really are facts. Moreover, if my response to the Quine-Putnam argument is acceptable, and if I am right that the Quine-Putnam argument is the only initially promising argument for the (face-value, non-factually-empty) truth of mathematics, then it follows that FBP-ists have no *argument* for the claim that their so-called "facts" really are facts. But unless fictionalists have an argument for the claim that these so-called "facts" really *aren't* facts — and more specifically, for the claim that our mathematical theories aren't true (in a face-value, non-factually-empty way) — we will be in a stalemate. And given the results that we've obtained so far, it's pretty clear that fictionalists *don't* have any argument here. To appreciate this, we need merely note that (a) fictionalists don't have any good *non-Ockham's-razor-based* argument here (for we've already seen that aside from the Ockham's-razor-based argument we're presently considering, there is no good reason for favoring fictionalism over FBP); and (b) fictionalists don't have any good *Ockham's-razor-based* argument here — i.e., for the claim that the platonist's so-called "facts" really aren't facts — because Ockham's razor cannot be used to settle disputes over the question of what the facts that require explanation *are*. That principle comes into play only after it has been agreed what these facts are. More specifically, it comes into play only in adjudicating between two explanations of an agreed-upon collection of facts. So Ockham's razor cannot be used to adjudicate between realism and anti-realism (whether in mathematics, or empirical science, or common sense) because there is no agreed-upon set of facts here, and in any event, the issue between realists and anti-realists is not which explanations we should accept, but whether we should suppose that the explanations that we eventually settle upon, using criteria such as Ockham's razor, are really *true*, i.e., provide us with accurate descriptions of the world.

Fictionalists might try to respond here by claiming that the platonist's appeal to the so-called "fact" of mathematical truth, or the so-called "fact" that $2+2=4$, is just a disguised assertion that platonism is true. But platonists can simply turn this argument around on fictionalists: if it is question begging for platonists simply to assert that mathematics is true, then it is question begging for fictionalists simply to assert that it's *not* true. Indeed, it seems to me that the situation here actually favors the platonists, for it is the fictionalists who are trying to mount a positive argument here and the platonists who are merely trying to defend their view.

Another ploy that fictionalists might attempt here is to claim that what we need to consider, in deciding whether Ockham's razor favors fictionalism over FBP, is not whether fictionalism accounts for all the *facts* that FBP accounts for, but whether fictionalism accounts for all the *sensory experiences*, or all the *empirical phenomena*, that FBP accounts for. I will not pursue this here, but I argue in my book (chapter 7, section 4.2) that fictionalists cannot legitimately respond to the above argument in this way.

Before we move on, it is worth noting that there is also a historical point to be made here. The claim that there are certain facts that fictionalism cannot account for is not an *ad hoc* device, invented for the sole purpose of staving off the appeal to Ockham's razor. Since the time of Frege, the motivation for platonism has always been to account for mathematical truth. This, recall, is precisely how I formulated the argument for platonism (or against anti-platonism) in section 2.2.1.

I now move on to condition (3) of Ockham's razor. In order to show that this condition isn't satisfied in the present case, I need to show that there are certain non-ontological respects in which FBP is simpler than fictionalism. My argument here is this: unlike fictionalism, FBP enables us to say that our scientific theories are true (or largely true) and it provides a uniform picture of these theories. As we have seen, fictionalists have to tell a slightly longer story here; in addition to claiming that our mathematical theories are fictional, they have to maintain that our empirical theories are, so to speak, *half* truths — in particular, that their nominalistic contents are true (or largely true) and that their platonistic contents are fictional. Moreover, FBP is, in this respect, more *commonsensical* than fictionalism, because it enables us to maintain that sentences like '$2+2=4$' and 'the number of Martian moons is 2' are true.

Now, I do not think that the difference in simplicity here between FBP and fictionalism is very substantial. But on the other hand, I do not think that the ontological parsimony of fictionalism creates a very substantial difference between the two views either. In general, the reason we try to avoid excess ontology is that ontological excesses tend to make our worldview more cumbersome, or less elegant, by adding unnecessary "loops and cogs" to the view. But we just saw in the preceding paragraph that in the case of FBP, this is not true; the immense ontology of FBP doesn't make our worldview more cumbersome, and indeed, it actually makes it less cumbersome. Moreover, the introduction of abstract objects is extremely uniform and non-arbitrary within FBP: we get *all* the abstract objects that there could possibly be. But, of course, despite these considerations, the fact remains that FBP does add a category to our ontology. Thus, it is less parsimonious than fictionalism, and so, in this respect, it is not as simple as fictionalism. Moreover, since the notion of an abstract object is not a commonsensical one, we can say that, in this respect, fictionalism is more commonsensical than FBP.

It seems, then, that FBP is simpler and more commonsensical than fictionalism in some ways but that fictionalism is simpler and more commonsensical in other ways. Thus, the obvious question is whether one of these views is simpler *overall*. But the main point to be made here, once again, is that there are no good *arguments* on either side of the dispute. What we have here is a matter of *brute intuition*: platonists are drawn to the idea of being able to say that our mathematical and empirical theories are straightforwardly true, whereas fictionalists are willing to give this up for the sake of ontological parsimony, but neither group has any *argument* here (assuming that I'm right in my claim that there are acceptable responses to all of the known arguments against platonism and fictionalism, e.g., the two Benacerrafian arguments and the Quine-Putnam argument). Thus, the dispute between FBP-ists and fictionalists seems to come down to a head-butt of intuitions. For my own part, I have *both* sets of intuitions, and overall, the two views seem *equally* simple to me.

3 CONCLUSIONS: THE UNSOLVABILITY OF THE PROBLEM AND A KINDER, GENTLER POSITIVISM

If the arguments sketched in section 2 are cogent, then there are no good arguments against platonism or anti-platonism. More specifically, the view I have been arguing for is that (a) there are no good arguments against FBP (although Benacerrafian arguments succeed in refuting all *other* versions of platonism); and (b) there are no good arguments against fictionalism (although Fregean arguments succeed in undermining all other versions of anti-platonism). Thus, we are left with exactly one viable version of platonism, *viz.*, FBP, and exactly one viable version of anti-platonism, *viz.*, fictionalism, but we do not have any good reason for favoring one of these views over the other. My first conclusion, then, is that we do not have any good reason for choosing between mathematical platonism and anti-platonism; that is, we don't have any good arguments for or against the existence of abstract mathematical objects. I call this the *weak epistemic conclusion*.

In the present section, I will argue for two stronger conclusions, which can be formulated as follows.

> *Strong epistemic conclusion*: it's not just that we *currently* lack a cogent argument that settles the dispute over mathematical objects — it's that we could *never* have such an argument.
>
> *Metaphysical conclusion*: it's not just that *we* could never settle the dispute between platonists and anti-platonists — it's that there is *no fact of the matter* as to whether platonism or anti-platonism is true, i.e., whether there exist any abstract objects.[38]

I argue for the strong epistemic conclusion in section 3.1 and for the metaphysical conclusion in section 3.2.

---

[38] Note that while the two epistemic conclusions are stated in terms of mathematical objects in particular, the metaphysical conclusion is stated in terms of abstract objects in general. Now, I actually think that generalized versions of the epistemic conclusions are true, but the arguments given here support only local versions of the epistemic conclusions. In contrast, my argument for the metaphysical conclusion is about abstract objects in general.

3.1 The Strong Epistemic Conclusion

If FBP is the only viable version of mathematical platonism and fictionalism is the only viable version of mathematical anti-platonism, then the dispute over the existence of mathematical objects comes down to the dispute between FBP and fictionalism. My argument for the strong epistemic conclusion is based on the observation that FBP and fictionalism are, surprisingly, very *similar* philosophies of mathematics. Now, of course, there is a sense in which these two views are polar opposites; after all, FBP holds that all logically possible mathematical objects exist whereas fictionalism holds that *no* mathematical objects exist. But despite this obvious difference, the two views are extremely similar. Indeed, they have much more in common with one another than FBP has with other versions of platonism (e.g., Maddian naturalized platonism) or fictionalism has with other versions of anti-platonism (e.g., Millian empiricism). The easiest way to bring this fact out is simply to list the points on which FBP-ists and fictionalists agree. (And note that these are all points on which platonists and anti-platonists of various other sorts do *not* agree.)

1. Probably the most important point of agreement is that according to both FBP and fictionalism, all consistent purely mathematical theories are, from a metaphysical or ontological point of view, equally "good". According to FBP-ists, all theories of this sort truly describe some part of the mathematical realm, and according to fictionalists, *none* of them do — they are all just fictions. Thus, according to both views, the only way that one consistent purely mathematical theory can be "better" than another is by being aesthetically or pragmatically superior, or by fitting better with our intentions, intuitions, concepts, and so on.[39]

2. As a result of point number 1, FBP-ists and fictionalists offer the same account of undecidable propositions, e.g., the continuum hypothesis (CH). First of all, in accordance with point number 1, FBP-ists and fictionalists both maintain that *from a metaphysical point of view*, ZF+CH and ZF+∼CH are equally "good" theories; neither is "better" than the other; they simply characterize different sorts of hierarchies. (Of course, FBP-ists believe that there actually exist hierarchies of both sorts, and fictionalists do not, but in the present context, this is irrelevant.) Second, FBP-ists and fictionalists agree that the question of whether ZF+CH or ZF+∼CH is correct comes down to the question of which is true in the intended parts of the mathematical realm (or for fictionalists, which would be true in the intended parts of the mathematical realm if there were sets) and that this, in turn, comes down to the question of whether CH or ∼CH is inherent in *our notion of set*. Third, both schools of thought allow that it *may* be that neither CH nor ∼CH is inherent in our notion of set and, hence, that there is no fact of the matter as

---

[39] In my book (chapter 8, note 3) I also argue that there's no important difference between FBP and fictionalism in connection with *in*consistent purely mathematical theories.

to which is correct. Fourth, they both allow that even if there is no correct answer to the CH question, there could still be good pragmatic or aesthetic reasons for favoring one answer to the question over the other (and perhaps for "modifying our notion of set" in a certain way). Finally, FBP-ists and fictionalists both maintain that questions of the form 'Does open question Q (about undecidable proposition P) have a correct answer, and if so, what is it?' are questions for *mathematicians* to decide. Each different question of this form should be settled on its own merits, in the above manner; they shouldn't all be decided in *advance* by some metaphysical principle, e.g., platonism or anti-platonism. (See my [2001] and [2009] and my book (chapter 3, section 4, and chapter 5, section 3) for more on this.)[40]

3. Both FBP-ists and fictionalists take mathematical theory at face value, i.e., adopt a realistic semantics for mathematese. Therefore, they both think that our mathematical theories are straightforwardly *about* abstract mathematical objects, although neither group thinks they are about such objects in a metaphysically *thick* sense of the term 'about' (see note 17 for a quick description of the thick/thin distinction here). The reason FBP-ists deny that our mathematical theories are "thickly about" mathematical objects is that they deny that there are *unique* collections of objects that correspond to the totality of intentions that we have in connection with our mathematical theories; that is, they maintain that certain collections of objects just happen to satisfy these intentions and, indeed, that *numerous* collections of objects satisfy them. On the other hand, the reason *fictionalists* deny that our mathematical theories are "thickly about" mathematical objects is entirely obvious: it is because they deny that there are any such things as mathematical objects. (See my book (chapters 3 and 4) for more on this.)

4. I didn't go into this here, but in my book (chapter 3), I show that according to both FBP and fictionalism, mathematical knowledge arises directly out of logical knowledge and that, from an epistemological point of view, FBP and fictionalism are on all fours with one another.

5. Both FBP-ists and fictionalists accept the thesis that there are no causally efficacious mathematical objects and, hence, no causal relations between mathematical and physical objects. (See my book (chapter 5, section 6) for more on this.)

6. Both FBP-ists and fictionalists have available to them the same accounts of the applicability of mathematics and the same reasons for favoring and rejecting the various accounts. (In this essay I said only a few words about

---

[40] I am not saying that every advocate of fictionalism holds this view of undecidable propositions. For instance, Field [1998] holds a different view. But his view is available to FBP-ists as well, and in general, FBP-ists and fictionalists have available to them the same views on undecidable propositions and the same reasons for favoring and rejecting these views. The view outlined in the text is just the view that *I* endorse.

the account of applicability that I favor (section 2.2.4); for more on this account, as well as other accounts, see my book (chapters 5–7).)

7. Both FBP-ists and fictionalists are in exactly the same situation with respect to the dispute about whether our mathematical theories are contingent or necessary. My own view here is that both FBP-ists and fictionalists should maintain that (a) our mathematical theories are logically and conceptually contingent, because the existence claims of mathematics — e.g., the null set axiom — are neither logically nor conceptually true, and (b) there is no clear sense of metaphysical necessity on which such sentences come out metaphysically necessary. (For more on this, see my book (chapter 2, section 6.4, and chapter 8, section 2).)

8. Finally, an imprecise point about the "intuitive feel" of FBP and fictionalism: both offer a neutral view on the question of whether mathematical theory construction is primarily a process of invention or discovery. Now, *prima facie*, it seems that FBP entails a discovery view whereas fictionalism entails an invention view. But a closer look reveals that this is wrong. FBP-ists admit that mathematicians discover objective facts, but they maintain that we can discover objective facts about the mathematical realm by merely inventing consistent mathematical stories. Is it best, then, to claim that FBP-ists and fictionalists both maintain an invention view? No. For mathematicians do discover objective facts. For instance, if a mathematician settles an open question of arithmetic by proving a theorem from the Peano axioms, then we have discovered something about the natural numbers. And notice that *fictionalists* will maintain that there has been a discovery here as well, although, on their view, the discovery is not about the natural numbers; rather, it is about our *concept* of the natural numbers, or our *story* of the natural numbers, or what *would* be true if there were mathematical numbers.

I could go on listing similarities between FBP and fictionalism, but the point I want to bring out should already be clear: FBP-ists and fictionalists agree on almost everything. Indeed, in my book (chapter 8, section 2), I argue that there is only *one* significant disagreement between them: FBP-ists think that mathematical objects exist and, hence, that our mathematical theories are true, whereas fictionalists think that there are no such things as mathematical objects and, hence, that our mathematical theories are fictional. My argument for this — i.e., for the only-one-significant-disagreement thesis — is based crucially on points 1 and 3 above. But it is also based on point 5: because FBP-ists and fictionalists agree that mathematical objects would be causally inert if they existed, they both think that the question of whether or not there do exist such objects has no bearing on the physical world and, hence, no bearing on what goes on in the mathematical community or the heads of mathematicians. This is why FBP-ists and fictionalists can agree on so much — why they can offer the same view of mathematical practice

— despite their bottom-level ontological disagreement. In short, both groups are free to say the same things about mathematical practice, despite their bottom-level disagreement about the existence of mathematical objects, because they both agree that it wouldn't matter to mathematical practice if mathematical objects existed.

If I'm right that the only significant disagreement between FBP-ists and fictionalists is the bottom-level disagreement about the existence of mathematical objects, then we can use this to motivate the strong epistemic conclusion. My argument here is based upon the following two sub-arguments:

(I) We could never settle the dispute between FBP-ists and fictionalists in a *direct* way, i.e., by looking *only* at the bottom-level disagreement about the existence of mathematical objects, because we have no epistemic access to the alleged mathematical realm (because we have access only to objects that exist within spacetime), and so we have no direct way of knowing whether any abstract mathematical objects exist.[41]

and

(II) We could never settle this dispute in an *indirect* way, i.e., by looking at the *consequences* of the two views, because they don't differ in their consequences in any important way, i.e., because the only significant point on which FBP-ists and fictionalists disagree is the bottom-level disagreement about the existence of mathematical objects.

This is just a sketch of my argument for the strong epistemic conclusion; for more detail, see chapter 8, section 2 of my book.

3.2 The Metaphysical Conclusion

In this section, I will sketch my argument for the metaphysical conclusion, i.e., for the thesis that there is no fact of the matter as to whether there exist any abstract objects and, hence, no fact of the matter as to whether FBP or fictionalism is true (for the full argument, see my book (chapter 8, section 3)). We can formulate the metaphysical conclusion as the thesis that there is no fact of the matter as to whether the sentence

(*) There exist abstract objects; i.e., there are objects that exist outside of spacetime (or more precisely, that do *not* exist *in* spacetime)

---

[41] This might seem similar to the Benacerrafian epistemological argument against platonism, but it is different: that argument is supposed to show that platonism is false by showing that even if we assume that mathematical objects exist, we could not know what they are *like*. I refuted this argument in my book (chapter 3), and I sketched the refutation above (section 2.1.1.5). The argument I am using here, on the other hand, is not directed against platonism or anti-platonism; it is aimed at showing that we cannot know (in any direct way) which of these views is correct, i.e., that we cannot know (in a direct way) whether there are any such things as abstract objects.

is true. Given this, my argument for the metaphysical conclusion proceeds (in a nutshell) as follows.

(i) We don't have any idea what a possible world would have to be *like* in order to count as a world in which there are objects that exist outside of spacetime.

(ii) If (i) is true, then there is no fact of the matter as to which possible worlds count as worlds in which there are objects that exist outside of spacetime, i.e., worlds in which (*) is true.

Therefore,

(iii) There is no fact of the matter as to which possible worlds count as worlds in which (*) is true — or in other words, there is no fact of the matter as to what the *possible-world-style truth conditions* of (*) are.

Now, as I make clear in my book, given the way I argue for (iii) — i.e., for the claim that there is no fact of the matter as to which possible worlds count as worlds in which (*) is true — it follows that there is no fact of the matter as to whether the *actual* world counts as a world in which (*) is true. But from this, the metaphysical conclusion — that there is no fact of the matter as to whether (*) is true — follows trivially.

Since the above argument for (iii) is clearly valid, I merely have to motivate (i) and (ii). My argument for (i) is based on the observation that we don't know — or indeed, have any idea — what it would be like for an object to exist outside of spacetime. Now, this is not to say that we don't know what *abstract objects* are like. That, I think, would be wrong. Of the number 3, for instance, we know that it is odd, that it is the cube root of 27, and so on. Thus, there is a sense in which we know what it is like. What I am saying is that we cannot imagine what *existence outside of spacetime* would be like. Now, it may be that, someday, somebody will clarify what such existence might be like; but what I think is correct is that no one has done this *yet*. There have been many philosophers who have advocated platonistic views, but I don't know of any who have said anything to clarify what non-spatiotemporal existence would really *amount* to. All we are ever given is a *negative* characterization of the existence of abstract objects — we're told that such objects do *not* exist in spacetime, or that they exist *non*-physically and *non*-mentally. In other words, we are told only what this sort of existence *isn't* like; we're never told what it *is* like.

The reason platonists have nothing to say here is that our whole conception of what existence *amounts* to seems to be bound up with extension and spatiotemporality. When you take these things away from an object, we are left wondering what its existence could *consist* in. For instance, when we say that Oliver North exists and Oliver Twist does not, what we *mean* is that the former resides at some particular spatiotemporal location (or "spacetime worm") whereas there is nothing in spacetime that is the latter. But there is nothing analogous to this in connection with abstract objects. Contemporary platonists do not think that the existence of 3 consists in there being something more encompassing than spacetime where 3 *resides*. My charge is simply that platonists have nothing substantive to say here, i.e., nothing substantive to say about what the existence of 3 consists in.

The standard contemporary platonist would respond to this charge, I think, by claiming that existence outside of spacetime is just like existence *inside* spacetime — i.e., that there is only *one* kind of existence. But this doesn't solve the problem; it just relocates it. I can grant that "there is only one kind of existence," and simply change my objection to this: we only know what certain *instances* of this kind are like. In particular, we know what the existence of concrete objects amounts to, but we do not know what the existence of abstract objects amounts to. The existence of concrete objects comes down to extension and spatiotemporality, but we have nothing comparable to say about the existence of abstract objects. In other words, we don't have anything more *general* to say about what existence amounts to than what we have to say about the existence of concrete objects. But this is just to say that we don't know what non-spatiotemporal existence amounts to, or what it might *consist* in, or what it might be *like*.

If what I have been arguing here is correct, then it would seem that (i) is true: if we don't have any idea what existence outside of spacetime could be like, then it would seem that we don't have any idea what a possible world would have to be like in order to count as a world that involves existence outside of spacetime, i.e., a world in which there are objects that exist outside of spacetime. In my book (chapter 8, section 3.3), I give a more detailed argument for (i), and I respond to a few objections that one might raise to the above argument.

I now proceed to argue for (ii), i.e., for the claim that if we don't have any idea what a possible world would have to be like in order to count as a world in which there are objects that exist outside of spacetime, then there is no fact of the matter as to which possible worlds count as such worlds — i.e., no fact of the matter as to which possible worlds count as worlds in which (*) is true, or in other words, no fact of the matter as to what the *possible-world-style truth conditions* of (*) are. Now, at first blush, (ii) might seem rather implausible, since it has an epistemic antecedent and a metaphysical consequent. But the reason the metaphysical consequent follows is that the ignorance mentioned in the epistemic antecedent is an ignorance of truth *conditions* rather than truth *value*. If we don't know whether some sentence is true or false, that gives us absolutely no reason to doubt that there is a definite fact of the matter as to whether it really *is* true or false. But when we don't know what the truth *conditions* of a sentence are, that is a very different matter. Let me explain why.

The main point that needs to be made here is that English is, in some relevant sense, *our* language, and (*) is *our* sentence. More specifically, the point is that *the truth conditions of English sentences supervene on our usage*. It follows from this that if our usage doesn't determine what the possible-world-style truth conditions of (*) are — i.e., doesn't determine which possible worlds count as worlds in which (*) is true — then (*) simply doesn't *have* any such truth conditions. In other words, (iia) If our usage doesn't determine which possible worlds count as worlds in which (*) is true, then there is no fact of the matter as to which possible worlds count as such worlds.

Again, the argument for (iia) is simply that (*) is *our* sentence and, hence, could obtain truth conditions only from our usage.[42]

Now, given (iia), all we need in order to establish (ii), by hypothetical syllogism, is

(iib) If we don't have any idea what a possible world would have to be like in order to count as a world in which there are objects that exist outside of spacetime, then our usage doesn't determine which possible worlds count as worlds in which (*) is true.

But (iib) seems fairly trivial. My argument for this, in a nutshell, is that if the consequent of (iib) were false, then its antecedent couldn't be true. In a bit more detail, the argument proceeds as follows. If our usage *did* determine which possible worlds count as worlds in which (*) is true — i.e., if it determined possible-world-style truth conditions for (*) — then it would also determine which possible worlds count as worlds in which there are objects that exist outside of spacetime. (This is trivial, because (*) just *says* that there are objects that exist outside of spacetime.) But it seems pretty clear that if our usage determined which possible worlds count as worlds in which there are objects that exist outside of spacetime, then we would have at least *some idea* what a possible world would have to be like in order to count as a world in which there are objects that exist outside of spacetime. For (a) it seems that if we have *no idea* what a possible world would have to be like in order count as a world in which there are objects that exist outside of spacetime, then the only way our usage could determine which possible worlds count as such worlds would be if we "lucked into" such usage; but (b) it's simply not plausible to suppose that we have "lucked into" such usage in this way.

This is just a sketch of my argument for the metaphysical conclusion. In my book (chapter 8, section 3), I develop this argument in much more detail, and I respond to a number of different objections that one might have about the argument. For instance, one worry that one might have here is that it is illegitimate to appeal to possible worlds in arguing for the metaphysical conclusion, because possible worlds are themselves abstract objects. I respond to this worry (and a

---

[42] One way to think of a *language* is as a function from sentence types to meanings and/or truth conditions. And the idea here is that *every* such function constitutes a language, so that English is just one abstract language among a huge infinity of such things. But on this view, the truth conditions of English sentences do *not* supervene on our usage, for the simple reason that they don't supervene on *anything* in the physical world. We needn't worry about this here, though, because (a) even on this view, which abstract language is *our* language will supervene on our usage, and (b) I could simply reword my argument in these terms. More generally, there are lots of ways of conceiving of language and meaning, and for each of these ways, the supervenience point might have to be put somewhat differently. But the basic idea here — that the meanings and truth conditions of *our* words come from *us*, i.e., from our usage and intentions — is undeniable.

number of other worries) in my book, but I do not have the space to pursue this here.

3.3 My Official View

My official view, then, is distinct from both FBP and fictionalism. I endorse the FBP-fictionalist interpretation, or picture, of mathematical theory and practice, but I do not agree with either of the metaphysical views here. More precisely, I am in agreement with almost everything that FBP-ists and fictionalists say about mathematical theory and practice,[43] but I do not claim with FBP-ists that there exist mathematical objects (or that our mathematical theories are true), and I do not claim with fictionalists that there do not exist mathematical objects (or that our mathematical theories are not true).

BIBLIOGRAPHY

[Armstrong, 1978] D. M. Armstrong. *A Theory of Universals*, Cambridge University Press, Cambridge, 1978.
[Ayer, 1946] A. J. Ayer. *Language, Truth and Logic*, second edition, Dover Publications, New York, 1946. (First published 1936.)
[Azzouni, 1994] J. Azzouni. *Metaphysical Myths, Mathematical Practice*, Cambridge University Press, Cambridge, 1994.
[Balaguer, 1992] M. Balaguer. "Knowledge of Mathematical Objects," PhD Dissertation, CUNY Graduate Center, New York, 1992.
[Balaguer, 1994] M. Balaguer. Against (Maddian) Naturalized Platonism, *Philosophia Mathematica*, vol. 2, pp. 97-108, 1994.
[Balaguer, 1995] M. Balaguer. A Platonist Epistemology, *Synthese*, vol. 103, pp. 303-25, 1995.
[Balaguer, 1996a] M. Balaguer. A Fictionalist Account of the Indispensable Applications of Mathematics, *Philosophical Studies* vol. 83, pp. 291-314, 1996.
[Balaguer, 1996b] M. Balaguer. Towards a Nominalization of Quantum Mechanics, *Mind*, vol. 105, pp. 209-26, 1996.
[Balaguer, 1998] M. Balaguer. *Platonism and Anti-Platonism in Mathematics*, Oxford University Press, New York, 1998.
[Balaguer, 1998b] M. Balaguer. Attitudes Without Propositions, *Philosophy and Phenomenological Research*, vol. 58, pp. 805-26, 1998.
[Balaguer, 2001] M. Balaguer. A Theory of Mathematical Correctness and Mathematical Truth, *Pacific Philosophical Quarterly*, vol. 82, pp. 87-114, 2001.
[Balaguer, 2008] M. Balaguer. Fictionalism in the Philosophy of Mathematics, *Stanford Encyclopedia of Philosophy*, http://plato.stanford.edu/entries/fictionalism-mathematics/.
[Balaguer, 2009] M. Balaguer. Fictionalism, Theft, and the Story of Mathematics, *Philosophia Mathematica*, forthcoming, 2009.
[Balaguer, in progress] M. Balaguer. Psychologism, Intuitionism, and Excluded Middle, in progress.
[Beall, 1999] JC Beall. From Full-Blooded Platonism to Really Full-Blooded Platonism, *Philosophia Mathematica*, vol. 7, pp. 322-25, 1999.
[Benacerraf, 1965] P. Benacerraf. What Numbers Could Not Be, 1965. Reprinted in [Benacerraf and Putnam, 1983, 272–94].
[Benacerraf, 1973] P. Benacerraf. 'Mathematical Truth, *Journal of Philosophy*, vol. 70, pp. 661-79, 1973.

---

[43]That is, I am in agreement here with the kinds of FBP-ists and fictionalists that I've described in this essay, as well as in my book and my [2001] and my [2009].

[Benacerraf and Putnam, 1983] P. Benacerraf and H. Putnam, eds. *Philosophy of Mathematics*, second edition, Cambridge University Press, Cambridge, 1983.
[Brouwer, 1948] L. E. J. Brouwer. Consciousness, Philosophy, and Mathematics, 1948. Reprinted in [Benacerraf and Putnam, 1983, 90–96].
[Burgess, 2004] J. Burgess. Mathematics and *Bleak House*, *Philosophica Mathemtica*, vol. 12, pp. 18–36, 2004.
[Carnap, 1934] R. Carnap. *Logische Syntax der Sprache*, 1934. Translated by A. Smeaton as *The Logical Syntax of Language*, Harcourt Brace, New York, 1937.
[Carnap, 1952] R. Carnap. Meaning Postulates, *Philosophical Studies*, vol. 3, pp. 65-73, 1952.
[Carnap, 1956] R. Carnap. Empiricism, Semantics, and Ontology, 1956. Reprinted in [Benacerraf and Putnam, 1983, 241–257].
[Chihara, 1990] C. Chihara. *Constructibility and Mathematical Existence*, Oxford University Press, Oxford, 1990.
[Colyvan and Zalta, 1999] M. Colyvan and E. Zalta. Review of M. Balaguer, *Platonism and Anti-Platonism in Mathematics*, *Philosophia Mathematica*, vol. 7, pp. 336-49, 1999.
[Curry, 1951] H. B. Curry. *Outlines of a Formalist Philosophy of Mathematics*, North-Holland, Amsterdam, 1951.
[Dedekind, 1888] R. Dedekind. *Was sind und was sollen die Zahlen?* 1888. Translated by W.W. Beman as "The Nature and Meaning of Numbers," in Dedekind, *Essays on the Theory of Numbers*, Open Court Publishing, Chicago, IL, 1901, pp. 31-115.
[Dehaene, 1997] S. Dehaene. *The Number Sense*, Oxford University Press, New Yor, 1997k.
[Dummett, 1973] M. Dummett. The Philosophical Basis of Intuitionistic Logic, 1973. Reprinted in [Benacerraf and Putnam, 1983, 97–129].
[Field, 1980] H. Field. *Science Without Numbers*, Princeton University Press, Princeton, NJ, 1980.
[Field, 1989] H. Field. *Realism, Mathematics, and Modality*, Basil Blackwell, New York, 1989.
[Field, 1993] H. Field. The Conceptual Contingency of Mathematical Objects, *Mind*, vol. 102, pp. 285-29, 19939.
[Field, 1998] H. Field. Mathematical Objectivity and Mathematical Objects, in C. MacDonald and S. Laurence (eds.) *Contemporary Readings in the Foundations of Metaphysics*, Basil Blackwell, Oxford, pp. 387-403, 1998.
[Frege, 1884] G. Frege. *Der Grundlagen die Arithmetic*, 1884. Translated by J.L. Austin as *The Foundations of Arithmetic*, Basil Blackwell, Oxford, 1953.
[Frege, 1893–1903] G. Frege. *Grundgesetze der Arithmetik*, 1893–1903. Translated (in part) by M. Furth as *The Basic Laws of Arithmetic*, University of California Press, Berkeley, CA, 1964.
[Frege, 1894] G. Frege. Review of Husserl's *Philosophie der Arithmetik*, in *Zeitschrift für Philosophie und phil. Kritik*, vol. 103, pp. 313-332, 1894.
[Frege, 1919] G. Frege. Der Gedanke, 1919. Translated by A.M. and M. Quinton as "The Thought: A Logical Inquiry," in Klemke, E. (ed.), *Essays on Frege*, University of Illinois Press, Urbana, IL, 1968, pp. 507-35.
[Frege, 1980] G. Frege. *Philosophical and Mathematical Correspondence*, University of Chicago Press, Chicago, 1980.
[Gödel, 1964] K. Gödel. What is Cantor's Continuum Problem? 1964. Reprinted in [Benacerraf and Putnam, 1983, 470–485].
[Hale, 1987] R. Hale. *Abstract Objects*, Basil Blackwell, Oxford, 1987.
[Hale and Wright, 1992] R. Hale and C. Wright. Nominalism and the Contingency of Abstract Objects, *Journal of Philosophy*, vol. 89, pp. 111-135, 1992.
[Hellman, 1989] G. Hellman. *Mathematics Without Numbers*, Clarendon Press, Oxford, 1989.
[Hempel, 1945] C. Hempel. On the Nature of Mathematical Truth, 1945. Reprinted in [Benacerraf and Putnam, 1983, 377–393].
[Hersh, 1997] R. Hersh. *What Is Mathematics, Really?* Oxford University Press, New York, 1997.
[Heyting, 1931] A. Heyting. The Intuitionist Foundations of Mathematics, 1931. Reprinted in [Benacerraf and Putnam, 1983, 52–61].
[Hilbert, 1899] D. Hilbert. *Grundlagen der Geometrie*, 1899. Translated by E. Townsend as *Foundations of Geometry*, Open Court, La Salle, IL, 1959.
[Hilbert, 1925] D. Hilbert. On the Infinite, 1925. Reprinted in [Benacerraf and Putnam, 1983, 183–201].

[Husserl, 1891] E. Husserl. *Philosophie der Arithmetik*, C.E.M. Pfeffer, Leipzig, 1891.
[Irvine, 1990] A. Irvine, ed. *Physicalism in Mathematics*, Kluwer Academic Publishers, Norwell, MA, 1990.
[Katz, 1981] J. Katz. *Language and Other Abstract Objects*, Rowman and Littlefield, Totowa, NJ, 1981.
[Katz, 1998] J. Katz. *Realistic Rationalism*, MIT Press, Cambridge, MA, 1998.
[Kitcher, 1984] P. Kitcher. *The Nature of Mathematical Knowledge*, Oxford University Press, Oxford, 1984.
[Lewis, 1986] D. Lewis. *On the Plurality of Worlds*, Basil Blackwell, Oxford, 1986.
[Locke, 1689] J. Locke. *An Essay Concerning Human Understanding*, 1689. Reprinted 1959, Dover, New York.
[Maddy, 1980] P. Maddy. Perception and Mathematical Intuition, *Philosophical Review*, vol. 89, pp. 163-196, 1980.
[Maddy, 1990] P. Maddy. *Realism in Mathematics*, Oxford University Press, Oxford, 1990.
[Maddy, 1990b] P. Maddy. Physicalistic Platonism. In [Irvine, 1990, 259–89].
[Maddy, 1997] P. Maddy. *Naturalism in Mathematics*, Oxford University Press, Oxford, 1997.
[Malament, 1982] D. Malament. Review of H. Field, *Science Without Numbers*, *Journal of Philosophy*, vol. 79, pp. 523-34, 1982.
[Mill, 1843] J. S. Mill. *A System of Logic*, Longmans, Green, and Company, London, 1843. (New Impression 1952.)
[Parsons, 1980] C. Parsons. Mathematical Intuition, *Proceedings of the Aristotelian Society*, vol. 80, pp. 145-168, 1980.
[Parsons, 1990] C. Parsons. The Structuralist View of Mathematical Objects, *Synthese*, vol. 84, pp. 303-46, 1990.
[Parsons, 1994] C. Parsons. Intuition and Number, in George, A. (ed.), *Mathematics and Mind*, Oxford University Press, Oxford, pp. 141-57, 1994.
[Plato, 1981] Plato. *The Meno* and *The Phaedo*. Both translated by G.M.A. Grube in *Five Dialogues*, Hackett Publishing, Indianapolis, IN, 1981.
[Priest, 2003] G. Priest. Meinongianism and the Philosophy of Mathematics, *Philosophia Mathematica*, vol. 11, pp. 3-15, 2003.
[Putnam, 1967a] H. Putnam. Mathematics Without Foundations, 1967. Reprinted in [Benacerraf and Putnam, 1983, 295–311].
[Putnam, 1967b] H. Putnam. The Thesis that Mathematics is Logic, 1967. Reprinted in [Putnam, 1979, 12–42].
[Putnam, 1971] H. Putnam. Philosophy of Logic, 1971. Reprinted in [Putnam, 1979, 323–357].
[Putnam, 1975] H. Putnam. What is Mathematical Truth? 1975. Reprinted in [Putnam, 1979, 60–78].
[Putnam, 1979] H. Putnam. *Mathematics, Matter and Method: Philosophical Papers Volume 1*, second edition, Cambridge University Press, Cambridge, 1979. (First published 1975.)
[Quine, 1948] W. V. O. Quine. On What There Is, 1948. Reprinted in [Quine, 1961, 1–19].
[Quine, 1951] W. V. O. Quine. Two Dogmas of Empiricism, 1951. Reprinted in [Quine, 1961, 20–46].
[Quine, 1961] W. V. O. Quine. *From a Logical Point of View*, second edition, Harper and Row, New York, 1961. (First published 1953.)
[Resnik, 1980] M. Resnik. *Frege and the Philosophy of Mathematics*, Cornell University Press, Ithaca, NY, 1980.
[Resnik, 1981] M. Resnik. Mathematics as a Science of Patterns: Ontology and Reference, *Nous*, vol. 15., pp. 529-550, 1981.
[Resnik, 1985] M. Resnik. How Nominalist is Hartry Field's Nominalism? *Philosophical Studies*, vol. 47, pp. 163-181, 1985.
[Resnik, 1997] M. Resnik. *Mathematics as a Science of Patterns*, Oxford University Press, Oxford, 1997.
[Restall, 2003] G. Restall. Just What *Is* Full-Blooded Platonism? *Philosophia Mathematica*, vol. 11, pp. 82-91, 2003.
[Rosen, 2001] G. Rosen. Nominalism, Naturalism, Epistemic Relativism, in J. Tomberlin (ed.), *Philosophical Topics* XV (Metaphysics), pp. 60-91, 2001.
[Routley, 1980] R. Routley. *Exploring Meinong's Jungle and Beyond*, Canberra: RSSS, Australian National University, 1980.
[Salmon, 1998] N. Salmon. Nonexistence, *Nous*, vol. 32, pp. 277-319, 1998.

[Shapiro, 1983] S. Shapiro. Conservativeness and Incompleteness, *Journal of Philosophy*, vol. 80, pp. 521-531, 1983.
[Shapiro, 1989] S. Shapiro. Structure and Ontology, *Philosophical Topics*, vol. 17, pp. 145-71, 1989.
[Shapiro, 1997] S. Shapiro. *Philosophy of Mathematics: Structure and Ontology*, Oxford University Press, New York, 1997.
[Steiner, 1975] M. Steiner. *Mathematical Knowledge*, Cornell University Press, Ithaca, NY, 1975.
[Thomasson, 1999] A. Thomasson. *Fiction and Metaphysics*, Cambridge University Press, Cambridge, 1999.
[van Inwagen, 1977] P. van Inwagen. Creatures of Fiction, *American Philosophical Quarterly*, vol. 14, pp. 299-308, 1977.
[Wittgenstein, 1956] L. Wittgenstein. *Remarks on the Foundations of Mathematics*, Basil Blackwell, Oxford. Translated by G.E.M. Anscombe, 1978.
[Wright, 1983] C. Wright. *Frege's Conception of Numbers as Objects*, Aberdeen University Press, Aberdeen, Scotland, 1983.
[Yablo, 2002] S. Yablo. Abstract Objects: A Case Study, *Nous* 36, supplementary volume 1, pp. 220-240, 2002.
[Zalta, 1983] E. Zalta. *Abstract Objects: An Introduction to Axiomatic Metaphysics*, D. Reidel, Dordrecht, 1983.
[Zalta, 1988] E. Zalta. *Intensional Logic and the Metaphysics of Intentionality*, Bradford/MIT Press, Cambridge, MA, 1988.
[Zalta and Linsky, 1995] E. Zalta and B. Linsky. Naturalized Platonism vs. Platonized Naturalism, *Journal of Philosophy*, vol. 92, pp. 525-55, 1995.

ARISTOTELIAN REALISM

James Franklin

1 INTRODUCTION

Aristotelian, or non-Platonist, realism holds that mathematics is a science of the real world, just as much as biology or sociology are. Where biology studies living things and sociology studies human social relations, mathematics studies the quantitative or structural aspects of things, such as ratios, or patterns, or complexity, or numerosity, or symmetry. Let us start with an example, as Aristotelians always prefer, an example that introduces the essential themes of the Aristotelian view of mathematics. A typical mathematical truth is that there are six different pairs in four objects:

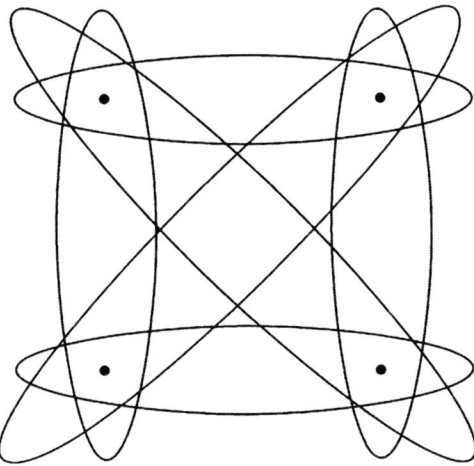

Figure 1. There are 6 different pairs in 4 objects

The objects may be of any kind, physical, mental or abstract. The mathematical statement does not refer to any properties of the objects, but only to patterning of the parts in the complex of the four objects. If that seems to us less a solid truth about the real world than the causation of flu by viruses, that may be simply due to our blindness about relations, or tendency to regard them as somehow less real than things and properties. But relations (for example, relations of equality between parts of a structure) are as real as colours or causes.

Handbook of the Philosophy of Science. Philosophy of Mathematics
Volume editor: Andrew D. Irvine. General editors: Dov M. Gabbay, Paul Thagard and John Woods.
© 2009 Elsevier B.V. All rights reserved.

The statement that there are 6 different pairs in 4 objects appears to be necessary, and to be about the things in the world. It does not appear to be about any idealization or model of the world, or necessary only relative to axioms. Furthermore, by reflecting on the diagram we can not only learn the truth but understand why it must be so.

The example is also, as Aristotelians again prefer, about a small finite structure which can easily be grasped by the mind, not about the higher reaches of infinite sets where Platonists prefer to find their examples.

This perspective raises a number of questions, which are pursued in this chapter.

First, what exactly does "structure" or "pattern" or "ratio" mean, and in what sense are they properties of real things? The next question concerns the necessity of mathematical truths, from which follows the possibility of having certain knowledge of them. Philosophies of mathematics have generally been either empiricist in the style of Mill and Lakatos, denying the necessity and certainty of mathematics, or admitting necessity but denying mathematics a direct application to the real world (for different reasons in the case of Platonism, formalism and logicism). An Aristotelian philosophy of mathematics, however, finds necessity in truths directly about the real world (such as the one in the diagram above). We then compare Aristotelian realism with the Platonist alternative, especially with regard to problems where Platonism might seem more natural, such as uninstantiated structures such as higher-order infinities. A later section deals with epistemology, which is very different from an Aristotelian perspective from traditional alternatives. Direct knowledge of structure and quantity is possible from perception, and Aristotelian epistemology connects well with what is known from research on baby development, but there are still difficulties explaining how proof leads to knowledge of mathematical necessity. We conclude with an examination of experimental mathematics, where the normal methods of science are used to explore a pre-existing mathematical realm.

The fortunes of Aristotelian philosophy of mathematics have fluctuated widely. From the time of Aristotle to the eighteenth century, it dominated the field. Mathematics, it was said, is the "science of quantity". Quantity is divided into the discrete, studied by arithmetic, and the continuous, studied by geometry [Apostle, 1952; Barrow, 1734, 10-15; *Encyclopaedia Britannica* 1771; Jesseph, 1993, ch. 1; Smith, 1954]. But it was overshadowed in the nineteenth century by Kantian perspectives, except possibly for the much maligned "empiricism" of Mill, and in the twentieth by Platonist and formalist philosophies stemming largely from Frege (and reactions to them such as extreme nominalism). The quantity theory, or something very like it, has also been revived in the 1990s, and a mainly Australian school of philosophers has tried to show that sets, numbers and ratios should also be interpreted as real properties of things (or real relations between universals: for example the ratio 'the double' may be something in common between the relation two lengths have and the relation two weights have.) [Armstrong, 1988; 1991; 2004, ch. 9; Bigelow, 1988; Bigelow & Pargetter, 1990, ch. 2; Forge, 1995; Forrest & Armstrong, 1987; Michell, 1994; Mortensen, 1998; Irvine, 1990, the "Sydney

School"]. The project has as yet made little impact on the mainsteam of northern hemisphere philosophy of mathematics.

The "structuralist" philosophy of Shapiro [1997], Resnik [1997] and others could naturally be interpreted as Aristotelian, if structure or pattern were thought of as properties that physical things could have. Those authors themselves, however, interpret their work more Platonistically, conceiving of structure and patterns as Platonist entities similar to sets.

2 THE ARISTOTELIAN REALIST POINT OF VIEW

Since many of the difficulties with traditional philosophy of mathematics come from its oscillation between Platonism and nominalism, as if those are the only alternatives, it is desirable to begin with a brief introduction to the Aristotelian alternative. The issues have nothing to do with mathematics in particular, so we deliberately avoid more than passing reference to mathematical examples.

"*Orange is closer to red than to blue.*" That is a statement about *colours*, not about the particular things that have the colours — or if it is about the things, it is only about them *in respect of their colour*: orange things are like red things but not blue things in respect of their colour. There is no way to avoid reference to the colours themselves.

Colours, shapes, sizes, masses are the repeatables or "universals" or "types" that particulars or "tokens" share. A certain shade of blue, for example, is something that can be found in many particulars — it is a "one over many" in the classic phrase of the ancient Greek philosophers. On the other hand, a particular electron is a non-repeatable. It is an individual; another electron can resemble it (perhaps resemble it exactly except for position), but cannot literally be it. (Introductions to realist views on universals appear in [Moreland, 2001, ch. 1; Swoyer, 2000].)

Science is about universals. There is perception of universals — indeed, it is universals that have causal power. We see an individual stone, but only as a certain shape and colour, because it is those properties of it that have the power to affect our senses. Science gives us classification and understanding of the universals we perceive — physics deals with such properties as mass, length and electrical charge, biology deals with the properties special to living things, psychology with mental properties and their effects, mathematics with quantities, ratios, patterns and structure.

This view is close to Aristotle's account of how mathematicians are natural scientists of a sort. They are scientists who study patterns or forms that arise in nature. In what way, then, do mathematicians differ from other natural scientists? In a famous passage at *Physics* B, Aristotle says that mathematicians differ from physicists (in the broad sense of those who study nature) not in terms of subject-matter, but in terms of emphasis. Both study the properties of natural bodies, but concentrate on different aspects of these properties. The mathematician studies the properties of natural bodies, which include their surfaces and volumes, lines, and points. The mathematician is not interested in the properties of natural bodies considered as the properties of natural bodies, which is the concern of the physicist. [*Physics* II.2, 193b33-4] Instead, the mathematician is interested in the properties of natural bodies that are 'separable in thought from the world of change'. But, Aristotle says, the procedure of separating these properties in thought from the world of change does not make any difference or result in any falsehood. [Aristotle, *Physics* II.2, 193a36-b35]

Science is also the arbiter of what universals there are. To know what universals there are, as to know what particulars there are, one must investigate, and accept the verdict of the best science (including inference as well as observation). Thus universals are not created by the meanings of words. On the other hand, language is part of nature, and it is not surprising if our common nouns, adjectives and prepositions name some approximation of the properties there are or seem to be, just as our proper names label individuals, or if the subject-predicate form of many basic sentences often mirrors the particular-property structure of reality.

Not everyone agrees with the foregoing. *Nominalism* holds that universals are not real but only words or concepts. That is not very plausible in view of the ability of all things with the same shade of blue to affect us in the same way — "causality is the mark of being". It also leaves it mysterious why we do apply the word or concept "blue" to some things but not to others. *Platonism* (in its extreme version, at least) holds that there are universals, but they are pure Forms in an abstract world, the objects of this world being related to them by a mysterious relation of "participation". (Arguments against nominalism appear in in [Armstrong, 1989, chs 1-3]; against Platonism in [Armstrong, 1978, vol. 1 ch. 7].) That too makes it hard to make sense of the direct perception we have of shades of blue. Blue things affect our retinas in a characteristic way because the blue is in the things themselves, not in some other realm to which we have no causal access. Aristotelian realism about universals takes the straightforward view that the world has both particulars and universals, and the basic structure of the world is "states of affairs" of a particular's having a universal, such as this table's being approximately square.

Because of the special relation of mathematics to complexity, there are three issues in the theory of universals that are of comparatively minor importance in general but crucial in understanding mathematics. They are the problem of uninstantiated universals, the reality of relations, and questions about structural and "unit-making" universals.

The Aristotelian slogan is that universals are *in re*: in the things themselves (as opposed to in a Platonic heaven). It would not do to be too fundamentalist about that dictum, especially when it comes to uninstantiated universals, such as numbers bigger than the numbers of things in the universe. How big the universe is, or what colours actually appear on real things, is surely a contingent matter, whereas at least some truths about universals appear to be independent of whether they are instantiated — for example, if some shade of blue were uninstantiated, it would still lie between whatever other shades it does lie between. One expects the science of colour to be able to deal with any uninstantiated shades of blue on a par with instantiated shades — of course direct experimental evidence can only be of instantiated shades, but science includes inference from experiment, not just heaps of experimental data, so extrapolation (or interpolation) arguments are possible to "fill in" gaps between experimental results. Other uninstantiated universals are "combinatorially constructible" from existing properties, the way "unicorn" is made out of horses, horns, etc. More problematic are truly "alien" universals, like nothing in the actual universe but perhaps nevertheless possible. However, these seem beyond the range of what needs to considered in mathematics — for all the vast size and esoteric nature of Hilbert spaces and inaccessible cardinals, they seem to be in some sense made out of a small range of simple concepts. What those concepts are and how they make up the larger ones is something to be considered later.

The shade of blue example suggests two other conclusions. The first is that knowledge of a universal such as an uninstantiated shade of blue is possible only because it is a member of a structured space of universals, the (more or less) continuous space of colours. The second conclusion is that the facts known in this way, such as the betweenness relations holding among the colours, are necessary. Surely there is no possible world in which a given shade of blue is between scarlet and vermilion?

At this point it may be wondered whether it is not a very Platonist form of Aristotelianism that is being defended. It has a structured space of universals, not all instantiated, into which the soul has necessary insights. That is so. There are three, not two, distinct positions covered by the names Platonism and Aristotelianism:

- (Extreme) Platonism — the Platonism found in the philosophy of mathematics — according to which universals are of their nature not the kind of entities that could exist (fully or exactly) in this world, and do not have causal power (also called "objects Platonism" [Hellman, 1989, 3], "standard Platonism" [Cheyne & Pigden, 1996], "full-blooded Platonism" [Balaguer, 1998; Restall, 2003]; "ontological Platonism" [Steiner, 1973])

- Platonist or modal Aristotelianism, according to which universals can exist and be perceived to exist in this world and often do, but it is an contingent matter which do so exist, and we can have knowledge even of those that are uninstantiated and of their necessary interrelations

- Strict this-worldly Aristotelianism, according to which uninstantiated universals do not exist in any way: all universals really are *in rem*.

It is true that whether the gap between the second and third positions is large depends on what account one gives of possibilities. If the "this-worldly" Aristotelian has a robust view of merely possible universals (for example, by granting full existence to possible worlds), there could be little difference in the two kinds of Aristotelianism. But supposing a deflationary view of possibilities (as would be expected from an Aristotelian), a this-worldly Aristotelian will have a much narrower realm of real entities to consider. The discrepancy is not a matter of great urgency in considering the usual universals of science which are known to be instantiated because they cause perception of themselves. It is the gargantuan and esoteric specimens in the mathematical zoo that strike fear into the strict empirically-oriented Aristotelian realist. Our knowledge of mathematical entities that are not or may not be instantiated has always been a leading reason for believing in Platonism, and rightly so, since it is knowledge of what is beyond the here and now. It does create insuperable difficulties for a strict this-worldly Aristotelianism; but it needs to be considered whether one might move only partially in the Platonist direction. There is room to move only halfway towards strict Platonism for the same reason as there is space in the blue spectrum between two instantiated shades for an uninstantiated shade. The non-adjacency of shades of blue is a necessary fact about the blue spectrum (as Platonism holds), but whether an intermediate shade of blue is instantiated is contingent (contrary to extreme Platonism, which holds that universals cannot be literally instantiated in reality). It is the same with uninstantiated mathematical structures, according to the Aristotelian of Platonist bent: a ratio (say), whether small and instantiated or huge and uninstantiated, is part of a necessary spectrum of ratios (as Platonists think) but an instantiated ratio is literally a relation between two actual (say) lengths (as Aristotelians think). The fundamental reason why an intermediate position between extreme Platonism and extreme Aristotelianism is possible is that the Platonist insight that there is knowledge of uninstantiated universals is compatible with the Aristotelian insight that instantiated universals can be directly perceived in things.

The gap between "Platonist" Aristotelianism and extreme Platonism is unbridgeable. Aristotelian universals are ones that could be in real things (even if some of them happen not to be), and knowledge of them comes from the senses being directly affected by instantiated universals (even if indirectly and after inference, so that knowledge can be of universals beyond those directly experienced). Extreme Platonism — the Platonism that has dominated discussion in the philosophy of mathematics — calls universals "abstract", meaning that they do not have causal powers or location and hence cannot be perceived (but can only be postulated or inferred by arguments such as the indispensability argument).

Aristotelian realism is committed to the reality of relations as well as properties. The relation being-taller-than is a repeatable and a matter of observable fact in the same way as the property of being orange. [Armstrong, 1978, vol. 2, ch. 19] The visual system can make an immediate judgement of comparative tallness, even if its internal arrangements for doing so may be somewhat more complex than those for registering orange. Equally important is the reality of relations between universals themselves, such as betweenness among colours — if the colours are real, the relations between them are "locked in" and also real. Western philosophical thought has had an ingrained tendency to ignore or downplay the reality of relations, from ancient views that attempted to regard relations as properties of the individual related terms to early modern ones that they were purely mental. [Weinberg, 1965, part 2; Odegard, 1969]

But a solid grasp of the reality of relations such as ratios and symmetry is essential for understanding how mathematics can directly apply to reality. Blindness to relations is surely behind Bertrand Russell's celebrated saying that "Mathematics may be defined as the subject where we never know what we are talking about, nor whether what we are saying is true" [Russell, 1901/1993, vol. 3, p.366].

Considering the importance of structure in mathematics, important parts of the theory of universals are those concerning structural and "unit-making" properties. A structural property is one that makes essential reference to the parts of the particular that has the property. "Being a certain tartan pattern" means having stripes of certain colours and widths, arranged in a certain pattern. "Being a methane molecule" means having four hydrogen atoms and one carbon atom in a certain configuration. "Being checkmated" implies a complicated structure of chess pieces on the board. [Bigelow & Pargetter, 1990, 82-92] Properties that are structural without requiring any particular properties of their parts such as colour could be called "purely structural". They will be considered later as objects of mathematics.

"Being an apple" differs from "being water" in that it structures its instances discretely. "Being an apple" is said to be a "unit-making" property, in that a heap of apples is divided by the universal "being an apple" into a unique number of non-overlapping parts, apples, and parts of those parts are not themselves apples. A given heap may be differently structured by different unit-making properties. For example, a heap of shoes consists of one number of shoes and another number of pairs of shoes. Notions of (discrete) number should give some account of this phenomenon. By contrast, "being water" is homoiomerous, that is, any part of water is water (at least until we go below the molecular level). [Armstrong, 2004, 113-5]

One special issue concerns the relation between sets and universals. A set, whatever it is, is a particular, not a universal. The set {Sydney, Hong Kong} is as unrepeatable as the cities themselves. The idea of Frege's "comprehension axiom" that any property ought to define the set of all things having that property is a good one, and survives in principle the tweakings of it necessary to avoid paradoxes. It emphasises the difference between properties and sets, by calling attention to the possibility that different properties should define the same set. In a classical (philosophers') example, the properties "cordate" (having a heart) and "renate" (having a kidney) are co-extensive, that is, define the same set of animals, although they are not the same property and in another possible world would not define the same set.

Normal discussion of sets, in the tradition of Frege, has tended to assume a Platonist view of them, as "abstract" entities in some other world, so it is not clear what an Aristotelian view of their nature might be. One suggestion is that a set is just the heap of its singleton sets, and the singleton set of an object x is just x's having some unit-making property: the fact that Joe has some unit-making property such as "being a human" is all that is needed for there to be the set {Joe}. [Armstrong, 2004, 118-23]

A large part of the general theory of universals concerns causality, dispositions and laws of nature, but since these are of little concern to mathematics, we leave them aside here.

3 MATHEMATICS AS THE SCIENCE OF QUANTITY AND STRUCTURE

If Aristotelian realists are to establish that mathematics is the science of some properties of the world, they must explain which properties. There have been two main suggestions, the relation between which is far from clear. The first theory, the one that dominated the field from Aristotle to Kant and that has been revived by recent authors such as Bigelow, is that mathematics is the "science of quantity". The second is that its subject matter is structure.

The theory that mathematics is about quantity, and that quantity is divided into the discrete, studied by arithmetic, and the continuous, studied by geometry, plainly gives an initially reasonable picture of at least elementary mathematics, with its emphasis on counting and measuring and manipulating the resulting numbers. It promises direct answers to questions about what the object of mathematics is (certain properties of physical and possibly non-physical things such as their size), and how they are known (the same way other natural properties of physical things are known). It was the quantity theory, or something very like it, that was revived in the 1990s by the Australian school of realist philosophers.

Following dissatisfaction with the classical twentieth-century philosophies of mathematics such as formalism and logicism, and in the absence of a general wish to return to an unreconstructed Platonism about numbers and sets, another realist philosophy of mathematics became popular in the 1990s. Structuralism holds that mathematics studies structure or patterns. As Shapiro [2000, 257-64] explains it, number theory deals not with individual numbers but with the "natural number structure", which is "a single abstract structure, the pattern common to any infinite collection of objects that has a successor relation, a unique initial object, and satisfies the induction principle." The structure is "exemplified by" an infinite sequence of distinct moments in time. Number theory studies just the properties of the structure, so that for number theory, there is nothing to the number 2 but its place or "office" near the beginning of the system. Other parts of mathematics study different structures, such as the real number system or abstract groups. (Classifications of various structuralist views of mathematics are given in [Reck & Price, 2000; Lehrer Dive, 2003, ch. 1; Parsons, 2004]). It is true that Shapiro [1997; 2004] favours an "*ante rem* structuralism" which he compares to Platonism about universals, and Resnik is also Platonist with certain qualifications [Resnik, 1997, 10, 82, 261]. But Shapiro and Resnik allow arrangements of physical objects, such as basketball defences, to "exemplify" abstract structures, thus allowing mathematics to apply to the real world in a somewhat more direct way that classical Platonism and so encouraging an Aristotelian reading of their work, while certain other structuralist authors place much greater emphasis on instantiated patterns. [Devlin, 1994; Dennett, 1991, section II]

The structuralist theory of mathematics has, like the quantity theory, some initial plausibility, in view of the concentration of modern mathematics on structural properties like symmetry and the purely relational aspects of systems both physical and abstract. It is supported by the widespread concentration of modern pure mathematics on "abstract structures" such as groups and topological spaces (emphasised in [Mac Lane, 1986] and [Corfield, 2003]; background in [Corry, 1992]).

The relation between the concepts of quantity and structure are unclear and have been little examined. The position that will be argued for here is that quantity and structure are different sorts of universals, both real. The sciences of them are approximately those called by the (philosophically somewhat unsatisfying) names of elementary mathematics and advanced mathematics. That is a more exciting conclusion than might appear. It means that the quantity theory will have to be incorporated into any acceptable philosophy of mathematics, something very far from being done by any of the current leading contenders. It also means that modern (post eighteenth-century) mathematics has discovered a completely new subject matter, creating a science unimagined by the ancients.

Let us begin with some examples, chosen to point up the difference between structure and quantity. This is especially necessary in view of the inability of supporters of either the quantity theory or the structure theory to provide convincing definitions of what properties exactly should count as quantitative or structural. (An attempt will be made later to remedy that deficiency, but the attempted definitions can only be appreciated in terms of some clear examples.)

The earliest case of a mathematical problem that seemed clearly not well described as being about "quantity" was Euler's example of the bridges of Königsberg (see Figure 2). The citizens of that city in the eighteenth century noticed that it was impossible to walk over all the bridges once, without walking over at least one of them twice. Euler [1776] proved they were correct.

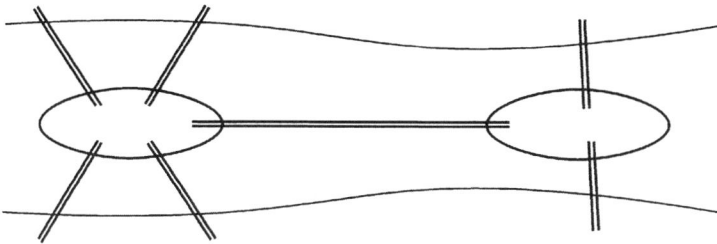

Figure 2. The Bridges of Königsberg

The result is intuitively about the "arrangement" or pattern of the bridges, rather than about anything quantitative like size or number. As Euler puts it, the result is "concerned only with the determination of position and its properties; it does not involve measurements." The length of the bridges and the size of the islands is irrelevant. That is why we can draw the diagram so schematically. All that matters is which land masses are connected by which bridges. Euler's result is now regarded as the pioneering effort in the topology of networks. There now exist large bodies of work on such topics as graph theory, networks, and operations research problems like timetabling, where the emphasis is on arrangements and connections rather than quantities.

The second kind of example where structure contrasts with quantity is symmetry, brought to the fore by nineteenth-century group theory and twentieth-century physics. Symmetry is a real property of things, things which may be but need not be physical (an argument, for example, can have symmetry if its second half repeats the steps of the first half in the opposite order; Platonist mathematical entities, if any exist, can be symmetrical.) The kinds of symmetry are classified by group theory, the central part of modern abstract algebra [Weyl, 1952].

The example of structure most discussed in the philosophical world is a different one. In a celebrated paper, Benacerraf [1965] observed that if the sequence of natural numbers were constructed in set theory, there is no principled way to choose which sets the numbers should be; the sequence

$$\varnothing, \{\varnothing\}, \{\{\varnothing\}\}, \{\{\{\varnothing\}\}\}, \ldots$$

would do just as well as

$$\varnothing, \{\varnothing\}, \{\varnothing, \{\varnothing\}\}, \{\varnothing, \{\varnothing\}, \{\varnothing, \{\varnothing\}\}\}, \ldots$$

simply because both form a 'progression' or 'ω-sequence' — an infinite sequence with a start, which does not come back on itself. He concluded that "Arithmetic is ... the science that elaborates the abstract structure that all progressions have in common merely in virtue of being progressions." The assertion that that is all there is to arithmetic is more controversial than the assertion that ω-sequences are indeed one kind of order structure, and that the study of them is a part of mathematics.

Now by way of contrast let us consider some examples of quantities which seem to have nothing inherently to do with structure. The universal 'being 1.57 kilograms in mass' stands in a certain relation, a ratio, to the universal 'being 0.35 kilograms in mass'. Pairs of lengths can stand in that same ratio, as can pairs of time intervals. (It is not so clear whether pairs of temperature intervals can stand in a ratio to one another; that depends on physical facts about the kind of scale temperature is.) The ratio itself is just what those binary relations between pairs of masses, lengths and time intervals have in common ("A ratio is a sort of relation in respect of size between two magnitudes of the same kind": Euclid, book V definition 3). A (particular) ratio is thus not merely a "place in a structure" (of all ratios), for the same reason as a colour is not merely a position in the space of all possible colours — the individual ratio or colour has intrinsic properties that can be grasped without reference to other ratios or colours. Though there is indeed a system or space of all ratios or all colours, with its own structure, it makes sense to say that a certain one is instantiated and a neighbouring one not. It is perfectly determinate which ratios are instantiated by the pairs of energy levels of the hydrogen atom, just as it is perfectly determinate which, if any, shades of blue are missing.

Discrete quantities arise differently from ratios. It is characteristic of 'unit-making' or 'count' universals like 'being an apple' to structure their instances discretely. That is what distinguishes them from mass universals like 'being water'. A heap of apples stands in a certain relation to 'being an apple'; that relation is the number of apples in the heap. The same relation can hold between a heap of shoes and 'being a shoe'. The number is just what these binary relations have in common. The fact that the heap of shoes stands in one such numerical relation to 'being a shoe' and another numerical relation to 'being a pair of shoes' (made much of by Frege [1884, §22, p. 28 and §54, p. 66]) does not show that the number of a heap is subjective or not about something in the world, but only that number is relative to the count universal being considered. (Similarly, the fact that the probability of a hypothesis is relative to the evidence for it does not show that probability is subjective, but that it is a relation between hypothesis and evidence.) Like a ratio, a number is not merely a position in the system of numbers. There is a perfectly determinate number of apples in a heap, independently of anything systematic about numbers (and independent of any knowledge about it, such as that obtained through counting).

The differing origins of continuous and discrete quantity led to some classical problems in Aristotelian philosophy of quantity. The distinction between the two kinds of quantity was reinforced by the discovery of the incommensurability of the diagonal (a significance somewhat obscured by calling it the irrationality of $\sqrt{2}$): there can exist a continuous ratio that is not the ratio of any two whole numbers. That only increased the mystery as to why some of the more structural features of the two kinds of ratios should be identical, such as the principle of alternation of ratios (that if the ratio of a to b equals the ratio of c to d, then the ratio of a to c equals that of b to d). Is this principle part of a "universal mathematics", a science of quantity in general (Crowley 1980)? Is there anything to be gained, philosophically or mathematically, by Euclid's attempt to define equality of ratios without defining a way of measuring ratios (Book V definition 5)? Genuine and interesting as these questions are, they will not be attacked here. The purpose of mentioning them is simply to indicate the scope of a realist theory of quantity.

Two tasks remain. The first is to indicate where in the body of known truths the sciences of quantity and of structure, respectively, lie. The second is to inquire whether there are convincing definitions of 'quantity' and 'structure', which would support proofs of their distinctness, or other mutual relations.

The theory of the ancients that the science of quantity comprises arithmetic plus geometry may be approximately correct, but needs some qualification. Arithmetic as the science of discrete quantity is adequate, though as the Benacerraf example shows, the study of a certain kind of order structure is reasonably regarded as part of arithmetic too. The distinction between cardinal and ordinal numbers corresponds to the distinction between pure discrete quantity and linear order structures. But geometry as the science of continuous quantity has more serious problems. It was always hard to regard shape as straightforwardly 'quantity' — it contrasts with size, rather than resembling it — though geometry certainly studies it. From the other direction, there can be discrete geometries: the spaces in computer graphics are discrete or atomic, but obviously geometrical. Hume, though no mathematician, certainly trounced the mathematicians of his day in arguing that real space might be discrete [Franklin, 1994]. Further, there is an alternative body of knowledge with a better claim to being the science of continuous quantity in general, namely, the calculus. Study of continuity requires the notion of a limit, as defined and made use of in the differential calculus of Newton and Leibniz, and made more precise in the real analysis of Cauchy and Weierstrass. On yet another front, there is another body of knowledge which seems to concern itself with quantity as it exists in reality. It is measurement theory, the science of how to associate numbers with quantities. It includes, for example, the requirement that physical quantities to be equated or added should be dimensionally homogeneous [Massey, 1971, 2] and the classification of scales into ordinal, linear interval and ratio scales ([Ellis, 1968, ch. 4]; many references in [Diez, 1997], conclusions for philosophy of mathematics in [Pincock, 2004]).

In summary, the science of quantity is elementary mathematics, up to and including the calculus, plus measurement theory.

That leaves the 'higher' mathematics as the science of structure. It includes on the one hand the subject traditionally called mathematical 'foundations', which deals with what structures can be made from the purely topic-neutral material of sets and categories, using logical concepts, as well as matters concerning axiomatization. On the other hand, most of modern pure mathematics deals with the richer structures classified by Bourbaki into algebraic, topological and order structures [Bourbaki, 1950; Mac Lane, 1986].

There is then the final question of whether there are formal definitions of 'quantity' and 'structure', which will exhibit their mutual logical relations. For 'quantity', one may loosely call any order structure a kind of quantity (in that it permits comparisons on a kind of scale), but a true or paradigmatic quantity should be a relation in a system isomorphic to the continuum, or to a piece of it (for example, the interval from 0 to 1, in the case of probabilities) or a substructure of it (such as the rationals or integers) [Hale, 2000, 106]. One might go so far as to allow fuzzy quantities by a family resemblance, as they share the properties of the continuum except for absolute precision.

It must be admitted that the difficulty of defining 'structure' has been the Achilles heel of structuralism. As one observer says, "It's probably not too gross a generalization to say that the main problems that have faced structuralism have been concerned with lack of clarity. After all, the slogans used to describe the view are nothing but highly evocative metaphors. In particular, philosophers have wondered: What is a structure?" [Colyvan, 1998, p. 653]. The matter is far from resolved, but one suggestion involves mereology. 'Structure' it is proposed, can be defined as follows.

A property S is *structural* if and only if "proper parts of particulars having S have some properties T ... not identical to S, and this state of affairs is, at least in part, constitutive of S." [Armstrong, 1978, vol. 2, 69] Under this definition, structural properties include such examples as "being a certain tartan pattern" [Armstrong, 1978, vol. 2, 70] or "being a baseball defence" [Shapiro, 1997, 74, 98] Plainly the reference in such properties to the parts having colours or being baseball players makes such structures not appropriate as objects of mathematics — not of pure mathematics, at least. Something more purely structural is needed. As Shapiro puts it in more Platonist language, a baseball defence is a kind of system, but the purer structure to be studied by mathematics is "the abstract form of a system, highlighting the interrelationships among the objects, and ignoring any features of them that do not affect how they relate to other objects in the system." [Shapiro, 1997, 74]; or again, "a position [in a pattern] ... has no distinguishing features other than those it has in virtue of being the particular position it is in the pattern to which it belongs." [Resnik, 1997, 203] These desiderata can be achieved by the following definition.

A property is *purely structural* if it can be defined wholly in terms of the concepts same and different, and part and whole (along with purely logical concepts).

To be symmetrical with the simplest sort of symmetry, for example, is to consist of two parts which are the same in some respect. To demonstrate that a concept is purely structural, it is sufficient to construct a model of it out of purely topic-neutral building blocks, such as sets — the capacities of set theory and pure mereology for construction being identical [Lewis, 1991, especially 112].

4 NECESSARY TRUTHS ABOUT REALITY

An essential theme of the Aristotelian viewpoint is that the truths of mathematics, being about universals and their relations, should be both necessary and about reality. Aristotelianism thus stands opposed to Einstein's classic dictum, 'As far as the propositions of mathematics refer to reality, they are not certain; and as far as they are certain, they do not refer to reality.' [Einstein, 1954, 233]. It is clear that by 'certain' Einstein meant 'necessary', and philosophers of recent times have mostly agreed with him that there cannot be mathematical truths that are at once necessary and about reality.

Mathematics provides, however, many *prima facie* cases of necessities that are directly about reality. One is the classic case of Euler's bridges, mentioned in the previous section. Euler proved that it was impossible for the citizens of Königsberg to walk exactly once over (not an abstract model of the bridges but) the actual bridges of the city.

To take another example: It is impossible to tile my bathroom floor with (equally-sized) regular pentagonal lines. It is a proposition of geometry that 'it is impossible to tile the Euclidean plane with regular pentagons'. That is, although it is possible to fit together (equally-sized) squares or regular hexagons so as to cover the whole space, thus:

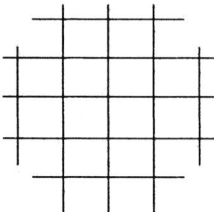

Figure 3. Tiling of the plane by squares

and

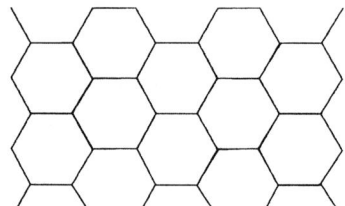

Figure 4. Tiling of the plane by regular hexagons

it is impossible to do this with regular pentagons:

No matter how they are put on the plane, there is space left over between them.

Now the 'Euclidean plane' is no doubt an abstraction, or a Platonic form, or an idealisation, or a mental being — in any case it is not 'reality'. If the 'Euclidean plane' is something that could have real instances, my bathroom floor is not one of them, and it may be that there are no exact real instances of it at all. It is a further fact of mathematics, however, that the proposition has 'stability', in the sense that it remains true if the terms in it are varied slightly. That is, it is impossible to tile (a substantial part of) an almost Euclidean-plane with shapes that are nearly regular pentagons. (The qualification 'substantial part of' is simply to avoid the possibility of taking a part that is exactly the shape and size of one tile; such a part could of course be tiled). This proposition has the same status, as far as reality goes, as the original one, since 'being an almost-Euclidean-plane' and 'being a nearly-regular pentagon' are as purely abstract or mathematical as 'being an exact Euclidean plane' and 'being an exactly regular pentagon'. The proposition has the consequence that if anything, real or abstract, does have the shape of a nearly-Euclidean-plane, then it cannot be tiled with nearly-regular-pentagons. But my bathroom floor does have, exactly, the shape of a nearly-Euclidean-plane. Or put another way, being a nearly-Euclidean-plane is not an abstract model of

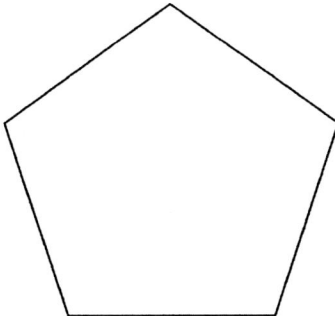

Figure 5. A regular pentagon, with which it is impossible to tile the plane

my bathroom floor, it is its literal shape. Therefore, it cannot be tiled with tiles which are, nearly or exactly, regular pentagons.

The 'cannot' in the last sentence is a necessity at once mathematical and about reality. (A further example in [Franklin, 1989])

That example was of impossibility. The next is an example of necessity in the full sense.

For simplicity, let us restrict ourselves to two dimensions, though there are similar examples in three dimensions. A body is said to be symmetrical about an axis when a point is in the body if and only if the point opposite it across the axis is also in the body. Thus a square is symmetrical about a vertical axis, a horizontal axis and both its diagonals. A body is said to be symmetrical about a point P when a point is in the body if and only if the point directly opposite it across P is also in the body. Thus a square is symmetrical about its centre. The following is a necessarily true statement about real bodies: All bodies symmetrical about both a horizontal and a vertical axis are also symmetrical about the point of intersection of the axes:

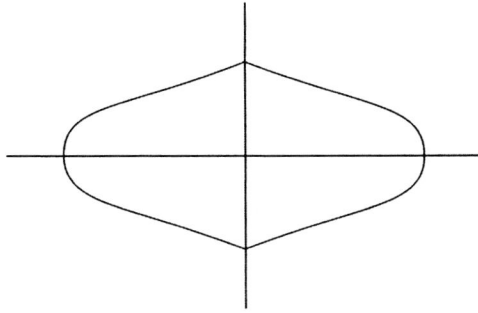

Figure 6. Symmetry about two orthogonal axes implies symmetry about centre

Again, the space need not be Euclidean for this proposition to be true. All that is needed is a space in which the terms make sense.

These examples appear to be necessarily true mathematical propositions which are about reality. It remains to defend this appearance against some well-known objections.

Objection 1.
The proposition $7 + 5 = 12$ appears at first both to be necessary and to say something about reality. For example, it appears to have the consequence that if I put seven apples in a bowl and then put in another five, there will be twelve apples in the bowl. A standard objection begins by noting that it would be different for raindrops, since they may coalesce. So in order to say something about reality, the mathematical proposition must need at least to be conjoined with some proposition such as, 'Apples don't coalesce', which is plainly contingent. This consideration is reinforced by the suspicion that the proposition $7 + 5 = 12$ is tautological, or almost so, in some sense.

Perhaps these objections can be answered, but there is plainly at least a *prima facie* case for a divorce between the necessity of the mathematical proposition and its application to reality. The application seems to be at the cost of introducing stipulations about bodies which may be empirically false.

The examples above are not susceptible to this objection. Being nearly-pentagonal, being symmetrical and so on are properties that real things can have, and the mathematical propositions say something about things with these properties, without the need for any empirical assumptions.

Objection 2.
This objection is perhaps in effect the same as the first one, but historically it has been posed separately. It does at least cast more light on how the examples given escape objections of this kind.

The objection goes as follows: Geometry does not study the shapes of real things. The theory of spheres, for example, cannot apply to bronze spheres, since bronze spheres are not perfectly spherical ([Aristotle, *Metaphysics* 997b33-998a6, 1036a4-12; Proclus, 1970, 10-11]). Those who thought along these lines postulated a relation of 'idealisation' variously understood, between the perfect spheres of geometry and the bronze spheres of mundane reality. Any such thinking, even if not leading to fully Platonist conclusions, will result in a contrast between the ideal (and hence necessary) realm of mathematics and the physical (and contingent) world.

It has been found that the problem was simply a result of the primitive state of Greek mathematics. Ancient mathematics could only deal with simple shapes such as perfect spheres. Modern mathematics, by studying continuous variation, has been able to extend its activities to more complex shapes such as imperfect spheres. That is, there are results not about particular imperfect spheres, but about the ensemble of imperfect spheres of various kinds. For example, consider all imperfect spheres which differ little from a sphere of radius one metre — say which do not deviate by more than one centimetre from the sphere anywhere. Then the volume of any such imperfect sphere differs from the volume of the perfect sphere by less than one tenth of a cubic metre. So imperfect-sphere shapes can be studied mathematically just as well as — though with more difficulty than — perfect spheres. But real bronze things do have imperfect-sphere shapes, without any 'idealisation' or 'simplification'. So mathematical results about imperfect spheres can apply directly to the real shapes of real things.

The examples above involved no idealisations. They therefore escape any problems from objection 2.

Objection 3.
The third objection proceeds from the supposed hypothetical nature of mathematics. Bertrand Russell's dictum, 'Pure mathematics consists entirely of assertions to the effect that, if such and such a proposition is true of anything, then such and such another proposition is true of that thing' [Russell, 1917, 75] suggests a connection between hypotheticality and lack of content. Even those who have not gone so far as to think that mathematics is just logic have generally thought that mathematics is not about reality, but only, like logic, relates statements which may happen to be about reality. Physicists, Einstein included, have been especially prone to speak in this way, since for them mathematics is primarily a bag of tricks used to deduce consequences from theories.

The answer to this objection consists fundamentally in a denial that mathematics is more hypothetical than any other science. The examples given above do not look hypothetical, but they could easily be cast in hypothetical form. But the fact that mathematical statements are often written in if-then form is not in itself an argument that mathematics is especially hypothetical. Any science, even a purely classificatory one, contains universally quantified statements, and any 'All As are Bs' statement can equally well be expressed hypothetically, as 'If anything is an A, it is a B'. A hypothetical statement may be convenient, especially in a complex situation, but it is just as much about real As and Bs as 'All As are Bs'.

No-one argues that

> All applications of 550 mls/hectare Igran are effective against normal infestations of capeweed

is not about reality nerely because it can be expressed hypothetically as

> If 550 mls/hectare Igran is applied to a normal infestation of capeweed, the weed will die.

Neither should mathematical propositions such as those in the examples be thought to be not about reality because they can be expressed hypothetically. Real portions of liquid can be (approximately) 550 mls of Igran. Real tables can be (approximately) symmetrical about axes. Real bathroom floors can be (nearly) flat and real tiles (nearly) regular pentagons [Musgrave, 1977, §5].

The impact of this argument is not lessened even if the process of recasting mathematics into if-then form goes as far as axiomatisation. Einstein thought it was. His quotation with which the section began continues as follows:

> As far as the propositions of mathematics refer to reality, they are not
> certain; and as far as they are certain, they do not refer to reality. It
> seems to me that complete clarity as to this state of things became
> common property only through that trend in mathematics which is
> known by the name of 'axiomatics'. [Einstein, 1954, 233]

Einstein goes on to argue that deductive axiomatised geometry is mathematics, is certain and is 'purely formal', that is, uninterpreted; while applied geometry, which includes the proposition that solid bodies are related as bodies in three-dimensional Euclidean space, is a branch of physics. Granted that it is a contingent physical proposition that solid bodies are related in this way, and granted that an uninterpreted system of deductive 'geometry' is possible, there remain two main problems about Einstein's conclusion that 'mathematics as such cannot predicate anything about ... real objects' [Einstein, 1954, 234].

Firstly, non-mathematical topics, such as special relativity, can be axiomatised without thereby ceasing to be about real things. This remains so even if one sets up a parallel system of 'purely formal axiomatised special relativity' which one pretends not to interpret.

Secondly, even if some of the propositions of 'applied geometry' are contingent, not all are, as the examples above showed. Doubtless there is a 'proposition' of 'purely formal geometry' corresponding to 'It is impossible to tile my bathroom floor with regular pentagonal tiles'; the point is that the modality, 'impossible', is still there when it is interpreted.

In theory this completes the reply to the objection that mathematics is necessary only because it is hypothetical. Unfortunately it does nothing to explain the strong feeling among ordinary users of mathematics, such as physicists and engineers, that mathematics is a kind of tool kit for getting one scientific proposition out of another. If an electrical engineer is accustomed to working out currents by reaching for his table of Laplace transforms, he will inevitably see this mathematical method as a tool whose 'necessity', if any, is because mathematics is not about anything, but is only a kind of theoretical juice extractor.

It must be admitted that a certain amount of applicable mathematics really does consist of tricks or calculatory devices. Tricks, in mathematics or anywhere else, are not *about* anything, and any real mathematics that concerns them will be in explaining why and when they work; this is a problem the engineer has little interest in, except perhaps for the final answer. The difficulty is to explain how mathematics can have both necessity and application to reality, without appearing to do so to many of its users.

The short answer to this lies in the mind's tendency to think of relations as not really existing. Since mathematics is so tied up with relations of certain kinds, its subject matter is easy to overlook. A familiar example of how mathematics applies in physics will make this clearer.

Newton postulated the inverse square law of gravitation, and derived from it the proposition that the orbits of the planets are elliptical. Let us look a little more closely at the derivation, to see whether the mathematical reasoning is in some way about reality or is only a logical device for deriving one scientific law from another.

First of all, Newton did not derive the shape of the orbits from the law of gravitation alone. An orbit is a path along which a planet moves, so there needs to be a proposition connecting the law of force with movement; the link is, of course,

$$\text{force} = \text{mass} \times \text{acceleration}.$$

Then there must be an assertion that net accelerations other than those caused by the gravitation of the sun are negligible. Ideally this should be accompanied by a stability analysis showing that small extra net forces will only produce small deviations from the calculated paths. Adding the necessary premises has not, however, introduced any ellipses. What the premises give is the local change of motion of a planet at any point; given any planet at any point with any speed, the laws give the force, and hence the acceleration — change of speed — that the planet undergoes. The job of the mathematics — the only job of the mathematics — is to add together these changes of motion at all the points of the path, and reveal that the resulting path must be an ellipse. The mathematics must track the path, that is, it must extract the global motion from the local motions.

There are two ways to do this mathematics. In this particular case, there are some neat tricks available with angular momentum. They are remarkable enough, but are still purely matters of technique that luckily allow an exact solution to the problem with little work. The other method is more widely applicable and is here more revealing because more direct; it is to use a computer to approximate the path by cutting it into small pieces. At the initial point the acceleration is calculated and the motion of the planet calculated for a short distance, then the new acceleration is calculated for the new position, and so on. The smaller the pieces the path is cut into, the more accurate the calculation. This is the method actually used for calculating planetary orbits, since it can easily take account of small extra forces, such as the gravitational interaction of the planets, which render special tricks useless. The absence of computational tricks exposes what the mathematics is actually doing — extracting global structure from local.

The example is typical of how mathematics is applied, as is clear from the large proportion of applied mathematics that is concerned one way or another with the solution of differential equations. Solving a differential equation is, normally, entirely a matter of getting global structure from local — the equation gives what is happening in the neighbourhood of each point; the solution is the global behaviour that results. [Smale, 1969] A good deal of mathematical modelling and operations research also deals with calculating the overall effects of local causes. The examples above all involve some kind of interaction of local with global structure.

Though it is notoriously difficult to say what 'structure' is, it is at least something to do with relations, especially internal part-whole relations. If an orbit is elliptical globally, its curvature at each point is necessarily that given by the inverse square law, and vice versa. In general the connections between local and global structure are necessary, though it seems to make the matter more obscure rather than less to call the necessity 'logical'. Seen this way, there is little temptation to regard the function of mathematics as merely the deducing of consequences, like a logical engine. It is easy to see, though, why mathematics has been seen as having no subject matter — the western mind has had enormous difficulty focussing on the reality of relations at all [Weinberg, 1965, section 2], let alone such abstract relations as structural ones. Nevertheless, symmetry, continuity and the rest are just as real as relations that can be measured, such as ratios of masses; bought and sold, such as interest rate futures; and litigated over, such as paternity.

Typically, then, a scientist will postulate or observe some simple local behaviour in a system, such as the inverse square law of attraction or a population growth rate proportional to the size of the population. The mathematical work, whether by hand or computer, will put the pieces together to find out the global effect of the continued operation of the proposed law – in these cases elliptical orbits and exponential growth. There are bad reasons for thinking the mathematics is just 'turning the handle' — for example it costs less than experiment, and many scientists' expertise runs to only simple mathematical techniques. But there are no good reasons. The mathematics investigates the necessary interconnections between the parts of the global structure, which are as real properties of the system studied as any other.

This completes the explanation of why mathematics seems to many to be just a deduction engine, or to be purely hypothetical, even though it is not.

Objection 4.
Certain schools of philosophy have thought there can be no necessary truths that are genuinely about reality, so that any necessary truth must be vacuous. 'There can be no necessary connections between distinct existences.'

Answer: The philosophy of mathematics has enough to do dealing with mathematics, without taking upon itself the refutation of outmoded metaphysical dogmas. Mathematics must be appreciated on its own terms, and wider metaphysical theories adjusted to take account of whatever is found.

Nevertheless something can be said about the exact point where this objection fails to make contact with the examples above. The clue is the word 'distinct'. The word suggests a kind of logical atomism, as if relations can be thought of as strings joining point particulars. One need not be F.H. Bradley to find that view too simple. It is especially inappropriate when treating things with internal structure, as is typical in mathematics. In an infinitely divisible thing like the surface of a bathroom floor, where are the point particulars with purely external relations? (The points of space, perhaps? But the relations between tile-sized parts of space and the whole space either have nothing to do with points at all or are properties of the whole system of relations between points.)

All the objections are thus answered. The conclusion stands, therefore, that the three examples are, as they appear to be, mathematical, necessary and about reality.

The thesis defended has been that *some* necessary mathematical statements refer directly to reality. The stronger thesis that *all* mathematical truths refer to reality seems too strong. It would indeed follow, if there were no relevant differences between the examples above and other mathematical truths. But there are differences. In particular, there are more things dreamed of in mathematics than could possibly be in reality. Some mathematical entities are just too big; even if something in reality could have the structure of an infinite dimensional vector space, it would be too big for us to know it did. Other mathematical entities seem obviously fictions from the way they are introduced, such as negative numbers. Statements about negative numbers can refer to reality in some way, since one can make true conclusions about debts by using negative numbers. But the reference is indirect, in the way that statements about the average wage-earner refer to reality, but not in the direct sense of asserting something about an entity, 'the average wage-earner'. Indirect reference of this kind is not in principle mysterious, though it needs to be explained in each particular case. So it can be conceded that many of the entities mentioned in mathematics are fictional, without any admission that this makes mathematics unique; minus-1 can be seen as like fictional entities elsewhere, such as the typical Londoner, holes, the national debt, the Zeitgeist and so on.

What has been asserted is that there are properties, such as symmetry, continuity, divisibility, increase, order, part and whole which are possessed by real things and are studied directly by mathematics, resulting in necessary propositions about them.

5 THE FORMAL SCIENCES

Aristotelians deplore the narrow range of examples chosen for discussion in traditional philosophy of mathematics. The traditional diet — numbers, sets, infinite cardinals, axioms, theorems of formal logic — is far from typical of what mathematicians do. It has led to intellectual anorexia, by depriving the philosophy of mathematics of the nourishment it would and should receive from the expansive world of mathematics of the last hundred years. Philosophers have almost completely ignored not only the broad range of pure and applied mathematics and statistics, but a whole suite of 'formal' or 'mathematical' sciences that have appeared only in the last seventy years. We give here a few brief examples to indicate why these developments are of philosophical interest to those pursuing realist views of mathematics.

It used to be that the classification of sciences was clear. There were natural sciences, and there were social sciences. Then there were mathematics and logic, which might or might not be described as sciences, but seemed to be plainly distinguished from the other sciences by their use of proof instead of experiment, measurement and theorising. This neat picture has been disturbed by the appearance in the last several decades of a number of new sciences, variously called the 'formal' or 'mathematical' sciences, or the 'sciences of complexity' [Pagels, 1988; Waldrop, 1992; Wolfram, 2002] or 'sciences of the artificial.' [Simon, 1969]

The number of these sciences is large, very many people work in them, and even more use their results. Their formal nature would seem to entitle them to the special consideration mathematics and logic have obtained. Not only that, but the knowledge in the formal sciences, with its proofs about network flows, proofs of computer program correctness and the like, gives every appearance of having achieved the philosophers' stone; a method of transmuting opinion about the base and contingent beings of this world into the necessary knowledge of pure reason. They also supply a number of concepts, like 'feedback', which permit 'in principle' explanatory talk about complex phenomena.

The oldest properly constituted formal science is perhaps operations research (OR). Its origin is normally dated to the years just before and during World War II, when multi–disciplinary scientific teams investigated the most efficient patterns of search for U–boats, the optimal size of convoys, and the like. Typical problems now considered are task scheduling and bin packing. Given a number of factory tasks, subject to constants about which must follow which, which cannot be run simultaneously because they use the same machine, and so on, one seeks the way to fit them into the shortest time. Bin packing deals with how to fit a heap of articles of given sizes most efficiently into a number of bins of given capacities. [Woolsey & Swanson, 1975]. The methods used rely essentially on search through the possibilities, using mathematical ideas to rule out obviously wrong cases. The diversity of activities in OR is illustrated by the sub-headings in the American Mathematical Society's classification of 'Operations research and mathematical science': Inventory, storage, reservoirs; Transportation, logistics; Flows in network, deterministic; Communication networks; Flows in networks, probabilistic; Highway traffic; Queues and service; Reliability, availability, maintenance, inspection; Production models; Scheduling theory; Search theory; Management decision–making, including multiple objectives; Marketing, advertising; Theory of organisations, industrial and manpower planning; Discrete location and assignment; Continuous assignment; Case–oriented studies. [*Mathematical Reviews*, 1990]

The names indicate the origin of the subject in various applied questions, but, as the grouping of actual applications into the last topic indicates, OR is now an abstract science. Plainly, a philosophy of mathematics that started with OR as its typical example would have a different — more Aristotelian — flavour than one starting with the theory of infinite sets.

Other formal sciences include control theory (noted for introducing the now familiar concepts of 'feedback' and 'tradeoff'), pattern recognition, signal processing, numerical taxonomy, image processing, network analysis, data mining, game theory, artificial life, mathematical ecology, statistical mechanics and the various aspects of theoretical computer science including proof of program correctness, computational complexity theory, computer simulation and artificial intelligence. Despite their diversity, it is clear they have in common the analysis of complex systems (both real systems and models of real systems). That is partly what accounts for their growing prominence since the computer revolution — computation can discover results about large systems by modelling them. But the role of proof in the formal sciences shows their commonality with mathematics. The general philosophical tendency of these sciences will therefore be to support a philosophy of mathematics that is structuralist (since the formal sciences deal with complexity, that is, a great deal of structure) and Aristotelian (since the structures are mostly realized fully in real world cases such as transportation networks or computer code).

The greatest philosophical interest in the formal sciences is surely the promise they hold of necessary, provable knowledge which is at the same time about the real world, not just some Platonic or abstract idealisation of it.

There is just one of the formal sciences in which a debate on precisely this question has taken place, and done so with a degree of philosophical sophistication. It is worth reviewing the arguments, as they address matters that are common to all the formal sciences. At issue is the status of proofs of correctness of computer programs. The late 1960s were the years of the 'software crisis', when it was realised that creating large programs free of bugs was much harder than had been thought. It was agreed that in most cases the fault lay in mistakes in the logical structure of the programs: there were unnoticed interactions between different parts, or possible cases not covered. One remedy suggested was that, since a computer program is a sequence of logical steps like a mathematical argument, it could be proved to be correct. The 'program verification' project has had a certain amount of success in making software error-free, mainly, it appears, by encouraging the writing of programs whose logical structure is clear enough to allow proofs of their correctness to be written. A lot of time and money is invested in this activity. But the question is, does the proof guarantee the correctness of the actual physical program that is fed into the computer, or only of an abstraction of the program? C. A. R. Hoare, a leader in the field, made strong claims:

> Computer programming is an exact science, in that all the properties of a program and all the consequences of executing it can, in principle, be found out from the text of the program itself by means of purely deductive reasoning. [Hoare, 1969]

The philosopher James Fetzer argued that the program verification project was impossible in principle. Published not in the obscurity of a philosophical journal, but in the prestigious *Communications of the Association for Computing Machinery,* his attack had effect, being suspected of threatening the livelihood of thousands. [Fetzer, 1988] Fetzer's argument relies wholly on the gap between abstraction and reality, and applies equally well to any case where a mathematical model is studied with a view to achieving certainty about the modeled reality:

> These limitations arise from the character of computers as complex causal systems whose behaviour, in principle, can only be known with the uncertainty that attends empirical knowledge as opposed to the certainty that attends specific kinds of mathematical demonstrations. For when the domain of entities that is thereby described consists of

> purely abstract entities, conclusive absolute verifications are possible; but when the domain of entities that is thereby described consists of non-abstract physical entities ... only inconclusive relative verifications are possible. [Fetzer, 1989]

It has been subsequently pointed out that to predict what an actual program does on an actual computer, one needs to model not only the program and the hardware, but also the environment, including, for example, the skills of the operator. And there can be changes in the hardware and environment between the time of the proof and the time of operation. In addition, the program runs on top of a complex operating system, which is known to contain bugs. Plainly, certainty is not attainable about any of these matters.

But there is some mismatch between these (undoubtedly true) considerations and what was being claimed. Aside from a little inadvised hype, the advocates of proofs of correctness had admitted that such proofs could not detect, for example, typos. And, on examination, the entities Hoare had claimed to have certainty about were, while real, not unsurveyable systems including machines and users, but written programs. [Hoare, 1985] That is, they are the same kind of things as published mathematical proofs.

If a mathematician says, in support of his assertion, 'my proof is published on page X of volume Y of *Inventiones Mathematicae*', one does not normally say — even a philosopher does not normally say — 'your assertion is attended with uncertainty because there may be typos in the proof', or 'perhaps the Deceitful Demon is causing me to misremember earlier steps as I read later ones.' The reason is that what the mathematician is offering is not, in the first instance, absolute certainty in principle, but necessity. This is how his assertion differs from one made by a physicist. A proof offers a necessary connection between premises and conclusion. One may extract practical certainty from this, given the practical certainty of normal sense perception, but that is a separate step. That is, the certainty offered by mathematics does depend on a normal anti-scepticism about the senses, but removes, through proof, the further source of uncertainty found in the physical and social sciences, arising from the uncertainty of inductive reasoning and of theorising. Assertions in physics, about a particular case, have two types of uncertainty: that arising from the measurement and observation needed to check that the theory applies to the case, and that of the theory itself. Mathematical proof has only the first.

It is the same with programs. While there is a considerable certainty gap between reasoning and the effect of an actually executed computer program, there is no such gap in the case Hoare was considering, the unexecuted program. A proof (in, say, the predicate calculus) is a sequence of steps exhibiting the logical connection between formulas, and checkable by humans (if it is short enough). Likewise a computer program is a logical sequence of instructions, the logical connections among which are checkable by humans (if there are not too many).

One feature of programs that is inessential to this reply is their being textual. So, one line taken by Fetzer's opponents was to say that not only could programs be proved correct, but so could machines. Again, it was admitted that there was a theoretical possibility of a perceptual mistake, but this was regarded as trivial, and it was suggested that the safety of, say, a (physically installed) railway signalling system could be assured by proofs that it would never allow two trains on the same track, no matter what failures occurred.

The following features of the program verification example carry over to reasoning in all the formal sciences:

- There are connections between the parts of the system being studied, which can be reasoned about in purely logical terms.

- The complexity is, in small cases, surveyable. That is, one can have practical certainty by direct observation of the local structure. Any uncertainty is limited to the mere theoretical uncertainty one has about even the best sense knowledge.

- Hence the necessity translates into practical certainty.

- Computer checking can extend the practical certainty to much larger cases.

Euler's example of the bridges of Königsberg, considered earlier, is an early example of network theory and an especially clear case for discussion. The number and importance of such examples has grown without bound, and it is time for more serious philosophical consideration of them.

6 COMPARISON WITH PLATONISM AND NOMINALISM

The main body of philosophy of mathematics since Frege has moved along a path unsympathetic to Aristotelian views. We collect here some comparisons of the present point of view with standard philosophy of mathematics and reply to some of the objections arising from it.

Frege set terms for the debate that were essentially Platonist. His language is Platonist about sets and numbers, and almost all subsequent philosophy of mathematics has either accepted Frege's views literally and hence embraced Platonism, or attempted to deploy broad-based nominalist strategies to undermine realism (Platonist or not) in general.

The crucial move towards Platonism in modern philosophy of numbers occurred in Frege's argument for the conclusion that numbers are not properties of physical things. From the Aristotelian point of view, there is a core of Frege's argument that is correct, but his Platonist conclusion does not follow. Frege argues, in a central passage of his *Foundations of Arithmetic*, that attributing a number to things is quite unlike attributing an ordinary property like 'green':

> It is quite true that, while I am not in a position, simply by thinking of it differently, to alter the colour or hardness of a thing in the slightest, I am able to think of the Iliad as one poem, or as 24 Books, or as

> some large Number of verses. Is it not in totally different senses that we speak of a tree as having 1000 leaves and again as having green leaves? The green colour we ascribe to each single leaf, but not the number 1000. If we call all the leaves of a tree taken together its foliage, then the foliage too is green, but it is not 1000. To what then does the property 1000 really belong? It almost looks as though it belongs neither to any single one of the leaves nor to the totality of them all; is it possible that it does not really belong to things in the external world at all? [Frege, 1884, §22, p. 28].

Frege's preamble in this passage is sound and his question "to what does the property 1000 really belong?" is a good one. The Platonist direction of his conclusion that numbers must be properties of something beyond the external world does not follow, because he has not included the Aristotelian option among those that make sense of the preamble. There are three possible directions to go at this point:

- An idealist or psychologist direction, according to which number is relative to how we choose to think about objects; Frege quotes Berkeley as taking that option but is firmly against it himself as it is unable to make sense of the objectivity of mathematics

- A Platonist direction, as Frege and his followers adopt, according to which number is either a self-subsistent entity itself or an objective property of something not in this world, such as a Concept (in Frege's non-psychological sense of that term) or an extension of a Concept (a set or function conceived Platonistically) [Frege, 1884, especially §72, p. 85]

- An Aristotelian direction, which Frege does not consider, according to which 1000 is not a property of the foliage simply but of the relation between the foliage and the universal 'being a leaf', while the foliage's being divided into leaves is a property of it "in the external world" as much as its green colour is.

When Frege returns to the issue later in the *Foundations*, he expresses himself in language that is interpretable at least as naturally from an Aristotelian as from a Platonist perspective:

> ... the concept, to which the number is assigned, does in general isolate in a definite manner what falls under it. The concept "letters in the word three" isolates the t from the h, the h from the r, and so on. The concept "syllables in the word three" picks out the word as a whole, and as indivisible in the sense that no part of it falls any longer under the same concept. Not all concepts possess this quality. We can, for example, divide up something falling under the concept "red" into parts in a variety of ways ... Only a concept which falls under it in a definite manner, and which does not permit an arbitrary division of

it into parts, can be a unit relative to a finite Number. [Frege, 1884, §54, p. 66]

On an Aristotelian view, Frege is here distinguishing correctly unit-making universals from others. The parallel he draws between them and a straightforward physical property like "red" is reason against his unargued Platonist understanding of "concepts". If red's being homoiomerous (true of parts) is compatible with red's being physical, it is unclear why being non-homoiomerous is in itself incompatible with being physical. Being large is not homoiomerous, in that the parts of a large thing are not all large, but that does not suggest that the property large is non-physical.

The degree of Frege's Platonism has been debated, as he does not emphasise the otherworldliness of the Forms and is content with the kind of reason that performs mathematical proofs as a means of knowledge of them (rather than requiring a mysterious intuition). But the emphasis here is not so much on the interpretation of Frege as on the effect of his forceful statements of Platonism on later work.

Frege's Platonism, in logic as much as in mathematics, has dominated the agenda of later analytic philosophy of logic, language and mathematics. It has led to a characteristic view of what counts as an adequate answer to questions in those areas, a view that Aristotelians (and often other naturalists) find inadequate.

Characteristic features of the philosophy of mathematics of the last hundred years that seem to Aristotelians to be mistakes or at least unfortunate biases in emphasis inspired by Frege include:

- Regarding Platonism and nominalism as mutually exhaustive answers to the question "Do numbers exist?", and hence taking a fundamentalist attitude to mathematical entities, as if they exist as "abstract" Platonist substances or not at all

- Resting satisfied that a concept (e.g. structure, the continuum) has been explained if it has been constructed out of some simple Platonist entities such as sets

- Feeling no need to ask for an account of what sets are

- Emphasising infinities and downplaying the role of small finite structures, the counting of small numbers and the measurement of finite quantities

- Regarding the problem of the "applicability of mathematics" or "indispensability of mathematics" as a question about the relation of some Platonist entities (e.g. numbers) and the physical world

- Regarding measurement as a relation between numbers and measured parts of the world

- Taking the epistemology of mathematics to be mysterious because requiring access to a Platonist realm.

We will examine how some of these issues have played out in the most prominent writings in the philosophy of mathematics in recent decades.

The assumption that the real alternatives in the philosophy of mathematics are Platonist realism or nominalism is pervasive in the philosophy of mathematics, as is clear from the survey of realism in Balaguer's chapter in this *Handbook*, as well as in standard works such as the *Routledge Encyclopedia of Philosophy*. In the introduction to this section, we found little non-Platonist realism to list, and that has not been taken with much seriousness by the mainstream of philosophy of mathematics.

The dichotomy also makes it too easy for nominalists to claim success if they analyse a concept without reference to numbers or sets. Hartry Field in *Science Without Numbers*, for example, proposed to "nominalize" basic mathematical physics. Typical of his strategy is his account of temperature, considered as a quantity that varies continuously over space. Temperature is often described in mathematical physics textbooks as a function (that is, a Platonist mathematical entity) from space-time points to the set of real numbers (the function that gives, for each point, the number that is the temperature at that point). Field rightly says that one can say what one needs to say about temperature without reference to functions or numbers. He begins with "a three-place relation [among space-time points] Temp-Bet, with y Temp-Bet xz meaning intuitively that y is a space-time point at which the temperature is (inclusively) between the temperatures of points x and z; and a 4-place relation Temp-Cong, with xy Temp-Cong zw meaning intuitively that the temperature difference between points x and y is equal in absolute value to the temperature difference between points z and w." He then provides axioms for Temp-Cong and Temp-Bet so as ensure they behave as congruence and betweenness should, and so that it is possible to prove a "representation theorem" stating that a structure $\langle A, \text{Temp-Bet}_A, \text{Temp-Cong}_A \rangle$ is a model of the axioms if and only if there is a function ψ from A to an interval of real numbers such that

a. for all x, y, z, y Temp-Bet$_A$ $xz \leftrightarrow \psi(x) \leq \psi(y) \leq \psi(z)$ or $\psi(z) \leq \psi(y) \leq \psi(x)$

b. for all x, y, z, w, xy Temp-Cong$_A$ $zw \leftrightarrow |\psi(x) - \psi(y)| = |\psi(z) - \psi(w)|$.

[Field, 1980, 56]

Since the clauses to the right of the double-arrows refer to numbers and functions while the terms to the left do not, Field can rightly claim to have dispensed with numbers and functions understood Platonistically. But is the result nominalist? It is all very well to write Temp-Bet and Temp-Cong as if they are atomic predicates, but they can only perform the task of representing facts about temperature if they really do "intuitively mean" betweenness and interval-equality of temperature, and if the axioms describe those relations as they hold of the real property of temperature (to a close approximation at least). In virtue of what, the Aristotelian asks, is Temp-Cong taken to be, say, transitive? It must be required because congruence of temperature intervals really is transitive. Field has not gone any way towards eliminating reference to the real continuous property, temperature.

The case of the "construction of the continuum" well illustrates the second problem with Platonist strategy, arising from its analysis of concepts via construction of them out of sets. According to Platonists, an obscure concept such as the continuum or "structure", or the meaning of sentences in natural language, is adequately explained if the concept is constructed out of some simpler Platonist entities such as sets or propositions that are taken to be so basic they need no further explanation. Aristotelian scepticism about this strategy focuses on two points: firstly, the alleged self-explanatoriness of these basic entities, and secondly, on how we know that the proposed construction in sets or propositions is adequate to the original concept we were trying to explicate — or rather (since the question is not fundamentally epistemological) what it is that would make the construction an adequate explanation. We treat the second problem here, and the first in the next section.

What account is to be given of why that particular set of sets of sets of ... is the (or a) correct construction of the explanandum, such as "the continuum"? We have an initial intuitive notion of the continuum as a continuous line, a universal that could be realised in real space (though whether real space is infinitely divisible is an empirical question, to which the answer is currently not known). [Franklin, 1994] There exists an elaborate classical construction of "the continuum" as a set of equivalence classes of Cauchy sequences of rational numbers, with Cauchy sequences and rational numbers themselves constructed in complex ways out of sets. What is it that makes that particular set an analysis of the original notion of the continuum? The Aristotelian has an answer to that question: namely that the notion of closeness definable between two equivalence classes of Cauchy sequences reflects the notion of closeness between points in the original continuum. "Reflects" means here an identity of universals: closeness is a universal literally identical in the two cases (and so satisfying the same properties such as the triangle inequality). The statement that closeness is the same in both cases is not subject to mathematical proof, because the original continuum is not a formalised entity. It can only be subject to the same kind of understanding as any statement that a portion of the real world is adequately modelled by some formalism, for example, that a rail transport system is correctly described as a network with nodes. The Platonist, however, does not have any answer to the question of why that construction models the continuum; the Platonist will avoid mention of real space as far as possible and simply rely on the tradition of mathematicians to call the set-theoretical construction "the continuum". The fact that Cantor constructed something with exactly the properties assigned by Aristotle to the continuum [Newstead, 2001] is important but unacknowledged in the Platonist story.

Similar considerations apply to all of the many constructions of mathematical concepts out of sets. There is some mathematical point to the exercise, mainly to demonstrate the consistency of the concepts (or more exactly, the consistency of the concepts relative to the consistency of set theory). But there is no philosophical point to them. The Aristotelian is not impressed by the construction of a relation as a set of ordered pairs, for example. To see that as an analysis of relations would make the same mistake as identifying a property with its extension. [Armstrong, 1978, vol. 1 ch. 4] The set of blue things is not the property blue, nor is it in any sense an "analysis" of the concept blue. It is the property blue that pre-exists and unifies the set (and supports the counterfactual that if anything else were blue, it would be a member of the set). Similarly the ordered pair (3,4) is a member of the extension of the relation "less than" because 3 is less than 4, not vice versa. The same remarks apply to, for example, the definition of a group as a set with a binary operation satisfying the associative, identity and inverse laws. That definition only has point because of pre-existing mathematical experience with groups of symmetries that do satisfy those laws, and the abstraction from those cases is what makes the abstract definition of a group a correct one. The case of groups is an instance of the more general Bourbakist notion of (algebraic or topological) "structure" as a set-theoretical construction. [Corry, 1992] Certainly if one has sets one can construct any number of sets of sets of sets ... of them, but the Aristotelian demands an answer as to why one such construction is an adequate analysis of symmetry groups and another an adequate analysis of topology. That answer must be in terms of one construction sharing a property with symmetry groups and another sharing a different property with topology. It is the shared property, as the mathematician using the sets as an analysis knows, that is the reason for the whole exercise. The philosopher with less mathematical experience is likely the make the mistake (in Aristotle's language) of confusing formal and material cause, that is, of thinking something is explained when one knows what it is made of. Constructing some structure or concept out of sets does not mean that the structure or concept is therefore about sets, for the same reason as an ability to construct the concept out of wood would not make the concept one of carpentry.

There is thus nothing to recommend the idea that if the philosophy of mathematics can explain sets, it can explain anything in mathematics since "technically, any object of mathematical study can be taken to be a set." [Maddy, 1992, 4] That gives a partial explanation of why mathematicians find standard philosophy of mathematics so irrelevant to their concerns. If mathematicians are studying the structures that can be constructed in sets while philosophers are discussing the material in which they are constructed, there is the same mismatch of concerns as if experts in concrete pouring set themselves up as gurus on architecture.

In any case, if some concept is constructed out of sets, that is only an advance, philosophically, if the Platonist conception of sets is clear. That is not the case. David Lewis exposes the unclarity of the concept in Cantor ('many, which can be thought of as one, i.e., a totality of definite elements that can be combined into a whole by a law') and in mathematics textbooks. [Lewis, 1991, 29-31] There is no explanation provided of the relation of singletons to their elements, for example. Philosophers, Lewis implies, have done even worse with the problem of what a set is than the writers of mathematics textbooks. They have simply ignored it. And when Aristotelians have offered an answer, such as David Armstrong's suggestion that the singleton set of an object x is the state of affairs of x's having some unit-making property, [Armstrong, 1991] Platonists have ignored it on the grounds that they do not need it. Since any analysis of the basic Platonist entities in terms of something non-Platonist (such as states of affairs) would threaten the whole Platonist edifice, Platonists must pretend that their basic building blocks are perfectly clear and have no need of analysis.

The Platonist mindset prefers to rush into the higher infinities and the technicalities associated with them, at the expense of achieving a correct philosophical view of the simpler finite cases first — cases such as counting small numbers, measuring small quantities, timetabling and the like. Philosophers of mathematics have been quick to accept that physics requires the full ontology of traditional real analysis, including the continuum conceived of an infinite set of points, and hence have conceived their task as essentially including an explanation of the role of infinities. But that does things in the wrong order. Firstly, the simple should in general be explained first and extended to the complex, so it is natural to ask first that we understand small numbers and counting before we ask about infinities. Secondly, the computer age has shown how to do most mathematics with finite means. A symbolic manipulation package such as Mathematica or Maple can do almost all mathematics needed for applications (and more pure mathematics than most mathematics graduates can do) but it is a finite object and manipulates only finite objects (such as formulas). It is possible to put forward with at least some degree of credibility an "ultrafinitist" philosophy that admits only finite numbers, [Zeilberger, 1991] which if not philosophically convincing is a sufficient reminder of how much of the mathematics one needs to do can be done in a strictly finite setting. Proposals that the universe (including space and time) is finite and can be adequately described by a discrete (though computationally intensive) mathematics in place of traditional real analysis [Wolfram, 2002, esp. 465-545] also cast doubt on whether infinities are really needed in applied mathematics.

Nowhere is the divergence between the Aristotelian and Platonist standpoints more obvious than in how they begin the problem of the applicability of mathematics. Even that description of the problem has a Platonist bias, as if the problem is about the relations between mathematical entities and something distinct from them in the "world" to which they are "applied". On an Aristotelian view, there is no such initial separation between mathematics and its "applications".

That undesirable assumed split between mathematical entities and their "applications" is first evident in accounts of measurement. Considering the fundamental importance of measurement as the first point of contact between mathematics and what it is about, it is surprising how little attention has been paid to it in the standard literature of the philosophy of mathematics. What attention there has been has tended to concentrate on "representation theorems" that describe the conditions under which quantities can be represented by numbers. "Measurement theory officially takes homomorphisms of empirical domains into (intended) models of mathematical systems as its subject matter", as one recent writer expresses it. [Azzouni, 2004, 161] That again poses the problem as essentially one about the association of numbers to parts of the world, which leads to a Platonist perspective on the problem. The Aristotelian insists that the system of ratios of lengths, for example, pre-exists in the physical things being measured, and measurement consists in identifying the ratios that are of interest in a particular case; the arbitrary choice of unit that allows ratios to be converted to digital numerals for ease of calculation is something that happens at the last step (similar in [Bigelow & Pargetter, 1990, 60-61]).

Fregean Platonism about logic and linguistic items has also contributed to a distorted view of the indispensability argument, widely agreed to be the best argument for Platonism in mathematics. It is obvious that *mathematics* (mathematical practice, mathematical statement of theories, mathematical deduction from theories) is indispensable to science, but the argument arises from more specific claims about the indispensability of reference to mathematical entities (such as numbers and sets), concluding that such entities exist (in some Platonist sense). As Quine put the argument,

> Ordinary interpreted scientific discourse is as irredeemably committed to abstract objects — to nations, species, numbers, functions, sets — as it is to apples and other bodies. All these things figure as values of the variables in our overall system of the world. The numbers and functions contribute just as genuinely to physical theory as do hypothetical particles. [Quine, 1981, 149-50]

As stated (and as further explained by Quine and Putnam) that argument implies an attitude to language both exceedingly reverent and exceedingly fundamentalist, an attitude that was only credible — in the mid-twentieth-century heyday of linguistic philosophy when it was credible at all — in the wake of Frege's Platonism about such entities as propositions and the objects of reference. Later more naturalist perspectives have not found it plausible that the language tail can wag the ontological dog in that way.

It is true that the careful defence of the indispensability argument by Colyvan is not so easily dismissed. Nevertheless it preserves the main features that Aristotelians find undesirable, the fundamentalism of the interpretation of reference to entities (if it cannot be paraphrased away) and the assumed Platonism of the conclusion. Colyvan does begin by redefining "Platonism" so widely as to include Aristotelian realism. [Colyvan, 2001, 4] That is not a good idea, because Plato and Aristotle do not bear the same relation as Cicero and Tully, and the name "Platonism" has traditionally been reserved for a realist philosophy that contrasts with the Aristotelian. But in any case Colyvan's discussion proceeds without further notice of that option. The strategies for the realist, he says, are either a mysterious perception-like "intuition" of the Forms, or an inference to mathematical objects as "posits" similar to black holes and electrons, which are not perceived but are posited to exist by the best physical theory. And he takes it for granted that the Platonism to which he believes the indispensability argument leads denies the "Eleatic principle" that "causality is the mark of being". The numbers, sets or other objects whose existence is supported by the indispensability argument are, he believes, causally inactive, in contrast to scientific properties like colours, and hence he argues that the Eleatic principle is false. [Colyvan, 2001, ch. 3] Cheyne and Pigden [1996], however, argue that any indispensability argument ought to conclude to entities that have causal powers, as atoms do: it is their causal power that makes them indispensable to the theory. 'If we are genuinely unable to leave those objects out of our best theory of what the world is like ... then they must be responsible in some way for that world's being the way it is. In other words, their indispensability is explained by the fact that they are causally affecting the world, however indirectly. The indispensability argument may yet be compelling, but it would seem to be a compelling argument for the existence of entities *with* causal powers.' At the very least, the existence of atoms *causally* explains the observations that led to their postulation. It is not clear what corresponds in the case of Platonic mathematical entities.

But surely there is something far-fetched in thinking of numbers as inferred hidden entities like atoms or genes. The existence of atoms is not obvious. It is only inferred from complex considerations about the ratios in which pure chemicals combine and from subtle observations of suspensions in fluids. On the other hand, a five-year-old understands all there is to know about why $2 + 2 = 4$. Kant's view that we understand counting thoroughly because we impose the counting structure on experience [Franklin, 2006] may be going too far, but he was right in believing that we do understand counting completely, and do not need inference to hidden entities or information on the web of total science to do so. It is the same with symmetry and any other mathematical structure realised in the world. It can be perceived in a single instance and understood to be repeated in another instance, without any extra-worldly form of symmetry needing to be inferred.

If the Platonist insists that the question was not about "applications" of numbers like counting by children but about the Numbers themselves, he faces the dilemma that was dramatised by Plato and Aristotle as the Third Man Argument. What good, Aristotle asks, is a Form of Man, conceived of as a separate entity from the individual men it is supposed to unify? What does it have in common with the men that enables it to perform the act of unifying them? Would not that require a "Third Man" to unite both the Form of Man and the individual men? An infinite regress threatens. [Plato, *Parmenides* 132a1-b21; Fine, 1993, ch. 15]. The regress exposes the uselessness of a Platonic form outside space and time and without causal power, even if it existed, in performing the role assigned to it. Either the individual men already have something in common that makes them resemble the Form of Man, in which case the Form is not needed, or they don't, in which case the Form has no power to gather them together and distinguish them from non-men. The same reasoning applies to the relation of numbers and sets (conceived of as Platonic entities) to counting and measurement. If a five-year-old can see by counting that a parrot aggregate is four-parrot-parted, and knows equally well how to count four apples if asked, no postulation of hidden other-worldly entities can add anything to the child's understanding, as it is already complete. The division of an apple heap into apple parts by the universal 'being an apple', and its parallel with the division of a parrot heap into parrot parts, is accomplished in the physical world; there is no point of entry for the supposed other-worldly entities to act, even if they had any causal power. Epistemologically, too, counting and measurement are as open to us as it is possible to be (self-knowledge possibly excepted), and again there is neither the need nor the possibility of intervention by other-worldly entities in our perception that a heap is four-apple-parted or that one tree is about twice as tall as another.

7 EPISTEMOLOGY

From an Aristotelian point of view, the epistemology of mathematics ought to be easy, in principle. If mathematics is about such properties of real things as symmetry and continuity, or ratios, or being divided into parts, it should be possible to observe those properties in things, and so the epistemology of mathematics should be no more problematic than the epistemology of colour. An Aristotelian point of view should solve the epistemology problem at the same time as it solves the problem of the applicability of mathematics, by showing that mathematics deals directly with properties of real things. [Lehrer Dive, 2003, ch. 3]

Plainly there are some difficulties with that plan, for example in explaining knowledge of some of the larger and more esoteric structures such as infinite-dimensional Hilbert spaces, which are not instantiated in anything observable. Nevertheless, it would be impressive if the plan worked for some simple mathematical structures, even if it did not work for all.

It would be desirable if an epistemology of mathematics could fulfill these requirements:

- Avoid both Platonist implausibilities involving contact with a world of Forms and logicist trivializations of mathematical knowledge

- At the lower level, be continuous with what is known in perceptual psychology on pattern recognition and explain the substantial mathematical knowledge of animals and babies

- At the higher level, explain how knowledge of uninstantiated structures is possible

- Explain the role of proof in delivering certainty in mathematics

- Explain the mental operation of "abstraction", which delivers individual mathematical concepts "by themselves".

If those requirements were met, there would be less motivation either to postulate Platonist intuition of forms, or to try to represent mathematics as tautologous or trivial so as not to have to postulate a Platonist intuition of forms.

Animal and infant cognition is not as well understood as one would wish, as experiments are difficult and inference from the observed behaviour problematic. Nevertheless it is clear in general terms that animals and babies, though they lack language, have high levels of generalization, memory, inference and inner experience. In particular, babies and animals share a numerical sense, as has become clear through careful experiments in the 1980s and 90s. To have any numerical ability (as opposed to just estimating sizes of heaps), a baby or animal must achieve three things:

- Recognition of objects against background — that is, cutting out discrete objects from the visual background (or discrete sounds from the sound stream) [Huntley-Fenner, Carey & Solimando, 2002]

- Identifying objects as of the same kind (e.g. food pellets, dots, beeps)

- Estimating the numerosity of the objects identified (the phraseology is intended to avoid the connotations of "counting" as possibly including reference to numbers or a pointing procedure, and exactitude of the answer).

Human babies can do that at birth. A newborn that sucks to get nonsense 3-syllable "words" will get bored, but perks up when the sounds suddenly change to 2-syllable words. [Bijeljac-Babic, Bertoncini & Mehler, 1993] Monkeys, rats, birds and many other higher animals can choose larger sets of food items, flee another group that substantially outnumbers their own, and with training press approximately the right number of times on a bar to obtain food. Babies and animals have an accurate immediate perception (called "subitization") of one, two and three items, and an inherently fuzzy estimate of larger sets — it is easy to tell the difference between 10 and 20 items, but not between 10 and 12. Various experiments, especially on the time taken to reach judgements, show that the reasons lie in an internal analog representation of numerosity; the persistence of this representation in adults is shown by such facts as that subjects presented with pairs of digits are slower at judging that 7 is greater than 5 than that 7 is greater than 2. None of these judgements involve anything like counting, in the sense of pairing off items with digits or numerals. [Review in Dehaene, 1997, chs 1-2; update in Xu, Spelke & Goddard, 2005]

There has been less research on the perception on continuous quantities. But infants of no more than six months can distinguish between the same and different heights of similar things side by side, and can be surprised if liquid poured into a container results in a grossly wrong final height of the liquid (though they are poor at judging quantities against a remembered standard). [Huttenlocher, Duffy & Levine, 2002] Four-year-olds can make some sense of the scaling of ratios needed to read a map. [Stea, Kirkman, Pinon, Middlebrook & Rice, 2004] Mature rats also have some kind of internal map of their surroundings. [Nadel, 1990]

But if animals are inept at counting beyond the smallest numbers, they are excellent at perceiving some other mathematical properties that require keeping an approximate running average of relative frequencies. The rat, for example, can behave in ways acutely sensitive to small changes in the frequencies of the results of that behaviour. [Review in Holland, Holyoak, Nisbett & Thagard, 1986, section 5.2] Naturally so, since the life of animals is a constant balance between coping adequately with risk or dying. Foraging, fighting and fleeing are activities in which animal evaluations of frequencies are especially evident. Those abilities require some form of counting, in working out the approximate relative frequency of a characteristic in a moderately large dataset (after identifying, of course, the population and characteristic).

Very recently, it has become clear that covariation plays a crucial role in the powerful learning algorithms that allow a baby to make sense of its world at the most basic level, for example in identifying continuing objects. Infants pay attention especially to "intermodal" information — structural similarities between the inputs to different senses, such as the covariation between a ball seen bouncing and a "boing boing boing" sound. That covariation encourages the infant to attribute a reality to the ball and event (whereas infants tend to ignore changes of colour and shape in objects). [Bahrick, Lickliter & Flom, 2004]

There is also much to learn on how the lower levels of the perceptual systems of animals and humans extract information on structural features of the world afforded by perception, for example, what algorithms are implemented in the visual system to allow inference of the curvature of surfaces, depth, clustering, occlusion and object recognition. Decades of work on visual illusions, vision in cats, models of the retina and so on has shown that the visual system is very active in extracting structure from — sometimes imposing structure on — the raw material of vision, but the overall picture of how it is done (and how it might be imitated) has yet to emerge. (A classic attempt is in [Marr, 1982].)

We have reached the furthest limits of what is possible in the way of mathematical knowledge with the cognitive skills of animals. According to traditional Aristotelianism, the human intellect possesses an ability completely different in kind from animals, an ability to abstract universals and understand their relations. That ability, it was thought, was most evident in mathematical insight and proof. The geometry of eclipses, Aristotle says, not only describes the regularities in eclipses, but demonstrates why and how they must take place when they do. [Aristotle, *Posterior Analytics*, bk II ch. 2] A true science differs from a heap of observational facts (even a heap of empirical generalizations) by being organised into a system of deductions from self-evidently true axioms which express the nature of the universals involved. Ideally, each deduction from the premises allows the human understanding to grasp why the conclusion must be true. Euclid's geometry conforms closely to Aristotle's model. [McKirahan, 1992] The Aristotelianism of the medieval scholastics argued that such an ability to grasp pure relations of universals was so far removed from sensory knowledge as to prove that the "active intellect" must be immaterial and immortal. [Kuksewicz, 1994]

Perhaps those claims were overwrought, but they were right in highlighting how remarkable human understanding of universals is and how different it is from sensory knowledge. Let us take a simple example.

Euclid defines a circle as a plane figure "such that all straight lines drawn from a certain point within the figure to the circumference are equal". That is not an arbitrary definition, or an abbreviation. A circle at first glance is not given with reference to its centre — perceptually (to an animal, for example) it is more like something "equally round all the way around". Understanding that Euclid's definition applies to the same object requires an act of imaginative insight. The genius of the definition lies in its suitability for use in proofs of the kind Euclid gives immediately afterwards, proofs which would be very difficult with the more obvious phenomenological definition of a circle. [Lonergan, 1970, 7-11]

We are ready to move toward the notion of proof. If we gain knowledge of $2 \times 3 = 3 \times 2$ not by rote but by understanding the diagram

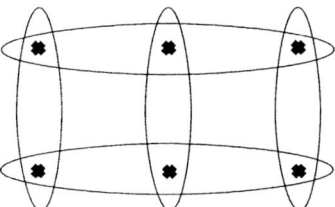

Figure 7. Why $2 \times 3 = 3 \times 2$

then we have fulfilled the Aristotelian ideal of complete and certain knowledge through understanding the reason why things must be so. We can also understand why the size of the numbers is irrelevant, and we can perform the same proof with more rows and columns, leading to the conclusion that $m \times n = n \times m$ for any whole numbers m and n. The insight permits knowledge of a truth beyond the range of actual or possible sensory experience, evidence again of the sharp difference in kind between sensory knowledge like subitization and intellectual understanding.

Consider six points, with each pair joined by a line. The lines are all coloured, in one of two colours (represented by dotted and undotted lines in the figure). Then there must exist a triangle of one colour (that is, three points such that all three of the lines joining them have the same colour).

Proof. Take one of the points, and call it O. Then of the five lines from that point to the others, at least three must have the same colour, say colour A. Consider the three points at the end of those lines. If any two of them are joined by a line of colour A, then they and O form an A-colour triangle. But if not, then the three points must all be joined by B-colour lines, so there is a B-colour triangle. So there is always a single-coloured triangle. ∎

There is nothing in this proof except what Aristotelian mathematical philosophy says there should be — no arbitrary axioms, no forms imposed by the mind, no constructions in Platonist set theory, no impredicative definitions, only the

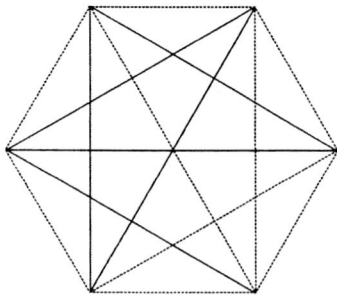

Figure 8. Six-point graph colouring

necessary relations of simple structural universals and our certain, proof-induced insight into them.

Unfortunately there is a gap in the story. What exactly is the relation between the mind and universals, the relation expressed in the crude metaphor of the mind "grasping" universals and their connection? "Insight" (or "eureka moment") expresses the psychology of that "grasp", but what is the philosophy behind it? Without an answer to that question, the story is far from complete. It is, of course, in principle a difficult question in epistemology in general, but since mathematics has always been regarded as the home territory of certain insight, it is natural to tackle the problem first in the epistemology of mathematics.

It is not easy to think of even one possible answer to that question. That should make us more willing to give a sympathetic hearing to the answer of traditional Aristotelianism, despite its strangeness. Based on Aristotle's dictum that "the soul is in a way all things", the scholastics maintained that the relation between the knowing mind and the universal it knows is the simplest possible: identity. The soul, they said, knows heat by actually *being hot* ("formally", of course, not "materially", which would overheat the brain).

That theory, possibly the most astounding of the many remarkable theses of the scholastics, can hardly be called plausible or even comprehensible. What could "being hot formally" mean? Nevertheless, it has much more force for the structural universals of mathematics than for physical universals like heat and mass. The reason is that structure is "topic-neutral" and so, whatever the mind is, structure could in principle be shared between mental entities (however they are conceived) and physical ones. While there seem insuperable obstacles to the thought-of-heat being hot, there is no such problem with the thought-of-4 being four-parted (though one will still ask what makes it the single thought-of-4 instead of four thoughts).

In fact, on one simple model of (some) mathematical knowledge, the identity-of-structure theory is straightforwardly true. If a computer runs a weather simulation, what makes it a simulation is an identity of structure between its internal model and the physical weather. The model has parts corresponding to the spatiotemporal parts of the real weather, and relations between the parts corresponding to the causal flow of the atmosphere. (The correspondence is very visible in an analog computer, but in a digital computer it is equally present, once one sees through the rather complicated correspondence between electronically implemented bit strings and spatiotemporal points.) That certainly does not imply that the structural similarity between mental/computer model and world is all there is to knowledge — that would be to accept thermostat tracking as a complete account of knowledge. In the weather model case, there must at least be code to generate and run the model and more code to interpret the model results, for example in announcing a cold front two days ahead. Nevertheless, it is clear that it is perfectly reasonable for structural type identities between knower and known to be an essential part of knowledge, and that that thesis does not require any esoteric view of the nature of the mind.

The possibility of mental entities having literally the same structural properties as the physical systems they represent has implications for the certainty of mathematical knowledge. If mental representations literally have the structural properties one wishes to study, one avoids the uncertainty that attends sense perception and its possible errors. The errors of the senses cannot intrude on the relation of the mind to its own contents, so one major source of error is removed, and it is not surprising if simple mathematical knowledge is accompanied by a feeling of certainty, predicated on the intimate relation between knower and known in this case. That is not to maintain that such knowledge is infallible just because of this close relation. In dealing with a complex mental model, especially, such as a visualized cube, the mind may easily become confused because the single act of knowledge has to deal with many parts and their complicated relations. A mental model of some complexity may even be harder to build and to compute with than one of similar complexity in wood — although experts at the mental abacus are very fast, most people find a physical abacus much easier to use. Nevertheless, the errors of perception are a large part of the reasons for our uncertainty about matters of fact, and the removal of that source of error for a major branch of knowledge is a matter of great epistemological significance.

8 EXPERIMENTAL MATHEMATICS AND EVIDENCE FOR CONJECTURES

If mathematical realism — whether Platonist or Aristotelian — is true, then mathematics is a scientific study of a world "out there". In that case, in addition to methods special to mathematics such as proof, ordinary scientific methods such as experiment, conjecture and the confirmation of theories by observations ought to work in mathematics just as well as in science. An examination of the theory and practice of experimental mathematics will do three things. It will confirm realism in the philosophy of mathematics, since an objectivist philosophy of science is premised on realism about the entities and truths that science studies. It will suggest a logical reading of scientific methodology, since the methods of science will be seen to work in necessary as well as contingent matter (so, for example, the need to assume any contingent principles like the 'uniformity of nature' will be called into question). And it will support the objective Bayesian philosophy of probability, according to which (some at least) probabilities are strictly logical — relations of partial implication between bodies of evidence and hypothesis.

Mathematicians often speak of conjectures as being confirmed by evidence that falls short of proof. For their own conjectures, evidence justifies further work in looking for a proof. Those conjectures of mathematics that have long resisted proof, as Fermat's Last Theorem did and the Riemann Hypothesis still does, have had to be considered in terms of the evidence for and against them. It is not adequate to describe the relation of evidence to hypothesis as 'subjective', 'heuristic' or 'pragmatic'; there must be an element of what it is rational to believe on the evidence, that is, of non-deductive logic. Mathematics is therefore (among other things) an experimental science.

The occurrence of non-deductive logic, or logical probability, in mathematics is an embarrassment. It is embarrassing to mathematicians, used to regarding deductive logic as the only real logic. It is embarrassing for those statisticians who wish to see probability as solely about random processes or relative frequencies: surely there is nothing probabilistic about the truths of mathematics? It is a problem for philosophers who believe that induction is justified not by logic but by natural laws or the 'uniformity of nature': mathematics is the same no matter how lawless nature may be. It does not fit well with most philosophies of mathematics. It is awkward even for proponents of non-deductive logic. If non-deductive logic deals with logical relations weaker than entailment, how can such relations hold between the necessary truths of mathematics?

Work on this topic has therefore been rare. There is one notable exception, the pair of books by the mathematician George Polya on *Mathematics and Plausible Reasoning*. [Polya, 1954; revivals in Franklin, 1987; Fallis, 1997; Corfield, 2003, ch. 5; Lehrer Dive, 2003, ch. 6] Despite their excellence, Polya's books have been little noticed by mathematicians, and even less by philosophers. Undoubtedly this is largely because of Polya's unfortunate choice of the word 'plausible' in his title — 'plausible' has a subjective, psychological ring to it, so that the word is almost equivalent to 'convincing' or 'rhetorically persuasive'. Arguments that happen to persuade, for psychological reasons, are rightly regarded as of little interest in mathematics and philosophy. Polya made it clear, however, that he was not concerned with subjective impressions, but with what degree of belief was *justified* by the evidence. [Polya, 1954, vol. I, 68] This will be the point of view argued for here.

Non-deductive logic deals with the support, short of entailment, that some propositions give to others. If a proposition has already been proved true, there is of course no longer any need to consider non-conclusive evidence for it. Consequently, non-deductive logic will be found in mathematics in those areas where mathematicians consider propositions which are not yet proved. These are of two kinds. First there are those that any working mathematician deals with in his preliminary work before finding the proofs he hopes to publish, or indeed before finding the theorems he hopes to prove. The second kind are the long-standing conjectures which have been written about by many mathematicians but which have resisted proof.

It is obvious on reflection that a mathematician must use non-deductive logic in the first stages of his work on a problem. Mathematics cannot consist just of conjectures, refutations and proofs. Anyone can generate conjectures, but which ones are worth investigating? Which ones are relevant to the problem at hand? Which can be confirmed or refuted in some easy cases, so that there will be some indication of their truth in a reasonable time? Which might be capable of proof by a method in the mathematician's repertoire? Which might follow from someone else's theorem? Which are unlikely to yield an answer until after the next review of tenure? The mathematician must answer these questions to allocate his time and effort. But not all answers to these questions are equally good. To stay employed as a mathematician, he must answer a proportion of them *well*. But to say that some answers are better than others is to admit that some are, on the evidence he has, more reasonable than others, that is, are rationally better supported by the evidence. This is to accept a role for non-deductive logic.

The area where a mathematician must make the finest discriminations of this kind — and where he might, in theory, be guilty of professional negligence if he makes the wrong decisions — is as a supervisor advising a prospective Ph.D. student. It is usual for a student beginning a Ph.D. to choose some general field of mathematics and then to approach an expert in the field as a supervisor. The supervisor then chooses a problem in that field for the student to investigate. In mathematics, more than in any other discipline, the initial choice of problem is the crucial event in the Ph.D.-gathering process. The problem must be

1. unsolved at present

2. not being worked on by someone who is likely to solve it soon

but most importantly

3. tractable, that is, probably solvable, or at least partially solvable, by three years' work at the Ph.D. level.

It is recognised that of the enormous number of unsolved problems that have been or could be thought of, the tractable ones form a small proportion, and that it is difficult to discern which they are. The skill in non-deductive logic required of a supervisor is high. Hence the advice to Ph.D. students not to worry too much about what field or problem to choose, but to concentrate on finding a good supervisor. (So it is also clear why it is hard to find Ph.D. problems that are also (4) interesting.)

It is not possible to dismiss these non-deductive techniques as simply 'heuristic' or 'pragmatic' or 'subjective'. Although these are correct descriptions as far as they go, they give no insight into the crucial differences among techniques, namely, that some are more reasonable and consistently more successful than others. 'Successful' can mean 'lucky', but 'consistently successful' cannot. 'If you have a lot of lucky breaks, it isn't just an accident', as Groucho Marx said. Many techniques can be heuristic, in the sense of leading to the discovery of a true result, but we are especially interested in those which give reason to believe the truth has been arrived at, and justify further research. Allocation of effort on attempted proofs may be guided by many factors, which can hence be called 'pragmatic', but those which are likely to lead to a completed proof need to be distinguished from those, such as sheer stubbornness, which are not. Opinions on which approaches are likely to be fruitful in solving some problem may differ, and hence be called 'subjective', but the beginning graduate student is not advised to pit his subjective opinion against the experts' without good reason. Damon Runyon's observation on horse-racing applies equally to courses of study: 'The race is not always to the swift, nor the battle to the strong, but that's the way to bet'.

It is true that similar remarks could also be made about *any* attempt to see rational principles at work in the evaluation of hypotheses, not just those in mathematical research. In scientific investigations, various inductive principles obviously produce results, and are not simply dismissed as pragmatic, heuristic or subjective. Yet it is common to suppose that they are not principles of *logic*, but work because of natural laws (or the principle of causality, or the regularity of nature). This option is not available in the mathematical case. Mathematics is true in all worlds, chaotic or regular; any principles governing the relationship between hypothesis and evidence in mathematics can only be logical.

In modern mathematics, it is usual to cover up the processes leading to the construction of a proof, when publishing it — naturally enough, since once a result is proved, any non-conclusive evidence that existed before the proof is no longer of interest. That was not always the case. Euler, in the eighteenth century, regularly published conjectures which he could not prove, with his evidence for them. He used, for example, some daring and obviously far from rigorous methods to conclude that the infinite sum

$$1 + \frac{1}{4} + \frac{1}{9} + \frac{1}{16} + \frac{1}{25} + \ldots$$

(where the numbers on the bottom of the fractions are the successive squares of whole numbers) is equal to the prima facie unlikely value $\pi^2/6$. Finding that the two expressions agreed to seven decimal places, and that a similar method of argument led to the already proved result

$$1 - \frac{1}{3} + \frac{1}{5} - \frac{1}{7} + \frac{1}{9} - \frac{1}{11} + \ldots = \frac{\pi}{4}$$

Euler concluded, 'For our method, which may appear to some as not reliable enough, a great confirmation comes here to light. Therefore, we shall not doubt at all of the other things which are derived by the same method'. He later proved the result. [Polya, 1954, vol. I, 18-21]

Even today, mathematicians occasionally mention in print the evidence that led to a theorem. Since the introduction of computers, and even more since the recent use of symbolic manipulation software packages like Mathematica and Maple, it has become possible to collect large amounts of evidence for certain kinds of conjectures. (Comments in [Borwein & Bailey, 2004; Epstein, Levy & de la Llave, 1992]) A few mathematicians argue that in some cases, it is not worth the excessive cost of achieving certainty by proof when "semirigorous" checking will do. [Zeilberger, 1993]

At present, it is usual to delay publication until proofs have been found. This rule is broken only in work on those long-standing conjectures of mathematics which are believed to be true but have so far resisted proof. The most notable of these, which stands since the proof of Fermat's Last Theorem as the Everest of mathematics, is the Riemann Hypothesis.

Riemann stated in a celebrated paper of 1859 that he thought it 'very likely' that

> All the roots of the Riemann zeta function (with certain trivial exceptions) have real part equal to $1/2$.

This is the still unproved Riemann Hypothesis. The precise meaning of the terms involved is not very difficult to grasp, but for the present purpose it is only necessary to observe that this is a simple universal proposition like 'all ravens are black'. It is also true that the roots of the Riemann zeta function, of which there are infinitely many, have a natural order, so that one can speak of 'the first million roots'. Once it became clear that the Riemann Hypothesis would be very hard to prove, it was natural to look for evidence of its truth (or falsity). The simplest kind of evidence would be ordinary induction: Calculate as many of the roots as possible and see if they all have real part $1/2$. This is in principle straightforward, though computationally difficult. Such numerical work was begun by Riemann and was carried on later with the results below:

Worker	Number of roots found to have real part $1/2$
Gram (1903)	15
Backlund (1914)	79
Hutchinson (1925)	138
Titchmarch (1935/6)	1,041

'Broadly speaking, the computations of Gram, Backlund and Hutchinson contributed substantially to the plausibility of the Riemann Hypothesis, but gave no insight into the question of why it might be true.' [Edwards, 1974, 97] The next investigations were able to use electronic computers, and the results were

Lehmer (1956)	25,000
Lehman (1966)	250,000
Rosser, Yohe & Schoenfeld (1968)	3,500,000
Te Riele, van de Lune *et al* (1986)	1,500,000,001
Gourdon (2004)	10^{13}

It is one of the largest inductions in the world.

Besides this simple inductive evidence, there are some other reasons for believing that Riemann's Hypothesis is true (and some reasons for doubting it). In favour, there are

1. Hardy proved in 1914 that infinitely many roots of the Riemann zeta function have real part $1/2$. [Edwards, 1974, 226-9] This is quite a strong consequence of Riemann's Hypothesis, but is not sufficient to make the Hypothesis highly probable, since if the Riemann Hypothesis is false it would not be surprising if the exceptions to it were rare.

2. Riemann himself showed that the Hypothesis implied the 'prime number theorem', then unproved. This theorem was later proved independently. This is an example of the general non-deductive principle that non-trivial consequences of a proposition support it.

3. Also in 1914, Bohr and Landau proved a theorem roughly expressible as 'Almost all the roots have real part very close to $1/2$'. This result 'is to this day the strongest theorem on the location of the roots which substantiates the Riemann hypothesis.' [Edwards, 1974, 193]

4. Studies in number theory revealed areas in which it was natural to consider zeta functions analogous to Riemann's zeta function. In some famous and difficult work, André Weil proved that the analogue of Riemann's Hypothesis is true for these zeta functions, and his related conjectures for an even more general class of zeta functions were proved to widespread applause in the 1970s. 'It seems that they provide some of the best reasons for believing that the Riemann hypothesis is true — for believing, in other words, that there is a profound and as yet uncomprehended number-theoretic phenomenon, one facet of which is that the roots ρ all lie on Re $s = 1/2$'. [Edwards, 1974, 298]

5. Finally, there is the remarkable 'Denjoy's probabilistic interpretation of the Riemann hypothesis'. If a coin is tossed n times, then of course we expect about $1/2 n$ heads and $1/2 n$ tails. But we do not expect *exactly* half of each. We can ask, then, what the average deviation from equality is. The answer, as was known by the time of Bernoulli, is $\sqrt{n}$. One exact expression of this fact is

> For any $\varepsilon > 0$, with probability one the number of heads minus the number of tails in n tosses grows less rapidly than $n^{1/2+\varepsilon}$. (Recall that $n^{1/2}$ is another notation for $\sqrt{n}$.)

Now we form a sequence of 'heads' and 'tails' by the following rule: Go along the sequence of numbers and look at their prime factors. If a number has two or more prime factors equal (i.e., is divisible by a square), do nothing. If not, its prime factors must be all different; if it has an even number of prime factors, write 'heads'. If it has an odd number of prime factors, write 'tails'. The sequence begins

2	3	4	5	6	7	8	9	10	11	12	13	14	15	16	17	...
		2^2		2×3		2^3	3^2	2×5		$2^2 \times 3$		2×7	3×5	2^4		
T	T		T	H	T			H	T		T	H	H		T	...

The resulting sequence is of course not 'random' in the sense of 'probabilistic', since it is totally determined. But it does look 'random' in the sense of 'patternless' or 'erratic' (such sequences are common in number theory, and are studied by the branch of the subject called misleadingly 'probabilistic number theory'). From the analogy with coin tossing, it is likely that

> For any $\varepsilon > 0$, the number of heads minus the number of tails in the first n 'tosses' in this sequence grows less rapidly than $n^{1/2+\varepsilon}$.

This statement is equivalent to Riemann's Hypothesis. Edwards comments, in his book on the Riemann zeta function,

> One of the things which makes the Riemann hypothesis so difficult is the fact that there is no plausibility argument, no hint of a reason, however unrigorous, why it should be true. This fact gives some importance to Denjoy's probabilistic interpretation of the Riemann hypothesis which, though it is quite absurd when considered carefully, gives a fleeting glimmer of plausibility to the Riemann hypothesis. [Edwards, 1974, 268]

Not all the probabilistic arguments bearing on the Riemann Hypothesis are in its favour. In the balance against, there are the following arguments:

1. Riemann's paper is only a summary of his researches, and he gives no reasons for his belief that the Hypothesis is 'very likely'. No reasons have been found in his unpublished papers. Edwards does give an account, however, of facts which Riemann knew which would naturally have seemed to him evidence of the Hypothesis. But the facts in question are true only of the early roots; there are some exceptions among the later ones. This is an example of the non-deductive rule given by Polya, 'Our confidence in a conjecture can only diminish when a possible ground for the conjecture is exploded.'

2. Although the calculations by computer did not reveal any counterexamples to the Riemann Hypothesis, Lehmer's and later work did unexpectedly find values which it is natural to see as 'near counterexamples'. An extremely close one appeared near the 13,400,000th root. [Edwards, 1974, 175–9] It is partly this that prompted the calculators to persevere in their labours, since it gave reason to believe that if there were a counterexample it would probably appear soon. So far it has not, despite the distance to which computation has proceeded, so the Riemann Hypothesis is not so undermined by this consideration as appeared at first.

3. Perhaps the most serious reason for doubting the Riemann Hypothesis comes from its close connections with the prime number theorem. This theorem states that the number of primes less than x is (for large x) approximately equal to the integral

$$\int_2^x \frac{dt}{\log t}$$

If tables are drawn up for the number of primes less than x and the values of this integral, for x as far as calculations can reach, then it is always found that the number of primes less than x is actually *less than* the integral. On this evidence, it was thought for many years that this was true for all x. Nevertheless Littlewood proved that this is false. While he did not produce an actual number for which it is false, it appears that the first such number is extremely large — well beyond the range of computer calculations. It gives some reason to suspect that there may be a very large counterexample to the Hypothesis even though there are no small ones.

It is plain, then, that there is much more to be said about the Riemann Hypothesis than, 'It is neither proved nor disproved'. Without non-deductive logic, though, nothing more can be said.

Another example is Goldbach's conjecture that every number except 2 is the sum of two primes, unproved since 1742, which has considerable evidence for it but is believed to be far from being solved. Examples where the judgement of experts that the evidence for a conjecture was overwhelming was vindicated by subsequent proof include Fermat's Last Theorem and the classification of finite simple groups. [Franklin, 1987]

The correctness of the above arguments is not affected by the success or failure of any attempts to formalise, or give axioms for, the notion of non-deductive support between propositions. Many fields of study, such as geometry in the time of Pythagoras or pattern-recognition today, have yielded bodies of truths while still resisting reduction to formal rules. Even so, it is natural to ask whether the concept *is* easily formalisable. This is not the place for detailed discussion, since the problem has nothing to do with mathematics, and has been dealt with mainly in the context of the philosophy of science. The axiomatisation that has proved serviceable is the familiar axiom system of conditional probability: if h (for 'hypothesis') and e (for 'evidence') are two propositions, $P(h|e)$ is a number between 0 and 1 inclusive expressing the degree to which h is supported by e, which satisfies

$$P(\text{not-}h|e) = 1 - P(h|e)$$
$$P(h'|h\&e) \times P(h|e) = P(h|h'\&e) \times P(h'|e)$$

While some authors, such as Carnap [1950] and Jaynes [2003] have been satisfied with this system, others (e.g. Keynes [1921] and Koopman [1940]) have thought it too strong to attribute an exact number to $P(h|e)$ in all cases, and have weakened the axioms accordingly. Their modifications are essentially minor.

Needless to say, command of these principles alone will not make anyone a shrewd judge of hypotheses, any more than perfection in deductive logic will make him a great mathematician. To achieve fame in mathematics, it is only necessary to string together enough deductive steps to prove an interesting proposition, and submit the results to *Inventiones Mathematicae*. The trick is finding the steps. Similarly in non-deductive logic, the problem is not in knowing the principles, but in bringing to bear the relevant evidence.

The principles nevertheless provide *some* help in deciding what evidence will be helpful in confirming the truth of a hypothesis. It is easy to derive from the above axioms the principle

If $h\&b$ implies e, but $P(e|b) < 1$, then $P(h|e\&b) > P(h|b)$.

If h is thought of as hypothesis, b as background information, and e as new evidence, this principle can be expressed as 'The verification of a consequence renders a conjecture more probable', in Polya's words. [Polya, 1954, vol. II, 5] He calls this the 'fundamental inductive pattern'; its use was amply illustrated in the examples above. Further patterns of inductive inference, with mathematical examples, are given in Polya.

There is one point that needs to be made precise especially in applying these rules in mathematics. If e entails h, then $P(h|e)$ is 1. But in mathematics, the typical case is that e does entail h, though this is perhaps as yet unknown. If, however, $P(h|e)$ is really 1, how is it possible in the meantime to discuss the (non-deductive) support that e may give to h, that is, to treat $P(h|e)$ as not equal to 1? In other words, if h and e are necessarily true or false, how can $P(h|e)$ be other than 0 or 1?

The answer is that, in both deductive and non-deductive logic, there can be *many* logical relations between two propositions. Some may be known and some not. To take an artificially simple example in deductive logic, consider the argument

If all men are mortal, then this man is mortal
All men are mortal
Therefore, this man is mortal

The premises entail the conclusion, certainly, but there is more to it than that. They entail the conclusion in two ways: firstly, by *modus ponens*, and secondly by instantiation from the second premise alone. More complicated and realistic cases are common in the mathematical literature, where, for example, a later author simplifies an earlier proof, that is, finds a shorter path from established facts to the theorem.

Now just as there can be two deductive paths between premises and conclusion, so there can be a deductive and non-deductive path, with only the latter known. Before the Greeks' development of deductive geometry, it was possible to argue

All equilateral (plane) triangles so far measured
   have been found to be equiangular
This triangle is equilateral
―――――――――――――――――――――――――――
Therefore, this triangle is equiangular

There is a non-deductive logical relation between the premises and the conclusion; the premises support the conclusion. But when deductive geometry appeared, it was found that there was also a deductive relation, since the second premise alone entails the conclusion. This discovery in no way vitiates the correctness of the previous non-deductive reasoning or casts doubt on the existence of the non-deductive relation.

That non-deductive logic is used in mathematics is important first of all to mathematics. But there is also some wider significance for philosophy, in relation to the problem of induction, or inference from the observed to the unobserved.

It is common to discuss induction using only examples from the natural world, such as, 'All observed flames have been hot, so the next flame observed will be hot' and 'All observed ravens have been black, so the next observed raven will be black'. This has encouraged the view that the problem of induction should be solved in terms of natural laws (or causes, or dispositions, or the regularity of nature) that provide a kind of cement to bind the observed to the unobserved. The difficulty for such a view is that it does not apply to mathematics, where induction works just as well as in natural science.

Examples were given above in connection with the Riemann Hypothesis, but let us take a particularly straightforward case:

The first million digits of π are random
―――――――――――――――――――――――――――――――――――
Therefore, the second million digits of π are random.

('Random' here means 'without pattern', 'passes statistical tests for randomness', not 'probabilistically generated'.)

The number π has the decimal expansion

$$3.14159265358979323846264338327950288419716939937\ldots$$

There is no apparent pattern in these numbers. The first million digits have long been calculated (calcultions now extend beyond one trillion). Inspection of these digits reveals no pattern, and computer calculations can confirm this impression. It can then be argued inductively that the second million digits will likewise exhibit no pattern. This induction is a good one (indeed, everyone believes that the digits of π continue to be random indefinitely, though there is no proof), and there seems to be no reason to distinguish the reasoning involved here from that used in inductions about flames or ravens. But the digits of π are the same in all possible worlds, whatever natural laws may hold in them or fail to. Any reasoning about π is also rational or otherwise, regardless of any empirical facts about natural laws. Therefore, induction can be rational independently of whether there are natural laws.

This argument does not show that natural laws have no place in discussing induction. It may be that mathematical examples of induction are rational because there are *mathematical* laws, and that the aim in natural science is to find some substitute, such as natural laws, which will take the place of mathematical laws in accounting for the continuance of regularity. But if this line of reasoning is pursued, it is clear that simply making the supposition, 'There are laws', is of little help in making inductive inferences. No doubt mathematics is completely lawlike, but that does not help at all in deciding whether the digits of π continue to be random. In the absence of any proofs, induction is needed to support the law (if it is a law), 'The digits of π are random', rather than the law giving support to the induction. Either 'The digits of π are random' or 'The digits of π are not random' is a law, but in the absence of knowledge as to which, we are left only with the confirmation the evidence gives to the first of these hypotheses. Thus consideration of a mathematical example reveals what can be lost sight of in the search for laws: laws or no laws, non-deductive logic is needed to make inductive inferences.

These examples illustrate Polya's remark that non-deductive logic is better appreciated in mathematics than in the natural sciences. [Polya, 1954, vol. II, 24] In mathematics there can be no confusion over natural laws, the regularity of nature, approximations, propensities, the theory-ladenness of observation, pragmatics, scientific revolutions, the social relations of science or any other red herrings. There are only the hypothesis, the evidence and the logical relations between them.

9 CONCLUSION

Aristotelian realism unifies mathematics and the other natural sciences. It explains in a straightforward way how babies come to mathematical knowledge through perceiving regularities, how mathematical universals like ratios, symmetries and continuities can be real and perceivable properties of physical and other objects, how new applied mathematical sciences like operations research and chaos theory have expanded the range of what mathematics studies, and how experimental evidence in mathematics leads to new knowledge. Its account of some of the more traditional topics of the philosophy of mathematics, such as infinite sets, is less natural, but there are initial ideas on how to rival the Platonist and nominalist approaches to those questions. Aristotelianism will be an enduring option in twenty-first century philosophy of mathematics.

BIBLIOGRAPHY

[Apostle, 1952] H. G. Apostle. *Aristotle's Philosophy of Mathematics*, University of Chicago Press, Chicago, 1952.
[Aristotle, Metaphysics] Aristotle, *Metaphysics*.
[Aristotle, Physics] Aristotle, *Physics*.
[Aristotle, Posterior Analytics] Aristotle, *Posterior Analytics*.
[Armstrong, 1978] D. M. Armstrong. *Universals and Scientific Realism* (Cambridge, 1978).

[Armstrong, 1988] D. M. Armstrong. Are quantities relations? A reply to Bigelow and Pargetter *Philosophical Studies* 54, 305-16, 1988.
[Armstrong, 1989] D. M. Armstrong. *Universals: An Opinionated Introduction* (Boulder, 1989).
[Armstrong, 1991] D. M. Armstrong. Classes are states of affairs, *Mind* 100, 189-200, 1991.
[Armstrong, 2004] D. M. Armstrong. *Truth and Truthmakers*, Cambridge University Press, Cambridge, 2004.
[Azzouni, 2004] J. Azzouni. *Deflating Existential Consequence: A Case for Nominalism* (Oxford, 2004).
[Bahrick et al., 2004] L. E. Bahrick, R. Lickliter and R. Flom. Intersensory redundancy guides the development of selective attention, perception, and cognition in infancy, *Current Directions in Psychological Science* 13, 99-102, 2004.
[Balaguer, 1998] M. Balaguer. *Platonism and Anti-Platonism in Mathematics*, Oxford, 1998.
[Barrow, 1734] I. Barrow. *The Usefulness of Mathematical Learning Explained and Demonstrated*, London, 1734. Repr. Frank Cass, London, 1970.
[Benacerraf, 1965] P. Benacerraf. What numbers could not be, *Philosophical Review* 74, 495-512, 1965.
[Bigelow, 1988] J. Bigelow. *The Reality of Numbers: A Physicalist's Philosophy of Mathematics*, Clarendon, Oxford, 1988.
[Bigelow and Pargetter, 1990] J. Bigelow and R. Pargetter. *Science and Necessity*, Cambridge University Press, Cambridge, 1990.
[Bijeljac-Babic et al., 1993] R. Bijeljac-Babic, J. Bertoncini and J. Mehler. How do four-day-old infants categorize multisyllabic utterances? *Developmental Psychology* 29, 711-21, 1993.
[Borwein and Bailey, 2004] J. M. Borwein and D. Bailey. *Mathematics by Experiment: Plausible Reasoning in the 21st Century* (Natick, MA, 2004).
[Bourbaki, 1950] N. Bourbaki. The architecture of mathematics, *American Mathematical Monthly* 57, 221-32, 1950.
[Carnap, 1950] R. Carnap. *Logical Foundations of Probability* (London, 1950).
[Cheyne and Pigden, 1996] C. Cheyne and C. R. Pigden. Pythagorean powers, *Australasian Journal of Philosophy* 74, 639-45, 1996.
[Colyvan, 1998] M. Colyvan. Review of Resnik (1997) and Shapiro (1997), *British Journal for the Philosophy of Science* 49, 652-656, 1998.
[Colyvan, 2001] M. Colyvan. *The Indispensability of Mathematics* (Oxford, 2001).
[Corfield, 2003] D. Corfield. *Towards a Philosophy of Real Mathematics*, Cambridge University Press, Cambridge, 2003.
[Corry, 1992] L. Corry. Nicolas Bourbaki and the concept of mathematical structure, *Synthese* 92, 315-48, 1992.
[Crowley, 1980] C. B. Crowley. *Universal Mathematics in Aristotelian-Thomistic Philosophy*, University Press of America, Washington, DC, 1980.
[Dedekind, 1887] R. Dedekind. The nature and meaning of numbers, in *Essays on the Theory of Numbers*, Chicago, 1887, 1901, repr. Dover, New York, 1963.
[Dehaene, 1997] S. Dehaene. *The Number Sense*, (New York, 1997).
[Dennett, 1991] D. Dennett. Real patterns, *Journal of Philosophy* 88, 27-51, 1991.
[Devlin, 1994] K. J. Devlin. *Mathematics: The Science of Patterns*, Scientific American Library, New York, 1994.
[Diez, 1997] J. A. Diez. A hundred years of numbers: An historical introduction to measurement theory 1887-1990. II, *Studies in History and Philosophy of Science* 28, 237-65, 1997.
[Edwards, 1974] H. M. Edwards. *Riemann's Zeta Function* (New York, 1974).
[Einstein, 1954] A. Einstein. *Ideas and Opinions*. (New York, 1954).
[Ellis, 1968] B. Ellis. *Basic Concepts of Measurement*, Cambridge University Press, Cambridge, 1968.
[Encyclopaedia Britannica, 1771] *Encyclopaedia Britannica*. 1st ed, Edinburgh, article 'Mathematics', vol. III pp. 30-1, 1771.
[Epstein et al., 1992] D. Epstein, S. Levy and R. de la Llave. About this journal, *Experimental Mathematics* 1 (1992), pp. 1-13, 1992.
[Euler, 1776] L. Euler. Solutio problematis ad geometriam situs pertinentis, 1776. Trans. in *Graph Theory 1736-1936*, ed. N. Biggs, E. Lloyd and R. Wilson, Clarendon, Oxford, 1976, pp. 3-8.
[Fallis, 1997] D. Fallis. The epistemic status of probabilistic proof, *Journal of Philosophy* 94, 165-86, 1997.

[Fetzer, 1988] J. H. Fetzer. Program verification: the very idea, *Communications of the Association for Computing Machinery* 31, 1048-1063, 1988.
[Fetzer, 1989] J. H. Fetzer. Program verification reprise: the author's response, *Communications of the Association for Computing Machinery* 32, 377-381, 1989.
[Field, 1980] H. Field. *Science Without Numbers: A Defence of Nominalism* (Princeton, 1980).
[Fine, 1993] G. Fine. *On Ideas: Aristotle's Criticism of Plato's Theory of Forms* (Oxford, 1993).
[Franklin, 1987] J. Franklin. Non-deductive logic in mathematics, *British Journal for the Philosophy of Science* 38, 1-18, 1987.
[Franklin, 1989] J. Franklin. Mathematical necessity and reality, *Australasian Journal of Philosophy* 67, 286-294, 1989.
[Franklin, 1994] J. Franklin. Achievements and fallacies in Hume's account of infinite divisibility, *Hume Studies* 20, 85-101, 1994.
[Franklin, 2006] J. Franklin. Artifice and the natural world: Mathematics, logic, technology, ch. 28 of *Cambridge History of Eighteenth Century Philosophy*, ed. K. Haakonssen (Cambridge, 2006), 817-53.
[Forge, 1995] J. Forge. Bigelow and Pargetter on quantities, *Australasian Journal of Philosophy* 73, 594-605, 1995.
[Forrest and Armstrong, 1987] P. Forrest and D. M. Armstrong. The nature of number, *Philosophical Papers* 16, 165-186, 1987.
[Frege, 1884] G. Frege. *The Foundations of Arithmetic*, 1884. Trans. J.L. Austin, 2nd revised ed, Oxford, 1980.
[Hale, 2000] B. Hale. Reals by abstraction, *Philosophia Mathematica* 8, 100-123, 2000.
[Hellman, 1989] G. Hellman. *Mathematics Without Numbers*, Oxford, 1989.
[Hoare, 1969] C. A. R. Hoare. An axiomatic basis for computer programming, *Communications of the Association for Computing Machinery* 12, 576-580, 1969.
[Hoare, 1985] C. A. R. Hoare. Programs are predicates, in C. A. R. Hoare and J. C. Stephenson (eds.), *Mathematical Logic and Programming Languages* (Englewood Cliffs, NJ: Prentice-Hall, 1985), 141-155.
[Holland et al., 1986] J. Holland, K. Holyoak, R. Nisbett and P. Thagard. *Induction* (Cambridge, Mass, 1986).
[Huntley-Fenner et al., 2002] G. Huntley-Fenner, S. Carey and A. Solimando. Objects are individuals but stuff doesn't count: perceived rigidity and cohesiveness influence infants' representations of small groups of discrete entities, *Cognition* 85, 203-21, 2002.
[Huttenlocher et al., 2002] J. Huttenlocher, S. Duffy and S. Levine. Infants and toddlers discriminate amount: are they measuring?, *Psychological Science* 13, 244-49, 2002.
[Irvine, 1990] A. Irvine. Introduction to A. Irvine, ed., *Physicalism in Mathematics* (Dordrecht, 1990).
[Jaynes, 2003] E. T. Jaynes. *Probability Theory: the logic of science* (Cambridge, 2003).
[Jesseph, 1993] D. M. Jesseph. *Berkeley's Philosophy of Mathematics*, University of Chicago Press, Chicago, 1993.
[Keynes, 1921] J. M. Keynes. *A Treatise on Probability* (London, 1921).
[Koopman, 1940] B. O. Koopman. The axioms and algebra of intuitive probability, *Annals of Mathematics*, 42, 269-92, 1940.
[Kuksewicz, 1988] Z. Kuksewicz. The potential and the agent intellect, ch. 29 of *The Cambridge History of Later Medieval Philosophy*, ed. N. Kretzmann et al (Cambridge, 1988).
[Lear, 1982] J. Lear. Aristotle's philosophy of mathematics, *Philosophical Review* 91, 161-92, 1982.
[Lewis, 1991] D. Lewis. *Parts of Classes*, Blackwell, Oxford, 1991.
[Lehrer Dive, 2003] L. Lehrer Dive. An Epistemic Structuralist Account of Mathematical Knowledge, PhD Thesis, University of Sydney, 2003.
[Lonergan, 1970] B. Lonergan. *Insight: A Study of Human Understanding* (3rd ed, New York, 1970).
[Mac Lane, 1986] S. Mac Lane. *Mathematics: Form and Function*, Springer, New York, 1986.
[Maddy, 1992] P. Maddy. *Realism in Mathematics* (Oxford, 1992).
[Marr, 1982] D. Marr. *Vision: A computational investigation into the human representation and processing of visual information* (San Francisco, 1982).
[Massey, 1971] B. S. Massey. *Units, Dimensional Analysis and Physical Similarity*, Van Nostrand Reinhold, London, 1971.

[McKirahan, 1992] R. D. McKirahan. *Principles and Proofs: Aristotle's Theory of Demonstrative Science* (Princeton, 1992).
[Michell, 1994] J. Michell. Numbers as quantitative relations and the traditional theory of measurement, *British Journal for the Philosophy of Science* 45, 389-406, 1994.
[Moreland, 2001] J. P. Moreland. *Universals* (Chesham, 2001).
[Mortensen, 1998] C. Mortensen. On the possibility of science without numbers, *Australasian Journal of Philosophy* 76, 182-97, 1998.
[Mundy, 1987] B. Mundy. The metaphysics of quantity, *Philosophical Studies* 51, 29-54, 1987.
[Musgrave, 1977] A. Musgrave. Logicism revisited, *British Journal for the Philosophy of Science* 28, 99-127, 1977.
[Newstead, 2000] A. G. J. Newstead. Aristotle and modern mathematical theories of the continuum, in D. Sfendoni-Mentzou, ed, *Aristotle and Contemporary Science*, vol. 2, Lang, New York, pp. 113-129, 2000.
[Odegard, 1969] D. Odegard. Locke and the unreality of relations, *Theoria* 35, 147-52, 1969.
[Pagels, 1988] H. R. Pagels. *Dreams of Reason: The Computer and the Rise of the Sciences of Complexity* (New York: Simon & Schuster, 1988).
[Parsons, 2004] C. Parsons. Structuralism and metaphysics, *Philosophical Quarterly* 54, 57-77, 2004.
[Peirce, 1881] C. S. Peirce. On the logic of number, *American Journal of Mathematics* 4, 85-95, 1881. Repr. in *Collected Papers,* ed. C. Hartshorne & P. Weiss, Belknap Press, Harvard, 1960, vol 3, pp. 158-70.
[Pincock, 2004] C. Pincock. A new perspective on the problem of applying mathematics, *Philosophia Mathematica* 12, 135-161, 2004.
[Plato, Parmenides] Plato, *Parmenides*.
[Polya, 1954] G. Polya. *Mathematics and Plausible Reasoning* (vol. I, *Induction and Analogy in Mathematics*, and vol. II, *Patterns of Plausible Inference*), Princeton, 1954.
[Quine, 1981] W. V. Quine. 'Success and limits of mathematization', in W. V. Quine, *Theories and Things* (Cambridge, Mass. 1981), 148-155.
[Proclus, 1970] Proclus. *Commentary of the First Book of Euclid's Elements*, trans. G. R. Morrow (Princeton, N.J., 1970)
[Reck, 2003] E. H. Reck. Dedekind's structuralism: an interpretation and partial defense, *Synthese* 137, 369-419, 2003.
[Reck and Price, 2000] E. Reck and M. Price. Structures and structuralism in contemporary philosophy of mathematics, *Synthese* 125, 341-383. 2000.
[Resnik, 1997] M. D. Resnik. *Mathematics as a Science of Patterns,* Clarendon, Oxford.
[Restall, 2003] G. Restall. Just what is full-blooded Platonism?, *Philosophia Mathematica* 11, 82-91, 2003.
[Russell, 1917] B. Russell. *Mysticism and Logic and Other Essays.* (London, 1917).
[Russell, 1901/1993] B. Russell. Recent work on the principles of mathematics, 1901. In *Collected Papers of Bertrand Russell*, vol. 3, G. H. Moore, ed. Routledge, London and New York, 1993.
[Shapiro, 1977] S. Shapiro. *Philosophy of Mathematics: structure and ontology,* Oxford University Press, New York, 1977.
[Shapiro, 2000] S. Shapiro. *Thinking About Mathematics: the philosophy of mathematics,* Oxford University Press, New York, 2000.
[Shapiro, 2004] S. Shapiro. Foundations of mathematics: metaphysics, epistemology, structure, *Philosophical Quarterly* 54, 16-37, 2004.
[Simon, 1981] H. A. Simon. *The Sciences of the Artificial* (Cambridge, MA: MIT Press, 1969; 2nd edn., 1981).
[Smale, 1969] S. Smale. What is global analysis?, *American Mathematical Monthly* 76, 4-9, 1969.
[Smith, 1954] V. E. Smith. *St Thomas on the Object of Geometry,* Marquette University Press, Milwaukee, 1954.
[Stea et al., 2004] D. Stea, D. D. Kirkman, M. F. Pinon, N. N. Middlebrook and J. L. Rice. Preschoolers use maps to find a hidden object outdoors, *Journal of Environmental Psychology* 24, 341-45, 2004.
[Swoyer, 2000] C. Swoyer. Properties, *Stanford Encyclopedia of Philosophy* (online), 2000.
[Sydney School, 2005] The Sydney School, 2005, manifesto, http://web.maths.unsw.edu.au/~jim/structmath.html

[Steiner, 1973] M. Steiner. Platonism and the causal theory of knowledge, *Journal of Philosophy* 70, 57-66, 1973.
[Waldrop, 1992] M. M. Waldrop. *Complexity: The Emerging Science at the Edge of Order and Chaos* (New York: Simon & Schuster, 1992).
[Weinberg, 1965] J.R. Weinberg. *Abstraction, Relation, Induction* (Madison, 1965).
[Weyl, 1952] H. Weyl. *Symmetry,* Princeton, Princeton University Press, 1952.
[Wolfram, 2002] S. Wolfram. *A New Kind of Science* (Champaign, Ill: Wolfram Media, 2002).
[Woolsey and Swanson, 1975] C. D. Woolsey and H. S. Swanson. *Operations Research for Immediate Application: a quick and dirty manual* (New York: Harper & Row, 1975).
[Xu et al., 2005] F. Xu, E. S. Spelke and S. Goddard. Number sense in human infants, *Developmental Science* 8, 88-101, 2005.
Mathematical Reviews (1990), Annual Index, Subject Index, p. S34.
[Zeilberger, 1993] D. Zeilberger. Theorems for a price: tomorrow's semi-rigorous mathematical culture, *Notices of the American Mathematical Society* 46, 978-81, 1993.
[Zeilberger, 2001] D. Zeilberger. 'Real' analysis is a degenerate case of discrete analysis. http://www.math.rutgers.edu/~zeilberg/mamarim/mamarimhtml/real.html

EMPIRICISM IN THE PHILOSOPHY OF MATHEMATICS

David Bostock

1 INTRODUCTION

Two central questions in the philosophy of mathematics are 'What is mathematics about?' and 'How do we know that it is true?' It is notorious that there seems to be some tension between these two questions, for what appears to be an attractive answer to the one may lead us into real difficulties when we confront the other.[1] (For example, it is a well-known objection to the Platonism of Frege, or Gödel, or indeed Plato himself, that if the objects of mathematics are as they suppose, then we could not know anything about them.) The subject of this chapter is empiricism, which is a broad title for one general style of answer to the question 'How do we know?' This answer is 'Like (almost?) all other knowledge, our knowledge of mathematics is based upon our experience'. The opposite answer, of course, is that our knowledge of mathematics is special because it is *a priori*, i.e. is not based upon experience. To defend that answer one would, naturally, have to be more specific about the nature of this supposed *a priori* knowledge, and about how it can be attained. Similarly, to defend the empiricist answer one must say more about just how experience gives rise to our mathematical knowledge, and — as we shall see — there are several quite different answers to this question which all count as 'empiricist'. These different answers to the question about how knowledge is acquired will usually imply, or at least very naturally suggest, different answers to the *other* question 'What is mathematics about?' For one can hardly expect to be able to explain how a certain kind of knowledge is acquired without making some assumptions about what that knowledge is, about what it is that is known, i.e. about what it is that is stated by the true statements of mathematics. But any answer to that must presumably involve an account of the 'mathematical objects' that such statements (apparently) concern. So we cannot divorce epistemology from ontology. The title 'empiricism' indicates one kind of answer to the epistemological question, but the various answers of this kind cannot be appraised without also considering their implications for the ontological question.

As I have just implied, there are different varieties of empiricism, and no one theory which is *the* empiricist theory of mathematical knowledge. Equally, there

---

[1] A classic exposition of this dilemma is [Benacerraf, 1973].

Handbook of the Philosophy of Science. Philosophy of Mathematics
Volume editor: Andrew D. Irvine. General editors: Dov M. Gabbay, Paul Thagard and John Woods.
© 2009 Elsevier B.V. All rights reserved.

is no one ontology which all empiricist theories subscribe to. Traditionally, the various ontological theories are classified as realist, conceptualist, and nominalist. The central claim of realism is that mathematics concerns objects (e.g. numbers) which exist independently of human thought. There are two main sub-varieties: the Platonic version adds that these objects are also independent of anything which exists in this physical world that we inhabit; the Aristotelian version holds that these objects, while not themselves physical objects in quite the ordinary sense, nevertheless depend for their existence upon the familiar physical objects that exemplify them. (In metaphorical terms, the Platonic theory claims that numbers exist 'in another world', and the Aristotelian theory claims that they exist 'in this world'.) By contrast with each of these positions, the central claim of conceptualism is that mathematics concerns objects (e.g. numbers) which exist only as a product of human thought. They are to be regarded merely as 'objects of thought', and if there had been no thought then there would have been no numbers either. Finally, the central claim of nominalism is that there are no such things as the abstract objects (e.g. numbers) that mathematics *seems* to be about. There are two main subvarieties. The traditional 'reductive' version adds that what mathematicians assert is nevertheless *true*, for what *seem* to be names of abstract objects are not really names at all. Rather, they have another role, for when mathematical statements are properly analysed it will be seen that they do not really concern such abstract objects as numbers were supposed to be. A different and more recent version of nominalism may be called the 'error' theory of mathematics, according to which mathematical statements are to be taken at face value, so they do purport to refer to abstract objects, but the truth is that there are no such objects. Hence mathematical statements are never true, though it is admitted that they may be very useful.

An empirical theory of mathematical knowledge is perhaps most naturally combined with Aristotelian realism in ontology, and this was Aristotle's own position. A more recent proponent of this kind of position is Penelope Maddy. But another kind of empiricist theory, due mainly to Quine and Putnam, requires an ontology which is much closer to Platonic realism. A very different empirical theory, hailing from Aristotle, but combined now with reductive nominalism, is to be found in John Stuart Mill, and in his disciple Philip Kitcher. As for the 'error' version of nominalism, which is due mainly to Hartry Field, that is a view according to which mathematical statements cannot be known at all, by any means, since they simply are not true. But it also supposes that there are *related* statements that are true, i.e. roughly those which 'reductive' nominalism invokes in its reduction. The question whether our knowledge of *these* truths is or is not empirical rather quickly leads to the more general question whether our knowledge of *logic* is empirical. In what follows I shall have more to say about each of the positions here mentioned.

I shall not further discuss the possibility of combining an empiricist view of how mathematical knowledge is acquired with a conceptualist view of the existence of mathematical objects. So far as I know, no one has ever proposed such a combination. And indeed it is natural to suppose that if mathematical objects exist only as a result of our own thinking, then the way to find out what is true of them is just to engage in more of that thinking, for how would experience be relevant? Yet this combination is not at once impossible, and one could say that the position adopted by Charles Chihara, which I do describe in what follows, is quite close to it.

Just how to understand the notion of 'empirical' (or '*a posteriori*') knowledge, as opposed to '*a priori*' knowledge, is a question that will occupy us from time to time as we proceed (particularly in section 4.1). For the time being I assume that the traditional description, 'empirical knowledge is knowledge that depends upon experience', is at least clear enough for the discussion to get started. But it may be useful to make two clarifications before we go any further.

First, the 'experience' in question is intended to be experience gained from our ordinary perception of the world about us, for example by seeing or hearing or touching or something similar. There are theories of mathematical knowledge which posit a quite different kind of 'experience' as its basis. For example Plato (at one time) supposed that our knowledge of abstract objects such as the numbers was to be explained by our having 'experienced' those objects *before* being born into this world, and while still in 'another world' (namely 'the intelligible world'), which is where those objects do in fact exist.[2] This kind of 'experience' is emphatically *not* to be identified with the familiar experience of ordinary perceptible objects that we enjoy in this world and, if mathematical knowledge is based upon it, then that knowledge does not count as 'empirical' in the accepted sense of the word. Perhaps no one nowadays would take this Platonic theory of 'recollection of another world' very seriously, except as a metaphor for what could be more literally stated in other terms. But there are broadly similar theories current today, for example Gödel's view that our knowledge of mathematics depends upon a special kind of experience which he called 'mathematical experience', and which he described as the experience of finding that the axioms of mathematics 'force themselves upon us as being true'.[3] Whether there is any such experience may of course be doubted, but even if there is still it would not count as showing that mathematical knowledge is a kind of 'empirical' knowledge. For the word 'empirical', as normally understood, refers only to the ordinary kind of experience (Greek: *empeiria*) that occurs in the perception of ordinary physical objects by means of the five senses. If, as some have supposed, there is also a rather different kind of 'experience' of other things — e.g. of mathematical truths, or logical truths, or (say) moral truths — that would not be counted as showing that knowledge based upon it — e.g. of mathematics, or logic, or morals — counted as 'empirical' knowledge. This may seem a somewhat arbitrary restriction upon what is to count as 'experience', and hence as 'empirical' knowledge. But the restriction is traditional, and I shall

---

[2] For Plato's theory of 'recollection', and his distinction between the perceptible world and the intelligible world, see his *Meno* (80d-86b), *Phaedo* (72e-77a), *Republic* (507a-518d) and *Phaedrus* (249b-c).

[3] [Gödel, 1947, 271].

observe it. It is 'empirical knowledge' in the traditional sense that is the subject of this chapter.

A quite different point that it is useful to mention here is this. Almost all philosophers would accept that very often we *first* come to know a mathematical truth as a result of experience. For example, one may come to know that $7 + 5 = 12$ by the experience of hearing one's teacher say so, or by the experience of putting together a collection of 7 apples with a collection of another 5 apples, counting the new collection so formed, and thus discovering that it is a collection of 12 apples. But those who deny empiricism — let us call them the 'apriorists' — will want to add that this initial knowledge, which *is* based upon experience, can later be superseded by a genuine *a priori* knowledge which is not so based. They may perhaps claim that this happens when one becomes able to see that the result of this particular experience of counting *must* also hold for any other like-numbered collections as well. Or they might say that genuine *a priori* knowledge arises only when one finds how to *prove* that $7 + 5 = 12$. But here we should notice that all proofs must start somewhere, so a proof could only yield *a priori* knowledge if the premises from which it starts are themselves known *a priori* to begin with. Pressing this line of thought will evidently lead one to focus on the axioms from which elementary arithmetic may be deduced, and the question becomes whether these are known *a priori* or known empirically (or perhaps not known at all — but let us set that possibility aside for the present). Once again the apriorist will no doubt concede that one may first come to know these axioms as a result of experience, for example the experience of reading a textbook on the subject, but he will insist that the knowledge could 'in principle' have been attained without any such experience. His claim is that (at least some?) mathematics *can* be known *a priori*, not that it actually is known in this way. Consequently, to provide a proper opposition to his position, the empiricist should be understood as claiming that *all* ways of acquiring mathematical knowledge *must* depend upon experience.

With so much by way of preamble, let us now consider the main varieties of empiricist theory that have been proposed.

2 ARISTOTLE

Much of Aristotle's thought developed in reaction to Plato's views, and this is certainly true of his philosophy of mathematics. Plato had held that the objects which mathematics is about — e.g. squares and circles in geometry, numbers in arithmetic — are not to be found in this world that we perceive. His main reason was that mathematics concerns ideal entities, and such ideals do not exist in this world. For example, geometry concerns perfect squares and perfect circles, but no actual physical circle ever is a *perfect* circle. As he believed, much the same applies to numbers, but this requires a little explanation. In Greek mathematics only one kind of number was officially recognised, and this was standardly explained by saying 'a number is a plurality of units'.[4] Plato took this to imply that in pure mathematics we are concerned with 'perfect' pluralities of 'perfect' units. These 'perfect units', he supposed, must be understood as exactly equal to one another in every way, and as divisible in no way at all. Moreover the 'perfect' number 4 (for example) was just four of such units, and was not also some other number as well.[5] But we see nothing in this world which fits these descriptions. Whatever perceptible things we take as units, they always will be further divisible, and they never will be perfectly equal to one another in *all* respects. Again, anything in this world which may be taken to be a plurality of four things may *also* be taken to be a plurality of some other number of things (e.g. as four complete suits of playing cards are also fifty-two individual cards[6]). So, in Plato's view, mathematics is about *perfect* numbers, and *perfect* geometrical figures, and such things do not exist in this world that we perceive. He therefore concluded that they must exist in 'another world', for mathematics could hardly be true if the things which it is about did not exist at all. This talk of 'two worlds' strikes us nowadays as wildly extravagant, and we would no doubt prefer Plato's other way of expressing his point, namely that the objects of mathematics (do exist and) are 'intelligible' but not 'perceptible'. But it is clear that Plato himself took the 'two worlds' picture quite seriously, and that Aristotle was right to understand him in this way.

So far I have been describing Plato's ontology, but his epistemology now follows in one quick step. Since the objects of mathematics do not exist in this world (i.e. are intelligible but not perceptible), we cannot find out about them by means of our experience of what is in this world. Rather, our knowledge of them must be attained by thought alone, thought which pays no attention to what can be perceived in this world. (As noted earlier,[7] this 'thought' was at one stage conceived as 'recollection' of our previous 'experiences' in the other world. It would seem that Plato later came to abandon this theory of 'recollection', but he always continued to think that mathematical knowledge is *not* gained by experience of this world.)

That is a quick sketch of the position that Aristotle aims to reject, and we can be quite sure of the main outline of the theory that he wishes to put forward in opposition. He holds that the objects that mathematics is about are the perfectly ordinary objects that we can perceive in this world, and that our knowledge of mathematics must be based on our perception of those objects. It may at first sight appear otherwise, but if so that is because in mathematics we speak in a very general and abstract way of these ordinary things, prescinding from many of the features that they do actually possess. For example, in mathematics we take no account of the changes that these objects do in fact undergo, but speak

---

[4] Note that on this explanation neither zero nor one is a number, and the number series begins with two. But in practice the series was generally counted as beginning with one. (However it was many centuries before zero was recognised as a number.)

[5] See e.g. Plato, *Republic* 523b-526a, *Philebus* 56c-e.

[6] This example is Frege's [1884, §22]. The passages just cited from Plato give no specific examples.

[7] See note 2.

of them as if they were things not subject to change (e.g. in Platonic language 'the square itself', 'the circle itself', 'the number 4 itself'). This does no harm, for their changes do not affect the properties studied in mathematics, but for all that it is these ordinary changeable things that we are speaking of (e.g. ordinary things that are square or circular, and pluralities of 4 quite ordinary objects, say the cows in a field). To take another instance, in geometry we do not mention the matter of which things are made, since it is not relevant to the study in question, but this does not mean that we are speaking of special things which are *not* made of matter; rather, they will be made of perfectly ordinary perceptible matter (and not some peculiar and imperceptible stuff called 'intelligible matter'). This much we can confidently attribute to Aristotle from what he does say, in the writings that have come down to us, but unfortunately we do not have any more detailed exposition of his own positive theory. Nor do we have any explicit response to the Platonic arguments just outlined, aiming to show that mathematics cannot be about the objects of this world. So I will supply a response on Aristotle's behalf.[8]

In a way, it is true that geometry idealises; it pays attention to *perfect* squares, circles, and so on, and not to the imperfect squares and circles that are actually found in this world. But, from our own perspective, we can easily see that there is not a serious problem here, for we are now quite familiar with scientific theories which 'idealise' in one way or another. For example, there is a theory of how an 'ideal gas' would behave — e.g. it would obey Boyle's law precisely — and this theory of 'ideal' gases is extremely helpful in understanding the behaviour of actual gases, even though no actual gas is an ideal gas. This is because the ideal theory simplifies the actual situation by ignoring certain features which make only a small difference in practice. (In this case, the ideal theory ignores the actual size of the molecules of the gas, and any attractive (or repulsive) force that those molecules exert upon one another.) But no one nowadays would be tempted to think that there must therefore be 'ideal gases' in some other world, and that the physicist's task must be to turn his back on this world and to try instead to 'recollect' that other world. That reaction would plainly be absurd. Something similar may be said of the idealisations in geometry. For example, a carpenter who wishes to make a square table will use the geometrical theory of perfect squares in order to work out how to proceed. He will know that in practice he cannot actually produce a *perfectly* straight edge, though he can produce one that is very nearly straight, and that is good enough; it obviously explains why the geometrical theory of perfect squares is in practice a very effective guide. Geometry, then, may perfectly well be regarded as a study of the spatial features — shape, size, relative position, and so on — of ordinary perceptible things. It does no doubt involve some 'idealisation' of these features, but that is no good reason for saying that it

---

[8] Books M and N of Aristotle's *Metaphysics* contain a sustained polemic against what he viewed as Platonic theories of mathematics. But most of the polemic concerns *details* — details that are often due not to Plato himself but to his successors — and the main arguments, which I have outlined above, are simply not addressed in those books, or anywhere else in Aristotle's surviving writings.

is not really concerned with such things at all, but with objects of a quite different kind which are not even in principle perceptible.

Let us turn to arithmetic. We, who have been taught by Frege, will of course think that Plato's arguments result only from a badly mistaken view of how numbers apply to things in this world. Frege claimed that a 'statement of number', such as 'Jupiter has 4 moons' or 'There are 4 moons of Jupiter' makes an assertion about a concept. That is, it says of the concept 'moon of Jupiter' that there are just 4 things that fall under it.[9] An alternative analysis, which (at first sight) does not seriously differ is that this statement says of the set of Jupiter's moons that it has 4 members. In either case, the thought is that '4' is predicated, not directly of a physical object, but of something else — a concept, a set — which has 4 physical objects that are instances or members of it. Once this indirectness is recognised, Plato's problems simply disappear. We see (i) that 'Jupiter has 4 moons' does not in any way require those moons to be indivisible objects; no doubt each moon does have parts, but since a mere part of a moon is not itself a moon this generates no problem. Again (ii) the statement does not imply that the 4 moons are 'equal' to one another in any way other than that each of them is a moon. And again (iii) the statement does not in any way imply that the matter which constitutes those 4 moons cannot *also* be seen as constituting some other totality with a different number of members. For example, it may be true both that there are 4 moons of Jupiter and that there are 10 billion billion molecules that are molecules of the moons of Jupiter. But this shows no kind of 'imperfection' in either claim, since the concepts (or sets) involved, given by 'moon of Jupiter' and 'molecule of a moon of Jupiter', are quite clearly different.

We cannot know quite how Aristotle himself would have responded to the two Platonic arguments just discussed, since no response of his is recorded in those of his writings that we now possess. I hope that it would have been something similar to what I have just been suggesting, but that is merely a pious hope. In any case, we can be sure that he endorsed the conclusions that these thoughts lead to, namely that such idealisations as are involved in geometry do not prevent the view that the actual subject-matter of geometry is ordinary (non-ideal) perceptible objects, and again that arithmetic applies straightforwardly to ordinary perceptible objects without any idealisation at all. There is therefore no obstacle to supposing that mathematics is to be understood as a (highly abstract) theory of the ordinary and familiar objects that we perceive. Finally, we add the expected step from ontology to epistemology: since mathematics is about the perceptible world, our knowledge of it must stem from the same source as all our other knowledge of this world, namely perception. Again, this step is one that Aristotle never argues, at least in the writings that have come down to us, but it must have seemed to him so obvious as to need no argument: *of course* knowledge of the perceptible world will be based upon our perception of that world. No doubt more needs to be said about just how one is supposed to 'ascend' from the initial perceptions of particular things, situations, and events to the knowledge of the first principles of a

---

[9][Frege, 1884, §54].

deductive science such as geometry. For Aristotle is convinced that every finished science will have its own first principles, and will proceed by deduction from them, even though in his own day — and for many centuries afterwards — geometry was the only major science that was so organised.[10] But his account of how to ascend to first principles is really so superficial that it is not worth discussing here.[11] So let us just say that this is another of the many areas in which Aristotle's view of mathematics needs, but does not get, further defence and elaboration.

There are many problems that would naturally arise if Aristotle had offered a more detailed account. But, since he does not, I postpone discussion of these until the next section, when we shall have a more detailed account to consider. Here I note just one problem that Aristotle did see himself, and did try to meet, namely over infinity. Even the simple mathematics that Aristotle was familiar with — i.e. what we now call elementary arithmetic and Euclidean geometry — quite frequently involves infinity, but it is not clear how that can be so if its topic is what we perceive. For surely we do not perceive infinity? Aristotle attacks this problem in his *Physics*, book III, chapters 4–8.

Geometry apparently involves infinity in two ways, (i) in positing an infinite space, and (ii) in assuming that a quantity such as length or area is infinitely divisible. To the first of these one might add, though in those days it was hardly a topic treated in mathematics, (iii) that time would appear to be infinitely extended too, both forwards and (at least according to Aristotle) backwards as well. Finally (iv) ordinary arithmetic apparently assumes the existence of an infinite plurality, because it assumes that there are infinitely many numbers. Let us take these points in turn.

(i) Aristotle simply denies that space is infinite in extent. On his account the universe is a finite sphere, bounded at its outer edge by the spherical shell which is the sphere of the fixed stars, and outside that there is nothing at all. In particular, there is not even any space, for space only exists within the universe. Now on the face of it this claim conflicts with the usual assumptions of geometry. For example, Euclid posits that any finite straight line may be extended for as far as you please in either direction, whereas Aristotle claims that there is a maximum length for any straight line, namely the length of a diameter of the universe.[12] Nevertheless he is clearly right to say (as he does at *Physics* $207^b 27$–34) that this does not deprive the geometers of their subject. It is true that some usual definitions would have to be altered; for example parallel lines could no longer be defined as lines

---

[10]The Greeks did add some others, though I would not call them 'major', e.g. Archimedes on the law of the lever.

[11]At different places he invokes either what he calls 'dialectic' or what he calls 'induction'. (I have summarised his discussion of these in my [2000, chapter X, sections 1–2].) But he shows no understanding of what we would regard as crucial, namely what is called 'inference to the best explanation'.

[12]Euclid is roughly one generation after Aristotle, so one cannot assume that Aristotle did know of Euclid's axioms in particular. But we can be sure that Euclid had his precursors, and that some axiomatisation of geometry was available in Aristotle's time. The details, however, are not known.

(in the same plane) which never meet, no matter how far (in either direction) they are extended. But it is quite easy to suggest an alternative definition. Moreover, wherever a proof would normally be given by assuming some extension to a given figure — an extension which may not be possible if the space is finite and the figure is large — we can always proceed instead by assuming some similar but smaller figure which can be extended in the required way. Aristotle's view does require a modification to ordinary Euclidean geometry, but it is an entirely minor modification.

(ii) On infinite divisibility his position is more complex. On the one hand he wishes to say (a) that *in a sense* a finite line is infinitely divisible, namely in the sense that, no matter how many divisions have been made so far, a further division would always (in principle) be possible. But he also wishes to say (b) that it is not possible for a finite line ever to have been infinitely divided, i.e. there cannot (even in principle) be a time when infinitely many divisions have been made. To explain his position in these simple terms, one must introduce an explicit mention of times, as I have just done, for Aristotle is smuggling in an assumption which he never does explicitly acknowledge, namely this: an infinite totality could exist only as the result of an infinite process being completed. But he then adds that infinite processes cannot be completed, and so infers that there are no infinite totalities. He has no objection to infinite processes as such; for example, there could perfectly well be an infinite process of dividing a finite line, with one more division made on each succeeding day, for an infinity of days to come. That is entirely conceivable. But (according to him) it is not conceivable that either this or any other infinite process should ever be *finished*. (His reason, I presume, is that one cannot come to the end of a process that has no end.)

He has another way of expressing his conclusion, by means of a distinction between 'actual' and 'potential' existence. For example, we may ask 'how many points are there on a finite line?' From Aristotle's perspective a point exists 'actually' only when it has in some way been 'actualised', which would happen if a division were made at that point, or if something else occurred at that point which in some way distinguished it from its neighbours (e.g. if a body in rectilinear motion changed its direction at that point). Until then the point exists only 'potentially'. So there is a 'potential infinity' of points on the line, but at any specified time there will be only finitely many that exist 'actually'.

Whether this position is defensible is a question that I must here set aside.[13] But in any case I think it is fair to say that it threatens no harm to the geometry of Aristotle's day. In the mathematical practice of that time, points and lines and planes were taken to be equally basic from an ontological point of view. Philosophers (including Aristotle) were attracted to the idea that a plane might be viewed as the limit of a solid, a line as the limit of a plane, and a point as the limit of a line. On this account, solids are the most basic of geometrical entities and points the least basic. Of course from a modern perspective it is usual to view solids, planes, and lines simply as sets of points, so that it is points that are

---
[13] I have discussed the point (and answered 'no') in my [1972/3].

the most basic entities. On this approach one must assume that infinitely many points do ('actually') exist if the subject is not to collapse altogether. But on the more ancient approach there seems to be no strong reason to say that there must ('actually') exist an infinity of points, so I think that we can once more say that Aristotle's proposals — though certainly unorthodox — again do not deprive the geometers of their subject.[14]

(iii) Whereas Aristotle believed that space is finite, he did not think the same of time. On the contrary, he supposed that the universe neither began to exist nor will cease to exist, and hence that time itself has no beginning and no end. The 'forwards' infinity of time is entirely compatible with the discussion that we have just given, for that simply means that time is an unending process which will never be completed, and Aristotle does not deny the existence of such processes. The 'backwards' infinity is much more difficult for him, for this appears to be an infinite process which (never started, but) *has* been completed. However, since he never discusses this point himself, I shall not do so for him. It would appear to be a problem for him, but one which concerns the nature of time rather than the nature of mathematics.[15]

(iv) Near the start of his discussion of infinity (*Physics* 203^b15–30) Aristotle cites a number of considerations that lead people to believe that there is such a thing as infinity, and one of these is that there appear to be infinitely many *numbers* (203^b22–5). He couples this with the idea that a geometrical quantity such as length is also infinite, in each case explaining the idea as due to the point that 'they do not give out in our thought'. The ensuing discussion then concentrates on geometrical magnitudes (as already explained), and we hear no more about the infinity of the natural numbers until the final summing up, which contains this claim: 'It is absurd to rely on what can be thought by the human mind, since then it is only in the mind, and not in the real world, that [these things] exist' (208^a14–16). Presumably this remark is intended to apply to the case of the numbers, mentioned initially but not explicitly treated anywhere else, save here. If so, then Aristotle's response is apparently this: it is true that the numbers do not give out 'in our thought', but they *do* give out in fact; and so there are only a finite number of numbers that 'actually' exist. Moreover, one can see that this position is forced upon him by his view that a number is simply (the number of) a plurality of ordinary perceptible objects. Since (on his account) the universe is finite in extent, and since no infinite division of a perceptible object can ever be completed, there can only be finitely many things to which numbers are applied.

---

[14] Aristotle argues with some force that a line *cannot* be regarded as made up out of points (*Physics* 231^a21–$^b18$). But this is not because he wishes to controvert anything that the mathematicians of his day asserted; rather, he is denying the 'atomist' claim that the smallest entities are both extended and indivisible.

[15] The infinite divisibility of time is treated in the same way as that of space. Thus, in the temporal stretch between now and noon tomorrow there will be only finitely many instants (i.e. points) of time that become 'actual'. This happens when something occurs at that instant which is not also occurring at all neighbouring instants. But however many do become 'actual', it is always possible that there should have been more.

So there are only finitely many numbers.[16]

This is a *shocking* conclusion. Ordinary arithmetic very clearly takes it for granted that the series of natural numbers has no end, since *every* number has a successor that is a number. But Aristotle commits himself to the view that this is not true, so (on his account) there must be some greatest number which has no successor. Unsurprisingly, he does not tell us which number this is, and one supposes that he would have to think that it is ever-increasing (for example as more 'divisions' are made, or as more days pass from some arbitrarily chosen starting point, or in other ways too). His two other claims that what ('actually') exists is only finite seem to me to be not obviously unacceptable, but the idea that there are only finitely many natural numbers is extremely difficult to swallow. And I do not find it much palliated by the defence that there is a *potential* infinity of natural numbers, since this only means that there *could* be more than there actually are (but still only a finite number). If empiricism in mathematics is committed to this claim, it is surely unappealing.

I add as a footnote that the infinity of the number series can be a problem not only for empiricists but also for other approaches to the philosophy of mathematics. For example conceptualists (such as the intuitionists), who hold that the numbers are our own 'mental mathematical constructions', are faced with the problem that on this account there will be infinitely many numbers only if there have been infinitely many such constructions, but this would appear to be impossible (if only because human beings have existed only for a finite time, and there must be some minimum time which every mental construction must take). Intuitionists pretend to respond to this problem by using Aristotle's terminology, and saying that the infinity of the number series is merely a 'potential' infinity, but not an 'actual' one.[17] This is a mere subterfuge, and it does not accord with their actual practice, either when doing mathematics or in explaining why they do it in their own (non-classical) way.[18] A reinterpretation of their position which seems to me to be forced upon them, by this and other considerations, is that a mathematical entity (such as a number) counts as existing so long as it is (in principle) *possible* that it should be constructed in our thought; and whether or not it has, at some time before now, actually been constructed is simply irrelevant. Moreover, this reinterpretation of the conceptualist's position would still allow him his basic thesis, that mathematical entities exist only because of human thought. But now

---

[16] Aristotle sometimes offers a further argument. If, as the Platonist supposes, the numbers exist independently of their embodiment in this world, then — he claims — there would have to be such a thing as the number of all those numbers, and this would have to be an infinite number. Since 'to number' is 'to count', it would then follow that one can count up to infinity (*Physics* III, 204^b7–10), and that there is a number which is neither odd nor even (*Metaphysics* M, 1083^b36–1084^a4). But both of these consequences are impossible.

[17] See e.g. [Dummett, 1977, 55–75].

[18] For example, they explain that quantification over numbers is quantification over an infinite domain, and for that reason quantifications over the numbers need not be (even in principle) decidable. But this explanation would collapse if they were to concede that the domain of the numbers is 'actually' a finite domain (though one that may be expected to grow as time goes on).

it is possible thought, rather than actual thought, that is what matters.

Could Aristotle have taken the same way out? Could he have said that, for a number to exist, what is required is that it be *possible* for there to be physical pluralities that have that number, and it does not matter whether there are actually any such pluralities? I think not. For he took it to be obvious that we find out about the numbers by perception because he supposed that numbers applied to perceptible pluralities of perceptible objects. But if we now modify this, and say instead that numbers apply to possible pluralities of possible objects, it will no longer seem obvious that perception is in any way relevant. It may seem very plausible to say that, when we are investigating what is actual, we cannot avoid relying on perception; but do we need perception at all if our topic is merely what is possible?

3   JOHN STUART MILL

Mill proposed his views on mathematics in conscious opposition to Kant (though in fact his own exposition scarcely mentions Kant at all). Kant in turn was reacting to his predecessors, and in particular to Hume. In order to set Mill's views in their context, I begin with a few brief remarks about this background.

Ever since Descartes, philosophers had paid much attention to what they called 'ideas', and which they construed as entities that exist only in minds. Hume's theory (which only makes more explicit the claims of his predecessors Locke and Berkeley) was that ideas are of two kinds, either simple or complex. Complex ideas may be deliberately created by us, put together from simpler ideas as their components, but the genuinely simple ideas can arise only as what Hume calls 'copies of impressions', where 'impressions' is his word for what occurs in the mind in a perception. All ideas, then, are derived directly or indirectly from perceptions, and this applies just as much to the ideas employed in mathematics as to any others. However, our *knowledge* of mathematics is special. Ordinary empirical knowledge Hume characterised as 'knowledge of matters of fact', and he contrasted this with 'knowledge of the relations of ideas', holding that mathematical knowledge was of the second kind. Thus the objects that mathematics is about — e.g. squares and circles, or numbers — are taken to *be* ideas, and the propositions of mathematics state relations between these ideas. Moreover, these relations can be discerned *a priori*, i.e. without relying on experience. That is, experience is needed to provide the ideas in the first place, but once they are provided we need no *further* recourse to experience in order to see the relations between them. This is taken to explain why the truths of mathematics are known with certainty, and cannot be refuted by experience.[19]

Kant agreed with a good part of this doctrine. He too thought that the truths of mathematics are necessary truths, known *a priori*, and not open to empirical

---

[19] A conveniently brief summary of Hume's position may be found in his *First Enquiry* [1748, section 20].

refutation. Moreover he does not dissent in any serious way from Hume's claim that these truths state relations between ideas, and become known when the mind attends to its own ideas. (Kant would say 'concept' rather than 'idea', and this is an important distinction, but not one that need concern us here.) However he did see a gap in Hume's account, which one can introduce in this way: just what relations are these, which are supposed to hold between our ideas (concepts), and just how are we able to discern them? It is here that he introduces his distinction between those necessary truths that are 'analytic' and those that are not.[20] One relation that may hold between ideas (concepts) is when one is *part* of another. (Let us understand this as including the case of an 'improper part', i.e. the case where the ideas are simply the same.) Kant saw no problem in our ability to discern this relation; it is done by *analysis* of our ideas (concepts), which Kant construes as a matter of anatomising a complex idea into its simpler parts. So truths which report this relation he calls 'analytic truths', and all others are contrasted as 'synthetic'. The question which lies at the heart of his *Critique of Pure Reason* is the question of how there can be *a priori* knowledge of truths that are not analytic but synthetic. And the discussion begins by claiming that such knowledge must somehow be possible, for the truths of mathematics are examples: they are 'one and all synthetic', but also known *a priori*.[21] It would be out of place in this chapter to pursue Kant's own investigations any further, though I do remark that the explanation of mathematical knowledge which he goes on to offer also leads him to say that the ideas with which mathematics is concerned are *not* derived from experience in the way that Hume had supposed.

The reasons that Kant offers for his two claims that mathematical truths are synthetic, and that they are known *a priori*, are not at all strong, and I pass over them. I think it likely that Kant did not argue very strongly because he took both claims to be uncontroversial. Certainly it was almost universally agreed amongst Kant's precursors that mathematical truths are known *a priori*, so he would not expect opposition to this. By contrast, the distinction between analytic and synthetic truths had not been applied to the case of mathematics by any of his precursors, and so no tradition was established on this point. But I think Kant took it to be simply obvious that, once the distinction was explained, everyone would agree that mathematical propositions could not be analytic. For analytic propositions are trivial and uninteresting truths, such as 'all men are men' or 'all men are animals' or 'no bachelor is married', whereas it is clear that the propositions of mathematics are much more interesting than these. Indeed, quite often we do not know whether a mathematical proposition is true or not, but it seems (at first sight) that analytic truths must be easy to discern, since the task of analysing a concept into its 'parts' is entirely straightforward.

For nearly two centuries following the publication of Kant's *Critique*, i.e. from 1781 to Quine's *Two Dogmas of Empiricism* in 1951, those philosophers who

---

[20] It is natural to call this 'Kant's distinction', though in fact it was drawn earlier by Leibniz and explained in a similar way. But Leibniz's use of it is so idiosyncratic that it is best ignored.
[21] Kant, *Critique of Pure Reason* [1781], Introduction.

thought of themselves as 'empiricists' felt that they had to face this dilemma: *either* show that our knowledge of mathematics is after all empirical knowledge, *or* admit that it is not, but explain it by showing how mathematical truths are really analytic truths. The second course was the one most usually taken, and in pursuit of this Kant's definition of 'analytic truth' has frequently been criticised, and various modifications have been proposed. (This road leads quite naturally to the logicist claim that mathematics is really no more than logic plus definitions.) There were not many who embraced the other horn of the dilemma, and argued that our knowledge of mathematics is after all empirical knowledge, in the traditional sense of 'empirical'. But amongst these there is one that stands out, namely John Stuart Mill. In one way he was absolutely right, as we can now see; in another, he was clearly badly wrong. That is, he was right about geometry and wrong about arithmetic, so I shall take each of these separately.

3.1 Mill on geometry[22]

Mill's main claim, stated at the outset of his discussion, is that 'the character of necessity ascribed to the truths of mathematics, and even (with some reservations ...) the peculiar certainty attributed to them, is an illusion'. Like almost all philosophers before Kripke's *Naming and Necessity* [1972], Mill runs together the ideas of necessary truth and *a priori* knowledge, so that his denial of necessity is at the same time a denial that knowledge of these truths is *a priori*. Indeed, his ensuing arguments are much more directly concerned with the nature of our knowledge than with the necessity or otherwise of what is known. And in fact they are mostly defensive arguments, claiming that the reasons given on the other side are not cogent.

The first is this. Some, he says, have supposed that the alleged necessity of geometrical truths comes from the fact that geometry is full of idealisations, which leads them to think that it does not treat of objects in the physical world, but of ideas in our minds. Mill replies that this is no argument, because the idealisations in question cannot be pictured in our minds either; for example, one may admit that there is no line in the physical world that has no thickness whatever, but the same applies too to lines in our imagination (section 1). In fact this response is mistaken, since it is perfectly easy to imagine lines with no thickness, e.g. the boundary between an area which is uniformly black and an area which is uniformly white. But that is of no real importance. It is clear that geometry *can* be construed as a study of the geometrical properties of ordinary physical objects, even if it does to some extent idealise, and I would say that it is better construed in this way than as a study of some different and purely mental objects. So here Mill parts company with the general tenor of the tradition from Descartes to Kant and beyond, and his revised (Aristotelian) ontology opens the path to his epistemology.

Mill very reasonably takes it for granted that geometrical knowledge is acquired

---

[22]See J. S. Mill, *System of Logic* [1843], book II, chapter V. The section references that follow are to this chapter.

by deduction, and that this deduction begins from axioms and definitions. In the present chapter he does not claim that there is any problem about our grasp of the deductions; that is, he accepts that if the premises were necessary truths, known *a priori*, then the same would apply to the conclusions. He also concedes (at least for the sake of argument) that there is no problem about the definitions, since they may be regarded as mere stipulations of ours, necessary truths and known *a priori*, just because we can know what we ourselves have stipulated. He is also prepared to grant that *some* of the propositions traditionally regarded as axioms might perhaps be rephrased as definitions, or replaced by definitions from which they would follow. (On this point he is somewhat over-generous to his opponents.) But he insists that the deductions *also* rely on genuine *axioms*, which are substantive assertions, not to be explained as concealed definitions. (The example that he most often refers to is: 'two straight lines cannot meet twice, i.e. cannot enclose a space'.)[23] So we can focus on the question of how axioms (such as this) are known (Sections 2–3).

His answer is that they are known only because they have consistently been verified in our experience. He concedes that it is not just that we never have experienced two distinct straight lines that meet twice, but also that we cannot even in imagination form a picture of such a situation. But he gives two reasons for supposing that this latter fact is not an *extra* piece of evidence. The first is just the counter-claim that what we can in this sense imagine — i.e. what we can imagine ourselves perceiving — is limited by what we have in fact perceived. (There is clearly at least some truth in this. To supply an example which Mill does not himself supply, we cannot imagine a radically new colour, i.e. a colour that falls quite outside the standard ordering of the colours that we do perceive. But that is no ground for saying that there could not *be* such a new colour, which might perhaps become available to our perception if human eyes develop a sensitivity to infrared light.) The second is that it is only because of our past experience, which has confirmed that spatial arrangements which we can imagine are possible, while those that we cannot imagine do not occur in fact, that we have any right to trust our imagination at all on a subject such as this. That is, the supposed connection between spatial possibility and spatial imaginability, which is here being relied upon, could not itself be established *a priori*. (Again, I supply an example which Mill does not: we can certainly imagine an Escher drawing, because we have seen them. But can we imagine the situation that such a drawing depicts? If so, the supposed connection between possibility and imaginability cannot be without exceptions.) For both these reasons Mill sets aside as irrelevant the claim that we cannot even picture to ourselves two straight lines meeting twice. The important point is just that we have never seen it (Sections 4–5).

But perhaps the most convincing part of Mill's discussion is his closing section 6 on the subject of conceivability. By this he means, not our ability to picture

---

[23] Strangely, Euclid's own text does not state this explicitly as a postulate, though he very soon begins to rely upon it. The gap was noted by his successors, and the needed extra postulate was added. For a brief history see Heath's commentary on Euclid's postulate I (pp. 195–6).

something, but our ability to see that it *might* be true. He concedes that we cannot (in this sense) even conceive of the falsehood of the usual geometrical axioms, but he claims that we cannot legitimately infer from this either that our knowledge of them is not based upon experience or that their falsehood is impossible. For what a person can conceive is again limited by what he has experienced, by what he has been brought up to believe, and by the weakness of his own creative thought. To substantiate these claims Mill cites several examples, from the history of science, of cases where what was once regarded as inconceivable was later accepted as true.

One of these is what we may call 'Aristotle's law of motion', which states that in order to keep a thing moving one has to keep applying force to it. It is clear that this seemed to Aristotle to be quite obviously true, and it is also clear why: it is a *universal* experience that moving objects will slow down and eventually stop if no further force is applied. Mill very plausibly claimed that for centuries no one could even conceive of the falsehood of this principle, and yet nowadays we do not find it difficult to bring up our children to believe in the principle of inertia. Another of Mill's examples is 'action at a distance', which *seems* to be required by the Newtonian theory of gravitational attraction. For example, it is claimed that the earth does not fly off from its orbit at a tangent because there is a massive object, the sun, which prevents this. But the sun is at a huge distance from the earth, and in the space between there is nothing going on which could explain how the sun's influence is transmitted. (To take a simple case, there is no piece of string that ties the two together.) The Cartesians could not believe this, and so felt forced into a wholly unrealistic theory of 'vortices'; Leibniz could not believe it, and said so very explicitly; interestingly, Newton himself could not — or anyway did not — believe it, and devoted much time and effort to searching for a comprehensible explanation of the apparent 'attraction across empty space' that his theory seemed to require.[24] But again we nowadays find it quite straightforward to explain the Newtonian theory to our children in a way which simply treats action at a distance as creating no problem at all. Of the several further examples that Mill gives I mention just one more, because it has turned out to be very apt, and in a way which Mill himself would surely find immensely surprising. He suggests that the principle of the conservation of matter (which goes way back to the very ancient dictum '*Ex nihilo nihil fit*') has by his time become so very firmly established in scientific thought that no serious scientist can any longer conceive of its falsehood. Moreover, he gives examples of philosophers of his day who did make just this claim of inconceivability. But of course we from our perspective can now say that this principle too turns out to be mistaken, for Einstein's $E = mc^2$ clearly denies it. Indeed, we from our perspective could add many more examples of how what was once taken to be inconceivable is now taken to be true; quantum theory would be a fertile source of such examples.

I am sure that when Mill was writing he did not know of the development that has conclusively proved his view of the axioms of geometry to be correct, namely

---

[24] But he never found an explanation that satisfied him, and so he remained true to the well-known position of his *Principia Mathematica*: on this question '*hypotheses non fingo*'.

the discovery of non-Euclidean geometries.[25] These deny one or more of Euclid's axioms, but it can be shown that if (as we all believe) the Euclidean geometry is consistent then so too are these non-Euclidean geometries. We say nowadays that the Euclidean geometry describes a 'flat' space, whereas the non-Euclidean alternatives describe a 'curved' space ('negatively curved' in the case of what is called 'hyperbolical' geometry, and 'positively curved' in the case of what is called 'elliptical' geometry). Moreover — and this is the crucial point that vindicates Mill's position completely — it is now universally recognised that it must count as an *empirical* question to determine which of these geometries fits actual space, i.e. the space of the universe that surrounds us. I add that as a matter of fact the current orthodoxy amongst physicists is that that space is not 'flat' but is 'positively curved', and so Euclid's axioms are not after all true of it. On the contrary, to revert to Mill's much-used example, in that geometry two straight lines *can* enclose a space, even though our attempts to picture this situation to ourselves still run into what seem to be insuperable difficulties.

For the curious, I add a *brief* indication of what a (positively) curved space is like as an appendix to this section. But for philosophical purposes this is merely an aside. What is important is that subsequent developments have shown that Mill was absolutely right about the status of the Euclidean axioms. There are alternative sets of axioms for geometry, and if we ask which of them is true then the *pure* mathematician can only shrug his shoulders and say that this is not a question for him to decide. He may say that it is not a genuine question at all, since the various axiom-systems that mathematicians like to investigate are not required to be 'true', and we cannot meaningfully think of them in that way. Or he may say (as the empiricist would prefer) that the question is a perfectly good question, but it can only be decided by an empirical investigation of the space around us, and — as a pure mathematician — that is not *his* task. In either case Mill is vindicated. The interesting questions about geometry are questions for the physicist, and not for the (pure) mathematician. Consequently they no longer figure on the agenda for the philosopher of mathematics.

Appendix: non-Euclidean geometry

Let us begin with the simple case of *two*-dimensional geometry, i.e. of the geometrical relations to be found simply on a surface. In this case it is easy to see what is going on. If the surface is a flat piece of paper, then we expect Euclid's axioms

---

[25] Mill tells us, in a final footnote to the chapter, that almost all of it was written by 1841. The first expositions of non-Euclidean geometry were due to Lobachevsky [1830] and Bolyai [1832], so in theory Mill could have known of them. But their geometry was the 'hyperbolical' one, in which it is still true that two straight lines cannot enclose a space, but another of Euclid's axioms is false, namely that there cannot be two straight lines, which intersect at just one point, and which are both parallel to the same line. Mill knows of this axiom, but does not take it as his main example, which he surely would have done if he had known that there is a consistent geometry which denies it. What he does take as a main example, namely that two straight lines cannot enclose a space, is false in 'elliptical' geometry, but that was not known at the time that Mill was writing. It is mainly due to Riemann (published 1867; proposed in lectures from 1854).

to hold for it, but if the surface is curved — for example, if it is the surface of a sphere — then they evidently do not. For in each case we understand a 'straight line' to be a line on the surface in question which is the shortest distance, as measured *over that surface*, between any two points on it. On this account, and if we think of our spherical surface as the surface of the earth, it is easy to see that the equator counts as a straight line, and so do the meridians of longitude, and so does any other 'great circle'. (You may say that such lines are not *really* straight, for between any two points on the equator there is shorter distance than the route which goes round the equator, namely a route through the sphere. But while we are considering just the geometry of a surface, we ignore any routes that are not on that surface, and on this understanding the equator does count as a straight line.)

Given this account of straightness, it is easy to see that many theses of Euclidean geometry will fail to hold on such a surface. For example, there will be no parallel straight lines on the surface (for, apart from the equator, the lines that we call the 'parallels' of latitude are not straight). Again, the sum of the angles of a triangle will always be greater than two right angles, and in fact the bigger the triangle the greater is the sum of its angles. (Think, for example, of the triangle which has as one side the Greenwich meridian of longitude, from the North pole to the equator, as another side a part of the equator itself, from longitude 0° to longitude 90°, and as its third side the meridian of longitude 90°, from the equator back to the North pole. This is an equilateral triangle, with three equal angles, but *each* of those angles is a right angle.) It is easy to think of other Euclidean theorems which will fail on such a surface. I mention just two. One is our old friend 'two straight lines cannot meet twice'; it is obvious that on this surface every two straight lines will meet twice, on opposite sides of the sphere. Another is that, unlike a flat surface, our curved surface is finite in area without having any boundary. Here is a simple consequence. Suppose that I intend to paint the whole surface black, and I begin at the North pole, painting in ever-increasing circles round that pole. Well, after a bit the circles start to decrease, and I end by painting myself into an ever-diminishing space at the South pole.

These points are entirely straightforward and easily visualised, but now we come to the difficult bit. We change from the two-dimensional geometry of a curved surface to the three-dimensional geometry of a genuine space, but also suppose that this space retains the same properties of curvature as we have just been exploring. A straight line is now the shortest distance in this *three*-dimensional space between any two points on it; i.e. it is genuinely a straight line, and does not ignore some alternative route which is shorter but not in the space: there is no such alternative route. But also the straight lines in this curved three-dimensional space retain the same properties as I have been saying apply to straight lines in a two-dimensional curved space. In particular, two straight lines can meet twice. So if you and I both start from here, and we set off (in our space ships) in different directions, and we travel in what genuinely are straight lines, still (if we go on long enough) we shall meet once more, at the 'other side' of the universe.

Again, the space is finite in volume, but also unbounded. So suppose that the volume remains constant and I have the magical property that, whenever I click my fingers, a brick appears. And suppose that I conceive the ambition of 'bricking in' the whole universe. Well, if I continue long enough, I will succeed. I begin by building a pile of bricks in my back garden. I continue to extend it in all directions, so that it grows to encompass the whole earth, the solar system, our galaxy, and so on. As I continue, each layer of bricks that I add will require more bricks than the last, until I get to the midpoint. After that the bricks needed for each layer will decrease, until finally I am bricking myself into an ever-diminishing space at the 'other end' of the universe. That is the three-dimensional analogue of what happens when you paint the surface of a sphere.

Well, imagination boggles. We say: that *could* not be what would actually happen. The situation described is just *inconceivable*. And I agree; I too find 'conception' extremely difficult, if not impossible. But there is no doubt that the mathematical theory of this space is a perfectly consistent theory, and today's physicists hold that something very like it is actually *true*.

Inconceivability is not a safe guide to impossibility.

3.2 *Mill on arithmetic*[26]

Mill's discussion of geometry was very much aided by the fact that geometry had been organised as a deductive science ever since Greek times. This allowed him to focus his attention almost entirely upon the status of its axioms. By contrast, there was no axiomatisation of arithmetic at the time when he was writing, and so he had no clear view of what propositions constituted the 'foundations' of the subject. He appears to have thought that elementary arithmetic depends just upon (a) the definitions of individual numbers, and (b) the two general principles 'the sums of equals are equal' and 'the differences of equals are equal' (Section 3). Certainly these are two basic assumptions which are made in the manipulation of simple arithmetical equations, though as we now see very well there are several others too. Mill claims that the two general principles he cites are generalisations from experience, which indeed they would be if interpreted as he proposed, i.e. as making assertions about the results of *physical* operations of addition and subtraction. To one's surprise he *also* says that the *definitions* of individual numbers are again generalisations from experience, and this is a peculiar position which (so far as I know) no one else has followed. But we may briefly explore it.

First we should notice his ontology. He opens his discussion (in Section 2) by rejecting what *he* calls 'nominalism', which he describes as 'representing the propositions of the science of numbers as merely verbal, and its processes as simple transformations of language, substitution of one expression for another'. The kind of substitution he has in mind is substituting '3' for '2 + 1', which the theory he is describing regards as 'merely a change in terminology'. He pours scorn upon such a view: 'The doctrine that we can discover facts, detect the hidden processes

---

[26] See J.S. Mill, *System of Logic* [1843, book II, chapter VI, sections 2–3].

of nature, by an artful manipulation of language, is so contrary to common sense, that a person must have made some advances in philosophy to believe it'. At first one supposes that he must be intending to attack what we would call a 'formalist' doctrine, which claims that the symbols of arithmetic (such as '1', '2', '3', and '+') have *no* meaning. But in fact this is not his objection, and what he really means to deny is just the claim that '3' and '2 + 1' have *the same meaning*.[27] We shall see shortly how, in his view, they differ in meaning.

He then goes on to proclaim himself as what *I* would call a 'nominalist': 'All numbers must be numbers of something; there are no such things as numbers in the abstract. *Ten* must mean ten bodies, or ten sounds, or ten beatings of the pulse. But though numbers must be numbers of something, they may be numbers of anything.' From this he fairly infers that even propositions about particular numbers are really generalisations; for example '2 + 1 = 3' would say (in Mill's own shorthand) 'Any two and any one make a three'. But it does not yet follow that these generalisations are known empirically, and the way that Mill tries to secure this further claim is, in effect, by offering an interpretation of the sign '+'.[28]

He says: 'We may call "three is two and one" a definition of three; but the calculations which depend on that proposition do not follow from the definition itself, but from an arithmetical theorem presupposed in it, namely that collections of objects exist, which while they impress the senses thus, $°_°°$, may be separated into two parts thus, $°$ $°$ $°$' (Section 2). I need only quote Frege's devastating response: 'What a mercy, then, that not everything in the world is nailed down; for if it were we should not be able to bring off this separation, and $2 + 1$ would not be 3!' And he goes on to add that, on Mill's account 'it is really incorrect to speak of three strokes when the clock strikes three, or to call sweet, sour and bitter three sensations of taste, and equally unwarrantable is the expression "three methods of solving an equation". For none of these is a [collection] which ever impresses the senses thus, $°_°°$.' It is quite clear that Mill's interpretation of '+' cannot be defended.[29]

One might try other ways of interpreting '+', so that it stood for an operation to be performed on countable things of any kind, which did not involve how they appear, or what happens when you move them around, or anything similar, but would still leave it open to us to say that arithmetical additions are established by experience. For example, one might suppose that '7 + 5 = 12' means something like: 'If you count a collection and make the total 7, and count another (disjoint) collection and make the total 5, then if you count the two collections together you will make the total 12'. That certainly makes it an empirical proposition, but of course one which is false, for one must add the condition that the counting is correctly done. But this then raises the question of whether the notion of *correct*

---

[27] Kant claimed that '7 + 5 = 12' was not analytic; you might say that Mill here makes the same claim of '2 + 1 = 3'.

[28] The position outlined in this paragraph is very similar to the position I attribute to Aristotle, except that Aristotle would have begun 'there *are* such things as numbers in the abstract, but they exist only in what they are numbers of'.

[29] G. Frege, *The Foundations of Arithmetic* [1884, trans. J.L. Austin 1959, 9–10].

counting can be explained in empirical terms, and the answer to this is not obvious.

Besides, there is another of Frege's objections to empiricism which becomes relevant here: what of the addition 7,000 + 5,000 = 12,000? Surely we do not believe that this is true because we have actually done the counting many times and found that it always leads to this result. So how could the empiricist explain this knowledge? Well, it is obvious that the answers to sums involving large numbers are obtained not by the experiment of counting but by calculating. If one thinks how the calculation is done in the present (very simple) case, one might say that it goes like this:

$$\begin{aligned} 7{,}000 + 5{,}000 &= (7 \times 1{,}000) + (5 \times 1{,}000) \\ &= (7+5) \times 1{,}000 \\ &= 12 \times 1{,}000 \\ &= 12{,}000 \end{aligned}$$

Here the first step and the last may reasonably be taken as simply a matter of definition (i.e. the definition of Arabic numerals); the third step depends upon the proposition $7 + 5 = 12$, which we take as already established, together with the principle that equals multiplied by equals yield equals; the second step is perhaps the most interesting, for it depends on the principle of distribution, i.e.

$$(x \times z) + (y \times z) = (x+y) \times z.$$

But how do we come to know that *that* general principle is true? And of course the same question applies to hosts of other general principles too, and not only to the two that Mill himself mentions (i.e. 'the sums of equals are equal' and 'the differences of equals are equal'). If the knowledge is to be empirical in the kind of way that Mill supposes, it seems that we can only say that we can run experimental checks on such principles where the numbers concerned are small, and then there is an inductive leap from small numbers to *all* numbers, no matter how large. But if that really is our procedure, then would you not expect us to be rather more tentative then we actually are on whether these principles really do apply to very large numbers?

Let us sum up. Frege's criticisms of Mill may be grouped under two main headings. (i) Arithmetical operations (such as addition) cannot simply be identified with physical operations performed on physical objects, even though they may share the same name (e.g. 'addition'). One reason is that arithmetical propositions are not falsified by the discovery that the associated physical operations do not always yield the predicted result (e.g. if 'adding' 7 pints of liquid to 5 pints of liquid yields, not 12 pints of liquid, but (say) an explosion).[30] Another reason

---

[30] I note incidentally that Hilbert's position is open to this objection. He believes that what he calls 'finitary arithmetic' does have real content, and to explain what that content is he takes it to be about operations on *numerals*. For example '2 + 1 = 3' says that if you write the numeral '||' and then after it the numeral '|', the result will be the numeral '|||'. But surely arithmetic would not be proved false if it so happened that, whenever you wrote one stroke numeral after another, the first one always altered (e.g. one of its strokes disappeared).

is that the arithmetical propositions may equally well be applied to other kinds of objects altogether, where there is no question of a physical addition. As Frege saw it, the mistake involved here is that of confusing the arithmetical proposition itself with what should be regarded as its practical applications. It will (usually) be an empirical question whether a proposed application of arithmetic does work or not, but arithmetic itself does not depend upon this. (ii) We are quite confident that the general laws of arithmetic apply just as much to large numbers as to small ones, but it is not easy to see how the empiricist can explain this. For on his account we believe them only because they have very frequently been verified in our experience, and yet the verification he has in mind would seem to be available only when the numbers concerned are manageably small. To these objections made by Frege, I add a third which (curiously) he does not make, but which we have seen bothered Aristotle: (iii) How can an empiricist account for our belief that there are infinitely many numbers? For, on the kind of account offered by both Aristotle and Mill, this belief seems in fact to be *false* (as is acknowledged by Aristotle, but overlooked by Mill).

In the next section I consider two attempts by post-Frege authors to re-establish empiricism in arithmetic, while yet bearing in mind these extremely powerful objections. These authors are of course familiar with modern axiomatisations of arithmetic, and so have a much better idea of just what the empiricist has to be able to explain. It may be useful if I here set out the usual axioms, which have become known as 'Peano's postulates'.[31]

1. 0 is a number.

2. Every number has one and only one successor, which is a number.

3. No two numbers have the same successor.

4. 0 is not the successor of any number.

5. (Postulate of mathematical induction:) Whatever is true of 0, and is true of the successor of any number when it is true of that number, is true of all numbers.

In the context of a second-order logic, these five axioms are sufficient by themselves, for in this context we can give explicit definitions of addition, multiplication, and so on, in terms of the vocabulary used here (i.e. '0' and 'successor'). But if the background is only a first-order logic then we shall need further axioms to introduce these notions. Using x' to abbreviate 'the successor of x', the standard axioms are these

$$x + 0 = x$$
$$x + y' = (x + y)'$$

---

[31] The axioms are in fact due to Dedekind [1888]. Peano [1901] borrowed them from him, with acknowledgement.

$$x \times 0 = 0$$
$$x \times y' = (x \times y) + x$$

The modern task for the empiricist is to show that there are empirical grounds for these axioms, and that there are not *a priori* grounds.

My next section concerns two philosophers who attempt to answer this challenge in roughly Mill's way; the following section moves to a completely different argument for empiricism.

4 MILL'S MODERN SUPPORTERS

Despite the shortcomings of Mill's approach to arithmetic, his main ideas are not without support from today's philosophers. In this section I choose two of them to discuss in some detail, namely Philip Kitcher and Penelope Maddy. The latter might not like to be described as one who 'supports Mill'; at any rate, in the work of hers that I shall mainly consider,[32] she refers to Mill's views on arithmetic only once, and that is in order to reject them (pp. 67–8, with notes). But the position that she puts forward seems to me to have at least some affinity with Mill's. On the other hand Philip Kitcher is very explicitly pro-Mill, and he himself calls his preferred theory 'arithmetic for the Millian'. So I shall take him first.

Kitcher offers two main lines of argument, which are not closely connected with one another. One is offered in support of the negative claim that arithmetical knowledge *could not* be *a priori*; the other is a positive account of how that knowledge is (or could be) empirically acquired. I shall treat these separately.

4.1 Kitcher against apriorism [33]

Two very traditional views on *a priori* knowledge are (i) that it is acquired independently of experience, and (ii) that it cannot (even in principle) be refuted by experience. Tradition apparently takes these two points to be connected, as if the second follows automatically from the first, and Kitcher does aim to explain the first in such a way that it will imply the second. I shall end by suggesting that this is a mistake, but let us first see what Kitcher's account is, and what conclusions he draws from it.

We must begin with a brief remark on the nature of knowledge in general. Kitcher takes it that x knows that P if and only if (i) it is true that P, (ii) x believes that P, and (iii) x's belief that P is 'warranted'. For the sake of argument, I very readily accept this description, taking the notion of a 'warrant' as just a way of referring to whatever it is that distinguishes knowledge from mere true belief. Kitcher hopes that he can avoid the old and much-disputed question of just what a warrant is in the general case, for he is going to give a proper definition

---

[32] P. Maddy, *Realism in Mathematics* [1990].
[33] See P. Kitcher, *The Nature of Mathematical Knowledge* [1984]. I refer throughout to the discussion in this book.

of what he calls an *a priori* warrant. This is clearly a reasonable procedure. But he does argue at the start for one general point about 'warrants', namely that a warrant attaches to 'the process by which x's belief was produced', and this is not exactly uncontroversial. We need not stickle on the word 'process'; others have preferred to speak of 'methods' by which beliefs are produced, and Kitcher would surely not object to this. Equally, we need pay no attention to the apparent suggestion that what matters is always how the belief was *first* acquired, for (as Kitcher notes, p. 17) the explanation of how I first came to believe something and the explanation of why I continue to believe it now may be very different, and it is the latter that matters. So I shall sometimes alter Kitcher's own terminology, and speak of 'methods by which a belief was acquired or is sustained', but this is not intended to indicate any significant disagreement with what Kitcher says himself. The substantive assumption is that what 'warrants' a (true) belief as knowledge is some feature of this 'process', or 'method', or whatever it should be called. I am happy to accept this assumption, and thereby to accept Kitcher's outline of what, in general, counts as knowledge (pp. 13–21). Let us press on to his account of what could make some knowledge *a priori*.

His main thought is that *a priori* knowledge is something that is knowledge, *and* would still continue to be knowledge whatever future experience turned out to be. In more detail, he accepts (as I do) the Humean point that some kinds of experience may be necessary in order to provide the 'ideas' which any belief requires. So the definition of *a priori* knowledge should explicitly allow for this, and set aside as irrelevant whatever experience was needed simply to entertain the relevant thought. Now suppose that someone, x, has the true belief that P, and we are asking whether it counts as *a priori* knowledge. Kitcher's definition is that it will do if and only if, given *any* course of experience for x, which was sufficiently rich to allow x to form the belief that P, if x had formed the belief by the same method as that by which his present belief is actually sustained, then that belief would have been both true and warranted (pp. 21–32).[34] Kitcher devotes some space to the question of what counts as 'the same method' ('the same process'), but for our purposes we can bypass this question. It will be enough to consider possible worlds in which x's experiences, thoughts, and beliefs are entirely as in this world *up to* the time at which we ask whether some true belief of his is known *a priori*; then we consider all possible variations in his *subsequent* experience, and ask whether the belief would still count as knowledge whatever his future experiences were.

Kitcher answers that future experience could upset *any* claim to mathematical knowledge, and so — according to the definition given — no such knowledge could count as *a priori*. His argument relies upon the thought that I could always experience what he calls a 'social challenge' to any of my mathematical beliefs. This is when *other people* tell me that I must be wrong. The challenge is especially powerful when these other people appear to be much better qualified than I am;

---

[34] I do not quote Kitcher's definition verbatim, because it introduces some special terminology which would need explanation. But I am confident that I do not misrepresent what he does say.

for example they are universally acknowledged as the experts in this field, and I myself recognise (what seem to me to be) their great achievements. Moreover they give reasons (of a general kind) which are at least convincing enough (to me) to make me acknowledge that my belief may be the result of a mistake on my part. Here is an example. My belief may rest upon what I take to be a proof, but the others may remind me that — as we all know — it is possible to make a mistake in a proof (especially a 'long' proof). Moreover, the others may show me what seems to be a countervailing proof, i.e. a proof that my result cannot be correct because it leads to a contradiction. We may suppose that I myself can see nothing wrong with this opposing proof (perhaps because I have been hypnotised not to notice what is actually an illicit step in the reasoning). So my situation is that I seem to have a proof that P (my own proof), but also a proof that not-P (given me by the others), and I cannot see anything wrong with either. In these circumstances the rational course for me to take must be to suspend judgment, and if I do continue to believe that P then that belief (though by hypothesis it is true) is no longer warranted. That is, there is a possible course of experience in which my belief would not count as knowledge, and so, even in the actual case (where I experience no such 'social challenge'), the belief does not count as *a priori* knowledge.

Kitcher wishes to apply a similar line of thought to my belief in the basic mathematical assumptions from which my proofs start. (He is — quite deliberately — non-committal on what exactly these basic assumptions are.) I shall not pursue the details of his discussion here, for indeed I think that the strategy that he himself follows is not the most convincing, and the argument could be very much strengthened.[35] But I simply summarise his general position. Suppose that I have some mathematical belief, say that 149 is a prime number. Kitcher will concede that this belief is true, and (in the present situation) warranted, and so counts as knowledge. He will also concede, at least for the sake of argument, that what I believe is necessarily true, and hence true in all possible worlds. But, he argues, it would not count as *knowledge* in all worlds, because we can envisage worlds in which I form the belief in just the same way as I do in this world, but then experience a powerful social challenge to it. This, he claims, would mean that in those worlds it would no longer count as a 'warranted' belief. So, by his definition of '*a priori* knowledge', I do not know it *a priori* even in this world. Clearly, the same line of thought could be applied to *any* of my beliefs, save for those very few about which Cartesian doubt is genuinely impossible.

I do not believe that any apriorist would be convinced by Kitcher's line of argument. He would certainly wish to maintain that the mere possibility of Kitcher's 'social challenges' is just irrelevant. But it will be useful to ask just how the argument is to be resisted. I think there are two main ways. One is to accept the proposed definition of *a priori* knowledge, but to take a special view about what

---

[35] Most people have no views on how they came to believe the basic assumptions, so one cannot get started on showing that the method they actually used may lead to mistakes. But Kitcher could rely on a course of experience that first *gives* them a view on this, and then later goes on to undermine it.

'warrants' a belief as knowledge which would prevent Kitcher's conclusion from following. The other is to reject the proposed definition. The second course seems to me to be the right one, but I begin with a brief explanation of the first.

Kitcher's line of argument depends upon the assumption that whether a belief of mine does or does not count as 'warranted' is affected by its relation to my other beliefs. For example, my recent calculation that 149 is a prime number gave me a true belief, and one that was acquired by a reliable method (i.e. simple calculation), and, on one (strongly 'externalist') account of what knowledge is, that is by itself enough to ensure that it is suitably 'warranted'. But Kitcher would not agree, for he thinks that if I had done exactly the same calculation in other circumstances, namely in circumstances in which I *also* had good reason to believe that 149 could not be a prime number, then the calculation would no longer warrant the belief. (This is an 'internalist' aspect to his thinking.) So one could resist Kitcher's conclusion by adopting the (strongly 'externalist') view that it simply does not matter what *other* beliefs I may have, for the question is just whether *this* belief was reached in a reliable way. But I myself would think that such 'strong externalism' is too strong,[36] and I would prefer a different line of objection.

Kitcher supposes that the traditional idea that *a priori* knowledge should be 'independent of experience' should be interpreted as meaning that such knowledge would still have been knowledge *however* experience had turned out to be. But this is surely not what we ordinarily mean by 'independence'. To take a simple example, my recent calculation that 149 is a prime number was independent of whatever might have been going on at the time on the other side of the road. That is to say, whatever it was that was actually occurring there did not have any effect upon my thought-processes at the time. (I was paying no attention to it; an ordinary causal account of why I thought as I did would have no reason to mention it.) Of course, this is not to say that what was happening there *could not* have influenced my calculation. No doubt, if a large bomb had exploded there, shattering my windows, then my calculation would certainly have been distracted, and probably never completed. Nevertheless we would normally say that, as things in fact were, there was no influence from one to the other, and we would feel that this justified the claim that each was 'independent' of the other. As a concession to Kitcher's way of thinking, we would accept that the *absence* of certain *possible* occurrences across the road was a necessary condition of my thinking proceeding as it did, but still we would not normally infer that my thought *depended* upon what did actually happen there.

---

[36] For an example of such 'strong externalism' see e.g. [Nozick, 1981, ch. 3]. On his account, whether a belief is warranted depends just on whether it was formed by a method that 'tracks the truth', for which we are given a counterfactual test which makes no mention of any other beliefs. Indeed Nozick explicitly claims that whether the belief is entailed by other beliefs of mine is simply irrelevant. I assume that he would say the same on whether the negation of the belief is entailed by other beliefs. (For a general account of the opposition between 'externalism' and 'internalism' see e.g. [Bonjour, 1980] and [Goldman, 1980]. I give no general account here, since for my present purposes it is not needed.)

To apply this point to *a priori* knowledge, and its claimed 'independence of experience', this should mean that, as things actually were, my experience played no role in the process that led me to acquire the belief (or, better, in whatever explains why I hold the belief now). As already admitted, this discounts any experience that was needed simply to provide an *understanding* of the relevant proposition. As we should now add, it also discounts the fact that an *absence* of distracting experiences was no doubt a necessary condition of my ever reaching the belief, for experience *could* have interfered with this in many ways. (To take a trivial example, the onset of a blinding headache whenever I tried to think of products of numbers greater than 10 would presumably have prevented me from ever calculating that 149 is prime.) We may put this by saying that the process which led to the belief could still have occurred in the absence of *all* experience (excepting — as always — whatever experience was needed to allow me even to have the thought in question). Provided that that condition is satisfied, then I would say that the belief counts as formed 'independently of experience'. I would add that if in addition the belief is true, and if it was reached by a method which counts as providing a suitable 'warrant' for it, then it will also count as knowledge that is 'independent of experience'. It therefore satisfies the traditional idea of what *a priori* knowledge is.

Kitcher could accept all of this except the last sentence, but that he must deny, for the account just given includes nothing that corresponds to the condition which he insists upon. He thinks that we should say, not only that the belief was formed by a method that is independent of experience, but also — if I may paraphrase somewhat loosely — that the belief's *warrant* should be independent of experience. (Hence no possible future experience could upset that warrant, and this is where his 'social challenges' become relevant.) But why should one feel the need for any such extra condition? I think the answer is that, without it, we do not capture another thought which really is part of the tradition, namely that a proposition which is known *a priori* is immune to refutation by experience. But what this requires is that the method of forming (or sustaining) the belief by itself guarantees the *truth* of that belief, and not that it guarantees its *warrantedness*. That is, the tradition does not require that future experience could not be such as to render the belief insufficiently warranted, but rather that future experience could not be such as to make the belief untrue. As Kitcher very explicitly concedes, his 'social challenges' do not show that the belief in question is not true, but they do create a situation in which it is not warranted. But it seems to me that the tradition is right to ignore such challenges, so the extra condition that is required should be concerned with truth and not with warrantedness.

One may wonder whether we do really need any such extra condition. After all, we have already said that the belief (formed independently of experience) must also be true. From this it follows that future experience will not in fact refute it. We have also added that the belief should be warranted, and — however the notion of a 'warrant' is understood — this surely implies that it is no accident that the belief is true. Why is this not enough?[37] Well, here again we cannot altogether avoid the question of what is to count as a 'warrant'. There are philosophers ('internalists') who think that a belief counts as warranted only if the believer himself can say what the warrant is, and why it counts as a warrant. On this approach, there surely is a further condition required, but it is easy to say what it is: the believer should *know* that and how his belief is warranted, and this knowledge in turn should also be 'independent of experience' in the way already explained. (It will be obvious that a regress, which appears to be vicious, is here threatened: to know that P one must also know that and how the belief that P is warranted; to know this in turn, one must also know that and how the belief that P is warranted is itself warranted; and so on.) But the opposite view ('externalism') is nowadays more popular, at least in the case of what appear to be the basic and 'foundational' beliefs, which are not themselves inferred from other beliefs. In this case the view is that in order to count as knowledge such beliefs must *be* (both true and) warranted, but it is not also required that the believer himself knows how they are warranted. He may have views on this question which are wholly mistaken, or — more probably — he may have no views at all. This, it seems, is the position that we must adopt if the basic truths of mathematics (and logic) are to be known at all, for the truth is that we simply do not know why we hold these beliefs. We do normally assume that the beliefs are warranted, but we cannot say how. So in this case too I think that, if such beliefs are to count as known *a priori*, an extra condition is needed: they must be true, and reached by a procedure which warrants them, *and* which does ensure their truth, no matter how experience turns out to be. That is, its efficacy as a warrant does not depend upon any contingent feature of this world, which we could become aware of only as a result of our experience. An example may help to clarify the point.

A method of forming beliefs which is presumably *a priori*, if any method is, is to see what one can imagine. Suppose that someone applies this method in a case, and in a manner, which most of us would say was inappropriate. He considers the proposition 'no new colour will ever be experienced', finds that he cannot imagine experiencing a new colour, and so concludes that the proposition is true. (Here, let us mean by 'a new colour' not something like Hume's missing shade of blue, which slots readily into the range of colours already perceived, but something that lies right outside that range.[38] And let us suppose that what is intended is perception *by human beings*, so that the possible experiences of bees or birds or Martians are simply not relevant.) I say that the method is not appropriately applied here, because we should distinguish between imagining a new colour-experience and imagining *that there should be* a new colour-experience. We cannot do the former, but it does not follow that we cannot do the latter. However, the example is of

---

[37] Several philosophers have argued, in response to Kitcher, that it is enough. E.g. [Edidin, 1984], [Parsons, 1986], [Hale, 1987, 129–37], [Summerfield, 1991], [BonJour, 1998, ch. 4], [Manfredi, 2000], [Casullo, 2003, ch. 2].

[38] Hence a fairly wide experience of colours will be necessary simply to provide understanding of 'a new colour'.

someone who does apply the method in this apparently inappropriate way.

We may easily suppose that the proposition in question is indeed true. We may also suppose that our subject's belief in it is, in a way, warranted, e.g. in this way. There is something about the nerve-cells responsible for human vision which does in fact limit their possible responses to visual stimuli. So for example if human beings were to evolve in such a way that their eyes became sensitive to infra-red light the effect would be that the existing range of perceived colours was preserved, though the external causes which give rise to it were shifted. That is, infra-red light would give us the experience that we now call 'seeing red', and consequently red light would give us the experience that we now call 'seeing orange', and so on throughout the spectrum. The suggestion is that there is a physical law which does confine the range of colours which humans can perceive to the range that is perceived now. So as a matter of fact the method of seeing what colours one can imagine is actually a very reliable guide to what colours could be perceived. Would the belief then be warranted? I presume that our subject knows nothing of the physical law here posited, for — if he did — that knowledge presumably could not be *a priori*. So far as *he* is concerned it is just his own powers of imagination that he is relying on. Consequently, from an 'internalist' point of view the belief is not warranted, since the subject cannot cite anything which warrants it. (He can say 'I cannot imagine a new colour', but has nothing at all to say when asked why that should be a good reason for supposing that there could not be any.) But I have already noted that an 'internalist' approach must generate a regress problem, so, let us now look at the question from a more 'externalist' perspective. If anything at all is to be known *a priori* then apparently there must be some things which are known *a priori* though the knower cannot himself cite any warrant for them. So let us come back to the example: is this one of them? By hypothesis our subject has a true belief, and I am presuming (for the sake of argument) that it is reached by a method which is in the relevant way 'independent of experience'. Moreover the method is, in this particular case, a reliable one, for there are physical laws which ensure its success. These laws are not known, or even suspected, by the subject, but from the externalist perspective that does not matter; they may all the same provide a 'warrant' for the belief. So it is knowledge, and reached by an *a priori* process. But should we therefore accept that it is *a priori* knowledge?

The intuitive answer, surely, is 'no'. For though the belief is formed by an *a priori* process, still that process does not *itself*, and of its *own* nature, guarantee any immunity from refutation by experience. What 'guarantees immunity' is only the physical laws that happen to hold in our world, and they could have been different. The same response would be appropriate to any other case of an 'external mechanism' which ensured the truth of a belief. (For example, if God were so friendly to me that whenever I dreamt that something would happen He ensured, for that reason, that it did happen.) What is needed, apparently, is the thought that the method of reaching the belief should *by itself* ensure the truth of that belief, without the aid of any external factors that could have been different. And this is what Kitcher's condition on '*a priori* warrants' was aiming for, though he wrongly puts it as the condition that the process should inevitably lead to knowledge, whereas I think he should have said just that it inevitably leads to truth.

As a result of this discussion, I suggest that *a priori* knowledge be defined thus: x knows *a priori* that P if and only if

(i) it is true that P;

(ii) x believes that P;

(iii) x's belief that P is acquired (or sustained) by a procedure which warrants it;

and

(iv) this procedure does not depend upon experience, in the sense that it could have occurred in the absence of all experience other than whatever was needed simply to allow x to have the thought that P;

(v) this procedure by itself guarantees that (if it is properly carried out)[39] the belief that it results in has to be true, whatever further experiences may be. (And, on this occasion, the procedure was correctly carried out.)

(Of course, condition (v) makes condition (i) superfluous, and presumably condition (iii) as well.) It seems to me that this definition represents the traditional conception better than Kitcher's does, and — if it is accepted — then the 'social challenges' that Kitcher's argument relies upon fall away as irrelevant. So I conclude that Kitcher has not shown that our ordinary mathematical knowledge *could not* be *a priori*. But the discussion also makes it clear how difficult the apriorist's position is, for what procedure could there be which would satisfy the conditions (iv) and (v) stated here? In my final section I shall try to argue in a different way that there are none. But meanwhile I come back to the other question: *could* it be that our present knowledge of mathematics — even such a simple area as our knowledge of elementary arithmetic — is generated by experience? Kitcher and Maddy both say 'yes', but their answer is open to serious objections.

4.2 Kitcher on arithmetic

Kitcher's general position on our knowledge of mathematics is that it has gradually evolved over the centuries, and that in practice the evolution works in this way. The knowledge that one generation has rests largely on the testimony of their teachers; they will of course try to extend that knowledge by their own efforts, but

---

[39] One might very naturally wish to maintain that (e.g.) the ordinary method of calculating whether a number is prime is bound to give the correct result *provided that* it is correctly carried out. But we all know that in practice slips are possible. The wording is intended to allow for that point.

still the extensions will be based upon what they were first taught; and knowledge based on testimony is, of course, empirical knowledge. The later stages of this evolution are quite well documented, and open to historical investigation, but there is no historical record of how it all began – i.e. of how men first learnt to count, to add, to multiply, and so on. So Kitcher offers a reconstruction of how elementary arithmetic might have started, taking as his model a way in which even nowadays small children may (at least in principle) gain arithmetical knowledge without relying on instruction from their elders. This, he supposes, is by noting what happens when they *manipulate* the world around them. ('To coin a Millian phrase, arithmetic is about "permanent possibilities of manipulation",' p. 108.)

Kitcher therefore presents his account as a theory of operations, thinking of these — at least in the first phase — as physical operations performed on physical objects, such as selecting certain objects by physically moving them and grouping them together in a place apart from the rest. These may be distinguished from one another as being 'one-operations', 'two-operations', 'three-operations', and so on, according to the number of objects selected by each. Another operation that is central to his account is a 'matching' operation, whereby one group of objects is matched with another, thereby showing that they have the same number. (This might be done, for example, by placing each fork to the left of one knife, and observing that as a result each knife was to the right of one fork.) His formal theory, however, takes 'matching' to be a relation, not between groups of objects, but between the selection-operations that generated those groups. He also makes use of a successor-relation: one selection-operation is said to 'succeed' another if it selects just one more object than the other does. And he adds too an addition-relation defined in this way: one selection-operation is the addition of two others when it selects just the objects that those two together selected, and those original two were disjoint (i.e. there was no object selected by both of them). He then presents us with a theory of such operations in this way. Let us abbreviate

Ux for x is a one-operation
Sxy for x is an operation that succeeds y
$Axyz$ for x is an addition on the operations y and z
Mxy for x and y are matchable operations.

Then the axioms of the theory are (pp. 113–4):

1. $\forall x (Mxx)$

2. $\forall xy \ (Mxy \rightarrow Myx)$

3. $\forall xyz \ (Mxy \rightarrow (Myz \rightarrow Mxz))$

4. $\forall xy \ ((Ux \ \& \ Mxy) \rightarrow Uy)$

5. $\forall xy \ ((Ux \ \& \ Uy) \rightarrow Mxy)$

6. $\forall xyzw \ ((Sxy \ \& \ Szw \ \& \ Myw) \rightarrow Mxz)$

7. $\forall xyz\, ((Sxy\ \&\ Mxz) \to \exists w(Myw\ \&\ Szw))$

8. $\forall xyzw\, ((Sxy\ \&\ Szw\ \&\ Mxz) \to Myw)$

9. $\forall xy\, \sim (Ux\ \&\ Sxy)$

10. $(\forall x\, (Ux \to \Phi x)\ \&\ \forall xy\, ((\Phi y\ \&\ Sxy) \to \Phi x)) \to \forall x\, (\Phi x)$, for all open sentences 'Φx' of the language

11. $\forall xyzw\, ((Axyz\ \&\ Uz\ \&\ Swy) \to Mxw)$

12. $\forall xyzwuv\, ((Axyz\ \&\ Szu\ \&\ Svw\ \&\ Awyu) \to Mxv)$.

Of these, axioms (1)–(7) state fairly obvious properties of the basic notions, axioms (8)–(10) state analogues to three of Peano's postulates (namely 'no two numbers have the same successor', '1 is not the successor of any number', and the principle of mathematical induction[40]), and (11)–(12) introduce analogues to the usual recursive equations for addition.[41] Let us pause here to take stock.

Kitcher has not forgotten Frege's crushing objection to Mill: 'what a mercy, then, that not everything in the world is nailed down'. He does think that arithmetical knowledge would have begun from people actually moving things around, and noting the results. But he is prepared to generalise from this starting point: 'One way of collecting all the red objects on the table is to segregate them from the rest of the objects, and to assign them a special place. We learn how to collect by engaging in this type of activity. However, our collecting does not stop there. Later we can collect the objects in thought without moving them about. We become accustomed to collecting objects by running through a list of their names, or by producing predicates which apply to them ... Thus our collecting becomes highly abstract' (pp. 110–111). This notion of an 'abstract collection' presumably meets Frege's objection that things do not have to be moved about for numbers to apply to them, and apparently it would also meet his objection that numbers apply also to all kinds of immovable things (e.g. sounds, tastes, questions). At any rate, Kitcher goes on to add that we can also learn to collect collectings themselves; in his view the notation '$\{\{a,b\},\{c,d\}\}$' should be viewed as representing three collectings, first the collecting of a and b, then the collecting of c and d, and finally the collecting of those two collectings (p. 111).[42]

But once the theory is generalised in this way, as it surely must be if it is claimed to be what *arithmetic* is really about, we must face anew the question 'how do

---

[40] Note incidentally that this axiom confines the domain to 'integral' selection-operations, excluding infinite selections, fractional selections, and so on.

[41] Kitcher notes that we could give a further explanation of multiplication in similar terms, and add a suitable pair of axioms for it. No doubt we could. But perhaps the most natural way of doing so would be by invoking a selection-operation *on selection-operations*. Thus an $n \cdot m$-selection is one that selects all the objects resulting from an n-selection of disjoint m-selections. But at this stage Kitcher does not appear to be contemplating operations which operate on other operations.

[42] Note that the notation that we ordinarily think of as referring to a set is taken by Kitcher to refer to an operation, i.e. an operation of collecting.

we know that it is true?' In the first phase, when the theory was understood as concerned with physical operations, our knowledge of it could only be empirical. For it seems obvious that only experience can tell us what happens to things when you move them about. But in the second phase no such physical activity is involved, but at most the mental activities of selecting, collecting, matching, and so on. To be sure, we do think that we know what the results of these activities will be, if they are correctly performed. But can this knowledge be understood as obtained simply by generalising from cases where the relevant mental activities have been performed? We must once again face Frege's question: what about large numbers? And if Kitcher should reply that his axiom (10), of mathematical induction, is what allows us to obtain results for *all* numbers, no matter how large, then we naturally ask: and how is *that* axiom known to be true?

The truth is that Kitcher never faces this question. He certainly begins by assuming that his axioms (1)–(12) are known empirically, and he seems to pay no attention to the possibility that, when we switch from physical collectings to 'abstract' collectings, the original empirical basis no longer applies. But he does think that there *is* a further development which is needed, and which leaves empiricism behind, and he describes this as introducing an 'idealisation'. (This further development is required by the addition of further axioms, which I come to shortly.) Using the expression 'an M-world' to describe a world which is in the relevant way 'ideal', he says: 'The usual theorems of arithmetic can be reinterpreted as sentences which are implicitly relativised to the notion of an M-world. The analogs of statements of ordinary arithmetic will be sentences describing the properties of operations in M-worlds. ("$2 + 2 = 4$" will be translated as "In any M-world, if x is a 2-operation and y is a 2-operation and z is an addition on x and y, then z is a 4-operation".) *These sentences will be logical consequences of the definitions of the terms they contain*' (p. 121, my emphasis). This apparently admits that, when the sentences are so interpreted, our knowledge of their truth is no longer empirical knowledge. But then one is inclined to respond: in the case of the example given, how is the status of our knowledge affected by whether the intended world is in some way ideal? If we know, simply by logic plus definitions, that in any M-world the addition of a 2-operation and a 2-operation is a 4-operation, then it would seem at first sight that by just the same means we also know that the same holds in any world whatever, including our own non-ideal world. But in fact this is not a fair criticism.

What Kitcher means is that it is simply *stipulated* that in a (relevantly) ideal world his axioms (1)–(12) are to be true. (So are some further axioms, which I shall come to shortly.) So when his paraphrase of '$2 + 2 = 4$' is relativised to an ideal world, it is 'true by definition' because of the definition *of an ideal world*. But that is just to say that it is a logical consequence of the axioms stated, plus — no doubt — perfectly straightforward definitions of '2-operation' and '4-operation'. But it has no implications on how we know the truth of the axioms. Yet Kitcher undoubtedly *does* think that we do know the truth of his axioms (1)–(12). It is because of that that he thinks that the addition of extra axioms (to come shortly) is a legitimate idealisation.[43] His assumption seems always to be that they are known simply as generalisations from experience, but this can surely be questioned. For example, consider axiom (1), which says that every selection-operation can be matched with itself. Do we really need experience, rather than just definitions of the terms involved, to assure us of that? Continue in this way through the other axioms. It seems to me that a likely first thought is that *all* of them follow simply from the definitions of the terms involved, until one comes to axiom (10), the principle of mathematical induction. If that is a consequence of any definition, it can only be a definition of what is to count as a selection-operation, and no such definition has actually been offered. But what is the alternative? *Surely* this principle *cannot* be regarded as a 'generalisation' of what experience will tell us about the small-scale selection-operations that we do actually perform? It seems a very obvious question. Kitcher pays it no attention whatever. I suspect that he would have done better to say that this axiom does not belong in his first group of axioms, but should be regarded as one of his second group, which introduce the 'idealisation'. So let us now turn to this second group of axioms.

Kitcher recognises the need to shift attention from our own world to an 'ideal' world — or, what comes to the same thing, from our own selection-operations to those of an 'ideal agent' — because the axioms (1)–(12) considered so far are not strong enough to allow us to deduce suitable analogues to Peano's postulates. This is because they do not yet include any existential claims. So, if we aim to obtain ordinary arithmetic, we must add something more, and Kitcher suggests this (p. 114):

13. $\exists x\, (Ux)$

14. $\forall x \exists y\, (Syx)$

15. $\forall xy\, \exists z\, (Azxy)$.

No doubt the proposed axiom (13) is entirely straightforward, and for present purposes we may simply set (15) aside.[44] For the obvious problem is with (14), which is needed to establish that every number has a successor, and which says that, for any selection-operation that has been (or will be?) performed, a selection-operation that succeeds it also has been (or will be?) performed. But we know that this is false of our world, and a sceptic might very naturally suggest that it is false of all other worlds too, for there is *no* possible world in which *infinitely* many selection-operations have been performed. Kitcher disagrees. He supposes

---

[43] Kitcher compares his 'idealisation' to the theory of an 'ideal' gas, but there is no real similarity between the two. The theory of an 'ideal' gas does not result from adding new axioms to a set of existing axioms which do accurately describe the behaviour of real gases.

[44] As is noted by Chihara [1990, p. 238–9], axiom (15) is not correctly formulated, given Kitcher's own informal explanations. It requires the condition that x and y are disjoint operations (i.e. no object is selected by both of them), and Kitcher has given us no way of even expressing this condition. But one expects that, if the axiom is formulated correctly, then its truth should follow by mathematical induction (axiom 10) from the recursive equations for addition (axioms 11–12) and the existential assumptions already given by axioms 13–14.

that, in a suitable 'ideal' world, 'the operation activity of an ideal subject' is not so restricted (p. 111). But I am sure that he has here taken a wrong turn, and in fact his own previous remarks explain why.

He has said: 'The slogan that arithmetic is true in virtue of human operations should not be treated as an account to rival the thesis that arithmetic is true in virtue of the structural features of reality ... [for] taking arithmetic to be about operations is simply a way of developing the general idea that arithmetic describes the structure of reality' (p. 109). I think that the moral of the previous paragraph is clear: operations may be limited in a way that 'reality' is not. So, rather than introduce an 'idealising' theory of operations, one should rather drop 'operations' altogether, and speak more directly of 'the structure of reality'. For example, if there are 7 cows in one field, and 5 in another, then there simply *are* 12 cows in the two fields together. For this to be so, it is not required that anyone has physically moved the two groups of cows, so as to amalgamate them both in the same field. Equally, it is not required that anyone has mentally selected first the 7 cows and then the 5 cows, and then has carried out a mental addition-operation on these two selections. There would still be 12 cows, whether or not *any* such operations had been performed, or would be performed by some posited 'ideal subject'. Operations of this kind are simply irrelevant to the truth of arithmetical propositions, and with this thought Kitcher's account of arithmetic may be dismissed.[45]

But the problem that led him to speak of 'idealisations' is a real one. How can an empiricist account for the infinity of the number series? Even if we forgo all talk of 'operations', and attempt a more direct account of 'the structure of reality', can an empiricist meet this challenge? My next subsection examines one attempt to do so.

4.3 Maddy on arithmetic[46]

Penelope Maddy's account of arithmetic is motivated quite differently from Philip Kitcher's, but their theories do have something in common. One feels that if Kitcher were to eliminate his mistaken stress on 'operations', he would end with a theory quite like Maddy's.

Maddy's main object is to defend 'realism' in the philosophy of mathematics. The two versions of realism that are most prominent today are that of Quine and Putnam on the one hand, and that of Gödel on the other. The former is an empiricist theory, and is my topic in the next section. I mention here only that Maddy rejects it on the ground that it does not account for the 'obviousness' of elementary arithmetic.[47] The latter is certainly not an empiricist theory in Gödel's

---

[45] Kitcher goes on (pp. 126–38) to give a 'Millian' account of the beginnings of set theory, and I shall not consider this. For some objections see e.g. [Chihara, 1990, 240–3].

[46] See P. Maddy, *Realism in Mathematics* [1990]. My page-references are to this work. I should note that her more recent book *Naturalism in Mathematics* [1997] has repudiated the 'realism' of the earlier book (but on quite different grounds from those which I put forward here). See her [1997, 130–60 and p.191n].

[47] The objection is cited from C. Parsons [1979/80, 101], and Maddy takes it to be so evidently

own presentation, but Maddy wishes to introduce some changes which turn it into one.

She emphasises that Gödel distinguishes two different reasons that we have for taking mathematical axioms to be true. In some (simple) cases we have an 'intuition' into these axioms, which Gödel describes by saying that 'they force themselves upon us as being true'.[48] In other (more recondite) cases an axiom may not strike us at once as 'intuitive', but we come to accept it as we discover its 'fruitfulness', e.g. how it yields simple proofs of results that otherwise could be shown only in a very roundabout way, how it provides solutions to problems hitherto insoluble, and so on.[49] Maddy takes over this 'two-tier' scheme, but applies it in a way that is quite different from Gödel's own intentions. This is because — like almost everyone else — she finds Gödel's appeal to 'intuition' very mysterious, and wishes to replace it by something more comprehensible. Gödel had *compared* his 'intuition' to sense-perception, but he had thought of it (as Plato did) as a special kind of 'mental perception' of abstract objects. Maddy wishes to say instead that it is just perfectly ordinary perception of familiar concrete objects, and she applies this view to elementary arithmetic in particular. Her thought is (I presume)[50] that if we ask how a simple mathematical truth such a '2 + 2 = 4' is known, then the answer is that we simply *see* that this is true in a quite literal sense of 'see', i.e. by visual perception. For example, we can simply *see* that 2 apples here and 2 apples there make 4 apples altogether. (I would expect Maddy to take the view that other forms of perception are also relevant; for example, the blind man will perceive that $2 + 2 = 4$ not by sight but by touch; but in fact it is exclusively visual perception that her discussion in chapter 2 concerns.) That is her answer to the epistemological problem that besets traditional Platonism. The answer, like Aristotle's answer to Plato, is one that brings numbers down from a Platonic 'other world' into 'this world', as is plainly required if it is to be ordinary perception that gives us our knowledge of them. Let us turn, then, to Maddy's preferred ontology.

Her basic thought is this: numbers are properties of sets; sets are perceptible objects; and we can simply see that an observed set has this or that number (i.e. this or that number of members). Some qualifications are needed at once. First, for the purposes of this part of her discussion, Maddy restricts attention to the 'hereditarily finite' sets. This means sets which are themselves finite sets, and such that any sets that they have as members are also finite sets, and in turn any members of *these* that are sets are finite sets, and so on. That is, in the

---

correct that she herself offers no further defence (p. 31). I shall differ from her (and from Parsons) on this point.

[48][Gödel, 1947, 484].

[49][*ibid.*, 477].

[50]I supply the example. So far as I have noticed Maddy herself gives no example of an arithmetical truth that we can simply *perceive* to be true. She makes this claim for some truths about sets, e.g. the axioms of pairing and union (67–8), but when it comes to numbers she is more concerned to maintain that simple perception can tell us (e.g.) that there are two apples on the table.

construction of such a set 'from the bottom up' no infinite set is ever needed. (In the terminology of ZF set theory, these are the sets of finite rank.) It is not that Maddy denies the existence of other sets, but she (very reasonably) thinks that other sets could hardly be said to be *perceptible* objects. (We know about them — if at all — only through axioms of Gödel's 'second tier', known because they are 'fruitful' but not simply through what Gödel called 'intuition', and what Maddy is regarding as just ordinary perception.) So she confines elementary arithmetic to the study of hereditarily finite sets.[51]

She does not say, and I imagine that she does not mean to say, that elementary arithmetic is confined to sets whose construction begins with *perceptible* individuals. Presumably it is only such sets that are themselves perceptible objects, but one would certainly expect arithmetic to apply too to imperceptible (but hereditarily finite) sets. Equally, she does not say, and I imagine that she does not mean to say, that elementary arithmetic is confined to 'hereditarily small' sets, i.e. to sets which have only a small number of members, and are such that their members in turn have only a small number of members, and so on. Again, it would seem that only such sets are perceptible,[52] but presumably arithmetic applies to larger (but finite) sets as well. What is left unexplained is how we know that arithmetic applies to these sets too, but perhaps Maddy would say that this falls under Gödel's 'second tier' of mathematical knowledge. That is to say, we accept axioms which extend the properties of perceptible sets to those which are (finite but) imperceptible, because we find such axioms 'fruitful'.[53] Admittedly, the one case of this sort that she does discuss leads her to a very strange idea. Arithmetic applies to finite sets of all kinds, whether or not they are also hereditarily finite. On this point Maddy says: 'When we demand that our numbers count more complicated, infinitary things, we are asking for more complicated numbers', and she adds in a footnote 'These new numbers are not more complicated in that they are infinite — I'm still talking about finite numbers — they are just more complicated in that the finite sets they number can have infinite sets in their transitive closures' (p. 100). I find this immensely puzzling. A set may have two apples as members; another set may have two infinite sets as members (e.g. the set of even numbers and the set of odd numbers); but we say of each of them that they have just two members. In what way is the number two that is predicated in the second case a 'more complicated number' than the number two that is predicated in the first case? I think that Maddy might like to reconsider this remark.

---

[51]'Knowledge of numbers is knowledge of sets, because numbers are properties of sets. Conversely, knowledge of sets presupposes knowledge of number ... From this perspective, arithmetic is part, perhaps the most important part, of the theory of hereditarily finite sets. Neither arithmetic nor this finite set theory enjoys epistemological priority; the two theories arise together' (p. 89).

[52]Perhaps this oversimplifies. For example, I may stand on a mountain-top and survey a forest that stretches for miles and miles all round me. Perhaps this may be counted as perceiving a rather large set of trees. But in such a case I do not also perceive how many members the set has.

[53]The idea that one good reason for accepting a proposed axiom is that it is 'fruitful' goes back at least to Russell [1907], if not earlier.

The last paragraph has simply applied to Maddy two points that were already made by Frege in his discussion of Mill, namely (i) that an empiricist account of how we find out about small numbers does not by itself explain how we know about large numbers, and (ii) that numbers apply to objects of *all* kinds, and not only to those that may be counted as perceptible. But we may (not unreasonably) assume that Maddy could find a way of responding to these points without abandoning her central claims. I now move on to the central claims, first noting that they also are in conflict with Frege's. Frege had argued that 'a statement of number contains an assertion about a concept' [1884, 59ff.]. Admittedly, it is not entirely clear how he wished us to understand his notion of a 'concept' when he first made this assertion,[54] but I think we can be quite confident that he did *not* regard a concept as a perceptible thing. Others since have wished to say (like Maddy) that a 'statement of number' contains an assertion about a set, and, at least at first glance, either view would seem to be defensible. *But* other authors have *not* at the same time supposed that sets are to be regarded as perceptible things; they would rather say that sets count as 'abstract objects', which they construe as implying that sets are not perceptible. It is (so far as I know) peculiar to Maddy that she thinks *both* that numbers are properties of sets *and* that sets are perceptible objects. She *begins* to see some of the difficulties involved in this combination of views in her chapter V, but she has not followed them through; and when we do follow them through it seems to me that her position becomes quite untenable.

Maddy thinks that, since sets are perceptible objects, they must have a location; in fact she takes it that a set is located where its members are located. This leads her to deny the existence of the null set, because if it were to exist it would have to be a perceptible object with no location, which seems to be a difficult conception (pp. 156–7).[55] In support of this proposal she cites various authors who have called the null set 'a fiction' or 'a mere notational convention' (p. 157n.). Given her account of what numbers are, this denies the existence of the number zero, but no doubt we can do elementary arithmetic without zero. Indeed, that is just how it *was* done for many centuries. So this first departure from present-day orthodoxy does not seem very serious in itself. But we have more to come.

In response to an objection that is forcefully put by Chihara,[56] she is also led

---

[54] His views on concepts appear to change between [1884] and later writings. In his [1884] it would seem that he does not construe concepts extensionally (cf. p. 80n), but after his [1892] he evidently does take it that concepts are the same if and only if the objects that they are true of are the same. (This view is very clearly put in his posthumous [1979, 118–25].) But of course he cannot say so in just these words without relying on the reader to 'meet him halfway', and 'not begrudge a pinch of salt', [1892, 54].

[55] One can see that there are no apples on the table. Should this be counted as perceiving the set of apples on the table, and perceiving that it has the number zero? And would this be an example of perceiving the null set? If so, it appears that 'the' null set would exist everywhere where there are no apples. By the same token, it would also exist everywhere where there are no bananas. So it would occupy every volume of space whatever. This conclusion is even stranger than the idea of a perceptible object which exists nowhere.

[56] [Chihara, 1982, 223], generously cited at length by Maddy on pp. 150–151. (Chihara develops the objection further in his [1990, ch. 10]. But this does not take account of Maddy's response

to identify an ordinary perceptible object (e.g. an apple) with its unit set (i.e. the set which has that apple as its one and only member). Chihara's objection is that if the unit set is taken to be a perceptible object, located exactly where its one member is located, then there is no way of distinguishing between the two. For example, the apple and its unit set *look* just the same as one another, and one can only presume that they taste just the same, smell just the same, and so on. Moreover, the set moves just as the apple does; it comes into existence and goes out of existence just as the apple does; and in all perceptible ways the two are indistinguishable. So what could differentiate between them? Maddy *accepts* this line of argument, and so identifies the individual with its unit set in such a case (pp. 150–153). Of course, any normal set theorist will reject it,[57] e.g. on the ground that the apple is not itself a set, and so has *no* members, whereas its unit set does have *one* member. And Frege, who thinks that numbers apply in the first place to concepts, rather than to sets, would find this identification quite intolerable, for it involves the claim that some concepts are objects, i.e. a concept under which just one object falls is identical with that one object.[58] But if Maddy is to maintain her claim that sets are perceptible objects, located where their members are located, it is not clear that there is any alternative that is open to her.[59] At any rate, she does accept that individuals and their unit sets are identical. But she does not seem to have realised where this proposal will lead to.

Suppose now that there are two apples on the table, and consider the set which has these two apples as members and no other members. How will *this* set differ from its unit set? All the same arguments apply: the two are located in exactly the same place; they look just the same; they move in the same way; and so on. It seems to me that Maddy is forced to say once more that the two are identical. Of course, a normal set theorist would deny this, e.g. on the ground that the set of two apples has two members whereas its unit set has only one member. But Maddy has been forced to set this ground aside before, and has no good reason for refusing to do so again.[60] Perhaps she might think that this concession too

---

in her [1990], whereas I fasten upon this response.

[57] As Maddy observes, Quine's set theories (NF and ML) accept this identification, but Quine's motive is simply technical simplicity (i.e. it is a way of making the usual axiom of extensionality apply not only to sets but to non-sets too). Quine certainly does *not* suppose that sets in general are perceptible objects. (I add, in parenthesis, that Quine can hardly be counted as a 'normal' set theorist.)

[58] Maddy says: 'What's the difference between a single object and its unit set? A "single object" already has an unambiguous number property' (p. 152). She means that both the apple and its unit set are *one*. This is a gross confusion, as anyone who has read Frege will immediately recognise.

[59] Maddy thinks that she has an alternative, as she could posit an *imperceptible* difference between the apple and its unit set. The difference, she suggests, might be that the apple is a 'concrete object' whereas the set is an 'abstract object'. I do not see how *this* distinction could be squared with her claim that sets are perceptible objects.

[60] She does refuse, i.e. she claims that the axiom of extensionality will prevent other sets being identified with their unit sets (p. 153). But why should she suppose that the axiom of extensionality applies to sets construed as *perceptible* objects? For she has *already* denied it in the case of individuals and their unit sets.

would do no noticeable harm, and we could accept that *every* hereditarily finite set was identical with its unit set, while still retaining most of the usual theory of hereditarily finite sets.

However, this line of thought is not yet exhausted. For suppose now that a, b, c, d are four distinct apples, and consider the two sets $\{\{a,b\},\{c,d\}\}$ and $\{a,b,c,d\}$. What is the difference between *them*? Obviously the normal set theorist will respond by saying (*inter alia*) that the first has two members whereas the second has four, so they could not possibly be the same set. But we have seen that Maddy is not entitled to *this* response, for she has to provide a *perceptible* difference between the two sets, and what could that be? As before, they are located in just the same place, they look just the same, they move in the same way, and so on. Suppose that this is conceded. Then of course we can generalise and say that whenever a set is built up by the set-operation, symbolised as $\{\ldots\}$, from a number of individuals, then — however often the notation indicates that this operation has been applied, and however complex is the structure thereby assigned to the resulting set — the resulting set *is* simply the same set as that which contains each of those individuals as members and no other members. In more technical terms, each hereditarily finite set is identical with the set of all the individuals in its transitive closure. This appears to be a consequence of the claim that the hereditarily finite sets are perceptible objects, and I do not see how Maddy could avoid it. But the consequence is disastrous.

We are back to what is (almost) the oldest problem in the book, namely Aristotle's problem of how there could be infinitely many numbers. For Maddy claims that numbers are properties of (hereditarily finite) sets, and (like Aristotle) she holds that these properties exist only if there do exist sets which have them. But we have seen that her claim that sets are perceptible objects has a consequence that there cannot be more (hereditarily finite) sets than there are ways of combining individuals, and if the individuals are finite in number then so also are the ways of combining them, and so also are the numerical properties which sets of them will have. In fact, if there are just n individuals then no hereditarily finite set will have more than n members, and so there will be no natural number greater than n. Could we accept this? Well, only if we are given a reason for supposing that there must be infinitely many individuals. But one cannot see how *perception* could provide such a reason.

In my opinion, another attempt at an empiricist theory of elementary arithmetic here bites the dust.

5 QUINE, PUTNAM AND FIELD

The previous empiricist proposals have all been prey to objections which are basically due to Frege. I now move on to a very different proposal which surely is not open to *these* objections. The proposal is essentially due to Quine, in various writings from his [1948] on, but it has become known as the Quine/Putnam theory because Hilary Putnam has expounded it (in his [1971]) at greater length than Quine himself ever did.

5.1 The indispensability of mathematics

In broad outline the thought is this. We can know that mathematics is true *because* it is an essential part of all our physical theories, and we have good ground for supposing that they are true (or, anyway, roughly true). Our reason for believing in physical theories is, of course, empirical; a theory which provides satisfying explanations of what we have experienced, and reliable predictions of what we will experience, should for that reason be believed. But all our physical theories make use of mathematics, and so they could not be true unless the mathematics that they use is also true. So this is a good reason to believe in the truth of the mathematics, and (it is usually held) a reason sufficiently strong to entitle us to claim *knowledge* of the mathematical truths in question. Knowledge grounded in this way is obviously empirical knowledge. That is the outline of what has come to be called 'the indispensability argument'. Let us expand it just a little.

It is the orthodox view nowadays that (most) physical theories should be construed 'realistically'. That is, these theories are presented as positing the existence of things which cannot plausibly be regarded as perceptible (e.g. atoms, electrons, neutrinos, quarks, and so on), and we should take such positings at face value. So if the theory is verified in our experience then that is a good reason for supposing that the entities which it posits really do exist. But, as we have said, today's physical theories all make heavy use of mathematics, and mathematics in its turn posits the existence of things (e.g. numbers) which are traditionally taken to be imperceptible. So we should take this too at face value, and accept that if mathematics is to be true then these entities must exist. But the argument just outlined gives us an *empirical* reason for supposing that mathematics is true, and we thereby have an *empirical* reason for supposing that the entities which it posits do really exist, even though they are not thought of as themselves perceptible entities. The ontology is Platonic, but the epistemology certainly is not.

A common objection is that it is only *some* parts of mathematics that could be justified in this way, by their successful application in physical theory (or in daily life), whereas the (pure) mathematician will think that all parts of his subject share the same epistemic status. Here I should pause to note that there certainly are many different areas of mathematics. In this chapter so far I have mentioned only the elementary geometry of squares, circles, and so on, and the elementary arithmetic of the natural numbers. These together *did* comprise *all* of mathematics at the time when Plato and Aristotle were writing, and ever since philosophers have tended to concentrate upon them. But of course many new subjects have been developed as the centuries have passed. From a philosophical point of view one might single out two in particular as demanding attention, namely the theory of the real numbers and the theory of infinite numbers. The first of these was very largely developed in response to the demands of physical theory: physics *needed* a good theory of real numbers, and the mathematicians did eventually find the very satisfying theory that we have today. (But they took a long time to do so.)⁶¹ By contrast, physical theory did not in any way require Cantor's development of the theory of infinite numbers, and the higher reaches of this theory still have no practical applications of any significance. In consequence, the indispensability argument that I have just sketched could provide a justification for saying that there really are those things that we call the real numbers, but it could not justify infinite numbers in the same way. But the (pure) mathematician is likely to object that each of these theories deserves the mathematician's attention, and that the same epistemic status (whatever that is) should apply to both.

This objection cuts no ice with proponents of the indispensability argument. They seriously do maintain that if there are only some parts of mathematics that have useful applications in science (or elsewhere), then it is only those parts that we have any reason to think true. Other parts should be regarded simply as fairy-stories. Putnam expresses the point in a friendly way: 'For the present we should regard [sets of very high cardinality] as speculative and daring extensions of the basic mathematical apparatus of science' [1971, p. 56]. His thought is that one day we might find applications for this theory, so we may accept that it is worth pursuing, even if there is now no reason to think it true. Quine is rather less friendly: 'Magnitudes in excess of such demands [i.e. the demands of the empirical sciences], e.g. $\beth_\omega$ or inaccessible numbers, I look upon only as mathematical recreation and without ontological rights'.⁶² They would say the same of any other branch of mathematics that has not found application in any empirically testable area.

I shall ask shortly just how much mathematics could receive the suggested empirical justification, but before I come to this I should like to deal with two other very general complaints. One I have already mentioned, namely the objection raised by Charles Parsons that this indispensability argument 'leaves unaccounted for precisely the *obviousness* of elementary mathematics'.⁶³ This seems to me a misunderstanding, which arises because the proponents of the argument do tend to speak of the applications of mathematics *in science* (and, especially, in physics). This is because they are mainly thinking of applications of the theory of *real* numbers, which one does not find in everyday life. Of course science applies the

---

⁶¹For a general history of mathematics see e.g. [Kline, 1972]. For the development of real number theory in particular I would suggest [Mancosu, 1996] for the period before Newton, and either [Boyer, 1949] or [Kitcher, 1984, ch. 10] for the period thereafter. The development was not completed until Dedekind's account of what real numbers are in his [1872] (or Cantor's different but equally good account, which was roughly contemporary).

⁶²[Quine, 1986, 400]. I should perhaps explain that '$\beth$' (pronounced 'beth') is the second letter of the Hebrew alphabet, and the beths are defined thus:

$\beth_0 = \aleph_0$ (= the smallest infinite cardinal)
$\beth_{n+1} = 2^{\beth_n}$
$\beth_\omega$ = the least cardinal greater than all the $\beth_n$, for finite n.

(The natural model for Russell's simple theory of types has cardinal $\beth_\omega$.) Inaccessible cardinals are greater than any that could be reached by the resources available in standard ZF set theory.

⁶³[Parsons, 1979/80, 101].

natural numbers too, but we do not have to wait until we learn what is called 'science' before we see that the natural numbers have many useful applications. Indeed, one's very first training in school mathematics is a training in how to use natural numbers to solve practical problems. (E.g., if I have 10p altogether, and each toffee costs 2p, how many can I buy?) It is hardly surprising that those who have undergone such training in their early childhood should find many propositions of elementary arithmetic just *obvious*, but that is no objection to the claim that the reason for supposing them to be true is the empirical evidence that they are useful. For that is exactly how such propositions were learnt in the first place.

A quite different objection is that this indispensability argument cannot show that our knowledge of (some parts of) mathematics is empirical, for it says nothing which would rule out the claim that there is *also* an *a priori* justification (perhaps for all mathematics, and not just some parts of it). This objection must be conceded. As I said right at the beginning, the apriorist's claim is that mathematical truths *can* be known *a priori*; it simply does not matter to this claim if (some of them) *can* also be known empirically. Of course, proponents of the indispensability argument do *believe* that no *a priori* justification can be given. In Quine's case, this is because he holds that there is absolutely no knowledge that is *a priori*, a claim that I shall consider in my final section. Others might wish to offer other reasons. For example, they might admit that some *a priori* knowledge is possible while still contending that one cannot have *a priori* knowledge *of existence*, and then pointing out that mathematics does claim the existence of innumerably many objects (e.g. of numbers of all kinds). In this chapter I cannot discuss the problems of apriorism, but it is clear that they do exist. Those who support the indispensability argument will usually suppose that these problems are insoluble, but the indispensability argument by itself gives no reason for thinking this.

I now move to a more detailed question: just how much mathematics could be justified on the ground that its applications cannot be dispensed with?

5.2 *How much mathematics is indispensable?*

All proponents of the indispensability argument will agree with this first step: we do not need any more than set theory, and the usual ZF set theory (or perhaps ZFC, which includes the axiom of choice) is quite good enough. One may happily admit that ordinary mathematics speaks of things (e.g. numbers) that are not sets. As we know, the numbers can be 'construed' as sets in various ways, but there are strong philosophical arguments for saying that numbers, as we actually think of them, cannot really *be* sets. (The best known is Benacerraf's argument, in his 'What Numbers Could Not Be' [1965].) But the reply is that, in that case, we can dispense with numbers as ordinarily thought of, for the sets with which they may be identified will do perfectly well instead. The physical sciences do not ask for more than numbers construed as sets, even if that is not how numbers are ordinarily construed. I think that this first step of reduction is uncontroversial.

But how much of ordinary set theory is indispensable? We have already said that its claims about infinite numbers seem to go well beyond anything that physics actually needs. One might say with some plausibility that physics requires there to be such a thing as the number of the natural numbers (i.e. $\aleph_0$), and perhaps that it also requires the existence of the number of the real numbers (i.e. $2^{\aleph_0}$), but it surely has no use for higher infinities than this. Yet this provokes a problem. For how can one stop the same principles as lead us from $\aleph_0$ to $2^{\aleph_0}$ from leading us higher still? Well, one suggestion that is surely worth considering is this: do we really need anything more than *predicative* set theory?

I cannot here give more than a very rough and ready description of what this theory is.[64] Historically, it began from a principle introduced by Poincaré [1905–6] called 'the vicious circle principle'. This was taken over by Russell [1908], and given various formulations by him, the most central one being this: whatever can be defined only by reference to all of a collection cannot itself be a member of that collection. Russell recommended this principle, partly because it seemed to provide a solution to a number of philosophical paradoxes, but also because it had what he called 'a certain consonance with common sense' (p. 59). Whether the principle is quite as effective as Russell supposed at solving his collection of paradoxes is a controversial question that I cannot enter into here.[65] In any case, we can certainly say nowadays that it is not the *only* known way of resolving these paradoxes. What is more interesting is Russell's claim that it conforms to 'common sense'. Since Gödel's discussion in his [1944], I think it has been very generally agreed that Russell's appeal to 'common sense' presupposes that common sense is basically 'conceptualist', i.e. it supposes that the objects of mathematics exist only because of our own mental activities. This approach leads very naturally to what is called 'constructivism' in the philosophy of mathematics, i.e. the claim that mathematical objects exist only if we can (in principle) 'construct' them. Seen from this perspective, Poincaré's 'vicious circle principle' seems very plausible, as it rules out what would indeed seem to be a kind of 'circularity' in an attempted 'construction'. (By contrast, from a more realist perspective, according to which sets exist quite independently of our ability to 'construct' them, the 'vicious circle principle' has absolutely no rationale.) Set theories which conform to this principle are called 'predicative' set theories.[66]

In practice, a set is taken to be constructible (in such theories) when and only when it has a 'predicative' definition, i.e. a definition that conforms to the vicious circle principle. Since (on this view) it is only constructible sets that exist, it follows that there cannot be more sets than there are definitions. But there cannot be more than denumerably many definitions (whether predicative or not), just because no learnable language can have more than denumerably many expressions (whether definitions or not). So the 'predicative universe' is denumerable. It contains all

---

[64] For more detail see McCarty's chapter in this volume.

[65] For some discussion see e.g. [Copi, 1971, ch. 3], and [Sainsbury, 1979, ch. 8].

[66] There is an accessible exposition in chapters IV and V of [Chihara, 1973]. This relies on earlier work by Wang, conveniently collected in his [1962].

the hereditary finite sets, for each of these can (in principle) be given a predicative definition, simply by listing its members. So, by a stratagem which I think is due to Quine,[67] the ordinary arithmetic of the natural numbers is available. It will also contain *some* infinite sets of these, which can be identified with real numbers in the usual way. But it cannot contain *all* the sets which a realist would recognise as built from the hereditarily finite sets, since (as Cantor showed) there are non-denumerably many of these. Consequently the full classical theory of the real numbers is not forthcoming, but a surprisingly large part of the classical theory can in fact be recovered by the predicativist.[68] It is quite plausibly conjectured (by [Chihara, 1973, 200–211]; and by [Putnam, 1971, 53–6]) that *all* the mathematics that is needed in science could be provided by a predicative set theory. (So far as I am aware, no one has tried to put this conjecture to any serious test.)

There is of course no reason why one who accepts the indispensability argument should also be a 'conceptualist' or 'constructivist' about the existence of mathematical entities. (Quine himself certainly was not.) So the fact — if it is a fact — that the indispensability argument will only justify a constructivist mathematics may be regarded as something of an accident. But it provokes an interesting line of thought, which one might wish to take further. The *intuitionist* theory of real numbers is even more restrictive than that which ordinary predicative set theory can provide. But is there any good reason for supposing that science actually *needs* anything more than intuitionistic mathematics? (Of course, the intuitionists themselves are not in the least bit motivated by the thought that they should provide whatever science wants. But perhaps, as it turns out, they do?)

More drastically still, one might propose that science does not *really need* any theory of real numbers at all. We all know that in practice no physical measurement can be 100% accurate, and so it cannot require the existence of a genuinely irrational number, rather than of some rational number that is close to it (for example, one that coincides for the first 100 decimal places). Discriminations finer than this simply cannot, in practice, be needed. Moreover, physical laws which are very naturally formulated in terms of real numbers can actually be reformulated (but in a more complex manner) in terms simply of rational numbers. The procedure is briefly illustrated in [Putnam, 1971, 54–3], who comments that 'A language which quantifies only over *rational* numbers, and which measures distances, masses, forces, etc., only by rational approximations ("the mass of a is $m \pm \delta$") *is*, in principle, strong enough to at least state [Newton's] law of gravitation.' I add that when the law is so stated we can make all the same deductions from it, but much more tediously.[69] The same evidently applies in other cases. At the cost of complicating our reasoning, our physics *could* avoid the real numbers altogether. If so, then there is surely no other empirical reason for wanting *any* infinite sets, and the indispensability argument could be satisfied just by positing the hereditarily finite sets. So we might next ask how many of these are strictly

---

[67][Quine, 1963, 74–7].
[68]There is a useful summary in [Feferman, 1964, part I].
[69]The general strategy is given in greater detail by [Newton-Smith, 1978, 82–4].

needed.

The answer would appear to be that we do not really need any sets at all, but only the natural numbers (or some other entities which can play the role of the natural numbers because they have the same structure, e.g. the infinite series of Arabic numeral types). Our scientific theories do apparently assume the existence of numbers, but they do not usually concern themselves with sets at all, and it has only seemed that sets have a role to play because the mathematicians like to treat real numbers as sets of rationals. But we have now said that real numbers could in principle be dispensed with, so that reason now disappears, and surely there is no other. It is true that the standard logicist constructions also treat rational numbers as sets (namely sets of pairs of natural numbers), but there is no need to do so. The theory of rational numbers can quite easily be reduced to the theory of natural numbers in a much more direct way, which makes no use of sets.[70] So apparently our scientific theories could survive the loss of all kinds of numbers except the natural numbers. But are even these really *needed*?

The most radical answer to the question which opened this subsection is that there are absolutely *no* 'mathematical objects' that are strictly indispensable for scientific (or other) purposes. This answer is proposed by Hartry Field in his *Science Without Numbers* [1980], and I have argued something similar in the final chapter of my [1979]. But before I come to discuss this claim directly it will be convenient to digress into a more general discussion of what is called 'nominalism' in the philosophy of mathematics. For Field certainly characterises his position as 'nominalism', but it is not the usual version of that theory.

5.3 Digression: Nominalism

Traditionally, nominalism is the doctrine that there are no abstract objects. It is called 'nominalism' because it starts from the observation that there are in the language words which appear to be names (*nomina*) of such objects, but it claims that these words do not in fact name anything. The *usual* version of the theory is that sentences containing such words are very often true, because they are not really names at all, but have another rôle. The sentences containing them are short for what could be expressed more long-windedly without using these apparent names. (To illustrate with a trivial example: one may say that abstract nouns are introduced 'for brevity' without supposing that the word 'brevity' is here functioning as the name of an abstract object. For (in most contexts) the phrase 'for brevity' is merely an idiomatic variant on 'in order to be brief', and this latter does not even look as if it refers to an abstract object.) A theory of this kind may be called 'reductive nominalism', for it promises to show how statements which apparently refer to abstract objects may be 'reduced' (without loss of meaning) to other statements which avoid this appearance.

---

[70]I have in mind a reduction in which apparent reference to and quantification over rational numbers is construed as merely a way of abbreviating statements which refer to or quantify over the natural numbers. For a brief account see e.g. [Quine, 1970, 75–6], or my [1979, 79–80].

One may be a nominalist about some kinds of abstract objects without being a nominalist about all of them. For example, one might feel that numbers, as construed by the Platonist, are incredible, and yet feel no such qualms about properties and relations of a more ordinary kind. (The thought might be that ordinary properties and relations are entities of a higher type than the objects they apply to, and this makes them acceptable, whereas the Platonist's numbers are not to be explained in this way.) Conversely, one might feel that numbers have to be admitted as objects, whereas ordinary properties and relations do not, since in their case it is usually quite easy to suggest a reductive paraphrase. Or, to take a quite different example, one might feel that numbers were highly problematic whereas numerals are entirely straightforward, though numerals (construed as types, rather than tokens) are presumably abstract objects. I shall be concerned here only with nominalism about numbers, and in this subsection I consider only the natural numbers. Can we say that the ordinary arithmetical theory of the natural numbers can be 'reduced' to some alternative theory, in which numerals no longer appear to be functioning as names of objects, and quantification over the numbers no longer appears as an ordinary first-order quantification over objects? Many have thought so.

The reduction which is most usually attempted is one which in effect replaces the number n by its associated numerical quantifier 'there are n objects x such that ...x...'. Something like this was surely what Aristotle was thinking of when he said that arithmetic should be viewed as a theory of quite ordinary objects, but one that is very general. It is also close to what Maddy has in mind when she claims that numbers should be taken to be properties of sets, for the relevant properties are those that we express by 'there are n members of ...'. It is also very natural to say that this is what the Millian theory would come to, when purged of Mill's own talk of the operations of moving things about, and of Kitcher's talk of less physical selection-operations. For the complaint, in both cases, is that numbers would still apply even in the absence of such operations, and it is the numerical quantifiers that apply them. As we know, Frege at one stage proposed exactly this reduction, but then went on to reject it, because he claimed that we must recognise numbers as objects (*Foundations of Arithmetic*, pp. 67–9). Those who disagree with him are likely to want to accept the reduction, and certainly it is the cornerstone of Russell's theory of natural numbers. For although he first takes numbers to be certain classes, his 'no-class' theory then eliminates all mention of classes in favour of what he calls the 'propositional functions' that define them; and in the case of the numbers these propositional functions just are the numerical quantifiers.

As is well known, Russell's theory runs into two main difficulties, and it will be useful to pause here for a brief reflection upon them. (i) Apparently some axiom of infinity is required, in order to ensure that each quantifier 'there are n ...' is true of something. We need to assume this in order to deduce, *via* the most natural definitions, that Peano's postulates do hold of the natural numbers, i.e. the numerical quantifiers. If we retreat to more complex definitions which introduce the idea of necessity,[71] then this axiom becomes the claim that each quantifier 'there are n ...' is *possibly* true of something. This at least has the advantage that, unlike Russell's axiom, it is certainly *true*. (For example, for each n, it is possible that there should have been just n apples in the universe.) But one cannot avoid all need for some such axiom without supplying enough entities of 'higher types' for the quantifiers to apply to, and this brings us to the second difficulty. (ii) Numerical quantifiers apply to (monadic) propositional functions of *all* types (or levels); indeed they even apply to propositional functions to which they themselves are arguments, as in

> There are 3 numerical quantifiers which come before the numerical quantifier 'there are 3 ...'.

But how can *any* consistent theory allow for that? Certainly, Russell's could not.

Frege was able to prove an axiom of infinity by taking numbers to be objects, and allowing them to 'apply to themselves' in just this way. (That is, he proved that each number n is the number of the numbers less than n.) One who does not wish to accept numbers as objects cannot proceed in this way, and will no doubt wish to point out that Frege is relying on a background logic that is inconsistent. It makes the impossible assumption that to every first-level concept there corresponds an object, in such a way that these objects are the same if and only if the concepts are equivalent. It is this that allows Frege to avoid something like Russell's (simple) theory of types, because what one might wish to say of an entity of higher type can always be said instead of its associated object.[72] Without such a reduction Frege's theory would be at least incomplete, because it would not cater for the fact that numbers can be applied to concepts of *every* level. From a technical point of view Frege could have achieved his deduction of arithmetic while avoiding inconsistency, if he had restricted his existential assumption to one specifically about numbers, namely that to every first-level concept there corresponds an object in such a way that these objects are the same if and only if the concepts can be correlated one-to-one. But there seems to be no rationale for restricting this assumption to first-level concepts only. Besides, although this assumption is certainly consistent, one may very well doubt whether it is true. For example, is there really any ground for supposing that there is an object (namely $Nx: x = x$) that is the number of *all* the objects that there are? For such reasons as these, one might feel that it would be premature to abandon all attempts at a reductive theory along something like Russell's lines.[73] Could Russell's difficulties be somehow met?

---

[71] For example, '3 is the next number after 2' might be rendered as
$$\Box \, \forall F (\exists_3 x (Fx) \leftrightarrow \exists x (Fx \,\&\, \exists_2 y (Fy \,\&\, y \neq x)))$$

[72] A natural language allows us to do exactly what Frege does, i.e. to exchange any predicate for an associated name (e.g. by prefixing the words 'the property of being ...', or 'the class of all ...', or simply by quoting the predicate), and taking it for granted that this name does name something.

[73] I add as a note that the usual set theory cannot do what we want. Since numerical quantifiers

In my [1980] I have proposed a solution, but I certainly have to admit that it introduces ideas which are not familiar, and which do not seem to be as 'clear and distinct' (in the Cartesian sense) as one would like. One cannot have very much confidence in it, and in fact a satisfying theory of the numerical quantifiers proves to be much more difficult to attain than one might at first have expected. Certainly it is a great deal more complex than the ordinary theory of the natural numbers that we all learn in early childhood. So the wholesale 'reduction' of the latter to the former might seem to be a somewhat dubious enterprise. Yet there are *some* reductions which seem to be very obviously available. For example, let the quantifiers 'there are 7 ...', 'there are 5 ...' and 'there are 12 ...' be defined in the obvious way, using the ordinary quantifiers $\forall$ and $\exists$ and identity. Then it seems very easy to suppose that '7 + 5 = 12' can be represented (in a second-level logic) as

$$\forall F \ (\exists G \ (\exists_7 x \ (Fx \ \& \ Gx) \ \& \ \exists_5 x \ (Fx \ \& \ \neg Gx)) \leftrightarrow \exists_{12} x \ (Fx)).$$

And of course, as logicians would desire, this proposition can be *proved* using only logic itself and the definitions indicated. But let us now step back and take a wider perspective.

As I said at the beginning of this subsection, the attempt to 'reduce' the theory of natural numbers to the theory of numerical quantifiers is certainly the one which has attracted the most attention. But it is not the only reduction worth considering. For example, Wittgenstein's *Tractatus* contains a different proposal, summed up as 'a number is the index of an operation' (*Tractatus* 6·021). The basic thought here is focused not on 'there is 1 ...', 'there are 2 ...', and so on, but rather on the series that begins with 'once', 'twice', 'thrice', understood as applied in this way. Starting from a given object, and operating on it 'twice' is first applying the operation once to the given object, and then applying the operation again to *what results from* the first operation. (Thus the instruction 'add 1 to 3 twice' is not obeyed by writing the *same* equation '3 + 1 = 4' twice over, but by successively writing the two different equations '3 + 1 = 4' and '4 + 1 = 5'.) I would myself prefer to generalise this a bit, on the ground that an 'operation' corresponds to a many-one relation (i.e. the relation between what is operated upon and what results from the operation), and the idea of a numerical index can be applied to all relations, not just those that represent operations. We define what are called the 'powers' of a relation in this way[74]

---

apply to sets of all ranks, they do not themselves determine sets. In a system such as NBG one may claim that each determines a proper class, but proper classes are not allowed to be members, either of sets or of other proper classes. So there is no set or class which has as members the proper classes corresponding to 'there are 0 ...', 'there is 1 ...', 'there are 2 ...'. Consequently we still cannot say, using the quantifier 'there are 3 ...' that *there are 3* quantifiers less than it. But this, which we cannot say, is surely true!

[74] If desired, one may add

$$xR^0 y \leftrightarrow x = y.$$

$$xR^1y \leftrightarrow xRy$$

$$xR^{n+1}y \leftrightarrow \exists z\, (xR^n z\ \&\ zRy).$$

Using this idea, '7 + 5 = 12' comes out very simply as

$$\forall R \forall xy\, (\exists z\, (xR^7 z\ \&\ zR^5 y) \leftrightarrow xR^{12}y).$$

Again, pure logic can obviously provide the proof. (Indeed in this case the logic can be even 'purer' than in the case of the numerical quantifiers, since we do not need to invoke the notion of identity.)[75] But again, when we pursue in detail the proposal that *all* of arithmetic be reduced in this way, we find exactly the same problems as before. If the definitions are given in the obvious way, then apparently we shall need something like an axiom of infinity; and what I call the 'type-neutrality' of the numbers will again cause problems. For we can consider the powers of a relation of any level whatever, and a logic that will allow us to do this is not easy to devise. (E.g. what uniform analysis of 'three times' will allow you to say (without quotation marks): if you start from the relation-index '0 times', and proceed *three times* from a relation-index to its successor, then you will reach the relation-index '3 times'?) A solution such as I have proposed for the numerical quantifiers is easily adapted to this case too, but of course the same objection still applies: the logic proposed is just too complicated to be that of ordinary arithmetic.

Once this line of thought is started, there is no end to it, for in our quite ordinary concerns the natural numbers are applied in *many* ways. We have mentioned so far their use as indices of operations ('double it twice'), or as powers of relations ('cousin twice removed'), and their use as cardinals ('there are two'). But obviously there are others. For example, the natural numbers are used as ordinals ('first', 'second', 'third') in connection with any (finite) series. They are also used in what I call 'numerically definite comparisons' such as 'twice as long' and 'three times as heavy'. One can of course suggest yet further uses (e.g. to state chances), but those I have mentioned will be quite enough to make my point. One can set out to 'reduce' the theory of the numbers themselves to the theory of any one of these uses. In each case one encounters essentially the same problems (infinity and type-neutrality), and if they can be solved in any one case then they can equally be solved in the other cases too. So there is nothing to choose between the various reductions on this score. Moreover, I do not believe that there is any other way of choosing between them either. So we are faced with a further application of Benacerraf's well known argument in his 'What Numbers Could Not Be' [1965]. Assume, for the sake of argument, that the technical problems with each of these proposed reductions can be overcome, so they are each equally possible. Moreover, there is nothing in our ordinary practice that would allow us

---

[75]This is *perhaps* the reason why the *Tractatus* prefers this reduction to Russell's, for the *Tractatus* does not allow the introduction of a sign for identity.

to choose between them, so they are each equally good. But they cannot *all* be right, for each proposes a different account of what the statements of ordinary arithmetic actually mean. Hence they must *all* be wrong. I find this argument very convincing.

What it shows is that *for philosophical purposes* no such reductive account of the natural numbers will do, for none preserves the meaning of the simple arithmetical statements that we began with. As Frege claimed (when commenting on Mill's proposed reduction) the theory of the numbers themselves must be distinguished from the theory of any of their applications [1884, 13]. Consequently a philosophical analysis must accept that the statements of arithmetic do (claim to) refer to, and quantify over, these things that we call numbers. So such things must exist if the statements are to be true. Note, however, that it does not follow that these reductions will not satisfy the demands *of science*. I began this discussion (p. 197) by noting that for scientific purposes it works perfectly well to construe numbers as sets, even though there are well-known philosophical objections to the claim that this is what numbers really are. Similarly, for scientific purposes it may work equally well to construe numbers as (say) numerical quantifiers, even if a similar philosophical objection applies. Of course, this assumes that the reduction can be made to work, and that is a controversial assumption, which I cannot here explore further. Part of the interest of Hartry Field's position is that it avoids this question, while still retaining a good part of what the reductive nominalist was trying to do.

Field is a nominalist, in that he does not believe that numbers exist as abstract objects, but not a reductive nominalist, in that he does not offer to reduce the statements of arithmetic to statements of some other kind which avoid referring to numbers as objects. Instead he grants that arithmetical statements do presuppose the existence of numbers as (abstract) objects, and for that reason claims that they are *not true*. What are true *instead* are those statements in which numbers are applied. So he does not offer a reduction, but rather claims that the reason why the arithmetical statements are accepted (even though they are not true) is that they are suitably related to the statements which apply numbers, and which genuinely are true.

5.4 *Doing Without Numbers*

Everyone will admit that ordinary arithmetic is very useful, not just in what is called 'science' but also in many aspects of everyday life. (That is why it is one of the first things that we learn at school.) The Quine/Putnam argument claims that it would not be useful unless it were true, and that this provides a good (empirical) reason for supposing that it must be true, and hence that all these infinitely many things called 'numbers' do actually exist. But the response is that a theory may perfectly well be useful even though it is *not* true. This is the central idea of Field's *Science Without Numbers*.

In the philosophy of science this is called an 'instrumentalist' view of theories.

The idea is that a scientific theory should be an effective 'instrument' for the derivation of predictions, but no more than this is required. And a theory may be a very efficient 'instrument' of this sort even though it is not true, nor even an approximation to the truth: it may be no more than a fairy tale. It has sometimes been claimed that *all* scientific theorising should be viewed in this way, but that is not a popular view these days. A common view (which I share) is that although a fairy tale may provide very useful predictions, it cannot provide *explanations* for why things happen as they do. In order to do that, a theory must also be true (or, at least, an approximation to the truth). But there are *some* cases of scientific theories which have deliberately been proposed *simply* as instruments of prediction, though that is not common. (A well-known historical example is Ptolemy's theory of planetary motion, which was the best theory available for 14 centuries or more. It is quite clear that Ptolemy himself did not suppose that the various mathematical devices used in his theory – epicycles, deferents, equant points, and so on – corresponded to anything that was really 'out there' in the heavens, and it seems to me that no one who really understood the theory could have thought that.[76] The theory did provide astonishingly good predictions of where in the night sky the planets would appear, but it offered no realistic explanation of why they move as they do.) To apply this idea to 'science without numbers', the suggestion is this. Arithmetic is a highly useful calculating device, but for this purpose it does not have to be true. We can perfectly well regard it as just a piece of fiction. So regarded, we cannot think of it as *explaining* anything, but (contrary to Quine) that does not matter. For whatever might be explained with the help of arithmetic can equally well be explained without it.

Field generalises this thought. He claims that whatever in science is explained with the help of any branch of mathematics *could* also be explained (but much more tediously) without it. But let us continue for a while just with the theory of the natural numbers, and the way that it may be applied by numerical quantifiers. Field offers this example (p. 22). Suppose that

(i) there are exactly twenty-one aardvarks;

(ii) on each aardvark there are exactly three bugs;

(iii) each bug is on exactly one aardvark.

The problem is: how many bugs are there? The *method* is: translate the problem into a problem in pure arithmetic, namely 'what is 21 × 3?'; calculate arithmetically that the answer is '63'; translate back into the language that we began with, concluding

(iv) there are sixty-three bugs.

Is that not a very convincing account of what we all learnt to do at school? But the point is that this detour through pure arithmetic was not in the strict

---

[76]For an account of Ptolemaic astronomy I recommend [Neugebauer, 1957, Appx.I].

sense *needed*. For, given standard definitions of the numerical quantifiers 'there are 3', 'there are 21', and 'there are 63', the result *could* have been reached just by applying first-order logic (with identity) to the premises. Of course this proof, if fully written out in primitive notation, would occupy several pages (and would scarcely be surveyable). But, in principle, it is available. This illustrates the claim that pure arithmetic is *useful*, but is never strictly *needed*.

Field's way of trying to put the claim more precisely is this. Suppose that we start with a 'nominalistic' theory, say the theory of the numerical quantifiers (but any other way of applying natural numbers would do instead). Suppose that we *add* to this theory the ordinary arithmetical theory of the natural numbers. For this addition to be of any use, of course, we must also add ways of translating between the one theory and the other; let us suppose this done. Then Field claims that the addition of ordinary arithmetic is a *conservative* addition, which means (as normally understood) that it does not allow us to prove any more results in the language of the numerical quantifiers than could have been proved (no doubt more tediously) without it. This is a very plausible claim. But to investigate it in detail we must first be much more precise about what to count as 'the theory of the numerical quantifiers'. For example, does it allow us to generalise over the numerical quantifiers? (I would say 'yes', but Field would say 'no'.) Does it also allow us to generalise over all properties of the numerical quantifiers? (Again, I would wish to say 'yes', for I think that what is called 'second-order' logic is in fact perfectly clear, and that there are very good reasons for wanting it; but Field would prefer to say 'no', for he would rather avoid second-order logic.)[77] These are no doubt details on which proponents of essentially the same idea might differ among themselves, but they do affect just what is to be meant by a 'conservative' addition.

First-order logic has a complete proof procedure, which is to say that in that logic the notions of semantic consequence (symbolised by $\models$) and syntactic consequence (symbolised by $\vdash$) coincide. Second-order logic is not complete in this way, and so the two notions diverge: whatever proof procedure is chosen there will be formulae which are not provable by that method but which are true in all (permitted) interpretations. So if a second-order logic is adopted as background logic, we have to be clear about what is to count as a 'conservative' addition to some original theory. In each case the idea is that statements of the original theory will be entailed by the axioms of the expanded theory only if they were already entailed by the axioms of the original theory, but we can take 'entailment' here either in its semantical or its syntactical sense. In the first case we are concerned with statements which have to be true if the original axioms are true, and in the second case with those that are provable from the original axioms. In the context of a second-order logic, there will be statements which have to be true (if the

---

[77] I present the case for second-order logic in my [1998]. [Field, 1980] does use a second-order logic in his 'nominalistic' construction of the Newtonian theory of gravitation (chs. 3–8), but then (ch. 9) he discusses first-order variants of this theory, and is evidently attracted by them. The preference for first-order theories is stated much more strongly in his [1985].

axioms are) but which are not provable (from those axioms). In this situation it seems to me that all that is required of an added 'Platonist' theory is that the addition be *semantically* conservative, which is enough to ensure that it cannot take us from 'nominalist' truths to 'nominalist' falsehoods. But if the addition allows us to *prove* more (as it may), then that should be regarded as just another way in which the addition could turn out to be useful.[78]

The relevance of this point may be seen thus. Suppose first (as I would prefer) that the proposed theory of the numerical quantifiers does allow us to quantify over these quantifiers, does allow us also to quantify over the properties of these quantifiers, and thereby allows us to prove (analogues of) Peano's postulates for these quantifiers. Then the theory will be a *categorical* theory, which means that all its models are isomorphic, and hence that any statement in the language of this theory is either true in all models of the axioms or false in all models.[79] It follows that the addition of any other theory *must* be conservative extension (in the semantic sense), provided only that the addition is consistent. Suppose, on the other hand, that the original theory of the numerical quantifiers is much more limited: it adds to ordinary first-order logic just the definition of each (finite) numerical quantifier, and adds no more than this. (This is apparently what Field himself envisages in this case, p. 21.) Then again the addition of any other theory must be a conservative extension (in either sense) unless it actually introduces an inconsistency. This is because the language of the original 'nominalistic' theory is now so limited that only very elementary arithmetical truths can be stated in it, and these can all be certified by the first-order theory of identity which is a complete theory. Either way, Field's claim is in this case vindicated: the addition of pure arithmetic to any theory which applies the natural numbers will be a conservative extension.

Of course, I have only argued for this in two particular cases. Both concern the numerical quantifiers, and the first was a very ambitious theory of these quantifiers while the second was extremely restricted. Obviously, there are intermediate positions which one might think worth considering. Also, I have not considered any of the other ways in which natural numbers may be applied, though we have noted that actually there are very many such ways. I think that we should come to the same conclusion in all cases, namely that the truth of pure arithmetic is not *required* for the explanation of any actual phenomenon. But I do not delay here to try to generalise the argument, for there are more difficult questions that

---

[78][Shapiro, 1983] points out the importance, for Field's programme, of distinguishing between semantic and syntactic conservativeness. But the point had in effect been anticipated by Field himself [1980, 104].

[79]When Dedekind (in 1888) discovered the axioms that are now called 'Peano's postulates', he proved that these axioms are categorical. The proof presupposes that the logic employed is a second-level logic, with the postulate of mathematical induction understood as quantifying over absolutely all properties of natural numbers, whether or not the vocabulary employed allows us to express those properties. This proof transfers quite straightforwardly to the theory of numerical quantifiers, provided again that we may quantify over absolutely all properties of those quantifiers.

lie ahead.

So far I have spoken only of the natural numbers, which in fact receive only scanty attention in Field's [1980] (i.e. only pp. 20–23), no doubt because he thinks that in this case his view encounters no serious problems. But many will want to say that it is in 'real science' that numbers become indispensable, and here it is mainly the real numbers that are relevant. Can the same idea be extended to their case? Field argues that it can, and most of his [1980] is devoted to this argument. Here the chief problem is to find a way of *formulating*, in a 'nominalistic' language, the theory to which the addition of the pure theory of real numbers is supposed to be a conservative extension.

The simplest case to begin with is that of (Euclidean) geometry. When our schoolchildren are introduced to this geometry, it is not long before they learn to speak in terms of the real numbers. For example, they learn that the area of a circle is πr^2, where 'π' is taken to be the name of a real number, and so is 'r', and the theorem is understood in some such way as this: multiplying the number π by the square of the number which measures the length of the radius of a circle gives the measure of the area of the circle, in whatever units were used to measure the length of the radius. Does this not presuppose the existence of the real numbers π and r^2? Well, it is certainly natural to say that, when the theorem is stated in these terms, it does have that presupposition.

But the existence of the real numbers cannot be *necessary* for Euclidean geometry, as can be argued in two ways. Field (p. 25) relies upon the point that a modern axiomatisation of geometry, such as is given in Hilbert's [1899], does not need to claim the existence of the real numbers anywhere in its axioms. So the basic assumptions say nothing of the real numbers, though in practice real numbers will quite soon be introduced, e.g. by showing that the axioms imply that the points on a finite line are ordered in a way which is isomorphic to the ordering of the real numbers in a finite interval. A line of argument which I prefer goes back to the ancient Greek way of doing geometry, which never introduces real numbers at any point, since the Greeks did not recognise the existence of such numbers. Nevertheless their techniques (with a little improvement) could be used to prove whatever we can prove today, though in a more longwinded fashion.[80] No doubt it does not matter which line of argument we prefer, since each leads to the same result: geometry does not strictly *need* the real numbers, though no doubt it is simplified by assuming them.

The case of geometry is no doubt a very simple case. Putnam [1971, 36] issues a challenge on the Newtonian theory of gravitation, where the basic law may be

---

[80] I give some account of this in my [1979, ch. 3]. An illustration may be useful here. The Greek version of the theorem on the area of a circle is: circles are to one another in area as are the squares on their radii (Euclid, book XII, proposition 2. But I have altered his 'diameter' to our 'radius', which is an entirely trivial change.) This proposition may be taken as saying that in any circle the ratio of the area of the circle to the area of the square on its radius is always the same, so we could introduce a symbol 'π' as a short way of indicating this ratio if we wished to. But for this purpose we would not have to suppose that ratios are themselves objects.

stated as

$$F = \frac{gM_a M_b}{d^2}.$$

Here F is the force, g is a universal constant, M_a is the mass of a, M_b is the mass of b, and d is the distance between a and b. On the face of it, *all* these symbols refer to real numbers. Can the law be restated without such a reference? Well, the answer is that it can, and the bulk of Field's [1980] is devoted to establishing this point. Others (including myself) might prefer a rather different way of effecting this elimination, but in the present context that is of no importance. What is crucial is just that *it can be done*. We do not *have* to call upon the existence of the real numbers in order to present Newtonian physics.[81]

It does not follow that all mention of the real numbers can be eliminated from *every* theory that the physicists have proposed or will propose. Since scientists nowadays have no scruples over presupposing the real numbers, their theories are usually formulated in a way which simply assumes the real numbers right from the start. Nevertheless, it seems to me that the two examples just considered do create quite a good *prima facie* case, and make it probable that sufficient effort could provide versions of other scientific theories which have been freed from this assumption.[82] Besides, there is a very general reason for saying that the existence of such abstract objects cannot really be *needed* in the explanation of why physical objects behave as they do. For who would suppose that, if the real numbers did not exist, then the behaviour of physical objects would be different (e.g. that apples would not fall from trees with the rate of acceleration that we call '32 ft/sec^2')?

I can put this more forcefully. We have seen earlier in this section that it is not obvious just how much mathematical theory today's physics does call for. In particular, there is the quite serious suggestion that a *predicative* set theory will provide all of the theory of real numbers that is actually needed. But in a predicative set theory there are no more sets than there are (predicative) definitions, which has led Charles Chihara to suggest that predicative sets might simply *be identified with* their definitions [1973, 185–9]. (This is, in his eyes, the first step of a nominalistic reduction of predicative set theory. The second stage claims that the existence of such defining formulae can in turn be reduced to the *potential* existence of actual defining activities on our part.) Taking this suggestion seriously, we reach the conclusion that the indispensability argument is claiming that physical objects would not behave as they do unless such definitions did exist (or, perhaps, do potentially exist). But that is recognisably absurd. The case can be made even more convincingly if we begin from the thought that no real numbers are strictly needed, and we could manage just with the natural numbers. For if

---

[81] I do not even sketch the elimination in this case, since it is somewhat complex. (But I remark that there are *categorical* theories of continuity, which can be applied not only to such abstract things as numbers but also to things of a more ordinary kind.)

[82] My own inclination would be to try in each case to 'translate' apparent references to the real numbers into the language of the Greek theory of proportion, which makes no such reference. I have given some account of this theory in my [1979, chs. 3–4.

that is all that is required then anything else that could play the role of the natural numbers would do instead, for example the Arabic numerals. But, again, it is obviously absurd to suppose that physical objects would not have behaved as they do if we had never invented the Arabic numerals. We may add that a theory which explains this behaviour in a satisfying way should be one that can be stated without assuming the existence of anything that does not affect the behaviour in question. It follows that it *must* be possible to formulate any good scientific theory in a way that does not assume the existence of such abstract objects as numbers are supposed (by the Platonist) to be, and so the indispensability argument will not justify the positing of these objects. We can certainly grant that the fictions which mathematicians explore are often very *useful* fictions, and they very much simplify our practical reasoning both in everyday life and in the advanced sciences. But that need not stop us regarding them as *fictions*.

So where does this leave empiricism?

6   LOGIC AND ANALYSIS

I set aside the question of how we know the truths of pure mathematics, because, according to the account given in the last section, there are no such truths. If pure mathematics is no more than a (very useful) fiction, then one can ask how we all come to believe it, but not how we know it. And the answer to that question is that, in practice, we believe it because we were taught to believe it, and — if we set aside the peculiar worries that philosophers have — we have found no reason to disbelieve it. For certainly the fiction, if that is what it is, is very useful. So the question shifts: how do we know that it is useful? And in broad outline the answer to that question must be that we are satisfied that *it works*, i.e. it never does lead us from true ('nominalistic') premises to false ('nominalistic') conclusions. But how do we know *that*?

If the argument of the last section is on the right lines, then the truths of the ('nominalistic') theories which apply pure mathematics can be known independently. How? Well, again, the answer must be that in practice our knowledge depends very largely on teaching, but it may be strengthened by our own experience in making use of these applications, e.g. in counting and in measuring. As I have noted already, the apriorist will not be concerned to deny this answer, but he will want to add that the knowledge *could*, in principle, be obtained a priori. How? Well, the only worthwhile explanation that has ever been offered is that the knowledge is obtained by combining two allegedly *a priori* resources: (i) *analysis* of the terms employed in these propositions, and (ii) *logic*.[83] So I take these in

---

[83]I set aside as wholly incredible Plato's alternative explanation (positing *recollection* of a previous existence) and Kant's explanation (that human beings cannot help *imposing* a certain form on their experience). Even if these accounts were accepted, they would at best show why we must believe these propositions, but not why they must be true. Some (e.g. Ayer in his *Language Truth & Logic* [1936, ch. 4]) have wanted to say that mere 'analysis' can by itself explain our knowledge of logic too. But it is well known that this view is open to many objections.

turn.

6.1 Analysis

Let us begin once more with the simplest and most familiar case, the application of the theory of natural numbers in numerical quantifiers. We 'analyse' the numerical quantifiers in terms of the ordinary quantifiers and identity, and then we think that 'logic alone' could (at least in principle) give us all the truths about these quantifiers. The topic of logic I postpone for the time being, but let us pause a little on the proposed analysis. There are two questions here, which in this case seem extremely simple, but which in other cases are somewhat more difficult.
(i) How do we know that (for example) 'there are two' may be analysed in terms of identity as 'there is one and one other', i.e.

$$\exists_2 x(Fx) \leftrightarrow \exists x(Fx \ \& \ \exists_1 y \ (Fy \ \& \ y \neq x))?$$

And (ii) How do we know that the notion of identity can be applied to the cases to which we wish to apply it? One is apt to be puzzled by both of these questions when they are seriously pressed, for in each case the proposition in question seems so *obviously* true. One says: *of course* 'two' is 'one and one other', and if you do not see that then you do not understand what 'two' means. One also says: *of course* the notion of identity applies to any objects whatever. There may in some cases be a difficulty about how a particular application should be understood — for example, philosophers have spilt much ink on what it is to be the same person — but one cannot doubt that the notion of identity does apply to this case, and to every other case too, whatever we are talking of. (As a matter of fact things are not quite so straightforward as this reply suggests, but there is no need to pursue that point here.)[84] So in each case we say 'That's obvious', but it is also clear on reflection that this response is not actually an answer to the question '*how* do you know?' It may suggest that the knowledge is *a priori*, but it certainly does not provide an argument for that claim.

Let us turn to a case which is slightly less straightforward, the application of the natural numbers as ordinals, as in 'the fourth house on the left'. In this case analysis will be needed to uncover the presuppositions, in particular the presupposition that we are dealing with a series, and to tell us what a series is. (This task is not altogether simple, since a series in the relevant sense may contain repetitions — e.g. the fourth house on the left may also be the fourteenth, if the road twists.) Given the appropriate analysis, it will then almost always be an empirical question whether what a particular application presumes to be a series really is one, e.g. in this case whether the phrase 'the houses on the left' does pick out a series of the appropriate kind. But one expects that logic alone should be able to tell us such

---

[84]Taking the domain to be what is on the table in front of us, we can surely say '$\exists x$ (x is water & $\exists y$ (y is water & $y \neq x$))'. But we cannot translate this into English as 'there are two waters here'. Why is this? Is there, perhaps, some deep metaphysical point that this feature of ordinary languages reveals?

general truths as 'in any series, the fourth term is the one that comes second after the second term'. Indeed, if such consequences were not deducible, that in itself would be a reason for saying that the proposed analysis must be wrong. But this does not yet seem to explain, by itself, how we can know that a proposed analysis is a correct analysis.

Let us move to a more difficult case, and one that genuinely is important for *science*, namely the use of numbers in such locutions as 'x is twice as long as y', and our knowledge of such truths as 'if x is twice as long as y, and y is three times as long as z, then x is six times as long as z'. Again, the apriorist will no doubt wish to say that such truths should be deducible by logic alone from a suitable analysis of the terms involved. But in this case the question of what counts as a correct analysis is thoroughly controversial. What are the conditions which a quantity has to satisfy if numbers are to be applied to it in this way? For example, everyone would say that 'twice as long as' makes perfectly good sense; some of us (including me) would say that 'twice as hot as' does not make sense; all of us would agree that 'twice as eloquent as' makes no sense at all. But what exactly is it that makes the difference? Well, as I say, this turns out to be a complex question, and various different answers to it have been proposed. I shall not attempt to explore it here.[85] Suffice it to say that in this case there genuinely are rival analyses, and it is not at all obvious how to choose between them. Perhaps there is some *a priori* method that would settle the question for us, but no one has any right to be confident that they have found it.

The question concerns not only the application of natural numbers but also the rational numbers and the real numbers, for they too are employed in what I call 'numerically definite comparisons'. (For example, the circumference of a circle is exactly π times as long as its diameter.) This is the primary application of real numbers in contemporary science, but of course it is justified only when the quantity concerned is also a continuous quantity, and that provokes another need for analysis: what is continuity? Well, since Dedekind's [1872] we have been fairly confident that we now know, but it was a long time before that analysis emerged, though it had been recognised ever since Aristotle that the notion of continuity is an important one.[86] Given a suitable analysis, it then becomes an empirical question whether a quantity such as length or time or mass is indeed continuous, but we expect logic to be able to deduce the properties which any continuous quantity must have, including those properties which are stated in terms of the real numbers. (For example, assuming that time is a continuous quantity, if x lasts $\sqrt{2}$ times as long as y, and y lasts $\sqrt{3}$ times as long as z, then x lasts $\sqrt{6}$ times as long as z, and the analysis of '$\sqrt{n}$ times' should provide the premises

---

[85] My own answer occupies the bulk of my [1979].

[86] For Aristotle's account see his *Physics*, book VI (with chapter 3 of book V). For the most part he is content to identify continuity just with infinite divisibility. But even he should have seen that infinite divisibility does not by itself imply divisibility in every ratio whatever, rational or *irrational*. He was an acute thinker, and I have never understood why he missed this point. But he did miss it, and so did everyone after him, for *centuries*.

from which this result could be deduced.)[87] But how such an analysis should be reached — indeed, how Dedekind's own analysis *was* reached — is a question that has no obvious answer. And if we are asked 'How do you know that this analysis is correct?' we are again at a loss for what to say.

When it is said that analysis and logic are two methods by which *a priori* knowledge can be attained, people usually have in mind *very simple* analyses. A typical example is 'a bachelor is an unmarried man', and here the proposition reached seems supremely obvious, and one only has to know what the word 'bachelor' means in order to see that it is true. If so, then the knowledge would apparently qualify as *a priori*, for we have said that whatever experience is needed in order to know what a word means is to be discounted. But mathematics is full of much more interesting analyses. I have mentioned three (i.e. the analysis of continuity, the analysis of numerically definite comparisons, and the analysis of the notion of a series). It is obvious that I could add many more. (Perhaps the most important was the analysis provided by Cauchy and Weierstrass of what was really going on in the so-called 'infinitesimal calculus'.) It clearly will not do to say that such analyses are immediately known by anyone who is familiar with the language being analysed, for we are well aware that for centuries they were not known. *Perhaps*, if we do come to know that such an analysis is correct, that knowledge will count as *a priori*; I have not argued against that suggestion directly; but I am highly sceptical. Just recall the attempt to say what *a priori* knowledge is that occupied us in section 4.1. I concluded there that, for knowledge to count as *a priori* in the traditional sense, we must stipulate that it is attained by a method which (is independent of experience, and) *by itself guarantees* that beliefs so reached must be true. But *are* there any such methods? It is surely very plausible to say that the methods (whatever they are), which we now think have led us to analyses that are correct, are just the same as the methods which, in the past, led to analyses which we now regard as incorrect (e.g. Aristotle's account of continuity, or the explanations first given — say by Leibniz and by Newton — of what was going on in their newly invented 'infinitesimal calculus'). But I here leave that as an open question, and move to the other: what about our knowledge of logic? Here I think that there is in fact a very strong reason for saying that this knowledge is *not a priori*.

6.2 *Logic*[88]

First off, one is apt to suppose that experience *could not* be relevant to such things as the correctness of *modus ponens*, the truth of the laws of non-contradiction and

---

[87] Dedekind very fairly complained in his [1872] that so far no one had ever given a proof that $\sqrt{2} \cdot \sqrt{3} = \sqrt{6}$ (p. 22). Of course a proof is easily available if we assume the axioms of Euclidean geometry and give a geometrical interpretation of the numbers involved, i.e. assuming that $\sqrt{2}$ is the length of the side of a square with area 2. But one of Dedekind's aims was to free the theory of real numbers from the assumptions (explicit or — quite often — tacit) of Euclidean geometry.

[88] This discussion is based on my [1990].

excluded middle, and so on. Why we believe this is not entirely evident, but I think that an ancient principle is very probably at work, namely this: we *cannot conceive* how experience might upset such claims. That may be true, but it does not establish the point in question, and to see this we have only to recall what Mill said about geometry: we *cannot conceive* a space that is not Euclidean, but it does not follow from this that our space *cannot be* non-Euclidean, for what we cannot conceive may nevertheless be true. Might not the same apply to logic? As a preliminary I note one relevant difference between Mill's account of geometry and what one might wish to say of logic, but also some important similarities.

The claim that we cannot 'conceive' a non-Euclidean space should presumably be understood as meaning that we cannot *picture* such a space to ourselves, i.e. that we cannot imagine what it would be like to *perceive* it. (It also means that we cannot imagine ourselves perceiving 'in one blow' some spatial feature that shows the space to be non-Euclidean. Of course we can imagine a whole series of perceptions which seems best interpreted on that hypothesis, but that is not what is intended.) However, in the case of logic, when we say that we cannot 'conceive' how one of the familiar principles might be false, we do not just mean that we cannot *picture* a falsifying situation. We mean that we can simply make no sense of this supposition at all, neither by forming pictures nor in any other way; we can find no way of understanding it whatever. This is a genuine difference between the two cases.

It has as a consequence that Mill's explanation of why we cannot do this 'conceiving' in geometry does not carry over to the case of logic. For Mill rather plausibly suggests that our ability to imagine ourselves perceiving this or that may well be limited by what we have actually perceived. (This is surely the right thing to say about imagining a new colour, as discussed earlier.) So if space has in fact always appeared as Euclidean in our perceptions so far, that could explain why we cannot picture it otherwise. But the case of logic is different, and if here again there is some contingent feature of the world which explains our inability to conceive otherwise, then it cannot just be this. For perception (in the literal sense) is scarcely relevant. Of course, there might be some other explanation. Perhaps it is 'human nature' to think in terms of the familiar logic (i.e., in materialist terms, perhaps our brains are so structured by our genes that they cannot step outside this way of thinking). Or perhaps the explanation can be provided by 'nurture' rather than 'nature', i.e. we have been so constantly brought up to think in this way that we have now become incapable of anything else. So I am not trying to suggest that this disanalogy must prevent what is recognisably the same general idea from being transferred from geometry to logic. But one must concede that the disanalogy exists.

But let us now attend to some features that genuinely are analogous. One should not expect there to be a particular experience, or a series of experiences, which by itself shows that a particular Euclidean axiom, e.g. the axiom of parallels, is false. One could describe what would appear to be a fairly direct refutation, e.g. finding a pair of lines, in the same plane, which kept everywhere the same distance between them, though only one of them was straight (i.e. followed always the shortest distance between any two points on it). But, as philosophers since Reichenbach [1927] have frequently observed, one *could* always account for such experiences in some other way. For example, one could suppose that our measurements of distance were at fault, because our measuring rods kept expanding and contracting as we moved them about, in ways that were unobservable and not predictable from our current theories of how rods come to change their length. This obviously opens up the possibility that it is *the latter* theories that are mistaken, and not the Euclidean postulate that we began with. Obviously this illustration is far too simple to be at all like the actual observations that have led to the rejection of Euclidean geometry, but it is perhaps enough to make clear the general position. One could not expect to be able to bring particular propositions of Euclidean geometry into direct confrontation with experiment, and one would not even expect the body of all such propositions to be testable by itself. One theory is put to the test only by relying on other theories, perhaps just the theory of the experimental apparatus employed, but perhaps in other and more general ways too. When what Quine calls a 'recalcitrant experience' is discovered, one knows that something has to be revised somewhere, but there will be a number of alternative revisions that might meet the case. The choice between them can only be made by considering which yields the best *total* theory, i.e. in the present case principally the theories of geometry and physics combined. And this choice is to be made by the ordinary scientific criteria for assessing rival theories, which include such things as economy, simplicity, predictive power, explanatory elegance, and so on. That is, in broad outline, how we have come to believe that Euclidean geometry is not after all the best theory of our space. Similarly, then, with the claim that logic too is open to empirical testing. What this requires is just the possibility of there being two different *total* theories, which differ from one another in employing different logics (and no doubt in other ways too), and where we can see that the choice between them should be made in accordance with the ordinary scientific criteria for assessing rival theories.

With so much by way of preliminaries, let us come to the argument. Both with geometry and with logic we begin with one side appealing to what we can conceive, and the other side replying that this is inconclusive, since what we cannot conceive may for all that still be true. The argument continues in the same way too. What vindicated Mill's position on geometry was (a) the discovery of non-Euclidean geometries, and (b) the eventual recognition that experience could provide a way of choosing between them, at least in the rather indirect fashion just outlined. I shall argue that the same line of thought applies to logic too, and with equal success. (I shall also simplify the discussion by considering only the simplest area of logic, namely propositional logic.)

There is no difficulty over the first step: we are nowadays entirely familiar with the idea that there are alternative logics. Besides the ordinary, two-valued, classical logic there is also a rival called 'relevance logic' which is pressed into service in so-called 'dialetheic logic'. In addition there is three-valued logic, many-valued logic, supervaluational logic, fuzzy logic, and of course there is the intuitionist logic that is now so often regarded as the 'right' logic for the 'anti-realist'. In each case it is fair to say that the rival logic aims to embody a conception that departs from the classical conception. Relevance logic seeks to provide an alternative to the classical conception of entailment (or following from), for it is held that the classical conception yields unintuitive consequences. But in other cases it is the conception of truth that is at issue. Dialetheic logic allows some propositions to be both true and false, in a somewhat desperate attempt to make sense of the semantic paradoxes. Obviously the classical conception of truth cannot allow this, and I think that very few of us are prepared to contemplate such a violent wrench to our ordinary concept. But other variations are more comprehensible. The classical conception of truth supposes that every proposition is determinately true or false, irrespective of our ability to recognise it as such, whereas the other logics mentioned deny this, but not always for the same reason. Thus many-valued logics, or logics which permit truth-value gaps (possibly closed off by supervaluations), are generally motivated by the thought that much of what we ordinarily say is uncomfortably vague, and vague propositions do not fit happily into the 'true/false' dichotomy. But it is not vagueness that motivates intuitionist logic, for indeed that logic has its original home in mathematics, which is an area of discourse that is less affected by vagueness than almost all others. Rather, the relevant feature in this case is that in mathematics we are constantly dealing with infinities of one kind or another, and it is here that truth, as classically conceived, most conspicuously diverges from our ability to recognise that truth, even 'in principle'. The intuitionist is unhappy with this gap, and so prefers to operate with a revised conception of truth, in which it is more or less equated with provability. Again, it is this different conception of truth that lies behind his different logic.

Obviously, this brief account of motivations is somewhat superficial, and much more could be said, but I shall not pursue it further. This is because the alternative logics mentioned so far have seldom been recommended on the ground that physical theory would benefit by changing to them.[89] Yet just this has been argued for another rival logic, namely quantum logic, so it is here that the empiricist claim about logic is best explored. I therefore set the others on one side, and will consider only quantum logic for the remainder of this discussion. Since this logic may not be familiar, it will be helpful if I begin with a brief description of it.

Formally speaking, quantum logic, like intuitionist logic, is a subsystem of classical logic, sharing many of its laws but not all. To put it briefly, intuitionist logic lacks the law of excluded middle, and consequently lacks some other laws too that would imply this one. By contrast, quantum logic retains excluded middle, but lacks the law of distribution in the form

$$P \wedge (Q \vee R) \vDash (P \wedge Q) \vee (P \wedge R).$$

---

[89] An exception is Reichenbach [1951], who recommended using three-valued logic in the interpretation of quantum theory.

Consequently it also lacks some other classical laws that would imply this. But we see why intuitionist logic lacks excluded middle only when we see that it is based upon a different conception of truth, and consequently a different account of the meaning of the logical connectives. (Very roughly, in classical logic they are 'truth-functors', but in intuitionist logic they are 'proof-functors'.) The case is exactly similar with quantum logic; different underlying conceptions are involved. I shall base what I have to say about this entirely upon Putnam's classic paper on the topic, namely 'Is logic empirical?'[90] This contains all the materials for explaining the different conceptions, though the fact that they are different is not something that Putnam himself wishes to stress.

Quantum logic is proposed as a way of dealing with the puzzles generated by quantum theory, which is concerned with the behaviour of very small things such as electrons. This behaviour is indeed puzzling, but the explanation offered by quantum theory seems at first sight even more puzzling. To put the point in a *very* simple way, in order to explain the behaviour of these things, the theory treats them as waves, spread out in space. But when we design an experiment to 'observe' what is going on, what we find is not a wave but a particle, i.e. something localised in one particular place. So it seems as if our observation itself changes the situation observed. Given a classical conception of truth, this is of course something which in principle makes sense, but it is very difficult to account for in any satisfying way. Putnam's proposal is, in effect, that we should change to a different notion of truth, in which this no longer makes sense: we should understand what *is* true as being indistinguishable from what *would* be 'observed' if tested for. In his own words, this is an 'idealised operational account', and he explains it thus: 'Let us pretend that to every physical property P there corresponds a test T such that something has P just in case it passes T (i.e. it *would* pass T, if T were performed)' (p. 195). This, as I have said, is not the classical understanding of what it is for (it to be true that) something *has* the property P, but it is what Putnam is proposing. To put it briefly, an elementary proposition (of quantum theory) is to count as true if and only if it would be verified if tested for.

Given this conception for elementary propositions, it is then quite natural to extend the notion to compound propositions in the way that Putnam does. The propositions $\neg P, P \wedge Q, P \vee Q$ are equally counted as true if those propositions would, as wholes, be verified if tested for. Thus $\neg P$ is explained as true if and only if the test for $\neg P$ would be satisfied. (The explanation is *not*: ... if the test for P would be not be satisfied.) Similarly $P \wedge Q$ is explained as true if and only if the test for $P \wedge Q$ would be satisfied. (The explanation is *not*: ... if the test for P would be satisfied and the test for Q would be satisfied.) Similarly again for $P \vee Q$. Once more, it is clear that this is not the classical account of the truth conditions of these compound propositions, but it is an account which harmonises perfectly well with the underlying conception of truth already mentioned. It needs to be supplemented, of course, with an account of what the tests are for $\neg P, P \wedge Q, P \vee Q$, and how they are related to the tests for P and Q. Putnam proceeds to

---

[90][Putnam, 1968].

give such an account.

For this purpose we again consider P and Q as properties ascribed to whatever quantum system is in question. Then it is a consequence of quantum theory itself that if there is a test for the property P there is also another test, T, such that everything passes either the test T or the test for P, and nothing passes both the test T and the test for P; so we take the test T as the test for $\neg P$. Equally, it is a consequence of the theory that, given a test for P and a test for Q there is also another test, T, which is the 'greatest lower bound' of these tests, in this sense: whatever passes T passes both the test for P and the test for Q, and for any further test T' such that whatever passes T' also passes both the test for P and the test for Q, it will hold that whatever passes T' also passes T. The test T is then taken to be the test for $P \wedge Q$. Similarly, there is a test which is the 'least upper bound' of the tests for P and Q, and this is taken to be the test for $P \vee Q$. That such tests do exist is of course an empirical claim, but one that is asserted by quantum theory. Finally, to obtain a 'logic' one adds that P entails Q (i.e. $P \models Q$) if and only if whatever passes the test for P also passes the test for Q, and the definitions just given then ensure that:[91]

$$P \vee Q \models R \text{ if and only if } P \models R \text{ and } Q \models R$$

$$R \models P \wedge Q \text{ if and only if } R \models P \text{ and } R \models Q$$

$$\models P \vee \neg P, \text{ and } P \wedge \neg P \models.$$

These laws are of course similar to the classical laws, but they do not imply the principle of distribution

$$P \wedge (Q \vee R) \models (P \wedge Q) \vee (P \wedge R).$$

On the contrary this principle is not valid in quantum logic, and again we invoke quantum theory to show this. For let P and Q be a pair of 'complementary' properties, such as the position and the momentum of a particle, for which the uncertainty principle holds. Let $P_1, P_2, \ldots, P_n$ be a finite list of propositions, each assigning a different position to the particle, specified with some precision but so that between them they exhaust all possible positions for the particle. Then we have as valid

$$\models P_1 \vee P_2 \vee \ldots \vee P_n .$$

For it must be the case that one of the disjuncts P_i would be verified if tested for, and the same therefore applies to the disjunction of them all. In the same way let $Q_1, Q_2, \ldots, Q_m$ be a similar list of propositions assigning different momenta to the particle, so that we equally have

---

[91] Other laws for negation can also be obtained from the claims of quantum theory, notably
$P \dashv\models \neg\neg P$, if $P \models Q$ then $\neg Q \models \neg P$.

$$\vDash Q_1 \lor Q_2 \lor \ldots \lor Q_m.$$

If the principle of distribution held, we should then validly infer

$$\vDash (P_1 \land Q_1) \lor (P_1 \land Q_2) \lor \ldots \lor (P_2 \land Q_1) \lor \ldots \lor (P_n \land Q_m).$$

But on the contrary if the propositions P_i and Q_j have been chosen so as to specify position and momentum with sufficient precision, then each conjunction $(P_i \land Q_j)$ will be what Putnam calls a 'quantum logical contradiction' since the uncertainty principle tells us[92] that nothing will pass a test for $(P_i \land Q_j)$, and so the theory tells us that our conclusion is false. Thus distribution fails.

Let this suffice as a description of what quantum logic is, and why it is as it is. Putnam's article goes on to make several claims about meaning, which I shall simply set aside. (For example, he claims that his account of truth remains a 'realist' one — which he takes to be virtue — whereas it seems clear to me[93] that this cannot be maintained. He also claims that his account of the logical connectives should not be seen as assigning them a new meaning, and again I certainly would not wish to defend this). There are besides all manner of difficulties in working out a version of quantum theory which adopts quantum logic consistently and throughout (not least because the *mathematics* employed in quantum theory is entirely classical), and I make no attempt to explore these issues. Nevertheless, the essence of Putnam's proposal seems to me to be clear enough. It suggests that we should, anyway for the purposes of quantum theory, adopt a non-classical conception of truth, in which it is more closely tied to verification, and in consequence a non-classical understanding of the familiar logical connectives, and therefore a non-classical logic. The rationale for this proposal is that it will remove what appears from the classical viewpoint to be an unanswerable puzzle, and in this way it will yield a simpler overall theory, one which is a better theory as judged by ordinary scientific criteria.

Now the merits of this proposal are highly controversial, and that is not an issue on which I offer any opinion. If I understand the current situation rightly, most of those who know what they are talking about hold that adopting quantum logic would not in fact simplify the puzzles which quantum theory seems to generate, and so the rationale suggested does not in practice work out. That may well be right. But what I wish to insist upon is that, from a purely philosophical point of view, the programme is not misconceived. It *could* turn out that exchanging one conception of truth for another did actually simplify our physics.

It cannot be denied that scientific progress does often require us simply to drop one scheme of concepts and exchange it for another. Moreover, one of the advantages that will be claimed for such a change of concepts is that questions

---

[92] One might quite naturally take the uncertainty principle as stating that there is no test for the conjunction, but Putnam's reasoning requires us to take it as stating that there is a test for it, but it is 'the contradictory test' which nothing passes.

[93] And to [Dummett, 1976].

which before had seemed puzzling or unanswerable will no longer arise. On the new way of thinking they simply disappear altogether. I illustrate this with a couple of examples from the Newtonian theory of motion, beginning with the question of 'absolute' rest and motion.

The puzzle arises in this way. First, it is built into Newton's theory that accelerations are 'absolute', and not merely 'relative' to some presupposed frame of reference, since the theory is that accelerations need forces to explain them, and forces are not in this way 'relative'. Next, since acceleration is defined as the rate of change of velocity, it then seems that velocity must be 'absolute' too, and there must be a difference in nature between one constant velocity and another (including velocity zero). This in turn seems to require that space also is 'absolute', in the sense that the same spatial position retains its identity over time, so that an object is (absolutely) at rest if it stays in what is (absolutely) the same position, and otherwise moving. Newton himself, of course, accepted this apparent consequence. But it generates a puzzle: how can we ever tell whether an object is (absolutely) at rest, or with what (absolute, but constant) velocity it is moving? And this puzzle bites, for it is apparently built into Newton's own theory that we cannot tell this, since no forces are required to explain the continuation of any constant velocity, including zero. It thus results that the conceptual scheme within which Newton operates generates a question which, according to his own theory, is unanswerable.

The scientifically accepted solution is to drop this conceptual scheme and substitute another. We shall no longer think of spatial positions as having a continuing identity over time, and the notion of a point *of space* will therefore disappear. Instead, we shall think of points *of space-time*, which of course cannot continue as 'the same point' from one time to another. Then being at rest may be explained as successively occupying a series of space-time points which stand in a certain relation to one another (namely, lying on the same 'geodesic'), and exactly the *same* explanation applies too to moving at a constant velocity. So, on the new way of thinking, there really is no difference between one constant velocity and another (including zero), and the old puzzle has simply disappeared. This in itself is an argument for changing from the old way of thinking to the new: it works better to think, not in terms of space and time separately, but in terms of space-time.

I add that this conceptual reform is available, and desirable, even within what is basically a Newtonian theory. But it becomes mandatory when we move from that theory to the theory of (special) relativity. My second example concerns the latter.

From the Newtonian point of view, events either are or are not simultaneous with one another 'absolutely', i.e. without any relativity to this or that frame of reference. (And even if we change from separate points of space, and of time, to joint points of space-time, still the absoluteness of simultaneity can be maintained as a relationship between such points). But subsequent empirical discoveries, especially concerning the behaviour of light, then lead once more to an unanswerable question. For if we retain what seem to be very natural assumptions on how to estimate the simultaneity of distant events, we find that the same pair of events will be counted as simultaneous by one observer, and as non-simultaneous by another, even though the only relevant difference between the two is that they are moving (with constant velocity) relative to one another. But again it is built into the theory that this cannot make any 'real' difference between them. So we ask 'are these two events "really" simultaneous or not?', and once more this question cannot be answered.

The scientifically accepted solution to this problem is again a conceptual reform. We should cease thinking of events as 'absolutely' simultaneous, and recognise that simultaneity is always relative to a point of view, i.e. to a particular frame of reference. Then all that can be said is that the two events are simultaneous relative to one 'observer', but not relative to the other, and that is *all* that can be said. On the new way of thinking, the old question of whether they are 'really' or 'absolutely' simultaneous simply cannot be raised, and that is an *advantage* for the new way, just because on the old way it could be raised but could not be answered. To put this in another way, the old question was a mistaken question, arising only because we were trying to interpret the world by means of a scheme of concepts that, as we now see, was not adequate for the task.

Examples could be multiplied, but the general point should already be quite clear. Experience certainly can show us that conceptual reform is required, that one way of thinking about the world is unprofitable, and is better replaced by another. In principle, I see *no* concept that is immune from this kind of revision in the light of experience, and that includes the classical concept of truth, and the associated logic of truth-functions. It really might turn out that this was better abandoned, in favour of an alternative conception and an alternative logic, since the new logic yielded a more satisfying theory overall. One particular way in which it could be more satisfying is that what had seemed, on the old way of thinking, to be quite unanswerable puzzles now simply disappear. No doubt there might be other ways too, but I need not enquire further into that, for this one way is enough to make the point, and it is the main consideration appealed to by those who advocate quantum logic. I am not claiming that their appeal is actually successful in this case, for certainly there is much that can be urged on the other side. But it serves perfectly nicely to illustrate how a change even in logic itself *could* turn out to make better sense of the world that we experience. This is not something that we can rule out *a priori*. There is no limit to the conceptual reform which, in the centuries to come, the 'tribunal of experience' might make desirable.

I add two brief footnotes. First, it should be clear that the argument generalises to show that there cannot be *any a priori* knowledge of the world that we experience. Indeed, we may generalise further: there cannot be *a priori* knowledge of any realm which exists, and has its own nature, independently of our way of thinking about it. For it may always turn out that our present way of thinking about it is not satisfactory, and is better replaced by another. I have concentrated upon the case of (very elementary) logic, because that is where most of us feel most resistance to this thought. But even in this case the resistance is, I have argued, misplaced.

Second, someone might wish to hold that there are certain *central* theses of classical logic which could never be abandoned — a plausible candidate might be the law of non-contradiction — even though we might perhaps be led to abandon more peripheral theses, such as the law of distribution, or of excluded middle, or some others. There are two obvious responses. (i) The task of drawing a line between those parts of classical logic which are central, and must always be retained, and other parts which might perhaps be abandoned, seems to me evidently impossible. Once it is admitted that a change in basic conceptions might lead us to abandon some parts, there is no principled way of saying just which parts could be thus affected. (ii) Supposing that this challenge could in some way be met, still the best that one could hope to do is to draw a line round those parts that seem central from our *present* perspective. But if in future we were to make alterations in the periphery beyond this line, then that would presumably have an effect upon what seemed to be central from the new point of view. So the boundaries of 'the centre' might themselves be expected to shift with each new reform. But if that is granted then clearly the line-drawing project cannot possibly succeed.

As I said before, there really is *no a priori* limit to the conceptual reforms that further experience may lead us to.

6.3 *Coda*

I add one brief concluding word. I believe that the empiricist approach to mathematics is correct, but that is because I believe that it is the correct approach to everything, including logic. However, if one wishes to stick to the apriorist view of logic, then I think one should accept it for mathematics too. There is no successful argument which shows that mathematics and logic are different in this respect.

Most of those who have adopted an empiricist attitude to our knowledge of mathematics have offered 'reductive' accounts of what mathematics is about. They have wanted to maintain that, when properly understood, mathematics concerns the ordinary observable features of quite ordinary objects in this world. This theme can be traced in Aristotle, in Mill, in Kitcher, and in Maddy. We have seen various difficulties in their different accounts. I have also offered (in section 5.3) a general argument against *any* such reduction. But suppose that we set these objections aside, and grant for the sake of argument that some such reduction could work. Then there is a question which *none* of those just mentioned have taken very seriously, namely: why should you suppose that these statements, to which mathematics is reduced, can be known only empirically? The fact that they concern 'ordinary observable features of quite ordinary objects' does not by itself ensure that *only* observation can establish their truth. Indeed, one of the several 'reductionist' theories (and perhaps the best one) is due to Russell, and *his* view was that the reduction showed that these statements can be established by logic alone. It seems to me that he was much more nearly right than other reductionists whom we have considered. But if the reductionist should be a logicist, then of course he should not also be an empiricist, unless he takes an empiricist view of logic itself.

Anyway, reductionism does not work. The statements of pure mathematics are *closely related* to those which the reductionist takes as their paraphrase, but the two cannot simply be identified. Nevertheless the reason why we all believe in the statements of pure mathematics is that they do generalise from, systematise, unify, and provide calculations that apply to those various statements in which numbers are applied. The generalisations (as I believe) cannot strictly be *needed* for the *explanation* of any actual physical phenomenon. The claim that physical phenomena would not be as they are, unless the posited abstract objects did really exist, is one that I find wholly incredible. So I reject the indispensability argument as put forward by Quine and by Putnam, preferring instead to construe pure mathematics in a purely instrumentalist way, as a convenient fiction that is very helpful for purposes of calculation, and helpful too as providing a vocabulary with which to express our scientific theories, but a fiction nonetheless. So far as all practical matters are concerned, we *could* (at least in principle) dispense with it, but only at a considerable cost in added intellectual labour. I therefore conclude that we do not have a reason for taking it to be true.

But others may think differently. They may suppose that, because the theory is so useful, it must be true. And what is it useful for? Well, as I have said, mainly for providing convenient calculations, which generalise, systematise, and unify a number of propositions in which we say that this theory is applied. These latter propositions form the data, to which the theory is responsible, for they can be known independently. How? Well, my answer is 'by analysis and logic', so if these can be known *a priori*, then *a priori* methods could provide the data on which this theory of pure mathematics is based. Then what about the epistemic status of the theory itself? I should think that if it counts as knowable at all then it must also count as knowable *a priori*, for it is known only because it fits so well these (supposedly) *a priori* data.

So I am inclined to conclude that, even if we accept the indispensability argument, there is *still* a good case for saying that if our knowledge of logic is (or could be) *a priori* then the same will apply to our knowledge of mathematics. Of course this is not a criticism of the chief exponents of the indispensability argument, namely Quine and Putnam. For they both believe (as I do) that our knowledge of logic could only be empirical.

No doubt there are many other considerations that could be adduced both for and against empiricism in the philosophy of mathematics. But my discussion must stop somewhere.

BIBLIOGRAPHY

[Aristotle,] Aristotle. Greek texts: Oxford Classical Texts (Oxford: Oxford University Press, various dates); many translations.

[Ayer, 1936] A. J. Ayer. *Language, Truth and Logic*. London: Victor Gollancz Ltd, 1936.

[Benacerraf, 1965] P. Benacerraf. What numbers could not be. *Philosophical Review*, 74: 47–73, 1965.
[Benacerraf, 1973] P. Benacerraf. Mathematical truth. *Journal of Philosophy*, 70: 661–79, 1973.
[Bolyai, 1832] J. Bolyai. The science of absolute space. Published as an appendix to W. Bolyai, *Tentamen Juventutem Studiosam in Elementa Matheseos*, Budapest, 1832.
[BonJour, 1980] L. BonJour. Externalist theories of empirical knowledge. In P. A. French, T. E. Uehling, H. K. Wettstein (eds.), *Midwest Studies in Philosophy 5: Studies in Epistemology*, University of Minnesota, pages 53–73, 1980.
[BonJour, 1998] L. BonJour. *In Defence of Pure Reason*, Cambridge: Cambridge University Press, 1998.
[Bostock, 1972/3] D. Bostock. Aristotle, Zeno, and the potential infinite. *Proceedings of the Aristotelian Society*, 73: 37–51, 1972/3.
[Bostock, 1979] D. Bostock. *Logic and Arithmetic*. Vol 2, Oxford: Clarendon Press, 1979.
[Bostock, 1980] D. Bostock. A study of type-neutrality. *Journal of Philosophical Logic*, 9: 211–296 and 363–414, 1980.
[Bostock, 1990] D. Bostock. Logic and empiricism. *Mind*, 99: 572–82, 1990.
[Bostock, 1998] D. Bostock. On motivating higher-order logic. In T. Smiley (ed.), *Philosophical Logic*, Oxford: Oxford University Press, pages 29–43, 1998.
[Bostock, 2000] D. Bostock. *Aristotle's Ethics*. Oxford: Oxford University Press, 2000.
[Boyer, 1949] C. B. Boyer. *The Concepts of the Calculus*, 1949. Re-published as *The History of the Calculus and its Conceptual Development*, New York: Dover, 1959.
[Casullo, 2003] A. Casullo. *A Priori Justification*. Oxford: Oxford University Press, 2003.
[Chihara, 1973] C. S. Chihara. *Ontology and the Vicious-Circle Principle*. Ithaca: Cornell University Press, 1973.
[Chihara, 1982] C. S. Chihara. A Gödelian thesis regarding mathematical objects: Do they exist? And can we perceive them? *Philosophical Review*, 91: 211–27, 1982.
[Chihara, 1990] C. S. Chihara. *Constructibility and Mathematical Existence*, Oxford: Oxford University Press, 1990.
[Copi, 1971] I. M. Copi. *The Theory of Logical Types*, London: Routledge & Kegan Paul, 1971.
[Dedekind, 1872] R. Dedekind. Continuity and irrational numbers, 1872. In his [1963].
[Dedekind, 1888] R. Dedekind. The nature and meaning of numbers, 1888. In his [1963].
[Dedekind, 1963] R. Dedekind. *Essays on the Theory of Numbers*. Tr. W. W. Beman, New York: Dover, 1963.
[Dummett, 1976] M. Dummett. Is logic empirical? Pages 269–89, 1976. (Reprinted in his *Truth and Other Enigmas*, London: Duckworth, 1978.)
[Dummett, 1977] M. Dummett. *Elements of Intuitionism*, Oxford: Oxford University Press, 1977.
[Edidin, 1984] A. Edidin. A priori knowledge for fallibilists. *Philosophical Studies*, 46: 189–97, 1984.
[Euclid,] Euclid. *The Elements*. Tr. and ed. T. L. Heath, 2nd ed, New York: Dover, 1956.
[Feferman, 1964] S. Feferman. Systems of predicative analysis. *Journal of Symbolic Logic*, 29, 1964. Reprinted in J. Hintikka (ed.), *Philosophy of Mathematics*. Oxford: Oxford University Press, pages 95–109, 1969.
[Field, 1980] H. H. Field. *Science Without Numbers*. Oxford: Blackwell, 1980.
[Field, 1985] H. H. Field. Comments and criticisms on conservativeness and incompleteness. *Journal of Philosophy*, 82: 239–60, 1985.
[Frege, 1884] G. Frege. *Foundations of Arithmetic*, 1884. Tr. J. L. Austin, Oxford: Blackwell, 1950.
[Frege, 1892] G. Frege. On concept and object. In his *Philosophical Writings*, tr. Geach and Black. Oxford: Blackwell, pages 42–55, 1892.
[Frege, 1893] G. Frege. *The Basic Laws of Arithmetic*, 1893. Tr. in part by M. Furth. Berkeley & Los Angeles: University of California Press, 1964.
[Frege, 1979] G. Frege. *Posthumous Writings*. H. Hermes, F. Kambartel, F. Kaulbach (eds.), tr. P. Long and R. White. Oxford: Blackwell, 1979.
[Gödel, 1944] K. Gödel. Russell's mathematical logic. In P. A. Schilpp (ed.), *The Philosophy of Bertrand Russell*. Evanston and Chicago: Northwestern University, 1944. Reprinted in P. Benacerraf and H. Putnam (eds.), *Philosophy of Mathematics; Selected Readings*. Cambridge: Cambridge University Press, 2nd ed 1983, pages 447–69, and cited from there.

[Gödel, 1947] K. Gödel. What is Cantor's continuum problem?. *American Mathematical Monthly*, 54: 515–25, 1947. Revised and expanded in P. Benacerraf and H. Putnam, (eds.), *Philosophy of Mathematics; Selected Readings*. Cambridge: Cambridge University Press, 2nd ed 1983, pages 470–85, and cited from there.

[Goldman, 1980] A. Goldman. The internalist conception of justification. In P. A. French, T. E. Uhling, H. K. Wettstein (eds.). *Midwest Studies in Philosophy 5: Studies in Epistemology*, University of Minnesota, pages 27–51, 1980.

[Hale, 1987] B. Hale. *Abstract Objects*. Oxford: Blackwell, 1987.

[Hilbert,] D. Hilbert. *Foundations of Geometry*. Leipzig, 1899.

[Hume, 1748] D. Hume. *Enquiry Concerning Human Understanding*. (Many editions), 1748.

[Kant, 1781/87] I. Kant. *Critique of Pure Reason*. (Many editions), 1781/87.

[Kitcher, 1984] P. Kitcher. *The Nature of Mathematical Knowledge*. Oxford: Oxford University Press, 1984.

[Kline, 1972] M. Kline. *Mathematical Thought from Ancient to Modern Times*. Oxford: Oxford University Press, 1972.

[Kripke, 1972] S. A. Kripke. Naming and Necessity. In G. Harman and D. Davidson (eds.), *Semantics of Natural Language*. Dordrecht: D. Reidel, 1972. Revised and enlarged edition as a book, Oxford: Blackwell, 1980, and cited from there.

[Lobatchevsky, 1830] N. I. Lobatchevsky. On the foundations of geometry. *Kazan Journal*, 1829/30.

[Maddy, 1990] P. Maddy. *Realism in Mathematics*. Oxford: Oxford University Press, 1990.

[Maddy, 1997] P. Maddy. *Naturalism in Mathematics*. Oxford: Clarendon Press, 1997.

[Mancosu, 1996] P. Mancosu. *Philosophy of Mathematics and Mathematical Practice in the Seventeenth Century*. Oxford: Oxford University Press, 1996.

[Manfredi, 2000] P. A. Manfredi. The compatibility of a priori knowledge and empirical defeasibility. *Southern Journal of Philosophy*, 38(Supplement): 159–77, 2000.

[Mill, 1843] J. S. Mill. *System of Logic*. (Many editions), 1843.

[Neugebauer, 1957] O. Neugebauer. *The Exact Sciences in Antiquity*. (2nd ed, Brown University Press, 1957.

[Newton, 1686] I. Newton. *Principia Mathematica*, 1686. Tr. Motte, rev. Cajori, Berkeley & Los Angeles: University of California Press, 1934.

[Newton-Smith, 1978] W. H. Newton-Smith. The underdetermination of theory by data. *Aristotelian Society Supplementary Volume*, 52: 71–91, 1978.

[Nozick, 1981] R. Nozick. *Philosophical Explanations*. Oxford: Oxford University Press, 1981.

[Parsons, 1986] C. Parsons. Review of Kitcher [1984]. *Philosophical Review*, 95: 129–37, 1986.

[Parsons, 1979/80] C. Parsons. Mathematical intuition. *Proceedings of the Aristotelian Society*, 80: 145–68. Reprinted in W. D. Hart (ed.), *The Philosophy of Mathematics*. Oxford, Oxford University Press, 1996 (and cited from there).

[Peano, 1901] G. Peano. *Formulaire de Mathématique*. Paris, 1901.

[Plato,] Plato. Greek texts: Oxford Classical Texts. Oxford: Oxford University Press, various dates; many translations.

[Poincaré, 1905-6] H. Poincaré. Les mathématiques et la logique. *Revue de Metaphysique et de Morale*, 13: 815–35, 14: 17–34, 294–317, 1905-6.

[Putnam, 1968] H. Putnam. Is logic empirical? In R. Cohen and M. Wartofsky (eds.), *Boston Studies in the Philosophy of Science 5*, Dordrecht: D. Reidel, 1968. Reprinted as 'The Logic of Quantum Mechanics', in his *Philosophical Papers vol I*, pages 174–97. Cambridge: Cambridge University Press, 2nd ed 1979, and cited from there.

[Putnam, 1971] H. Putnam. *Philosophy of Logic*. London: George Allen & Unwin, 1971. Reprinted in his *Philosophical Papers vol I*, pages 323–57. Cambridge: Cambridge University Press, 2nd ed 1979.

[Quine, 1948] W. V. Quine. On what there is. 1948. Reprinted in his [1980], pages 1–19.

[Quine, 1951] W. V. Quine. Two dogmas of empiricism. 1951. Reprinted in his [1980], pages 20–46.

[Quine, 1963] W. V. Quine. *Set Theory and its Logic*. Cambridge, Mass: Harvard University Press, 1963.

[Quine, 1970] W. V. Quine. *The Philosophy of Logic*. Englewood Cliffs, New Jersey: Prentice-Hall, 1970.

[Quine, 1980] W. V. Quine. *From a Logical Point of View*. Cambridge, Mass: Harvard University Press, 2nd ed 1980.

[Quine, 1986] W. V. Quine. Response to my critics. In L. E. Hahn and P. A. Schilpp (eds.), *The Philosophy of W. V. Quine*. La Salle, Illinois: Open Court, 1986.

[Reichenbach, 1927] H. Reichenbach. Tr. M. Reichenbach and J. Freund, *The Philosophy of Space and Time*, 1927. (New York: Dover, 1957).

[Reichenbach, 1951] H. Reichenbach. *The Rise of Scientific Philosophy*. Berkeley, California, 1951.

[Riemann, 1867] B. Riemann. *Über die Hypothesen, welche der Geometrie zugrunde liegen*. Darmstadt: Wissenschaftliche Buchgesellschaft, 1867.

[Russell, 1907] B. Russell. The regressive method of discovering the premises of mathematics. 1907. In D. Lackey (ed.), *Essays in Analysis by Bertrand Russell*. London: George Allen & Unwin, pages 272–83, 1973.

[Russell, 1908] B. Russell. Mathematical logic as based on the theory of types. 1908. Reprinted in R. C. Marsh (ed.), *Russell: Logic and Knowledge; Essays 1901-50*. London: George Allen & Unwin, 1956.

[Sainsbury, 1979] R. M. Sainsbury. *Russell*. London: Routledge & Kegan Paul, 1979.

[Shapiro, 1983] S. Shapiro. Conservativeness and incompleteness. *Journal of Philosophy*, 80, 1983. Reprinted in W. D. Hart (ed.), *The Philosophy of Mathematics*. Oxford: Oxford University Press, pages 225–34, 1996.

[Summerfield, 1991] D. M. Summerfield. Modest a priori knowledge. *Philosophy and Phenomenological Research*, 51: 39–66, 1991.

[Wang, 1962] H. Wang. *A Survey of Mathematical Logic*. Amsterdam: North-Holland, 1962.

[Wittgenstein, 1921] L. Wittgenstein. *Tractatus Logico-Philosophicus*. 1921. Tr. D. F. Pears and B. F. McGuinness, London: Routledge & Kegan Paul, 1961.

A KANTIAN PERSPECTIVE ON THE PHILOSOPHY OF MATHEMATICS

Mary Tiles

One of the most distinctive and original aspects of Kant's philosophy is the way in which it exploits the connection between the repeated application of a rule, law, or function and order, regularity, structure or form.[1] Kant distinguishes the rational (and thereby also moral) being from the non-rational on the basis of its capacity to act not merely according to a rule (or law) but according to its conception of the rule [Kant, 1959, 29, Ak.IV 412]. This is the capacity on which the possibility of logic, mathematics, scientific knowledge and morality depend. It links the dynamic, the temporal, the realm of action and process with the static, the spatial and quasi-spatial structures, the realm of representation and theoretically articulated knowledge. It connects thought with action and action to thought via the thought of action. Equally importantly Kant saw reason as issuing its own imperatives with regard both to thought and action Reason, with its demand for unifying principles, for ultimates, for unconditioned starting points dictates the form that theoretical and practical understanding should take.[2] It is

---

[1] Cassirer [1955, 79ff] too stresses this in his reading of Kant.

[2] Although Kant is frequently read as assuming a fixed universal rational capacity with its own innate principles, given once and for all, this is not well supported in his texts. Humans living in society find themselves endowed with language, with the capacity to communicate thoughts, to dispute about principles and to place demands upon one another in the name of reason. Kant's view of the origin of our capacity for thought is arguably more sociological than individualistically psychological.

> We do admittedly say that, whereas a higher authority may deprive us of freedom of *speech* or *writing*, it cannot deprive us of freedom of *thought*. But how much and how accurately would we *think* if we did not think, so to speak, in community with others to whom we *communicate* our thoughts and who communicate theirs to us! We may therefore conclude that the same external constraint which deprives people of the freedom to *communicate* their thoughts in public also removes their freedom of *thought*, the one treasure which remains to us amidst all the burdens of civil life, and which alone offers us a means of overcoming all the evils of this condition. [Kant, 1991, 247]

Kant points out that our presumption that objective knowledge can be distinguished from subjective opinion rests on an assumption of a basic uniformity in human capacities such that through communication they can come to agreement in judgment of matters of fact as well as on an assumption that there is a matter of fact (an object) about which to agree.

> The touchstone whereby we decide whether holding a thing to be true as conviction or mere persuasion is therefore external, namely the possibility of communicating it and find it to be valid for all human reason... [Kant, 1965, A820/B848].

thus also reason with its drive for unity and completion, which pushes us to think beyond finite limits and to postulate an infinite. In his discussion of the ways in which such a drive seems to bring reason into conflict with itself (to spawn internal contradictions) a key strategy is to provide freedom of philosophic movement by (a) rejecting the views of both rationalist and empiricist philosophy while (b) insisting on the need to acknowledge the distinctive roles of the (empirically) real and the (rationally) ideal by preserving the space between them. This seems to be a space that philosophers with reductionist or foundationalist tendencies seem to find hard to keep open.

The infinite was one of the central foci of philosophical and mathematical angst motivating a remarkable period, which extended from the latter half of the nineteenth century through the first half of the twentieth century. During this period philosophical mathematicians and mathematical philosophers sought ways to legitimize the use of infinitistic methods in mathematics and guarantee their freedom from internal contradiction. And even though those debates died down once mathematicians became satisfied that they had a sufficiently secure basis (in axiomatic set theory coupled with first order predicate calculus) to continue, the infinite has remained a locus of unresolved philosophical problems and of open mathematical questions (such as the status of the Continuum Hypothesis). If Kant's basic analysis of the source and nature of the drive to move beyond the finite is sound, then there could be much to gain, philosophically, by approaching the philosophy of mathematics and its commitments to the infinite from a Kantian perspective. Of course, many have argued that Kant was fundamentally misguided, that it is precisely his view of mathematical judgments as synthetic *a priori* in nature that proves to be the Achilles heel of his whole critical enterprise and that this vulnerability was exposed with the advent of non-Euclidean geometry. Nevertheless, many of these arguments miss their target because they treat Kant's claim about the status of mathematics as if it were an answer to questions raised in the kind of philosophical framework that Kant was rejecting.

1 MATHEMATICS, SCIENCE OF FORMS

Kant's claim that mathematical knowledge is synthetic *a priori* actually has two components. One is that mathematics can claim to give *a priori knowledge* of (universally applicable to) objects of possible experience because it is the science of the forms of intuition (space and time which are conditions under which all objects of experience are made known to us). The other is that the way in which mathematical knowledge is gained is through the synthesis (construction) of objects corresponding to its concepts, not by the analysis of concepts. The basis of its knowledge is distinguished both from that of general (formal) logic and from that of the empirical sciences. It can start with axioms and definitions and proceed thence to derive theorems only to the extent that its definitions result not only in concepts but also in (pure) intuitions of objects corresponding to them. In other words, the concepts are rules for constructing a form (structure) in pure intuition (i.e., out of nothing, no material).

There are thus two general theses here, between which there has been considerable confusion. These theses correspond to different questions Kant is trying to answer. One is "How is mathematical physics possible?" which is related to the broader question of how synthetic *a posteriori* knowledge is possible. Another has to do with the scope and limits of scientific knowledge, which arises out of the following conflict: on the one hand, we suppose that the world we live in is a world which is completely knowable in a manner that conforms to our ideals of what complete knowledge would be like, and, on the other, we suppose that we live in a world in which we are significant causal agents confronted with real choices (choices that make a difference, and whose outcome is not already known or knowable by an omniscient being (whether hypothetical or real).

Kant's account of the distinctive role and nature of mathematics forms a crucial part of his way of addressing both these questions. He argued that reason and mathematics are responsible for setting up ideals of complete knowledge and ideas corresponding to them. While the ideas are not even in principle applicable to the empirical world — i.e., the empirical world cannot be completely known as having the kind of fixed, fully determinate structures required for fully rationally articulated and demonstrated knowledge — the corresponding ideals have nevertheless an important practical, regulative function. Further he recognizes that we cannot have the ideals without the ideas, so it is not possible to go along with radical empiricists, such as Hume, who thought we could banish all non-empirically grounded ideas. In articulating this position Kant argues that the price of mathematical certainty is recognition that its possibility is grounded in the fact that the mathematical edifice is a human construction (but not an arbitrary one) and that its necessary employment in our empirical dealings with the world licenses neither the metaphysical claim that there is realm of entities existing independently of all human beings which is in itself mathematically structured, nor the claim that there is a realm of independently existing mathematical objects.

If pure mathematics is the study of the possible structures of manifolds, natural numbers are fundamental in that they are both measures and markers of discrete plurality. — a plurality is a plurality of units (individuals). But, if we follow the line of discussion indicated above, objects (units) are "given" (grounded) not conceptually, but practically or operationally. Whether one wants to call this intuition or not, it is that elusive interface between theoretical representation and practical application. This interface (the application of a concept to experience), Kant argues, always goes through the mediation of a schema, linked to a method or procedure for its application. In the case of *a priori* categorical concepts these have to be pure schema (products of pure productive imagination). Kant's argument is that our presumption that the categorical concepts (unity, plurality, causality, etc.) have empirical application is already a presumption that the world of possible objects of experience is one to which basic mathematical concepts necessarily apply. This is because the schemata of the concepts are already at work in constituting our (schematic) conception of the possible object of experience as a determinate unit extended in time and/or space.

So Kant's account of mathematics as a source of synthetic *a priori* knowledge has two closely interwoven, but distinguishable parts. One is an account of the nature of and necessity for empirical applications of mathematics (where it contributes to providing synthetic *a priori* knowledge of empirical objects). The other is an account of the distinctively constructive nature of pure mathematical objects (forms), concepts, and reasoning and of the need to recognize the status of these as products of idealization which are not to be encountered in the empirical world.

The development and application of non-Euclidean geometry in Einstein's theories of relativity does not fundamentally disrupt this picture, but it bears more directly on the first of the two Kantian theses than on the second. In fact it serves to underscore Kant's message that the mathematical forms in which we write our causal laws themselves have implications for the geometrical structure attributed to space and time (or space-time). The truths of Euclidean geometry are not deniable within Newtonian mechanics since they are built into its causal structure. Equally, Newton's three laws of motion assume the role of synthetic *a priori* truths, structuring the theoretical framework he brought to bear to organize and explain empirical phenomena. This does not, however, make them immune to revision. (Detailed discussion of this is given in the Appendix.)

The more general plank of the Kantian position is a point about the role of relations in constituting a world of individual empirical objects, and about mathematics as the provider of the theory not just of pure relational structures but also of the identification, individuation and definition of objects within them. For this Kant drew heavily on Leibnizian ideas, while at the same time being highly critical of Leibnizian metaphysics.

2 INDIVIDUAL OBJECTS — WHY MATHEMATICS CANNOT BE REDUCED TO LOGIC

The rise of Newtonian mechanics as a paradigm for the kind of knowledge to be sought by science, and of the mix of experimental and mathematical methods by which it could be achieved, represented a change in the "object" of scientific knowledge.[3] Scientific understanding was no longer focused on the structure of genus and species, on knowledge of essences, expressed in terms of *conceptual* relations, but on knowledge of the laws according to which the world of individual objects (including events) is regulated, ordered and structured. As Leibniz repeatedly emphasized, it is impossible to capture the particularity of an individual object with a finite number of predicates, so complete conceptual knowledge of individual objects is beyond our grasp. In addition, he argued that between any two spatially distinguishable objects there is always some qualitative difference

---

[3]For further elaboration on this way of describing such changes and the role of mathematics in them, see Bachelard [1934, especially Chapter VI].

— however small. His thesis of the identity of indiscernibles then becomes an infinitistic (or second order) principle — two objects are identical if and only if they have all the same properties ($a = b \leftrightarrow \forall \Phi(\Phi(a) \leftrightarrow \Phi(b))$). As Hume had earlier argued, such a concept of identity can neither be derived from experience, nor will any (finite amount of) experience ever fully justify an application of it. Yet, identity and unity are presupposed in all talk and thought of objects. They have the status of categories, *a priori* concepts, presupposed by the logical forms of judgment (whether expressed in thought or in language).

Kant, and subsequent neoKantians, have thus argued that the possibility of scientific knowledge gained by experience, of a world of individual objects, is conditional upon presupposing empirically applicable means of identifying and individuating the objects under investigation.[4] This requires that they be identified and individuated in terms of their relations to one another, not in terms of purely intrinsic qualities. The most universal frameworks within which we do this are those of space and time whose founding relations (again identified by Leibniz) are those of succession and co-existence. But to be able to use space and time as frameworks for the individuation and identification of empirical objects and events they need to be established as reference frames (they need empirical measures and the presupposed mathematical structures that come with them). In furthering the argument that the structures required here cannot be logical, conceptual structures, Kant argues for the distinctness of the part-whole relation for concepts and the part-whole relation for (extended) objects. The relational complexity of parts in a physical whole is of a different order from that of the conceptual part-whole relation. (E.g., whereas whatever can be truly predicated of the genus (whole), can be truly predicated of the species (part), a spatially asymmetrical object (a spiral snail shell) does not necessarily have only spatially asymmetrical parts). At the very least, it has to be granted that the logic of relations is distinct from that of concepts and has an important role to play in the articulation of empirical, scientific knowledge. The application of concepts to objects presupposes that their identity and individuality is given in a relational reference frame, a frame that plays a constitutive role in relation to the objects identifiable within it.

Mathematics, as the pure theory of manifolds and their possible (relational) structures is thus presupposed in any knowledge of objects, and in any logic which includes the forms of knowledge of objects as well as concepts, since it presupposes objects as given, as identifiable and capable of individuation in some manner. Equally, mathematics is dependent on logic for the expression of its knowledge and for the theory of the forms of its judgments and principles of its reasoning. Thus in insisting that knowledge requires both intuitions and concepts, Kant is also insisting that it requires both mathematics and logic to articulate its forms.

Some logicists have followed Frege [1953, §104–106] in wanting to preserve the idea that mathematical knowledge is knowledge of abstract objects, and not merely knowledge of what the logical consequences of a set of axioms are. Frege insisted both that arithmetic is knowledge of numbers as objects and that this knowledge

---

[4]This issue is explored at much greater length in [Tiles, 2004].

can be obtained by reasoning from definitions according to laws of logic. (He retained a Kantian view of the status of geometry [Frege, 1971, 14].) To uphold this view it is necessary to believe that it is possible to define numbers as objects. The much more recent neologicist program launched by Wright and Hale, which is neo-Fregean rather than neo-Russellian, attempts to show that fundamental mathematical theories, such as arithmetic and analysis, can be founded in *abstraction principles*. These are principles that have the form $(\alpha)(\beta)(O(\alpha) = O(\beta) \leftrightarrow \alpha \equiv \beta)$, where $\equiv$ is an equivalence relation on entities of the type, over which the variables α and β range, and O is a function from entities of that type to objects, [Hale, 2002, 304]. But the quantifiers here are assumed to range over individual entities (to which concepts apply); they presume a manifold, and in so doing already presuppose the founding concepts of arithmetic.[5]

Although Frege extended logic to include the logic of relations, he did so by assimilating relations to concepts, so the distinction is marked now only as the distinction between one- and many-place predicates (note the use of numbers to express this.) In so doing he fails to recognize any constitutive role for relations. His attempts to secure an *absolute* reference for numbers through use of abstraction principles fails, as he himself recognized [Frege, 1893, §10; Frege, 1903, Appendix], for although he specified numbers as classes, he cannot define what it is to be a class and hence secure reference to classes as unique objects.

We should further note that in this respect at least the developments in formal logic have not fundamentally changed the situation. First order theories satisfiable only in infinite domains cannot secure a unique interpretation of their "objects" nor can they ensure categoricity (the isomorphism of all structures satisfying the axioms.)[6] Logic requires the identity of indiscernibles to assure uniqueness and this is a second order (infinitistic) principle since it requires quantification over all predicates of the language in question. The quantifiers in first order logic presume a "manifold" of individual objects as given. Even if identity is added as a primitive "logical" relation, there are no first order axioms that can prevent its interpretation as an equivalence relation, rather than a "true" identity, relation. The Kantian approach to identity would say that it is not a logically grounded relation (since it is presupposed by all the logical functions of judgment), rather it is pragmatically grounded; the functions and purposes of our representation systems (discursive frameworks) determine what we count as identity for the purposes at hand. So those functions and purposes play a constitutive role in relation to objects represented.

---

[5]This is basically also Hilbert's [1925] argument against logicists, [Hilbert, 1967, 192].
[6]For further implications see, for example, [Quine, 1969].

3 FORMAL RULES — WHY MATHEMATICS CANNOT BE REDUCED TO MANIPULATION OF MARKS ON PAPER

Pragmatists and neo-Kantians have argued for a kind of reverse application of the principle of identity of indiscernibles — identity is pragmatically determined. It is grounded in our practices, founded on establishing relations among objects, and has no *ultimate* justification. Moreover, understanding of this relation is not and cannot be, conceptual; the basis lies in practice, in what we do and the practical standards we enforce through training.[7] We adopt measurement standards (standard objects or standard procedures) and count these as invariant — there is no further standard against which to check (they are conventions). This gives us units that we presume to be identical in the relevant respect. We do this to the point where it seems that (as a result of other comparisons) there is reason to recognize differences and adopt a different standard. This might be one place where one has to agree that objects have to be given in intuition — in a kind of cognition which is non-conceptual and which has no foundation in the pure nature of things, but only in the rules we succeed in setting up to govern our transactions with the world and each other. Such rules are not adopted arbitrarily, they are there to facilitate certain functions and must be rejected when they fail (whether because of internal incoherence or because of inapplicability to the situations in which we attempt to use them).

This is as much as to say that mathematics is formal in the sense that it is the science of possible forms of intuition, not of its possible content. Are we then arguing for formalism? And haven't formalists claimed to have an account of mathematics that eliminates all reliance on intuition? Some formalists have indeed (mistakenly as I shall argue) made this claim but the most notable proponent of what has been called formalism, Hilbert, did not.

True, one can build machines to operate according to rules that *we* interpret as rules of logic or calculation; this is done by translating rules of logic or of arithmetic into causal operating principles of mechanisms. We might also be able to train humans to operate according to those same rules without having any comprehension that they might be rules *for* calculating or reasoning. There is a sense in which *they* too would not be calculating or reasoning because they attach to their performance none of the consequences, none of the potential applications, of their activity, even if others might.[8]

However, we again see the importance of Kant's claim that the mark of rational agents is their ability to act not only in accordance with a rule, but also in accordance with their conception of the rule. The idea of a pure formal calculus, an uninterpreted notation, is that of a system generated by a set of rules for pro-

---

[7]This is the burden of many of Wittgenstein's discussions of rules and rule following, especially, for example, [Wittgenstein, 1963, paragraphs 206–289].

[8]Cf. Getting people to sign their names on pieces of paper that contain text they have not been able to read (perhaps in a foreign language). They are signing their names ignorant of the consequences, while others know that the consequences are that they have just made a confession, or signed away rights to their property.

ducing sequences of marks on paper, where it is possible to specify an algorithm (another rule in the guise of an effective procedure) that will determine whether any given sequence has or has not been produced in accordance with the rules. But can a rational agent ever knowingly play within a *pure* uninterpreted system?

To do so it has to be able to recognize and distinguish the various marks (use concepts classifying them) in order to be able to obey the rules for producing strings and for transforming one string into another. Rules introduce normativity; there are constraints on formation and transformation of strings (some are admissible, others are not) but they also introduce generality. Any rational agent able to follow such a rule, as an explicitly formulated rule, has to have grasped that it is to apply to every presentation of a particular type of mark, or sequence of marks. Concrete marks must thus be read as tokens of a type. In this way recognition of identity is built into application of the rule — these are just two sides of the same coin. Further, any rational agent will realize that the production of a particular sequence of marks (token) will be representative of all other productions issuing from the same sequence of rule applications (the same procedure of construction). That is, grasp of a generative rule (of token production) already presupposes an advance from token (concrete object) to a type (abstract object). There is no further abstraction principle required here; repeatedly applicable formal rules for construction and recognition of abstract objects are indissolubly linked. Rules that are rules for the production (or construction) of objects determine the character of the product (are constitutive) in just those ways that make it possible to tell from the product whether it was or was not constructed according to the rules. Thus the kind of rule thought to characterize a formal system immediately traverses the gap between the particular and the universal, token and type, precisely by being purely formal. In this way concrete marks cannot remain without signification; they symbolically signify the types of which they are tokens. Rules of this kind thus characterize types of processes and a type of structure generated by those processes, and this type of structure can be characterized and known through reflection on active participation in the production of symbols that signify beyond themselves.

This is what makes meta-mathematics, proof theory, etc. possible. Formal languages and formal systems become objects of *mathematical* study and indeed are constituted as mathematical objects in much the manner that Kant describes. The resort to formal systems does not eliminate reliance on "intuition" — on the grasp of rules as rules for constructing objects (and simultaneously defining concepts of them). As objects of study formal systems and their components are no less abstract and no less the subject of mathematical investigation than numbers, points, or sets; numerals are no less abstract than numbers.

4 RULES AND FORMS OF REPRESENTATION — HILBERTIAN FORMALISM

Although Hilbert's name is that most frequently invoked when mention is made of formalism as a philosophy of mathematics, it is important to remember that Hilbert never took the view that mathematics was just an empty game of formal rules. He acknowledges the extent to which arithmetic and geometry have some basis in practices of counting, measuring, computing and for using numerals and diagrams to facilitate indirect measurement, in other words, planning and practical (artisanal) reasoning generally. These rules, being pragmatic in origin are justified if they have been found to work.

The question that philosophers and mathematicians want to answer though is why do their rules work? Can we be assured that they always will work? If we reason using these rules can we be sure they will never lead us astray, especially if they go via ideal elements? Are there other and better ways of "modeling" the situations in which we are interested?

Plato, Aristotle and Euclid set the pattern for answering these kinds of question, and, however one is going to answer, it requires establishing the practices of mathematical representation on a more rigorously rational footing. Exactly what it means to do this has of course been a continuing subject of debate, both philosophical and mathematical. What are the appropriate standards of rigor? Nevertheless, proceeding by analysis to reach basic concepts and basic assumptions — finding secure starting points from which rational reconstruction can proceed — is a common theme. This making rigorous through analysis, explicit definition and axiomatization has always been a matter of reworking something already given to which the definition, axiomatization or formalization is held accountable. A formal arithmetic that cannot be related back to ordinary arithmetic has no right to be called an arithmetic. This is why Hilbert was at pains to distinguish between those statements in a formalized axiomatic theory of arithmetic that had finitary significance (significance not limited to the role of the symbol in the system) and those that, because they invoked ideal elements, did not. Ideal elements could have no empirical interpretation but represent limits, completions or totalizations of those components that do have finitary significance.[9] Finitary arithmetic is thus synthetic in Kant's sense, namely, that in order to understand its statements as asserting something true or false, and in order to determine their truth value, it is necessary to look beyond the formal definitions available in a formalized arithmetic, to something which is instead grounded in the construction of numerals as objects and in their use as numerals (to record the results of counting, measuring or calculating).

In the case of geometry Hilbert interpolates geometrical diagrams between material objects and their mathematical representations. The intuitions on which

---

[9]So, for example, whereas the claim $\exists n = mP(n)$ could be finitarily significant, the claim $\exists n P(n)$ would not be in general because there is no guarantee that one could reach a determination of its truth value in a finite number of steps.

theoretical geometry is founded are derived from practices of drawing diagrams to represent spatial situations (in architecture, in map making and surveying, in astronomy). These practices already perform the "abstraction" of separating what is spatial or structural from what is material.[10] The diagrams don't represent material or qualitative characteristics. Geometers differ from architects in that they aren't interested in what bit of land a map represents, or on the practicality of the methods by which it is produced. They are however interested in being able to answer questions such as "If we assume a piece of land to have particular specified dimensions and topography, can we be sure that the methods used to construct and interpret the map are such as to be able accurately to move from map to terrain and back again?" In other words do these methods have empirical objectivity? But note what happens in doing this — to judge the methods objectively valid we have to assume the objects of representation already have geometrically represented spatial characteristics. In this way the formal characteristics of the representational practice become constitutive not only of representations (as themselves constructed objects) but also of objects as represented; objects which are only ever known as represented in some way or other.

Geometrical diagrams come to have a double reading — as (potential) representations of empirical objects, and as tokens of abstract types — types of figures that are drawn (constructed) in specified ways, where the operation of construction too has a double reading — literally the drawing of a diagram but also abstractly the non-material construction of a pure figure. Geometry thus requires a move from the drawing and use of diagrams (particular empirical representations of empirical situations) to the abstract (universal) form via a *method* of construction (schema). At the same time it imposes a secondary (symbolic) reading on the diagram (a reference to a non-empirical, ideal object). It involves reasoning from construction of an object (representation) according to a general method of construction that becomes definitive of the concept. Euclid's geometry, for example, limited its field of study to and its methods by reference to what can be achieved using straight edge and compass construction (straight lines and circles). The first three postulates are postulates about possible operations.

1. To draw a straight line from any point to any point.

2. To produce a finite straight line continuously in a straight line.

3. To describe a circle with any centre and distance. (See [Heath, 1926, 154].)

Descartes in his geometry (see [Descartes, 1925]) had to argue for an extension of its subject matter to allow other kinds of construction so that figures such as conic sections become legitimate geometrical objects. He did not however present his theory axiomatically.

Geometry is important because it reveals the extent to which even our conventions for representing finite spaces, finite figures and the continuous movement of

---

[10]Thus Aristotle [1984] remarks that "while geometry investigates natural lines, but not *qua* natural, optics investigates mathematical lines, but not *qua* mathematical." p. 331, 193b20-24.

(rigid) objects within finite spaces implicitly introduce the infinite and in more than one way. The infinite lies coiled within the concept of the homogeneous continuity of a line or of space (infinite divisibility); it is there in the definition of parallel lines and their use in facilitating comparison of angles and ratios. It is there in our presumption that objects represented have determinate lengths, areas, etc., that can be ever more accurately approximated by empirical measurements. It is there in the conception of points, lines and planes as limits, as pure boundaries lacking volume or area or anything that could make them possible objects of experience. It also reveals how it takes the analytic effort involved in axiomatization to reveal what exactly are the assumptions on which our accepted methods rest. And the repeated reconceptualizations of the subject show that analysis back to what are considered simple starting points (simple constructions, simple objects, simple concepts, and defining statements about their relations) changes in response to changes in the broader field of mathematics and in the practical, representational, demands placed on it by other fields (rational mechanics, theory of perspective, and optics, fluid mechanics, etc.) Thus Hilbert said

> The use of geometrical symbols as a means of strict proof presupposes the exact knowledge and complete mastery of axioms which lie at the foundations of those figures; and in order that these geometrical figures may be incorporated in the geometrical features of mathematical symbols, a rigorous axiomatic investigation of their conceptual content is necessary. Just as in adding two numbers, one must place the digits under each other in the right order so that only the rules of calculation, i.e. the axioms of arithmetic, determine the correct use of the digits, so the use of geometrical symbols is determined by the axioms of geometrical concepts and their combinations. [Hilbert, 1900, 79]

For Hilbert rigorization through axiomatization is a process in which familiar concepts are reforged, rather than eternal truths intuited (contra Frege) or purely arbitrary rules set up (contra hard-headed formalists).

5 AXIOMATIZATION AND STRUCTURES — CHANGING THE OBJECT OF MATHEMATICS

Hilbert's own axiomatization of geometry was given to make more rigorous a far more extensive corpus of geometrical practices than those of the geometry of Euclid's *Elements*. It included the practices of analytic geometry where algebraic and geometrical representations are combined and where geometric conclusions are based on algebraic reasoning. Hilbert stated his goal as being

> ... to establish for geometry a *complete* and as *simple as possible* set of axioms and to deduce from them the most important geometric theorems in such a way that the meaning of the various groups of axioms, as well as the significance of the conclusions that can be drawn from the individual axioms comes to light. [Hilbert, 1971, 2]

The controversial aspect of his approach (in which he disagreed strongly with Frege) was that he did not treat axioms as the expression of truths about space conceived as having its own, intrinsic and determinate structure, but as combining to define the structure of Euclidean space by more precisely determining the primitive concepts and relations required to characterize this structure as well as clarify the meaning of these primitive concepts. His approach reflects the changed conception of geometry as no longer focused solely on spatial figures and the establishment of their geometrical characteristics and interrelations, but as recognizing that any such study makes more fundamental presuppositions about the nature of the space of which these spatial objects are determinations (or limitations). Specifically (as Leibniz and Kant had already been urging) this means presuppositions about its structuring relations (relations of coexistence). Hilbert's axioms are divided into three groups. Each of the first three groups aims to characterize the structural properties of a single relation: I — incidence, II — order, III — congruence. Group IV consists simply of an axiom of parallels, and Group V contains two continuity axioms. Before presenting the axioms he gives a "definition":

> Consider three distinct sets of objects. Let the objects of the **first** set be called *points* and be denoted by $A, B, C, \ldots$, Let the objects of the **second** set be called *lines* and be denoted by $a, b, c, \ldots$; let the objects of the **third** set be called *planes* and be denoted by $\alpha, \beta, \chi, \ldots$; the points and lines and planes are called ... the elements of the space. [Hilbert, 1971, 3]

The elements are thus merely presumed to belong to distinct sets with notationally distinguished variables to range over each. The axioms have to do the work of filling out these concepts that are defined only in relation to one another. However, the whole project would fail were it not possible to recapture as theorems standard geometrical theorems expressed using our antecedent understanding of the terms point, line and plane. Yet although the axiomatization is aimed at providing a more rigorous and complete analysis of antecedent concepts, it is equally important that other sets of objects, with other relations, can satisfy the axioms. Hilbert uses such "models" as diagnostic tools for probing the properties of his axioms. By showing that the real numbers can be used to provide a model for all the axioms, he shows them to be consistent, relative to the theory of real numbers. And, conversely, that sets of real numbers can provide numerical substitute representations for the space of experience or for diagrams.

By showing that all axioms except the last continuity axiom have a model in the field of algebraic numbers, he shows that this last axiom is independent of the rest (cannot be proved from them).[11] In other words, with his axiomatization of geometry Hilbert also illustrated the utility of the methods of model theory for investigating axiom systems, but model theory itself needs somewhere from which

---
[11] Hilbert requires axioms to be consistent and mutually independent. Of course, the question of how or whether consistency can be proved in an absolute fashion was to be the problem posed in Hilbert's program.

to draw models. Along with the use of axiom systems to characterize relational structures came the need for a theory of systems of objects (manifolds) to provide the modeling tools. Hence the idea that set theory is the foundational theory for mathematics — all the rest of mathematics can be reduced to set theory and proved consistent relative to it.

6 DOES SET THEORY PROVIDE A PURE THEORY OF MANIFOLDS?

Another plank of anti-Kantian views of mathematics is thus the claim that with the arithmetization of analysis, Hilbert's axiomatization of geometry, Peano's axiomatization of arithmetic and the demonstration that axiomatic set theory can provide a foundation for (almost) all of mathematics, reliance on intuition has been eliminated from mathematics. However, in this case, because there are significant questions about sets (perhaps most notably the Cantor's Continuum Hypothesis) that have been proved not to be decidable on the basis of the most widely accepted axioms (those of Zermelo-Fraenkel), traditional rationalist and empiricist forms of dogmatic realism have re-emerged as ways to save the view that the axioms of set theory do express truths about sets, with some form of intuition as the source of at least the basic concept of set. Platonists [Brown, 1999; Gödel, 1964] appeal to non-empirical intuition; empirical realists to empirical intuition [Maddy, 1990]. Realist positions (whether empiricist or rationalist) take the notion of object (and thus unity and identity) as given; Kantian idealist positions do not. So the disagreement here is not over the need for some sort of appeal to intuition, but over the nature of that appeal (or the account of the role of intuition).

It has been presumed moreover that any broadly Kantian account of mathematics must follow the path of Brouwer and his intuitionist and constructivist successors, in repudiating (Zermelo–Fraenkel) set theory altogether because of its deployment of infinististic methods and because, in order to play its role *vis-à-vis* the rest of mathematics it must assume the existence of actually infinite sets. It might be more profitable to leave that as an open question for the time being while we pursue a little further the theme of mathematics as a study of the *apriori* forms of manifolds of intuition (pluralities of objects).

First, however, let us note that, while set theory does indeed play the role of providing the models (domains of objects) for the first order axiomatic theories used to characterize and define the kinds of structures to be studied by the mathematician, it can play this role only because the notion of set is not the logician's notion of class (extension of a predicate). It thus constitutes a response to the Kantian argument, rehearsed above, about the need to distinguish the part-whole relation for objects from that for concepts. The founding relation in set theory is not that of part and whole, but that between a set and an individual member of the set, where the set is again an individual object that may belong to further sets. But the membership relation (unlike the inclusion relation) is not transitive; '$a \in b$ and $b \in c$' does not entail '$a \in c$'. This means that the set theoretic universe is a universe of individual objects some of which have very significant levels of internal complexity. But axiomatic set theory is itself written in the language of first order logic and thus still *presupposes* a domain of individuals as its domain of quantification; it thus still presupposes the notion of a manifold — a plurality of *individual* objects.[12] Moreover, to perform its foundational role it has to countenance actually infinite sets.

7 ORDINAL, CARDINAL AND TWO KINDS OF INFINITE

Kant argued that when one examines the cognitive underpinning of the two basic orders structuring all of our experience — those of coexistence and succession — they are seen to be interdependent. Our ability to think about temporal succession depends on having some atemporal means of representing it. If each moment just slips by unmarked, it is as if it had never existed (as for victims of Alzheimer's disease); any cognition of events and of temporal sequences thus requires a way of (re)presenting the sequence in the order of co-existents. Equally, for us to recognize an order among coexistents also takes time — takes the integration of several cognitive acts. The use of numbers as measures and markers of plurality similarly requires this integration. Even the most basic (cardinal) representation, where a notch is made in a tally stick, or a knot placed on a quipu rope in the process of counting — say, a herd of cows — can be thought of as a transfer from the order of succession to that of co-existence. The successive registering of a cow in a herd by a notch in a stick leads to a cumulation of notches, the "permanent" record capable of direct comparison with last year's, for example. Yet even that comparison might take time, might require a process of counting whose result is recorded in a numeral. Our representational conventions are there in part to facilitate the cumulation of successive structure into something that can be grasped "all at once" and unpacked if need be.[13] In counting using numerals, it is the operation of adding one (of ticking off another object) whose repetitions are recorded (counted) by each successive numeral. But the point of numerals is that via their conventional serial ordering the numeral (ordinal number) reached carries with it information about the size (cardinal number) of the collection counted (the tally stick marks can be dispensed with). This opens up the path to using methods other than direct counting as a way to calculate size. Repetitions are by definition doing the same thing again (there is no further court of appeal here — going up the number sequence in the conventional way while ticking off objects is just what we mean by counting). But if we want to give a schema of the counting process so

---

[12]It could be retorted that axiomatic set theory doesn't need any primitive objects, it only needs the null (or empty) set. Here I would agree with the arguments in Mayberry [2000, 76–7] that the founding mathematical concept of set is that of a plurality of objects which is itself an object, and that it should be regarded as distinct from the logical notion "extension of a concept". On this view the empty set is a kind of ideal (or conventional) object, introduced as a counterpart to 0 as a kind of operational closure (a limit case).

[13]Descartes [1931, Rule XI] makes a similar remark about the need to continually run through a proof (given as a succession of deductive steps) until one can capture the whole in a single intuition.

that we can start to discuss and verify methods of computation, we need to resort to tallying — using counters or symbols — treated as identical units — markers of each (identical repetition). In order words, for the foundations of the finite, natural numbers, ordinal and cardinal (successive and co-existent) orders are both equally fundamental. Axiomatic set theory recognizes this in the Von-Neumann ordinals

$$\emptyset, \{\emptyset\}, \{\emptyset, \{\emptyset\}\}, \{\emptyset, \{\emptyset\}, \{\emptyset, \{\emptyset\}\}\}, \ldots x, x \cup \{x\} \ldots$$

where each finite ordinal is the set of all its predecessors and is the result of adding one object (a set) to its predecessor. So any given number includes in its set theoretic structure (and symbolic, notational structure) the marks of its place in the order of succession and of its representation as a standard coexistent plurality whose number is measured by the number appearing in that place in the order of succession.

So far we have clearly been talking only about finite pluralities. Now Kant had also argued in addition that the manifold in which the order of co-existence is given is that of continuous (not discrete) magnitude. It is the order in which all other magnitudes have to be represented in order to be become objects of cognition — and this supposition is justified only by assuming that division can be made anywhere.[14] However, analytic geometry, with its algebraic methods and use of co-ordinate systems presupposes that the continuum can be modeled arithmetically (by numbers). Hilbert's axiomatization of geometry gives a sense in which that is true; the real numbers can provide a model for his axioms, but in doing this the mathematician now needs to think in terms of infinite manifolds — ones which are not just potentially infinite, in the manner of the natural numbers, but actually infinite — an infinite order of coexistence. One of the problems associated with allowing infinite sets as pluralities is that the cardinal and ordinal aspects of the normal measures of plurality (numbers) seem to part company. The Continuum Hypothesis[15] represents an attempt to reunite them in a particular way, but the concept of "set" captured by the Zermelo-Fraenkel axioms is indeterminate in this regard. Yet, Hilbert's case for the consistency of his axioms for Euclidean geometry rests on the consistency of the theory of real numbers, the theory that provides his model of the axioms. If crucial questions about the continuum remain unanswered within axiomatic set theory, with what justice can it be claimed to have provided a foundation for geometry that supplants any appeal to geometric intuition? In order to think about how this situation should be described from a Kantian perspective it is necessary to recall earlier debates about the infinite.

In the history of Western philosophy one can find an ongoing debate about the infinite (see [Moore, 2001]). The debate is about what our concept of the infinite can be and whether it has any legitimate cognitive employment. Once mathe-

---

[14]Indeed Descartes [1931, Rules XII and XVIII] assumed that any ratio of magnitudes can be represented as a ratio of line lengths, and Bostock [1979, 238 ff.] show that theory of real numbers can be developed as a theory of a system of ratios capable of realizing any ratio of magnitudes.

[15]Namely, $2^{\aleph_0} = \aleph_1$.

matics, with the development of calculus, began more centrally to concern itself with the nature and structure of the infinite, the historical debate, not unnaturally found reflection within a more mathematical setting. Although there are many variations, the basic opposition is between the infinite negatively conceived as the continual lack of completion of a series, such as that of the natural numbers, for which we cannot posit a last member without contradiction, and the infinite positively conceived as a maximum, ideal, perfect, all encompassing unity by reference to which the finite is defined by limitation and so is conceived as imperfect or deficient in some respect. Note that the former would be an infinite associated with an order of succession, the latter an infinite order of co-existence.

Empiricists, wishing to ground all knowledge in experience, and recognizing the finiteness of human beings, naturally can see no legitimate cognitive role for the infinite positively conceived since it can have no basis in experience. Instead they insist that the only legitimate infinite is the potential infinite (which is only ever actually finite). All other infinitistic talk is strictly meaningless and can have no legitimate claim to express knowledge. Rationalists on the other hand (Descartes providing perhaps the clearest example (Meditation III, [Descartes, 1931b, 166])) argue for the primacy of the positive conception of the infinite as a necessary ground for deployment of the finite-infinite distinction, and, as Cantor did, for the actual completed infinite whole as a necessary ground for thinking the concept of potential infinity to have any application.[16]

In the Antinomy of Pure Reason, Kant [1965] lays out four contested applications of the idea of the infinite as completion of a series, giving the (rationalist) argument for and (empiricist) argument against the thought that objects corresponding to the completions exist. His point in doing this is to say that the arguments from both perspectives should claim our attention (indeed they bring reason into conflict with itself); but that neither can be conclusive. The empiricist fails to recognize that it is reason, with its demands and idealizations that is the source of the conception of a series which has no end, a series without limit. The rationalist, on the other hand, fails to recognize that rational ideals are not empirically realizable and does not recognize that to assume they are empirically realizable is to come into conflict with the conception of the world as open to, but never exhausted by, empirical investigation. In other words, this conception of the objectivity of the empirical world as known only by empirical means, but never exhausted by them, is one framed by reason (by our conception of the goals of scientific knowledge as theoretically organized, empirically grounded understanding), but filled in only by experience.

There is a continual mismatch between the idea of (coexistent) totality (set by reason) and that of the (successively generated) potential infinite. The complete totality is always either too large (cannot be reached from below) or too small — once a boundary has been set it can be surpassed [Kant, 1965, A489/B517]. One might note the analogy here with the attempt to find a recursive decision procedure for a formal system of arithmetic. All recursively defined sets of sentences are either

---

[16]See [Hallett, 1984, 25].

too small (they leave out some theorems) or too large (they include some non-theorems); similarly the recursively enumerable set of theorems is always either too small or too large in relation to the set of true sentences of arithmetic. (See, for example, [Hofstadter, 1999, 71].)

It is along these same lines that Kant sums up his discussion by extracting the general structure underlying all four of the antinomies.

> All the dialectical representations of totality, in the series of conditions for a given conditioned, are throughout of the same character. The condition is always a member of the series of conditions along with the conditioned, and so is *homogeneous* with it. On such a series the regress was never thought as completed, or if it had to be so thought, a member, in itself conditioned, must have been falsely supposed to be a first member, and therefore to be unconditioned; the object, that is, the conditioned, might not always be considered merely according to its magnitude, but at least the series of its conditions was so regarded. Thus arose the difficulty — a difficulty which could not be disposed of by any compromise but solely cutting the knot — that reason made the series either too long or too short for the understanding, so that the understanding could never be equal to the prescribed idea. (A529/B557)

We can further note that reason in its logical guise — concerned with the order of concepts — introduces the infinite in two forms (linked to Kant's two dynamical antinomies): (a) with the series generated by inference and the demand for regression to first principles (or *summa genera*), and (b) with the conception of the individual object (*infima species*) as determinate through the law of excluded middle/bivalence and the assumption that any predication made of an object must be true or false). This makes concepts of objects (unitary totalities) dividers of the field of possible predicates. The determinate individual object as co-ordinate with a given infinite totality of concepts is one for which there can be no contingency and no freedom. Reason in its mathematical guise — concerned with the order of objects — also introduces the infinite in two forms: (a) with the series of natural numbers, and (b) with the concept of a "homogeneous" continuum subject to infinite division.

Structurally there are just two senses of infinite here — one the potential infinity of a series for which there is a starting point and a "rule" for progressing from any given term to the next, and the other the infinity of possible divisions in a whole such that the products of division are assumed always to be further divisible. (The ability to treat a formal system of logic as a generalized arithmetic links the two serial conceptions, and the one-one correspondence between the real numbers and the set of all subsets of the natural numbers links the concepts of infinite division. Indeed, the infinite binary tree can be read as generating either real numbers, and thus is linked to the process of division of a line, or of generating definitions of species (*per genus et differentiam*) as directed by Plato's method of division.)

Of the two senses of infinite, the first is clearly modeled by the *sequence* of natural numbers and their ordinal features, the second is modeled by a continuous line segment whose homogeneity (scale invariance) implies that there can be a one-one correspondence between every point on the line and every point in a proper subset of it (divide the line in half, take a parallel line the length of the half, project the whole line into that and then reflect it back into itself with a perpendicular map.

Although the points on a finite line segment can be linearly ordered, they cannot be counted off while preserving that order, because between any given point and another there are always intermediate points. So if there were a concept of number with immediate application to such a totality of points it would be that of cardinal rather than ordinal number. The general mark of an infinite totality, whether ordered or not, is that there is a one-one correspondence between it and a proper subset of itself.

At one level this is not at all a paradoxical property, and is exploited in the construction of numeral systems that will compactly record (or give a way of referring to) large numbers. We define a recursive function each repetition of which moves up the natural numbers not one at a time, but many (say 10) at a time. Yet this function, in counting the number of repetitions also sets up a one-one correspondence between all the natural numbers and those divisible by 10. It is paradoxical if we think that the existence of a one-one correspondence (independently of counting) gives a measure of size, in the sense of number of members. For our finitely conditioned common sense tells us that if one set N has all the members that also belong to another T and some more in addition, then N has more members than T. For this reason our finitely grounded common sense says that even if there are infinite totalities, part of what we mean by infinite is that they are immeasurably large — the concept of number has no application here.

There the matter might have been left had it not been for Cantor's proof (see [Hallett, 1984, 75]) that there can be no one-one correspondence between the totality of natural numbers and the totality of real numbers coupled with his interpretation of this result as indicating that there are different *sizes* of infinite set (i.e., that it is possible to extend the concept of number into the realm of the infinite). To interpret the proof in this way one must, of course, admit that the concept of size can sensibly be extended to apply to infinite totalities. What Cantor actually demonstrated was that a contradiction results if one supposes that the real numbers can be enumerated (that there is a one-one mapping from the natural numbers onto the real numbers). This is proved by showing that any given enumeration of the real numbers must be incomplete because there is a (diagonal) method which, given that enumeration, uses it as the basis for defining a real number not included in the original enumeration. This same method can be used to show there can be no one-one correspondence between the set of all subsets of the natural numbers and the natural numbers and is also exploited in the proof of Gödel's first incompleteness theorem for formalized arithmetic. The basic method (applicable in many contexts) is one that demonstrates the incommensurability between the kind of totalization postulated by reason with its demands for maximal completeness and that accompanying the uniformity of the products of successive generation.

Kant has already pointed out that the limit of a infinitely repeated process — a limit postulated by reason in thinking the completion of the process and of the series as a determinate completed whole (totality) — if countenanced at all, must (if consistency is to be preserved) be treated as being different in kind from the terms of the series which generated it. It must, that is to say, be regarded as inaccessible by repetition of the process, and as incommensurate with the terms of the series it limits (not measurable by them). In extending the concept of number into the infinite Cantor observed this principle, distinguishing in the case of ordinals between limit numbers and successor numbers, and sacrificing the complete co-ordination between ordinal and cardinal numbers that occurs in the finite case. The Continuum Hypothesis is then an attempt to locate (measure) the set of real numbers (set of subsets of the natural numbers) within the "numerical" order of cardinalities of sets of infinite ordinal numbers. But the idea that one can talk of number here, whether cardinal or ordinal presupposes that there is a totality with a determinate "number" of members.

Since the extension of number concepts into the infinite requires a distinction be drawn between ordinal and cardinal concepts, debate ensued about which concept is the "founding" concept. Those eager to extended the concept into the infinite argued for the priority of cardinality since the definition of cardinal number, using the principle of abstraction (sets A and B have the same cardinality if and only if there exists a one-one correspondence between their members), need make no reference to whether the sets in question are finite or infinite. Those resisting the extension insisted on the priority of the ordinal concept and of number as generated in a potentially infinite series.[17]

---

[17]Of this situation Cassirer makes the following comments, which indicate again that more is at stake than just how to answer some questions about the concept of number. After noting [Cassirer, 1950, 59] that there are two trends in foundations of the theory of numbers, one that starts from cardinal, the other from ordinal aspects of number, he goes on to say

> It must seem strange indeed as first sight that a problem concerning pure mathematics, and wholly confined to it, should excite so much vehemence and such argumentation. From a purely mathematical standpoint it seems to make little difference whether one starts out from the cardinals or the ordinals in thinking of number, for it is clear that every deduction of the number concept must take both into account. Number is cardinal and ordinal all in one; it is the expression of the "how many", as well as the determination of the position of a member in an ordered series. As the two factors are inseparable and really strictly correlative,

The cardinal concept of number is clearly parasitic on a presumed, given field of objects and collections of them, so to take this as mathematically foundational denies mathematics any constitutive role *vis-à-vis* the presupposed manifold of objects. But then what can consistently be said about mathematical objects? Are they already members of the presupposed field (out there waiting to be discerned and described)? If so, how do we come to know anything about them? If not, then what is the semantic function of those mathematical terms that look (grammatically) as if they name objects? These questions are all familiar within post-Fregean philosophies of mathematics, based on the conception of knowledge as accurate representation of an independently given realm of objects.

Intuitionists and constructivists take ordinal numbers to be constructs, and take the generation of the natural numbers in sequence as the founding paradigm of what it is to construct mathematical objects. Only products of methods of construction recognized as having legitimacy are granted the status of objects of mathematical study and investigation, and that definitely does not license treating infinite series on the same basis as finite series. This approach then challenges the legitimacy of any mathematics that is dependent on Cantor's "extension" of numbers into the infinite, or the use of set theory as a theory of actually infinite totalities. Is this really where Kant would leave one?

As Cassirer put it,

> Epistemologically two fundamental views stood opposed and their differences far transcended the sphere of pure mathematics. For what was at stake was no longer the concept of the object of mathematics but the universal question of how knowledge is actually related to "objects" and what conditions it must fulfill in order to acquire "objective" meaning. [Cassirer, 1950, 61]

Establishing a view of the nature and role of mathematics was crucial to the debate between philosophical traditions. This is why Russell and Reichenbach (see Appendix) invested heavily in having a conclusive refutation of Kant's view of mathematics because that is the lynch pin around which their philosophies turn. If mathematics cannot be reduced either to logic or to a body of analytic truths its nature and status will continue to present problems for any empiricist philosophy. Cassirer put the point slightly more generally as follows:

> The crucial question always remains whether we seek to understand the function by the structure or the structure by the function, which one we

---

philosophical criticism was right insisting that it was fruitless to argue over which of these two functions of number is primary and which is dependent on the other and merely follows by implication. ... The ordinal theory had to do justice to the plurality of actual number, just as the cardinal theory had to show how numbers that were defined independently of one another could be arranged in a fixed series. As a matter of fact both theories had distinguished mathematicians behind them; on one side were arrayed Dedekind and Peano, alongside of Helmholtz and Kronecker; on the other, Cantor, Frege and Russell. [Cassirer, 1950, 60]

choose to "base" upon the other. This question forms the living bond
connecting the most diverse realms of thought with one another....
For the fundamental principal of critical thinking, the principle of the
"primacy" of the function over the object, assumes in each special field
a new form and demands a new and dependent explanation. [Cassirer,
1955, 79]

Accounts which start from thinking that possession of knowledge is a matter of having accurate representations (whether mental or linguistic) presume that this requires having an external relation of correspondence between the representation and its object; requiring thus an external relation (reference) of name to object and predicate to concept. Objects (particulars) are thus presumed as given, they are there to be designated and described. Accounts that think of knowledge in functional (or pragmatic) terms work in the opposite direction. The object is not treated as given but as an unknown, as the goal of knowledge not its starting point. Here the first philosophical questions are not as to the nature of these objects, but as to how knowledge of them is possible, what are the means by which we can come to know them.

In other words, following Kant's lead, instead of starting from the object as the known and given, we have to begin with the laws of reason and understanding. It is this opposition in methodological orientation that we will need to pursue into its more detailed consequences for logic and mathematics.

8 INTUITION AND THE THEORY OF PURE MANIFOLDS

What are the foundations of any theory of manifolds of intuition (pluralities of objects) and what, if any, are their founding "intuitions"? The foundations of the manifolds of space and time were argued by Leibniz to be relations of succession and co-existence. Here Kant and subsequent Kantians have agreed, except that they have not interpreted this as claiming that space and time are concepts which can be extracted from given objects that happen to stand in these relations, but as claiming that relations of coexistence and succession are constitutive in relation to spatio-temporal objects and that, in addition, space and time must be conceived as "objects" (relational structures). Spatio-temporal objects are not objects merely contingently capable of standing in spatio-temporal relations to one another but are the objects that they are because of their mode of space-time occupancy and the characteristics of the space-time within which they appear.

Kant makes this point in terms of enantiomorphs. His argument is repeated and strengthened by Nerlich [1976, ch.3] as making the point that one of the ways in which spaces can differ from one another is over the ways in which objects are regarded as intrinsically distinct or indistinct within them. Such investigations are mathematical investigations (prompted by empirical observations.) Global characteristics of the space (rather than the individual objects in it) play a role in determining what is possible or impossible for objects within it. (Thus, for example, spatially separated objects are necessarily distinct material objects, however qualitatively similar they may be; temporally separated ticks of a clock are necessarily distinct events, etc., as well as the fact that a right hand cannot fit into a left hand glove; spirals and helices come in right handed and left handed forms.)

Conversely our causal views about what is possible and impossible determine the mathematical features attributed to the space of objects subject to those causal relations. The requirement for space and time as forms of intuition is that they must provide the basis for individuation and identification of possible objects of experience — for the possibility of setting up reference frames for fixing spatio-temporal location (and those conditions can be explored by the mathematician). In addition these frames have to be such that they can at least in theory be established by physical devices as allowed for by the basic principles of physics. Kant had assumed the requirement was for a unitary, universal frame. Einstein argued that the physical theory which accords best with experimental observations is one that makes establishment of a unitary universal frame impossible. Reference frames (and thus reference) are established locally, but with rules for translating from one to another.

It is this core conception of forms of relation as being able to play a constitutive role in respect of objects, rather than taking objects as simple givens, that marks off Kantian and neo-Kantian approaches from those which take as part of their framework the assumption that semantics can be separated from syntax, or the world (of objects) can be separated from language, leaving for philosophy the task of determining how they relate. When one starts with an emphasis on function, on the practical, form and content are never given independently; even though form does not fully determine content, there is no determinate content without form.

What is significant here is the recognition of two levels of thought about manifolds: (i) as built up from component objects (a sequential process — with a resultant collection whose identity is given by the components and mode of composition), and (ii) a manifold as a structured whole whose parts have their identity only as parts of the whole, with the conditions for the existence of parts and their relation to one another being founded in the structure of the whole, which is in turn characterized by axioms governing the basic structuring relations. The latter is the mode of investigation to be found in topology, category theory and universal algebra.

From the latter, holist perspective the function of axioms is to define by limiting possibilities, not to specify or identify the parts distinguished by the structure:

> .... what matters in mathematics, and to a very great extent in physical science, is not the intrinsic nature of our terms, but the logical nature of their inter-relations. (Russell 1919, p.59)

A pure, homogeneous continuum is only potentially a manifold — and is in one sense completely structureless. It is the ground over which any structure may be imposed, or in which all coherent structures are realizable (and is the counterpart of Aristotle's prime matter, being similarly an abstract object never empirically realized). Its potential (but wholly uncharacterisable) parts could be thought to provide the domain of quantification for any first order axiomatization of a relational structure, where the axioms themselves limit possibilities by requiring certain kinds of relations between parts always to be present. A Euclidean continuum is one in which only a limited range of geometrical possibilities can be realized, and yet it retains other characteristics of homogeneity — infinite divisibility, in infinitely many ways, and the similarity of products of division to the whole. The fewer the axiomatically imposed restrictions, the more possibilities, but less structure; the more structure, the more limited the possibilities. A structure is completely characterized when every realization of it is isomorphic to every other. This doesn't necessarily mean that there are no transformations of the structure onto itself that are isomorphisms but are distinct from the identity transformation. That is, it doesn't necessarily mean that the structure *alone* serves to constitute its elements as objects in the sense of being able to provide a definite description of each that would guarantee the application of the law of excluded middle to statements involving that definite description. (For example, a square has to have four corners with certain relationships between them, but there is nothing further that would distinguish one corner from another.)

Equally clearly, the question of consistency is crucial for axiomatisation viewed as definition. How is one to be assured that what is defined is a (logically) possible structure? It would seem, only by showing that it is realizable over at least one domain of objects given independently of the axioms whose consistency is in question. But if this domain in turn has only an axiomatic characterization the question of consistency is only deferred. Hilbert placed two conditions on axiom systems: they should be consistent, and the axioms had to be mutually independent of one another, i.e., for a system S and any axiom A in S, neither A nor its negation should be derivable from the remaining axioms of S. This is one way to assure consistency, for if axioms are successively added under this condition the resulting collection will be consistent. But proving independence can be a far from simple matter (think of how long it took to prove the independence of the parallel postulate or of the Axiom of Choice), and since it usually has to go via the construction of models, it ends up being no simpler to resolve than the question of consistency. At bottom, it would seem there is a need either to acknowledge the homogeneous (structureless) continuum as a legitimate starting point for construction (by limitation) or for a domain of objects given independently of all axioms, i.e., given either constructively or at least as a collection constituted by objects identified independently. Work on the foundations of mathematics, because concerned to secure the foundations of differential and integral calculus and analysis, and their seeming presupposition that the continuum can be arithmetised, has all tended to see the latter as the only available route.

9 MANIFOLDS AS AGGREGATES

Part of the formalist approach was to eliminate Kantian appeals to imagination by thinking of numbers in terms of their representations (numerals). The serial definition of the natural numbers then reduces to a formal definition of what is to be counted a numeral.

> 0 is a numeral.
> If t is a numeral, then t' is a numeral.
> Only expressions containing (constructed from) one occurrence of '0' followed by a string of '$'$'s are numerals.

As was argued earlier, numerals are just as abstract as numbers. However, this approach has the merit, important from the foundational perspective, of seeming to offer some assurance of the existence of an unending (potentially infinite) supply of objects (numerals). The way in which Frege had tried to *prove* the existence of infinitely many natural numbers was part of what was responsible for the inconsistency in his system. Russell, realizing Frege's error, had to invoke an axiom of infinity that asserts there exist infinitely many individuals, and axiomatic set theory has to include an axiom asserting the existence of an infinite set.[18] Dedekind too needed to argue for the existence of an infinite system.[19]

The potentially infinite collection of formally defined numerals (types of marks on paper) serves merely the function of translating the temporal serial operation of addition of a stroke to the cumulative co-existent series of its results, and while it cannot persuade those who object to the transition from potential to actual infinity of the existence of an actual totality of numerals, it does come with an effective criterion for deciding whether any given collection of marks is or is not a numeral. Since the rules given are rules for constructing objects (numerals) there is no question of their consistency or otherwise in the logical sense, only in the practical sense — can they be followed? Since there is only one rule of construction, a rule to be repeatedly applied, and it only adds to the results of previous construction (never subtracts) there is no room for practical conflict. This I think is why Hilbert felt justified in assuming the consistency of the finitary part of arithmetic and also further bolsters the view that the existence of a potentially infinite series of symbols, together with an effective criterion for establishing whether any given symbol belongs to the series, is assumed in all uses of formal systems, whether of logic or pure computation.

But *all* this gives is a "manifold" of objects serially ordered by the complexity (in this case length measured in discrete units) of their construction. It does not give numerals in the sense of signs for numbers unless we already presume to understand the *function* of numbers in counting and in assessing the size of collections of

---

[18] There is a set x, such that $\emptyset \in x$, and such that $\forall y\ (y \in x \rightarrow y \cup \{y\} \in x)$.

[19] His argument, which few would find convincing, does not show the existence of a potentially infinite series, but purports to prove the existence of a set which can be put in one-one correspondence with a proper part of itself. His exemplar is the totality of things that can be the object of one's thought, [Dedekind, 1963, 64].

discrete objects. We have either to count or compare the number of /s in two given numerals in order to say which comes before the other in the series, or whether they are two tokens of the same numeral. So a "mechanical" constructive "intuition" is required for recognizing numerals as objects, and an "intuition" based on a grasp of the function of numerals is required to read numerals as symbols signifying numbers. "Intuition" here is merely used to mark the non-logical contribution of practical understanding, based on the creation and manipulation of signs as objects, to filling out the concept of number.

However, if we push the direction of this reliance on concretely manifest numerals for providing us with assurance of the existence of a potentially infinite series a little harder, it too can run into trouble. From the perspective of applied mathematics, and particularly of the numerical methods used in real (very definitely finite) computers we may be led in a direction that would question our right to assurance about the infinity of the series. It would mean adopting much the same stance as that which led Einstein to realize that in setting up reference frames one needs to take physics into account. The result might be that just as we now discuss non-Euclidean geometries and the relationships between them, we have to distinguish a variety of non-"Euclidean" number systems. For one might insist (as in [Rotman, 1993]) that there is a real difference between the imagined pure seriality of the intuitionists, in which the number series results from the iteration of the same operation (which in turn licenses the thought of a series which can never come to and end and the principle of complete induction) and any theory of marks on paper, however idealized or abstract.

The serial construction of numerals makes each numeral different from the next; it coexists with all preceding numerals and is differentiated from them by its length. The addition of each new numeral to the series of numerals is thus not merely a repetition of the same operation. Each numeral is formed *in the same way* from its predecessor, but its addition to the series is the addition of a new, distinct, and *longer* member. Recognition of it as a new, distinct numeral and of its place in the serial order must thus already invoke counting as a means of size comparison (the function of numerals as signs for numbers). It also means that it gets harder and harder to add new numerals (consumes more resources — paper, disc space, memory.) In this case, since we do not know what exactly the limits of our resources are or may be in the future we cannot put any once-and-for-all fixed upper bound on the series of natural numbers, but we know that there always will be an upper limit — as we approach the limit it just gets harder and harder to add new numbers. This then is a not a potentially infinite series, but an indefinitely long finite one.

Rotman has sketched some of the consequences. One of these is that the integers would not be closed in any standard sense under arithmetic operations. The point beyond which a function ceases to be defined is lower, the "faster" the function climbs up the numbers. This view would also have implications for rational numbers and the division of a continuum — this too could not be conceived as the potentially infinite repetition of the same operation, but the successive creation of something different and more complex. This approach represents a way to bring the mathematical structure of forms of intuition (representation) better in to line with what is empirically realizable — i.e., bringing about a better coordination of what is represented as possible for an object of experience and what our experience-based theories tell us is (and is not) possible. It displaces some of the idealizations projected onto the empirical world by ideas of reason suggesting an inappropriately exact conception of what we should aspire to by way of knowledge — of what can be made objective. In this regard it would be hard for a Kantian to resist the thought that these are ideas as worthy of further exploration as non-Euclidean geometries.

9.1 Aggregates as Manifolds

The (reductionist) tendency has been to assume that arguments, such as that given above, to the effect that holism on its own cannot be enough, not only indicate the need for something besides axiomatic holism, but the need for it to be reduced to a theory of aggregates or sets. Reductionism requires one of the concepts of manifold to be reducible to the other. Only one can indicate the "right" way to come to know and understand, provide the "right" foundation for building our castle of knowledge. I take it that the lesson to be learned from Kant is that there is no justification for this assumption, and that to proceed as if there are nothing but manifolds given one way or the other is to be taken in by an illusion of reason. Rather one should recognize the distinctive functions of these forms of representation and the kinds of knowledge associated with each together with the fact that they are not independent. The very notion of a unitary manifold (exactly what Kant insists on for the forms of intuition) constituting an object with multiple parts/constituents already requires both to be in play. The demand for unity imposes the necessity of a holistic conception of the manifold and its structure (which in turn limits possibilities for its constituents.) Recognition that this unity is a manifold (has many descriminable objects as constituents) requires thought about how those objects are given, how their relations are determined and how they can in virtue of those relations be aggregated into complex units.

We have just argued that the holist approach (from the whole manifold down to parts) needs supplementation from the aggregative approach. But equally the aggregative approach needs supplementation from the holist, systems view, even in the simplest case of a finite collection. The serially given, in order to be recognized as a plurality, as a collection, must be postulated as a system of coexisting elements. But then the information necessary to considering such an individual object as a unit (an object) is the information that this is all there are — this is what makes it a (determinate) whole — the specification of when it is complete. The whole then has properties in its own right, based on its components, and possibly the way they were put together. Equally the components "acquire" new properties, based on their relation to the whole and to all other components as the other components of that whole. This is why the question of what sets exist is significant even if one supposes the universe of individual objects (objects that are not sets) to be given. The axioms of set theory stipulate which sets exist and provide for there to be enough for most mathematical purposes and yet, as with all first order axiomatizations, they do not uniquely determine or settle questions of set existence.

One key question for set existence was, as we have seen, whether infinite sets exist — in what sense these can be complete objects in their own right. If sets (aggregates of objects) have their identity fully determined by the objects that belong to them (satisfy the axiom of extensionality) then the membership of any set must be assumed to be determinate (any object either does or does not belong to it). Should it be thought that for this reason every set has a determinate number of elements — even if it is not finite? The standard answer (following Cantor) has been 'Yes'. Since the existence of one-one correspondences between sets allows for definition of an equivalence relation (same cardinality) and ordering in respect of cardinality, it is appropriate to extend the concept of number into the infinite. The first exemplar of an infinite set (the smallest) is the set of natural numbers; another, larger, is the set of all subsets of the natural numbers. But regarding it this way does have the counterintuitive consequence that, because any infinite set is such that there is a one-one correspondence between the whole set and a proper part of itself, it will have the same cardinal number as a set that contains "fewer" elements than it does. This might equally be seen as indication that it is a mistake to think that there can be any infinite sets (objects whose identity is *fully* determined by their members), since an infinite set would be such that the "number" of its elements doesn't depend crucially on all of them being present. The postulation of a totality as an infinite set thus still represents a way of thinking from the top down, as it were, (the principle defining the whole), and not from the bottom (members) up. If sets, as aggregates of their members, do have their characteristics determined by their members, then some connection has to be retained between specification from below, by members, and from above as a system of objects.

Only by coming from this dual perspective can one do full justice to the concept of set as one object — a completed unit whose identity is given by the axiom of extensionality — and to the difference between a set and its disaggregated members. Coming from this perspective one might insist that the only sets there are, are those that can be numbered, namely those that are finite in the sense that if you take any element away from the set the remaining elements together form a smaller totality. If an element can be taken away from a set without affecting its "size" it would seem to imply that size is not a determinate characteristic of a set, because it is not determined by the set's identity (having just the members it does.)

Mayberry [2000] explores the consequences of developing set theory and arithmetic, as founded in set theory, on this basis.[20] He proposes an axiom that says

---

[20]Here one should be careful to note that Mayberry adamantly repudiates *all* appeals to metaphors of construction and generation. He works strictly from the direction of seeking axioms

that all sets are finite in the sense that there can be no one-one correspondence between a set and a proper part of itself. Such an axiom makes no presupposition about the generation of sets. Nonetheless something akin to the principle of induction,[21] can be proved for sets, without countenancing the collection of all sets as itself a set (although again this system is formulated using quantifiers ranging over the domain of sets so there must be questions raised about whether, or the extent to which, this totality of sets is presupposed as determinate.)

One of the interesting features of the theory so developed is that it has to recognize simply infinite systems of different lengths and that no simply infinite system measures the totality of Euclidean (i.e., finite) sets (p. 382). As Mayberry conjectures this may mean that we have to recognize that in the absence of postulating infinite sets, we cannot assume that even all the non-infinite sets can be measured against a single scale of cardinal numbers. What is in question is whether every simply infinite system measures every other (p. 385). They all have the same global structure because they all satisfy the axioms for a simply infinite system, but their local structure is tied to their ordering relation (successor function). This situation prevents closure under addition, multiplication, etc. Mayberry's exploration of these matters and of the differences between sets so conceived, and as conceived under the standard assumption of the acceptability of the Cantorian hierarchy of infinite sets, gives a clear sense of the way in which taking the rationally projected ideal realm of Cantorian set theory for the mathematically real has unwarrantedly closed off important questions and lines of investigation. Without dismissing work in Cantorian set theory it is nonetheless necessary to adopt a critical attitude toward it, recognizing that it is a construct whose objective validity (applicability in relation to the world of possible experience through provision of the framework of mathematical representations of empirical objects) needs investigation and cannot be taken for granted.

Once we cease to take it for granted that set theory is inevitably the theory of the hierarchy of Cantorian infinite sets, many interesting questions foreclosed by this assumption are opened up for fresh investigation. Some of those listed by Mayberry (pp. 387–95) are the following:

> What global logic can be used for set theory?
> What is the connection between the arithmetic of arithmetical
>   functions and relations and that of simply infinite systems?
> How should real numbers be defined?
> How can we introduce analytical methods in a natural way so that
>   these discrete geometries have appropriate "continuity" and
>   "smoothness" properties?
> How might this geometry relate to and impact the mathematics of
>   quantum theory?

---

that are true of an independently existent domain.

[21]Something Mayberry calls the Principle of one point extension induction [Mayberry, 2000, 278].

What are the implications for logic if model theory is restricted to using Euclidean set theory?

10 MAXIMA, MINIMA — TOTALITIES AND QUANTIFIERS

Now, while both Rotman and Mayberry present excellent critical analyses of the way in which mathematics has foundations in the theory of sets, their approaches are still foundationalist and reductivist. Mayberry is looking for a once and for all grounding (in the order of coexistence) in propositions that are self-evidently true and he presumes that this grounding goes from the bottom up, as it were, from objects to their aggregation in sets. Rotman is repudiating the mathematics of the infinite and rejecting the claims of set theory to be foundational; instead he starts from the successive order of iterated constructive operations. As we have seen, the Kantian position suggests that both may be folorn quests, that there is no ultimate grounding of mathematics in truths, but only in practical principles and constructive definitions. Equally there is no definitive priority to be given to the constructive order of succession and the static order of coexistence. Any knowledge of objects as complex and of their complexity requires both. Even if we do treat the natural number series as indefinite in length, rather than potentially infinite, we still need to be able to answer questions about how to read quantification over the natural numbers. Moreover, both approaches (as Rotman and Mayberry acknowledge) have to face up to their implications in relation to continuity, the concept which really pushed the infinite into mainstream mathematics. These implications may indeed be very interesting but it is also possible that they will reveal the impossibility of completely recapturing the functions of this concept from the finitary bases from which they start. This could open the way to acknowledgment that an alternative is to recognize, with Kant, that there are two founding "intuitions" required by our forms of intuition (structures within which objects can be identified and individuated) each of which has to be manifest in both the dynamic order of succession and in the static order of coexistence in order to yield simultaneous construction of an object and recognition of that object in a concept. The concept is constructed with the object as a conception of the rule or procedure of construction. The two intuitions/concepts would be identity (or the repetition of an operation giving rise to an aggregate of units) and continuity (or the flowing uniformity of unimpeded motion) giving rise to the homogeneously extended continuum. Neither of these is given in experience; both are imposed through our representations as a matter of pragmatic necessity, as a way of fixing the level of detail we want to discriminate (the scale at which we are going to constitute our objects).

The function of the continuum is to be the ground within which structure can be characterized, objects identified and interrelated. In this sense it takes over the role of the absolutely infinite — the infinite within which the finite is revealed by limitation (or division). Because Kant uses both continuity and identity as primitive intuitions the scope of mathematics recognized in a Kantian framework is not as limited as would be suggested by intuitionism or constructivism. It is a framework recognizing two poles, continuity and identity, along with their corresponding ideal objects — the unitary continuum (a potential manifold) and plurality of units (a potentially unitary object). Indivisible units are postulated as limits of division of the continuum, suggesting its resolution into an aggregate of discrete objects. The infinite totality of natural numbers is postulated as the limit of the aggregation of discrete units into a single system, but the minimal infinity of the natural numbers (the smallest possible instantiation of the Peano axioms) and the maximal infinity of the continuum are functionally distinct, not merely distinct in cardinality.

Treating the continuum as a maximum gives no recipe for proving universally quantified statements about it on the basis of what can be proved of its members individually. In the case of real numbers, infinite decimals or subsets of the natural numbers, it says that nothing can be excluded and that the continuum as a set of elements (limits of division) is placed beyond all *determination* as a field of limitless possibilities which constructive explorations can never exhaust. This would be to side with those who suggest there are grounds for thinking Cantor's continuum hypothesis should not be regarded as correct. Cantor was attempting to characterize the structure of the continuum from below, as an aggregate of identifiable elements using infinitistic assumptions and seeking to identify a *minimal* structure that would serve (making the cardinality of the continuum the next smallest after that of the natural numbers). This conflicts with the epistemological function of the continuum as maximal.

The natural numbers function to recognize the finite plurality of distinguished objects as well as the possibility of indefinite hierarchical organization of units which are themselves composed of units without end (as the continuum assures is possible). The interaction of the two concepts sets up cognitive goals bringing the methods of investigation of each to bear on the other. The axiomatic method is brought to bear on arithmetic; numbers are thought of as a structured system of objects. Algebra allows arithmetic methods to be extended into geometry and suggests that the continuum can be given a discrete numerical representation. The gulf between the finite definiteness of discrete magnitudes represented by natural numbers and the maximal infinity of the continuum is the space within which mathematical exploration of possible structures, their properties and interrelation, occurs. The other challenge is in bridging the transition from operational, procedural, rules to conceptual characterization within the static order of coexistence. The challenge goes both ways — the function which generates (has as its range) a recursively enumerable set, does not immediately disclose how to determine the objects in that set. This may or may not be effectively decidable. Gödel's first incompleteness theorem is an illustration of the fact that this is not always possible. The problems encountered in proving that algorithms really do compute the functions intended, or really do execute the intended operations, is critical and non-'trivial. Similarly the ability of go from an analytic function to a computer model based on being able to compute values (find solutions to equations) is similarly non-trivial — and in the case of the n-body problem, intractable by analytic means.

The order of understanding and of formal logic is sequential. Questions of the relation of knowledge to its possible object belong to the sphere of reason and of transcendental logic, which in its attempts to unify and systematize drives the quest for ever more encompassing and more detailed characterizations (the two imperatives of modern science) by totalizing what is sequentially given as if it were of an order of coexistents. Equally from this perspective the use of limits, whether minima or maxima come as imperatives rather than as descriptions of what is antecedently the case, and they do reach beyond the bounds of formal logic.

11 WHAT IS A KANTIAN APPROACH?

The burden of the forgoing discussion has been to illustrate that a Kantian approach to the philosophy of mathematics, by being non-foundationalist and non-reductivist, is also more open to the view of mathematics as an evolving subject. If mathematics is concerned with our forms of representation, it has both internal and external drivers for development — demands from the increasing numbers of contexts in which those forms are deployed and from its own internal attempts to bridge the gap between knowledge founded in constructive methods (order of succession and rule understanding) and knowledge founded in axiomatic methods, in ideal completions and totalizations (order of coexistence, principles and reason).

The basic epistemological insight is the need to insist, on multiple levels, that there is a necessity for dual approaches:

- dynamic succession — static coexistence

- ordinal — cardinal

- successive construction of objects according to a rule — successive division of a whole according to a principle

- definition by construction of a complex object — axiomatic characterization of a relational structure

In each case both components are necessary; neither can be reduced to the other, nor will there be a meeting in the middle,[22] even though there can be ongoing

---

[22] Which is why a logic, such as first order predicate calculus, for which a completeness theorem can be proved cannot provide a sufficient basis for the characterization of mathematical objects or mathematical reasoning about them. Second order logic (whose claims to being logic are disputed) at least recognizes two, very different realms of "objects" with its two domains of quantification — over the referents of predicate symbols (whatever those are) and over individual objects. But if the domain of second order quantification is interpreted maximally (as having to be non-denumerable) the logic is not complete — there will be valid sentences that are not provable.

mutual elucidation and elaboration. For each one of the pair there is a supplement required from the other direction; the supplement which is the missing content or "intuition" preventing mathematics from being a collection of analytic truths. The realm of the ideal remains ideal, the projection of practical principles the need for which comes from outside mathematics itself (the product of a synthesis of intuitions coupled with intuition of the synthesis) in the practical need we have for forms of representation of objects as a condition of the possibility of any knowledge of objects through experience. The need for mathematical forms is thus an *a priori* universal necessity. The justification for any given representational form is practical not logical; nevertheless the implementation of practical rules is creative, whether in mathematics or in law.

Laws create rights, and obligations, as well as crimes of various kinds, and even create entities such as corporations. The transition from being able to follow a law to being able to discern the structures created by its implementation is not straightforward, and is not a logically deductive process, but it nonetheless has objective standards of proof without any guarantee that all possibilities will be either forbidden or required. However, the standard of justification for a rule or law itself isn't that of correct description, but its appropriateness to the task at hand.

The core value behind the kind of critical, non-dogmatic, philosophy that Kant urged is the need continually to go back to re-examine principles (and co-ordinate ideals), subjecting them to critical analysis and modification as required. The necessity emanating from these principles is that of practical necessity (obligation to have principles, which in turn constrain possibilities), not of theoretical necessity (eternal truth):

> Reason must not, therefore, in its transcendental endeavours, hasten forward with sanguine expectations, as though the path which it has traversed directly to the goal, and as though the accepted premises could be so securely relied upon that there can be no need of constantly returning to them and of considering whether we may not perhaps, in the courses of the inferences, discover defects which have been overlooked in the principles, and which render it necessary either to determine these principles more fully or to change them entirely, [Kant, 1965, A736 B 764].

APPENDIX

A NON-EUCLIDEAN GEOMETRY AND EINSTEIN'S RELATIVITY THEORIES

The death knell for Kant's position on the nature of mathematics was asserted by Russell and others to have been sounded by (i) the success of Einstein's theories of relativity, in which non-Euclidean geometries find application to the physical (spatio-temporal) world, (ii) developments in logic and the development of a logic of relations in particular, (iii) the arithmetization of analysis produced by Weierstrass, Dedekind and others, and (iv) Hilbert's axiomatization of Euclidean geometry. The combined effect of (ii)–(iv) provided the basis on which Russell claimed, that thanks to the progress of symbolic logic especially as treated by Peano, that

> This part of the Kantian philosophy is now capable of a final and irrevocable refutation ... The fact that Mathematics is Symbolic Logic is one of the greatest discoveries of our age; and when this fact has been established, the remainder of the principles of mathematics consist in the analysis of Symbolic Logic itself, [Russell, 1903, 4–5].

As we now know, the heroic efforts of Frege, Whitehead, Russell and Carnap to demonstrate that mathematics can be reduced to the new formal logic, and that its application in physics is a matter simply of logical deduction, failed. Their efforts did, however, contribute to the demonstration that set theory can, in principle, provide a "foundation" for most of mathematics, but, as Quine [1963] argued in detail, set theory does not reduce to logic although reasoning within axiomatic set theory can be formalized in classical first order predicate calculus. From a foundational point of view this still leaves open questions about the status of the axioms of set theory and of sets as founding "objects" for mathematics and it is on this topic that much twentieth century philosophy of mathematics has focused.[23]

But if Russell was wrong about the power of the new symbolic logic and accompanying axiomatic methods to reveal the analytic character of all mathematical propositions, the only remaining basis for rejecting a broadly Kantian position out of hand would be Einstein's demonstration of the applicability of non-Euclidean geometries.[24] Reichenbach [1949] gives perhaps the most trenchant statement of the anti-Kantian, logical positivist/logical empiricist reading of the significance of Einstein's work. His argument is that Kant asserts that there are synthetic *a priori* statements that are absolutely necessary and that amongst these are the truths of Euclidean geometry. But since "propositions contradictory to them have been developed and employed for the construction of knowledge" (p. 307), these principles must now be considered *a posteriori* empirical hypotheses, verifiable through experience only. Reichenbach goes on to say:

> It is the philosophy of empiricism, therefore, to which Einstein's relativity belongs.... Einstein's empiricism is that of modern theoretical physics, the empiricism of mathematical construction, which is so devised that it connects observational data by deductive operations and

---

[23] Debate has continued with Bennett [1966; 1974] reasserting the demise of Kantian position, while others such as Brittan [1978], Parsons [1980], and Holland [1992] have sought to rescue it in various ways.

[24] Clearly claims about the foundational role of set theory are also likely to be problematic for a Kantian view of mathematics and will be taken up below. However, since they do not involve claiming analytic status for mathematical truths they presumably allow that they are synthetic. The question then becomes how to understand this status.

> enables us to predict new observational data ... the enormous amount of deductive method in such physics can be accounted for in terms of analytic operations alone .... The method of modern science can be completely accounted for in terms of an empiricism which recognizes only sense perception and the analytic principles of logic as sources of knowledge, [Reichenbach, 1949, 309-10].

Reichenbach here states clearly the central tenet of what came to be logical atomism and logical positivism. The idea that sense-data/observation forms the objective foundation for scientific knowledge and that all further organization of this data is purely logical. All empirical claims should be reducible, through logical analysis, to their observational content, there is no empirical content added by logical (and hence mathematical) structure. Otherwise stated — the only necessity is logical necessity. In line with the tradition of Humean empiricism, Reichenbach reveals that his argument here is part of a campaign against metaphysics — against the philosopher who claims to know truth from intuition or any "super-empirical" source.

> There is no separate entrance to truth for philosophers. The path of the philosopher is indicated by that of the scientist: all philosophy can do is to analyze the results of science, to construe their meaning and stake out their validity. Theory of knowledge is theory of science, [Reichenbach, 1949, 310].

(Reichenbach seems somehow to have forgotten that Kant too was preoccupied with dismissing the claims of dogmatic metaphysics, with arguing that our cognitive claims are limited to the domain of possible experience. Equally Kant was concerned to reveal the inadequacies of any purely empiricist philosophy.)

In the same volume in which Reichenbach's article was published, Einstein himself remarked:

> The theoretical attitude here advocated is distinct from that of Kant only by the fact that we do not conceive of the "categories" as unalterable (conditioned by the nature of the understanding) but as (in the logical sense) free conventions. They appear to be *a priori* only in so far as thinking without the positing of categories and of concepts in general would be as impossible as breathing in a vacuum, [Einstein, 1949, 674]

Einstein's mention of the "categories" is significant. The categories are not specifically mathematical concepts, but they are the concepts whose application within the spatio-temporal world of possible experience yields synthetic *a priori* knowledge of that world, including its geometry. Crucial amongst the categories is the concept of causality. What changes from Newtonian to Einsteinian physics is the mathematical form assumed by fundamental causal laws. So in this sense the category has been reinterpreted. But that this category should play a constitutive role *vis à vis* the world investigated by physics has not changed and has not been shown to be a "free convention".

In mathematical physics the mathematical form of its causal laws, coupled with the assumption that space and time do not of themselves have causal properties, has implications for the geometry attributed to space-time.[25] It was because Maxwell's laws of electro-dynamics did not obey the same invariance conditions (were not invariant under the same (Gallilean) group of spatio-temporal transformations as the laws of classical Newtonian mechanics) that Einstein, imposing the very Kantian requirement of unity in our representation of physical reality was led to suggest an alternative geometry for space-time in the theory of general relativity. This is in complete accord with Kant's argument that the structure of space and time must be determined by causal relationships, since space and time as pure intuitions have no determinate structure and are not possible objects of experience.

The way in which causal assumptions interact with assumptions about the geometry of space-time is illustrated in an article by Robertson, to which Einstein refers the reader [Robertson, 1949]. Robertson illustrates how the question "Is space really curved?" is not a question that can be settled by any simple observation. The import of the question has to go via a clarification of what it means mathematically and empirically for space to be curved. His account can be summarized as follows.

Mathematically speaking, a geometry is taken to be defined by a set of axioms involving the concepts point, angle, and a unique relation called "distance" between pairs of points. The only constraint on the axioms is that they form a consistent set. Theorems have to be derivable from the axioms. Mathematicians then ask what distinguishes Euclidean geometry from other geometries. It can be characterized by the group of translations and rotations under which distance relations are invariant; it is a congruence geometry, or the space comprising its elements is homogeneous and isotropic. The intrinsic relations between points and other elements of a configuration are unaffected by the position or orientation of the configuration. What is notable is that only in such a space can the traditional concept of rigid body be maintained. In other words all our assumptions about the ways material objects can be moved around and measured (all of which contribute to their identity criteria) are valid only if space is assumed to have a congruence geometry. However, Euclidean geometry is not the only congruence geometry. Hyperbolic, spherical and elliptical geometries are too. Each of them is characterized by a real number K ($K = 0$ for Euclidean space), which can be interpreted as the "curvature" of the space. How might this "curvature" be detectable through measurement? One such gauge is the measure of the sum of the internal angles of a triangle, another is the ratio between the surface and the volume of a sphere.[26]

---

[25] Excellent discussions of the interplay between geometry and physics in their mutual development can be found in Gray [1999].

[26] $S = 4\pi r^2(1 - Kr2/3 + \ldots), V = 4/3\pi r^3(1 - Kr2/5\ldots)$.

Robertson then gives an example to illustrate both the interconnection between measurement and choice of geometry and of the role of universality in such considerations. He describes an experiment with a flat (by normal Euclidean standards) metallic plate, which is heated so that the temperature across it is not uniform (it is constrained so it cannot buckle). Measurements are taken across it using a metal ruler that is allowed to reach thermal equilibrium with the region of the plate measured before a reading is taken. Robertson argues that the geometry revealed by these measurements will in general not be a congruence geometry and that it will be hyperbolic if heat flow is constant through the plate. Do we say the plate is flat or not? The real question is whether we accept measurement by the ruler that has been allowed to reach thermal equilibrium with the plate. If we do the role of heat in "causing" expansion or the ruler will disappear. Since the ruler gives the standard by which sameness of distance is judged, it cannot be allowed to have changed in length; thus there will be no change to explain. However, if we require our rulers to yield invariable results then the latter system of measurement doesn't work. If we changed the metal from which the ruler was made we would get different results. Because the point of a system of measurement is that it should yield invariable results we opt for judging the plate to be Euclideanly flat, and then explain deviations in measurement results as the effect of heat on the ruler. In the case of general relativity, however, the force involved (gravitation) is assumed to be universal — the gravitational and inertial masses of any body are asserted to be rigorously proportional for all matter.

The point is that even if there are choices here, they are interconnected and subject to non-empirical constraints. Measurement practices, essential to the possibility of any science being both experimental and mathematical, require invariance assumptions together with causal assumptions about there being an explanation for variations in measurement results (these assumptions are required to underwrite the objectivity of measurement; i.e., to underwrite the validity of the claim that what is being measured is a feature of the empirical real object of measurement and not a product of the measuring instrument (or observer). This is one way of restating a key part of Kant's argument against empiricists; the possibility of experimental mathematical physics rests on assumptions about the identity and difference of its possible objects. Such assumptions are constitutive of the identity of those objects and so yield necessary *a priori* truths about them, but these truths are not such as could be revealed by logical analysis of concepts. In other words, there are no bare particulars (intuitions), particular objects are always objects to which concepts already apply and between which there are already relations. The role of synthetic *a priori* truths is that they do hold necessarily within the domain of objects for which they play a constitutive role.

Another, lengthy and sustained, Kantian reflection on the impact of Einstein's theories is provided by Cassirer [1923]. In commenting on the fact that relativistic physics denies the possibility of establishing a universal frame of temporal reference he says:

The 'dynamic unity of temporal determinations' is retained as a postulate; but it is seen that we cannot satisfy this postulate if we hold on to the laws of the Newtonian mechanics, but that we are necessarily driven to a new and more universal and more concrete form of physics. The objective determination shows itself thus to be essentially more complex that the classical mechanics had assumed, which believed it could literally grasp with its hands the objective determination in its privileged systems of reference. That a step is thereby taken beyond Kant is incontestable, for he shaped his "Analogies of Experience" essentially on the three fundamental Newtonian laws: the law of inertia, the law of proportionality of force and acceleration, and the law of equality of action and reaction. But in this very advance the doctrine that it is the "rule of understanding" that forms the pattern of all our temporal and spatial determinations is verified anew. In the special theory of relativity, the principle of the constancy of the velocity of light serves as such a rule; in the general theory of relativity this principle is replaced by the more inclusive doctrine that all Gaussian coordinate systems are of equal value for the formulation of natural laws. It is obvious that we are not concerned here with the expression of an empirically observed fact, but with a principle that the understanding uses hypothetically as a norm of investigation in the interpretation of experience ... [Cassirer, 1923, 415]

Cassirer goes on to explain the difference between the space-time of the physicist and the *a priori* "forms of intuition". "What the physicist calls "space" and "time" is for him a concrete measurable manifold, which he gains as the *result* of coordination, according to law, of the particular points; for the philosopher, on the contrary, space and time signify nothing else than forms and *modi*, and thus presuppositions of this coordination itself. They do not result for him from the coordination, but they are precisely this coordination and its fundamental directions. It is coordination from the standpoint of coexistrency and adjacency or from the standpoint of succession, which he understands by space and time as "forms of intuition" [Cassirer, 1923, 417]. These forms are *a priori* in that no physics (science of change and the changeable) can lack the form and function of spatiality and temporality in general.

Empiricist philosophers such as Reichenbach might be prepared to admit that physics cannot do without the *concepts* of space and time. What is distinctive of the Kantian position is its insistence that the cognitive basis of our thought of the world of experience as spatio-temporally structured cannot be purely conceptual and cannot be derived from experience. This is what is meant by saying that space and time are *a priori* forms of intuition and is the basis of the claim that mathematics, as the science of the possible pure structures of these forms, is not part of logic (which deals only with concepts). Its truths, established *a priori*, are therefore not analytic (not revealed by the analysis of concepts). Russell's claim was that, whereas there was some justice in Kant's position, given the primitive state of logic at the time he was writing, subsequent developments, especially those incorporating the logic of relations and the development of set theory have rendered it unnecessary to move beyond the structures afforded by logic to account for mathematical knowledge or its applications. This is the claim that has come to seem to be almost beyond question by those working within analytic philosophy, for to confront it requires challenging assumptions from which that way of doing philsophy takes its whole orientation.

So even if the death-knell for a broadly Kantian view on the nature of mathematics was sounded prematurely, it was nonetheless heard and believed to have signaled the end for such an approach. Kant's critical questioning focused on the seeking the conditions for the possibility of mathematical physics, whereas philosophy of mathematics from the late nineteenth century on has focused more on the epistemological and ontological foundations of pure mathematics, seeming to assume, for the most part, that the uses of mathematics in science have nothing to contribute to these investigations. Mathematics has changed significantly since the eighteenth century, and so have the sciences. We now have not only to think of mathematical physics, but also of mathematical biology and of the ubiquity of mathematics in the many disciplines that have acquired scientific status since Kant's time. Acknowledging these changes, is an approach to philosophy of mathematics that is broadly Kantian in spirit likely to be fruitful? Or, was Russell right to consign Kant's approach to the scrapheap of history? Clearly I think Russell was too hasty.

BIBLIOGRAPHY

[Aristotle, 1984] Aristotle. *Physics* in J. Barnes (ed.) *The Complete Works of Aristotle*, revised Oxford translation, Princeton: Princeton University Press, 1984.
[Bachelard, 1934] G. Bachelard. *Le nouvel esprit scientifique*. Paris: Presses Universitaires de France, 1934.
[Benacerraf and Putnam, 1964] P. Benacerraf and H. Putnam, eds. *The Philosophy of Mathematics: Selected Readings*, Englewood Cliffs, NJ: Prentice-Hall, 1964. 2nd edition 1983, Cambridge: Cambridge University Press.
[Bennett, 1966] J. Bennett. *Kant's Analytic*, Cambridge: Cambridge University Press, 1966.
[Bennett, 1974] J. Bennett. *Kant'sDialectic*, Cambridge: Cambridge University Press, 1974.
[Bostock, 1979] D. Bostock. *Logic and Arithmetic Vol. II: Rational and Real Numbers*, Oxford: Oxford University Press, 1979.
[Brittan, 1978] G. Brittan, Jr. *Kant's Theory of Science*, Princeton: Princeton University Press, 1978.
[Brown, 1999] J. Brown. *Philosophy of Mathematics: an introduction to the world of proofs and pictures*, London and New York: Routledge, 1999.
[Cassirer, 1923] E. Cassirer. "Einstein's Thoery of Relativity" in *Substance and Function & Einstein's Theory of Relativity*, Chicago: Open Court, 1923. Reprinted, 1953 New York: Dover.
[Cassirer, 1950] E. Cassirer. *The Problem of Knowledge:Philosophy Science and History Since Hegel*, trans. W. H. Woglom and C. W. Hendel, New Haven and London: Yale University Press, 1950. First published in this English edition.
[Cassirer, 1955] E. Cassirer. *The Philosophy of Symbolic Forms, Vol.I: Language*, trans. Ralph Manheim. New Haven, CT: Yale University Press, 1955. First published 1923 as *Philosophie der symbolischen Formen: Die Sprache*, Berlin: Bruno Cassirer.

[Dedekind, 1963] R. Dedekind. *Essays on the Theory of Numbers*, trans. W. W. Beman, New York: Dover, 1963. First published 1893 as *Was sind und was sollen die Zahlen?* Braunschweig: Vieweg.

[Descartes, 1925] R. Descartes. *The Geometry of René Descartes*, trans. and ed. D. E. Smith and M. L. Latham, Chicago, IL and London: Open Court, 1925. Translation of *La géometrie* published as an appendix to *Discours de la methode*, 1637.

[Descartes, 1931a] R. Descartes. "Rules for the Direction of the Mind" in *The Philosophical Works of Descartes*, trans. and ed. E. S. Haldane and G. R. T. Ross, Cambridge: Cambridge University Press, 1931. 2nd edition, New York: Dover, 1955.

[Descartes, 1931b] R. Descartes. "Meditations on First Philosophy" in *The Philosophical Works of Descartes*, trans. and ed. E.S. Haldane and G.R.T. Ross, Cambridge: Cambridge University Press, 1931. 2nd edition, New York: Dover, 1955.

[Einstein, 1949] A. Einstein. "Reply to Criticisms" in Schilpp, 1949.

[Frege, 1893] G. Frege. *Grundgesetze der Arithmetik, begriffsschriftlich abgeleitet, Band I*, Jena: Verlag Hermann Pohle, 1893. Partial English translation Frege, 1964.

[Frege, 1903] G. Frege. *Grundgesetze der Arithmetik, begriffsschriftlich abgeleitet, Band II*, Jena: Verlag Hermann Pohle, 1903. Appendix appears in English in Frege, 1964.

[Frege, 1953] G. Frege. *The Foundations of Arithmetic*, trans. J. L. Austin., Oxford: Blackwell, 1953. First published 1884 as *Die Grundlagen der Arithmetik*, Breslau: Keobner.

[Frege, 1964] G. Frege. *The Basic Laws of Arithmetic: Exposition of the System*, partial translation of Frege 1893, by M. Furth, Berkeley and Los Angeles, CA: University of California Press, 1964.

[Frege, 1971] G. Frege. *On the Foundations of Geometry and Formal Theories of Arithmetic*, trans. And ed. E.H. Kluge, London & New Haven, CT: Yale University Press, 1971.

[Gödel, 1964] K. Gödel. "What is Cantor's Continuum Problem?" in Benacceraf and Putnam, 1964.

[Gray, 1999] J. Gray, ed. *The Symbolic Universe: Geometry and Physics 1890-1930*, Oxford: Oxford University Press, 1999.

[Hale, 2002] B. Hale. "Real Numbers, Quantities, and Measurement" *Philosophia Mathematica Volume Ten*, 304-320, 2002.

[Hallett, 1984] M. Hallett. *Cantorian set theory and limitation of size*, Oxford Logic Guides: 10, Oxford: Oxford University Press, 1984.

[Heath, 1926] T. L. Heath. *The Thirteen Books of Euclid's Elements,* trans. T. L.Heath, Cambridge: Cambridge University Press, 1926. 2nd edition reprinted New York: Dover 1956.

[Hilbert, 1967] D. Hilbert. "On the Infinite" in van Hiejenhoort 1967. Also reprinted in Benacerraf & Putnam 1964 and 1983.

[Hilbert, 1970] D. Hilbert. "The Future of Mathematics", Chapter X of C. Reid *Hilbert*, Berlin: Springer-Verlag, 1970.

[Hilbert, 1971] D. Hilbert. *Foundations of Geometry*, La Salle, IL: Open Court, 1971. Originally published as *Grundlagen der Geometrie*, Stuttgart: Teubner, 1899.

[Hofstadter, 1999] D. Hofstadter. *Gödel, Escher, Bach: an Eternal Golden Braid*, 20th anniversary edition, New York: Basic Books, 1999.

[Holland, 1992] R. A. Holland. "A Priority and Applied Mathematics" *Synthese* 92, 349-370, 1992.

[Kant, 1991] I. Kant. "What is Orientation in Thinking?" in *Kant's Political Writings*, (2nd. Edition) introduction and notes by Hans Reiss, trans. H. B. Nisbet. Cambridge, UK: Cambridge University Press, 1991. First published October 1786 in *Berlinische Monatsschrift*, *VIII*, 304-30.

[Kant, 1965] I. Kant. *Critique of Pure Reason*, trans. Norman Kemp Smith. New York: St. Martin's Press, 1965. First edition, 1781, second edition published 1787 as *Kritik der reinen Vernunft*, Riga: Johann Friedrick Hartknoch.

[Kant, 1959] I. Kant. *Foundations of the Metaphysics of Morals*, trans. Lewis White Beck, Indianapolis and New York: Bobbs-Merrill Company Inc, 1959.

[Maddy, 1900] P. Maddy. *Realism in Mathematics*, Oxford: Oxford University Press, 1900.

[Mayberry, 2000] J. Mayberry. *The Foundations of Mathematics in the Theory of Sets*, Cambridge: Camrbidge University Press, 2000.

[Moore, 2001] A. Moore. *The Infinite* 2nd edition, London and New York: Routledge, 2001.

[Nerlich, 1976] G. Nerlich. *The Shape of Space*, Cambridge: Cambridge University Press, 1976.

[Parsons, 1980] C. Parsons. "Mathematical Intuition" *Proceedings of the Aristotelian Scoiety*, 80 pp. 145-68, 1980. Reprinted in *The Philosophy of Mathematics*, W.D.Hart ed., 1996, Oxford: Oxford University Press.
[Quine, 1963] W. V. Quine. *Set Theory and its Logic*, Cambridge, MA: Harvard University Press, 1963.
[Quine, 1969] W. V. Quine. "Ontological Relativity" in *Ontological Relativity and Others Essays*, New York: Columbia University Press, 1969.
[Riechenbach, 1949] R. Reichenbach. "The Philosophical Relevance of the Theory of Relativity" in Schilpp, 1949.
[Robertson, 1949] H. P. Robertson. "Geometry as a branch of Physics" in Schilpp, 1949.
[Rotman, 1993] B. Rotman. *Ad Infinitum: The Ghost in Turing's Machine*, Stanford: Stanford University Press, 1993.
[Russell, 1903] B. Russell. *Principles of Mathematics*, Cambridge: Cambridge University Press, 1903.
[Russell, 1919] B. Russell. *Introduction to Mathematical Philosophy*, London: Allen & Unwin, 1919.
[Schilpp, 1949] P. A. Schilpp, ed. *Albert Einstein: Scientist-Philosopher, Library of Living Philosophers Vol. VII*, Evanston, IL: The Library of Living Philosophers Inc, 1949.
[Tiles, 1991] M. Tiles. *Mathematics and the Image of Reason*, London: Routledge, 1991.
[Tiles, 2004] M. Tiles. "Kant: From General to Transcendental Logic" in *Handbook of the History of Logic, Vol.3: The Rise of Modern Logic from Leibniz to Frege*, ed. Dov M. Gabbay and John Woods, Amsterdam-Boston-Heidelberg-London-New York-Oxford-Paris-San Diego-San Francisco-Singapore-Sydney-Tokyo: Elsevier North Holland, 2004.
[van Heijenoort, 1967] J. Van Heijenoort. *From Frege to Gödel*, Cambridge, MA: Harvard University Press, 1967.
[Wang, 1986] H. Wang. *Beyond Analytic Philosophy*, Cambridge, MA: MIT Press, 1986.
[Wittgenstein, 1963] L. Wittgenstein. *Philosophical Investigations*, Oxford, UK: Blackwell, 1963.

LOGICISM

Jaakko Hintikka

1 WHAT IS LOGICISM?

Logicism can be characterized as the doctrine according to which mathematics is, or can be understood as being, a branch of logic. Historically speaking, logicism became a major position in the late nineteenth century. The most prominent representatives of this view have been Gottlob Frege (1848–1925), Bertrand Russell (1872–1970), and Rudolf Carnap (1891–1970). The term "logicism" did not gain currency until the late twenties, largely through Fraenkel [1928] and Carnap [1929]. Another formulation says that according to logicism mathematics can be reduced to logic.

A more detailed statement is given in a classical paper by C.G. Hempel [1905-1997], (see [Hempel, 1945]). According to Hempel the logicist thesis means that

(a) All concepts of mathematics, i.e., of arithmetic, algebra, and analysis, can be defined in terms ... of pure logic.

(b) All the theorems of mathematics can be deduced from those definitions by means of the principles of logic (including the axioms of infinity and choice).

Such characterizations leave a large number of loose ends, however. For one thing, it is not clear precisely what is supposed to be reduced to precisely what. Hempel's formulation speaks of a deduction of mathematical theorems from the principles of logic. This presupposes that mathematical theorems and logical principles are commensurate at least to the extent that the former can be deduced from the latter. But mathematical and logical systems are not in fact commensurate in a natural and widely accepted perspective. Mathematical theorems deal with what is true in a certain structure, for instance, in the structure of natural numbers or in that of real numbers. In contrast, logical principles deal with logical truths. These are not a subclass of truths simpliciter, that is truths in some one structure. They are truths in every possible structure. They can be considered empty or "tautological", just because they do not exclude any possibilities. How could mathematical truths possibly be deduced from them? The difference is among other things illustrated by the fact that the inference rules used in systematizing the two kinds of truth can be different. For instance, some actually used inference rules in logic that preserve logical truth but do not preserve ordinary truth.

This problem did not bother early logicists like Frege and Russell, for whom logical truths were simply the most general truths about the world. But as soon as one is forced to distinguish between logical truth and truth simpliciter, a logicist is in for a serious difficulty. Hence the very conception of logical truth presupposed by Frege and Russell points to a difficulty in the logicist position.

These problems lead us to the two fundamental questions on which any examination and evaluation of logicism crucially depends: What is mathematics? What is meant by logic? It is important to realize that the meaning and the reference of both of these crucial terms has changed in the course of history. This makes the force of the term "logicism" also dependent on the historical context in which it is being applied.

2 WHAT IS MATHEMATICS?

One important change in the meaning of mathematics was beginning to take place at the very time logicism first became an important movement in the philosophy of logic through the efforts of Frege and Russell. According to the earlier view, mathematics has two subject matters, number and space. The two most basic parts of mathematics are therefore arithmetic and geometry. Admittedly, the concept of number was generalized so as to include real numbers and complex numbers. Accordingly, arithmetic was extended to infinitesimal or "higher" analysis. Yet, in spite of this tremendous growth of mathematics, someone like Leopold Kronecker (1823–1891) could still maintain that natural numbers are at the bottom of all mathematics.

Slowly the scope and function of mathematics began to change. The study of number and space was transformed into a study of structures which may be instantiated in arithmetic as well as in algebra and in geometry, and, perhaps, altogether outside the realms of number and space. Not only were analogies discovered between geometry and algebra, analogies which had already been exploited in analytic geometry. The structures now being studied were more general than either algebra or geometry. They could be realized in yet different material. For instance, group structures became crucial both in algebra and in geometry, as witnessed by the Galois theory in algebra and by Felix Klein's (1849–1925) Erlanger Program. But groups could be found also outside mathematics. Perhaps the most important single step in this generalization process was Bernhard Riemann's (1826–1866) introduction of the idea of manifold. Manifolds were not in themselves geometrical any more than algebraic or analytical, even though different geometries could be thought of as special cases of such structures. The most abstract structures studied in the "new mathematics" were sets. The genesis of the set theory in the hands of Georg Cantor (1845–1918) and others was thus a crucial step in the development of the new conception of mathematics. It is revealing of the antecedents of set theory that the Riemannian term "manifold" (*Mannigfaltigkeit*) was initially applied to sets, too.

Philosophical formulations of this new conception of mathematics are found among other places in Edmund Husserl (1859–1938) who at one point characterized mathematics as the science of theoretical systems in general [Husserl, 1983; Schumann and Schumann 2001, 91]. Husserl refers to the old mathematics as *Quantitätsmathematik*.

The new conception of mathematics is sometimes called conceptual or abstract mathematics. As was already indicated, the most important pioneer of this conception was probably Bernhard Riemann, (see [Laugwitz, 1996]). Its development is often characterized as being motivated by a search of greater rigor. This is not the whole story, and philosophically the new role of mathematics as a tool of conceptual analysis is a more interesting one. In fact, one service that an abstract mathematics could render was to analyze and define different concepts originally formed intuitively rather than logically. For instance, in the theory of surfaces developed by C.F. Gauss (1777–1853) and Riemann, mathematicians could explicitly define concepts like curvature which originally were formed intuitively. A.L. Cauchy (1789–1857), Karl Weierstrass (1815–1897) and others showed how to define the basic concepts of analysis, such as convergence, continuity, differentiation, etc. In Cantor's set theory, the very notions of cardinal and ordinal number were extended to infinite numbers.

This development of abstract mathematics means that mathematics and logic were spontaneously converging at the time when logicism began its career. In a sufficiently general historical perspective, the genesis of logicism is but one particular manifestation of this general development. However, the first major figures of logicism, Frege and Russell, formulated their project by reference to the earlier conception of mathematics. Frege sought to define the concept of number and to show that when this definition is taken into account, all mathematical truths become logical truths. By mathematical truths Frege meant in the first place arithmetical truths. He exempts geometry completely from his treatment. Logical truths were considered by Frege analytic in contradistinction to Kant, who had considered mathematical truths synthetic *a priori*.

There is another related development in the nature of mathematics that is relevant to the motivation and prospects of logicism. It is the proliferation of mathematics into ever more numerous independent theories. Perhaps this is a consequence of the idea of mathematics as the study of all different kinds of structures. The multiplicity of different kinds of structures necessitates a similar multiplicity of mathematical theories.

In contrast, logic is usually thought of as one unified discipline. Admittedly, recent decades have seen a host of different "nonclassical logics" and "philosophical logics" making their appearance all the way from modal logics to nonmonotonic logics to quantum logic. Much of this multiplication, be it with or without necessity, is nevertheless irrelevant to any attempted reductions of mathematics to logic. The reason is that the logic involved in such reductions is mostly old-fashioned classical logic. For instance, one can barely find more than a couple of applications of modal logic to the foundations of mathematics. Admittedly, the intuitionistic logic of Heyting (see, e.g., [Heyting, 1956]) is closely related to the modal logic known as S4. However, the father of intuitionism, L.E.J. Brouwer, did not accept this logic as a representation of his ideas. In any case, intuitionistic ideas seem to be best implemented along different lines. (See sec. 13 below.)

3 WHAT IS THE LOGIC OF LOGICISM?

But it is not only the changing fortunes of the idea of mathematics that matter in discussing logicism. In a historical perspective, there have also been important changes in what is included in the purview of logic. Probably the most important such change is associated with the contrast between general concepts (universals) and notions of particulars. From Aristotle on, logic, being a matter of reason, was taken to deal with universals. In contrast, only sense-perception was considered appropriate for dealing with particulars. Even for thinkers like Frege for whom logical truths were still truths about reality, they were the most general truths of that kind.

However, there are modes of apparently logical reasoning that seem to involve the use of particular representatives of general concepts. From our contemporary vantage point, they are rules of instantiation. Even though the rules of modern logic can, formally speaking, be formulated seemingly without explicit instantiation rules, in a deeper perspective the rules of existential and universal instantiation are the mainstays of first-order logic.

The explicit formulation of instantiation rules as central tools of logic is a recent development which involves such techniques as natural deduction, Gentzen's sequent calculus, Beth's semantical *tableaux* and what are known as tree methods. Under different names instantiation rules also played a role in much earlier discussions. (For the history of these methods see Judson Webb [2004].) Aristotle already used certain modes of reasoning of this kind in his logical theory under the title *ekthesis* (exposition). Because they involved particulars, such rules were not purely logical. Accordingly, Aristotle tried to dispense with *ekthesis* in his logical theory, but could not do so completely. Alexander Aphrodisias later declared that the use of *ekthesis* involves an appeal to sense-perception and hence is not purely logical.

In mathematical reasoning instantiation rules are likewise crucially important. In axiomatic geometry, instantiations play a role, partly in the use of what looks like particular figures exemplifying theorems and problems, partly in the form of the auxiliary constructions that introduce new geometrical objects — apparently particular objects — into the figures so "constructed". The part of a Euclidean proposition in which former kinds of instantiations are used was even called by the same name *ekthesis* as instantiation in logic. Instantiations of the second kind are fairly obviously indispensable, for typically theorems could not be proved without suitable auxiliary constructions. And *ekthesis* was in turn indispensable, for it introduced the figure which was amplified by auxiliary constructions. It is fairly obvious that from our modern point of view both *ekthesis* and so-called auxiliary constructions can be thought of as applications of purely logical instantiation rules.

In their historical situation, it was nevertheless natural for early theorists of mathematics to think of instantiation rules as representing typically mathematical but not logical modes of reasoning. This idea was systematized by Kant into his theory of the mathematical method as being based on appeals to intuitions. By "intuitions" (*Anschauungen*) Kant by definition meant particular representatives of general concepts. Hence appeals to intuition in mathematical proofs amounted for Kant to instantiations (cf. here [Hintikka, 1969]).

It was the use of "constructions" in the form of *ekthesis* and auxiliary constructions that made mathematical truths synthetic for Kant. (The force of the term in his philosophy of mathematics is thus reminiscent of the meaning of "synthetic" in synthetic geometry.) In contrast, logical truths were for Kant based on the law of contradiction and hence analytic.

From a historical point of view, this brings out a crucial presupposition of the rise of logicism. The logicist position was not viable in the first place until the purview of logic was tacitly widened so as to include the uses of instantiation procedures illustrated by *ekthesis* and auxiliary constructions.

This happened as a part of the creation of modern logic by Frege and others. They did not just create *ex nihilo* the new structure called modern logic, in one possible formulation. They unwittingly (or in the case of C.S. Peirce (1839–1914), perhaps wittingly,) expanded the scope of what counts as logic. They made the use of instantiation methods not only a part of modern logic, but arguably its central part. Without this extension of the scope of logic, logicism would not have any plausibility whatsoever.

Once all this is understood, it can be seen that logicist theses need not be incompatible with the theses of earlier philosophies of logic and mathematics, when they are interpreted in the light of these changes in the conception of logic. In particular, logicism is compatible with Kant's claim that instantiation rules are the root of the mathematical method.

Frege seems to have harbored some apprehensions as to whether he was really contradicting Kant, (see [Frege, 1884, sec. 88]). At one point he speaks of the idea that in analytical reasoning the conclusions are contained in the premises. But contained in what sense? Like a plant in a seed or like a building-block in a house? Yet he does not qualify his claim that mathematical truths are analytic.

This tacit widening of the scope of logic is especially important to keep in mind when it comes to mathematical inferences (cf. here [Hintikka, 1982]). Even traditional logicians who thought of mathematical theorems as being proved by synthetic methods usually attributed this synthetic character exclusively to the use of *ekthesis* and of constructions in mathematical reasoning. The rest of a geometrical proposition, including the part called *apodeixis* where inferences are drawn is purely logical and purely analytic. Even Kant [1787, 14] acknowledged that all the inferences (*Schlüsse*) of mathematicians proceed according to the "principle of contradiction" and are therefore logical and analytic "as required by the nature of all apodeictic certainty". Hence it is not surprising to find nineteenth-century German thinkers refer to mathematical reasoning as being logical. The chances are that they did not think that they were necessarily contradicting Kant, and likewise they most likely were not prepared to embrace logicism.

The pioneers of modern logic were not aware of extending the concept of logic so as to comprehend what were earlier thought of as being characteristically mathematical. It is often said that the rise of contemporary logic originated as an attempt to apply mathematical methods in logic. However, this is not the whole story. There are unmistakable applications or at best hopes of applications in the other direction.

The different traditions of emerging symbolic logic were all more or less knowingly preparing the logicist case in that the intended applications of the new logic prominently included the foundations of mathematics, (see [Peckhaus, 1997, 307–308]). This includes not only the British tradition that was primarily oriented toward logic, but to some extent also the tradition of Charles S. Peirce and Ernst Schröder (1841–1902). Indeed, the unformalized logic which was employed by Weierstrass and others and which has become known as the epsilon-delta technique is part of the logic of quantifiers developed by Peirce and Frege and further studied by Schröder. This informal logicization of mathematics seems to be what is often intended by references to a quest of a rigor in the foundation of mathematics. As the example of set theory shows, the result of logical analyses of mathematical concepts sometimes led to greater uncertainties rather than directly to enhanced rigor.

As a matter of historical fact, Peirce rejected logicism in the sketchy and programmatic form in which he found it in Dedekind, (see *Collected Papers* 4.239, and cf. [Haack, 1993]). Yet Peirce makes it perfectly clear that his work on his iconic logic was calculated to enhance our understanding of, and capacity to carry out, mathematical reasoning (*Collected Papers* 4.428–429).

4 FREGE THE FIRST LOGICIST

However, when Frege first conceived the program of logicism, the development of modern logic had not yielded a system of logic which he could use as a target of a reduction of mathematics to logic. Hence he had to create such a logic himself. A preliminary result was published under the telling title *Begriffsschrift* (concept-notation) (1879). This logic, which will be discussed below, is essentially a higher-order logic of quantification, complicated by Frege's distinction between concepts and their extensions. The basic ideas of the reduction of mathematics to logic were outlined in *Die Grundlagen der Arithmetik* (Foundations of arithmetic) in 1884. Frege accepts the view of mathematics as the study of numbers and of space, that is as comprehending arithmetic (with its ramifications in analysis and elsewhere) and geometry. He exempts geometry from his reduction. Hence the basic part of Frege's project was a reduction of arithmetic to logic. Again, the crucial step is that reduction was the definition of number in what Frege took to be purely logical terms. Frege's insight was that the notion of the equinumerosity (equicardinality) of two sets can be characterized purely logically. Hence a number could be defined as the class of all equinumerous sets. Actually, Frege chose a slightly more complicated definition and defined a number as the class of all concepts whose extension are equinumerous.

Frege undertook to carry out the project that he had explained in the *Grundlagen* in explicit detail in his monumental work *Grundgesetze der Arithmetik* (Fundamental laws of arithmetic, two volumes, 1893 and 1903). Alas, just as Frege was reading the proofs of the second volume of the *Grundgesatze*, he received a letter from Bertrand Russell, pointing a contradiction in Frege's axiomatic system of logic [Russell, 1902]. This was the famous paradox of Russell's. In Frege's system, it arises by considering the concept "object that is the extension of some concept under which it does not fall". Does its extension fall under it or not? Either answer is easily seen to lead to an impossibility. In Frege's system, this argument is sanctioned by his assumptions. Hence his system is inconsistent.

Does that mean that Frege's project failed? And if so, what does that imply concerning the prospects of logicism?

It turned out that the contradiction could not be eliminated in any straightforward way from Frege's particular system. However, arguably the same problems arise in competing approaches, for instance in axiomatic set theory (see below). Hence it is not at all clear that the failure of Frege's project tells against logicism in particular. In order to see what there is to be said, a closer look at the presuppositions of Frege's logic is in order. Russell's paradox is only the proximate cause of Frege's difficulties. The real reasons for them lie much deeper. Frege's logicism will stand or fall with his logic.

5 FREGE'S LOGIC OF QUANTIFIERS

There are in fact several deeper flaws in Frege's logic. The crucial novelty of this logic is that it prominently is the logic of (existential and universal) quantifiers. This might not seem much of a novelty, for quantifier words like *some* and *every* are used already in Aristotle's syllogisms. However, the modern conception of quantifier is grounded on the assumption that they refer to some given domain of values of quantified variables over which the variables of quantification range. That is to say, quantifiers range over a given "universe of discourse" of particular objects, usually referred to as individuals. This idea is foreign to Aristotle, and it was developed only by the British nineteenth-century logicians. Frege swallowed the ranging-over idea completely. For him, quantifiers are higher-order predicates which tell whether a lower-order predicate is nonempty or exceptionless.

What Frege missed is an important fact concerning quantifiers. Their meaning is not exhausted by the "ranging over" idea. They serve another important function. On the first-order level, the only way in which we can express the actual (material) dependence of a variable (say y) on another variable (say x) is by means of the formal dependence of the quantifier (Q_2y) to which it is bound on the quantifier (Q_1x) to which the independent variable is bound.

Now in the logic of Frege and Russell, and in most of the logics of their successors, the formal dependence of quantifier $(Q_2 y)$ on the quantifier $(Q_1 x)$ is expressed by its occurring in the (syntactical) scope of $(Q_1 x)$, that is, within the parentheses that follow it:

$$(Q_1 x)(-(Q_2 y)(-)-)$$

Such scopes are assumed in Frege's and Russell's logic to be nested, that is, to exhibit a tree structure. Accordingly, the scope relation is antisymmetric and transitive, and can only serve to express similar modes of dependence. But this means that not all possible patterns of dependence relations can be expressed by means of Frege's logic. For instance, symmetrical patterns or branching patterns cannot be so expressed. An instance of the latter is the Henkin quantifier structure

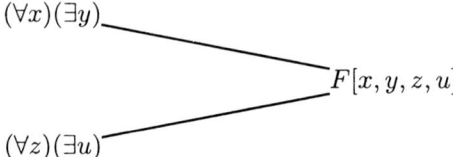

In other words, Frege's logic of quantifiers has a shortcoming that limits its expressive power. And this shortcoming has direct implications for Frege's logicist project. With some qualifications, it can be said that Frege defined number as the set of all equicardinal sets, that is, of all sets with the same number of members. But the equicardinality of two sets α and β cannot be defined in Frege's logic on the first-order level, even though it can be so defined when the restrictions that affect his logic are removed. Hence Frege had to use second-order logic, that is a logic in which quantifiers can range, not only over individuals, but over sets of individuals, or over properties and relations of individuals. Then the equicardinality can be expressed by saying that there exists a one-to-one relation that maps α on β and vice versa.

6 DEFINING REAL NUMBERS

Similar conclusions would have emerged if someone had tried to use the first-order part of Frege's logic as a tool in defining numbers other than natural numbers. Sharper logical tools are perhaps not needed for the introduction of negative numbers or rational numbers. But things are more difficult in the theory of real numbers. Different ways of defining them in terms of rational numbers were explored by different mathematicians, most prominently by Weierstrasss, Cantor and Richard Dedekind (1831–1916). This work has a much more direct impact on mathematical practice than questions of how to define natural numbers, the reason being that real-valued functions were at that time the true bread-and-butter subject of a working mathematician. Whatever definition of real numbers is adopted, they must have as a consequence of the definition of the properties that are needed in analysis. For instance, any set of real numbers so defined must have a real number as its least upper bound. Most of the logicians and mathematicians in the late nineteenth century did not, with the partial exception of Dedekind, relate the problems of defining real numbers to the logicist program. Nevertheless, in a systematic perspective finding such definitions is a major challenge to a logicist. The difficulty of this task was brought home to mathematicians by the criticism of classical mathematics by intuitionists like L.E.J. Brouwer (1881-1966) and Hermann Weyl (1885-1955).

7   FREGE'S HIGHER-ORDER LOGIC

Thus in the light of hindsight it can be seen why Frege had to build a higher-order logic, in other words a logic in which the values of the variables of quantification could be higher-order entities, perhaps sets or properties and relations. This choice between properties and sets involves a choice between extensions of concepts and concepts themselves as values of variables. Now not only set theory but, as Frank Ramsey (1903-1930) noted in 1925, practically all modern mathematics deals with extensions. Formally speaking, they traffic in extensions of predicates and what used to be called relations-in-extension. But Frege did not think that we can speak of extensions directly, without considering the concepts whose extensions they are. Extensions were for him only a special kind of particular objects. Hence the same logic of quantification applies to them as applies to ordinary individuals. What remains to be determined in order for us to have a higher-order logic are the identity conditions of extensions. As these conditions, Frege assumes what look like the natural ones. They are formulated as the two parts of his Basic Law V. They say that two concepts have the same extension if and only if the same individuals fall under them. In the same Basic Law, Frege also assumed that each simple or complex predicate of his formal language expresses a concept.

Natural or not, this basic law quickly led to the contradiction Russell pointed out to him in his famous letter dated June 16, 1902. Moreover, the difficulty turned out to be impossible to eliminate in any simple way. Thus Frege's grand logicist project failed. But where did it leave logicians and mathematicians?

8   AXIOMATIC SET THEORY VS. LOGICISM

In the light of hindsight it can be said that the most important repair operation in the foundations of mathematics was the axiomatization of set theory. It involved discarding Frege's use of concepts altogether and building up a theory of extensions (sets, classes) only. Even though the fact was not appreciated by the first axiomatic set theorist, Ernest Zermelo (1871-1953), such an alternative to higher-order logic is not likely to make sense only if the logic used in it is first-order logic.

Axiomatic set theory came to be considered widely as the natural medium of mathematical reasoning and theorizing. Such a view implies a rejection of the logicist thesis, for set theory does not reduce to the logic it presupposes, which is normally assumed to be the traditional unamended first-order logic. First-order set theory requires additional assumptions, in the first place various assumptions of set existence.

A comparison with axiomatic set theory reveals in fact an important weakness in Frege's treatment of higher-order logic and *a fortiori* in his logicism. Frege thought that extensions (classes) were simply objects of a certain kind. What is peculiar to them is merely how they are obtained from concepts. This is what Frege's Basic Law V was calculated to tell us. However, there is no hope that this law could give us all that we need for the purposes of mathematics. Even if something like this law had not led into contradictions, Frege would have needed some rules for higher-order entities, rules that do not apply to other kinds of objects but which apply to them in virtue of their being the higher-order objects that they are. The axiom of choice is a typical example of such laws.

In axiomatic set theory, such higher-order laws take the form of axioms that are assumed over and above the first-order logic that is being used. But this axiomatic treatment does not give us any reason to think that such laws are logical and not essentially mathematical.

Set theory is accordingly considered in our days almost universally as a mathematical rather than logical theory. The widespread reliance on axiomatic set theory as the *lingua franca* of mathematics has therefore led to a perception of logicism as a defense of a lost cause. This rejection of logicism nevertheless cannot be considered as a *fait accompli*. Axiomatic set theory faces much greater difficulties than has been realized, (cf. here [Hintikka, 2004]). In a perfectly natural sense, some theorems of first-order axiomatic set theory are even false, (see sec. 12 below).

Historically speaking, axiomatic set theory was created as a response to other kinds of difficulties. For Frege, as was seen, extensions were simply certain kinds of objects. The treatment of set theory on the first-order level is but a codification of that idea. However, such a treatment of sets and their members on the same level easily leads to problems. Plausible-looking assumptions were seen to lead to outright contradictions, known as paradoxes of set theory. Zermelo's axiomatization was calculated to restrict the assumptions made in set theory so as to weed out all inconsistencies and yet to give the resulting theory enough power to serve all of mathematics.

9 PRINCIPIA MATHEMATICA AND ITS AFTERMATH

This is a delicate task, and some more logically minded mathematicians and philosophers preferred another idea which preserved the logicist program. Not unexpectedly, this idea was to stratify the set-theoretical universe, that is, to treat sets and their members always on a different level. This is essentially the idea of higher-order logic that Frege already tried to implement. After experimenting with different approaches it is also the idea Bertrand Russell ended up embracing. Together with A.N. Whitehead, he tried to show how to reconstruct all of mathematics on this basis in their monumental work *Principia Mathematica* (1910–1913).

Did they succeed better than Frege in trying to carry out the logicist project? Their higher-order logic was stratified into levels in the same way as Frege's logic. Different levels of the hierarchy are called different types. Quantifiers of a given type can only range over entities of the same or a lower type. Moreover, there is a more refined distinction between what are known as ramified types. Did this logic work out? Russell and Whitehead had the benefit of knowing the intensive discussion of the paradoxes which had come up not only in Frege's higher-order logic but in the original naïve set theory itself. Many things can be said and have been said of the system of *Principia Mathematica*, but in a sufficiently deep philosophical perspective it can and perhaps should be discussed in the first place by reference to one aspect of logicism. This is the close relation of logicism to the quest of a *lingua franca* of mathematics, (see sec. 2 above). A purely logical language is presumably universal, the most general language that there is, at least for the purposes of mathematics. But if all mathematics can be done in such a purely logical language, then so must be the metatheory of any mathematical theory and ultimately the metatheory of this very universal mathematical language.

Even though the logicians who have stressed the importance of such metatheory do not seems to have pointed it out, this stress is very much in keeping with the kind of mathematical practice to which the development of abstract, conceptual mathematics gave rise. Even a typical axiomatic theory in conceptual mathematics, for instance, group theory, does not consist mainly or even primarily of deductions of theorems from axioms. Most of it is in present-day terminology metatheory, for instance, classifying of different kinds of groups or proving representation theorems. If mathematics is to be reduced to logic, the logical language to which it is reduced hence must include its own metatheory.

No further metalanguage should therefore be needed to discuss what goes on in this universal logical language to which mathematics could be reduced. But set-theoretical languages are not likely to satisfy this requirement. For instance, we must be able to speak in such a language of what is and is not definable in it. If such a language allows only a countable number of definitions, there must exist sets indefinable in the language, for then provably exists uncountably many sets. But if our language enables us to speak of what is definable in it, we can for instance define in it the least undefinable ordinal, which would involve a contradiction.

Hence questions of definability are crucial for the logicist program. Another problem concerning definability was the crux of the project of Russell and Whitehead. Consider a set s definable by means of quantifiers ranging over a class to which s is itself is supposed to belong. Such definitions are called impredicative. They seem to involve a kind of vicious circle, and Russell attributed a number of paradoxes to the use of impredicatively defined sets.

The ramified hierarchy of Russell's and Whitehead's is an attempt to rule out all impredicativities. It was supposed to be the crucial element of their logicist project. However, the ramified theory of types ran into formidable complexities. Its upshot is that the logical status of a set could depend crucially on the way it is defined. In this sense, the logic of *Principia Mathematica* is not purely extensional. What is worse, Russell and Whitehead could carry our their overall project only by making assumptions which do not have much theoretical justification. The most important assumption is known as the axiom of reducibility. Strangely enough, it eliminated some of the very complexities that the ramified hierarchy was calculated to introduce.

As was indicated, in 1925 Frank Ramsey proposed to replace the system of *Principia Mathematica* by an extensional one which dispensed with the ramified hierarchy and with the axiom of reducibility. The result was a version of higher-order logic known as the simple theory of types.

Most mathematicians and logicians nevertheless preferred set theory to the theory of types as a medium of mathematical theorizing, hence in effect disregarding logicism. This is partly due to the greater flexibility of set-theoretical foundations and their closeness to the usual mathematical symbolism. This preference may nevertheless have tacit deeper reasons.

10 LOGICISM VS. METAMATHEMATICS

What are they? Whatever the merits of a theory of types or a higher-order logic are or may be, it is not obvious that they can provide a vindication of logicism. For one thing, it is no longer clear that all mathematics can be done in such a logic, the reason being that its own metatheory, which is a legitimate subject of mathematical investigation, is apparently impossible to develop in the logical theory itself. We would, for instance, have to quantify over all types, which is blatantly impossible in type theory itself.

Other aspects of the metatheory of logic point in the same direction. The ramified theory of types was partly motivated as a way of avoiding the so-called semantical paradoxes of higher-order logic and set theory. They arise when one tries to discuss the metatheory of a logical language in the same language, for instance discussing what is or is not definable in it. Ramsey's elimination of ramified types can be said to be based on giving up the project of such self-applied theory. This metatheory is typically mathematical in nature, often called in fact metamathematics. But that meant that in the resulting theory you could no longer deal with its own metamathematics. Hence some parts of mathematics could not be reduced to it.

This point is related to the reasons for which the main architect of contemporary metamathematics, the great German mathematician David Hilbert (1862–1943) did not accept logicism. For according to him, some mathematics is needed already in the theory of purely formal logic. Hence logic and mathematics have to be built together, without trying to reduce one to the other.

In spite of these difficulties, logicism continued to find supporters. For instance, at the historical meeting in Königsberg in 1930 logicism was considered as one of the main currents in the foundations of mathematics represented by an invited main speaker. The others were Hilbert's metamathematics, represented by John von Neumann (1903-1957), intuitionism, represented by A. Heyting (1898–1980) and Wittgenstein's philosophy of mathematics, represented by Friedrich Waismann (1896–1969). Logicism was represented by one of the central figures of Vienna Circle, Rudolf Carnap (1891–1970). Carnap had literally had a vision of a universal language in which we could, among other things, reconstruct mathematics and also speak of itself. Alas, this project faltered on the impossibility results of Kurt Gödel (1906–1978) and Alfred Tarski (1902–1983). In particular, Tarski's famous undefinability theorem seemed to shatter all reasonable hopes for a universal mathematical language and thereby to logicism. Tarski proved that the crucial metalogical concept of truth can be defined for a first-order language (of the received Frege-Russell sort) only in a richer metalanguage. Thus the result of Carnap's efforts, *Logische Syntax der Sprache* (*Logical Syntax of Language*, 1934) failed to produce a universal language which would have vindicated the logicist position. In spite of this, some of the other logical positivists continued to support logicism, among others C.G. Hempel quoted earlier.

Gödel's and Tarski's results changed radically the entire question of the truth of logicism. In so far as first-order logic is thought of as the logic of our actual discourse, this seems to end all hope of a kind of universal language that is apparently needed for logicism.

Gödel's result had different kinds of consequences. He showed that as basic parts as elementary arithmetic must be incomplete in the sense that in any axiomatization of elementary arithmetic there must be sentences that are true but unprovable. Since arithmetical truth can easily be captured by means of higher-order logic, it follows that higher-order logic must likewise by unaxiomatizable.

11 THE TRANSFORMATION OF LOGICISM

How are these results relevant to the nature and prospects of logicism? What they show is not so much that logicism is wrong but that its earlier formulations do not make any sense, that is, that the way logicism was earlier conceived is inappropriate. Earlier logicians and philosophers typically construed the reduction of mathematics to logic as a reduction of the axioms of arithmetic (or whatever other part of mathematics is at issue) to an axiomatic system of logic. Now it turns out that even elementary arithmetic is not axiomatizable. Furthermore, since higher-order logics are not axiomatizable, they do not offer any axiomatic systems to reduce mathematics to. What remains axiomatizable is first-order logic, but it is by itself woefully inadequate as a medium of nontrivial mathematics. This might seem to end all hopes of carrying out a logicist program. However, what emerges is the need of reinterpreting the very claims of logicism. They cannot be construed as claiming the reducibility of mathematical concepts or theories to logical concepts or logical systems. Such claims make little sense in the light of the change in our conception of mathematics noted above. If mathematics is not the study of certain particular numerical and geometrical structures but a study of structures of all different kinds, a reduction of one system to another has little relevance to the realities of the relation of mathematics to logic. The tasks of logic and mathematics are beginning to look very similar. What distinguishes them will be a difference between the conceptual tools used. A reduction of mathematics to logic will be essentially a reduction of the methods of reasoning (proof) used in mathematics to the modes of reasoning codified in logic. This relation of the two is in any case what matters to mathematical practice, the focal point of which is often considered to be theorem-proving.

This shift of emphasis from axiomatic reductions of mathematics to logic to comparisons of mathematical and logical modes of reasoning is thus in keeping with the development noted earlier of conceptual (abstract) mathematics and can be considered part of this development. It did not come about suddenly, either. For instance, Peirce's project of understanding better our modes of mathematical reasoning in logical terms can be taken to be a part of the same general project.

At first sight, this shift of perspective nevertheless does not seem to matter very much to the problem of logicism. Set theory in its axiomatic form can, from this point of view, be thought of as an inventory of modes of inference acceptable in mathematics. (Of course this refers to modes of inference that go beyond first-order logic, for an axiomatic set theory uses itself first-order logic.) This role of set theory as a theory of mathematical modes of inference may sound strange, for set theory in its axiomatic form is like any axiomatic theory a theory of some domain of entities, the set-theoretical universe, not a theory of forms of valid inference. But this distinction perhaps does not make much difference. For instance, the axiom of choice, which codifies a mathematical inference pattern *par excellence*, appears in set theory as one of its axioms. An important indication of how the difference between ways of looking at set theory can be overcome is the flourishing research program known as reverse mathematics, (see here, e.g., [Simpson, 1999]). It is was created principally by Harvey Friedman (born 1948). In it, the difficulty of a mathematical proof is measured by the sets that have to exist according to axiomatic set theory in order for the proof to go through. Hence the study of forms of mathematical inference, that is, according to this view, set theory, is itself a mathematical rather than logical theory.

Also, the nonaxiomatizability of higher-order logic might perhaps be taken to count against its ability to serve as a medium of logical proofs. For we do not have any longer an exhaustive method of deciding which proof steps are valid or not, as we had in Frege-Russell logic.

Yet arguably we should look at the relation to these results to the idea of logicism in a different way. As was indicated earlier, there are serious difficulties in the idea of set theory as a depository of valid modes of mathematical inference. For one thing, are all the modes of inference sanctioned by axiomatic set theory valid? It has been well known that there are counterintuitive theorems in axiomatic set theory. They have nevertheless been about a very large set concerning which our intuitions can only be expected to be shaky. However, it can be shown (see [Hintikka, 2004]) that such counterintuitive theorems can pertain to relatively "small" sets and that the intuitions which are being violated concern our pretheoretical notion of truth rather than sets *per se*. It is not even difficult to give an indication of what such false theorems say. In the same (qualified) sense in which the famous Gödelian sentence says, "I am not provable", the new paradoxical sentence says, "My Skolem functions do not exist." A moment's reflection shows that the existence of the Skolem functions for a given sentence S is the natural truth condition for S. In the sense appearing from these remarks, there are false theorems in axiomatic set theory, which therefore is a poor guide to valid mathematical inferences.

12 CORRECTING FREGE'S THEORY OF QUANTIFICATION

Are we therefore driven back to higher-order logic? The answer depends on how much can be done in first-order logic. Now it was noted earlier that the received first-order logic that goes back to (a fragment of) Frege's and Russell's logic does not fully satisfy its job description, in that there are patterns of dependence and independence between quantifiers not expressible in it. Now this shortcoming is corrected in what is known as independence-friendly (IF) first-order logic. (For it, see [Hintikka, 1996].) It is obtained from the received first-order logic by merely allowing a quantifier to be independent of another one even when it occurs in the syntactical scope of the latter. IF first-order logic is obviously our genuine basic logic, free from the unnecessary limitations of the Frege-Russell quantification theory. How does its discovery affect the prospects of logicism?

First, it reinforces the reinterpretation of logicism as claiming that mathematical modes of reasoning can all be interpreted as logical ones. For IF first-order logic is not axiomatizable in the same way ordinary first-order logic is. Hence there is no rock-bottom axiom system of logic to which mathematical axioms systems could be reduced. Accordingly, the only natural sense of reduction here is for mathematical modes of inference to be reduced to the semantically valid logical inferences.

Is such a reduction possible, as the reconstructed logicist thesis claims? At first, this may seem unlikely, for deductively IF logic is in certain respects weaker than the received first-order logic. For one thing, the negation used in it is a strong (dual) negation which does not obey the law of excluded middle. However, IF logic has expressive capabilities that the Frege-Russell logic does not have. Among other things, the equicardinality of two sets can be expressed by its means, as can such mathematically crucial notions as the infinity of a set, topological continuity and a suitable formulation of the axiom of choice. In general, a great deal of what has been taken to be characteristically mathematical reasoning can now be carried out in logic, viz., IF first-order logic. One important thing that this means is that the tacit reasons that forced Frege to resort to higher-order logic are weakened.

On the other hand, the absence of the law of excluded middle from IF logic suggests that it can serve as an implementation of intuitionistic ideas.

13 REDUCTION TO THE FIRST-ORDER LEVEL

Indeed, second-order logic can in a sense be dispensed with altogether. Even though not all mathematical reasoning can be carried out in IF first-order logic, this logic can be extended and strengthened while still remaining on the first-order level in the sense that all quantification is over individuals (particular members of the domain). By itself, IF first-order logic is equivalent to Σ_1^1 fragment of second-order logic. (This is the logic of sentences which have the form of a string of second-order existential quantifiers followed by a first-order formula.) It can nevertheless be extended by adding to it a sentence-initial contradictory negation. This adds to it the strength of Π_1^1 second-order logic. In order to extend IF logic further, a meaning must be associated to contradictory negation also when it occurs in the scope of quantifiers. This can be done, but it involves a strongly infinitary rule which involves the possibly infinite domain of individuals as a closed totality and which is tantamount to an application of the law of excluded middle to propositions of a complexity. This complexity can be thought of as a measure of the nonelementary (infinitistic) character of the application. If no limits are imposed on this complexity, we obtain a logic which is as strong as the entire second-order logic but is itself a first-order logic in the sense of involving only quantification over individuals.

14 LOGICISM VINDICATED?

This development can be taken to constitute a qualified vindication of re-interpreted logicism. For virtually all normal mathematical reasoning can be carried out in second-order logic. (This logic is here and throughout this article naturally understood as having the standard semantics in the sense of Henkin [1950].) As was pointed out earlier, the character of second-order logic as involving quantification over higher-order entities has prompted doubts as to its status as a logic and not as a mathematical theory, as "set theory in sheep's clothing", to use Quine's phrase. Now it turns out that in principle no quantification over higher-order entities is needed. All reasoning codified in terms of second-order logic can in principle be carried out in terms which obviously are purely logical. Admittedly, the reconstruction of second-order logic on the first-order level involves strongly infinitary assumptions, but this was only to be expected. In conjunction with the problems affecting the main rival of higher-order logic as a codification of mathematical reasoning, axiomatic set theory, this development strengthens the reconstructed logicist position.

This conclusion is reinforced by other considerations. Earlier, it was seen that a failure of logical languages to deal with their own metatheory was considered an objection to logicism. The force of such objections is reduced by the fact that some aspects of the metatheory of an IF first-order language L can be expressed in the same language. In particular, if L is rich enough to enable a formulation of its own syntax, then the concept of truth can be defined for L in L itself, (see

[Hintikka, 1996]). This shows the limitations of Tarski's impossibility result, both in itself and as a basis of objections to logicism.

More generally speaking, even when the metatheory of a language cannot for some reason be formulated in the same language, it does not necessarily follow that the modes of reasoning needed in the metatheory must be stronger than those involved in the theory itself. Thus an interpretation of logicism as claiming a reduction of mathematical modes of reasoning to logic eliminates a class of objections to it. The reason may be that those kinds of reasoning are applied to more demanding cases.

When logicism is construed as a thesis about the relation of mathematical modes of inference, the problems caused by the incommensurability of logical and *de facto* truth also disappear.

It is not clear, either, that the presumed advantages of axiomatic set theory in the foundations of mathematics cannot be duplicated by means of second-order logic reconstructed as an infinitistic first-order logic. An example may be offered by the reverse mathematics mentioned earlier. There the demands of a mathematical proof are measured by the sets that have to exist for the proof to go through. But suppose that a set **s** with a definiens $D[x]$ and with the explicit definition

$$(\forall x)(x \varepsilon s \leftrightarrow D[x])$$

is proved to exist. Then the principle of excluded middle can be applied to the definiens $D[x]$. The complexity of $D[x]$ can then be read as a measure of the nontriviality of the same step in a mathematical argument as relied on the existence of **s**, as is suggested by the use of the complexity of applications of *tertium non datur* as a natural measure of the nontriviality of a logical argument.

The possibility of construing mathematical reasoning as moving (in the last analysis) always on the first-order level removed several obstacles from the path of logicism. It was seen that the quantification over higher-order entities is a crucial difficulty for logicists. But since everything now happens at the first-order level, all problems concerning the existence or nonexistence of higher-order entities disappear. We do not have to search for those principles of peculiarly higher-order reasoning that set theory is supposed to catch but which it cannot ever fully completely exhaust.

Instead of a search for stronger set-theoretical axioms, mathematicians now face the problem of discovering new and more powerful principles of logical proof. This problem remains because not even the new basic logic, IF first-order logic, is not axiomatizable in one fell swoop. But this problem concerns the existence and nonexistence of different structures of particular objects (individuals). Such structures are much easier to have intuitions about and to experiment with in thought than complexes of higher-order entities. And the study of such structures in general belongs as much and more to logic than to mathematics.

Furthermore, the predicativity or impredicativity of definitions ceases to be an issue. Since there is no quantification over predicates, no definition of a predicate can involve a totality to which it itself belongs. The same holds for all other kinds of higher-order entities. Definitions of individuals by means of quantifiers will admittedly involve a totality to which the defined individual belongs, viz., the range of quantifiers. But this is simply the given universe of discourse, an appeal to which does not introduce any vicious circles.

For similar reasons, problems concerning the definitions of real numbers (cf. section 6 above) are dissolved into the unavoidable perennial problem of finding better and better principles of logical reasoning. Thus once again new developments in logic have changed the prospects of logicism.

An especially interesting suggestion concerning the relations of logic and mathematics that ensues from these different results is the creative component in mathematics and in logic. This creative component cannot be restricted to mathematics as distinguished from logic, as used to be generally thought. For instance, it would not be appropriate to locate it in the search of further axioms of set theory, as for instance Gödel seems to have thought. The most basic core area of logic is deductively incomplete, which means that we have to go on searching for deductive axioms already there. And these logical truths are all we need in our mathematics. In this deep sense, all that is needed in mathematics can already be done in logic. And in this same sense, the basic idea of logicism seems to be vindicated.

BIBLIOGRAPHY

Much of the literature on the foundations of mathematics in general is relevant to logicism. Only such literature is listed here that deals specifically with the issues dealt with in this article.

[Benacerraf, 1995] P. Benacerraf. Frege: The Last Logicist, in [Demopoulos, 1995, 41–67].
[Boolos, 1990] G. Boolos. The Standard of Equality of Numbers. In *Meaning and Method: Essays in Honor of Hilary Putnam*, Cambridge University Press, Cambridge, pp. 261-277, 1990.
[Bostock, 1979a] D. Bostock. *Logic and Arithmetic*, Vol. 1, Oxford University Press, 1979.
[Bostock, 1979b] D. Bostock. *Logic and Arithmetic*, Vol. 2, Oxford University Press., 1979
[Carnap, 1929] R. Carnap. *Abriss der Logistik mit besonderer Berücksichtigung der relationstheorie und ihrer Anwendungen*, Verlag Julius Springer, Wien, 1929.
[Carnap, 1930–31] R. Carnap. Die Mathematik als Zweig der Logik, *Blätter für deutsche Philosophie* vol.4, Berlin, pp. 298-310, 1930–31.
[Carnap, 1983] R. Carnap, The Logicist Foundations of Mathematics. In *Philosophy of Mathematics*, 2nd ed., Paul Benacerraf and Hilary Putnam, editors, Cambridge University Press, Cambridge, 1983 (original 1939).
[Carnap, 1934] R. Carnap. *The Logical Syntax of Language*, Routledge and Kegan Paul, London, 1934.
[Church, 1962] A. Church. Mathematics and Logic. In *Logic, Methodology and Philosophy of Science, Proceedings of the 1960 international Congress*, Stanford California, pp. 181-186, 1962.
[Coffa, 1995] A. Coffa. Kant, Bolzano and the Emergnce of Logicism. In [Demopoulos, 1995, 41–67].
[Demopoulos, 1995] W. Demopoulos, ed. *Frege's Philosophy of Mathematics*, Harvard University Press, Cambridge, Mass, 1995.
[Egidi, 1966] R. Egidi. Aspetti della crisi interna del logicismo, *Archivio di Filosofia*, vol. 66, pp. 109-119, 1966.
[Fraenkel, 1928] A. Fraenkel. *Einleitung in die Mengenlehre*, Springer, Berlin, 1928.
[Frege, 1879] G. Frege. *Begriffsschrift, eine der arithmetischen nachgebildete Formelsprache des reinen Denkens*, Louis Nebert, Halle a. S, 1879.

[Frege, 1884a] G. Frege. *Die Grundlagen der Arithmetik: eine logisch-mathematische Untersuchung über den Begriff der Zahl*, W. Koebner, Breslau, 1884.
[Frege, 1884b] G. Frege. *Grundgesetze der Arithmetik*, Vol. I, Verlag Hermann Pohle, Jena, 1884.
[Frege, 1903] G. Frege. *Grundgesetze der Arithmetik*, Vol. II, Verlag Hermann Pohle, Jena, 1903.
[Gödel, 1967] K. Gödel. On Formally Undecidable Propositions of *Principia Mathematica* and Related Systems I. In *From Frege to Gödel*, Jean van Heijenoort, editor, Harvard University Press, Cambridge, Mass., pp. 596-616, 1967 (original 1931).
[Gödel, 1990] K. Gödel. Russell's Mathematical Logic. In his *Collected Works*, 4 vols., Oxford University Press, New York, 1986–2003, vol. 2, pp. 119-143, 1990 (original 1944).
[Grattan-Guinness, 1979] I. Grattan-Guinness, On Russell's Logicism and its Influence, 1910-1930. In *Wittgenstein, der Wiener Kreis und der Kritische Rationalismus, Akten des Dritten Internationalen Wittgenstein Symposiums*, Vol. 13, H. Berghel, A. Hübner, and E. Köhler, editors, Wien, pp. 275-280, 1979.
[Haack, 1993] S. Haack. Peirce and Logicism; Notes towards an Exposition, *Transactions of the Charles S. Peirce Society*, vol. 29, no.1, pp. 33-56, 1993.
[Hempel, 1945] C. G. Hempel. On the Nature of Mathematical Truth, *The American Mathematical Monthly* vol. 52, pp.543-556, 1945.
[Henkin, 1950] L. Henkin. Completeness in the Theory of Types, *Journal of Symbolic Logic* vol. 14, pp. 159-166, 1950.
[Heyting, 1983] A. Heyting. The Intuitionist Foundations of Mathematics. In Paul Benacerraf and Hilary Putnam, editors, *Philosophy of Mathematics*, Cambridge University Press, 1983, (original 1931).
[Hetying, 1956] A. Heyting. *Intuitionism: An Introduction*, North-Holland, Amsterdam, 1956.
[Hintikka, 1969] J. Hintikka. 'On Kant's Notion of Intuition (Anschzuung). In *The First Critique: Reflections on Kant's Critique of Pure Reason*, T. Penelhum and J. J. MacIntosh, editors, Wadsworth, Belmont, CA, pp. 38-53, 1969.
[Hintikka, 1982] J. Hintikka. Kant's Theory of Mathematics Revisited. In *Essays on Kant's Critique of Pure Reason*, J. N. Mohanty and Robert W. Shehan, editors, University of Oklahoma, Norman, pp. 201-215, 1982.
[Hintikka, 2001] J. Hintikka. Post-Tarskian Truth, *Synthese*, vol.126, pp. 17-36, 2001.
[Hintikka, 2004] J. Hintikka. Independence-friendly Logic and Axiomatic Set Theory, *Annals of Pure and Applied Logic*, vol. 126, pp. 313-333, 2004.
[Husserl, 1983] E. Husserl. *Studien zur Arithmetik und Geometrie. Texte and dem Nachlass.* (Husserliana vol. 21), Martinus Nijhoff, Dordrecht, 1983. (See especially the material concerning Husserl's "Doppelvortrag".)
[Kant, 1787] I. Kant. *Kritik der reinem Vernunft*, Zwarte Auflage (B), Johann Friedrich Hartknoch, Rigam 1787.
[Laugwitz, 1999] D. Laugwitz. *Bernard Riemann, 1826-1866: Turning Points in the Conception of Mathematics*, Abe Shenitzer, translator, Birkhauser, 1999 (original 1996)
[Musgrave, 1977] A. Musgrave. Logicism Revisited, *British Journal for the Philosophy of Science*, vol. 28, pp. 99-127, 1977.
[Peckhaus, 1997] V. Peckhaus. *Logik, Mathesis Universalis und allgemeine Wissenschaft*, Akademie Verlag, Berlin, 1977.
[Peirce, 1931-58] C. S. Peirce. *Collected Papers*, Vols. 1-6, Charles Hartshorne and Paul Weiss, editors, and Vols. 7-8, A. W. Burks, editor, Harvard University Press, Cambridge, Mass, 1931–58.
[Putnam, 1967] H. Putnam. The Thesis that Mathematics is Logic. In *Bertrand Russell, Philosopher of the Century: Essays in His Honour*, R. Schoenman, editor, London, Boston, Toronto, pp. 273-303, 1967.
[Quine, 1966] W. V. O. Quine. Ontological Reduction and the World of Numbers, in W. V. O. Quine, *The Ways of Paradox and Other Essays*, Random House, New York, 212-220, 1966.
[Radner, 1975] M. Radner. Philosophical Foundations of Russell's Logicism, *Dialogue* vol. 14, pp. 241-253, 1975.
[Ramsey, 1978] F. P. Ramsey. The Foundations of Mathematics, in *Foundations*, D.H. Mellor, editor, Humanities Press, Atlantic Highlands, N. J., pp. 152-212, 1978 (original 1925).
[Russell, 1967] B. Russell. Letter to Frege. In *From Frege to Gödel*, Jean van Heijenoort, editor, Harvard University Press, Cambridge, Mass., pp. 124-125, 1967, (original 1902).

[Schumann and Schumann, 2001] E. Schumann and K. Schumann, eds. Husserls Manuskripte zu seinem Göttinger Doppelvortrag von 1901, *Husserl Studies* vol. 17, pp. 87-123, 2001.

[Shaw, 1916] J. B. Shaw. Logistic and the Reduction of Mathematics to Logic, *Monist* vol. 26, pp. 397-414, 1916

[Simpson, 1999] S. G. Simpson. *Subsystems of Second-Order Arithmetic*, Springer, Berlin, 1999.

[Steiner, 1975] M. Steiner. *Mathematical Knowledge,* Cornell University Press, Ithaca, NY, 1975.

[Tarski, 1956] A, Tarski. The Concept of Truth in Formalized Languages. In *Logic, Semantics, Metamathematics: Papers from 1923 to 1938*, Clarendon Press, Oxford, 1956.

[von Neumann, 1983] J. von Neumann. The Formalist Foundations of Mathematics. In *Philosophy of Mathematics*, 2nd ed., Paul Benacerraf and Hilary Putnam, editors, Cambridge University Press, Cambridge 1983 (original 1931).

[Webb, 2006] J. Webb. Hintikka on Aristotelian Constructions, Kantian Intuitions, and Peircean Theorems. In *The Philosophy of Jaakko Hintikka* (Library of Living Philosophers), Randall Auxier, editor, Open Court, LaSalle, Illinois, 2006.

[Whitehead and Russell, 1910–13] A. N. Whitehead and B. Russell. *Principia Mathematica*, 3 vols, Cambridge University Press, Cambridge, 1910–13.

FORMALISM

Peter Simons

Formalism is a philosophical theory of the foundations of mathematics that had a spectacular but brief heyday in the 1920s. After a long preparation in the work of several mathematicians and philosophers, it was brought to its mature form and prominence by David Hilbert and co-workers as an answer to both the uncertainties created by antinomies at the basis of mathematics and the criticisms of traditional mathematics posed by intuitionism. In this prominent form it was decisively refuted by Gödel's incompleteness theorems, but aspects of its methods and outlook survived and have come to inform the mathematical mainstream. This article traces the gradual assembly of its components and its rapid downfall.

1 PRELIMINARIES

1.1 Problem of Definition

Formalism, along with logicism and intuitionism, is one of the "classical" (prominent early 20th century) philosophical programs for grounding mathematics, but it is also in many respects the least clearly defined. Logicism and intuitionism both have crisply outlined programs, by Frege and Russell on the one hand, Brouwer on the other. In each case the advantages and disadvantages of the program have been clearly delineated by proponents, critics, and subsequent developments. By contrast, it is much harder to pin down exactly what formalism is, and what formalists stand for. As a result, it is harder to say what clearly belongs to formalist doctrine and what does not. It is also harder to say what count as considerations for and against it, with one very clear exception. It is widely accepted that Gödel's incompleteness theorems of 1931 dealt a severe blow to the hopes of a formalist foundation for mathematics. Yet even here the implications of Gödel's results are not unambiguous. In fact many of the characteristic methods and aspirations of formalism have survived and have even been strengthened by tempering in the Gödelian fire. As a result, while few today espouse formalism in the form it took in its heyday, a generally formalist attitude still lingers in many aspects of mathematics and its philosophy.

1.2 Hilbert

As Frege and Russell stand to logicism and Brouwer stands to intuitionism, so David Hilbert (1862–1943) stands to formalism: as its chief architect and proponent. As Frege and Russell were not the first logicists, so Hilbert was not the first formalist: aspects of Hilbert's formalism were anticipated by Berkeley, and by Peacock and other nineteenth century algebraists [Detlefsen, 2005]. Nevertheless, it is around Hilbert that discussion inevitably centers, because his stature and authority as a mathematician lent the position weight, his publications stimulated others, and because it was his energetic search for an adequate modern foundation for mathematics that focussed the energies of his collaborators, most especially Paul Bernays (1888–1977), Wilhelm Ackermann (1896–1962) and to some extent John von Neumann (1903–1957). As admirably recounted by Ewald [1996, 1087–9], Hilbert tended to focus his prodigious mathematical abilities on one area at a time. As a result, his concentration on the foundations of mathematics falls into two clearly distinct periods: the first around 1898–1903, when he worked on his axiomatization of geometry and the foundational role of axiomatic systems; and the second from roughly 1918 until shortly after his retirement in 1930. The latter period coincided with a remarkable flowering of mathematical talent around Hilbert at Göttingen, and must be considered formalism's classical epoch. It was brought to an abrupt end by Gödel's limitative results and by the effects of the National Socialist *Machtergreifung*, which emptied Germany in general and Hilbert's Göttingen in particular of many of their most fertile mathematical minds. In the foundations of mathematics, Hilbert's own writings are not as crystalline in their clarity as Frege's, and his successive adjustments of position combine with this to rob us of a definitive statement of formalism from his pen.

1.3 Working Mathematicians

Despite the consensus among mathematicians and philosophers of mathematics alike that Hilbert's program in its fully-fledged form was shown to be unrealizable by Gödel's results, many of Hilbert's views have survived to inform the views of working mathematicians, especially when they pause from doing mathematics to reflect on the status of what they are doing. While their weekday activities may effectively embody a platonist attitude to the objects of their researches, surprisingly many mathematicians are weekend formalists who happily subscribe to the view that mathematics consists of formal manipulations of essentially meaningless symbols according to strictly prescribed rules, and that it is not truth that matters in mathematics as much as interest, elegance, and application. So whereas formalism is widely (whether wisely is another matter) discounted among philosophers of mathematics as a viable philosophy or foundation for the subject, and is often no longer even mentioned except in passing, it is alive and well among working mathematicians, if in a somewhat inchoate way. So formalism cannot be written off simply as an historical dead end: something about it seems to be right enough to convince thousands of mathematicians that it, or something close to it, is along the right lines.

2 THE OLD FORMALISM AND ITS REFUTATION

2.1 Contentless Manipulation

As mentioned above, formalism did not begin with Hilbert, even in Germany. In the latter part of the 19th century several notable German mathematicians professed a formalist attitude to certain parts of mathematics. In conformity with Kronecker's famous 1886 declaration "*Die ganzen Zahlen hat der liebe Gott gemacht, alles andere ist Menschenwerk*",[1] Heinrich Eduard Heine (1821–1881), Hermann Hankel (1839-1873), and Carl Johannes Thomae (1840–1921) all understood theories of negative, rational, irrational and complex numbers not as dealing with independently existing entities designated by number terms, but as involving the useful extension of the algebraic operations of addition, multiplication, exponentiation and their inverses so as to enable equations without solution among the natural (positive whole) numbers to have solutions. In this way whereas an expression like '$(2+5)$' unproblematically stands for the number 7, an expression like '$(2-5)$' has sense not by denoting a number -3 but as part of the whole collection of operations regulated by their characteristic laws such as associativity, commutativity, and so on. Such symbols may be manipulated algebraically in a correct or incorrect manner without having to correspond to their own problematic entities. The rules of manipulation on their own suffice to render the expressions significant.

In his 'Die Elemente der Functionenlehre' Heine wrote,

> "To the question what a number is, I answer, if I do not stop at the positive rational numbers, not by a conceptual definition of number, for example the irrationals as limits whose *existence* would be a presupposition. When it comes to definition, I take a purely formal position, in that I call certain tangible signs numbers, so that the existence of these numbers is not in question." [Heine, 1872, 173]

and Hankel writes in his *Theorien der komplexen Zahlensysteme*

> "It is obvious that when $b > c$ there is no number x in the series 1, 2, 3, ... which solves the equation $[x + b = c]$: in that case subtraction is *impossible*. But nothing prevents us in this case from *taking* the difference $(c - b)$ as a sign which solves the problem, and with which we can operate exactly as if it were a numerical number from the series 1, 2, 3, ..." [Hankel, 1867, 5].

Thomae's *Elementare Theorie der analytischen Funktionen einer komplexen Veränderlichen* is particularly candid about this method, which he calls 'formal arithmetic'. He considered that non-natural numbers could be

---

[1] Reported in [Weber, 1893].

> "viewed as pure schemes without content [whose] right to exist [depends on the fact] that the rules of combination abstracted from calculations with integers may be applied to them without contradiction."

It was Thomae's fate to have Gottlob Frege as a colleague in Jena. Frege's criticisms of the formalist position prompted Thomae to extend his introduction in the second edition in justification:

> "The formal conception of numbers sets itself more modest limits than the logical. It does not ask what numbers are and what they are for, but asks rather what we require of numbers in arithmetic. Arithmetic, for the formal conception, is a game with signs, which may be called empty, which is to say that (in the game of calculating) they have no other content than that which is ascribed to them regarding their behaviour in certain rules of combination (rules of the game). A chess player uses his pieces similarly: he attributes certain properties to them which condition their behaviour in the game, and the pieces are merely the external signs of this behaviour. There is inded an important difference between chess and arithmetic. The rules of chess are arbitrary; the system of rules for arithmetic is such that by means of simple axioms the numbers may be related to intuitive manifolds and as a consequence perform essential services for us in the knowledge of nature. [...] The formal theory lifts us above all metaphysical difficulties; that is the advantage it offers." [Thomae, 1898, 1.]

2.2 Frege's Critique

Frege was the old formalism's most trenchant and effective critic. In *Die Grundlagen der Arithmetik* (*Foundations of Arithmetic*), Sections 92–103, entitled "Other Numbers", he takes issue with those who would introduce new numbers simply to provide solutions to equations that were previously insoluble, as had Hankel and others, and as had been standardly practiced and preached by many mathematicians, including Gauss. Frege is unimpressed. Simply introducing new signs to do new things is inadmissible, since they could be introduced to perform contradictory tasks:

> "One might as well say: there are no numbers among those known hitherto that simultaneously satisfy the equations
>
> $$x + 1 = 2 \text{ and } x + 2 = 1;$$
>
> but nothing prevents us from introducing a sign that solves the problem." (Section 96.)

While ordinary numbers would yield a contradiction if they solved both equations, what is to say new numbers would also entail a contradiction? We could introduce them and see what happened. Frege does not admit free creation:

"Even the mathematician can no more arbitrarily create anything than the geographer: he can only discover what is there, and give it a name." (Ibid.)

Since contradictions do not always show themselves easily, the "try and see" attitude will not suffice. The only way to show a theory consistent is to produce an object that satisfies it: a model.[2] The unwitting irony of these remarks would not emerge until 1902, when Russell showed Frege that his own system contained a hidden contradiction.

A year after *Grundlagen*, Frege published in 1885 a short essay, "On Formal Theories of Arithmetic", which dealt again with the issues, though it did so without naming adherents to the formalist position. Contrasting formalism with his own logicist view, he criticises the formalists' theory of definition of numbers as either circular in presupposing the consistency of what is defined, which supposes the signs signify something after all, or else as impotent to secure the truth of the propositions that formal manipulations are supposed to underwrite. He also points out that the formalists are not thoroughgoing in their attitude, since they do not offer a formal theory of the positive integers: "usually one does not feel a need to justify the most primitive of numbers." [Frege, 1984, 121].

Russell's contradiction prevented Frege from completing his program of showing how all of arithmetic and analysis is logical in nature. The foundations of analysis were discussed in Part III, "The Real Numbers", of *Grundgesetze der Arithmetik* (*Basic Laws of Arithmetic*), volume II, published in 1903. Russell's Antinomy overshadows this second volume, and prevented the formal continuation, but before Frege introduced his own theory of real numbers he criticised in prose other extant theories, as he had done other theories of natural numbers in *Grundlagen*. The earlier book's wit and light touch are here replaced by protracted, sarcastic and tedious schoolmasterly lecturing of others, most particularly Thomae. Cutting away the redundant verbiage, Frege's criticisms come down to three further points. Firstly, the formalists are excessively cavalier about the distinction between signs and what they signify, ascribing properties of the one to the other and vice versa. Since they *identify* numbers with signs, this is to be expected. Secondly, for this reason, they are unable to distinguish between statements made *within* a formal context and statements made *about* a formal context. For example, when we say that a king and two knights cannot force checkmate, we have stated a well-known theorem of chess. But we have made a statement *about* chess, not a statement *within* chess. Chess positions and chess pieces do not have meanings: they are what they are, but do not state or say anything [Frege, 1903, Section 91]. By contrast, a mathematical statement has a meaning and states something. To suppose that a theory about the signs of arithmetic is a theory about numbers is to confuse statements within the language of arithmetic, *arithmetical statements*, with statements about the language of arithmetic, *meta-arithmetical statements* (the terminology is modern, not Frege's). Finally the major difference between

---

[2] Ibid., § 95.

mathematical theories with content (such as arithmetic and analysis) and mere games is that mathematical theories may be applied outside mathematics: "It is application alone that raises arithmetic up above a game to the rank of a science." [Frege, 1903, Section 91].

Frege's major critical points — the importance of the sign/object distinction; the requirement of consistency; the difference between statement and metastatement; and the importance of application; lack of thoroughgoing application of the program — carried the day in the argument against the earlier formalists. They were however to be consciously noticed and incorporated into the more sophisticated kind of formalism put forward by Hilbert.

3 THE NEW AXIOMATICS

3.1 Hilbert's Grundlagen

In 1899 Hilbert published his *Grundlagen der Geometrie*. This work was radically innovative in a number of ways. It established the basic pattern for axiomatic systems from that time on in modern mathematics. Although the subject matter — Euclidean geometry — was not new, Hilbert's way of treating it was. Axioms in Euclid and in the subsequent tradition were statements considered self-evidently true. In Hilbert this status is put aside. Axioms are simply statements which are laid down or postulated, not because they are seen to be true, but for the sake of investigating what follows logically from them. The choice of axioms is of course not arbitrary: the aim is to find axioms from which the normal theorems of geometry follow. Further, these axioms should be as few and simple as possible, they should contain as few primitive terms as possible, and they should be independent, that is, no one should be derivable from the remainder. Further, where Euclid postulated that certain constructions could be carried out, Hilbert stated the existence of certain geometrical objects.

3.2 Implicit Definition and Contextual Meaning

Hilbert's axiomatization constituted an advance in rigor over Euclid, since it did not depend on having separate suites of definitions, such as "a point is that which has no part"; postulates, such as "To draw a straight line from any point to any point"; and common notions, such as "The whole is greater than the part". In Hilbert, everything is set out in a system of 21 axioms (one was later shown to be redundant). There are three primitive concepts, *point, line* and *plane*, and seven primitive relations: a ternary relation of *betweenness* linking points, three binary relations of *incidence* and three of *congruence*. Important axioms include Euclid's Parallels Axiom, and the Archimedean Continuity Axiom. Speaking in anticipation of later developments, the last means the system is not one of first order (where only individual points, lines and planes are quantified over) but second-order, where is is necessary to quantify over classes or properties of elements. In the course of his study, Hilbert lays stress on ensuring that the axioms are consistent, by producing a countable arithmetical model for them. Of course this only shows consistency *relative to* arithmetic, not absolute consistency. He showed that any two models are isomorphic, that is, in current terminology, that his axiom system is categorical. He also demonstrates the independence of axioms, again by using models, allowing different interpretations of the primitive terms.

The fact that the words 'point', 'line' and 'plane' are chosen for the three basic kinds of element is a concession to tradition. Their employment is inessential. As early as 1891, Hilbert remarked after hearing a lecture on geometry by Hermann Wiener that "Instead of 'points, lines, planes' we must always be able to say 'tables, chairs, beer mugs'." This distinguishes his approach to axioms from that of his predecessors and contemporaries. It is not required that the primitive terms have a fixed and determinate meaning. Rather, Hilbert regards them as being given meaning by the axioms in which they occur. He describes the axioms as affording an *implicit definition* of the primitive terms they contain, in terms of one another and the various logical components making up the remainder of the axioms.

The most important innovation in Hilbert's approach was, as Bernays put it later, to dissociate the status of axioms from their epistemological status. Axioms are no longer assumed to be true, as guaranteed by self-evidence or intuition. The approach is more liberal, and more experimental. A certain number of axioms are put forward, and their logical interrelations and consequences investigated. The enterprise takes on a hypothetical character rather than the categorical character traditionally assumed. The greater freedom this allows (and Hilbert constantly emphasized the mathematician's creative freedom) comes at a price however, since the loss of intuitive or evident guarantees of truth means the consistency of the axioms can no longer be taken for granted. This turns out to be the crux of the issues facing the new formalism later.

3.3 *Dispute with Frege*

Hilbert's work prompted a reaction from Frege, who wrote to him objecting to his treatment of axioms, definitions and geometry. Frege's part in their exchange of letters was published by Frege after Hilbert discontinued the correspondence, and when Korselt replied on behalf of Hilbert, Frege criticised him too. The exchange is illuminating both for what it reveals about the issues and for what it tells us about the relative positions of Hilbert and Frege in the German mathematical community.

Frege's view of axiom systems is staunchly Euclidean. Axioms are truths which are intuitively self-evident. Their being individual truths entails their being propositions having a determinate meaning (sense), in all their parts. Their being severally true guarantees their consistency with one another without need of a consistency proof. Definitions on the other hand are stipulations endowing a new sign with meaning (sense) on the basis of the pre-existing meanings of all the terms of the definiens. Hilbert's procedure of taking axioms not to be fully determinate in all their parts, and in considering that they severally define the primitive terms occurring in them, mischaracterizes both axioms and definitions, and unnecessarily blurs the distinction between them. For Frege it also blurs the important epistemological distinction between the truths of geometry, whose validating intuitions are geometric in nature, and so synthetic *a priori*, and the truths of arithmetic, which according to Frege are analytic, following from logic and suitable definitions.

Frege's positive characterization of Hilbert's position is illuminating. The conjunction of the axioms with the primitive terms 'point', 'line', 'between' etc. taken as distinct *free variables* gives an open sentence in several first-order variables, so a second-order open sentence. The question of consistency then becomes the question whether this open sentence can be satisfied. Hilbert's position is subtly different from this. Using modern terminology, we could say his view is that his axioms contain *schematic* first-order variables, so that valid inferences from them are schematic inferences after the fashion now familiar in first-order predicate logic, rather than subclauses in a true second-order logical conditional as they would be for Frege. The axioms and their consequences hold not just for a single system of things, the points of space, as Frege would have it, but for *any* system of things that satisfies the axioms. Consistency though would amount to the same thing: there can be a model.

However, this is precisely *not* how Hilbert saw the issue. In correspondence with Frege he writes 29 December 1899 (Simpson's translation):

> You write "From the truth of the axioms follows that they do not contradict one another". It interested me greatly to read this sentence of yours, because in fact for as long as I have been thinking, writing and lecturing about such things, I have always said the very opposite: if arbitrarily chosen axioms together with everything which follows from them do not contradict one another, then they are true, and the things defined through the axioms exist. For me that is the criterion of truth and existence. The proposition 'every equation has a root' is true, or the existence of roots is proved, as soon as the axiom 'every equation has a root' can be added to the other arithmetical axioms without it being possible for a contradiction to arise by any deductions. This view is the key not only for the understanding of my [*Foundations of Geometry*], but also for example my recent [*Über den Zahlbegriff*], where I prove or at least indicate that the system of all real numbers *exists*, while the system of all Cantorean cardinalities or all Alephs — as Cantor himself states in a similar way of thinking but in slightly different words — *does not exist*.

This is the clearest statement by Hilbert of a position which has become notorious: the view that, in mathematics, consistency is existence. It is clear why Frege could not accept Hilbert's view. For Hilbert, non-Euclidean geometry can be treated in just the same axiomatic way as Euclidean geometry, so since all three are consistent (relative to one another), all three are true and their objects exist. But for Frege they cannot all be true because they are mutually inconsistent: if one is true (Euclidean geometry, for Frege), the others are false, and their objects do not exist.

In Hilbert, truth is not absolute in the way it is for Frege. To say that the theorems of a system of geometry are true is for Hilbert to say that they follow logically from the axioms (assuming always the axioms are consistent). Finally, for Hilbert the axioms are subject to different interpretations, which he employs in independence proofs, whereas for Frege they must have a fixed meaning and cannot be reinterpreted. On these matters, while Frege makes his points clearly, it is he rather than Hilbert who is out of step with subsequent mathematical developments. Hilbert's treatment of axiom systems has become orthodoxy.

Hilbert did not continue the correspondence, bring unwilling to publish it, no doubt irritated by Frege's schoolmasterly and patronising tone, and after Frege published his part, the cudgels were taken up by Alwin Korselt, who attempted to mediate between the two positions. The result was another polemical piece by Frege against Korselt, in a much testier tone even than before.

3.4 The Axioms of Real Numbers

In 1900 Hilbert published a short memoir called 'On the Concept of Number'. In this he assembled into an axiom system a number of principles about real numbers which he had mentioned in the *Grundlagen*, characterizing the real numbers axiomatically as an ordered Archimedean field which is maximal, i.e., cannot be embedded in a larger such field. This was in effect the first axiomatization of the reals. He contrasts this axiomatic method with what he calls the *genetic* method, which is the successive introduction of extensions to the natural numbers, such as is found in Dedekind. His preference for the axiomatic method is clearly stated: "Despite the high pedagogic and heuristic value of the genetic method, for the final presentation and the complete logical grounding of our knowledge the axiomatic method deserves the first rank." (*vide* [Ewald, 1996, 1093].)

4 THE CRISIS OF CONTENT

4.1 Logicism's Waterloo and other Paradoxes

At the same time as Hilbert was proposing his axiomatization, Frege was, so he supposed, crowning his logicism program by showing how to derive the principles of the real numbers from purely logical principles, and establishing the existence of the real numbers by producing a model based on sequences of natural numbers, taken as already established as existing as a matter of logic in the previous volume of *Grundgesetze*. This was the task that Frege set himself in the third part, 'The Real Numbers', of his monumental *Basic Laws of Arithmetic*. The thrust of Frege's approach unified two strands in previous thinking about the foundations of mathematics. One was his own logicism, which went back to Leibniz, and which he shared, in many respects, with his older contemporary Dedekind and (unknown to him at this stage) his younger contemporary Russell. According to logicism, the principles of mathematics — or as Frege less ambitiously believed, arithmetic and analysis — are logical in nature, and can be demonstrated to follow from logical principles alone. The second strand was the idea, going back to Gauss and Dirichlet, and also shared with Dedekind, that the arithmetic of finite numbers may in some way serve as the basic mathematical theory for grounding "higher" theories such as analysis. In order to vindicate his view, Frege had not only been inspired to create the first comprehensive modern system of logic; he had also been led to introduce a kind of entity called *value-ranges*, a species of abstract object whose existence is demanded by logic, and which includes, as a special case, the extensions of concepts, which Frege called *classes*. Numbers, according to Frege, are particular extensions of concepts, and so are classes in this sense.

The concept of number had in the preceding period been subject to an unprecedented development and enlargement by Georg Cantor. In his revolutionary works, Cantor, building on tentative beginnings by Bolzano, had begun to work with the general notion of a class or set, and had established that sets with infinitely many members need not all have the same size (cardinality), or number of elements. In particular the size of the continuum, that of all numbers on a continuous line, is greater than the size of the set of all finite natural numbers. Cantor's second proof of this result in 1891 uses a device now called the method of diagonalization; this was quickly generalized to show that for any size of set, another of greater size can be shown to exist, namely the set of all subsets of the former set (its power set), so that there is no greatest number. The theory of transfinite numbers to which this led was the most radical extension of the domain of arithmetic since its very beginning. However the very generality of the notion of size or cardinality of a set led to that curious result: there could not be a largest set, because if there were, by the diagonalization argument, there would have to be one larger still, contradicting the original assumption that there was a largest. Hence there could be no such set as the set of all things, for it would by definition have the largest cardinality. While this conclusion undercut an infamous attempt by Dedekind to prove that there is at least one infinite set, it did not give Cantor much concern. For theological reasons he was quite happy to accept that there were pluralities of things too numerous to be collected together into a set: he called them "inconsistent totalities".

The same indifference could not apply to Frege, whose logical system required him to quantify over all objects, including all sets, and for whom sets were included among the objects. Bertrand Russell, like Frege working with the idea of all objects, discovered in 1901 by considering Cantor's proof that there is no greatest cardinal number that a similar curious result could be derived concerning sets: according to logical assumptions he shared with Frege about the existence of sets, the set of all sets which are not elements of themselves would have to be an element of itself and also not an element of itself. Russell communicated this result to Frege in 1902, about a year after he had discovered it. Frege, disconcerted, hastily concocted a patched repair to his logical system for the publication of the second volume of *Basic Laws* in 1903, but the repair was unsuccessful,[3] as Frege must soon have realised, since he thenceforth gave up publishing about the foundations of mathematics, and declared that the contradiction showed set theory to be impossible. Russell's Paradox was also independently discovered by Ernst Zermelo at about the same time, but unlike Russell, Zermelo did not think it worth mentioning in a publication.

Russell's Paradox, though the clearest and most damaging, was but one of a cluster of paradoxes which had begun to infest post-Cantorian mathematics, starting with Cesare Burali-Forti's argument in 1897 that there could not be a greatest ordinal number. Cantor's result that there could be no greatest cardinal number followed in 1899. The general atmosphere conveyed by the rash of paradoxes coming to light was that modern mathematics was in a crisis. What had precipitated it was a matter for debate. Uncritical assumptions about the infinite, especially the uncountable infinite, or the assumption of the existence of objects not directly constructed, or the uncritical application of logical principles in an unrestricted context were three not unconected potential sources of the difficulties. All of these potential sources were to be confronted in the "classical" phase of formalism. The paradoxes also dramatically highlighted the importance of ensuring that mathematical theories are consistent.

4.2 Self-Restriction

Reactions to the paradoxes varied. Russell pressed forward with the attempt to maintain logicism, blocking the paradoxes by stratifying entities into logical types. Expressions of entities of different type could not be substituted for one another on pain of producing ungrammatical nonsense. Russell diagnosed the paradoxes as arising through vicious circles in definition, whose use was strongly criticised by Henri Poincaré. To avoid impredicative definitions, that is, those where the object defined is in the domain of object quantified over in the *definiens*, the types were themselves typed, or ramified, into infinitely many orders. However, this ramification, while it avoided impredicativity, did not allow standard mathematical laws to be derived, so the ramification was effectively neutralized by an axiom of reducibility, according to which every defined function is extensionally equivalent to one of lowest order in the type. The logical system Russell and Whitehead produced, under the influence of Peano and Frege, was the first widely recognised system of mathematical logic. The motivations for its complications were largely philosophical. By contrast, Hilbert's Göttingen colleague Ernst Zermelo produced for mathematical purposes (deriving Cantor's principle that every set can be well ordered from the axiom of choice) a surprisingly straightforward axiomatic ver-

---

[3] This was first shown by Lesniewski: cf. Sobocinski 1949. Lesniewski showed that Frege's repair entails the unacceptable result that there is only one object. But Frege certainly must have realised fairly soon that the repair was also too restrictive to allow him to prove that every natural number has a successor, a crucial theorem of number theory.

sion of set theory which retained most of Cantor's results, but by weakening the conditional set existence principles did not allow the formation of the paradoxical Russell set. Mathematicians showed themselves generally unwilling to accept the complications of the type system, and set theory quickly became the framework of choice for the then rapidly developing discipline of topology. Zermelo's achievement was a twofold vindication of the value of working with axiomatic systems as Hilbert had proposed: it largely silenced critics of set theory who had regarded it as a piece of mathematical extravagance, and it apparently avoided inconsistency, though that was (and is) still unproven.

Cantor's extension of arithmetic into the transfinite had been staunchly opposed by Leopold Kronecker, who propounded the principle that all mathematical objects were to be constructed from the finite integers. Kronecker's insistence on constructing mathematical objects was seconded for more philosophical reasons by L. E. J. Brouwer, who first used the terms 'formalism' and 'intuitionism' in 1911. By 1918, Brouwer had rejected the uncountably infinite as well as unrestricted employment of the law of excluded middle, in particular its use in infinite domains. Similar and at the time more influential views were put forward by Hilbert's former student Hermann Weyl in his 1918 book *The Continuum*, developing a logical account of analysis which used only predicative principles, and avoided using the axiom of choice or proofs by *reductio ad absurdum*. Coming from a former Göttingen student, Weyl's book and his 1921 essay 'On the New Crisis in the Foundations of Mathematics' took the challenge of Brouwer's arguments directly to the doors of the Göttingen mathematicians, declaring, "Brouwer, that is the revolution." It was their response, particularly that of Hilbert and his assistant Paul Bernays, that ushered in the intense but short-lived classical period of formalism.

5 THE CLASSICAL PERIOD

5.1 *Preparations*

The first outward response to the challenge of Brouwer and Weyl came in the form of two papers published in 1922: Hilbert's 'The New Grounding of Mathematics' and Bernays' 'Hilbert's Significance for the Philosophy of Mathematics'. However, as Wilfried Sieg has emphasized, these papers emerged from a richer matrix of work in progress, and not merely as a response to the intuitionist challenge. After a period of over ten years in which Hilbert had concentrated on functional analysis and, under the influence of Hermann Minkowski, on the mathematics of physics, he returned to foundational issues. In 1917 he delivered a lecture course 'Principles of Mathematics', for which Paul Bernays, newly recruited to Göttingen from Zurich, produced lecture notes. Notes from these and subsequent lectures, later reworked by Hilbert's student Wilhelm Ackermann, became the basis for Hilbert and Ackermann's classic 1928 book *Mathematical Logic* (*Grundzüge der mathematischen Logik*), the first modern textbook of the subject. In the lectures,

Hilbert, availing himself of the developments since Whitehead and Russell's *Principia mathematica*, gave a modern formulation of mathematical logic in what has become the standard form, separating propositional from predicate calculus, and first-order from higher-order predicate calculus. Metamathematical questions are posed such as whether the various systems of axioms are consistent, independent, complete, and decidable. Although Hilbert soon distanced himself from the foundationally suspect axioms of infinity and reducibility, for the first time he and the Göttingen school had a precise logical instrument with which to approach the revisionary challenge to mathematics posed by intuitionism.

5.2 Hilbert's Maximal Conservatism

Brouwer himself had pointed out that adopting the constructive viewpoint of intuitionism meant foregoing acceptance of such mathematical results as that every real number has an infinite decimal expansion. It soon became clear that the intuitionistic program, at this stage not cast in the form of an alternative logic, would involve a large-scale rejection of many well-established mathematical results as genuinely false. In addition, Brouwer's rejection of completed infinities meant that Cantor's transfinite revolution was to be repudiated wholesale. In time, this threatened loss of contentual mathematics was to cost Brouwer even the support of Weyl.

Short of inconsistency, Hilbert was not prepared to accept restrictions on what mathematics can be accepted. His goal indeed was, as it had been earlier, to provide an epistemologically respectable foundation for *all* mathematics, and that included not just traditional number theory, analysis, and geometry, but also the newly added regions of set theory and transfinite number theory. His program was thus conservative, in the sense of wishing to conserve accepted mathematical results, in contradistinction to the revisionism of Brouwer, Weyl and Poincaré. And his conservatism was *maximal*, in that *any* consistent mathematical theory was acceptable, whether or not the patina of time-honored acceptance clung to it. What was new was the way in which mathematics, including the new mathematics of the infinite, was to be defended. Hilbert decided to break radically with foundational attempts by Dedekind, Frege and Russell, and to beat the intuitionists at their own game.

5.3 Finitism

The sticking point in establishing the consistency of geometry, analysis and number theory had always been the infinite. Any attempt to transmit consistency from finite cases to all cases by a recursive procedure, such as that sketched by Hilbert in 1905, was subject to Poincaré's criticism that the consistency of inductive principles was being assumed, so that a vicious circularity was involved. Hilbert adopted a distinction and a strategy to circumvent this. The distinction was between reasoning *within* some part of mathematics, represented by an axiomatic system, and reasoning *about* the axiomatic system itself, considered as a collection of symbol-combinations. Any mathematical proof, even one using transfinite induction, is itself a finite combination of symbols. Provided the notion of proof can be regimented uniformly, a procedure which advances in mathematical logic since Frege gave reason to think could be done, then provided conceptions of logical derivation and consistency could be formulated which did not depend on the content of a mathematical theory but only on the graphical form of its formulas, as a formula A and its negation $\sim A$ differ only by the presence of the negation symbol, the question of consistency could be tackled by examination of the formulas themselves. A consistency proof for a given mathematical theory, suitably formalized, would show that from the given finite collection of axioms, each a finite combination of symbols, no pair of formulas could be logically derived which differed solely in that one was the negation of the other.

The reasoning about a mathematical system was *metamathematics*. In so far as such reasoning, aimed at establishing consistency of a system, considered only the shapes and relationships of formulas and their constituent signs, not what they are intended to mean or be about, it is concerned only with the *form* or *syntax* of the formulas. The theory of the formulas themselves however is not formal in this way: it has a content; it is about formulas! Poincaré's accusation of circularity could be circumvented provided any inductive principles used in reasoning about formulas are themselves acceptable: the status of formulas within the theory (as suspicious because inductive) now becomes irrelevant, because their meaning is disregarded.

Hilbert signals this turn to the sign as a radical break with the past:

> the objects of number theory are for me — in direct contrast to Dedekind and Frege — the signs themselves, whose shape can be generally, and certainly recognized by us [...] The solid philosophical attitude that I think is required for the grounding of pure mathematics — as well as for all scientific thought, understanding and communication — is this: *In the beginning was the sign.* [Hilbert, 1922, 202; Mancosu, 1998, 202]

Formulas are essentially simply finite sequences or strings of primitive symbols, so the kind of reasoning applied to them could be expected to be not essentially more complex than the kind of reasoning applied to finite numbers. Hilbert and Bernays called such reasoning "finitary". The exact principles and bounds of finitary reasoning were nowhere spelled out, but the expectation was that combinatorial methods involving only finitely many signs could be employed to demonstrate in finitely many steps in the case of a consistent system that no pair of formulas of the respective forms A and $\sim A$ could be deduced (derived) from the axioms. This hope — for hope it was — turned out to be unrealizable.

Formalism's finitism was not simply an exercise in hair-shirt self-denial. Brouwer's and Weyl's criticisms of classical mathematical reasoning stung the formalists into a more extreme response. While intuitionists rejected certain forms of inference, and also uncountable infinities, they were prepared to use countably infinite sequences. Finitism went further in its rejection of infinitary toools, and looked to achieve its results using only finitely many objects in any proof. This was the point of the turn to symbols. It is possible to formulate many a short quantified sentence of first-order logic using just one binary relation, such that these sentences cannot be true except in an infinite domain. The infinite is then "tamed" by any such sentence. If formalized theories of arithmetic, analysis etc. could be shown consistent using finitely many finitely long sentences in finitely many steps, then even the uncountable infinities of real analysis that intuitionism rejected would be "tamed", and by sterner discipline than the intuitionists themselves admitted. Finitism was thus in part an exercise in one-upmanship.

5.4 Syntacticism and Meaning

Consistency of a formal theory (essentially, a set of formulas, the axioms, with their consequences) can be defined in terms of the lack of any pair of formulas A and $\sim A$ of the theory, both of which derive from the axioms. This characterization depends solely on the graphical fact that the two formulas are exactly alike (type-identical) except that one has an additional sign, the negation sign, at the front. The process of proof or derivation is likewise so set up that the rules apply solely in virtue of the syntactic form of the formulas involved, for example *modus ponens* consists in drawing a conclusion B from two premises A and $A \to B$, no matter what the formulas A and B look like *in concreto*. Likewise other admissible proof steps such as substitution and instantiation can be described in purely syntactic terms, though with somewhat more effort. This metamathematical turn was in many respects the most radically revolutionary part of formalism: it consisted in treating proofs themselves not (simply) as the vehicles of mathematical derivation, *but as mathematical objects in their own right*. It is ironic indeed that while the general idea of formalization was well understood by the formalists, the implications of the formal nature of proof only became apparent when Gödel showed in detail how to encode these formal steps in arithmetic itself, which was precisely what set up the proof that there could be no finite proof of arithmetic's consistency.

Nevertheless, the oft-repeated charge that according to formalists mathematics is a game with meaningless symbols is simply untrue. The metamathematics that deals with symbols is meaningful, even though it abstracts from whatever meaning the symbols might have. And in the case of an axiomatic system like that for Euclidean geometry, the axioms (provided, as ever, that they are consistent) themselves limit what the symbols can mean. Though in general they do not fix the meanings unambiguously, this very constraining effect gives the symbols a schematic kind of meaning, which it is the task of the mathematician to tease out by her inferences. That is the point of Hilbert's infamous view that the axioms constitute a kind of implicit definition of the primitive signs they contain. While for several reasons the word 'definition' aroused antipathy, the point is that the meaning is as determinate as the axioms constrain it to be, and no more. The "objects" discussed and quantified over in such a theory are considered only from the point of view of the structure of interrelationships that they embody, which is what the axioms describe.

6 GÖDEL'S BOMBSHELL

In their 1928 *Grundzüge der theoretischen Logik*, Hilbert and Ackermann formulated with admirable clarity the interesting metamathematical questions that needed to be answered. Is first-order logic complete, in the sense that all valid statements and inferences can be derived in its logical system? Are basic mathematical theories such as those of arithmetic and analysis, expressed in the language of first- or higher-order predicate logic, consistent? Hilbert had already begun to take steps along the way of showing the consistency of parts of natural number theory and real number theory, in papers in the early 1920s. The aim was to work up to the full systems, including quantifiers for the "transfinite" part, as Hilbert termed it. Ackermann tried unsuccessfully in 1924 to show the consistency of analysis, while Johann von Neumann in 1927 gave a consistency proof for number theory where the principle of induction contains no quantifiers. When Kurt Gödel in his 1930 doctoral dissertation proved the completeness of first-order predicate calculus, it appeared that the ambitious program to show the consistency of mathematics on a finite basis was nearing completion, and that number theory, analysis and set theory would fall in turn. In 1930 Gödel also started out trying to prove the consistency of analysis, but in the process discovered something quite unexpected: that it is possible to encode within arithmetic a true formula which, understood as being about formulas, "says" of itself that it cannot be proved. The formal theory of arithmetic was incomplete.

This in itself was both unexpected and disappointing, but Gödel's second incompleteness theorem was much more devastating to the formalist program, since it struck at the heart of attempts to show portions of formalized mathematics to be consistent. Gödel showed namely that in any suitable formal system expressively powerful enough to formulate the arithmetic of natural numbers with addition and multiplication, if the system is consistent, then it cannot be proved consistent using the means of the system itself: it contains a formula which can be construed as a statement of its own consistency and this formula is unprovable if and only if the system is consistent. Therefore any proof of consistency of the system can only be made in a system which is proof-theoretically *stronger* than the system whose consistency is in question. The idea of the formalists had been to demonstrate, given some system whose consistency is not straightforwardly provable (such as arithmetic with only addition or only multiplication as an operation), that despite its apparent strength it could be shown by finite formal methods that it is consistent. Gödel's Second Incompleteness Theorem showed to the contrary that no system of sufficient strength, and therefore questionable consistency, could be shown consistent except by the use of a system with greater strength and *more* questionable consistency. The formalist goal was destined forever to recede beyond the capacity of "acceptable" systems to demonstrate. Gödel himself offered a potential loophole to formalists, by suggesting that perhaps there were finitary methods that could not be formalized within a system. However, this loophole was not exploited, and the effect was simply to highlight the unclarity of the concept 'finitary', which has continued to resist clear explication. Other aspects of Gödel's proofs which have remained controversial concern the question in what sense the formula "stating consistency" of the system in the system in fact does state this.

It is usual to portray Gödel's incompleteness theorems as a death-blow to formalism. They certainly closed off the line of giving finitistic consistency proofs for systems with more than minimal expressive power. However they were if anything more deadly to logicism, since logicism claimed that all mathematics could be derived from a given, fixed logic, whereas Gödel showed that any logical system powerful enough to formulate Peano arithmetic — which included in particular second-order predicate logic, set theory, and Russell's type theory — would always be able to express sentences it could be shown were not provable in the system and yet which could be seen by metamathematical reasoning to be true. Logicists aside, most mathematicians were fairly insouciant about this: many had not believed logicism's claims in the first place.

The effect on formalism was more immediate but also ultimately more helpful. Hilbert's dream had proved untenable in its most optimistic form, but interest shifted to investigating the relative strengths of different proof systems, to seeing what methods could be employed beyond the finitary to showing consistency, to investigating the decidability of problems, and in general to further the science of metamathematics. Like a river in spate, formalism was obstructed by the Rock of Gödel, but it soon found a way to flow around it.

7 THE LEGACY OF FORMALISM

7.1 Proof Theory

Hilbertian metamathematics initiated the treatment of proofs as mathematical objects in their own right, and introduced methods for dealing with them such as structural induction. In the 1930s a number of advances by different logicians and mathematicians, principally Herbrand, Gödel, Tarski and Gentzen, showed that there were a number of perspectives from which proofs could be investigated as mathematical objects. Probably the most important was the development of the sequent calculus of Gentzen, which allowed precise formulations of statements and proofs about what a given system proves. In Gentzen's treatment, the subject-matter of the formulas treated is irrelevant: what matters are the structural principles for manipulating them. Proof theory was to go on to become one of the most important pillars of mathematical logic.

7.2 Consistency Proofs

The first post-Gödelian consistency proof was due to Gentzen [1936], who showed that Peano arithmetic could be proved consistent by allowing transfinite induction up to ε_0, an ordinal number in Cantor's transfinite hierarchy. Later results by Kurt Schütte and Gaisi Takeuti showed that increasingly powerful fragments of mathematics, suitable for formulating all or nearly all of "traditional" mathematics, could be given transfinite consistency proofs. Any sense that the consistency of ordinary mathematics is under threat has long since evaporated.

7.3 Bourbakism

Hilbert's attitude to axiom systems, revolutionary in its day, has become largely unquestioned orthodoxy, and informs the axiomatic approach not just to geometry and arithmetic but all parts of (pure) mathematics. The reformulation of pure mathematics as a plurality of axiomatic theories, carefully graded from the most general (typically: set theory) to the more specific, propagated by the Bourbaki group of mathematicians, effectively took Hilbert's approach to its limit. As to the entities such theories are "about", most commentators adopt a structuralist approach: mathematics is concerned not with any inner or intrinsic nature of objects, but only with those of their characters which consist in their interrelationships as laid down by a given set of axioms. While this stresses the ontology of mathematics more than the formalists did, it is an ontology which is informed by and adapted to the changes in thinking about the axiomatic method which drove formalism. Not all mathematics is done in Bourbaki style, nor is it universally admired or followed, but the organisational work accomplished by the Bourbakist phase is of permanent value to an increasingly sprawling discipline.

8 CONCLUSION

In the "classical" form it briefly took on in the 1920s, formalism was fairly decisively refuted by Gödel's incompleteness theorems. But these impossibility results spurred those already working in proof theory, semantics, decidability and other areas of mathematical logic and the foundations of mathematics to increased activity, so the effect was, after the initial shock and disappointment, overwhelmingly positive and productive. The result has been that, of the "big three" foundational programs of the early 20th century, logicism and intuitionism retain supporters but are definitely special and minority positions, whereas formalism, its aims adjusted after the Gödelian catastrophe, has so infused subsequent mathematical practice that these aims and attitudes barely rate a mention. That must count as a form of success.

BIBLIOGRAPHY

General remarks: There is an extensive secondary literature on formalism. For the "horse's mouth" story the papers of Hilbert are indispensable, but be warned that though attractively written, they are often quite difficult to pin down on exact meaning, and his position does change frequently. Of modern expositions, for historical background and the long perspective it is impossible to beat Detlefsen 2005, while for more detailed accounts of the shifting emphasis and tendencies within formalism as it developed, the works by Sieg, Mancosu, Peckhaus and Ewald are valuable signposts. Sieg is editing the unpublished lectures of Hilbert, so we can expect further detailed elaboration and clarification of the twists and turns of the classical period and the run up to it.

[Detlefsen, 2005] M. Detlefsen. Formalism. In S. Shapiro, ed., *The Oxford Handbook of the Foundations of Mathematics*. Oxford: Oxford University Press, 2005, 236-317.

[Ewald, 1996] W. Ewald. *From Kant to Hilbert. A Source Book in the Foundations of Mathematics*. Oxford: Clarendon Press, 1996. 2 volumes. Chapter 24 (pp. 1087-1165) consists of translations of seven papers by Hilbert.

[Frege, 1884] G. Frege. *Die Grundlagen der Arithmetik*, 1884. Translation: *Foundations of Arithmetic*. Oxford: Blackwell, 1951.

[Frege, 1893/1903] G. Frege. *Grundgesetze der Arithmetik*. Jena: Pohle, 1893/1903. (A complete English translation is in preparation.)

[Frege, 1971] G. Frege. *On the Foundations of Geometry and Formal Arithmetic*. Ed. and tr. E.-H. W. Kluge. New Haven: Yale University Press, 1971.

[Frege, 1984] G. Frege. *Collected Papers*. Oxford: Blackwell, 1984.

[Gödel, 1931] K. Gödel. Über formale unentscheidbare Sätze der *Principia Mathematica* und verwandter Systeme. *Monatshefte für Mathematik und Physik* **38**: 173-98, 1931. Translation: On formally undecidable propositions of *Principia Mathematica* and related systems, in J. van Heijenoort, ed., *From Frege to Gödel: A Source Book in Mathematical Logic, 1879-1931*. Cambridge, Mass.: Harvard University Press, 1967, 596-616.

[Hallett, 1995] M. Hallett. Hilbert and Logic. In: M. Marion and R. S. Cohen, *Quebec Studies in the Philosophy of Science*, Vol. 1, Dordrecht: Kluwer, 135-187, 1995.

[Heine, 1872] E. Heine. Die Elemente der Functionenlehre. *Journal für die reine und angewandte Mathematik* **74**, 172-188, 1872.

[Hilbert, 1899] D. Hilbert. Grundlagen der Geometrie. In *Festschrift zur Feier der Enthüllung des Gauss-Weber Denkmals in Göttingen*. Leipzig: Teubner, 1899. Translation *The Foundation of Geometry*, Chicago: Open Court, 1902.

[Hilbert, 1900] D. Hilbert. Über den Zahlbegriff. *Jahresbericht der deutschen Mathematiker-Vereinigung* **8**, 180-94, 1900. Translation: On the Concept of Number. In Ewald 1996, 1089-95.

[Hilbert, 1918] D. Hilbert. Axiomatisches Denken. *Mathematische Annalen* **78**, 405-15, 1918. Translation: Axiomatic Thought. In Ewald 1996, 1105-14.

[Hilbert, 1922] D. Hilbert. Neubegründung der Mathematik. *Abhandlungen aus dem mathematischen Seminar der Hamburger Universität* **1**, 157-77, 1922. Translation: The New Grounding of Mathematics. In Ewald 1996, 1115-33.

[Hilbert, 1923] D. Hilbert. Die logischen Grundlagen der Mathematik. *Mathematische Annalen* **88**, 151-65, 1923. Translation: The Logical Foundations of Mathematics. In Ewald 1996, 1134-47.

[Hilbert, 1931] D. Hilbert. Die Grundlegung der elementaren Zahlentheorie. *Mathematische Annalen* **104**, 485-94, 1931. Translation: The Grounding of Elementary Number Theory. In Ewald 1996, 1148-56.

[Mancosu, 1998] P. Mancosu, ed. *From Brouwer to Hilbert. The Debate on the Foundations of Mathematics in the 1920s*. Oxford: Oxford University Press, 1998.

[Mancosu, 1998a] P. Mancosu. Hilbert and Bernays on Metamathematics, in Mancosu 1998, 149-188, 1988

[Moore, 1997] G. H. Moore. Hilbert and the emergence of modern mathematical logic, *Theoria* (Segunda Epoca), **12**, 65-90, 1997.

[Peckhaus, 1990] V. Peckhaus. *Hilbertprogramm und kritische Philosophie*, Göttingen: Vandenhoek & Ruprecht, 1990.

[Sieg, 1988] W. Sieg. Hilbert's program sixty years later, *Journal of Symbolic Logic*, **53**: 338-348, 1988.
[Sieg, 1990] W. Sieg. Reflections on Hilbert's program, in W. Sieg (ed.), *Acting and Reflecting*. Dordrecht: Kluwer, 1990.
[Sieg, 1999] W. Sieg. Hilbert's Programs: 1917-1922, *Bulletin of Symbolic Logic*, **5**: 1-44, 1999.
[Sobocinski, 1949] B. Sobocinski. L'analyse de l'antinomie Russellienne par Lesniewski. *Methodus* I, 94-107, 220-228, 308-316; II, 237-257, 1949.
[Thomae, 1880] J. Thomae. *Elementare Theorie der analytischen Funktionen einer komplexen Veränderlichen*. Halle, 1880. 2nd ed. 1898.
[Weber, 1893] H. Weber. Leopold Kronecker. *Jahresberichte der deutschen Mathematikervereinigung* **2**, 5-31, 1893.

CONSTRUCTIVISM IN MATHEMATICS

Charles McCarty

1 INTRODUCTION: VARIETIES OF CONSTRUCTIVISM

Constructivism in mathematics is generally a business of practice rather than principle: there are no significant mathematical axioms or attitudes characteristic of constructivism and statable succinctly that absolutely all constructivists, across the spectrum, endorse. Instead, one finds sparsely shared commitments, indefinite orientations and historical precedents. For instance, some constructivists demand that legitimate proofs of crucial existential theorems, perhaps those concerning natural numbers, be constructive in that there be available admissible means for educing, from the proofs, specific, canonically-described instances of the theorems proven. Let $\mathbb{N}$ be the set of natural numbers and $\Phi(x)$ a predicate of natural numbers. A rough, preliminary expression of the notion 'constructive proof,' when it comes to statements about natural numbers, is

> P is a constructive proof of $\exists x \in \mathbb{N}.\Phi(x,y)$ when there is to hand mathematical means Θ such that one can operate in a recognizable fashion with Θ on P and perhaps values m of y so that the result $\Theta(P,m)$ both describes appropriately a natural number n and yields a constructive proof that $\Phi(n,m)$.

As characterization of or condition for constructive proof, the immediately preceding cries out, at least, for recursive unwinding. It is demonstrable that a preponderance of the weight carried by this brief statement rests on the relevant meanings, yet to be explained, of such qualifiers as 'appropriately.' Mapping out divergences and disagreements over those meanings aids in demarcating some of the several varieties of constructivism now either extant or remembered, and in illuminating the variation one detects when surveying the now relatively unpopulated field of constructive mathematics. One finds among brands of constructivism Brouwerian intuitionism, Markovian constructivism, Errett Bishop's new constructivism, predicativism, and finitism, to mention only the most prominent features in the landscape.

Here is a patently nonconstructive proof. Let S be a statement of a mathematical problem as yet unsolved, e.g., Riemann's Hypothesis or Goldbach's Conjecture. Define the natural number function f to be constantly 0 if S is true, and constantly 1 otherwise. Obviously, if S is true, then f is a total, constant function and, in case S is false, f is a (different) total, constant function. In any case, f is total and constant. Even when statement S is known, this (admittedly contrived) proof of f's constancy is nonconstructive because it yields no specific indication which function f is and which natural number is the constant output of f. Short of learning whether S is true or false, one cannot know what that function and number are. A famous constructive proof is Euclid's proof of the conclusion that, given any natural number n, there exists a prime number strictly greater than n. The reasoning informs us that we can always locate, by searching if desired, a prime number between n and $n! + 2$. The proof itself implicitly includes a means Θ, as above, for finding an instance of the theorem proven.

A conventional mathematician may call for a constructive proof of an existence theorem for a variety of reasons, among them a desire to compute numerical solutions, without thereby becoming a constructivist. Dyed-in-the-wool constructivists may be distinguished by their insistence that all proofs of such theorems be constructive. Constructivists of the Brouwerian and Markovian stripes set the existence of a constructive proof of $\exists n.\Phi(n)$ as a necessary and sufficient condition for the truth of $\exists n.\Phi(n)$. As the sample nonconstructive proof may suggest (a suggestion to be examined more closely in a moment), those demanding that all proofs of existence theorems be constructive may be obliged to abridge or revise conventional thinking about the validity of logical and mathematical rules. A natural target of possible adjustment would be the *tertium non datur* or TND — that every instance of the scheme

$$\phi \vee \neg \phi$$

is true. Many constructivists (but not Russellian predicativists, for example) reject or avoid this law. Some constructivists maintain that TND is provably invalid, that not every instance of the scheme is true.

One way or another, all versions of constructivism described herein advance markedly nonstandard theses in logic or mathematics. Brouwerian intuitionists and Markovian constructivists demand that constructively correct mathematics and logic alter and extend, in nontrivial fashions, ordinary mathematics and logic, both groups endorsing anticlassical mathematical laws. Bishop-style constructivists, finitists, and predicativists favor more or less strict limitations on allowable rules, definitions and proofs, but do not generally look to extend the reach of conventional mathematics by adjoining anticlassical principles. When formalized, the mathematical and logical claims characteristic of the latter kinds of constructivism give rise to proper subtheories of familiar arithmetic, analysis, and set theory.

1.1 Crucial Statements

Some constructivists maintain that only certain *crucial* statements of existence be proved constructively. For example, predicativists who accepted the doctrines expressed in Whitehead and Russell's *Principia Mathematica* [Whitehead and Russell, 1910-13] would have demanded that a stricture on existential claims apply, in the first instance, to uses of comprehension principles for class existence, and that the class specifications in those principles be predicative. A Markovian constructivist holds that every existence statement concerning natural numbers is crucial, and, if such a parametrized existential statement as $\exists y \in N$. $\Phi(x,y)$ is true, there will be an abstract Turing machine that computes, from each number m, an $f(m)$ such that $\Phi(m, f(m))$ is also true. Intuitionists who are fans of Brouwer claim to prove, from principles of intuitionistic analysis, that the Markovian is mistaken in calling for Turing machine computations to register existential theorems of arithmetic.

1.2 Appropriateness

Without some nontrivial constraint on specifications, a call for constructive existence proofs governing natural number statements could be answered by citing a simple fact of conventional mathematics: if $\exists x.\Phi(x)$ holds over the natural numbers, then there will always be a unique least n such that $\Phi(n)$ is true and one can employ the μ-term $\mu x.\Phi(x)$ to pick out that least number. Therefore, constructivists who are fussy about natural number existence will likely avoid unrestricted use of μ-terms; in fact, Brouwerian intuitionists claim to prove that the classical least number principle

> Every nonempty set of natural numbers has a least member

is false.

Constructivists of different breeds also disagree over what it takes for a specification of a crucial mathematical object to be appropriate. A Whitehead–Russell predicativist rejected class specifications that violate the Vicious-Circle Principle, that is, those that involve quantification or other reference in the *definiens* to a class that contains the *definiendum* or is presupposed by it. Hilbertian finitists would have maintained that, in the final analysis, an appropriate specification for a natural number is a tally numeral. Bishop's constructivists take standard numerals in base ten as canonical representations for natural numbers.

The appropriateness of a specification therefore makes demands on the sorts of notations that are available. For certain brands of constructivism, as more admissible notations are devised, more existential claims will be deemed constructive and, hence, there will be more numbers. Consequently, it has been natural for such constructivists to think of the extent of their mathematical universes as time-dependent and growing in synch with the collection of appropriate notations. Henri Poincaré , a predicativist, conceived of the universe of classes in this way. He was a champion of the potentially infinite: the collection of all classes and that of all real numbers are never fixed and complete, but continually expanding as more members get defined.

1.3 Constructions

What are constructively legitimate means for specifying, given a parameter, a mathematical object? Often, the means will be operational or functional, and the relevant operations or functions will be constructions or computable functions, perhaps in the sense of Turing, perhaps in some other sense. A desire that proofs of crucial existence statements yield computations that can be carried out in principle, a desire plainly visible in the writings of such constructivists as Bishop and Leopold Kronecker, at times motivated mathematicians to adopt constructivism. This desire is also manifested in the work of computer scientists who look to extract implementable algorithms from constructive proofs, as in [Constable et. al., 1986].

Many constructivists would embrace this version of the Axiom of Choice.

> If $\forall x \in \mathbb{N} \exists y \in \mathbb{N}.\Phi(x,y)$, there is a computable (in some sense) function or construction f over the natural numbers such that $\forall x \in \mathbb{N}.\Phi(x, f(x))$.

In conventional set theory, the Axiom of Choice (often in the presence of other axioms) requires that, when $\forall x \in A \exists y.\Phi(x,y)$ obtains, there is a function f on the set A such that $\forall x \in A.\Phi(x, f(x))$. Generally, constructivists understand the term 'construction' in the above display more restrictively than they do the words 'conventional function.' It should not now come as a surprise to the reader that different constructivists construe 'construction' in different ways. Intuitionists have it that constructions are functions, perhaps partial, given by computing recipes that humans can follow in principle. Normally, intuitionists refuse to identify humanly computable functions with Turing computable functions. Finitists ask that recipes for constructions be required (finitistically, of course) to yield an output on any given input. Other constructivists agree with the Markovians that all constructive operations are to be governed by explicit rules, formulated within a delimited language, such as the replacement rules giving Markov algorithms.

The claim in the foregoing display is to be distinguished from expressions of the Church–Turing Thesis operative in classical theories of computability, viz., that every natural number function computable in principle by a human is also computable by a Turing machine. Plainly, if the word 'computable' in the constructive Axiom of Choice means 'Turing computable,' that statement cannot without further ado be adjoined to conventional arithmetic. For example, where $\Psi(x,y)$ is a reasonable definition in elementary arithmetic of the graph of the characteristic function for the halting problem, $\forall x \in \mathbb{N} \exists y \in \mathbb{N}.\Psi(x,y)$ is true and conventionally provable. However, there is no Turing computable f such that $\forall x \in \mathbb{N}.\Phi(x, f(x))$ is true, on pain of solving the halting problem effectively.

1.4 Constructive Logics and Proof Conditions

Let P be any mathematical statement. Define the predicate $\Phi^P(n)$ over the natural numbers so that Φ^P holds exclusively of 0 when P is true and holds exclusively of 1 otherwise. (Please note that this *definition* of a *predicate* does not itself imply TND and would be admissible in all but the most restrictive forms of constructive mathematics.) Obviously, if P holds, there is a natural number m such that $\Phi^P(m)$

and most constructivists would accept this conclusion. If P fails, there is also a natural number m (this time it is 1) such that $\Phi^P(m)$. Therefore, if one were to assume TND, it would follow that $\exists n.\Phi^P(n)$.

Now, were this existence claim crucial and this little proof treated as constructive, as explained above, there would have to be available an appropriate specification of a natural number n such that $\Phi^P(n)$. From that specification, one should be able to tell whether $n = 0$ or not. If the former obtains, P holds. If the latter, not-P or, in symbols, $\neg P$. Hence, we can tell from the specification which of P or $\neg P$ is true. Since there will always be propositions P like Riemann's Hypothesis whose truth-values are still dark to us, (a strict reading of) the call for constructive existence proofs appears to rule TND out of bounds. TND seems to lead inferentially from premises that are provable constructively to conclusions that are not. Therefore, as earlier suggested, it would seem that a thoroughgoing constructivist about natural number existential statements will be called to reject laws conventionally thought logically valid.

In this connection, it is essential to remember that the impact on logic of the requirement that proofs be constructive is not always restrictive. Existential claims will feature in mathematical arguments not only as final or intermediate conclusions, as in the preceding example, but also as initial assumptions or as antecedents of conditionals. In these cases, constructive proofs of existence for premises would insure that mathematicians possess extra information that would normally not be available in a conventional setting.

Some constructivists would have us pare logic down by allowing only those inferences to be constructively valid that satisfy conditions set on constructive proofs of statements more generally. On such accounts, the truth conditions of a mathematical statement are to be given in terms of its constructive proof conditions, and an inference is to be allowed just in case it preserves constructive provability. Markovian constructivists and Brouwerian intuitionists often set restrictions like the following on disjunction $\vee$ and universal number quantification $\forall x \in \mathbb{N}$.

> A constructive proof of $A \vee B$ is an appropriately specified natural number less than 2 plus another constructive proof. If the specified number is 0, then the latter proof proves A. If it is 1, it proves B.
>
> A constructive proof of $\forall x \in \mathbb{N}.\Phi(x)$ is a construction f such that, for any natural number n, $f(n)$ is a constructive proof of $\Phi(n)$.

Constructivists who accept these conditions will likely refuse the quantified TND. Let $\Phi(x)$ be an undecidable predicate defined over the natural numbers. (This time, 'decidable predicate' need mean no more than 'predicate with a construction for its characteristic function.') Then, this instance of quantified TND

$$\forall x \in \mathbb{N}.(\Phi(x) \vee \neg\Phi(x))$$

cannot be proved constructively. If it were, there would have to be a construction f such that, for each $n \in \mathbb{N}$, $f(n)$ yields an appropriate specification for a natural number less than 2. From this specification, one should be able to tell, using suitable constructive means, whether $\Phi(n)$ holds or not, for each number n. However, this conclusion contradicts the assumption that $\Phi(x)$ is undecidable.

The idea of using conditions set on constructive proofs to determine which logical principles are to count as constructively correct is an old one: it goes back at least to the writings of Brouwer's student Arend Heyting [Heyting, 1934] who was among the first to give constructive proof conditions for mathematical statements across the board.

1.5 Formalizations of Constructive Logic and Mathematics

Except for finitists and predicativists, the constructivists here considered look upon the formal logics deriving from the foundational work of Heyting [Heyting, 1930a-c] as fair representations of the basic principles of reasoning allowed in mathematics. In their standard natural deduction formulations, the rules of Heyting's propositional logic deviate from the familiar rules of formal logic in one respect only: the scheme of negation elimination (aka the negative form of *reductio ad absurdum*)

$$\text{If } \Delta; \neg\phi \vdash \bot, \text{ then } \Delta \vdash \phi$$

is replaced by the rule *ex falso quodlibet*

$$\text{If } \Delta \vdash \bot, \text{ then } \Delta \vdash \phi.$$

Here, $\bot$ stands for any formal contradiction, ϕ for any formula, and Δ for any finite set of formulae. Simple metamathematical arguments show that Heyting's propositional logic, also called 'intuitionistic propositional logic,' will not derive the obvious formalization of TND. In fact, it is not difficult to prove that intuitionistic propositional logic possesses the *disjunction property*: whenever $\phi \vee \psi$ is a theorem, then so is either ϕ or ψ individually.

Heyting's intuitionistic predicate logic can be described as resulting from making the very same replacement of the negative *reductio* rule by *ex falso quodlibet* in a natural deduction formulation of conventional predicate logic. Elementary metamathematical considerations prove that TND is not derivable here either, and that the logic manifests the disjunction and the existence properties for closed formulae. A formal logic (over a particular formal language) has the *existence property* when, if $\vdash \exists x \phi(x)$, then $\vdash \phi(t)$, for some closed term t of the language.

2 CONSTRUCTIVISM IN THE 19TH CENTURY: DU BOIS-REYMOND AND KRONECKER

2.1 Paul du Bois-Reymond

David Paul Gustav du Bois-Reymond was born in Berlin on 2 December 1831. He died on 7 April 1889, while passing through Freiburg on a train. When he died, Paul du Bois-Reymond had been a professor of mathematics at Heidelberg, Freiburg, Berlin and Tübingen. His elder brother was the world-famous physicist, physiologist and essayist Emil du Bois-Reymond. Paul started his scientific career by studying medicine and physiology at Zürich. While there, Paul collaborated on important research concerning the blindspot of the eye. Later, he turned to mathematical physics and pure mathematics, first tackling problems of partial differential equations. He went on to do impressive work in the areas now known as analysis, topology and foundations of mathematics, becoming one of Georg Cantor's greatest competitors in the last subject. In his lifetime and for some decades afterwards, Paul du Bois-Reymond was widely recognized as a leading opponent and critic of efforts to arithmetize analysis, as confirmed in the section *Du Bois-Reymond's Kampf gegen die arithmetischen Theorien* [*Du Bois-Reymond's battle against arithmetical theories*] of Alfred Pringsheim's article for the *Encyklopädie der mathematischen Wissenschaften* [*Encyclopedia of Mathematical Sciences*] [Pringsheim, 1898-1904].

Both Paul and his brother Emil took large roles in the *Ignorabimusstreit*, a spirited public debate over agnosticism in the natural sciences. Emil's 1872 address to the Organization of German Scientists and Doctors, *Über die Grenzen des Naturerkennens* [*On the limits of our knowledge of nature*] [E. du Bois-Reymond, 1886], both sparked the debate and baptized the controversy, for it closed with the dramatic pronouncement,

> In the face of the puzzle over the nature of matter and force and how they should be conceived, the scientist must, once and for all, resign himself to the far more difficult, renunciatory doctrine, '*Ignorabimus*' [We shall never know]. [E. du Bois-Reymond, 1886, 130]

Emil argued that natural science is inherently incomplete: there are fundamental and pressing questions concerning physical phenomena, especially the ultimate natures of matter and force, to which science will never find adequate answers.

Argument and counterargument in the press and learned journals over Emil's scientific agnosticism and attendant issues continued well into the 20th Century. This was the *Ignorabimus* against which Hilbert so often railed; Hilbert's denunciation of it loomed large in the Problems Address [Browder, 1976, 7] as well as in his final public statement, the Köningsberg talk of 1930 [Hilbert, 1935, 378-387]. The latter concluded with a direct reference to Emil's lecture: "[I]n general, unsolvable problems don't exist. Instead of the ridiculous Ignorabimus, our solution is, by contrast, We must know. We will know." [Hilbert, 1935, 387] The words "We must know. We will know" are inscribed on Hilbert's burial monument in Göttingen.

Paul du Bois-Reymond's 1882 monograph *Die allgemeine Functionentheorie* [*General Function Theory*] [P. du Bois-Reymond, 1882] and his posthumously published *Über die Grundlagen der Erkenntnis in den exakten Wissenschaften* [*On the Foundations of Knowledge in the Exact Sciences*] [P. du Bois-Reymond, 1966] were devoted expressly to planting in the realm of pure mathematics the banner of agnosticism first unfurled by Emil. In those writings, Paul enunciated a skeptical philosophy of mathematics that was no simple paraphrase in mathematical terms of his brother's views. As a critic of arithmetization and logicism (and in other respects as well), Paul du Bois-Reymond was a true precursor of the intuitionist Brouwer. In *General Function Theory*, he drew a clear distinction between infinite and potentially infinite sets and, recognizing that a call for the existence of potential but nonactual infinities makes demands on logic, questioned the validity of TND. In his article *Über die Paradoxon des Infinitärcalcüls* [*On the paradoxes of the infinitary calculus*] [P. du Bois-Reymond, 1877], he explained that the denial of the validity of TND is required for a satisfactory understanding of mathematical analysis.

Paul du Bois-Reymond may also have been the first to conceive of lawless sequences, real-number generators the successive terms of which are not governed by any predetermined rule or procedure, and to attempt to demonstrate that they exist. In illustrating the idea, du Bois-Reymond imagined sequences whose terms are given by successive throws of a die:

> One can also think of the following means of generation for an infinite and lawless number: every place [in the sequence] is determined by a throw of the die. Since the assumption can surely be made that throws of the die occur throughout eternity, a conception of lawless number is thereby produced. [P. Du Bois-Reymond, 1882, 91]

Readers are encouraged to compare the foregoing with the explanation of lawless sequence in terms of dice-throwing given by contemporary intuitionists A. Troelstra and D. van Dalen in their [Troelstra and van Dalen, 1988, 645ff].

Du Bois-Reymond believed that information about the physical world could be so encoded in certain sequences that, if such a sequence were governed by a law, a knowledge of that law would yield us predictions about the universe that would otherwise be impossible to make. Were we aware of laws for developing such sequences, he reasoned, we would be able to answer correctly questions about the precise disposition of matter at any point in space and at any time in the past. He wrote,

> If we think of matter as infinite, then a constant like the temperature of space is dependent on effects that cannot be cut off at any decimal place. Were its sequence of terms to proceed by a law of formation, then this law would contain the history and picture of all eternity and the infinity of space. [P. du Bois-Reymond, 1882, 91-92]

Du Bois-Reymond's reflections on lawless sequences prefigured not only Brouwer's later thoughts on choice sequences but also Brouwer's arguments for so-called weak counterexamples, themselves forerunners of reduction arguments in classical recursion theory.

A goodly part of du Bois-Reymond's *General Function Theory* takes the form of a dialogue between two imaginary philosophical characters, Idealist and Empirist.

The Idealist championed a conception of the number continuum on which its constituent real numbers are generally transcendent and have infinitesimal numbers among them. The Empirist restricted mathematics to those real numbers and relations on them open to geometrical intuition. According to du Bois-Reymond, the literary artifice of a fictional debate between Idealist and Empirist corresponds to a natural and permanent duality in human mathematical cognition. He maintained that our current and future best efforts at foundational studies, philosophy of mathematics, and philosophy of mind will discern only these two distinct, mutually inconsistent outlooks on the foundations of mathematics, and no final decision between idealism and empirism will ever be reached. No knockdown mathematical argument will be devised for preferring one over the other. Now or later, a choice between them will be a matter of scientific temperament. In anticipation of the absolute incompleteness arguments of Finsler [1926] and Gödel [1995], du Bois-Reymond came to believe that mathematicians have to cope with absolute undecidability results, meaningful questions of mathematics answers to which depend entirely upon the outlook — Idealist or Empirist — adopted. The Idealist answers the question one way, the Empirist another. Since no conclusive mathematical argument will ever decide between the two, du Bois-Reymond concluded that no final decision will be forthcoming.

2.2 Leopold Kronecker

Leopold Kronecker was born on 7 December 1823 in what is now Liegnica, Poland, and died on 29 December 1891 in Berlin. While studying at the gymnasium in Leignica, Kronecker was taught by the noted algebraist Ernst Kummer. Kronecker matriculated at Berlin University in 1834, where he studied with Dirichlet and Steiner. The young Kronecker later followed his former teacher Kummer to Breslau, where the latter had been awarded a chair in mathematics. In 1845, Kronecker completed his PhD on algebraic number theory at Berlin under Dirichlet, and returned home to enter into the family banking business. In due course, Kronecker became independently wealthy and in no need of a university post to support his mathematical research. He returned to Berlin during 1855, where Kummer and Karl Weierstrass were shortly to join the faculty. 1861 saw Kronecker elected to the Berlin Academy on the recommendation of Kummer, Borchardt and Weierstrass. As a member of the Academy, Kronecker began to lecture at the university on his ongoing mathematical work. He was also elected to the Paris Academy and to foreign membership in the British Royal Society. In 1883, Kronecker was appointed to take up the chair in mathematics left vacant upon Kummer's retirement.

To Kronecker is attributed the remark

> God created the integers, all else is the work of man.

It would seem that this saying was first associated with him, at least in print, in Heinrich Weber's memorial article *Leopold Kronecker* [Weber, 1893]. There, Weber wrote,

> Some of you will remember the expression he used in an address to the 1886 meeting of Berlin Natural Scientists, namely, "Dear God made the whole numbers; all else is the work of men." [Weber, 1893, 19]

According to Pringsheim [Pringsheim, 1898-1904, 58, fn 40], Kronecker was the first to use the expression *Arithmetisierung* [arithmetization] in the foundational context, the arithmetization he desired being much more stringent than that sought by his colleague Weierstrass. In those mathematical fields to which Kronecker devoted his closest attention, branches of number theory and algebra, he insisted that all mathematical claims be reducible to statements about the natural numbers, and that all mathematical operations be resolvable into numerical calculations of finite length. Kronecker set out his program for constructivizing analysis via arithmetization in an article *Über den Zahlbegriff* [*On the number concept*] [Kronecker, 1887]. There he wrote,

> So the results of *general* arithmetic also belong properly to the special, ordinary theory of numbers, and all the results of the profoundest mathematical research must in the end be expressible in the simple forms of the properties of integers. [Kronecker, 1887, 955]

For Kronecker, a definition of a notion would be acceptable only if it could be checked, in a finite number of steps, whether or not an arbitrary object satisfies the definition. Kronecker seems to have been among the earliest mathematicians to call into question nonconstructive existence proofs, as well as assertions and theories dependent upon them. He rejected the concept of arbitrary, non-arithmetic sequences or sets of rational numbers as they featured in the foundational schemes of Heine, Dedekind and Cantor. Accordingly, he refused the least upper bound principle, the use of arbitrary irrational numbers, and Weierstrass's proof of the Bolzano-Weierstrass Theorem. Kronecker criticized Lindemann's 1882 proof of the transcendentality of π on the grounds that transcendental numbers do not exist, and rejected Weierstrass's proof that there exist continuous but nowhere differentiable functions. In keeping with this outlook, he opposed the publication in *Crelle's Journal* of Georg Cantor's papers on set theory and the foundations of analysis.

3 INTUITIONISM AND L. E. J. BROUWER

The notoriety of 20th Century intuitionism seems permanently linked to that of its progenitor, mathematician and philosopher L.E.J. Brouwer, who introduced his intuitionism to the mathematical community in a series of revolutionary articles on the foundations of set theory and analysis, the most influential of which were published between 1907 and 1930. Luitzen Egbertus Jan Brouwer was born on 27 February 1881 in Overschie, later part of Rotterdam. On 21 December 1966, he was hit by a car and died in Blaricum, also in the Netherlands. Brouwer entered high school at the age of nine and completed his high school studies at fourteen, before attending gymnasium for two years. Beginning in 1897, Brouwer worked under Diederik Korteweg and Gerrit Mannoury at the University of Amsterdam, proving original results concerning four-dimensional space that were published by the Dutch Royal Academy of Science in 1904. In addition to topology and the foundations of mathematics, the student Brouwer was interested in the philosophy of mathematics, mysticism, and German idealism. He recorded his ruminations on these topics in a treatise *Leven, Kunst en Mystiek [Life, Art and Mysticism]* [Brouwer, 1905].

Written under Korteweg's supervision, Brouwer's doctoral dissertation *Over de grondlagen der wiskunde [On the Foundations of Mathematics]* [Brouwer, 1907] contributed in creative fashion to the debate between logicists like Russell and antilogicists like Poincaré over logically and mathematically suitable foundations. The dissertation reflected two mental attitudes that would inform Brouwer's entire intellectual life: a strong desire to subject widely accepted foundations for mathematics to trenchant criticism, and a love for geometry and topology. It was in the article *De onbetrouwbaarheid der logische principes [The Unreliability of the Logical Principles]* [Brouwer, 1908] that he first gave the critical ideas of his graduate work a new and startling direction: Brouwer there claimed to show that TND is inappropriate for mathematical use. With justice, one can assert that, in [Brouwer, 1908], mathematical intuitionism in the 20th Century was born and, with it, a great part of contemporary logic, mathematics, and philosophy.

Over the next decade, Brouwer undertook original research in two domains. He continued his penetrating studies of the logical foundations of mathematics, laying down the basis for intuitionistic mathematics, and he put great effort into topological problems from the list Hilbert presented during his Problems Address to the 1900 International Congress of Mathematicians at Paris [Browder, 1976]. In 1908, Brouwer spoke before the International Congress of Mathematicians in Rome on topology and group theory. In April of the following year, he was appointed *privaat docent* in the University of Amsterdam. He delivered an inaugural lecture on 12 October 1909, entitled *Het wezen der meetkunde [The Nature of Geometry]* in which he outlined his research program in topology and listed unsolved problems he planned to attack. A few months later, Brouwer made an important visit to Paris, meeting Poincaré, Hadamard and Borel.

Brouwer was elected to the Dutch Royal Academy of Sciences in 1912 and, in the same year, was appointed extraordinary professor of set theory, function theory and axiomatics at the University of Amsterdam. His professorial inaugural address was published as *Intuitionism and formalism* [Brouwer, 1912]. He would succeed Korteweg as professor *ordinarius* the next year. David Hilbert had written a letter of recommendation for Brouwer that helped him acquire the chair.

In this period, Brouwer proved theorems of tremendous significance in topology. His fixed-point theorem states that a continuous function from the closed unit ball into itself will always hold at least one of the ball's points fixed. Later, he extended the theorem to balls of any finite dimension. He constructed the first correct definition of 'dimension' and proved a theorem on its invariance. He also formulated the concept of degree of a mapping and generalised the Jordan curve theorem to n dimensions.

In 1919, Hilbert endeavored to entice Brouwer away from Amsterdam with a mathematics chair in Göttingen. Over his lifetime, Brouwer would receive a number of academic offers, including positions in Germany, Canada, and the United States, but he would never leave the Netherlands permanently. Brouwer served on the editorial board of *Mathematische Annalen* from 1914 until 1928, when Hilbert, over the objections of Einstein and Carathéodory, had him ejected from the board. After his retirement in 1951, Brouwer traveled, lecturing in South Africa, Canada, and the United States. His list of academic honors includes election to the Royal Dutch Academy of Sciences, the Preussische Akademie der Wissenschaften in Berlin, the Akademie der Wissenschaften in Göttingen and the British Royal Society. The University of Oslo awarded Brouwer an honorary doctorate in 1929 and Cambridge University in 1954. He was named knight of the Order of the Dutch Lion in 1932.

In 1918, Brouwer began the systematic reconstruction of mathematics in the intuitionistic fashion with his paper *Begründung der Mengenlehre unabhängig vom logischen Satz vom ausgeschlossenen Dritten. Erster Teil, Allgemeine Mengenlehre [Foundation of set theory independent of the logical law of the excluded middle. Part One, general set theory]* [Brouwer, 1918]. In a lecture of 1920, published as [Brouwer, 1921], Brouwer gave a negative answer to the question, "Does every real number have a decimal expansion?" He proved there that the assumption that every real number has a decimal expansion leads immediately to an unacceptable consequence of TND,

$$\forall P \forall n \in \mathbb{N}(P(n) \vee \neg P(n)).$$

In his *Beweis dass jede volle Funktion gleichmässigstetig ist [Proof that every total function is uniformly continuous]* [Brouwer, 1924], he claimed to show that every function defined on all the real numbers is uniformly continuous on every closed, bounded interval. Now known as Brouwer's Uniform Continuity Theorem, this result is paradigmatic of his intuitionism.

Brouwer took a predicative form of the Dedekind-Peano Axioms for arithmetic to hold of the natural numbers: 0 is not a successor, the successor function is one-to-one, and mathematical induction obtains for all predicatively specifiable properties of numbers. Of course, use of TND and other nonintuitionistic logical laws is not permitted in proofs from those axioms. Brouwer rejected proofs by induction featuring impredicative property specifications, i.e., those containing unbounded quantification over all or some classes or properties of numbers. (Contemporary intuitionists are often willing to consider relatively unrestricted induction principles). A predicative inductive argument shows that equality between natural numbers is decidable, that is, equality satisfies TND:

$$\forall n, m \in \mathbb{N}(n = m \vee n \neq m).$$

Primitive recursive functions, with addition, multiplication and exponentiation among them, are defined as usual and retain in intuitionism all their decidable features.

Intuitionists refuse induction in its 'downward' or 'least number' variant, since the assumption that every inhabited set of natural numbers contains a least number leads immediately to TND. To say that a set A of natural numbers is *inhabited* is to say that $\exists n.\ n \in A$. This is strictly stronger intuitionistically than the claim that A is nonempty, viz., $\neg \forall n \in \mathbb{N}. \neg n \in A$. To see that least number induction fails intuitionistically, let S be any mathematical statement and consider the set A^S of numbers where $n \in A^S$ if and only if

either ($n = 0$ and S) or $n = 1$.

A^S certainly exists and it is inhabited, since 1 is a member in any case. However, if A^S has a least member a, it is either 0 or 1. If the former, S is true. If the latter, since a is the *least* member of A^S, S must be false. Therefore, were least number induction true, TND would be valid. Least number induction is acceptable to intuitionists when the set A at issue is both inhabited and decidable. Indeed, a simple inductive argument shows that the decidability of a set of natural numbers is equivalent to the truth, for it, of the least number principle.

With simple intuitionistic set theory, the elementary theories of the integers and rational numbers can be worked out along familiar lines. For example, the equality relation between integers and that between rational numbers are provably decidable; this follows immediately from the decidability of equality over the natural numbers and the usual definitions of integers and rational numbers.

It is characteristic of Brouwer's outlook that the ontology of his intuitionism is, by comparison with that of Markovian constructivism, liberal. For Brouwer as for Dedekind before him, mathematics was a 'free creation,' independent of language, metaphysics or the needs of empirical science. This sentiment Brouwer expressed in his dissertation. In particular, mathematics did not wait upon the provision of appropriate notations for mathematical objects thus denoted to exist. On the whole, Brouwer treated entities as legitimate if their existence is apparent to an inner intuition of the continuous passage of time, displayable as either a discrete, finite unfolding construction, a choice sequence of those, or a set of the foregoing. In addition to the natural numbers, Brouwerian intuitionists were happy to countenance sets of a number of sorts (provided they were predicatively defined), abstract proofs both finitary and infinitary, constructive sequences of numbers governed by a rule or algorithm, and nonconstructive or *choice sequences* that may resist any rule-governance.

Natural number functions governed by rules for computations are often called 'lawlike' by intuitionists. It would have offended Brouwer's anti-linguistic outlook to have demanded that such rules be computable by Turing machines or be specifiable in some closely delimited but alternative fashion. Consequently, Church's Thesis in the form

> Every total function from the natural numbers $\mathbb{N}$ into $\mathbb{N}$ is Turing computable

is not accepted by intuitionists, despite the fact that it is consistent with a great deal of the formalized mathematics commonly deemed Brouwerian, including a formulation of Brouwer's Continuity Theorem. (This intuitionistic Church's Thesis should not be confused with the Church-Turing Thesis of classical recursion theory.) In his [1952], American logician Stephen Kleene proved, in effect, that Church's Thesis is inconsistent with the Fan Theorem, the intuitionistic version of the compactness of Cantor space (*vide infra*). Some intuitionists, the present author among them, have expressed a favorable attitude toward a form of Church's Thesis weakened by the insertion of a double negation, namely,

> For any total function f from $\mathbb{N}$ into $\mathbb{N}$ it is false that all Turing machines fail to compute f.

The latter principle plays a signal and salutary role in a proof that, unlike the conventional mathematician, the intuitionist need not countenance Tarskian nonstandard models of arithmetic [McCarty, 1988].

To Brouwer's thinking, there were two intuitionistic analogues to the conventional notion of set or class: species and spreads. The former were predicatively-defined mathematical properties individuated extensionally: species are the same when they have the same members, irrespective of the means by which their correlative properties are expressed. With his interest in topology, Brouwer laid much emphasis on the notion of spread. Spreads were either species of sequences constrained so that their individual terms obey a computable law, a 'spread law,' or species of sequences obtained from those via a continuous mapping. It is in the latter sense that sequences generating real numbers comprise a Brouwerian spread.

In defining real numbers, intuitionists can follow Cantor and Meray, and take reals to be Cauchy sequences of rational numbers under the equivalence relation of co-convergence. They can go with Dedekind and take real numbers to be Dedekind cuts, sets of rational numbers that are proper, inhabited, located, possessing no greatest member, and not bounded below. Brouwer himself took a Cantorean course and let real numbers be sequences of nested intervals of rational numbers. Conventionally as well as intuitionistically, these Cantorean and Dedekindian accounts of real number are provably order isomorphic provided that one adopts an Axiom of Dependent Choice that some intuitionists find acceptable, namely,

$$\forall n \in \mathbb{N} \exists m \in \mathbb{N}. \Phi(n, m) \rightarrow \exists f \forall n \in \mathbb{N}. \Phi(n, f(n)).$$

It is possible to show that the Cantorean or sequential reals represent intuitionistically a proper subset of the collection of Dedekind cuts, unless Dependent Choice adopted. It is here worth mentioning that the general Axiom of Choice for sets A and B,

$$\forall x \in A \exists y \in B. \phi(x, y) \rightarrow \exists f \forall x \in A. \phi(x, f(x)),$$

is disprovable intuitionistically, as Diaconescu [1975] has shown.

Such basic properties of addition and multiplication on the rational numbers as associativity and commutativity extend to the real numbers, given either Cantor's or Dedekind's approach. However, not all the expected properties carry over. For example, Brouwerian intuitionists prove, as a corollary to Brouwer's Uniform Continuity Theorem, that equality over the real numbers is undecidable. Let r be a fixed real number. Were real number equality decidable, the discontinuous function that maps r into 1 and any other real number into 0 would be a total function.

For real numbers, intuitionists recognize more than one notion of distinctness. Real numbers are weakly distinct when their Cauchy sequences fail to co-converge. They are strongly distinct or *apart* when there exists a positive rational number that separates the tails of their Cauchy sequences. Plainly, if two real numbers are apart, they are weakly distinct. The converse does not hold generally, unless Markov's Principle (*vide infra*) is adopted. Following Heyting, intuitionists have studied abstract apartness relations, where an abstract apartness relation # is anti-reflexive, symmetric, and has the property that if $a\#b$, then, for any c from the field of #, either $c\#a$ or $c\#b$. It is consistent with strong formalizations of intuitionistic set theory to assume that every set carrying an apartness relation is the functional image of a set of natural numbers [McCarty, 1986].

Brouwer constructed *weak counterexamples* to such theorems of classical mathematics as that every real number is either rational or irrational, and that the set of all real numbers is linearly ordered. In each case, he showed that the statement implies an intuitionistically unacceptable instance of the law of the excluded third. For example, in the first case, one generates in stepwise fashion the decimal expansion of $\sqrt{2}$ and, when the nth term appears, one checks to see if $2n$ is the sum of two primes. If it is not, one terminates the expansion. If it is, one continues to generate terms. Clearly this decimal expansion defines a real number. If that number is rational, then Goldbach's Conjecture is false. If it is irrational, Goldbach's Conjecture is true. Therefore, if every real number is either rational or irrational, then the famous Conjecture is either true or false. According to Brouwer, since neither the truth nor the falsity of Goldbach's Conjecture is presently known, it is not correct intuitionistically to assert that every real number is either rational or irrational. This unconvincing manner of reasoning can be made convincing by endorsing intuitionistic principles of continuity or a weak form of Church's Thesis.

In his early papers on intuitionism, Brouwer worried that a number continuum consisting entirely of lawlike sequences would not be adequate for real analysis. (Bishop's constructive mathematics and Kleene's work on realizability have since proven such a worry ill-founded.) This encouraged Brouwer to introduce choice sequences into intuitionistic analysis. Earlier conceived by Paul du Bois-Reymond, the notion of choice sequence is often considered Brouwer's greatest contribution to intuitionism and to constructivism generally. At this point, Brouwer's mathematical creativity extended beyond the constructivism adumbrated in the opening sections of the present article, at least on a strict interpretation. For example, Brouwer was willing to countenance lawless sequences among choice sequences, i.e., sequences of natural numbers not generated according to any rule, recipe or law, and was willing to assert that such sequences exist, even though, by the nature of the case, no one would be able to specify an individual lawless sequence.

For sequences s of natural numbers, Brouwerian intuitionists accept a principle of Local Continuity, LC:

> Let Φ be an extensional predicate. Assume that $\forall s \exists n \in \mathbb{N}.\Phi(s,n)$. Each sequence s has a finite initial segment u and a natural number m such that, for any sequence t containing u, $\Phi(t,m)$.

LC informs us that infinite sequences of natural numbers can only be related to individual numbers by relating them as neighborhoods; all sequences that are approximately the same get assigned a common number by Φ. Brouwer [1918] first enunciated LC and Heyting [1930c] gave it a fully explicit formulation. LC for sequences is anticlassical: one easily sees that $\neg$TND follows from it. If one assumes that $\forall s \in S(\forall n.\ s(n) = 0 \vee \exists n.\ s(n) \neq 0)$, then a discontinuous function is definable over all sequences that maps the constantly 0 sequence into 0 and all others into 1. Hence, from LC, one obtains

$$\neg \forall s \in S(\forall n.\ s(n) = 0 \vee \exists n.\ s(n) \neq 0).$$

In attempting a justification for LC, intuitionists treat sequences as choice sequences, the successive terms of which are conceived to appear one-by-one in a way relatively or wholly unregulated by rules or other constraints. Therefore, all the intuitionist may know of a choice sequence at a time is the initial segment of it consisting of those of its terms that have already appeared. Hence, because a mathematical relation or operation is thought to apply to an infinite sequence at some specific time or other, its action can depend only upon the terms of the sequence that are available at or before that time. So, the relation or operation affects all the sequences in some neighborhood in the same way.

The most celebrated intuitionistic result of the early days was Brouwer's Uniform Continuity Theorem: that every real-valued function of a real variable total over a closed, bounded interval is uniformly continuous on that interval. Brouwer's proof of it is importantly different from the proof of Čeitin's Theorem in Markovian constructivism [Čeitin, 1959] or the related Kreisel-Lacombe-Shoenfield Theorem [Kreisel et al., 1959] in recursion theory. For one thing, Brouwer's argument relied upon the Fan Theorem; Čeitin's does not. The Fan Theorem is an intuitionistic analogue to König's Lemma (or the compactness of Cantor space) in conventional mathematics and states that a tree is finite if its branching is uniformly finitely bounded and every one of its branches is finite. Brouwer endeavored to prove the Fan Theorem by proving the Bar Theorem, a principle of induction on well-founded trees. Here, a *bar* for an initial segment t of a sequence is a set B of initial segments such that every sequence containing t also contains a member of B. A set of segments S is *hereditary* if it contains a segment whenever it contains all its immediate descendants. The Bar Theorem asserts that, if a collection S of initial segments is decidable and contains an hereditary bar of the empty segment, then the empty segment belongs to S. The cogency of Brouwer's argument for the Bar Theorem remains a subject of scholarly disagreement [Dummett, 1977, 94-104][van Atten, 2004, 40-63].

Brouwer's most controversial contribution to intuitionism, even in the eyes of his fellow intuitionists, was his theory of the creative subject. In papers beginning with [1948], Brouwer allowed the definition of choice sequences based on the activity of an idealized mathematician or creative subject. By such means, Brouwer obtained (strong) counterexamples to theorems of classical analysis, e.g., that a linear equation with nonzero real coefficients has a real solution. In his [1954], Brouwer advanced a principle governing arguments using the creative subject, a scheme now known as the 'Brouwer-Kripke Scheme': for every mathematical statement S there is a function f_S on the natural numbers such that

$$S \leftrightarrow \exists n \in \mathbb{N}.f_S(n).$$

For example, it is easy to see that, in the presence of LC, the Brouwer-Kripke Scheme implies the falsity of Markov's Principle in the form

$$\forall f(\neg\neg\exists n.f(n) = 0 \to \exists n.f(n) = 0).$$

4 HEYTING AND FORMAL INTUITIONISTIC LOGIC

Brouwer's approach tended to be nonaxiomatic and highly informal. His student and, later, colleague Arend Heyting favored careful, axiomatic formulations and, working in that style, extended the reach of intuitionistic mathematics into formal logic, geometry, algebra, and the theory of Hilbert spaces. Heyting was born in Amsterdam on 9 May 1898 and died of pneumonia while on vacation in Lugano, Switzerland, on 9 July 1980. At the University of Amsterdam, Heyting studied mathematics under Brouwer, earning his tuition by tutoring high school students in the evenings. He received his masters degree *cum laude* in 1922. His doctoral dissertation, written under Brouwer's direction and entitled *Intuitionistic Axiomatics of Projective Geometry*, was successfully examined in 1925 and also earned for Heyting the designation *cum laude*. Heyting was working as a teacher at two secondary schools in Enschede when he entered a prize essay competition sponsored by the Dutch Mathematical Association *Het Wiskundig Genootschap*. In 1928, Heyting won with an article containing a formalization of Brouwer's intuitionistic logic and set theory; he published the revised and expanded essay as his [1930a-c]. To some extent, Heyting's formalization was anticipated by A. Kolmogorov [1925] and V. Glivenko [1928].

The first paper of the series [1930a-c] contains an axiomatization of intuitionistic propositional logic. Heyting there proves, using truth tables, that the axioms are independent. The second paper presents axioms for intuitionistic predicate logic with decidable identity and arithmetic. Like later intuitionistic logics, such as that of [Scott, 1979], Heyting's predicate logic allows terms to be undefined. Heyting's axioms for arithmetic are essentially those of Dedekind-Peano Arithmetic. The last paper contains principles for species, spreads and choice sequences, including a formulation of LC.

Heyting became a *privaat-dozent* at the University of Amsterdam in 1936, a reader the following year, and a professor in 1948. He retired from university work in 1968. Throughout his career, Heyting acted at home and abroad as spokesperson for and expositor of intuitionism and Brouwer's ideas; his elegant *Intuitionism: An Introduction* [Heyting, 1956] remains the classic introduction to the subject. As a representative of intuitionism, Heyting lectured on the topic *The intuitionist foundations of mathematics* to the second conference on the Epistemology of the Exact Sciences, held at Königsberg, 5-7 September 1930 [Heyting, 1931]. At the same conference, Rudolf Carnap and John von Neumann spoke on logicism and formalism, respectively. Kurt Gödel made there a brief presentation of his proof of the completeness theorem for predicate logic, and gave the first public announcement of his incompleteness theorems [Gödel, 1931].

In his monograph [1934], Heyting proposed a proof-theoretic or proof-conditional treatment of the intuitionistic logical signs in terms of constructions and informal proofs, abstractly considered. He believed this to express the special significance Brouwer attached to the signs. The treatment proceeds, as does Tarski's definition of satisfaction, by recursion on the structures, determined by quantifiers and connectives, of formulae. These are the clauses governing $\vee, \rightarrow, \neg$ and quantification $\exists$ and $\forall$ over natural numbers:

$\vee$ A proof of $\Phi \vee \Psi$ consists of a natural number $n < 2$ and a further proof P such that, if n is 0, then P proves Φ and, if n is 1, then P proves Ψ.

$\rightarrow$ A proof of $\Phi \rightarrow \Psi$ affords a construction Θ that will convert any proof P of Φ into a proof $\Theta(P)$ of Ψ.

$\neg$ A proof of $\neg \Phi$ affords a construction that converts a proof of Φ into a proof of $0 = 1$.

$\exists n$ A proof of $\exists n \in \mathbb{N}.\Phi(n)$ consists of a natural number m and a proof of $\Phi(m)$.

$\forall n$ A proof of $\forall n \in \mathbb{N}.\Phi(n)$ affords a construction Θ such that, for any natural number m, $\Theta(m)$ is a proof of $\Phi(m)$.

A statement Φ is said to be true when there exists a proof, in this sense, of it.

With clause [$\vee$], the decidability of equality between natural numbers underwrites an ability in principle to determine effectively, whenever a disjunction is true, which of its two disjuncts is true. The assumption of such an ability is implicit in Brouwer's arguments for weak counterexamples. If we agree that there can be no proofs of $0 = 1$, then clause [$\neg$] is equivalent to the statement that any construction counts as a proof of $\neg \Phi$ as long as there are no proofs of Φ. The provision of number m in clause [$\exists n$] is intended to attach the constructive proof condition permanently to the existential number quantifier.

The constructions mentioned in clauses [→], [¬], and [∀n] are operations that are computable, in some reasonable sense, on proofs. Let $\Phi(n)$ define an undecidable set of natural numbers. Then, there can be no proof of $\forall n \in \mathbb{N}.(\Phi(n) \vee \neg\Phi(n))$. Were there such a proof, clauses [∀n] and ∨ together would guarantee the existence of a computable operation $\Theta(n)$ mapping the natural numbers into the set $\{0,1\}$ and such that $\Theta(n) = 0$ if and only if $\Phi(n)$ is true. Therefore, it must be the case that $\neg\forall n \in \mathbb{N}.(\Phi(n) \vee \neg\Phi(n))$

Working independently of Heyting, Kolmogorov [1932] defined a structurally similar relation between intuitionistic statements and problems rather than proofs or constructions. Consequently, Heyting's proof-theoretic treatment is often called (e.g., by Troelstra and van Dalen in their [1988]) the Brouwer-Heyting-Kolmogorov or BHK interpretation. Academic disputes have arisen over the correctness of some of the clauses. For instance, scholars have worried about the impredicativity of the [→] clause; in laying down what it means for a construction to prove a conditional statement $\Phi \to \Psi$ the *definiens* quantifies over all proofs: 'converts any proof of Φ into a proof of Ψ.' Following the lead of G. Kreisel [1962b], others have endeavored, by adding restrictions to clauses [→] and [∀n], to insure that the proof relation defined à la Heyting between constructions and statements be decidable.

5 MARKOVIAN OR RUSSIAN CONSTRUCTIVISM

The founder of the school of Russian constructivism was Andrei A. Markov (1903-1979), whose lectures on constructive mathematics in the years 1948 and 1949 inspired his colleagues to adopt a relatively strict constructive outlook. His father, also named 'Andrei A. Markov,' introduced Markov chains. In addition to Markov, prominent members of the Russian school are Nikolai A. Shanin (b. 1919) and Grigorij S. Čeitin (b. 1936). Relative to intuitionism and predicativism, Markovian constructivism is exacting in its requirements: mathematical objects are admissible only if fully encodable as words on a finite alphabet. As Markov held, such encodings must be potentially realizable as concrete notations for words [Markov, 1971]. Natural and rational numbers are treated as finite words; the collections of natural and rational numbers are deemed potentially, but not actually, infinite [Shanin, 1968, 10]. Constructions are given as algorithms for Turing machines, these also conceived as words over a finite alphabet. Sets (of potentially realizable objects) are identified with their specifications: a set is a description, often expressible as a formula in first-order arithmetic with one free variable. What Oliver Aberth wrote of computable analysis holds of Markovian constructivism as well:

> [It] may be informally described as an analysis wherein a computation algorithm is required for every entity employed. The functions, the sequences, even the numbers of computable analysis are defined by means of algorithms." [Aberth, 1980, 2]

Markovian constructivists often accept a form of Church's Thesis or CT:

If $\forall n \in \mathbb{N}.\exists m \in \mathbb{N}.\Phi(n,m)$, there is an index e for a Turing machine such that $\forall n \in \mathbb{N}.\Phi(n, \{e\}(n))$.

Here, '$\{e\}(n)$' is short for 'the output of machine e on input n.' The above schema is to be distinguished from the Church-Turing Thesis familiar from standard computability theory, the claim that every humanly computable function is computable by a Turing machine.

Markovians may also endorse a stronger computability principle, Extended Church's Thesis or ECT. One expression of ECT is

If $\forall n \in \mathbb{N}(\neg\Psi(n) \to \exists m \in \mathbb{N}.\Phi(n,m))$, there is a Turing machine index e such that $\forall n \in \mathbb{N}(\neg\Psi(n) \to \Phi(n, \{e\}(n)))$.

The negation in the antecedent $\neg\Psi(x)$ is essential. ECT is strictly stronger than CT over intuitionistic arithmetic. Forms of ECT arise naturally in the study of axiomatizations of Kleene's number realizability.

With an argument akin to one sketched above, it is easy to see that CT derives, in first-order arithmetic, negations of instances of quantified TND. Let $\exists m.\, T(n,n,m)$ be a standard self-halting predicate. Assume that

$$\forall n(\exists m.\, T(n,n,m) \vee \neg\exists m.\, T(n,n,m)).$$

By applying CT to this assumption, one infers that the characteristic function of the halting problem is Turing computable. Since it is provable constructively (even on a narrow rendering of constructivity) that the halting problem is recursively unsolvable, it follows that

$$\neg\forall n(\exists m.\, T(n,n,m) \vee \neg\exists m.\, T(n,n,m)).$$

Hence, the TND is demonstrably invalid, given CT. The Markovian interpretation of constructivity, if it is seen to require CT, enjoins upon Russian constructivists anticlassical principles of logic.

Both the constructivists of Bishop's school and some Brouwerian intuitionists refuse to adopt an official logic. By contrast, the Markovians explicitly endorse a version of Heyting's formal intuitionistic logic. Shanin [1958] gave the signs of the logic a reading associated with what he called a 'deciphering algorithm.' On that algorithm, constructive existence is attached semantically to the existential quantifier $\exists$: the holding of a claim $\exists n \in \mathbb{N}.\Phi(n,m)$ marks the existence of an algorithm, implementable on a Turing machine, the action of which is described correctly by $\Phi(n,m)$. In this and other regards, Shanin's interpretation bears a clear resemblance to Kleene's realizability for arithmetic. In the name of this interpretation, mathematical principles characteristic of Markovian constructivism, such as CT and MP, are endorsed. MP or Markov's Principle is the statement

$$\forall n \in \mathbb{N}(\Phi(n) \vee \neg\Phi(n)) \to (\neg\forall n \in \mathbb{N}.\, \neg\Phi(n) \to \exists n \in \mathbb{N}.\, \Phi(n)).$$

Since $\neg\forall x\neg\phi$ is provably equivalent in Heyting's logic to $\neg\neg\exists x\phi$, one can say that MP allows double negations to be dropped from existential claims of arithmetic whenever the matrix of the claim satisfies TND. In constructive computability theory, MP implies that a subset of $\mathbb{N}$ is recursive provided both it and its complement are recursively enumerable. In Markovian real analysis, MP guarantees that a pair of real numbers are apart if and only if they are unequal [Troelstra and van Dalen, 1988, 205]. In intuitionistic metamathematics, (a form of) MP is equivalent to the completeness of Heyting's formal predicate logic with respect to Kripke or Beth models [Kreisel, 1962a][McCarty, 1996]. MP is consistent with the strong formal set theory IZF, intuitionistic Zermelo-Fraenkel, as realizability interpretations for set theory show [Beeson, 1985, 162].

The adoption of MP is illuminated, but hardly justified, by the algorithmic reading of logical signs Shanin proposed. On that reading, $\forall n \in \mathbb{N}(\Phi(n) \vee \neg\Phi(n))$ requires that the extension of Φ be decidable: there is a Turing machine M computing a total function that, on any natural number input n, outputs 0 when $\Phi(n)$ holds of n and 1 otherwise. Now, one imagines that M is run on the sequence 0, 1, 2..., successively searching for an input on which M outputs 0. The assumption that $\neg\forall n \in \mathbb{N}.\neg\Phi(n)$ implies that it is impossible for this search to fail. M cannot then output 1 for every number input. On these assumptions, the consequent of MP has it that M, as a search machine, will eventually locate a number on which M outputs 0. Such a result is not provable intuitionistically.

As do Brouwerian intuitionists, Russian constructivists recognize as legitimate a variety of conceptions of real number. For example, a real number can be an *FR-sequence*, where 'FR' stands for 'Fundamental sequence with Regulator of convergence.' Such a sequence consists of two algorithms, the first giving a standard Cauchy sequence with rational number terms, the second providing a modulus of convergence for the first. A real number is an *F-sequence* when it is a Cauchy sequence of rational terms, with or without modulus. Markov introduced the latter concept into Russian constructivism in the article [Markov, 1958].

In Russian constructivism, the most celebrated theorem of real analysis is that of Čeitin: every total function from the real numbers into the real numbers is pointwise continuous [Čeitin 1959]. Its proof requires MP. It seems that Čeitin obtained his result prior to the related theorem proved by Kreisel, D. Lacombe, and J. Shoenfield, and named for them [Kreisel *et al.*, 1959]. Markov had already obtained the weaker result that total real-valued functions on the reals cannot be discontinuous [Markov, 1958]. In a narrowly constructive mathematics, Čeitin's Theorem does not imply that every real-valued function over a closed, bounded interval is uniformly continuous. I.D. Zaslavskiĭ (b. 1932) had already shown that there exist functions that are continuous pointwise on the closed unit interval $[0,1]$ but are not uniformly continuous there [Zaslavskiĭ, 1955]. By employing a result proved by Ernst Specker [1949], Markovian constructivists show that there exist continuous functions that are unbounded on closed intervals.

6 BISHOP'S NEW CONSTRUCTIVISM

The monograph *Foundations of Constructive Analysis* [Bishop, 1967] by Errett A. Bishop (1928-1983) banished any lingering doubts that an elegant and meaningful form of mathematical analysis can be developed in a thoroughly constructive fashion. Bishop was a mathematical child prodigy, having studied textbooks belonging to his father, a mathematics professor in Wichita, Kansas. He matriculated at the University of Chicago in 1944, and earned there both BS and MS degrees in 1947. For his PhD, granted by the University of Chicago in 1954, he researched spectral theory under Paul Halmos. From 1954 until 1965, Bishop taught at Berkeley. At the time of his death, he was professor of mathematics at the University of California, San Diego. New constructivists, among them Fred Richman and Douglas Bridges (b. 1945), followed Bishop's lead in rebuilding mathematics, especially analysis, in the style of [Bishop, 1967].

As in Brouwer's intuitionism, Bishop's constructive universe is open-ended. Bishop wrote, "Constructive mathematics does not postulate a pre-existent universe, with objects lying around waiting to be collected and grouped into sets, like shells on a beach." [Bishop, 1985, 11] Ordinary base ten notation is canonical for natural numbers. According to Bishop, "Every integer can be converted in principle to decimal form by a finite, purely routine process." [Bishop, 1985, 8] In addition to numbers, new constructivists accept abstract operations, functions, proofs and sets. They do not require, as do Markovians, that all admissible objects be coded as natural numbers or finite strings over an alphabet. No special notations are adopted for sets or operations. A set A is given once a recipe is available for constructing elements of A and conditions are at hand for determining when elements of A are provably equal. Any function on a set A must preserve the A-equality of A-elements. For Bishop, a real number is a triple $\langle r, p, s \rangle$ wherein r and s are operations and p is a proof that s is a modulus of convergence for the sequence r of natural numbers.

Bishop and his followers ask that mathematical operations and constructions be humanly computable in principle. No limit is set ahead of time on the manners in which recipes can be conveyed linguistically. Therefore, new constructivists do not endorse CT or ECT. However, they allow that CT, ECT, and MP are together consistent with their mathematics; logicians have proven formalizations of these consistent with systems, such as the set theories B [Friedman, 1977] and IZF [Beeson,1985], that can represent Bishop's mathematics. It follows that new constructivists will not be able to show that there are total discontinuous real-valued functions over the reals. In consequence, a constructive theory of measure and integration cannot be established by them in a straightforward fashion using the standard doctrine of real-valued step functions.

As did Kronecker, Bishop held that numerical meaning is the only legitimate meaning discoverable in mathematics and that his version of constructive mathematics accords with the numerical meanings of mathematical statements. Such meanings can be displayed explicitly by realizing constructive theorems as computer programs, perhaps as R. Constable and colleagues have done in the NuPRL Project [Constable et. al., 1986]. Bishop encouraged an analogy between suitable formalizations of new constructivism and high-level languages for specification and programming. The idea was that, in the future, proofs in the formalism would be compiled more or less directly into implementable code [Bishop, 1985, 14-15].

Bishop did not recommend a development of constructive mathematics in isolation from conventional mathematics, but a careful and elaborate recreation of the latter. He deemed this project "the most urgent task of the constructivist." [Bishop, 1970, 54] In practice, its completion often requires that extra hypotheses be added to the statements of conventional theorems so that their proofs are possible using strictly constructive reasoning. For example, the classical least upper bound theorem states that every nonempty set of real numbers that is bounded above has a least upper bound. With extra hypothesis, Bishop's version of the theorem is that every inhabited set of real numbers that is *order located* and has an upper bound has a least upper bound. Here, a set A of real numbers is order located if, for any real numbers r and s with $r < s$, either s is an upper bound for A or there exists a real number $x \in A$ such that $r < x$. In fact, A being order located is also a necessary condition for A having a least upper bound [Bishop and Bridges, 1985, 37].

The reasoning accepted by Bishop's new constructivists is formalizable in Heyting's first-order predicate logic. Their mathematics can be captured in intuitionistic Zermelo-Frankel set theory IZF or the weaker CZF [Aczel, 1978]. Harvey Friedman has argued [Friedman, 1977] that the work of the new constructivists can also be formalized within the even weaker set theory B, which is provably conservative over the elementary intuitionistic formal arithmetic HA or Heyting Arithmetic [Beeson, 1985, 321].

7 PREDICATIVISM

Predicativism manifests itself as a restriction of mathematics, primarily set theory, class theory or analysis, imposed upon the linguistic means by which supposed higher-order entities such as logical classes, real numbers and infinitary functions are defined by intension or abstraction. Roughly put, a specification of a class C via a class abstract $\{x : \Phi(x)\}$ is *impredicative* when $\Phi(x)$ contains a variable ranging over elements of either a collection that has C as a member or another class requiring C for its proper definition. When the abstract $\Phi(x)$ fails to contain such a variable, the specification is *predicative*, and can be allowed by predicativists. Therefore, a class C is deemed to be predicatively well-defined by $\{x : \Phi(x)\}$ when all variables in $\Phi(x)$ are restricted to ranging over collections of classes or other entities already known to be well-defined predicatively prior to the moment of C's definition. In standard Zermelo-Fraenkel set theory, impredicativity is ubiquitous: the familiar definition of the set ω of natural numbers as the $\subset$-least inductive set,

$$\omega = \{x : \forall y \ (y \text{ is inductive} \rightarrow x \in y)\},$$

is impredicative since ω is itself an inductive set.

Although the committed predicativist does not call for any large restriction in classical logic beyond predicative limits imposed upon schemes of comprehension or definitions of classes and sets, predicativism does count as a species of constructivism as delimited *infra*. Quantification over numbers or finite strings is not often thought subject to predicativistic restriction. (However, Nelson [1986] argued that, because the definition of natural number via inductive sets is impredicative, a consistent predicativist should call for a predicative arithmetic in which bounded number quantifications play the starring role.) Normally, existential quantification over classes $\exists x.\Phi(x)$ is crucial for predicativists and a proof of such a quantified statement is admissible if the proof gives a predicatively specifiable class C such that $\Phi(C)$ is also admissibly provable. One can think of a class within a predicative universe of classes as an abstract collection constructed through a well-founded process of definition involving conventional class operations like union, intersection, and relative complement, as well as quantification, subject to predicative restrictions on variables.

Henri Poincaré (1854-1912) and, after him, Bertrand Russell (1872-1970) initially advanced the idea that use by mathematicians of impredicative definitions is fallacious and that mathematics should be reconstructed so that all definitions are strictly predicative in form. Mathematician and physicist Poincaré was cousin to Raymond Poincaré, prime minister and president of France. After working as a mining engineer, he completed a doctorate in mathematics in 1879 under the direction of Charles Hermite, submitting a dissertation on differential equations. From 1886 until his death, Poincaré held chairs at the Sorbonne and the École Polytechnique. He introduced into complex analysis the study of automorphic functions, discovered and developed basic ideas of algebraic topology, including that of homotopy group, and made major contributions to the fields of analytic functions, number theory, and algebraic geometry. In physics, he receives joint credit, with Einstein and Lorentz, for discovering the special theory of relativity. In addition to his membership in the Académie Francaise and the Académie des Sciences, he was a corresponding member of scientific societies in Amsterdam, Berlin, Boston, Copenhagen, Edinburgh, London, Munich, Rome, Stockholm, St. Petersburg, and Washingston.

Poincaré's views on the foundations of mathematics inspired Brouwer and the Dutch intuitionists; he emphasized that only potential, rather than completed or actual infinities exist, and that the principle of induction over the natural numbers is known to us exclusively by the exercise of an intuition irreplaceable by deduction from logical axioms. Poincaré anticipated Bishop in requiring that, to be adequate, infinitary mathematics must retain a clear numerical meaning; he wrote,

> Every theorem concerning infinite numbers or particularly what are called infinite sets, or transfinite cardinals, or transfinite ordinals, etc., etc., can only be a concise manner of stating propositions about finite numbers. [Poincaré, 1963, 62]

Poincaré thought to see a fallacy of impredicativity (which he explicitly called 'a vicious circle') underlying both foundational paradoxes such as Richard's and what he believed a questionable extension, by Georg Cantor, of mathematics into the transfinite. Reinforcing Poincaré's objections to impredicativity was his vision of the existence of mathematical entities as time-dependent and, accordingly, variable over time: a mathematical object can exist only if it is properly defined, and comes to exist once it is defined. Therefore, collections of mathematical objects are not immutable when it comes to members. A mathematical object like a real number may be a logical class C that does not exist unless and until it has been specified using an abstract. Hence, the collection of real numbers is constantly growing in size as more real numbers are defined.

It was in terms of the growth of classes and their membership relations over time that Poincaré first explicated the terms 'predicative' and 'impredicative' as applied to classes. For him, a class is predicative when it is so defined that its membership is stable: any new members that get added to the class remain permanently in the class. The use of an abstract defining C and featuring a variable construed to range over an infinite collection D containing C makes the false assumption that C already exists and is well-defined. Furthermore, since membership in C is determined with reference to all members of D, C's impredicative definition is likely to be a cause of instability, membership in C being dependent upon elements of D that are not yet defined.

Poincaré objected in his [1963] to Zermelo's proof of the Well-Ordering Theorem on the grounds that the set-theoretic operation of arbitrary union, as exploited by Zermelo, is impredicative. By today's standards, he was right on the last point: the union $\bigcup x$ of a set x is given by the abstract

$$\{z : \exists y \, (z \in y \wedge y \in x)\}$$

in which the unrestricted bound set variable y is intended to range over all sets, including $\bigcup x$. For example, in standard set theory, the union $\bigcup n$ of a nonzero (von Neumann) natural number n is always a member of n itself.

Russell formulated the demand that all logical classes be predicatively defined in his Vicious Circle Principle:

> I recognise, however, that the clue to the paradoxes is to be found in the vicious-circle suggestion; I recognise further this element of truth in M. Poincaré's objection to totality, that whatever in any way concerns *all* or *any* or *some* (undetermined) of the members of a class must not be itself one of the members of a class. In M. Peano's language, the principle I wish to advocate may be stated: "Whatever involves an apparent variable must not be among the possible values of that variable." [Russell, 1906, 198]

By Russell's lights, classes are naturally organized into a noncumulative hierarchy of orders or types in such a way that "any expression containing an apparent [bound] variable is of higher type than that variable." [Russell, 1908] Here, the means to implement a ban upon impredicativity is to conceive of classes falling of their own accord into mutually disjoint types or orders or levels, and to adopt as an official means of expression and deduction a many-sorted formal system with variables for such classes indexed with symbols for those types, orders or levels. In such a system, a quantifier with a bound variable carrying type index α can only be replaced, in universal instantiation or existential generalization, by a variable or parameter carrying that same index α. Formal systems of this kind include the ramified type theory of *Principia Mathematica* [Whitehead and Russell, 1910-1913] and that of Hao Wang's systems Σ [Wang, 1964] [Chihara, 1973].

One thinks of a standard model of a predicative type theory as a subuniverse of the standard model of simple type theory over the natural numbers, but having classes further divided or *ramified* into a sequence of orders or levels indexed by natural numbers or constructive ordinal numbers. Classes of natural numbers on level 1 are those that are specifiable using variables ranging only over natural numbers. Classes of level 2 are those specifiable using variables ranging exclusively over natural numbers or classes of level 1. Classes of level 3 have, in their specifications, variables ranging only over natural numbers or classes of levels 1 and 2, and so on. Every class is required to exist in some level. Such requirement is intended to rule impredicativity out. Predicativists often imagine that the levels are formed by some kind of definitional process evolving in discrete stages over time, with new classes and levels appearing on the bases of classes and levels already in existence. Of this process, Wang wrote, "[N]ew objects are only to be introduced stage by stage without disturbing the arrangement of things already introduced or depending for determinedness on objects yet to be introduced at a later stage." [Wang, 1964, 640] Consequently, latter-day predicativists often followed Poincaré in thinking that the mathematical universe of classes expands over time as new definitions and specifications become available.

If real numbers are classes of rational numbers, e.g., Dedekind cuts, then in ramified analysis, there is no single class containing all real numbers and there may be 'new' real numbers appearing at every level from some point onward. Therefore, if one cleaves strictly to the ramified conception, there can be no single variable that ranges over all real numbers. As Russell came to realize, ramification appears to block all satisfying formulations of Dedekind's Theorem that every nonempty collection of cuts that is bounded above has a least upper bound. The least upper bound of a nonempty class of order α of cuts, defined as it is by a union over all cuts of order α, must be a cut of order at least $\alpha + 1$. To circumvent such drawbacks to his type theory, Russell reluctantly adjoined to his system the controversial Axiom of Reducibility: in the above terms, the assumption that every class of any level is coextensive with some class of level 1.

Hermann Weyl (1885-1955), a distinguished mathematician and a leading student of David Hilbert, championed the cause of predicativism in his monograph *Das Kontinuum* [Weyl, 1918]. Weyl studied mathematics and physics, first at Munich and later under Hilbert at Göttingen. After obtaining his doctorate, Weyl took up a professorial post at the Swiss Federal Institute of Technology in Zürich.

He later replaced Hilbert at Göttingen, before emigrating to America — to the Advanced Institute at Princeton — in 1933. As mathematician and physicist, Weyl made notable contributions not only to the foundations of mathematics but also to the theories of integral and differential equations, geometric function theory, differential topology, analytic number theory, gauge field theory, group theory, and quantum mechanics. In his [1918], Weyl sidestepped the technical issues besetting Russell's formulation of ramified analysis by constructing a predicative version of analysis using strictly arithmetic comprehension, that is, taking the natural numbers as given and permitting classes of numbers at level 1 only. He wrote,

> A "hierarchical" [ramified] version of analysis is artificial and useless. It loses sight of its proper object, i.e., number. ... Clearly, we must take the other path — that is, we must restrict the existence concept to the basic categories (here, the natural and rational numbers) and must not apply it in connection with the system of properties and relations (or the sets, real numbers, and so on, corresponding to them). [Weyl, 1918, 32]

By such means, Weyl was able to obtain a sequential version of Dedekind's Theorem. For that, he treated real numbers not as cuts but as Cauchy sequences, and used strictly level 1 definitions to prove that every Cauchy sequence of real numbers has a real number as its limit.

Close metamathematical study of predicativity, using the formal tools forged by Gödel, Kleene, Tarski and others, began anew in the 1950s with efforts to extend the hierarchy of arithmetically definable sets into the transfinite. Prominent here are the contributions of Wang [1954], Lorenzen [1955], and Kreisel [1960]. More recently, Solomon Feferman (b. 1928) has been largely responsible for the detailed proof-theoretic study of the depth and extent of predicative mathematics, writing a number of papers (some published with coworkers) from his classic [1964] up through the retrospective [2005]. Independently of Kurt Schütte [1965], Feferman identified the precise upper bound on the predicatively provable ordinal numbers.

8 FINITISM

Finitists demand that mathematicians avoid all reference, explicit or implicit, to infinite totalities, including the totality of natural numbers. Sometimes, as was the case with the finitism of David Hilbert [Hilbert, 1926], this avoidance is allied with nominalism and a desire, epistemically motivated, to replace abstract notions with notations that are relatively concrete and physically realized. In addition to the finitism of Hilbert, one should count Skolem's primitive recursive arithmetic [1923] and Yessenin-Volpin's ultra-intuitionism [1970] among influential versions of finitism in the 20th Century.

Only natural numbers or items simply and fully encodable as natural numbers count as finitistically acceptable. The natural numbers are not deemed to constitute a completed infinite totality, but are permitted as concrete, readily visualizable notations. Hilbert [1926] insisted that finitistic talk of operations on numbers or symbols (the only sort of talk permitted in his metamathematics for proof theory) must be understood entirely in terms of performable manipulations upon the intuitable forms of strings of tally marks. Hilbert wrote,

> The subject matter of mathematics is, in accordance with this theory, the concrete symbols themselves whose structure is immediately clear and recognizable. [Hilbert, 1926, 142]

Finitists maintain that the customary use of unbounded existential quantification over natural numbers in mathematics commits the user to the existence of completed infinite totalities and, hence, existential claims require, for their full legitimacy, finitistic reconstrual, perhaps by the imposition of explicit numerical bounds on all arithmetic quantifiers. Such completely bounded quantifications over the numbers are usually finitistically admissible without further ado. According to Hilbert, thoroughly finitistic statements express mathematical propositions that are contentual. Since the kinds of claims recognized as finitistic are so narrowly delimited as always to be decidable, classical logic reigns in finitistic mathematics, whether in the style of Hilbert or of Skolem.

Hilbert proposed to construe some unbounded numerical quantifications as 'incomplete statements.' The completion of an unbounded existential statement with primitive recursive matrix includes the provision of a correct bound on the existential quantifier. In the case of universal quantification $\forall n.\Phi(n)$ with $\Phi(x)$ primitive recursive, completion requires a finitistic proof of each instance of the free-variable scheme $\Phi(x)$. Hilbert treated the claims of analysis and set theory that do not admit finitistic reconstrual as ideal. These ideal statements lack denotative meaning but can be manipulated by the deductive apparatus of a theory containing them.

One can speak of constructions or operations in finitistic mathematics, but they do not constitute an absolutely infinite collection of functions, and are not conceived as bearing with them infinite domains or ranges of input and output values. An operation on natural numbers is finitistic whenever it can be seen as a transformation that can be carried out on concrete signs so that there is an explicit, uniform, humanly calculable bound on the number of steps required to complete the transformation in any given case. Natural candidates for numerical functions that fit this bill are the primitive recursive functions, since these can be defined as computation routines that never require of the computer an unbounded search. Whether the finitistic functions of Hilbert include — either conceptually or historically — more than the primitive recursive functions has been a subject of some dispute among the cognoscenti. W. Tait [1981] argued that Hilbertian finitistic mathematics is to be identified with primitive recursive mathematics, but see his [2002] and [2005a] for qualification.

Thoralf Skolem (1887-1963), a pioneer in model theory and set theory, and professor of mathematics at Oslo, admitted as finitistic only those assertions whose truth or falsity is determinable in a finite number of steps via calculations that are primitive recursive. Hence, he allowed simple equations between primitive recursive terms and statements obtainable from such equations via combinations with sentential connectives and bounded quantifiers.

R. L. Goodstein (1912-1985), who studied under L. Wittgenstein and J.E. Littlewood, developed, in his [1957] and [1961], finitistic mathematics further in the fashion established by Skolem. Goodstein examined various conceptions of real number, among them the notion of a primitive recursive Cauchy sequence attached to a primitive recursive modulus of convergence. Here, as in Hilbert's and Skolem's versions of finitism, nonclassical axioms are not accepted, so there can be no proof of any theorem such as Čeitin's or Brouwer's Continuity Theorem that flouts laws of classical analysis. Since finitistic mathematics is also intuitionistically correct, there can be no finitistic theorem that contradicts Brouwer's intuitionism. According to Goodstein, a Skolemite finitist cannot prove that any bounded, monotonically increasing sequence of rational numbers has a real number as its limit.

The ultra-intuitionism of A. S. Yessenin-Volpin is substantially different from the finitistic outlooks just described. Yessenin-Volpin (b. 1924) is the son of the Russian poet Sergei Esenin, once the husband of Isadora Duncan, and Nadezhda Volpin, a writer and translator. In 1949, he was arrested by the Soviet authorities for his poetry, and was committed to a mental institution. In 1950, he was exiled to Khazakhstan. Yessenin-Volpin emigrated to the United States in 1972. In his mathematical work, he rejected both the standard notion of a natural number system closed under the successor operation and the ideas for a primitive recursive mathematics set out by Skolem. Numbers are not thought to be potentially realizable in terms of concrete notations; only those numbers that are feasible, literally displayable, are to be accepted. Following P. Bernays [1935], scholars refer to views like Yessenin-Volpin's as 'strict finitism.' C. Kielkopf [1970] and other investigators have thought to see in the writings of Wittgenstein, principally his [1956], an endorsement of a form of strict finitism.

BIBLIOGRAPHY

[Aberth, 1980] O. Aberth. *Computable Analysis.* New York, NY: McGraw-Hill International Book Company. xi+187, 1980.

[Aczel, 1978] P. Aczel. *The type-theoretic interpretation of constructive set theory.* A. Macintyre et al. (eds.) Logic Colloquium '77. Amsterdam, NL: North-Holland Publishing Company. pp. 55-66, 1978.

[Beeson, 1985] M. Beeson. *Foundations of Constructive Mathematics.* Berlin, DE: Springer-Verlag, 1985.

[Benacerraf and Putnam, 1964] P. Benacerraf and H. Putnam, eds. *Philosophy of Mathematics. Selected Readings.* First Edition. Englewood Cliffs, NJ: Prentice-Hall, Inc, 1964. Second Edition. Cambridge, UK: Cambridge University Press. viii+600, 1983.

[Bernays, 1935] P. Bernays. *Sur le platonism dans les mathématiques.* L'enseignement Mathématique. Volume 34. pp. 52-69, 1935. Reprinted C. Parsons (tr.) *On platonism in mathematics.* [Benacerraft and Putnam 1964]. pp. 274-286, 1964.

[Bishop, 1967] E. Bishop. *Foundations of Constructive Analysis.* New York, NY: McGraw-Hill. xiii+370, 1967.

[Bishop, 1970] E. Bishop. *Mathematics as a numerical language*. A. Kino et al. (eds.) Intuitionism and Proof Theory. Proceedings of the Summer Conference at Buffalo, N.Y. 1968. Amsterdam, NL: North-Holland Publishing Company. pp. 53-71, 1970.

[Bishop, 1985] E. Bishop. *Schizophrenia in contemporary mathematics*. M. Rosenblatt (ed.) Errett Bishop: Reflections on Him and His Research. Contemporary Mathematics. Volume 39. Providence, RI: American Mathematical Society. pp. 1-32, 1985.

[Bishop and Bridges, 1985] E. Bishop and D. S. Bridges. *Constructive Analysis*. Grundlehren der mathematischen Wissenschaften. Volume 279. Berlin, DE: Springer-Verlag. xii+477, 1985.

[Bridges and Richman, 1987] D. S. Bridges and F. Richman. *Varieties of constructive mathematics*. London Mathematical Society Lecture Notes Series. Volume 97. Cambridge, UK: Cambridge University Press. x+149, 1987.

[Brouwer, 1905] L. E. J. Brouwer. *Leven, Kunst en Mystiek*. Delft, NL: Waltman. 99 pp., 1905. Reprinted W.P. van Stigt (ed.) *Life, art and mysticism*. The Notre Dame Journal of Formal Logic. Volume 37. 1996. pp. 381-429, 1996.

[Brouwer, 1907] L. E. J. Brouwer. *Over de grondslagen der wiskunde*. [*On the Foundations of Mathematics.*] Amsterdam, NL: University of Amsterdam dissertation. 183 pp, 1907.

[Brouwer, 1908] L. E. J. Brouwer. *De onbetrouwbaarheid der logische principes*. [*The unreliability of the logical principles.*] Tijdschrift voor Wijsbegeerte. Volume 2. pp. 152-158, 1908.

[Brouwer, 1909] L. E. J. Brouwer. *Het wezen der meetkunde*. [*The Nature of Geometry.*] Inaugural Lecture as *Privaat Docent*. Amsterdam, NL. 23 pp, 1909.

[Brouwer, 1913] L. E. J. Brouwer. *Intuitionism and formalism*. A. Dresden (tr.) Bulletin of the American Mathematical Society. Volume 20. pp. 81-96, 1913. Reprinted [Bernacerraf and Putnam 1964]. pp. 66-77, 1913.

[Brouwer, 1918] L. E. J. Brouwer. *Begründung der Mengenlehre unabhängig vom logischen Satz vom ausgeschlossenen Dritten. Erster Teil, Allgemeine Mengenlehre.* [*Foundation of set theory independent of the logical law of the excluded middle. Part One, general set theory.*] Verhandelingen der Koninklijke Akademie van wetenschappen te Amsterdam. First Section. Volume 12. Number 5. pp. 1-43, 1918.

[Brouwer, 1921] L. E. J. Brouwer. *Besitzt jede reelle Zahl eine Dezimalbruchentwichlung?* [*Does every real number have a decimal expansion?*] Mathematische Annalen. Volume 83. pp. 201-210, 1921.

[Brouwer, 1924] L. E. J. Brouwer. *Beweis dass jede volle Funktion gleichmässigstetig ist.* [*Proof that every total function is uniformly continuous.*] Koninklijke Nederlandse Akademie van Wetenschappen. Proceedings of the Section of Sciences. Volume 27. pp. 189-193, 1924.

[Brouwer, 1948] L. E. J. Brouwer. *Essentieel negatieve eigenschappen.* [*Essentially negative properties.*] Indagationes Mathematicae. Volume 10. pp. 322-323, 1948.

[Brouwer, 1954] L. E. J. Brouwer. *Points and spaces*. Canadian Journal of Mathematics. Volume 6. pp. 1-17, 1954.

[Browder, 1976] F. E. Browder, ed. *Mathematical Developments arising from Hilbert Problems*. Proceedings of Symposia in Pure Mathematics. Volume XXVIII. Providence, RI: American Mathematical Society. xii+628, 1976.

[Čeitin, 1959] G. S. Čeitin. *Algoritmičéskié operatory v konstruktivnyk polnyk séparabélnyk métričéskyk prostranstvak.* [*Algorithmic operators in constructive complete separable metric spaces.*] Doklady Akadérnii Nauk SSR. Volume 128. pp. 49-52, 1959.

[Chihara, 1973] C. Chihara. *Ontology and the Vicious-Circle Principle*. Ithaca, NY: Cornell University Press. xv+257, 1973.

[Constable, 1986] R. Constable et al. *Implementing Mathematics with the NuPRL Proof Development System*. Englewood Cliffs, NJ: Prentice-Hall. x+299, 1986.

[Diaconescu, 1975] R. Diaconescu. *Axiom of choice and complementation*. Proceedings of the American Mathematical Society. Volume 51. pp. 175-178, 1975.

[Du Bois-Reymond, 1886] E. Du Bois-Reymond. *Über die Grenzen des Naturerkennens.* [*On the limits of the knowledge of nature.*] Reden von Emil Du-Bois Reymond. Erste Folge. Leipzig: Verlag von Veit und Comp. viii+550, 1886.

[Du Bois-Reymond, 1877] P. Du Bois-Reymond. *Über die Paradoxon des Infinitärcalcüls.* [*On the paradoxes of the infinitary calculus.*] Mathematische Annalen. Volume 10. pp. 149-167, 1877.

[Du Bois-Reymond, 1966] P. Du Bois-Reymond. *Über die Grundlagen der Erkenntnis in den exakten Wissenschaften.* [*On the Foundations of Knowledge in the Exact Sciences.*] Sonderausgabe. Darmstadt, DE: Wissenschaftliche Buchgesellschaft. vi+130, 1966.

[Du Bois-Reymond, 1882] P. Du Bois-Reymond. *Die allgemeine Funktionentheorie.* [*General Function Theory.*] Tübingen, DE: H. Laupp. xiv + 292, 1882.

[Dummett, 1977] M. Dummett. *Elements of Intuitionism.* Oxford, UK: Clarendon Press. First Edition. xii+467, 1977. Second Edition, 2000.

[Feferman, 1964] S. Feferman. *Systems of predicative analysis.* The Journal of Symbolic Logic. Volume 29. pp. 1-30, 1964.

[Feferman, 2005] S. Feferman. *Predicativity.* S. Shapiro (ed.) The Oxford Handbook of Philosophy of Mathematics and Logic. Oxford, UK: Oxford University Press. pp. 590-624, 2005.

[Finsler, 1926] P. Finsler. *Formale Beweise und Entscheidbarkeit.* [*Formal proofs and decidability.*] Mathematische Zeitschrift. Volume 25. pp. 676-682, 1926. Reprinted S. Bauer-Mengelberg (tr.) [van Heijenoort, 1967]. pp. 438-445, 1967.

[Friedman, 1977] H. Friedman. *Set-theoretic foundations for constructive analysis.* Annals of Mathematics. Volume 105. pp. 1-28, 1977.

[Glivenko, 1928] V. Glivenko. *Sur la logique de M. Brouwer.* Academie Royale de Belgique. Bulletin de la Classe des Sciences. Volume 5. Number 14. pp. 225-228, 1928.

[Gödel, 1930] K. Gödel. *Über die Vollständingkeit des Logikkalküls.* [*On the completeness of the calculus of logic.*] [Gödel, 1986]. pp.124-125, 1930.

[Gödel, 1931] K. Gödel. *Diskussion zur Grundlegung der Mathematik.* [*Discussion on the foundation of mathematics.*] [Gödel 1986]. pp. 200-203, 1931.

[Gödel, 1986] K. Gödel. *Collected Works. Volume I. Publications 1929-1936.* S. Feferman et al. (eds.) Oxford, UK: Oxford University Press. xvi+474, 1986.

[Gödel, 1995] K. Gödel. *Some basic theorems on the foundations of mathematics and their implications.* Collected Works. Volume III. Unpublished Essays and Lectures. S. Feferman et. al. (eds.) Oxford, UK: Oxford University Press. pp. 304-323, 1995.

[Goodstein, 1957] R. Goodstein. *Recursive Number Theory.* Studies in Logic and the Foundations of Mathematics. Amsterdam, NL: North-Holland Publishing Company. XII+190, 1957.

[Goodstein, 1961] R. Goodstein. *Recursive Analysis.* Studies in Logic and the Foundations of Mathematics. Amsterdam, NL: North-Holland Publishing Company. viii+138, 1961.

[Heyting, 1930a] A. Heyting. *Die formalen Regeln der intuitionistischen Logik.* [*The formal rules of intuitionistic logic.*] Sitzungsberichte der Preussischen Akademie der Wissenschaften, Physikalisch-mathematicsche Klasse. pp. 42-56, 1930.

[Heyting, 1930b] A. Heyting. *Die formalen Regeln der intuitionistischen Mathematik I.*[*The formal rules of intuitionistic mathematics I.*] Sitzungsberichte der Preussischen Akademie der Wissenschaften, Physikalisch-mathematicsche Klasse. pp. 57-71, 1930.

[Heyting, 1930c] A. Heyting. *Die formalen Regeln der intuitionistischen Mathematik II.* [*The formal rules of intuitionistic mathematics II.*] Sitzungsberichte der Preussischen Akademie der Wissenschaften, Physikalisch-mathematicsche Klasse. pp. 158-169, 1930.

[Heyting, 1931] A. Heyting. *Die intuitionistische Grundlegung der Mathematik.* [*Intuitionistic foundations of mathematics.*] Erkenntnis. Volume 2. pp. 106-115, 1931. Reprinted *The intuitionist foundation of mathematics.* [Benacerraf and Putnam, 1964]. pp. 42-49, 1964.

[Heyting, 1934] A. Heyting. *Mathematische Grundlagenforschung. Intuitionismus. Beweistheorie.* [*Research in the Foundations of Mathematics. Intuitionism. Proof Theory.*] Berlin, DE: Springer Verlag. iv+73, 1934.

[Hilbert, 1926] Hilbert, D. [1926] *Über das Unendliche.* [*On the infinite.*] Mathematische Annalen. Volume 95. pp. 161-190, 1926 Reprinted E. Putnam and G. Massey (trs.) *On the infinite.* [Benacerraf and Putnam, 1964]. pp. 134-151, 1964.

[Hilbert, 1935] D. Hilbert. *Gesammelte Abhandlungen.* [*Complete Works.*] Dritter Band. Berlin, DE: Verlag von Julius Springer. vii+435 pp, 1935.

[Kielkopf, 1970] C. Kielkopf. *Strict Finitism: An Examination of Ludwig Wittgenstein's Remarks on the Foundations of Mathematics.* Studies in Philosophy. The Hague, NL: Mouton, 192 pp, 1970.

[Kleene, 1952] S. Kleene. *Recursive functions and intuitionistic mathematics.* L. Graves et. al. (eds.) Proceedings of the International Congress of Mathematicians. August 1950. Cambridge, Mass. Providence, RI: American Mathematical Society. pp. 679-685, 1952.

[Kolmogorov, 1925] A. Kolmogorov. *On the principle of the excluded middle.* [van Heijenoort, 1967]. pp. 414-437, 1925.

[Kolmogorov, 1932] A. Kolmogorov. *Zur Deutung der intuitionistischen Logik.* [*Toward an interpretation of intuitionistic logic.*] Mathematische Zeitschrift. Volume 35. pp. 58-65, 1932.

[Kreisel, 1960] G. Kreisel. *La predicativité*. Bulletin de la Société Mathématique de France 88. pp. 371-391, 1960.

[Kreisel, 1962a] G. Kreisel. *On weak completeness of intuitionistic predicate logic*. The Journal of Symbolic Logic. Volume 27. pp. 139-158, 1962.

[Kreisel, 1962b] G. Kreisel. *Foundations of intuitionistic logic*. E. Nagel et. al. (eds.) Logic, Methodology and Philosophy of Science I. Stanford, CA: Stanford University Press. pp. 198-210, 1962.

[Kreisel et al., 1959] G. Kreisel, D. Lacombe and J. Shoenfield. *Partial recursive functions and effective operations*. A. Heyting (ed.) Constructivity in Mathematics. Proceedings of the Colloquium held at Amsterdam 1957. Studies in Logic and the Foundations of Mathematics. Amsterdam, NL: North-Holland Publishing Company. pp. 290-297, 1959.

[Kronecker, 1887] L. Kronecker. *Über den Zahlbegriff*. Philosophische Aufsätze, Eduard Zeller zu seinem fünfzigjährigen Doctorjubiläum gewidmet. Leipzig, DE: Fues. pp. 261-274, 1887. Reprinted W. Ewald (tr.) *On the concept of number* W. Ewald (ed.) From Kant to Hilbert: A Source Book in the Foundations of Mathematics. Volume II. 1996. Oxford, UK: Clarendon Press. pp. 947-955, 1996.

[Lorenzen, 1955] P. Lorenzen. *Einführung in die operative Logik und Mathematik.* [*Introduction to Operative Logic and Mathematics.*] Berlin, DE: Springer-Verlag. VII+298, 1955.

[Markov, 1958] A. A. Markov. *On constructive functions*. American Mathematical Society Translations (2). Volume 29. pp. 163-196, 1963. Translated from Trudy Matematicheskogo Instituta imeni VA Steklova. Volume 52. pp. 315-348, 1958.

[Markov, 1971] A. A. Markov. *On constructive mathematics*. American Mathematical Society Translations. Series 2. Volume 98. pp. 1-9, 1971. Translated from Trudy Matematicheskogo Instituta imeni VA Steklova. Volume 67. pp. 8-14, 1962.

[Martin-Löf, 1984] P. Martin-Löf. *Intuitionistic Type Theory*. Naples, IT: Bibliopolis. iv+91, 1984.

[McCarty, 1986] C. McCarty. *Subcountability under realizability*. Notre Dame Journal of Formal Logic. Volume 27. Number 2. April 1986. pp. 210-220, 1986.

[McCarty, 1988] C. McCarty. *Constructive validity is nonarithmetic*. The Journal of Symbolic Logic. Volume 33. Number 4. December 1988. pp. 1036-1041, 1988.

[McCarty, 1996] C. McCarty. *Undecidability and intuitionistic incompleteness*. The Journal of Philosophical Logic. Volume 25. pp. 559-565, 1996.

[Nelson, 1986] E. Nelson. *Predicative Arithmetic*. Mathematical Notes. Volume 32. Princeton, NJ: Princeton University Press. viii+190, 1986.

[Poincaré, 1963] H. Poincaré. *The Logic of Infinity*. J. Bolduc (tr.) Mathematics and Science: Last Essays. New York, NY: Dover Publications, Inc. pp.45-64, 1963.

[Pringsheim, 1898-1904] A. Pringsheim. *Irrationalzahlen und Konvergenz unendlicher Prozesse.* [*Irrational numbers and the convergence of infinite processes.*] W. F. Meyer (ed.) Encyklopädie der mathematischen Wissenschaften. Erster Band in zwei Teilen. Arithmetik und Algebra. [Encyclopedia of the Mathematical Sciences. First Volume in Two Parts. Arithmetic and Analysis.] Leipzig, DE: Druck und Verlag von B.G. Teubner. pp. 47 - 146, 1898-1904.

[Russell, 1906] 68] B. Russell. *Les paradoxes de la logique*. Revue de Métaphysique et la Morale. Volume 14. September 1906. pp.627-650, 1906. Reprinted *On 'insolubilia' and their solution by symbolic logic*. D. Lackey (ed.) Bertrand Russell. Essays in Analysis. New York, NY: George Braziller. pp. 190-214, 1973.

[Russell, 1908] B. Russell. *Mathematical logic as based on a theory of types*. American Journal of Mathematics. Volume 30. pp.222-262, 1908. Reprinted [van Heijenoort, 1967]. pp. 150-182, 1967.

[Schütte, 1965] K. Schütte. *Predicative well orderings*. J. Crossley and M. Dummett (eds.) Formal Systems and Recursive Functions. Amsterdam, NL: North-Holland Publishing Company. pp. 279-302, 1965

[Shanin, 1958] N. A. Shanin. *On the constructive interpretation of mathematical judgments*. American Mathematical Society Translations. Series 2. Volume 23. pp.108-189, 1958. Translated from *O konstruktiviom ponimanii matematicheskikh suzhdenij*. Trudy Ordena Lenina Matematicheskogo Instituta imeni V.A. Steklova. Akademiya Nauk SSSR. Volume 52. pp. 226-311, 1958.

[Shanin, 1968] N. A. Shanin. *Constructive real numbers and constructive function spaces*. E. Mendelson (tr.) Translations of Mathematical Monographs. Volume 21. Providence, RI: American Mathematical Society. iv+325, 1968.

[Skolem, 1923] T. Skolem. *Begründung der elementaren Arithmetik durch die rekurrierende Denkweise ohne Anwendung scheinbarer Veränderlichen mit unendlichem Ausdehnungsbereich.* Skrifter utgit av Videnskapsselskapet I Kristiania, I. Matematisk-naturvidenskabelig Klasse 6. pp. 1-38, 1923. Reprinted S. Bauer-Mengelberg (tr.) *The foundations of elementary arithmetic established by means of the recursive mode of thought, without the use of apparent variables ranging over infinite domains.* [van Heijenoort, 1967]. pp. 302-333, 1967.

[Specker, 1949] E. Specker. *Nicht konstruktiv beweisbare Sätze der Analysis. [Nonconstructively provable sentences of analysis.]* The Journal of Symbolic Logic. Volume 14. pp. 145-158, 1949.

[Tait, 1981] W. Tait. *Finitism.* The Journal of Symbolic Logic. Volume 78. Number 9. pp. 524-546, 1981. Reprinted [Tait, 2005b]. pp. 21-41, 2005.

[Tait, 2002] W. Tait. *Remarks on finitism.* W. Sieg et. al. (eds.) Reflections on the Foundations of Mathematics. Essays in honor of Solomon Feferman. Assocation for Symbolic Logic. Lecture Notes in Logic. Natick, MA: A.K. Peters, Ltd. pp. 410-419, 2002. Reprinted [Tait, 2005b]. pp. 43-53, 2005.

[Tait, 2005a] W. Tait. *Appendix to Chapters 1 and 2.* [Tait, 2005b]. pp. 54-60, 2005.

[Tait, 2005b] W. Tait. *The Provenance of Pure Reason. Essays in the Philosophy of Mathematics and Its History.* Oxford, UK: Oxford University Press. 2005. viii+332, 2005.

[Troelstra, 1981] A. Troelstra. *Arend Heyting and his contribution to intuitionism.* Nieuw Archief voor Wiskunde. Volume XXIX. pp. 1-23, 1981.

[Troelstra and van Dalen, 1988] A. S. Troelstra and D. van Dalen. *Constructivism in Mathematics. An Introduction.* Volume I. Studies in Logic and the Foundations of Mathematics. Volume 121. xx+342+XIV. Volume II. Studies in Logic and the Foundations of Mathematics. Volume 123. xvii+345-879+LII. Amsterdam, NL: North-Holland, 1988.

[Turing, 1936-37] A. M. Turing. *On computable numbers with an application to the Entscheidungsproblem.* Proceedings of the London Mathematical Society. Series 2. Volume 42. pp. 230-265, 1936-37.

[van Atten, 2004] M. van Atten. *On Brouwer.* London, UK: Thomson/Wadsworth. 95 pp, 2004.

[van Dalen, 1999] D. van Dalen. *Mystic, Geometer, and Intuitionist. The Life of L. E. J. Brouwer. Volume I: The Dawning Revolution.* Oxford, UK: The Clarendon Press. xv+440, 1999.

[van Dalen, 2005] D. van Dalen. *Mystic, Geometer, and Intuitionist. The Life of L. E. J. Brouwer. Volume II: Hope and Disillusion.* Oxford, UK: The Clarendon Press. x+946, 2005.

[van Heijenoort, 1967] J. van Heijenoort. (ed.) *From Frege to Gödel: A Source Book in Mathematical Logic, 1879-1931.* Cambridge, MA: Harvard University Press. xi+665, 1967.

[Wang, 1954] H. Wang. *The formalization of mathematics.* The Journal of Symbolic Logic. Volume 19. pp. 241-266, 1954.

[Wang, 1964] H. Wang. *A Survey of Mathematical Logic.* Amsterdam, NL: North-Holland Publishing Company. x+651, 1964.

[Weber, 1893] H. Weber. *Leopold Kronecker.* Jahresberichte der Deutschen Mathematiker Vereinigung. Volume II. pp. 5-31, 1893.

[Weyl, 1918] H. Weyl. *Das Kontinuum. Kritische Untersuchungen über die Grundlagen der Analysis.* Leipzig, DE: Von Veit. vi+83, 1918. Reprinted S. Pollard and T. Bole (trs.) *The Continuum. A Critical Examination of the Foundation of Analysis.* New York, NY: Dover Publications, Inc. xxvi+130, 1994.

[Whitehead and Russell, 1910–13] A. N. Whitehead and B. Russell. *Principia Mathematica.* Volumes I, II and III. Cambridge, UK: Cambridge University Press, 1910-13.

[Wittgenstein, 1956] L. Wittgenstein. *Remarks on the Foundations of Mathematics.* G. von Wright et. al. (eds.) G. Anscombe (tr.) New York, NY: The Macmillan Company. xix+204, 1956.

[Yessenin-Volpin, 1970] A. S. Yessenin-Volpin. *The ultra-intuitionistic criticism and the antitraditional program for foundations of mathematics.* A. Kino et al. (eds.) Intuitionism and Proof Theory. Proceedings of the Summer Conference at Buffalo. New York. 1968. Amsterdam, NL: North-Holland Publishing Company. pp. 3-45, 1970.

[Zaslavskiĭ, 1955] I. D. Zaslavskiĭ. *The refutation of some theorems of classical analysis in constructive analysis.* [in Russian] Uspehi Mat. Nauk. Volume 10. pp. 209-210, 1955.

FICTIONALISM

Daniel Bonevac

Fictionalism, in the philosophy of mathematics, is the view that mathematical discourse is in some important respect fictional: mathematical objects such as π or $\emptyset$ have the same metaphysical status as Sherlock Holmes or Macbeth. Admittedly, this definition is vague. But it hard to do better without ruling out the positions of some philosophers who consider themselves fictionalists. Hartry Field [1980, 2], who is largely responsible for contemporary interest in fictionalism as a strategy in the philosophy of mathematics, defines it as the view that there is no reason to regard the parts of mathematics that involve reference to or quantification over abstract entities such as numbers, sets, and functions as true. I do not adopt that definition here, for some fictionalists (e.g., Stephen Yablo [2001; 2002; 2005]) think of fictional discourse as in some important sense true. Zoltan Szabo [2003] treats fictionalism about Fs as the belief that 'There are Fs' is literally false but *fictionally* true. This rules out the possibility that fictional statements lack truth values and introduces a concept of fictional truth that only some fictionalists endorse.

A number of writers, influenced by Kendall Walton [1978; 1990; 2005], treat fictionality as a matter of attitude: a discourse is fictional if its participants approach it with a cognitive attitude of pretense or make-believe. As John Burgess, Gideon Rosen [1997], and Jason Stanley [2001] point out, however, there is little evidence that participants in mathematical discourse approach it with such an attitude. One might expect, moreover, that philosophers who diverge in their accounts of mathematics would also diverge in their attitudes toward mathematical discourse: fictionalists might approach it with an attitude of make-believe, while realists approach it with an attitude of discovery. To avoid a choice between relativism and trivial falsity, therefore, fictionalists about mathematics had better not take attitude as definitive of fictionality.

Mark Kalderon [2005] defines fictionalism at a higher level of abstraction: "The distinctive commitment of fictionalism is that acceptance in a given domain of inquiry need not be truth-normed, and that the acceptance of a sentence from an associated region of discourse need not involve belief in its content" (2). This seems to make fictionalism a close relative of noncognitivism. But mathematics is unquestionably a realm of inquiry that admits rational discourse and determination; indeed, it seems a paradigm of rational inquiry. It does not express emotion; it does not issue commands. Taking this into account, the fictionalist, we might say, sees some kinds of discourse that count as rational inquiry as aiming at something other than truth and as accepted in a sense weaker than belief. In what follows, I will usually speak of *success* rather than acceptance to make it clear that the relevant notion of acceptance is one of being accepted by the participants in a discourse as a successful contribution to that discourse. Fictional sentences do not aim at truth (whether or not they in some sense achieve it) and succeed in playing their roles in discourse even if they are not believed — or, at any rate, could play their roles even if they were not believed.

1 KINDS OF FICTIONALISM

To understand what fictionalists mean to assert about mathematics, consider a simple instance of fictional discourse, from Nathaniel Hawthorne's *Twice Told Tales*:

> I built a cottage for Susan and myself and made a gateway in the form of a Gothic Arch, by setting up a whale's jawbones.

Hawthorne's utterance, conceived as fictional, succeeds, even if there are no cottage, gateway, whale, and jawbones related in the way described. Conceived as nonfictional, in contrast, the nonexistence of any of those entities would prevent the utterance from succeeding. In ordinary, nonfictional discourse, pragmatic success requires truth, and truth, for existential sentences, at least, requires that objects of certain kinds exist. In fiction, pragmatic success and existence come apart. Existential utterances can succeed even if no objects stand in the relations they describe. The pragmatic success of a fictional discourse, that is, is independent of the existence of objects to which the discourse is ostensibly committed (or, perhaps, would be ostensibly committed if it were asserted as nonfiction).

The fictionalist about mathematics, then, maintains that the pragmatic success of mathematical discourse is independent of the existence of mathematical objects. Mathematics can do whatever it does successfully even if there are no such things as numbers, sets, functions, and spaces.

Not everyone who holds this view, however, counts as a fictionalist. Fictionalism is a variety of mathematical *exceptionalism*, the view that the success of mathematical statements is exceptional, depending on factors differing from those upon which the success of ordinary assertions depends. There are many other versions of exceptionalism: (a) Reductionists, for example, maintain that mathematical statements are exceptional in that they are about something other than what they appear to be about; they translate into statements with different ontological commitments. Those statements they take to be more metaphysically revealing than the originals. (See [Link, 2000].) (b) Supervenience theorists dissent from the translation thesis, but contend nevertheless that the success of mathematical statements depends on facts about something other than what they appear to be about, namely, nonmathematical entities. (c) Logicism (e.g., that of Russell [1918] or Hempel [1945]) maintains that the success of mathematical statements depends solely on logic — classically, because mathematical truths translate into truths of logic. (d) Putnam's [1967] deductivism (or if-thenism) maintains that the success of mathematical statements is determined not according to the truth conditions of the statements themselves but instead according to those of associated conditionals. (e) Hellman's [1989] modal structuralism maintains that mathematics, properly understood, makes no existence claims, but speaks of all possible realizations of structures of various kinds. Resnik [1997] advances another version of structuralism. (f) Chihara [1990] replaces traditional existential assertions with constructibility theorems. (g) Schiffer's [2003] account of pleonastic propositions and entities that result from "something-from-nothing transformations" (see also Hofweber [2005a; 2005b; 2006; forthcoming]) might offer a foundation for an account of mathematical statements without the need to invoke extra-mental mathematical entities. (Note, however, that fictional entities are just one kind of pleonastic entity.) These generally do not count as fictionalist accounts of mathematics in the contemporary sense — though they are all in a sense anti-realist, though Schiffer's might count as a generalization of fictionalism, perhaps, and though, as we shall see, several (including, especially, the reductionist account) might have been considered fictionalist throughout considerable stretches of philosophical history.

What distinguishes fictionalism from these kinds of exceptionalism, all of which are kinds of anti-realism? (1) It does not necessarily endorse the thesis that mathematical statements, properly understood, are true. Some fictionalists hold that fiction is in a sense true and that mathematics is true in the same or at least in an analogous sense. But fictionalism per se carries no such entailment. And that by virtue of which mathematical statements are true is not the same as that by virtue of which ordinary statements about midsize physical objects, for example, are true. It seems fair to summarize this, as Kalderon does, by saying that mathematics is not truth-normed; it aims at something other than truth. Reductionists, supervenience theorists, logicists, deductivists, etc., in contrast, all take mathematical statements (properly understood or translated) as true and as aiming at truth. (2) The success of mathematics is independent of belief. The reductionist, supervenience theorist, deductivist, modal structuralist, constructibility theorist, etc., all believe that mathematical statements, as true, are worthy objects of belief, even if their surface forms are misleading. The fictionalist, however, sees belief as inessential to the success of mathematical statements. (3) Fictional entities, if it is proper to speak of them at all, are products of free creative activity. A fictionalist about mathematics maintains that the same is true of its entities. More neutrally, we might say that according to the fictionalist mathematical statements, like those in fiction, are creative products. Among the facts by which their success is determined are facts about human creative activity. (4) Fiction is nevertheless in some sense descriptive. It describes objects and events. It differs from ordinary descriptive discourse only in that the nonexistence of its objects and the nonoccurrence of its events does not detract from its success.

If fictionalists agree about that much, they disagree about much else. I shall classify fictionalists here according to their views on three issues: truth, interpretation, and elimination.

1.1 Truth

To say that mathematics is not truth-normed is not to say that it is not truth-evaluable. Mathematics might aim at something other than truth but nevertheless be evaluable as true or false. Indeed, that seems to be just the status of fiction. A work of fiction aims at something other than truth. But we can still ask whether the sentences it comprises are true or false. In nonfiction, success requires truth, which in turn requires the existence of objects. If success in fiction is to be independent of the existence of objects, however, either success must be independent of truth, or truth must be independent of the existence of objects.

Fictionalist strategies, then, divide naturally into two kinds, depending on their attitude toward truth. According to one sort, Hawthorne's sentence succeeds despite its literal falsity. Although fictionalists and their critics sometimes speak this way — see Szabo [2003] — it seems not quite right; *literal* contrasts most naturally with *figurative*, but the literal/figurative distinction does not line up neatly with the nonfiction/fiction distinction. It would be more accurate to say that fiction is not *realistically* true. Yablo [2001; 2005] advances a version of fictionalism he calls *figuralism*, which maintains that mathematical statements are analogous specifically to fictional, figurative language. On his view, the kind of truth appropriate to mathematics does contrast naturally with literal truth. But it seems contentious to build such a view into our terminology from the beginning. According to the other, Hawthorne's sentence is in some sense true, despite the nonexistence of the objects it describes, precisely because it occurs in a fictional context.

Correspondingly, some fictionalists, such as Field [1980], contend that mathematics succeeds without being true. Others contend that mathematics can be true even if the objects it seems to describe do not exist. Among the latter are some whose views come close to Putnam's; they contend that mathematics is true in a purely deductive, "if-then" sense, mathematical truth being simply truth in a story (the standard set-theoretic hierarchy story, for example), much as fictional truth is truth in a story (the *Twice Told Tales*, for example). Whether such views remain fictionalist depends on the details. Others hold that mathematics, and fiction in general, are true in a more full-blooded sense.

1.2 Interpretation

Burgess and Rosen [1997] distinguish *hermeneutic* from *revolutionary* nominalists. Various writers draw the same distinction among fictionalists. Hermeneutic fictionalists about mathematics maintain that we *do* interpret mathematics as fictional; revolutionary fictionalists, that we *should*. As Yablo puts it: "Revolutionary nominalists want us to stop talking about so-and-so's; hermeneutic nominalists maintain that we never started" [2001, 85].

Both versions of fictionalism face serious problems. Most mathematicians and scientists, not to mention nonspecialists, do not appear to interpret mathematics fictionally. The various schemes for reinterpreting mathematics that fictionalists or nominalists have proposed, moreover, are complex and difficult to use; it is hard to see what sort of mathematical, scientific, or practical advantage they could possess that would justify that assertion that we ought to reinterpret mathematics in accordance with them. Indeed, they seem to have severe *dis*advantages by those measures. The revolutionary fictionalist (e.g., [Leng, 2005]) would presumably claim philosophical (specifically, epistemological) advantages. But, even if those claims can be sustained, it is hard to see why philosophical advantages should outweigh mathematical, scientific, or practical disadvantages.

Fortunately for fictionalism, Burgess and Rosen's distinction is not exhaustive. Fictionalists can argue, not that we *do* or *should* interpret mathematics fictionally, but that we *can*. Imagine a fictionalist about ancient Greek gods and goddesses, for example. Fictionalism about Zeus, Hera, etc., seems an entirely reasonable position. Such a person might contend that ancient Greeks adopted an attitude of make-believe toward the gods, but might not; he or she might contend that they should have adopted such an attitude, but, again, might not. (Belief in the gods might have been crucial to social stability, for example.) The fictionalist in this instance holds that we can best understand ancient Greek religion by interpreting it in fictional terms.

Alternatively, think of Russell's [1918] view of ontology as the study of what we must count as belonging to the basic furniture of the universe. A fictionalist interpretation of a discourse, such as Russell's no-class theory, on his view shows that we do *not* have the count the objects the discourse ostensibly discusses as among the basic furniture of the universe. Whatever attitude about the objects to which such a discourse is ostensibly committed we happen to have, and whatever attitude about them might be best for us practically, scientifically, psychologically, or linguistically, the theoretical viability of a fictionalist attitude is enough to show that we are capable of avoiding ontological commitment to them. From Russell's perspective, fictionalism pays ontological dividends even if no one does or should (except perhaps in a technical ontological sense) adopt a fictionalist attitude toward the pertinent discourse.

Similarly, a fictionalist about mathematics can hold that we can best understand the ontological commitments forced upon us by mathematics by interpreting it in fictional terms. The view implies nothing at all about how mathematicians themselves do or ought to interpret their subject. Mathematics can accomplish its purposes, according to the fictionalist, even if there are no mathematical objects. To show this, the fictionalist needs to (a) specify the purposes of mathematics that must be accomplished, and (b) demonstrate the possibility of accomplishing them with a theory that makes no commitment to mathematical entities. Field, for example, takes the application of mathematics in physical science as the purpose of mathematics that the fictionalist must explain, and attempts to show the physical application of mathematical theories can be understood without appeal to distinctively mathematical objects.

In addition to hermeneutic and revolutionary fictionalism, then, we should distinguish *deflationary* fictionalism, which maintains that there is no need for a substantive metaphysics or epistemology for mathematics. There is no deep mystery, the deflationary fictionalist insists, about how we know that Dr. Watson was Holmes's associate. Neither is there a deep mystery about how we know that $\pi > 3$. We are not committed to the existence of Dr. Watson, Holmes, π, or 3. Nor must we postulate any strange faculty of intuiting objects with which we stand in no causal relation. Mathematics serves its function without any such assumption.

Mark Balaguer expresses the spirit of deflationary fictionalism concisely:

> ... we use mathematical-object talk in empirical science to help us accurately depict the nature of the physical world; but we could do this even if there were no such things as mathematical objects; indeed, the question of whether there exist any mathematical objects is wholly irrelevant to the question of whether we could use mathematical-object talk in this way; therefore, the fact that we do use mathematical-object talk in this way does not provide any reason whatsoever to think that this talk is true, or genuinely referential. [Balaguer, 1998, 141]

We do not need to show that people actually adopt an attitude of make-believe in mathematics. All we need is to show is that they could use mathematics just as successfully if they did.

Deflationary fictionalism bears some resemblance to the indifferentism propounded by Eklund [in press]. On that view, the ontological commitments of our statements are nonserious features of them, features that are beside the point of the statements. In general, Eklund argues, "we do not commit ourselves either to its literal truth or to its truth in any fiction; we are, simply, non-committed." Think of a picture of content similar to that of Stalnaker [1978], in which statements in a context function to restrict the class of possible worlds that constitutes that context. Speakers in nonphilosophical contexts may not be interested in ruling out possibilities that differ only metaphysically. Their statements have a content, therefore, that is properly understood as metaphysically neutral. The similarity between this view and deflationary fictionalism emerges in this passage: "It can be that I do not in fact make, say, mathematical statements in a fictional spirit, but when I come to realize this is a possibility I can also realize that doing so would all along have satisfied all of my conversational aims" [Eklund, in press]. Deflationary fictionalism needs to show only that a fictional interpretation of a discourse is possible, and establishes thereby that the discourse is ontologically neutral.

1.3 Elimination

What does it take to show that mathematics can fulfill its purposes even if there are no mathematical objects? The oldest tradition historically falling under the heading of fictionalism maintains that the fictionalist must reduce mathematical to nonmathematical objects. William of Ockham, for example, contends that universals are *ficta*; they do not exist in any real sense. Everything that really exists is particular. He seems willing to discard some talk of universals as incorrectly assuming their real existence. Most discourse invoking universals, however, he seeks to reinterpret. *Socrates has wisdom*, for example, is true by virtue of the fact that Socrates is wise. It *seems* to be committed to the existence of a universal, wisdom, but in fact requires the existence of nothing more than Socrates. Nominalists (sometimes calling themselves fictionalists) and others have employed reductive strategies in a wide variety of contexts. Consider, for example, David Hume's account of necessary connection, Bertrand Russell's "no-class" theory, and Rudolf Carnap's [1928] phenomenological construction of the world.

Opponents of nominalist reinterpretation such as Burgess, Rosen, and Hofweber argue as follows: *Socrates has wisdom* can be interpreted as *Socrates is wise* only if they are equivalent. But, by the nominalist's own lights, the former entails the existence of universals, while the latter does not. So, they cannot be equivalent, and the nominalistic interpretation fails. Nominalists, in response, tend to deny that interpretation requires equivalence *tout court*; it requires only equivalence relative to the purposes of the kind of discourse in question.

Broadly speaking, it seems fair to say, nominalism takes its inspiration from something like empiricism. The purpose of discourse in the broadest sense is to account for our experience. Call sentences *A* and *B experientially equivalent* if and only if they have exactly the same implications for experience. (We might suppose that there is an experiential language E adequate to and restricted to describing experience such that A and B are experientially equivalent just in case, for any sentence C of E, $A \models C \Leftrightarrow B \models C$.) The claim is then that *Socrates is wise* and *Socrates has wisdom* are experientially equivalent; they have exactly the same implications for our experience. They differ in their ostensible commitments — that is to say, their prima facie commitments, what they seem to be committed to independent of any considerations about the possibility of translation, paraphrase, or elimination, etc. So, they are not equivalent all things considered. But they are empirically equivalent. The upshot, according to the nominalist: we would suffer no decline in our ability to describe our experience if we were to assert *Socrates is wise* in place of *Socrates has wisdom*. We have no reason, therefore, to take on the additional ontological commitments of the latter.

We might think in model-theoretic terms, as follows. M and N are elementarily equivalent with respect to E if and only if they agree on every sentence C of E: $M \models C \Leftrightarrow N \models C$. Suppose that M is a platonistic model including abstracta as part of its domain and that N is a nominalistic model with a domain consisting solely of concreta. If M and N are elementarily equivalent with respect to E, then they satisfy exactly the same experiential sentences. So, we may safely replace M with N, avoiding M's worrisome platonistic commitments, without adversely affecting our ability to account for our experience.

Experiential equivalence in this sense is weaker than reduction. Suppose that for every platonistic model M of our best theory of the world there is an experientially equivalent nominalistic model N of that theory. It does not follow that the theory reduces to one having only nominalistically acceptable entities as ostensible commitments, unless there is a function from platonistic to experientially equivalent nominalistic models meeting stringent criteria (see e.g., [Enderton, 1972]). So, there is plenty of logical space in which fictionalists may adopt a strategy based on experiential equivalence without committing themselves to reductionism. Mounting an *argument* that every platonistic model has an associated experientially equivalent nominalistic model without specifying such a function, on the other hand, presents a challenge. (See the discussion of Field below.) Fictionalists must steer a path between the Scylla of reductionism and the Charybdis of deductivism.

2 MOTIVATIONS FOR FICTIONALISM IN THE PHILOSOPHY OF MATHEMATICS

Mathematics does not appear to be a species of fiction. Why, then, insist on the possibility of construing it fictionally? Fictionalists fall into two camps. Just as debaters advance cases either on the basis of needs or on the basis of comparative advantage, so fictionalists argue for fictional interpretations either because alternative interpretations raise philosophical puzzles or because fictionalist interpretations simply provide better explanations for mathematical success.

2.1 Benacerraf's Dilemma

The traditional argument for fictionalism is that nonfictional interpretations of mathematics raise insuperable philosophical difficulties. The *locus classicus* of the argument, though presented with a different intent, is Paul Benacerraf's "Mathematical Truth" [1973]. Benacerraf argues that we can devise a successful semantics or a successful epistemology for mathematics, but not both. We cannot reconcile the demands of an account of mathematical truth with the demands of an account of mathematical knowledge.

> ... accounts of truth that treat mathematical and nonmathematical discourse in relevantly similar ways do so at the cost of leaving it unintelligible how we can have any mathematical knowledge whatsoever; whereas those which attribute to mathematical propositions the kinds of truth conditions we can clearly know to obtain, do so at the expense of failing to connect these conditions with any analysis of the sentences which shows how the assigned conditions are conditions of their *truth*. [Benacerraf, 1973, 662]

Benacerraf makes two assumptions. First, he assumes that we should maintain a unified Tarskian semantics for mathematical as well as nonmathematical discourse. Second, he assumes a causal theory of knowledge. The first assumption implies that mathematical objects exist; mathematical discourse succeeds only to the extent that it is true, and it is true only to the extent that the objects over which it quantifies exist. The second implies that we can have mathematical knowledge only by causally interacting with mathematical objects. But that, evidently, is what we do not and cannot do. (Some abstract objects are nevertheless *dependent* on concrete objects and events, as Szabo (2003) observes. Stories may be abstract but depend on concrete people and events; concrete events may in turn depend on them. So, it is a mistake to think of all abstracta as causally isolated. It seems doubtful that enough mathematical objects could be dependent in this sense to ground mathematical knowledge. But see Maddy [1990; 1992; 1997].)

The fading popularity of causal theories of knowledge may make Benacerraf's dilemma seem like something of a period piece, no longer compelling a choice between semantic and epistemological adequacy. There are various ways, however, of weakening these assumptions. Here I will present just one (developed at length in [Bonevac, 1982]). So long as (a) pragmatic success requires truth, (b) truth is to be explained in terms of reference and satisfaction, and (c) we must have epistemic access to the objects we take our discourse to be about (as Benacerraf puts it, we must have "an account of the link between our cognitive faculties and the objects known" (674)), we face the same problem. These assumptions are weaker than Benacerraf's in several respects. They do not assume that a single semantic theory must apply to mathematical and nonmathematical language, as well as all other forms of discourse. Most crucially, they do not assume a causal theory of knowledge. Demanding epistemic access requires that there be a relation between us as knowers and the objects of our knowledge that allows for the possibility of an empirical cognitive psychology. This might be causal, but it need not be. The central idea is motivated by a naturalized epistemology:

> In short, our ability to have knowledge concerning the objects assumed to

exist must itself be capable of being a subject for empirical, and preferably physiological, investigation. [Bonevac, 1982, 9]

I have summarized this by demanding an *empirically scrutable* relationship between ourselves and the objects postulated by theories we accept. It must be possible to explain our knowledge of those objects in a naturalized epistemology. One may spell out the required relationship differently: Field, for example, simply says that our knowledge must be explicable. It should not be chalked up to coincidence:

> The key point, I think, is that our belief in a theory should be undermined if the theory requires that it would be a huge coincidence if what we believed about its subject matter were correct. But mathematical theories, taken at face value, postulate mathematical objects that are mind-independent and bear no causal or spatio-temporal relations to us, or any other kinds of relations to us that would explain why our beliefs about them tend to be correct; it seems hard to give any account of our beliefs about these mathematical objects that doesn't make the correctness of the beliefs a huge coincidence. [Field, 1989, 7].

Mathematicians are reliable; surely that fact needs to be explained.

We might distinguish two kinds of mathematical theories: *existential* theories, which postulate the existence of mathematical entities such as $\emptyset$, π, or the exponentiation function, and *algebraic* theories, which do not, but instead speak only of objects related in certain ways, e.g., as groups, rings, fields, and so on. A structuralist or deductivist analysis of the latter seems natural, though they too make existence claims, which might be seen as derivative from the claims of existential theories or as *sui generis* and needing separate explanation. Benacerraf's dilemma seems most acute for existential theories, such as arithmetic, analysis, and set theory, which postulate the existence of numbers, sets, and functions. Classical mathematics, in Quine's words, "is up to its neck in commitments to an ontology of abstract entities" [Quine, 1951, 13]. But how can we know about such entities? Must our epistemology of mathematics remain nothing but "a mysterious metaphor" [Resnik, 1975, 30]?

Since the argument from Benacerraf's dilemma has fallen under widespread attack, let me try to spell it out somewhat more explicitly, restricting it to existential theories and keeping its assumptions as weak as possible:

1. Some existential mathematical theories — arithmetic and set theory, for example — are successful.

2. Mathematics is successful only to the extent that it is true.

3. An adequate theory of truth for mathematics must be continuous with the theory of truth for the rest of language.

4. An adequate theory of truth in general must be Tarskian, proceeding in terms of reference and satisfaction.

5. Any Tarskian theory interprets existential sentences as requiring the existence of objects in a domain.

6. Existential mathematical theories contain existential sentences.

7. Therefore, mathematical objects exist.

8. We know about objects only by standing in an explicable epistemic relation to them.

9. We stand in no explicable epistemic relation to mathematical objects.

10. Therefore, we cannot know anything about mathematical objects — even that they exist.

We can thus conclude that mathematical objects exist — a conclusion which, having been rationally justified, appears to be known. But we can also conclude that we cannot have any such knowledge. Any account of mathematics must confront the problem this poses. The fictionalist focuses on the first three of the above premises. Most deny the second, maintaining that mathematics may be successful without being true. Some deny the first, holding that all mathematical theories can be given an algebraic interpretation. And some deny the third, seeking a non-Tarskian semantics that can apply to fictional and mathematical discourse in a way that frees them from ontological commitment and associated epistemological difficulties. We interpret sentences in and about works of fiction in ways that do not seem to commit us to fictional characters in a way that raises serious metaphysical and epistemological difficulties. We have no trouble, for example, explaining how it is possible to know that Sherlock Holmes is a detective. That suggests to some that our semantics for fiction is non-Tarskian.

2.2 Yablo's Comparative Advantage Argument

The traditional argument just outlined faces an obvious problem, even if it is not quite "dead and gone,", as Yablo [2001, 87] says. It rests on a thesis requiring epistemic access to objects to which we make ontological commitments. Epistemic access need not be spelled out in terms of empirical scrutability or even explicability. Say simply that the objects of our commitments must exhibit property P. The problem then arises because mathematical objects lack P. But an opponent can turn this argument on the fictionalist by using the fact that mathematical objects lack P to refute the premise that the objects of our commitments must exhibit P. Mathematical objects, that is, can be used as paradigm cases undermining any epistemology a fictionalist or other anti-realist might use (see, e.g., [Hale, 1994]). It is hard to see how to resolve the resulting impasse.

Yablo argues for fictionalism on different grounds. The traditional argument focuses solely on the problems facing a platonistic account of mathematics, effectively granting that otherwise a platonistic theory would be preferable. But that, Yablo, insists, ignores the real advantages of a fictionalist approach while allowing the false advantages of platonism.

Yablo begins with the Predicament: "One, we find ourselves uttering sentences that seem on the face of it to be committed to so-and-so's — sentences that could not be true unless so-and-so's existed. But, two, we do not *believe* that so-and-so's exist" (72). We do not have to insist that so-and-so's are unknowable; it is enough to observe that it is coherent to speak of them without really believing in them. Nonplatonists adopt this attitude toward mathematical and other abstract objects, but we adopt it in all sorts of everyday contexts as well, in speaking of sakes, petards, stomach-butterflies, etc. Quine outlines three ways out of the predicament: reduction, elimination, and acceptance. Yablo argues that fictionalism constitutes a fourth way.

Let's begin with platonism. Waive epistemological objections to abstracta. What does the platonist explain by invoking them? Presumably, the objectivity of certain kinds of discourse. Nominalists have long thought that "explaining" the truth of *Socrates is wise* by pointing out that Socrates has wisdom is no explanation at all. Yablo sharpens the argument by constructing a dilemma. Consider our conception of the numbers, for example. It is either determinate (in the sense that it settles all arithmetical questions) or indeterminate. If it is determinate, "Then the numbers are not needed for objectivity. Our conception draws a bright line between true and false, whether anything answers to it or not" (88). If it is indeterminate, how do we manage to pick out one of the many possible models of our conception as the intended model — that is, as the numbers? In short, the numbers themselves are either unnecessary or insufficient for objectivity. (For a perceptive treatment of determinacy in mathematics, see [Velleman, 1993].)

By itself, this argument seems little better than the Benacerraf-inspired argument we have been considering. A platonist who believes in determinacy will surely insist that our conception of the numbers manages to settle all mathematical questions precisely because it is the conception of a determinate reality. Like a photograph that settles a set of questions (about who won the race, say) because it is the photograph of an event or state of affairs (the finish of the race), our conception of mathematical objects may settle questions because it is a conception of those objects. Such a conception does not demonstrate that the objects are not needed to account for the objectivity of mathematics any more than the photograph would show that the event or state of affairs is not needed to account for the objectivity of our judgment about who won the race. In short, we may need the objects themselves to account for the determinacy as well as the objectivity of our conception.

What of a platonist who believes in indeterminacy? Yablo's question — how then do we pick out one model as the intended model? — bears some similarity to Benacerraf's question of how we manage to refer to or have knowledge of mathematical entities without having any causal contact with them. Yablo seems to be asking for an explanation of our ability to pick out the objects of mathematics. But this seems to have exactly the status of Field's request for an explanation of the correctness of our mathematical beliefs.

The platonist, of course, might also say that we cannot pick out the intended model; the indeterminacy of our mathematical conceptions may entail the indeterminacy of reference in mathematics, which is arguably a conclusion of Benacerraf [1965]. This may do no more to disrupt mathematical discourse than the indeterminacy of reference in general disrupts our discourse about rabbits [Quine, 1960]. The upshot of Quine's arguments seems to be that it is indeterminate whether the natives (or, by [Quine, 1969], we) are speaking of rabbits, undetached rabbit parts, rabbithood, etc. But they are speaking of something with rabbit-like characteristics. Similarly, it may be indeterminate whether mathematicians are speaking of categories, sets, classes, numbers, etc., but clear that they are speaking of something with abstract characteristics. (It is perhaps enough to observe that, whatever they are speaking about, they are committing themselves to infinitely many of them.)

Let's turn to the positive portion of Yablo's argument. Yablo's fictionalist account, he contends, in contrast to a platonistic account, succeeds in explaining a great deal. It explains why numbers are "thin," lacking any hidden nature; for mathematical objects, nominal and real essence coincide. It explains why numbers are indeterminate with respect to identity relations involving nonnumbers; it is determinately true that $0 \times 2 = 0$, and determinately false that $0 \times 2 = 2$, but indeterminate whether $2 = \{\{\emptyset\}\}$ or $2 = \{\emptyset, \{\emptyset\}\}$. It explains why applied arithmetical statements are translucent; we immediately see through them to recognize their implications for concreta. It explains why people are impatient with any objection to mathematics on ontological grounds, since the ontological status of its objects makes no difference to what a mathematical theory is communicating. It explains why mathematical theories are excellent representational aids, applying readily to the world. And it explains why mathematical assertions strike us as necessary and a priori; they do not depend on the actual existence or contingent circumstances of their objects.

3 A BRIEF HISTORY OF FICTIONALISM

The problem fictionalism addresses is ancient: the problem of knowing the forms. Plato hypothesized the existence of forms, universals standing outside the causal order but explaining its structure. If they remain outside the causal order, however, how is it possible to know anything about them? Aristotle, arguably, had formulated a version of Benacerraf's dilemma in his criticisms of the theory of forms in *Metaphysics* I, 9: "if the Forms are numbers, how can they be causes?" Fictionalism might be read into some forms of ancient skepticism and into an alternative outlined (but not accepted) by Porphyry, that universals are *nuda intellecta*. But the earliest philosopher to endorse a fictionalist strategy is probably Roscelin. The father of nominalism, Roscelin sought to solve the problem posed by knowing the forms by denying their existence. He maintained that abstract terms are *flatus vocis*, puffs of the voice, empty noises, reflecting Boethius's thought that *Nihil enim aliud est prolatio (vocis) quam aeris plectro linguae percussio*. It is not clear exactly what position Roscelin meant to endorse by this claim; we do not know whether he thought abstract terms were empty of meaning or empty of reference. If the latter, which is in any case a more plausible position to hold, Roscelin might reasonably be considered a fictionalist about universals. Here are some highlights in the history of fictionalism, designed to illustrate some important types of fictionalism. (For an alternative history that fills in many gaps in the following, see Rosen [2005].)

3.1 William of Ockham: Reductive Fictionalism

The earliest fictionalism of which we have any detailed record is probably William of Ockham's. Ockham explicitly speaks of universals as *ficta*, and makes it clear that, in his view, they do not exist; everything that exists is particular. Whether we interpret this as a form of nominalism, as most traditional commentators have, or as a form of conceptualism, as some more recent scholars have, it is in any case the thesis that sentences seemingly referring to universals can succeed in their linguistic function even though nothing in mind-independent reality corresponds to a universal. (See, for example, [Boehner, 1946; Adams, 1977; 1987; Tweedale, 1992; Spade, 1998; 1999a; 1999b].)

It is important to note that, from Ockham's point of view, one must do more to break the success-existence link than interpret discourse involving abstract terms as fictional. Ockham assumes that sentences involving abstract terms play a role in discourse, and that, to make the case the case that such terms need not be taken seriously from an ontological point of view, one must explain how it is possible for them to play such a role. Ockham initiates one of the traditional strategies for doing so, arguing that abstractions are shorthand for expressions that fill the same semantic and pragmatic role but without invoking the existence of objects lying outside the causal order. *Socrates exemplifies wisdom* is an inefficient way of saying that Socrates is wise (Ockham 1991, 105ff); *Courage is a virtue* is an efficient way of saying that courageous people and courageous actions are *ceteris paribus* virtuous. To say *Socrates and Plato are similar with respect to whiteness*, Ockham says, is just to say that Socrates is white and Plato is white (572). In general, every true sentence containing abstract terms translates into a true sentence containing only concrete terms, and vice versa: "it is impossible for a proposition in which the concrete name occurs to be true unless the [corresponding] proposition in which its abstract [counterpart] occurs is true" (432).

Ockham thus outlines the *reductive* strategy: we may break the link between success and existence by arguing that the use of abstract terms, for example, is unnecessary. Anything that can be said with abstract terms can be said without them. A language without abstract terms could thus in principle fill every role in discourse that a language with abstract terms can fill. Of course, it might do so inefficiently; there may be good practical reasons to use abstract language. Nevertheless, its in-principle eliminability shows that the ontological commitments such language seems to force upon us are also in principle eliminable. If abstract language is unnecessary, then so is a commitment to abstract entities. Note that the reductive strategy breaks the chain from success to existence not by denying the truth of the sentences with troublesome ontological commitments but by denying that they form an essential part of an accurate description of the world. Though the sentences may make commitments to objects to which we stand in no empirically scrutable relation, *we* even in asserting them make no such commitment, because we can treat them as optional abbreviations for sentences making no such commitment.

Though Ockham considers reductionism a version of fictionalism, and though Bentham, Russell, and others have agreed, he is probably wrong to do so. At any rate, a reductive fictionalism seems not to take advantage of anything distinctive of fictionalism. (1) The analogy with fiction is not very strong [Burgess, 2004]. On a reductive account, sentences ostensibly making commitments to abstract objects are better understood as being about something else. But fictional statements, it seems plausible to claim, are not about something other than fictional characters. Sherlock Holmes is a detective by virtue of the fact than Sir Arthur Conan Doyle described him that way. But it would be strange to say that *Sherlock Holmes is a detective* should be translated, from an ontological point of view, into *Sir Arthur Conan Doyle described Sherlock Holmes as a detective*, and not only because the latter still seems to refer to Holmes. It would be even stranger to say that the former sentence is really about descriptions. In any case, such translations are not recursive in the way that reductive translations ought to be. (2) However this may be, reductive theories interpret purported truths about objectionable entities as truths about acceptable entities. They translate sentences with ostensible commitments to abstracta into sentences without such commitments. The criterion for the translation's success is truth preservation. Not only does the discourse thus emerge as truth-normed, though in an unexpected way; the discourse still commits one to objects, even if not to the objects one initially took it as being about. Existential sentences translate into other existential sentences. Hawthorne's sentence in *Twice Told Tales*, viewed as fictional, however, seems to force no commitments at all.

3.2 Jeremy Bentham: Instrumentalist Fictionalism

Jeremy Bentham is perhaps the first philosopher to have advocated fictionalism explicitly. His theory of fictions, which exerted a powerful influence on Bertrand Russell at an early stage of his thought, appears over the course of seven of Bentham's works and still receives surprisingly little attention. But Bentham thought of it as one of his chief achievements, something upon which most of his other philosophical conclusions depend.

Bentham distinguishes fictitious entities not only from real entities but also from *fabulous* entities, "supposed material objects, of which the separate existence is capable of becoming a subject of belief, and of which, accordingly, the same sort of picture is capable of being drawn in and preserved in the mind, as of any really existing object" [Bentham, 1932, *xxxv-xxxvi*]. Fabulous objects, in other words, are nonexistent objects of existent kinds (legendary kings, for example, or countries such as Atlantis or El Dorado) or nonexistent objects of nonexistent kinds (dragons, elves, or the Loch Ness Monster); they would pose no particular metaphysical or epistemological problems if they were to exist. Ordinary fiction thus introduces, primarily, fabulous objects by way of referring expressions that would, if they were to denote at all, denote substances.

Fictitious objects, in contrast, do not "raise up in the mind any corresponding images" (*xxxvi*); their names function grammatically as if they were names of substances, but would not refer to substances even if their referents were to exist. We speak as if such names refer to real objects, "yet in truth and reality existence is not meant to be ascribed" (12). "To language, then — to language alone — it is, that fictitious entities owe their existence; their impossible, yet indispensable, existence" (15). Every language must speak of a fictitious object as existing — in that sense, their existence is indispensable — "but without any such danger as that of producing any such persuasion as that of their possessing, each for itself, any separate, or, strictly speaking, any real existence" (16).

Distinguishing fabulous from fictitious objects might serve as an argument against fictionalism, for it suggests that the objects of mathematics, for example, are not very closely analogous to ordinary fictional entities. But Bentham nevertheless applies his theory of fictions directly to mathematics. Quantity, he contends, is the chief subject matter of mathematics, and is fictitious. "The ink which is in the ink-glass, exists there *in* a certain quantity. Here *quantity* is a fictitious substance — a fictitious receptacle — and in this receptacle the ink, the real substance, is spoken of as if it were lodged" (*xxxviii*). Pure mathematics, Bentham holds, "is neither useful nor so much as true" (*Works* IX, 72). This makes it sound as if Bentham were breaking the success-existence chain by denying the truth of mathematics in the way that most contemporary fictionalists do. In fact the picture is more complicated. "A general proposition which has no individual object to which it is truly applicable is not a true one" (*Works* VIII, 163), Bentham says, denying that certain branches of pure mathematics are true. But he concedes that geometry can be given an interpretation in which it applies, for example, to all spherical bodies, in which case it is properly viewed as a true or false theory of those physical entities. If such entities are "capable of coming into existence, it may be considered as having a sort of potential truth" (*Works* VIII, 162).

So, in Bentham's view, there are three kinds of mathematical theories: the theories of applied mathematics, which are empirical theories of the world; theories with potential truth, which would be true if their objects, which are capable of existing, were to exist; and theories with no kind of truth at all, of objects that are not even capable of existing. Corresponding to these three kinds of theories are three strategies for reconciling the semantics and epistemology of mathematics: the *empirical* strategy, the strategy of John Stuart Mill and W. V. O. Quine, which holds that mathematics is an empirical theory and is thus in no sense fictional; the *modal* strategy, the strategy of Cantor, Poincaré, and Chihara, which holds that mathematics is a theory of possible objects and is thus fictional in the sense that fiction, too, describes possible objects; and the *instrumentalist* strategy, the strategy of Hartry Field, which holds that mathematics is not true but rather an instrument or, as Bentham would say, a successful system of contrivances for a practical purpose.

There is considerable merit, Bentham maintains, in the reductive strategy, which locates the success of mathematics in "mere *abbreviation* ... nothing but a particular species of *short-hand*" (183), but ultimately, he insists, that is not enough; "Newton, Leibnitz, Euler, La Place, La Grange, etc., etc. — on this magnificent portion of the field of science, have they been nothing more than so many expert *short-hand writers*?" (37) He emphasizes, therefore, the view of mathematics as a system of contrivances. It is not clear what Bentham means by *contrivance*, but he does give examples: "the conversion of algebraic method into geometrical ... the method of *fluxions* ... and the *differential and integral calculus*" (169). The success of such contrivances consists in their satisfying two conditions. First, there must be kinds of empirical circumstances to which the theory with its attendant fictions would be applicable. Second, there must be some advantage to using such a theory in such circumstances. Mathematics is not true, but it succeeds to the extent that it serves as a successful instrument in reasoning about empirical states of affairs, specifically, in proceeding from the known to the unknown. Mathematics introduces objects — infinite collections, for example, limits, derivatives, integrals, tangent lines, etc. — that cannot be reduced to nonmathematical entities, but which prove useful in deducing conclusions about nonmathematical entities.

3.3 C. S. Peirce: Representational Fictionalism

The fictionalism of nominalists from Roscelin to Bentham has been motivated by an epistemological attitude we might construe as at least a proto-empiricism. Two other forms of fictionalism stem from rather different perspectives: Peirce's pragmatism and Vaihinger's Kantianism.

C. S. Peirce developed a view of mathematics that might be counted as a variety of fictionalism. "Mathematics," Peirce says, "has always been more or less a trade" [1898, 137]. We can understand the nature of mathematics only by understanding "what service it is" to other disciplines. Its purpose, he writes, is to draw out the consequences of hypotheses in the face of complexity:

> An engineer, or a business company (say, an insurance company), or a buyer (say, of land), or a physicist, finds it suits his purpose to ascertain what the necessary consequences of possible facts would be; but the facts are so complicated that he cannot deal with them in his usual way. He calls upon a mathematician and states the question. . . . He [the mathematician] finds, however in almost every case that the statement has one inconvenience, and in many cases that it has a second. The first inconvenience is that, though the statement may not at first sound very complicated, yet, when it is accurately analyzed, it is found to imply so intricate a condition of things that it far surpasses the power of the mathematician to say with exactitude what its consequence would be. At the same time, it frequently happens that the facts, as stated, are insufficient to answer the question that is put. [1898, B 137–38, CP 3.349]

Think of a child measuring a line segment as, say, 5 cm long. Almost certainly, the measurement is an approximation; the line is slightly shorter or longer. Similarly with a surveyor's angle. An economist predicting a firm's profits seems to be in a bit better position, since money comes in discreet units. But the complexity of predicting profits is immense, since many factors influence income and expenditure and do so in complicated ways. A physicist calculates the trajectory of a projectile using approximate values for force, velocity, distance, etc., and ignoring the effects of friction, the moon's gravitational pull, and so on. In short, the real world is overwhelmingly complicated. This provides the chief motivation but also the chief difficulty for applying mathematics to real-world problems.

Essential to the application of mathematics to the world, therefore, is idealization. Like the economic principles linking variables such as population growth, personal income, the inflation rate, etc. to income and expenditures, the physical laws governing the motion of projectiles successfully relate quantities in an idealized, simplified version of the real situation, not the real situation itself. The world is full of friction, vagueness, imprecision —

in short, noise. The noise itself can be studied and classified. But it cannot be eliminated. One can deduce consequences only by abstracting from it:

> Accordingly, the first business of the mathematician, often a most difficult task, is to frame another simpler but quite fictitious problem (supplemented, perhaps, by some supposition), which shall be within his powers, while at the same time it is sufficiently like the problem set before him to answer, well or ill, as a substitute for it. This substituted problem differs also from that which was first set before the mathematician in another respect: namely, that it is highly abstract. [1898, B 138, CP 349]

This "skeletonization" or "diagrammatization" serves "to strip the significant relations of all disguise" (B 138, CP 349) and thus to make it possible to draw consequences.

In idealizing a problem in this way, Peirce writes,

> The mathematician does two very different things: namely, he first frames a pure hypothesis stripped of all features which do not concern the drawing of consequences from it, and this he does without inquiring or caring whether it agrees with the actual facts or not; and, secondly, he proceeds to draw necessary consequences from that hypothesis. (1898, B 138, CP 349-350)

A pure hypothesis "is a proposition imagined to be strictly true of an ideal state of things" [1898, B 137, CP 348].

The principles used by the engineer, the economist, or even the physicist are not like that; "in regard to the real world, we have no right to presume that any given intelligible proposition is true in absolute strictness" [1898, B 137, CP 348]. The real world is complicated; principles tends to hold of it only *ceteris paribus*, or in the absence of any complicating factors. Some laws of nature may be simple enough to be easily intelligible while also holding absolutely, as Galileo thought, but we have no right to expect that to be the case in general. In some areas, "the presumption in favor of a simple law seems very slender" [1891, 318]. "We must not say that phenomena are perfectly regular, but that their degree of regularity is very high indeed" (unidentified fragment, 1976, xvi); "The regularity of the universe cannot reasonably be supposed to be perfect" ("Sketch of a New Philosophy," 1976, 376). Peirce offers an evolutionary account of laws that explicitly rejects the assumption that the ultimate laws of nature are simple and hold without exception:

> This supposes them [laws of nature] not to be absolute, not to be obeyed precisely. It makes an element of indeterminacy, spontaneity, or absolute chance in nature. Just as, when we attempt to verify any physical law, we find our observations cannot be precisely satisfied by it, and rightly attribute the discrepancy to errors of observation, so we must suppose far more minute discrepancies to exist owing to the imperfect cogency of the law itself, to a certain swerving of the facts from any definite formula. [1891, 318]

In sum, "There are very few rules in natural science, if there are any at all, that will bear being extended to the most *extreme cases*" [1976, 158]. Mathematics, however, abstracts away from complications, using principles that hold absolutely of idealized states of affairs. Since idealization or diagrammatization involves two components, the construction of a pure hypothesis or ideal state of affairs and the derivation of consequences from it, so mathematics can be defined in two ways, as the science "of drawing necessary conclusions" or "as the study of hypothetical states of things" (1902, 141). The latter conception is essentially fictionalist. They are equivalent, Peirce sometimes maintains, because the fact "that mathematics deals exclusively with hypothetical states of things, and asserts no matter of fact whatever" alone explains "the necessity of its conclusions" [1902, 140].

Peirce's account of the nature of mathematics comprises, then, the following theses:

1. The real world is too complicated to be described correctly by strictly universal principles.

2. At best, laws governing the real world hold *ceteris paribus* or in the absence of complicating factors.

3. In applying mathematics to the real world, we

    (a) build idealized models that abstract away from many features of the world but highlight others;

    (b) deduce consequences of assumptions using the 'pure hypotheses,' i.e., strictly universal principles holding in those models; and

    (c) use those consequences to derive consequences for the real world.

On this Peircean theory, which I will term *representational fictionalism*, the applicability of mathematics to the real world is no mystery; mathematical theories are designed to be applicable to the world. Mathematics is the science of constructing idealized models into which aspects of the real world can be embedded and using rules to derive consequences from the models that, *ceteris paribus vel absentibus*, apply to the real world. That means that measurement and, more generally, the representability of features of the world in mathematics is essential to mathematical activity. There is no need to worry about the interpretability of mathematics in concrete terms; the issue is the interpretability of our theories of the concrete in mathematical terms. In Peirce's terms, "all [combinations] that occur in the real world also occur in the ideal world.... [T]he sensible world is but a fragment of the ideal world" (1897, 146). Consequently, "There is no science whatever to which is not attached an application of mathematics" [1902, CP 112].

There are, in first-order languages, at least, two equivalent ways of thinking about the interpretability of theories of the real world in mathematics. We might interpret theories of the concrete in mathematics by translating nonmathematical language into mathematical language. More naturally, we might represent nonmathematical objects and relations mathematically by mapping them into mathematical objects and relations. We apply mathematics by embedding nonmathematical structures into mathematical ones. It is natural to identify mathematical theories, therefore, by the structures (e.g., the natural numbers, the integers, the reals) they describe, and to think of mathematical theories as describing intended models. It is likewise natural to think of mathematics as a science of structure or patterns; to apply mathematics to a real-world problem is to find a mathematical structure that encompasses the relevant structure of the real-world situation.

One may think of mathematics, consequently, as a universal container — a body of theory into which any actual or even possible concrete structure could be embedded. There may be a single mathematical theory — set theory or category theory, perhaps — general enough to serve by itself as a universal container. But this is not essential to the Peircean theory. What matters is that mathematics collectively be able to play this role. The more general a mathematical theory is, of course, the more useful it is, and the more basic it can be taken to be within the overall organization of mathematics.

Because mathematics strives to be a universal container in which any concrete structure can be embedded, mathematical theories tend to have infinite domains. This accounts for the dictum that mathematics is the science of the infinite. But mathematics does not study the infinite for its own sake; it studies the infinite because it studies structures into which other finite and infinite structures can be embedded.

The application of mathematics does not require that it be true. In some ways, Peirce's account of mathematics and its application presages Field's (1980) account. But there are important differences. For Peirce, there is no reason to assume that the real world can be described in nonmathematical terms. We map certain structures into mathematical structures so readily that we may lack any nonmathematical language for describing them. Also, for Peirce, mathematics need not be conservative; it may be possible to use a mathematical model to derive conclusions concerning a real-world problem that are false in the real world. This may happen because the mathematical theory contains structure that goes beyond the structure of the corresponding situation in the real world. The density and continuity of the real line, for example, may permit us to obtain conclusions about an actual concrete line segment that are false. It may also happen because of the complicated character of the real world. The real world is unruly, but the 'ideal world' into which it is embedded is rule-governed. Inevitably, some things that hold of the idealized model will not hold in the real world. That is why a *ceteris paribus* proviso is needed.

3.4 Hans Vaihinger: Free-range Fictionalism

From a very different epistemological standpoint is the fictionalism of Hans Vaihinger, who elaborated his particular form of Kantianism (called "Positivistic Idealism" or "Idealistic Positivism") in *The Philosophy of As-If*, the chief thesis of which is that "'As if', i.e. appearance, the consciously-false, plays an enormous part in science, in world-philosophies and in life" (*xli*). Vaihinger seems to think of fictionalism in attitudinal terms; he holds that "we operate intentionally with consciously false ideas." "Fictions," he maintains, "are known to be false" but "are employed because of their utility" (*xlii*). Fictions are artifices, products of human creative activity and as such "*mental structures*" (12). They act as "accessory structures" helping human beings make sense of "a world of contradictory sensations," "a hostile external world." Vaihinger, inspired by Schiller's phrase, "In error only is there life," finds fictions throughout our mental life, and interprets Kant's *Critique of Pure Reason* as showing that many of the concepts of metaphysics and ethics, not to mention ordinary life, are fictional. The same is true of mathematics.

How do we identify something as a fiction, particularly when there is disagreement about the attitude one ought to take toward it? The paradigms of a demonstration of fictional status, for Vaihinger, are Kant's antinomies of pure reason. In short, the mark of fiction is contradiction. Reality is entirely consistent; anything that successfully describes it must be consistent. But fiction need not obey any such constraint. Of course, a writer may and typically does seek consistency in a work of fiction, the better to describe a world that could be real. But consistency is inessential, and, in philosophically interesting cases, highly unusual. We resort to fictions to make sense of an otherwise contradictory experience, and characterize them to meet contradictory goals in generally contradictory ways.

In the case of mathematics, specifically,

> The fundamental concepts of mathematics are space, or more precisely empty space, empty time, point, line, surface, or more precisely points without extension, lines without breadth, surfaces without depth, spaces without content. All these concepts are contradictory fictions, mathematics being based upon an entirely imaginary foundation, indeed upon contradictions. (51)

Most fictionalists have thought of consistency as the only real constraint on fiction and, so, on mathematical existence. But Vaihinger, impressed perhaps with the inconsistencies in the theories of infinitesimals, infinite series, functions, and even negative numbers that prompted the great rigorization projects of Cauchy, Dedekind, Peano, and others in the nineteenth century, and certainly taking as an exemplar the theory of limits, held that

> The frank acknowledgment of these fundamental contradictions has become absolutely essential for mathematical progress. (51)

The point, from his perspective, is not to recognize contradictions in order to get rid of them but rather to understand the fictional nature of the objects supposedly being described.

> There is therefore no object in trying to argue away the blatant contradictions inherent in this concept [of pure absolute space]. To be a true fiction, the concept of space should be self-contradictory. Anyone who desires to "free" the concept of space from these contradictions, would deprive it of its characteristic qualities, that is to say, of the honour of serving as an ideal example of a true and justified fiction. (233)

In addition to the items already listed, Vaihinger counts as contradictory theses that a circle is an ellipse with a zero focus; that a circle is an infinite-sided polygon; and that a line consists of points. He would no doubt have taken Russell's paradox, the Burali-Forti paradox, Richard's paradox, and so on as lending strong support to his thesis. (*The Philosophy of As-If* was published in 1911, but he wrote Part I, in which he developed his fictionalism, in 1877.)

Mathematics is an instance of the method of abstract generalization, "one of the most brilliant devices of thought" (55), but one which easily generates contradictions and in general produces fictions. "The objects of mathematics are artificial preparations, artificial structures, fictional abstractions, abstract fictions" (233); "they are contradictory constructs, a nothing that is nevertheless conceived as a something, a something that is already passing over into a nothing. And yet just these contradictory constructs, these fictional entities, are the indispensable bases of mathematical thought" (234).

It is tempting to see Vaihinger's idealistic fictionalism as quaint, a product of nineteenth-century German idealism that has little relevance to contemporary discussions of fictionalism. As Vaihinger's analysis of the Leibniz-Clarke correspondence shows, however, that is not so. The fact that we might imagine everything in space displaced some distance to the right, or everything in time displaced some period into the future or past, but find the results of those thought experiments indistinguishable from the actual state of affairs shows, according to Vaihinger, that absolute space and time are fictions. The contradiction here, like the contradiction he finds in the idea of an extensionless point, a limit, and infinitely large or infinitely small quantity, etc., is not strictly speaking logical but epistemological. There is no contradiction in the idea of everything being displaced two feet to the right, or everything have been created two minutes before it actually was; but such hypotheses also seem pointless, for there would be no way to tell whether they were true or not. Vaihinger does not elaborate the exact nature of the contradiction this situation entails.

On one interpretation, absolute space and time imply the possibility of different states of affairs that are in principle (and not merely as a result of limitations of our own cognitive capacities) not empirically distinguishable. The contradiction is thus not really self-contradiction, though Vaihinger sometimes describes it in those terms. It is a contradiction with the positivistic part of his positivistic idealism. But that again describes it too narrowly, for the contradiction is with a thesis that can seem appealing for reasons independent of positivism, namely, that the objects a theory postulates ought to be in principle empirically scrutable. The objects of mathematics are not logically incoherent but epistemically incoherent; their inaccessibility makes them philosophically objectionable even if practically indispensable.

On another interpretation, the contradiction stems from our inability to distinguish isomorphic structures. It is characteristic not only of mathematics but of all discourse, Vaihinger believes, that isomorphic structures are indiscernible. Any concept, proposition, or entity that depends for its sense on the discernibility of isomorphic structures is fictional; its contradictory character lies in its presumption of the discernibility of indiscernibles. Vaihinger's view, so construed, bears an interesting relation to contemporary structuralist accounts of mathematics. The structuralist, from his perspective, is essentially correct, but misses part of the story. Mathematics, properly understood, is structuralist in the sense that its objects can be nothing more than roles in a certain kind of structure. But it purports to be something else; it presents its objects as if they were substances analogous to concrete objects in the causal order. The argument of Benacerraf 1965, that mathematics can characterize its objects only up to isomorphism even within the realm of mathematics itself, Vaihinger would no doubt take as demonstrating the fictional nature of mathematics.

4 SCIENCE WITHOUT NUMBERS

Hartry Field single-handedly revived the fictionalist tradition in the philosophy of mathematics in *Science Without Numbers* (Field 1980). Field begins with the question of the applicability of mathematics to the physical world, something about which most earlier versions of fictionalism had little to say. (Peirce's representational fictionalism is an obvious exception.) Field offers an extended argument that "it is not necessary to assume that the mathematics that is applied is true, it is necessary to assume little more than that mathematics is consistent" (*vii*). Since "*no* part of mathematics is true ... no entities have to be postulated to account for mathematical truth, and the problem of accounting for the knowledge of mathematical truths vanishes" (*viii*).

Field's fictionalism is plainly of the instrumentalist variety; mathematics is an instrument for drawing nominalistically acceptable conclusions from nominalistically acceptable premises. He shows that, for mathematics to be able to perform this task, it need not be true. It must, however, be *conservative*: Anything nominalistic that is provable from a nominalistic theory with the help of mathematics is also provable without it. Roughly, $\Gamma \cup M \models A \Leftrightarrow \Gamma \models A$, where M is a mathematical theory and Γ and A make no commitment to mathematical entities. Usually, a purely nominalistic proof would be far less efficient than a platonistic proof. Mathematics is thus practically useful, and perhaps even heuristically indispensable, since we might never think of certain connections if confined to a purely nominalistic language. But mathematics is theoretically dispensable; anything we can do with it can be done without it.

Field gives a powerful argument for the conservativeness of mathematics, though there is a limitation that points in the direction of representational fictionalism. (For an excellent discussion of the technical aspects of Field's work, see [Urquhart, 1990].) Let $ZFU_{V(T)}$ be Zermelo-Fraenkel set theory with urelements, with the vocabulary of a first-order, nominalistic physical theory T appearing in instances of the comprehension axiom schema. Let T^* be T with all quantifiers restricted to nonmathematical entities. If T can be modeled within ZF, then $ZF \vdash Con(T)$ (the consistency statement for T), and so $ZFU_{V(T)} + T^*$ is interpretable in $ZFU_{V(T)}$, indeed, in ZF. Interpretability establishes relative consistency, so, if ZF is consistent, $ZFU_{V(T)} + T^*$ is consistent. Let T be $N \cup \{\neg A\}$. Then, if $ZFU_{V(T)} + N^* + \neg A^*$ is inconsistent — that is, if $ZFU_{V(T)} + N^* \vdash A^*$ — then, if ZF is consistent, $ZF \nvdash Con(N^* + \neg A^*)$. Hence, $N^* + \neg A^*$ cannot be modeled in ZF, and so $N^* \vdash A^*$. On the assumption that ZFU is strong enough to model any mathematical theory that might usefully be applied to the physical world, this yields the conclusion that mathematics is conservative. Mathematics is conservative, that is, with respect to theories that are interpretable within it.

Field treats that assumption as safe, given set theory's foundational status in mathematics. It is not, however, unassailable; second-order set theory and ZFU + Con(ZFU) are two theories that cannot be modeled in ZFU. Second-order set theory, unlike ZF, is categorical; all its models are isomorphic, which means that the continuum hypothesis, for example, is determinately true or false within it. (We do not know which.) Perhaps nothing physical will ever turn on the truth of the continuum hypothesis. We know, however, that whether certain empirical testing strategies are optimal depends on the continuum hypothesis [Juhl, 1995], so empirical consequences should not be ruled out. Similarly, ZFU + Con(ZFU) might seem to add nothing of physical relevance to ZFU. But there may well be undecidable sentences of ZFU that do not involve coding but have real mathematical and even physical significance, just as there are undecidable sentences of arithmetic with real mathematical significance (Paris and Harrington 1977).

The difficult part of Field's argument lies in showing that we do not need mathematics to state physical theories in the first place. Given nominalistic premises, we can use mathematics without guilt in deriving nominalistic conclusions. But why should we have confidence that we can express everything we want to say in nominalistically acceptable form? Malament [1982] argues, for example, that Field's methods cannot apply to quantum field theory. It is hard to evaluate that allegation without making a full-blown attempt to rewrite quantum field theory in nominalistically acceptable language (but see [Balaguer, 1996; 1998]). There are similar problems involving relativity; the coordinate systems of Riemannian and differential geometries cannot be represented by benchmark points as Euclidean geometry can, and such geometries have not been formalized as Hilbert, Tarski, and others have formalized Euclidean geometry [Burgess and Rosen, 1997, 117–118]. Even if all of current science could be so rewritten, however, there seems to be no guarantee that the next scientific theory will submit to the same treatment (see [Friedman, 1981; Burgess, 1983; 1991; Horgan, 1984; Resnik, 1985; Sober, 1993]). Consequently, as impressive as Field's rewriting of Newtonian gravitation theory is — and it *is* impressive — it is hard to know how much confidence one should have in the general strategy without a recipe for rewriting scientific theories in general.

There is an obverse worry as well. Modern theories of definition (as in, for example, [Suppes 1971]) generally have criteria of eliminability and noncreativity or, in Field's language, conservativeness. That is, an expression is definable in a theory if (1) all occurrences of it can be eliminated unambiguously without changing truth values and if (2) one cannot prove anything in the remainder of the language by using the expression that one could not also prove without it. For an n-ary predicate R, these conditions hold if and only if the theory contains a universalized biconditional of the form $\forall x_1...x_n(Rx_1...x_n \leftrightarrow A(x_1...x_n))$, where A is an expression with n free variables that does not contain R. Now suppose that R is a mathematical predicate. Field's program requires that it be eliminable — in the strong sense of being replaceable *salva veritate* by nominalistically acceptable expressions — and conservative. It seems to follow that R is definable in nonmathematical terms. Generalizing to all mathematical expressions, the worry is that mathematics meets Field's conditions if and only if mathematics is reducible to nominalistically acceptable theories. As Stanley [2001] puts the objection, "a pretense analysis turns out to be just the method of paraphrase in disguise" (44).

Does Field's instrumentalist fictionalism collapse into a reductive fictionalism? To understand what Field is doing, we must understand the ways in which his approach falls short of a reduction of mathematics to nonmathematical theories. The key is the expressability of physical theories in nominalistically acceptable terms. In three respects, we might interpret Field's eliminability requirement as weaker than that invoked in theories of definition.

First, the eliminability need not be uniform. We must be able to write physical theories, for example, in forms that make no use of mathematical language and make no commitment to mathematical objects. But that might fall short of a translation of a standard physical theory into nominalistic language, for we might substitute different nominalistic language for the same mathematical expression in different parts of the theory. Presumably we can specify the contexts in which the mathematical expression is replaced in a given way. But we may not be able to express that specification within the language of the theory itself.

Second, even if we were to have a reduction of a physical theory expressed mathematically into a physical theory expressed nominalistically, we would not necessarily have a reduction of mathematics to a nominalistic theory. We might, for example, know how to express the thought that momentum is the integral of force in nominalistic terms without being able to translate *integral* in any context whatever. Any definitions of mathematical terms that emerge from this process, in short, might be contextual definitions, telling us how to eliminate mathematical expressions in a given context without telling us how to define them in isolation.

For an analogy, consider Russell's theory of descriptions. Russell stresses that he gives us, not a definition of *the*, but instead a contextual definition of a description in the context of a sentence. We can represent *The F is G* as $\exists x \forall y((Fy \leftrightarrow x = y) \& Gx)$, but we have no translation of *the* or even *the F* in isolation. By using the lambda calculus, we can provide such definitions; we might define *the F*, for example, as $\lambda G \exists x \forall y((Fy \leftrightarrow x = y) \& Gx)$ and *the* as $\lambda F \lambda G \exists x \forall y((Fy \leftrightarrow x = y) \& Gx)$. But such definitions are not expressible in the original first-order language. Moreover, an analogous strategy for mathematics does not seem particularly plausible. It may be true that integration is a relation that holds between momentum and force, for example, but it is hardly the only such relation, or even the only such mathematical relation. Even if it were, characterizing it in those terms would not be very helpful in eliminating integration from theories about work or electrical force, not to mention volume or aggregate demand.

Third, given the potentially contextual nature of nominalistic rewritings of physical theories, the best we might hope for within our original language is a translation of a mathematical expressions into universalized infinite disjunctions. The context dependence of the rewriting, in short, presents a problem similar to that of multiple realizability. Mathematical concepts are multiply realizable in physical theories, and we might not be able to do better than to devise an infinite disjunction expressing the possible realizations. That would yield a reduction of mathematics not to a nominalistically acceptable theory but instead to an infinitary extension of that theory. As I have argued elsewhere [1991; 1995], this would be equivalent to showing that mathematics supervenes on nominalistically acceptable theories.

At most, then, Field's fictionalism commits him to the claim that mathematics supervenes on theories of the concrete. That would be enough, however, to cast doubt on his claim to be advancing a version of fictionalism, for it would commit him to the claim that mathematics is true and moreover true by virtue of exactly what makes concrete truths true.

To evaluate this objection, we need to characterize Field's method more precisely. Field [1980, 89–90], presents his method as including the following steps:

1. We begin with a physical theory T expressed using mathematics. We define a nominalistic axiom system S any model of which is homomorphically embeddable in R^4. This is equivalent to requiring that S reduce to $\mathfrak{Th}(R^4)$: $S \leq \mathfrak{Th}(R^4)$.

2. We expand S to $S' \supseteq S$ by adding statements that express a nominalistic physical theory in a language making no commitments to mathematical entities. This expanded theory is such that $S' + M \models T$ and $T + M \models S'$, where $M \supseteq \mathfrak{Th}(R^4)$. If S' and T are both finitely axiomatizable by, say, $\&S'$ and $\&T$, then, $M \models \&S' \leftrightarrow \&T$.

3. Let A be a nominalistically stable physical truth. By the conservativeness of M, $T + M \models A \Rightarrow S' + M \models A \Rightarrow S' \models A$. Any nominalistically stable truth that follows from ordinary physical theory follows from a nominalistically stated theory. So, "mathematical entities are theoretically dispensable" [Field, 1980, 90].

The crucial step here is the second. Not only is it the difficult step from a technical point of view, the topic of Field's central chapters; it is the point at which a prospective nominalist is likely to become faint of heart. What justifies confidence that such an S' exists?

To answer this question, we need to examine Field's method in those central chapters. He begins, addressing step (1), by using Hilbert's axiomatization of geometry to provide a theory of space-time interpretable in $\mathfrak{Th}(R^4)$. Hilbert's approach proceeds by way of representation theorems, which show that the structure of phenomena under certain operations and relations is the same as the structure of numbers or other mathematical objects under corresponding mathematical operations and relations. A representation theorem for theories T and T' shows, in general, that any model of T can be embedded in a model of T'. (Generally, we would want to show in addition that the embedding is unique or at any rate invariant under conditions, something Field proceeds to do.) A representation theorem for T and T' thus establishes that $T \leq T'$. Throughout his treatment of Newtonian gravitation theory — that is, throughout his treatment of step (2) — Field employs the same method. He seeks "a statement that can be left-hand side of the representation theorem" (71), where the right-hand side is given by the mathematically expressed physical theory. In short, he seeks a nominalistic theory whose models are embeddable into models of the standard physical theory. Just as, in step (1), we need a theory $S \leq \mathfrak{Th}(R^4)$, so, in step (2), we need a theory $S' \leq T + M$. That fact suffices to justify the claim that $T + M \models S'$.

What, however, justifies the claim that $S' + M \models T$? What, that is, justifies Field's claim that "the nominalistic formulation of the physical theory in conjunction with standard mathematics yields the usual platonistic formulation of the theory" (90)? We must choose the nominalistic statements that expand S to S' in such a way that "if these further nominalistic statements were true in the model then the usual platonistic formulation . . . would come out true" (90). Our nominalistic statements must give "the full invariant content" of any physical law (60). And why should we believe that this can be done? The thought, inspired by the theory of measurement (e.g., [Krantz et al., 1971]), is that there must be intrinsic features of a physical domain by virtue of which it can be represented mathematically. Those intrinsic features can be expressed nominalistically. If the intrinsic features of the objects — and, thus, the statements we build into S' — were insufficient to entail the mathematical description of them resulting from such representation, then the platonistic physical theory of those objects would unjustifiably attribute to them a structure they do not in fact have. So, any mathematical representation of physical features of objects must, if justifiable, be entailed by intrinsic features of those objects. But that implies that $S' + M \models T$.

We are now in a position to understand in what respect Field's strategy falls short of establishing the supervenience of mathematics on a theory of concreta. Nothing in the above implies that $M \leq S'$ or even that there is some nominalistically stable theory $S'' \supseteq S'$ such that $M \leq S''$. We do, nevertheless, get the result that $M \models \&S' \leftrightarrow \&T$ if S' and T are finitely axiomatizable. Given mathematics, that is, we can demonstrate not only the reducibility of the nominalistic theory S' to our ordinary physical theory T, which is required for the representation theorem underlying the applicability of mathematics to the relevant physical phenomena, but also the equivalence of S' and T *modulo* our mathematical theory. To fall back on the account of reduction in Nagel [1961]: reducibility is equivalent to definability plus derivability. Since $M \models \&S' \leftrightarrow \&T$, we have derivability, but not definability. We can define the expressions of our nominalistic language in terms of the mathematical language of our standard physical theory, but not necessarily vice versa.

None of this is surprising, given Field's outline of his method. But carrying it out often gives rise to the temptation to think that we have definability as well. Consider, for example, his treatment of scalar quantities, as represented by a function $T : R^4 \mapsto R$ representing temperature, gravitational potential, kinetic energy, or some other physical quantity. Suppose that we have a bijection $\phi : D_S \mapsto R^4$ and a representation function ψ from a scalar quantity into an interval, each unique up to a class of transformations. Field observes that $T = \psi \circ \phi^{-1}$. He concludes (emphasis in original): "*This suggests that laws about T (e.g. that it obeys such and such a differential equation) could be restated as laws about the interrelation of ϕ and ψ; and since ϕ and ψ are generated by the basic [nominalistic] predicates ... it is natural to suppose that the laws about T could be further restated in terms of these latter predicates alone*" (59–60). In practice, then, Field often operates by translating mathematical into nonmathematical expressions. The result is not a wholesale reduction of mathematics to a theory of concreta. But it is a reduction of fragments of mathematics employed in a physical theory to something nominalistically acceptable.

Stewart Shapiro observes that, since Field provides a model of space-time isomorphic to R^4, we can duplicate basic arithmetic within Field's theory of intrinsic relations among points in space-time. But that allows us to duplicate the Gödel construction and so devise a sentence in that theory that holds if and only if it is not provable in the theory. Within the theory S of space-time, that is, we can construct a sentence G such that $S \vdash G \leftrightarrow \neg Pr[G]$, where Pr is the space-time correlate of the provability predicate and $[G]$ is the code of G. As usual, we can show that G is equivalent to $Con(S)$, the consistency statement for S, thus showing that $S \nvdash Con(S)$. So, some mathematical truths are not derivable from S.

Gödel's incompleteness theorems are widely recognized as having dealt a serious blow to Hilbert's program, which depended on being able to show the consistency of infinitary mathematics by finite means. Shapiro rightly recognizes the analogy between Hilbert's program and Field's strategy of using the conservativeness of mathematics to justify mathematical reasoning. It is not clear, however, whether the analogy is strong enough to generate a serious problem for Field. (For discussion, see [Shapiro, 1983a; 1983b; 1997; 2000; Field, 1989; 1991].) It does follow, it seems, that Field cannot demonstrate the conservativeness of mathematics by strictly nominalistic reasoning. We need to employ mathematics to prove its own conservativeness. But Field denies that this is troublesome, for he sees his project as a *reductio* of the assumption that mathematical reasoning is indispensable.

Still, we can show in set theory that G and $Con(S)$ are true, even though neither can be demonstrated in Field's space-time theory. That seems to show that mathematics allows us to prove some truths about space-time that Field cannot capture. We might put the point simply by saying that if space-time is isomorphic to R^4, the theory of space-time is not axiomatizable. It is, like arithmetic, essentially incomplete. How heavily this counts against Field's program seems to depend on how adequately he can account for set-theoretic reasoning in metamathematics, something no one has investigated in any detail. For any illuminating discussion of the issues, see [Burgess and Rosen, 1997, 118–123].

Field's program has encountered other, less technical objections. An excellent source for discussion of the issues is Irvine [1990], which contains discussions of Field's work by many of the leading figures in the philosophy of mathematics. Field [1989] develops Field's view further, partly in response to various criticisms. It diverges in complex ways from Field 1980 and will not be discussed in any depth here.

Bob Hale and Crispin Wright [1988; 1990; 1992] observe that Field's theory, especially as developed in Field 1989, takes consistency as a primitive notion, and raise two objections on that basis. First, both standard mathematical theories and their denials are consistent and, as conservative, cannot be confirmed or disconfirmed directly. So, they conclude, Field should be agnostic with respect to the existence of mathematical objects. Field [1993] points out that this ignores his argument that mathematics is dispensable, which constitutes indirect evidence against the existence of mathematical objects. (If mathematics were indispensable in physics, that would constitute indirect evidence in favor of their existence; Field accepts and, indeed, starts from the Quine [1951]–Putnam [1971] indispensability argument. See Colyvan [2001].) Second, on Field's view, the existence of mathematical objects is conceptually contingent. But what could it be contingent *on*? Hale and Wright argue that Field needs an answer. Without one, the claim of conceptual contingency is not only empty but incoherent. Field [1989; 1993] responds by denying the principle that, for every contingency, we need an account of what it is contingent on. It is conceptually possible that God exists; it is conceptually possible that God does not exist. But we have no account of what the existence of God might depend on, nor can we even imagine such an account. The same is true of many conceptual contingencies: the existence of immaterial minds, the n-dimensionality of space-time, the values of fundamental physical constants, and the amount of matter in the universe, for example.

Yablo [2001] raises three additional problems for Field:

- The problem of real content: What are we asserting when we say that $2 + 2 = 4$? According to Field, we are saying something false, since there are no objects standing in those relations. A fictionalist carrying out Field's program may be well aware of that. We might say that we are *quasi-asserting* that 2 + 2 = 4 without really asserting it. Is there anything we are really asserting? I take it that Field's answer is no. We do not have to be viewed as asserting mathematical statements at all. What we are doing is closer to *supposing* them. It is not clear why this generates a problem, or why we must be asserting anything at all.

- The problem of correctness: I assert that 2 + 2 is 4, not 5, even though there may be a consistent and thus conservative theory of the numbers according to which 2 + 2 = 5. (Let 4 and 5 switch places in the natural number sequence, for example. On some theories of mathematics, this suggestion makes no sense. But I take it that Field's is not one of them.) How can I distinguish correct from incorrect assertions about the numbers? If the purpose of a mathematical theory, however, is to characterize up to isomorphism a model into which we can embed aspects of reality, or even a class of such models, we can distinguish correct from incorrect characterization of that model, even if there is some arbitrariness about which model we use.

- The problem of pragmatism: Fictionalists seem to assert sentences, put forward evidence for them, attempt to prove them, get upset when people deny them, and so on — all of which normally accompany belief. How, then, does the fictionalist's attitude toward mathematical utterances fall short of belief? This raises the complex issue of trying to delineate an account of acceptance such as that of Bas van Fraassen [1980]; see Horwich [1991]. Placing this in the context of success in mathematical discourse, however, may make the problem easier. We need an account of what constitutes mathematical success. Arguably, the instrumentalist and representational aspects of Field's fictionalism provide a detailed answer.

Yablo [2002] raises an additional objection. Mathematics, if true at all, seems to be true necessarily. Most properties of mathematical objects seem to be necessary. Being odd, for example, seems to be a necessary property of 3. The relations between mathematical objects — that $\pi > 3$, for example, or, in set theory, that $\forall x \; \emptyset \in \wp(x)$ — appear to hold necessarily. Field thinks he has an explanation: the conservativeness of mathematics entails the applicability of mathematics in any physical circumstance. Conservativeness, he quips, is "necessary truth without the truth" [1989, 242]. Yablo objects to this explanation. First, any physical circumstance has a correlate with numbers; any circumstance with numbers has a correlate without them. If the first leads us to treat mathematics as necessary, why doesn't the second lead us to treat it as impossible? This asymmetry, however, seems easy to explain. The conservativeness of mathematics implies that, in any physical circumstance, it is safe to assume mathematics and use it in reasoning about physical situations and events. We cannot conceive of a situation in which mathematical reasoning fails. That any circumstance with numbers has a correlate without them seems to have no corresponding implication.

Second, mathematical truths seem necessary on their own. For Field, however, mathematical *theories* are conservative. That $\pi > 3$, however, seems necessary *simpliciter*, not merely relative to a theory. It is hard to evaluate this objection, since we learn about mathematical objects in the context of a theory; $\pi > 3$ is a necessary truth *of arithmetic*.

Third, inconsistent statements can both be conservative. But they cannot both be necessarily true. Both the axiom of choice and its negation are conservative over physics, presumably, but they cannot both be necessary (unless they are taken as holding of different parts of the domain of abstracta, as in the Full-Blooded Platonism of Balaguer [1998]). But this seems to relativize mathematics. It conflicts with our sense that $\pi > 3$, period, not merely relative to a certain quite specific mathematical theory — that is, not just relative to set theory, say, but relative to ZFC. Again, however, it is not clear how much force this objection has against Field's view. If a class of theories such as variants of set theory are all conservative over physics, and there is no other basis for choosing among them, it seems plausible to say of the sentences (e.g., the axiom of choice or the continuum hypothesis) on which they disagree not that they are necessarily true but that they are neither true nor false. Imagine a work of fiction in which there are variant readings in various manuscripts. (This is actually the case with *The Canterbury Tales*, for example; there are about eighty different versions.) A supervaluation over the variants seems the only reasonable policy. This may complicate Field's picture slightly, but only slightly; it explains our sense of necessity with respect to those sentences on which the appropriate conservative theories agree while also explaining our unwillingness to assert or deny those on which they disagree.

5 BALAGUER'S FICTIONALISM

Mark Balaguer [1998] develops versions of platonism and fictionalism bearing some affinity with Vaihinger's free-range fictionalism. He does so to show that "*platonism and anti-platonism are both perfectly workable philosophies of mathematics*" (4, emphasis in original). Platonists might feel vindicated, since they see the burden of proof as being on the anti-realist; anti-realists might conclude that they were right all along to insist that a commitment to abstracta is unnecessary. Balaguer himself, however, concludes that there is no fact of the matter.

To see what Balaguer takes fictionalism to accomplish, it is useful to begin with his preferred version of platonism, Full-Blooded Platonism (FBP). FBP's central idea is that "*all possible mathematical objects exist*" (5). For abstract objects, possibility suffices for existence. He observes that this solves many of the philosophical problems faced by platonism. How is it possible, for example, to know anything about mathematical entities? It's easy; any consistent set of axioms describes a realm of abstract entities. No causal contact, empirically scrutable relationship, or explanation of how knowledge is not coincidental is required. FBP is committed to abstract entities, but the entities themselves play no role in our explanation of our knowledge of them. How is it possible to refer to a particular entity? Again, it's easy (at least up to isomorphism); we simply need to make clear which kind of abstract entity we have in mind. Benacerraf [1965] points out that numbers are highly indeterminate, and seem to be nothing but places in an ω-sequence. Balaguer finds this easy to explain; in characterizing a kind of abstract object, we do not speak of a unique collection of objects. Since any possible collection falling under the kind exists, the theory speaks of all of them.

Fictionalism shares the advantages of FBP. Just as authors may tell stories however they like, mathematicians may characterize mathematical domains however they like, provided the characterization they give is consistent. Any fiction gives rise to a collection of fictional characters. Fictionalism is thus full-blooded in just the way FBP is, but without FBP's commitment to abstracta.

Shapiro [2000] complains that consistency is itself a mathematical notion; Balaguer's project is in danger of circularity. Balaguer responds that his concept of consistency is primitive. So long as consistency can be understood pre-mathematically — which seems plausible, since it seems a logical rather than specifically mathematical notion, although theories of logical relationships of course use mathematics — Balaguer's project avoids circularity. An alternative response would be to free both fiction and mathematics from the constraint of consistency, as Vaihinger does, perhaps employing a paraconsistent logic as the underlying logic of mathematics. Given the continuing utility of inconsistent mathematical theories such as naive set theory, this option may have other attractions as well. See, for example, [Routley and Routley, 1972; Meyer, 1976; Meyer and Mortensen, 1984; Mortensen, 1995; Priest, 1994; 1996; 1997; 2000].

Balaguer's preferred version of fictionalism is Field's, which may seem disappointing, since it is highly constrained by the need to nominalize scientific theories and thus quite different in spirit from FBP. But he develops the underlying picture in a pragmatist fashion reminiscent of Peirce, bringing out more fully the representational aspects of Field's fictionalism. His key premise: "Empirical theories use mathematical-object talk only in order to construct *theoretical apparatuses* (or *descriptive frameworks*) in which to make assertions about the physical world" (137). We do not deduce features of the physical world from features of mathematical objects alone; we understand the structure of physical states of affairs in terms of related mathematical structures. That is to say, we model features of the physical world mathematically, mapping intrinsically physical features into mathematical models and using our knowledge of those models to infer features of the physical world.

The reductionist thus has the picture exactly backwards. We should think of mathematics not as something that reduces to the nonmathematical but as something, by design, to which the nonmathematical reduces. Galileo famously defined mathematics as the language in which God has written the universe. From the pragmatist point of view, this is not so remarkable; we create mathematics to be a language in which the relationships we find in the physical universe can be expressed. Mathematics is a container into which any physical structure might be poured, a coordinate system on which any physical structure might be mapped. This is why mathematical domains are typically infinite. Field assumes a physical infinity, an infinite collection of space-time points. But, if mathematics is designed to be a "universal solvent," a theory to which any theory of physical phenomena reduces, and if there is no known limit to the size of physical structures, we need mathematical domains to be infinite if they are to serve their purpose reliably.

6 YABLO'S FIGURALISM

Yablo [2001; 2002; 2005] develops an alternative to Field's fictionalism which he refers to as figuralism or, sometimes, "Kantian logicism." Yablo draws an analogy with figurative speech. "The number of Fs is large iff there are many Fs," for example, he likens to "your marital status changes iff you get married or ...," "your identity is secret iff no one knows who you are," and "your prospects improve iff it becomes likelier that you will succeed." In every case, a figure of speech quantifies over an entity we do not need to take to be a real constituent of the world. We speak of marital status, identity, prospects, stomach butterflies, pangs of conscience, and the like not by describing a distinct realm of objects but instead describing a familiar realm in figurative ways. The unusual entities serve as representational aids. The point of the figurative discourse is not to describe them but to use them to describe other things. They may describe them truly; on this view, the fictional character of the figurative description does not contradict its truth.

Mathematical discourse similarly invokes a realm of specifically mathematical entities, typically using them as representational aids to describe nonmathematical entities. We use statements about numbers, for example, to say things about objects; "the number of asteroids is greater than the number of planets" holds if and only if there are more asteroids than planets. Statements of pure arithmetic, such as $2 + 2 = 4$, express logical truths (in this case, that $(\exists_2 x F x \,\&\, \exists_2 y G y \,\&\, \neg \exists z (Fz \& Gz)) \rightarrow \exists_4 (Fx \vee Gx)$; see, e.g., Hodes (1984)). Statements of pure set theory are logically true over concrete combinatorics, that is, are entailed by basic facts (for example, identity and distinctness facts) about concrete objects. Figuralism thus explains why mathematics is true necessarily and a priori. It also explains why mathematics is absolute rather than relative to a particular theory (until, at any rate, one reaches the frontiers of set theory).

Yablo refers to his figuralist view as a kind of fictionalism, specifically, relative reflexive fictionalism. It is reflexive to recognize that fictional entities can function in two ways, as representational aids (in applied mathematics, for example) or as things represented (e.g., in pure mathematics). They may even, in self-applied contexts, function in both ways, as when a fictionalist, speaking ontologically, says that the number of even primes is zero (on the ground that there are no numbers, and *a fortiori* no even primes).

This example brings out the need for a *relative* reflexive fictionalism. The relativity is not to a particular mathematical theory but to a perspective. Rudolf Carnap [1951] would draw the contrast as one between internal and external questions. Internally, the number of even primes is one; externally, by the fictionalist's lights, that number is zero. Yablo draws it in terms of engaged and disengaged speech. However the distinction is to be drawn, we must distinguish the perspective of those engaged in the relevant language game from the perspective of those talking about the game.

Stanley [2001], directing himself primarily at Yablo's theory, argues that "hermeneutic fictionalism is not a viable strategy in ontology" (36). He advances five objections against Yablo's fictionalism and against hermeneutic fictionalism as such. First, a fictionalist account does not respect compositionality: "there is no systematic relationship between many kinds of sentences and their real-world truth conditions" (41). Second, we move rapidly from pretense to pretense; is there any systematic way of understanding how we do it? Third, pretense is an attitude that evidently must be inaccessible to the person engaged in it, since people do not generally think of themselves as approaching mathematics in an attitude of make-believe. Fourth, the claim that people have the same psychological attitude toward games of make-believe and mathematics is empirically implausible. Fifth, fictionalism splits the question of a speaker's believed ontological commitments from what our best semantic theory postulates in the domains of the models it uses to interpret the speaker's discourse. But what should interest us is the latter, which is the key to explaining a speaker's actual commitments.

The first two objections pertain to compositionality, and may well afflict certain versions of fictionalism, including Yablo's. Yablo does not spell out any compositional way of generating the real-world truth conditions of various mathematical sentences. It is hard to say whether he thinks compositionality is not necessary or whether he takes Frege, Russell, and others as having already spelled it out (with implications he might not like; see [Rosen, 1993]), though he seems, in response to Stanley's criticism, to deny the need for compositionality in any strong sense. But hermeneutic fictionalism is after all a theory of what we *mean* in a certain realm of discourse. It does not seem unreasonable to demand that hermeneutic fictionalism meet the criteria we impose on any other semantic theory of what we mean. As Stanley interprets it, at any rate, hermeneutic fictionalism is by definition — as hermeneutic — a *one-stage* theory. It is a semantic theory intended to characterize what sentences within a certain part of language mean and to show, in the process, by virtue of that semantic theory, that those sentences lack the ontological commitments they seem to have on the basis of an analogy with other kinds of discourse.

From Yablo's perspective, this is a misconstrual. Faced with a sentence in a work of fiction such as that from the *Twice Told Tales*, we do not use an alternative semantics; we interpret the sentence as we always would, but recognize that it is not a literal description of reality. In short, Yablo seems to intend his theory as a *two-stage* theory. The first stage is that of semantic interpretation, and it proceeds in standard fashion. The second involves a recognition that the context is fictional, which leads us to reinterpret the seeming ontological commitments of the discourse.

It is not clear that Yablo can escape the objection so easily. First, he may have to sacrifice his claim to the "hermeneutic" moniker; a two-stage theory, arguably, is no longer a theory of meaning, but a supplement to a theory of meaning. Second, most fiction does not seem to fit his analysis, for most fiction is not figurative. Hawthorne does not in any obvious way invoke a gate made from a whale's jawbone as a representation of something else. Third, and most seriously, a demand for compositionality is not out of place even in the second stage of a two-stage theory. Reductionists have set out to accomplish their ontological aims by interpreting one theory in another in a fully recursive fashion. It is not clear that an *ad hoc* interpretive scheme that cannot be cast in compositional form deserves to be taken seriously.

Does the demand for compositionality tell against other versions of fictionalism in the philosophy of mathematics? A theory like Field's or the one I sketch in the last section of this paper provides a compositional semantics for mathematical sentences; there is no asystematicity about how such sentences are to be interpreted. It seems plausible, moreover, that mathematics constitutes a language game of its own — consider, for example, learning to count — and fictionalists such as Yablo can explain why it is in fact easy for us to switch into and out of that game.

The second two objections address questions of psychological attitude, and, again, may apply to hermeneutic versions of fictionalism, particularly if one defines fictionalism in terms of an attitude of pretense or make-believe. A deflationary fictionalist, however, advances a view that remains independent of the attitude a speaker takes toward his or her discourse. These objections thus miss deflationary fictionalism entirely.

Finally, the fifth objection is straightforwardly metaphysical, and raises an interesting issue about the significance of semantic theory for ontology. It tells against one-stage theories, but not against their two-stage counterparts. Any two-stage theorist must draw a distinction between two senses of ontological commitment. Semantic theory is a good guide to our *ostensible commitments*, the commitments we prima facie seem to have. That is Stage I; that is where ontology starts. But it is not where it ends. Semantic theory is not a reliable guide to our *real commitments*, the commitments we are bound to embrace once, at Stage II, the relations among various theories and discourses is taken into account. Someone who reduces everyday talk of medium-size physical objects to talk of atomic simples, microparticles, sense data, or object-shaped gunk avoids real commitment to ordinary objects as a distinct ontological kind. The deflationary fictionalist similarly avoids real commitment to objects to which he or she is ostensibly committed by a discourse that can successfully be interpreted as fictional.

7 SEMANTIC STRATEGIES

The fictionalist's goal is to resolve Benacerraf's dilemma by breaking the chain of reasoning that leads from the success of mathematics to its truth and then to the existence of mathematical objects. Most fictionalists have chosen to break the chain at the first link, denying that mathematics is true. But it is also possible to question the second link, granting that mathematics is true while denying that its truth requires countenancing distinctively mathematical objects. Reductionism is of course one attempt to accomplish this. Reinterpreting mathematics so that its existential sentences might be true without the existence of abstract objects is another.

7.1 Truth in a Fiction

The simplest strategy is probably to think of each sentence of a fiction as preceded by a fiction operator F meaning something like "it is true in the story that." It is harder to make this thought precise, however, than one might initially suppose. There is an obvious worry about compositionality akin to that Dever [2004] raises against modal fictionalism. Since that raises some technical issues, however, and tracks debates concerning modal fictionalism quite closely — see, for example, [Brock, 1993; Nolan, 1997; 2005; Nolan and Hawthorne, 1996; Divers, 1999; Kim, 2002; 2005; Rosen, 1990; 1995; 2003] — let me concentrate on another problem.

What is a story? How can we give a semantics for the F operator? In ordinary cases of fiction, this is not difficult to do; a story is a finite sequence of sentences. So, we can understand Fp as having the truth condition $S \models p$ (or equivalently, if S is first-order, $\models \& S \rightarrow p$). Notice, however, that this immediately collapses fictionalism into deductivism. It analyzes the (fictional) truth of a statement as being the logical truth of an associated conditional. (Actually, the story as articulated by the author also certainly needs supplementation with frame axioms to derive what are intuitively consequences of the story, but I will ignore this complication.) Mathematical stories (e.g., first-order Peano arithmetic or ZF) are often infinite stories. In the context of a first-order language, or for that matter in any compact logic, Fp will still be true if and only if an associated conditional is true, since $S \models p$ if and only if there is a finite $S_0 \subseteq S$ such that $S_0 \models p$.

We might alternatively think of a story as a model or class of models: we could count Fp as true if and only if, for every model $M \in K$, $M \models p$. But it would be hard to distinguish this version of fictionalism from platonism. How do we pick out the relevant model or model class? How do we know anything about it? Is our knowledge of it a coincidence? Any argument the fictionalist might use against platonism — even Yablo's comparative advantage case — might be turned against this version of fictionalism.

The only viable version of this simple semantic approach to fictionalism that remains truly fictionalist seems to be to second-order. Assume that we tell our mathematical stories in a second-order language. Since we can replace axiom schemata with higher-order axioms, we can safely assume that our stories are finite. We might then say that Fp holds if and only if $S \models p$, where S is the appropriate mathematical story and $\models$ is a second-order relation. This yields a theory like those of Hellman [1989], Shapiro [1997; 2000], and the later Field [1989]. Set aside the Quinean worry that the use of second-order logic smuggles mathematics in through the back door. Strikingly, a second-order approach too turns out to be a version of deductivism. It is not surprising that Hellman and Shapiro do not consider their views fictionalist.

7.2 Constructive Free-range Fictionalism

The obvious semantic route to fictionalism, then, in fact leads away from it. Is there another way of construing the semantic strategy?

Reading quantification in mathematics (and perhaps not only in mathematics) as substitutional rather than objectual, for example, might offer a way of accepting existential sentences in mathematics as true without being forced to recognize the existence of numbers, sets, functions, and other abstracta. I explored this possibility in a series of papers in the 1980s [Bonevac, 1983; 1984a; 1984b]. While this seemed to some a "wild strategy" (Burgess and Rosen 1997), Kripke [1976] had already cleared away the most serious objections to interpreting quantifiers substitutionally. Burgess and Rosen nevertheless argue that any strategy based on a nonstandard reading of the quantifiers is bound to fail, since "'ontological commitment' is a technical term, introduced by a stipulative definition, according to which, nearly enough, ontological commitment just *is* that which ordinary language quantification, in regular and paradigmatic cases, expresses" (204). It is far from clear that this is correct. First, ontological considerations have figured in philosophical discussion at least since the time of Plato; figures throughout philosophical history have debated questions of ontological commitment without using those words. Second, Quine [1939; 1951; 1960] treats his thesis that to be is to be a value of a variable as a substantive claim, not as a stipulative definition. Third, as Szabo [2003] emphasizes, it is not clear that ordinary language existential expressions are univocal, a point already emphasized in Parsons [1980].

It may seem that a substitutional strategy nevertheless succeeds too easily and much too broadly. If substitutional quantification avoids ontological commitment, and, as Benacerraf's semantic continuity requirement seems to demand, ordinary language quantification is substitutional, then it would seem that ordinary language quantification avoids ontological commitment, which is absurd, since commitment is defined in those very terms.

But substitutional quantification does not avoid commitment; it transfers the ontological question to the level of atomic sentences. The strategy means not to avoid metaphysical questions or assume that nothing at all requires the existence of objects but only to shift metaphysical questions from quantified and specifically existential sentences to quantifier-free sentences and their truth conditions. On a substitutional approach, the interesting metaphysical problem arises at the atomic level — why count those atomic sentences as true? — and there no longer seems to be any reason to assume that such a question must have a uniform answer that applies no matter what the atomic sentences happen to be about. In "regular and paradigmatic cases" ordinary language quantification expresses ontological commitment because, in such cases, the truth values of atomic sentences are determined in standard Tarskian fashion and so depend on the existence of objects. On my view, in short, not only is it true that ordinary quantification carries ontological commitment in paradigm cases, but there is an explanation of which cases count as paradigmatic and why.

Nevertheless, a central idea behind this strategy is independent of substitutional quantification. Mathematical domains are generally infinite; one cannot assume that there are enough terms in the language to serve the purposes of a traditional substitutional account. One must therefore think in terms of extensions of the language, counting an existential sentence true if it is possible to add a term to the language serving as a witness to the sentence. That, however, introduces a modal element to the theory, which can be isolated from a substitutional interpretation of the quantifiers. Here I shall present the revised-semantics strategy in a form that emphasizes its modal character, relying in part on unpublished work that Hans Kamp and I did some years ago but setting aside its substitutional features.

Two sets of considerations in addition to the substitutional considerations just outlined motivate the revised semantics for quantification that I am about to present. The first concerns the mathematician's freedom to introduce existence assumptions, which seems analogous to the freedom of an author to introduce objects in a work of fiction. Georg Cantor [1883] wrote that "the very essence of mathematics is its freedom." David Hilbert [1980] saw consistency as the only constraint on mathematical freedom: "[I]f the arbitrarily given axioms do not contradict one another with all their consequences, then they are true and the things defined by the axioms exist. This for me is the criterion of truth and existence." Henri Poincaré similarly maintained that "A mathematical entity exists provided there is no contradiction implied in its definition, either in itself, or with the proposition previously admitted" [Poincaré, 1952]; "In mathematics the word *exist* can have only one meaning; it signifies exemption from contradiction." This suggests that existence, in mathematics as in fiction, is tied to possibility. Something of a certain kind exists if it is possible to find or construct a thing of that kind.

The second motivation stems from thinking of mathematical objects and domains as constructed. We might think of fictional entities as introduced, not all at once, but as a work of fiction (or a series of works of fiction, such as Doyle's Sherlock Holmes stories or Whedon's *Buffy the Vampire Slayer* and *Angel* episodes) unfolds. Similarly, we might think of mathematical objects and mathematical domains as constructed over time or, more generally, over a series of creative mathematical acts. We can assert that a mathematical entity of a kind exists if and only if it is possible to construct an object of that kind.

This might suggest that *exists* means something quite different in mathematics from what it means in ordinary contexts. Poincaré [1952], indeed, maintained that "the word 'existence' has not the same meaning when it refers to a mathematical entity as when it refers to a material object." This flies in the face of Benacerraf's semantic continuity requirement, that language used in both mathematical and nonmathematical contexts must be given a uniform semantics covering both. What follows attempts to do just that: provide a uniform semantics for mathematical and nonmathematical language that explains the difference in existence criteria not in terms of meaning but in terms of ontology — in terms, that is, of the kinds of objects and models under consideration.

7.3 Definitions

The semantics I shall present follows the pattern of Kripke semantics for intuitionistic logic [Kripke, 1965]. A Kripke model $\mathfrak{K}$ for language L is a quadruple $< K, \leq, D, \Vdash >$, where $< K, \leq >$ is a poset, D is a monotonic function from K to inhabited sets, called *domains*, and $\Vdash$ is a relation from K to the set of atomic formulas of $L^* = L \cup D(K) (= \{D(k) : k \in K\})$ such that (a) $k \Vdash R^n(d_1...d_n) \Rightarrow d_j \in D(k)$ for $1 \leq j \leq n$ (existence), and (b) $k \Vdash R^n(d_1...d_n)$ and $k \leq k' \Rightarrow k' \Vdash R^n(d_1...d_n)$ (persistence).

We may extend $\Vdash$ to all formulas by inductive clauses for compound formulas in several ways. One familiar method is intuitionistic:

$k \Vdash_i A \& B \Leftrightarrow k \Vdash_i A$ and $k \Vdash_i B$
$k \Vdash_i A \vee B \Leftrightarrow k \Vdash_i A$ or $k \Vdash_i B$
$k \Vdash_i A \rightarrow B \Leftrightarrow \forall k' \geq k$ if $k' \Vdash_i A$ then $k' \Vdash_i B$
$k \Vdash_i \neg A \Leftrightarrow \forall k' \geq k \; k \nVdash_i A$
$k \Vdash_i \exists x A(x) \Leftrightarrow \exists d \in D(k) \; k \Vdash_i A(d)$
$k \Vdash_i \forall x A(x) \Leftrightarrow \forall k' \geq k \forall d \in D(k') \; k' \Vdash_i A(d)$

Another is classical:

$k \Vdash_c A \& B \Leftrightarrow k \Vdash_c A$ and $k \Vdash_c B$
$k \Vdash_c A \vee B \Leftrightarrow k \Vdash_c A$ or $k \Vdash_c B$
$k \Vdash_c A \rightarrow B \Leftrightarrow k \nVdash_c A$ or $k \Vdash_c B$
$k \Vdash_c \neg A \Leftrightarrow k \nVdash_c A$
$k \Vdash_c \exists x A(x) \Leftrightarrow \exists d \in D(k) \; k \Vdash_c A(d)$
$k \Vdash_c \forall x A(x) \Leftrightarrow \forall d \in D(k) \; k \Vdash_c A(d)$

For now, however, I want to focus on another possibility. Suppose we take seriously the thought that mathematical objects, like fictional objects, are constructed — are characterized in stages of creative acts, as Kripke models for intuitionistic logic seem to reflect.

We might be tempted to adopt intuitionistic logic as that appropriate to mathematical and fictional reasoning alike. Yet there is something odd about intuitionistic logic in this connection, something that the thought of the objects of the domain as constructed does not itself justify or explain. Intuitionistic logic treats the quantifiers asymmetrically. It is well-known that, in intuitionism, one cannot define the quantifiers as duals of each other. In fact, one could not do so even if negation were given a classical rather than intuitionistic analysis. The Kripke semantics brings out the reason why: the universal quantifier is essentially forward-looking, taking into account future stages of construction, while the existential quantifier is not. The existential quantifier, one might say, ranges over *constructed* objects, while the universal quantifier ranges over *constructible* objects. Nothing about taking the domain as constructed in stages seems to require that. Intuitionistic logic, in short, is not Quinean: to be cannot be construed as being the value of a variable, for being the value of a variable has no univocal meaning.

Intuitionistic logic rests on two independent theses expressing quite different kinds of constructivism. One is the metaphysical thesis that the domain consists of constructed objects; the other, the epistemological thesis that only constructive proofs can justify existence statements. The former seems compatible with fictionalism, and perhaps even to be entailed by it. The latter, however, has no obvious link to fictionalism, and in fact seems inconsistent with the freedom of existential assertion that Cantor, Poincaré, Hilbert, and Balaguer have found central to mathematical practice.

To be clear about this, we need to distinguish the perspective of the author of a work of fiction from the perspective of someone talking about the fiction — and, similarly, the perspective of the creative mathematician form the perspective of someone talking about the mathematical theory. What justifies me in saying that there are vampires with souls in the universe of *Buffy the Vampire Slayer* is Josh Whedon's construction or, better, stipulation of them. What justifies Josh Whedon in saying so is an entirely different matter, and seems to be nothing more than logical possibility. Just so, what justifies me in saying that every Banach space has a norm is Banach's stipulation, but what justified Stefan Banach in saying so is something else, and again seems to be nothing more than logical possibility. (Perhaps, as Vaihinger suggests, not even logical possibility is required; we could develop a version of the theory using a paraconsistent logic, by, for example, thinking of the semantics as relating sentences to truth values.)

In thinking of mathematics as in some sense fictional, it is important to focus on the perspective of the author or more generally creator of a fictional work rather than the perspective of the reader or viewer. The latter perspectives are derivative. I am justified in saying that there are ensouled vampires in the Buffyverse by virtue of Whedon's stipulation of them. I am justified in saying that all Banach spaces have norms by virtue of Banach's stipulation of them. In short, my epistemological perspective with respect to the universe of *Buffy* or Banach spaces is derivative. It requires no epistemic contact with vampires or Banach spaces. Given the stipulations of Whedon and Banach, it seems fairly easy to explain my knowledge of the relevant domains; I have contact not with the objects but with the stipulations.

The philosophically interesting question concerns their justification in making those stipulations in the first place. If they, as the authors of the relevant fictions, were indeed free, their acts of creation evidently required nothing like a constructive existence proof. One might be tempted to think that the author's act of creation is itself a construction of the kind demanded in intuitionism and other forms of constructive mathematics. But that would be a mistake. The author is not at all like "the constructive mathematician [who] must be presented with an algorithm that constructs the object x before he will recognize that x exists" [Bridges and Richman, 1987]. The author can stipulate whatever he or she pleases.

For fictionalist purposes, then, it makes sense to isolate the thought that mathematical objects and domains are constructed — and the related thought that existence in mathematics amounts to constructibility — from the further thought that existence claims can be justified only by the completion of certain kinds of constructions. Kripke semantics lends itself naturally to constructivism in the metaphysical sense. To free it from an insistence on constructive proof, however, and to bring it into line with Quine's understanding of the role of quantification as ranging over a domain, we might reasonably adapt the intuitionistic truth clauses to treat the quantifiers symmetrically, as ranging over constructible objects:

$k \Vdash A \& B \Leftrightarrow k \Vdash A$ and $k \Vdash B$
$k \Vdash A \vee B \Leftrightarrow k \Vdash A$ or $k \Vdash B$
$k \Vdash A \rightarrow B \Leftrightarrow \forall k' \geq k$ if $k' \Vdash A$ then $k' \Vdash B$
$k \Vdash \neg A \Leftrightarrow \forall k' \geq k \ k \nVdash A$
$k \Vdash \exists x A(x) \Leftrightarrow \exists k' \geq k \exists d \in D(k') \ k' \Vdash A(d)$
$k \Vdash \forall x A(x) \Leftrightarrow \forall k' \geq k \forall d \in D(k') \ k' \Vdash A(d)$

As usual, say that A is *valid* at k in $\mathfrak{K} \Leftrightarrow k \Vdash A$, and that A is valid in $\mathfrak{K} \Leftrightarrow \forall k \in K \ k \Vdash A$ (written $\mathfrak{K} \Vdash A$). $\Sigma \Vdash A \Leftrightarrow \forall \mathfrak{K}$ if $\mathfrak{K} \Vdash B$ for all $B \in \Sigma$ then $\mathfrak{K} \Vdash A$. A is *valid* ($\Vdash A$) iff $\emptyset \Vdash A$.

Let $\mathfrak{K}_k$ be $< K', \leq', D', \Vdash'>$, where $K' = \{k' : k' \geq k\}$, $\leq' = \leq \restriction k'$, $D' = D \restriction K'$, and $\Vdash' = \Vdash \restriction (K' \times$ the atomic formulas of L^*). It is easy to show the following:

$$k \Vdash A \Leftrightarrow \mathfrak{K}_k \Vdash' A \Leftrightarrow k \Vdash' A.$$

Validity in $\mathfrak{K}$ is just validity in $\mathfrak{K}$'s bottom node, if there is one. If A is built up from just disjunction and conjunction — the connectives receiving the same truth conditions in all three logics so far defined — then $\Vdash A \Leftrightarrow \Vdash_i A \Leftrightarrow \Vdash_c A$. Monotonicity fails for formulas containing existential quantifiers. That is, in intuitionistic logic, $k \Vdash_i A$ and $k' \geq k$ imply $k' \Vdash A$. That holds in the logic I have defined only for formulas without $\exists$. On such formulas, it agrees entirely with intuitionistic logic. In general, however, it is weaker. Every valid formula is valid in intuitionistic logic. The reverse, however, fails.

Some intuitionistically valid formulas that fail:

1. $\exists x \forall y A \rightarrow \forall y \exists x A$

2. $(\exists x A \& \exists y B) \rightarrow \exists x \exists y (A \& B)$ (where x does not occur in B and y does not occur in A)

3. $\forall x \forall y (A \vee B) \rightarrow (\forall x A \vee \forall y B)$ (where x does not occur in B and y does not occur in A)

The first of these may seem surprising, but, in speaking of fictional characters, its failure has some plausibility. *There are characters who treat everyone badly* does not appear to entail *Everyone is treated badly by some character*. We naturally read the former sentence as meaning that some characters treat every character in their fiction badly; the latter, however, seems to speak of everyone in or outside a work of fiction. The second and third formulas above in effect mix stories.

In general, what Quine calls "rules of passage" are not valid. Also, note that despite the symmetry of the quantificational truth clauses, the quantifiers are not duals of each other, because of the intuitionistic understanding of negation.

7.4   Settled Models

Say that $\Re$ is *settled* iff for all $k, k' \geq k$, if A is in the language of k, then $k' \Vdash A \Rightarrow k \Vdash A$. In settled models, that is, further stages of construction make no difference to truth values. It is easy to see that, in settled models, sentential connectives behave classically.

Even in settled models, however, the quantifiers do not behave classically. The above formulas remain invalid. A sound and complete system for settled models has, in addition to the rules

MP: $\vdash A, \vdash A \to B \Rightarrow \vdash B$

UG: $\vdash Ay/x \Rightarrow \vdash \forall xA$ (where y does not occur in A)

a complete set of schemata for classical sentential logic, and schemata defining &, $\vee$, and $\exists$ in terms of the other connectives, the following quantificational axiom schemata:

UI: $\forall xA \to At/x$, where t is a term free for x in A

CBV: $\forall xA \to \forall yAy/x$, where y is a variable free for x in A

DIS: $\forall x(A \to B) \to (\forall xA \to \forall xB)$

AMAL: $(\forall xA \& \forall yB) \to \forall x\forall y(A \& B)$, where x does not occur free in B and y does not occur free in A

BVQ: $A \to \forall xA$, where A is a basic formula and x does not occur free in A

The proof proceeds by defining a tableau system analogous to those for intuitionistic logic.

7.5   Minervan Constructions

The logic of quantification that emerges from this conception is thus weaker than intuitionistic logic and, even on the class of settled models, weaker than classical logic. Is there a restricted class of models on which the quantifiers behave classically? If so, we might see classical logic as stemming from a more general semantics plus ontological considerations. In fact, two rather different sets of ontological assumptions yield classical logic. One is metaphysically especially well-suited to ordinary physical objects; the other, to mathematical objects. Assume throughout that we are within the class of settled models. It is simplest to assume that we have replaced the intuitionistic clauses for sentential connectives with their classical counterparts.

We may think of the domain as constructed in two different senses. In the collective sense, the domain may be constructed in stages if objects are added to it gradually, at different stages of construction. In the distributive sense, the domain may consist of objects which themselves are constructed gradually, in stages. Call a construction (or the objects issuing from it) *minervan* if it produces elements that are complete as soon as they emerge, and *marsupial* if it produces elements that are fledgling at first and achieve maturity gradually as the structure unfolds. Fictional objects are typically marsupial; they are defined gradually throughout a work of fiction, and might develop in a variety of ways. Earlier stages of construction do not determine how later stages must go. Mathematical objects might be construed similarly. But they might also be construed according to the minervan conception. The properties of 2, π, or e appear to be determined as soon as they are introduced — though it might of course take a very long time for those properties to be known, articulated, and understood.

Interestingly, under certain conditions, each conception leads to a classical logic of quantification. Consider first the minervan conception. We might think of our ourselves as constructing a well-defined domain of objects. The domain expands in stages. Every object introduced, however, is fully defined as soon as it is introduced, and retains its identity across subsequent stages of construction. The domain, in other words, is constructed in the collective sense only; each object is fully characterized upon its introduction.

Formally, we can capture this conception by restricting ourselves to *Berkeley models*, models $\mathfrak{K}$ that are *strong nets* in the sense that, for all $k, k' \in K$ there is a $k'' \in K$ such that $k \cup k' \subseteq k''$. (The name is inspired by Bishop Berkeley's thought that our constructions are approximations of ideas in the mind of God.) Nodes in a strong net are compatible with each other; they must agree about the objects they have in common.

By induction on the complexity of formulas, we can show that, if $\mathfrak{K}$ is a settled strong net, then, for any sentence A of L and $k, k' \in K$, if $L(A) \subseteq L(k)$ and $k \leq k'$, $k \Vdash A \Leftrightarrow k' \Vdash A$. It follows that A holds in all settled strong nets iff A is classically valid.

The real significance of this result lies not in its application to mathematical objects, which, if viewed as fictional, seem to be construed more naturally as marsupial, but in its application to physical objects. If objects are minervan, then the constructive aspects of the semantics make little difference, and the logic that emerges is classical. That explains how the criteria of existence in mathematical and nonmathematical contexts can appear quite different, even though *exists* has a univocal meaning in both. It explains, in short, how it is possible for the fictionalist to satisfy Benacerraf's semantic continuity requirement, giving a uniform semantics for mathematical and nonmathematical language.

7.6 *Marsupial Constructions*

Marsupial constructions, too, under certain conditions yield classical logic. If we think of objects as constructed gradually in stages, we might naturally think of later stages of development as to some extent undetermined by earlier stages. The construction might therefore develop one or more relational structures. Suppose that moreover we think of the objects as "thin" from an ontological point of view — as described purely qualitatively, having no further hidden nature or essence [Azzouni, 1994; Yablo, 2001]. We might, for example, think of the names in our theory as mere pegs in a relational grid. As in Benacerraf [1965], for example, we might think of the numeral '2' not as standing for a determinate object but as marking a place in a relational grid of order type ω. I will call the models that make this conception precise *structuralist models*. Such models seems especially well suited to mathematical theories, which at best characterize their objects up to isomorphism.

Let f be a function from objects to objects such that the domain of f includes $D(k)$. Extend f to formulas by letting $f(A)$ be the result of replacing each designator of x in A with a designator of $f(x)$. Node $f(K)$ is such that, for every atomic sentence A of L, $f(k) \Vdash f(A) \Leftrightarrow k \Vdash A$. If f is a one-one function from $D(k_1)$ onto $D(k_2)$, say that k_1 and k_2 are equivalent modulo f ($k_1 \approx_f k_2 \Leftrightarrow k_2 = f(k_1)$).

$\mathfrak{K}$ has the *universal understudy property* iff the following holds for $k_1, k_2, k_3 \in K$ such that $k_1 \leq k_3$ and $k_1 \approx_f k_2$ for some f: If $C \subseteq D(K)$ is disjoint from $D(k_3)$ and g is any one-one function from $D(k_3) - D(k_1)$ onto C, then there is a $k_4 \in K$ such that $k_3 \approx_{f \cup g} k_4$ and $k_2 \leq k_4$. In models with this property, any group of objects not already in a node may play the roles of a group of that node's objects. In this sense, anything in the universe of the model but playing no role in a node may serve as "understudy" for anything that is playing a role. The objects of such a model are extremely versatile; they can play any role the model provides. That reflects well the idea that the objects in such a model are mere pegs, having no identity independent of the bundle of properties they instantiate.

Say that $\mathfrak{K}$ is *inexhaustible* iff for each $k_1, k_2 \in K$ there is a subset C of $D(K)$ such that $C \cap D(k_1) = \emptyset$ and $|C| = |D(k_2)|$. Given the assumption that $K \neq \emptyset$, this condition implies the weaker property that, for all $k \in K$, there is a $c \in D(K)$ such that $c \notin D(k)$. If $k \in K$ and n is a natural number, there is a set $C \subseteq D(K)$ disjoint from $D(k)$ which has cardinality n.

If $\mathfrak{K}$ is inexhaustible and has the universal understudy property, it also has the *existential understudy property*: for all $k_1, k_2, k_3 \in K$ such that $k_1 \leq k_2$ and $k_1 \leq k_3$, there exist a set $C' \subseteq D(K)$ such that $C' \cap (D(k_3) - D(k_1)) = \emptyset$, a one-one function f from $D(k_3) - D(k_1)$ onto C', and a $k_4 \in K$ such that $k_2 \approx_g k_4$, where g is the union of f and the identity function on $D(k_1)$. If a model has this property, each collection of objects playing roles in a node has a team of "understudies" not contained in the node who can take over the roles played by members of the collection — and in fact do so on another node of the model.

Call any inexhaustible model with the understudy properties *permutable*.

A model $\mathfrak{K}$ is a *weak net* iff, whenever $k_1, k_2 \in K$ and $k_1 \upharpoonright (D(k_1) \cap D(k_2)) = k_2 \upharpoonright (D(k_1) \cap D(k_2))$, there is a $k_3 \in K$ such that $k_1 \cup k_2 \subseteq k_3$. Any two nodes that agree on their common domain, that is, are together subsumed within some further node. In a weak net, nodes function as giving (partial) information about their objects; whenever they agree on the objects appearing on both, the model combines the information to yield a more complete picture of the objects in question. But there is no requirement that nodes agree on their common objects. The model may develop alternative and incompatible portraits of the objects under construction.

A *structuralist model*, then, is a permutable weak net. By induction on the complexity of formulas, we can show that, if $\mathfrak{K}$ is a settled structuralist model, then, for any sentence

A of L and $k, k' \in K$, if $L(A) \subseteq L(k)$ and $k \leq k'$, $k \Vdash A \Leftrightarrow k' \Vdash A$. It follows that A holds in all settled structuralist models iff A is classically valid. This explains how it is possible to maintain both that mathematical objects are constructed and that classical logic is appropriate to mathematical reasoning. It also explains how a unified semantics for mathematical and nonmathematical language can apply successfully to both and yield classical logic when applied to both, even though the semantics itself is nonclassical.

7.7 Open Models

We have so far developed two conceptions under which a symmetric, constructive semantics for quantification yields classical first-order logic. But our constructive fictionalism is not yet free-range. Neither conception reflects Cantor's idea that the essence of mathematics is its freedom; neither reflects Poincaré's thought that existential assertions in mathematics require nothing more than consistency. To capture those notions, we need the concept of an *open model*, an model, analogous to a canonical model in modal logic, in which all possibilities — or, at least, all possibilities consistent with certain constraints — are realized.

Let W be a set of infinite cardinality κ. The *open model of cardinality κ* for W, $O_W = \langle K, \leq, D, \Vdash \rangle$, is generated from $K = \{k : \exists U \subseteq W(|U| < \kappa \ \& \ D(k) = U)\}$, where $k \leq k' \Leftrightarrow D(k) \subseteq D(k')$. It is straightforward to show that O_W is an inexhaustible weak net with the understudy properties — in short, a structuralist model. It follows that $\mathfrak{Th}(O_W)$ is a first-order theory.

Say that model $\mathfrak{K}$ for L is *reductively complete* iff for all $k \in K$ and every L' such that $L \subseteq L' \subseteq L(k)$, $k \upharpoonright L' \in K$. In reductively complete models, altering the quantificational clauses to ones considering only nodes extending the current one by the addition of a single object would make no difference, provided that for any sentence A of L_k and $k \leq k' \in K$, $k \Vdash A \Leftrightarrow k' \Vdash A$.

We can use this fact to show (somewhat tediously) that if L contains no individual constants and O is an open model of cardinality κ, $\mathfrak{Th}(O)$ is decidable. Moreover, $\mathfrak{Th}(O)$ is an $\forall \exists$ theory, axiomatized by the set $\mathfrak{A}(O)$ of axioms of the form $\forall x_1...x_n \exists y_1...y_m F(\vec{x}, \vec{y})$, where $n \geq 0, m \geq 1, F(\vec{x}, \vec{y})$ is a consistent conjunction of basic formulas built from predicates of L and variables from among $x_1...x_n, y_1...y_m$, and in each conjunct there is at least one occurrence of one of the ys.

It is possible to generalize this result in two different directions. First, suppose that $\mathfrak{K}$ is a reductively complete structuralist model: a reductively complete permutable weak net. Then $\mathfrak{Th}(\mathfrak{K})$ is a first-order theory axiomatized by $\mathfrak{A}(\mathfrak{K}))$, consisting of

(i) all axioms of the form $\forall \vec{x}(G(\vec{x}) \rightarrow \exists y G'(\vec{x}, y))$, where for some $\vec{c}, c, G(\vec{c}), G'(\vec{c}, c)$ are diagrams of some $k, k' \in K$ such that $D(k')$ extends $D(k)$ by a single object, and

(ii) all axioms of the form $\forall \vec{x}, y(G(\vec{x}) \rightarrow \bigvee_i G_i(\vec{x}, y))$, where for some such $\vec{c}$, c, $G(\vec{c})$, $G'(\vec{c}, c)$, $G_1(\vec{c}, c)$, ..., $G_n(\vec{c}, c)$ are diagrams of all the $k' \in K$ extending k by a single object. If $\mathfrak{K}$ is not reductively complete, but has an analogous property with respect to finite extensions, then $\mathfrak{Th}(\mathfrak{K})$ is similarly axiomatizable, and will in fact be decidable iff $\mathfrak{A}(\mathfrak{K}))$ is recursively enumerable.

Second, and more immediately relevant to mathematics, suppose that P is a decidable set of purely universal sentences of L — sentences, from a metaphysical point of view, carrying no ontological commitment. Then we can characterize the *P-open model of cardinality* κ for W, $O_{P,W} = \langle K, \leq, D, \Vdash \rangle$, generated from $K = \{k : \exists U \subseteq W(|U| < \kappa \ \& \ D(k) = U \ \& \ $ the diagram of k is consistent with $P)\}$, where $k \leq k' \Leftrightarrow D(k) \subseteq D(k')$. It is straightforward to show that $O_{P,W}$ is an inexhaustible weak net with the understudy properties. It follows that $\mathfrak{Th}(O_{P,W})$ is a classical first-order theory.

Suppose, for example, that L consists of a single nonlogical two-place predicate $<$ characterized as a strict linear order by the purely universal axioms $\forall x, y(x < y \rightarrow \neg y < x)$, $\forall x, y, z((x < y \ \& \ y < z) \rightarrow x < z)$, and $\forall x, y(x < y \lor y < x \lor x = y)$. Among the theorems of the theory of the $<$-open model of cardinality $\aleph_0$ for N would be $\forall x \exists y \ x < y$, $\forall x \exists y \ y < x$, and $\forall x, y(x < y \rightarrow \exists z(x < z \ \& \ z < y))$. We thus get a theory of a dense linear order extending infinitely in both directions, even though every stage k in the structure has a finite domain.

If every mathematical theory could be analyzed as the theory of a P-open model of some cardinality, we could stop the account here, and have, perhaps, a version of modal structuralism that would capture a variety of fictionalist insights. Whether it would deserve to be called a version of fictionalism is unclear. Unlike other forms of structuralism, it does take seriously the thought that the structures in question are products of human creative activity governed by no constraints other than those applying to fiction. How many mathematical theories might be analyzed as theories of P-open models remains an open question.

It appears, however, that some mathematical theories are fictionalist in a stronger sense. They appear not to be analyzable as theories of P-open models. Peano arithmetic, for example, assumes the existence of zero as the sole natural number without a predecessor. Set theory assumes the existence of the null set and of an infinite set. Geometry, on Hilbert's axiomatization, assumes the existence of two points lying on a line, three points not lying on a line, and four points not lying in a plane. It is possible to account for some such theories in terms of Q-open models in which, among the axioms of Q, there are not only purely universal sentences but also (a) pure existentials (needed, for example, in the case of geometry) and (b) definitions of one or more constants (needed, for example, in the case of arithmetic). Suppose, for example, we define 0 by means of the formula $\forall x(x = 0 \leftrightarrow \neg \exists y \, S \, yx)$, where S is the successor relation, and stipulate that $\forall x, y, z((S \, xy \ \& \ S \, yz) \rightarrow y = z)$ and $\forall x, y, z((S \, yz \ \& \ S \, xz) \rightarrow x = y)$. The theory of the relevant open model then includes $\forall x \exists y S \, xy$ and $\forall x(x \neq 0 \rightarrow \exists y S \, yx)$. The induction schema corresponds to a pure universal in a second-order language, and so can perhaps in principle be included in the axiom set Q. For that reason, in fact, it is easier to analyze second-order arithmetic as the theory of a Q-open model than it is to analyze first-order arithmetic in similar fashion.

For the same reason, it is easier to attempt an analysis of second-order set theory. Whether set theory or geometry can be understood as the theory of a Q-open model is a large question I cannot discuss here in detail. The general strategy for set theory would be to use arithmetic or the theory of dense linear order to define an infinite set, thus justifying the axiom of infinity; to use Q to define unions, pair sets, and power sets, and to express a second-order abstraction axiom; and then to view axioms of sum sets, pair sets, and power sets as theorems.

However the details of this might go, the philosophical moral appears to be that certain mathematical theories, especially existential mathematical theories such as Peano arithmetic and set theory, if capable of being given a fictional interpretation, are fictional in two senses. They are fictional in the sense that they speak of objects constructible given the general criteria of construction governing fiction. They are also fictional in the sense that they require the postulation of an object — zero, in the case of the theory of natural numbers, or the null set, in the case of set theory — that accords with those criteria but the existence of which cannot be viewed as a logical truth. The semantic fictionalism I am outlining is irresistably fictionalist; it does not collapse into deductivism or reductionism.

7.8 Modal Translations

That both minervan and marsupial conceptions of objects — Berkeley models and structuralist models — yield classical first-order logic raises a number of questions. First, under what conditions does the semantics I have sketched yield first-order logic? We have seen two sets of sufficient conditions; what are necessary conditions? What effect do the understudy properties and the weak net condition have on their own? Are there conditions that might be imposed on the semantics to yield intuitionistic logic? How do strong nets and structuralist models behave when not restricted to the class of settled models?

I can only begin to address such question here. Some light is shed on them by considering a modal translation of formulas. Kurt Gödel [1933] showed that intuitionistic logic translates into modal logic in such a way that a formula is valid intuitionistically iff its translation is valid in S4. Almost exactly the same translation works for the logic I have developed here. We may define the translation of a formula recursively as follows:

$A^\pi = \Box A$, if A is atomic
$(\neg A)^\pi = \Box \neg A^\pi$
$(A \& B)^\pi = A^\pi \& B^\pi$
$(A \vee B)^\pi = A^\pi \vee B^\pi$
$(A \to B)^\pi = \Box(A^\pi \to B^\pi)$
$(\forall x A)^\pi = \Box \forall x A^\pi$
$(\exists x A)^\pi = \Diamond \exists x A^\pi$

This is identical to the Gödel translation except for its treatment of existential quantification. Let MS4 indicate quantified S4 with domains that may increase in accessible worlds, and CS4 indicate quantified S4 with constant domains (and similarly for other logics). MS4 thus validates the converse Barcan formula, $\Box \forall x A \to \forall x \Box A$, and CS4 validates in addition the Barcan formula, $\forall x \Box A \to \Box \forall x A$. It is straightforward to establish the following facts:

A is valid on the class of all models iff A^π is valid in MS4.
A is valid on the class of all settled models iff A^π is valid in MS5.
A is valid on the class of all settled Berkeley models iff A^π is valid in CS5.
A is valid on the class of all settled structuralist models iff A^π is valid in CS5.

A is valid on the class of all weak nets iff A^π is valid in MS4.2.

S4.2, with characteristic axiom $\Diamond\Box A \supset \Box\Diamond A$, characterizes the class of reflexive, transitive, and convergent models. (R is convergent iff $\forall x, y, z((xRy \wedge xRz) \supset \exists w(yRw \wedge zRw))$.) It points the way to the schema $\exists x\forall yA \to \forall y\exists xA$, which holds in all weak nets.

It is tempting to think that there might be a constraint that would yield intuitionistic logic. Brouwer [1913; 1949] and Heyting [1956] suggest such a possibility, contending that intuitionistic logic is required specifically for infinite mathematical constructions. Dummett [1973], in contrast, denies that intuitionism can be given any such ontological foundation, insisting that it rests solely on a semantical thesis about the nature of truth. It is easy, in the context of the view I have developed, to confirm Dummett's perspective: There is no class C of models such that A is true in C iff A is intuitionistically valid. Any constraint on models that would permit the existential quantifier to behave intuitionistically, so that $k \Vdash \exists xA \Leftrightarrow \exists d \in D(k) \, k \Vdash A(d)$, would force the universal quantifier to behave classically. The asymmetry of intuitionistic logic's treatment of the quantifiers cannot be removed by any ontological constraint on the class of models.

Burgess and Rosen object to semantic views that overtly or covertly introduce modality, as mine arguably does, on the ground that they trade one obscurity for another. Indeed, a number of fictionalist views — perhaps all of them — can be seen as trading ontology for ideology. Fictionalists often add modal notions, and can legitimately stand accused of eliminating commitments to abstracta and other troublesome entities at the expense of a commitment to possibilia. Some fictionalists see this as an advance, since possible objects or states of affairs need to be invoked for other kinds of modals. Others see it as at best an intermediate step, and argue for a modal fictionalism that employs the same strategy to eliminate commitment to possible worlds. Modal fictionalism has its own problems that I cannot discuss here. In any case, any philosopher of mathematics who takes seriously the need to explain the necessity of mathematics (or, more neutrally, perhaps, its applicability to hypothetical and counterfactual situations) encounters the problem of accounting for modal operators. So, this raises issues that are not unique to fictionalism or even especially forceful with respect to it.

It highlights, however, an issue that faces fictionalism in general. Metaphysical and epistemological problems arise when a discourse seems to commit us to empirically inscrutable objects or facts. Fictionalism addresses those problems by construing the discourse as fictional, as capable of succeeding despite the nonexistence of such objects or facts. One can go on to ask why, in the absence of such constraints, that discourse, as opposed to other possible competitors, succeeds. Attempts to respond by citing a fact threaten to collapse the fictionalist project into reductive or modal accounts. Only accounts that allow all possible discourses to count as successful seem to promise a theory that is stably fictional.

BIBLIOGRAPHY

[Adams, 1977] M. Adams. Ockham's Nominalism and Unreal Entities, *Philosophical Review* 87, 144–176, 1977.

[Adams, 1987] M. Adams. *William Ockham*. 2 vols., Notre Dame, Ind.: University of Notre Dame Press, 1987. (2nd rev. ed., 1989.)

[Azzouni, 1994] J. Azzouni. *Metaphysical Myths, Mathematical Practice*. Cambridge: Cambridge University Press, 1994.
[Balaguer, 1998] M. Balaguer. *Platonism and Anti-Platonism in Mathematics*. New York : Oxford University Press, 1998.
[Benacerraf, 1965] P. Benacerraf. What Numbers Could Not Be, *Philosophical Review* 74, 47–73, 1965.
[Benacerraf, 1973] P. Benacerraf. Mathematical Truth, *Journal of Philosophy* 70, 661–679, 1973.
[Benacerraf and Putnam, 1983] P. Benacerraf and H. Putnam. *Philosophy of Mathematics*. Cambridge: Cambridge University Press, 1983.
[Bentham, 1932] J. Bentham. *Bentham's Theory of Fictions*. C. K. Ogden, ed. London: Ams Press, 1932.
[Boehner, 1946] P. Boehner. The Realistic Conceptualism of William Ockham. *Traditio* 4, pp. 307–35, 1946.
[Bonevac, 1982] D. Bonevac. *Reduction in the Abstract Sciences*. Indianapolis: Hackett, 1982.
[Bonevac, 1983a] D. Bonevac. Freedom and Truth in Mathematics, *Erkenntnis* 20, 93–102, 1983.
[Bonevac, 1983b] D. Bonevac. Quantity and Quantification, *Noûs* 19, 229–248, 1983.
[Bonevac, 1984a] D. Bonevac. Mathematics and Metalogic, *Monist* 67, 56–71, 1984.
[Bonevac, 1984b] D. Bonevac. Systems of Substitutional Semantics, *Philosophy of Science* 57, 631–656, 1984.
[Bonevac, 1991] D. Bonevac. Semantics and Supervenience, *Synthèse* 87, 331–361, 1991.
[Bonevac, 1995] D. Bonevac. Reduction in the Mind of God, in *Supervenience: New Essays*. E. Savellos and U. Yalcin (ed.). New York: Cambridge University Press, 1995.
[Bridges and Richman, 1987] D. S. Bridges and F. Richman. *Varieties of constructive mathematics*. Cambridge: Cambridge University Press, 1987.
[Brock, 1993] S. Brock. Modal Fictionalism: A Response to Rosen, *Mind* 102, 147–150, 1993.
[Brouwer, 1913] L. E. J. Brouwer. Intuitionism and Formalism, 1913. Reprinted in [Benacerraf and Putnam, 1983, 77–89].
[Brouwer, 1949] L. E. J. Brouwer. Consciousness, Philosophy, and Mathematics, 1949. Reprinted in [Benacerraf and Putnam, 1983, 90–96].
[Burgess, 1983] J. Burgess. Why I am Not a Nominalist, *Notre Dame Journal of Formal Logic*, 23, 93–105, 1983.
[Burgess, 1991] J. Burgess. Synthetic Physics and Nominalist Reconstruction, in C. Wade Savage and P. Ehrlich (eds.), *Philosophical and Foundational Issues in Measurement Theory*. Hillsdale: Lawrence Erlbaum Associates, 119–38, 1991.
[Burgess and Rosen, 1997] J. Burgess and G. Rosen. *A Subject with No Object*. Oxford: Clarendon Press, 1997.
[Burgess, 2004] J. Burgess. Mathematics and *Bleak House*, *Philosophia Mathematica* 2004 12(1):18-36, 2004.
[Cantor, 1883] G. Cantor. Über unendliche, lineare Punktmannigfaltigkeiten, *Mathematische Annalen* 21, 545-591, 1883.
[Carnap, 1928] R. Carnap. *Der Logische Aufbau der Welt*. Leipzig: Felix Meiner Verlag, 1928. English translation by Rolf A. George, 1967. *The Logical Structure of the World: Pseudoproblems in Philosophy*. University of California Press.
[Chihara, 1990] C. Chihara. *Constructibility and Mathematical Existence*. Oxford: Oxford University Press, 1990.
[Colyvan, 2001] M. Colyvan. *The Indispensability of Mathematics*. New York: Oxford University Press, 2001.
[Dever, 2003] J. Dever. Modal Fictionalism and Compositionality, *Philosophical Studies* 114, 223-251, 2003.
[Divers, 1999] J. Divers. A Modal Fictionalist Result, *Nous* 33, 317-346, 1999.
[Dummett, 1973] M. Dummett. The Philosophical Basis of Intuitionistic Logic, 1973. Reprinted in [Benacerraf and Putnam, 1983, 97-129].
[Eklund, forthcoming] M. Eklund. Fiction, Indifference, and Ontology, *Philosophy and Phenomenological Research*, forthcoming.
[Enderton, 1972] H. B. Enderton. *A Mathematical Introduction to Logic*. New York: Academic Press, 1972.
[Field, 1980] H. Field. *Science Without Numbers*. Princeton: Princeton University Press, 1980.
[Field, 1989] H. Field. *Realism, Mathematics, and Modality*. Oxford: Basil Blackwell, 1989.
[Field, 1991] H. Field. Metalogic and Modality, *Philosophical Studies* 62, 1-22, 1991.
[Field, 1993] H. Field. The Conceptual Contingency of Mathematical Objects, *Mind* 102, 285-299, 1993.
[Field, 1998] H. Field. Mathematical Objectivity and Mathematical Objects, in C. MacDonald and S. Laurence (ed.), *Contemporary Readings in the Foundations of Metaphysics*. Oxford: Basil Blackwell, 389-405, 1998.
[Friedman, 1981] M. Friedman. Review of Hartry Field's *Science Without Numbers: A Defence of Nominalism*. *Philosophy of Science* 48, 505-506, 1981.
[Gödel, 1933] K. Gödel. An interpretation of the intuitionistic propositional calculus', in *Collected Works*, 1. Oxford: Oxford University Press, 144–195, 1933.
[Hale, 1994] R. Hale. Is Platonism Epistemologically Bankrupt? *Philosophical Review* 103, 299-325, 1994.

[Hale and Wright, 1992] R. Hale and C. Wright. Nominalism and the Contingency of Abstract Objects, *Journal of Philosophy* 89, 111-135, 1992.
[Hellman, 1989] G. Hellman. *Mathematics Without Numbers*. Oxford: Clarendon Press, 1989.
[Hempel, 1945] C. G. Hempel. On the Nature of Mathematical Truth, reprinted in Benacerraf and Putnam 1983, 377-393, 1945.
[Heyting, 1956] A. Heyting. *Intuitionism: An Introduction*. Amsterdam: North Holland, 1956.
[Hilbert, 1926] D. Hilbert. On the Infinite, *Mathematische Annalen* 95, 161-190, 1926; English translation by E. Putnam and G. Massey in [Benacerraf and Putnam, 1983, 183-201].
[Hodes, 1984] H. Hodes. Logicism and the Ontological Commitments of Arithmetic, *Journal of Philosophy* 81, 123-149, 1984.
[Hofweber, 2005a] T. Hofweber. A Puzzle about Ontology, *Nous* 39, 256-283, 2005.
[Hofweber, 2005b] T. Hofweber. Number Determiners, Numbers, and Arithmetic *Philosophical Review*, 2005.
[Hofweber, 2006] T. Hofweber. Schiffer's new theory of propositions *Philosophy and Phenomenological Research*, 2006.
[Hofweber, forthcoming] T. Hofweber. Innocent Statements and their Metaphysically Loaded Counterparts, forthcoming.
[Horgan, 1984] T. Horgan. Science Nominalized, *Philosophy of Science* 51, 529-549, 1984.
[Horgan, 1984] T. Horgan. Science Nominalized, *Philosophy of Science* 51, 529-549, 1984.
[Horwich, 1991] P. Horwich. On the Nature and Norms of Theoretical Commitment, *Philosophy of Science*, Vol. 58, No. 1 (Mar., 1991) , pp. 1-14, 1991
[Irvine, 1990] A. Irvine. *Physicalism in Mathematics*. Norwell: Kluwer, 1990.
[Juhl, 1995] C. Juhl. Is Gold-Putnam diagonalization complete? *Journal of Philosophical Logic* 24, 117-138, 1995.
[Kalderon, 2005] M. Kalderon. *Fictionalism in Metaphysics*. Oxford: Clarendon Press, 2005.
[Kim, 2002] S. Kim. Modal Fictionalism Generalized and Defended, *Philosophical Studies* 111, 121-146, 2002.
[Kim, 2005] S. Kim. Modal fictionalism and analysis, in [Kalderon, 2005].
[Krantz et al., 1971] D. Krantz, R. D. Luce, P. Suppes, and A. Tversky. *Foundations of Measurement*. New York: Academic Press, 1971.
[Kripke, 1965] S. Kripke. Semantical analysis of intuitionistic logic, in *Formal Systems and Recursive Functions*. J. Crossley and M. A. E. Dummett, eds. Amsterdam: North Holland, 92-130, 1965.
[Kripke, 1976] S. Kripke. Is there a problem about substitutional quantification? In G. Evans and J. McDowell, editors, *Truth and Meaning*, pages 325 — 419. Oxford: Clarendon Press, 1976.
[Leng, 2005] M. Leng. Revolutionary Fictionalism: A Call to Arms, *Philosophia Mathematica*, October 1, 2005; 13(3): 277 - 293.
[Link, 2000] G. Link. Reductionism as Resource-Conscious Reasoning, *Erkenntnis* 53, 173-193, 2000.
[Maddy, 1990] P. Maddy. *Realism in Mathematics*. Oxford: Oxford University Press, 1990.
[Maddy, 1992] P. Maddy. Indispensability and Practice, *Journal of Philosophy* 89, 275-289, 1992.
[Maddy, 1997] P. Maddy. *Naturalism in Mathematics*. Oxford University Press, 1997.
[Malament, 1982] D. Malament. Review of Field's *Science Without Numbers*, *Journal of Philosophy* 79, 523-534, 1982.
[Meyer, 1976] R. Meyer. Relevant Arithmetic, *Bulletin of the Section of Logic of the Polish Academy of Sciences*, 5, 133-137, 1976.
[Meyer and Mortensen, 1984] R. Meyer and C. Mortensen. Inconsistent Models for Relevant Arithmetics, *The Journal of Symbolic Logic*, 49, 917-929, 1984.
[Mortensen, 1995] C. Mortensen. *Inconsistent Mathematics*. Dordrecht: Kluwer, 1995.
[Nagel, 1961] E. Nagel. *The Structure of Science*. New York: Harcourt Brace and Company, 1961.
[Nolan, 1997] D. Nolan. Three Problems for 'Strong' Modal Fictionalism, *Philosophical Studies* 87, 259-275, 1997.
[Nolan, 2005] D. Nolan. Fictionalist attitudes about fictional matters, in Kalderon 2005.
[Noland and Hawthorne, 1996] D. Nolan and J. Hawthorne. Reflexive Fictionalisms, *Analysis* 56-23-32, 1996.
[Paris and Harrington, 1977] J. Paris and L. Harrington. A mathematical incompleteness in Peano arithmetic, *Handbook of Mathematical Logic*, ed. J.Barwise. Amsterdam: North-Holland, 1977.
[Parsons, 1980] T. Parsons. *Nonexistent Objects*. New Haven: Yale University Press, 1980.
[Peirce, 1878] C. S. Peirce. How to Make Our Ideas Clear, *Popular Science Monthly*, 1878; reprinted in J. Buchler (ed.), *Philosophical Writings of Peirce* (New York: Dover, 1955), 23-41, itself a republication of *The Philosophy of Peirce: Selected Writings* (London: Routledge and Kegan Paul, 1940); and in C. Hartshorne and P. Weiss (ed.), *Collected Papers of Charles Sanders Peirce*, Volume 5. Cambridge: Belknap Press of Harvard University Press, 1933: 388-410.

[Peirce, 1891] C. S. Peirce. The Architecture of Theories, *Monist*; reprinted in Buchler, 315-323, 18891.
[Peirce, 1896a] C. S. Peirce. Lessons from the History of Science, 1896. In C. Hartshorne and P. Weiss (ed.), *Collected Papers of Charles Sanders Peirce*, Volume 1: Principles of Philosophy. Cambridge: Belknap Press of Harvard University Press, 1960: 19-49.
[Peirce, 1896b] C. S. Peirce. The Regenerated Logic, *Monist* 7, 19-40, 1896; reprinted in C. Hartshorne and P. Weiss (ed.), *Collected Papers of Charles Sanders Peirce*, Volume 3: *Exact Logic*. Cambridge: Belknap Press of Harvard University Press, 1933: 266-287.
[Peirce, 1897] C. S. Peirce. The Logic of Relatives, *Monist*; reprinted in Buchler, 146-147; and in Hartshorne and Weiss, Volume 3, 288-345, 1897.
[Peirce, 1898] C. S. Peirce. The Logic of Mathematics in Relation to Education, *Educational Review*; 209-216, 1898; reprinted in Buchler, 135-139, and in Hartshorne and Weiss, Volume 3, 346-359.
[Peirce, 1902] C. S. Peirce. Minute Logic, manuscript; reprinted in Buchler, 139-145, 1902.
[Peirce, 1905a] C. S. Peirce. Issues of Pragmaticism, *Monist*; reprinted in Buchler, 290-294, 300-301, 1905.
[Peirce, 1905b] C. S. Peirce. What Pragmatism Is, *Monist*, 1905; reprinted in Buchler, 251-267.
[Peirce, 1976] C. S. Peirce. *The New Elements of Mathematics, Volume IV: Mathematical Philosophy*, C. Eisele (ed.). The Hague: Mouton, 1976.
[Poincaré, 1952] H. Poincaré. *Science and Hypothesis*. New York: Dover, 1952.
[Priest, 1994] G. Priest. Is Arithmetic Consistent? *Mind*, 103, 337-49, 1994.
[Priest, 1996] G. Priest. On Inconsistent Arithmetics: Reply to Denyer, *Mind*, 105, 649-59, 1996.
[Priest, 1997] G. Priest. Inconsistent Models of Arithmetic; I Finite Models, *Journal of Philosophical Logic*, 26, 223-35, 1997.
[Priest, 2000] G. Priest. Inconsistent Models for Arithmetic: II, The General Case, *The Journal of Symbolic Logic*, 65, 1519-29, 2000.
[Putnam, 1967] H. Putnam. Mathematics without Foundations, *Journal of Philosophy* 64, 5–22, 1967; reprinted in P. Benacerraf and H. Putnam 1983,295-311.
[Quine, 1939] W. V. O. Quine. A Logistical Approach to the Ontological Problem. *The Ways of Paradox* (Cambridge: Harvard University Press. 1976), 198, 1939.
[Quine, 1951] W. V. O. Quine. On What There Is, in *From a Logical Point of View*, New York: Harper & Row, 1-19, 1951.
[Quine, 1960] W. V. O. Quine. *Word and Object*. Cambridge: MIT Press, 1960.
[Quine, 1969] W. V. O. Quine. *Ontological Relativity and Other Essays*. New York: Columbia University Press, 1969.
[Resnik, 1975] M. Resnik. Mathematical Knowledge and Pattern Cognition, *Canadian Journal of Philosophy*, 5, 25-39, 1975.
[Resnik, 1985] M. Resnik. How Nominalist is Hartry Field's Nominalism? *Philosophical Studies* 47, 163-181, 1985.
[Resnik, 1997] M. Resnik. *Mathematics as a Science of Patterns*. Oxford: Oxford University Press, 1997.
[Rosen, 1990] G. Rosen. Modal Fictionalism, *Mind* 99, 327-354, 1990.
[Rosen, 1993] G. Rosen. The Refutation of Nominalism (?), *Philosophical Topics* 21, 149-186, 1993.
[Rosen, 1995] G. Rosen. Modal Fictionalism Fixed, *Analysis* 55, 67-73, 1995.
[Rosen, 2003] G. Rosen. A Problem for Fictionalism for Possible Worlds, *Analysis* 53, 71-81, 2003.
[Rosen, 2005] G. Rosen. Problems in the History of Fictionalism, in Kalderon 2005.
[Routley and Routley, 1972] R. Routley and V. Routley. The Semantics of First Degree Entailment, *Nous*, 6, 335-359, 1972.
[Russell, 1918] B. Russell. The Philosophy of Logical Atomism in *Logic and Knowledge*, ed. R.C. Marsh. London: Allen and Unwin, 1956, 177-281, 1918.
[Schiffer, 2003] S. Schiffer. *The Things We Mean*. Oxford: Clarendon Press, 2003.
[Shapiro, 1983a] S. Shapiro. Conservativeness and Incompleteness, *Journal of Philosophy* 80, 521-531, 1983.
[Shapiro, 1983b] S. Shapiro. Mathematics and Reality, *Philosophy of Science* 50, 523-548, 1983.
[Shapiro, 1997] S. Shapiro. *Philosophy of Mathematics*. New York: Oxford University Press, 1997.
[Shapiro, 2000] S. Shapiro. *Thinking About Mathematics*. Oxford: Oxford University Press , 2000.
[Sober, 1993] E. Sober. Mathematics and Indispensability, *Philosophical Review* 102, 35-57, 1993.
[Spade, 1998] P. Spade. Three Versions of Ockham's Reductionist Program. *Franciscan Studies* 56, pp. 335-46, 1998.
[Spade, 1999] P. Spade, ed. *The Cambridge Companion to Ockham*. New York: Cambridge University Press, 1999.
[Spade, 1999a] P. Spade. Ockham's Nominalist Metaphysics: Some Main Themes. In Spade [1999], 100-117, 1999.

[Stalnaker, 1972] R. Stalnaker. Pragmatics. In Donald Davidson and Gilbert Herman, eds., *Semantics of Natural Language*, 380-397. Dordrecht and Boston: D. Reidel, 1972.
[Stanley, 2001] J. Stanley. Hermeneutic Fictionalism, *Midwest Studies in Philosophy*, XXV, 36-71, 2001.
[Suppes, 1971] P. Suppes. *Introduction to Logic*. New York: Van Nostrand, 1971.
[Szabo, 2003] Z. Szabo. Nominalism, in M. J. Loux and D. Zimmerman eds., *Oxford Handbook of Metaphysics*. Oxford: Oxford University Press, 11-45, 2003.
[Tweedale, 1992] M. M. Tweedale. Ockham's Supposed Elimination of Connotative Terms and His Ontological Parsimony. *Dialogue* 31, pp. 431-44, 1992.
[Urquhart, 1990] A. Urquhart. The Logic of Physical Theory, in *Physicalism in Mathematics*, ed. A.D. Irvine, Kluwer 1990, 145-154
[Vaihinger, 1911] H. Vaihinger. *Die Philosophie des als ob. System der theoretischen, praktischen und religiosen Fiktionem der Menschheit auf Grund eines idealistichen*. Berlin: Reuther und Reichard, 1911; translated into English as *The Philosophy of As-If, a system of the theoretical, practical and religious fictions of mankind*, New York: Harcourt, Brace and Company, 1924.
[van Fraassen, 1980] B. C. van Fraassen. *The Scientific Image*. New York : Oxford University Press, 1980.
[Velleman, 1993] D. Velleman. Constructivism Liberalized, *Philosophical Review* 102, 59-84, 1993.
[Walton, 1978] K. Walton. Fearing Fictions, *The Journal of Philosophy*, 1978.
[Walton, 1990] K. Walton. *Mimesis as Make-believe: on the Foundations of the Representational Arts*. Cambridge: Harvard University Press, 1990.
[Walton, 2005] K. Walton. Metaphor and prop oriented make-believe, in Kalderon 2005.
[William of Ockham, 1991] William of Ockham. *Quodlibetal Questions*. New Haven: Yale University Press, 1991.
[Wright, 1988] C. Wright. Why Numbers Can Believably Be: A Reply to Hartry Field, *Revue Internationale de Philosophie* 42, 425-473, 1988.
[Yablo, 2001] S. Yablo. Go Figure: A Path Through Fictionalism, *Midwest Studies in Philosophy* XXV, 72-99, 2001.
[Yablo, 2002] S. Yablo. Abstract Objects: A Case Study, in Andrea Bottani, Massimiliano Carrara, and Pierdaniele Giaretta (ed.) *Individuals, Essence and Identity: Themes of Analytic Metaphysics*. Amsterdam: Kluwer, 2002.
[Yablo, 2005] S. Yablo. The myth of seven, in Kalderon 2005.

SET THEORY FROM CANTOR TO COHEN

Akihiro Kanamori

Set theory is an autonomous and sophisticated field of mathematics, enormously successful not only at its continuing development of its historical heritage but also at analyzing mathematical propositions and gauging their consistency strength. But set theory is also distinguished by having begun intertwined with pronounced metaphysical attitudes, and these have even been regarded as crucial by some of its great developers. This has encouraged the exaggeration of crises in foundations and of metaphysical doctrines in general. However, set theory has proceeded in the opposite direction, from a web of intensions to a theory of extension *par excellence*, and like other fields of mathematics its vitality and progress have depended on a steadily growing core of mathematical proofs and methods, problems and results. There is also the stronger contention that from the beginning set theory actually developed through a progression of *mathematical* moves, whatever and sometimes in spite of what has been claimed on its behalf.

What follows is an account of the development of set theory from its beginnings through the creation of forcing based on these contentions, with an avowedly Whiggish emphasis on the heritage that has been retained and developed by the current theory. The whole transfinite landscape can be viewed as having been articulated by Cantor in significant part to solve the Continuum Problem. Zermelo's axioms can be construed as clarifying the set existence commitments of a single proof, of his Well-Ordering Theorem. Set theory is a particular case of a field of mathematics in which seminal proofs and pivotal problems actually shaped the basic concepts and forged axiomatizations, these transmuting the very notion of set. There were two main junctures, the first being when Zermelo through his axiomatization shifted the notion of set from Cantor's range of inherently structured sets to sets solely structured by membership and governed and generated by axioms. The second juncture was when the Replacement and Foundation Axioms were adjoined and a first-order setting was established; thus transfinite recursion was incorporated and results about all sets could established through these means, including results about definability and inner models. With the emergence of the cumulative hierarchy picture, set theory can be regarded as becoming a theory of well-foundedness, later to expand to a study of consistency strength. Throughout, the subject has not only been sustained by the axiomatic tradition through Gödel and Cohen but also fueled by Cantor's two legacies, the extension of number into the transfinite as transmuted into the theory of large cardinals and the investigation of definable sets of reals as transmuted into descriptive set theory. All this can be regarded as having a historical and mathematical logic internal to set theory, one that is often misrepresented at critical junctures in textbooks (as will be pointed out). This view, from inside set theory and about itself, serves to shift the focus to those tensions and strategies familiar to mathematicians as well as to those moves, often made without much fanfare and sometimes merely linguistic, that have led to the crucial advances.

1 CANTOR

1.1 Real Numbers and Countability

Set theory had its beginnings in the great 19th-Century transformation of mathematics, a transformation beginning in analysis. Since the creation of the calculus by Newton and Leibniz the function concept had been steadily extended from analytic expressions toward arbitrary correspondences. The first major expansion had been inspired by the explorations of Euler in the 18th Century and featured the infusion of infinite series methods and the analysis of physical phenomena, like the vibrating string. In the 19th-Century the stress brought on by the unbridled use of series of functions led first Cauchy and then Weierstrass to articulate convergence and continuity. With infinitesimals replaced by the limit concept and that cast in the ϵ-δ language, a level of deductive rigor was incorporated into mathematics that had been absent for two millenia. Sense for the new functions given in terms of infinite series could only be developed through carefully specified deductive procedures, and proof reemerged as an extension of algebraic calculation and became basic to mathematics in general, promoting new abstractions and generalizations.

Working out of this tradition Georg Cantor[1] (1845–1918) in 1870 established a basic uniqueness theorem for trigonometric series: If such a series converges to zero everywhere, then all of its coefficients are zero. To generalize Cantor [1872] started to allow points at which convergence fails, getting to the following formulation: For a collection P of real numbers, let P' be the collection of limit points of P, and $P^{(n)}$ the result of n iterations of this operation. *If a trigonometric series converges to zero everywhere except on a P where $P^{(n)}$ is empty for some n, then all of its coefficients are zero.*[2]

It was in [1872] that Cantor provided his formulation of the real numbers in terms of fundamental sequences of rational numbers, and significantly, this was for the specific purpose of articulating his proof. With the new results of analysis to be secured by proof and proof in turn to be based on prior principles the regress led in the early 1870s to the appearance of several independent formulations of the real numbers in terms of the rational numbers. It is at first quite striking that the real numbers came to be developed so late, but this can be viewed as part of the expansion of the function concept which shifted the emphasis from the continuum taken as a whole to its extensional construal as a collection of objects. In mathematics objects have been traditionally introduced only with reluctance, but a more arithmetical rather than geometrical approach to the continuum became necessary for the articulation of proofs.

The other well-known formulation of the real numbers is due to Richard Dedekind [1872], through his cuts. Cantor and Dedekind maintained a fruitful correspondence,

---

[1] Dauben [1979], Meschkowski [1983], and Purkert-Ilgauds [1987] are mathematical biographies of Cantor.
[2] See Kechris-Louveau [1987] for recent developments in the Cantorian spirit about uniqueness for trigonometric series converging on definable sets of reals.

especially during the 1870s, in which Cantor aired many of his results and speculations.[3] The formulations of the real numbers advanced three important predispositions for set theory: the consideration of infinite collections, their construal as unitary objects, and the encompassing of arbitrary such possibilities. Dedekind [1871] had in fact made these moves in his creation of ideals, infinite collections of algebraic numbers,[4] and there is an evident similarity between ideals and cuts in the creation of new numbers out of old.[5] The algebraic numbers would soon be the focus of a major breakthrough by Cantor. Although both Cantor and Dedekind carried out an arithmetical reduction of the continuum, they each accommodated its antecedent geometric sense by asserting that each of their real numbers actually corresponds to a point on the line. Neither theft nor honest toil sufficed; Cantor [1872: 128] and Dedekind [1872: III] recognized the need for an *axiom* to this effect, a sort of Church's Thesis of adequacy for the new construal of the continuum as a collection of objects.

Cantor recalled[6] that around this time he was already considering infinite iterations of his P' operation using "symbols of infinity":

$$P^{(\infty)} = \bigcap_{n}^{\infty} P^{(n)}, P^{(\infty+1)} = P^{(\infty)'}, P^{(\infty+2)}, \ldots P^{(\infty \cdot 2)}, \ldots P^{(\infty^2)}, \ldots P^{(\infty^\infty)}, \ldots P^{(\infty^{\infty^\infty})}, \ldots$$

In a crucial conceptual move he began to investigate infinite collections of real numbers and infinitary enumerations for their own sake, and this led first to a basic articulation of size for the continuum and then to a new, encompassing theory of counting. Set theory was born on that December .1873 day when Cantor established that *the real numbers are uncountable.*[7] In the next decades the subject was to blossom through the prodigious progress made by him in the theory of transfinite and cardinal numbers.

The uncountability of the reals was established, of course, via *reductio ad absurdum* as with the irrationality of $\sqrt{2}$. Both impossibility results epitomize how a *reductio* can compel a larger mathematical context allowing for the deniability of hitherto implicit properties. Be that as it may, Cantor the mathematician addressed a specific *problem*, embedded in the mathematics of the time, in his seminal [1874] entitled "On a property of the totality of all real algebraic numbers". After first establishing this property, the countability of the algebraic numbers, Cantor then established: *For any (countable) sequence*

---

[3]The most complete edition of Cantor's correspondence is Meschkowski-Nilson [1991]. Excerpts from the Cantor-Dedekind correspondence from 1872 through 1882 were published in Noether-Cavaillès [1937], and excerpts from the 1899 correspondence were published by Zermelo in the collected works of Cantor [1932]. English translations of the Noether-Cavaillès excerpts were published in Ewald [1996: 843ff.]. An English translation of a Zermelo excerpt (retaining his several errors of transcription) appeared in van Heijenoort [1967: 113ff.]. English translations of Cantor's 1899 correspondence with both Dedekind and Hilbert were published in Ewald [1996: 926ff.].

[4]The algebraic numbers are those real numbers that are the roots of polynomials with integer coefficients.

[5]Dedekind [1872] dated his conception of cuts to 1858, and antecedents to ideals in his work were also entertained around then. For Dedekind and the foundation of mathematics see Dugac [1976] and Ferreirós [2007], who both accord him a crucial role in the development of the framework of set theory.

[6]See his [1880: 358].

[7]A set is *countable* if there is a bijective correspondence between it and the natural numbers $\{0, 1, 2, \ldots\}$. The exact date of birth can be ascertained as December 7. Cantor first gave a proof of the uncountability of the reals in a letter to Dedekind of 7 December 1873 (Ewald [1996: 845ff]), professing that ". . . only today do I believe myself to have finished with the thing . . .".

of reals, every interval contains a real not in the sequence. Cantor appealed to the order completeness of the reals:

Suppose that s is a sequence of reals and I an interval. Let $a < b$ be the first two reals of s, if any, in I. Then let $a' < b'$ be the first two reals of s, if any, in the open interval (a, b); $a'' < b''$ the first two reals of s, if any, in (a', b'); and so forth. Then however long this process continues, the (non-empty) intersection of these nested intervals cannot contain any member of s.

By this means Cantor provided a new proof of Joseph Liouville's result [1844, 1851] that there are transcendental numbers (real non-algebraic numbers) and only afterward did Cantor point out the uncountability of the reals altogether. This presentation is suggestive of Cantor's natural caution in overstepping mathematical sense at the time.[8]

Accounts of Cantor's work have mostly reversed the order for deducing the existence of transcendental numbers, establishing first the uncountability of the reals and only then drawing the existence conclusion from the countability of the algebraic numbers.[9] In textbooks the inversion may be inevitable, but this has promoted the misconception that Cantor's arguments are non-constructive.[10] It depends how one takes a proof, and Cantor's arguments have been implemented as algorithms to generate the successive digits of new reals.[11]

1.2 Continuum Hypothesis and Transfinite Numbers

By his next publication [1878] Cantor had shifted the weight to getting bijective correspondences, stipulating that two sets have the same *power* [Mächtigkeit] *iff* there is such a correspondence between them, and established that the reals $\mathbb{R}$ and the n-dimensional spaces $\mathbb{R}^n$ all have the same power. Having made the initial breach in [1874] with a negative result about the *lack* of a bijective correspondence, Cantor secured the new ground

---

[8]Dauben [1979: 68ff] suggests that the title and presentation of Cantor [1874] were deliberately chosen to avoid censure by Kronecker, one of the journal editors.

[9]Indeed, this is where Wittgenstein [1956: I,Appendix II, 1-3] located what he took to be the problematic aspects of the talk of uncountability.

[10]A non-constructive proof typically deduces the existence of a mathematical object without providing a means for specifying it. Kac-Ulam [1968: 13] wrote: "The contrast between the methods of Liouville and Cantor is striking, and these methods provide excellent illustrations of two vastly different approaches toward proving the *existence* of mathematical objects. Liouville's is purely *constructive*; Cantor's is purely *existential*." See also Moore [1982: 39]. One exception to the misleading trend is Fraenkel [1930: 237][1953: 75], who from the beginning emphasized the constructive aspect of diagonalization.

The first non-constructive proof widely acknowledged as such was Hilbert's [1890] of his basis theorem. Earlier, Dedekind [1888: §159] had established the equivalence of two notions of being finite with a non-constructive proof that made an implicit use of the Axiom of Choice.

[11]Gray [1994] shows that Cantor's original [1874] argument can be implemented by an algorithm that generates the first n digits of a transcendental number with time complexity $O(2^{n^{1/3}})$, and his later diagonal argument, with a tractable algorithm of complexity $O(n^2 \log^2 n \log \log n)$. The original Liouville argument depended on a simple observation about fast convergence, and the digits of the Liouville numbers can be generated much faster. In terms of 2.3 below, the later Baire Category Theorem can be viewed as a direct generalization of Cantor's [1874] result, and the collection of Liouville numbers provides an explicit example of a co-meager yet measure zero set of reals (see Oxtoby [1971: §2]). On the other hand, Gray [1994] shows that every transcendental real is the result of diagonalization applied to *some* enumeration of the algebraic reals.

with a positive investigation of the *possibilities* for having such correspondences.[12] With "sequence" tied traditionally to countability through the indexing, Cantor used "correspondence [Beziehung]". Just as the discovery of the irrational numbers had led to one of the great achievements of Greek mathematics, Eudoxus's theory of geometrical proportions presented in Book V of Euclid's *Elements* and thematically antecedent to Dedekind's [1872] cuts, Cantor began his move toward a full-blown mathematical theory of the infinite.

Although holding the promise of a rewarding investigation Cantor did not come to any powers for infinite sets other than the two as set out in his [1874] proof. Cantor claimed at the end of [1878: 257]:

> Every infinite set of reals either is countable or has the power of the continuum.

This was the *Continuum Hypothesis* (CH) in the nascent context. The conjecture viewed as a primordial question would stimulate Cantor not only to approach the reals *qua* extensionalized continuum in an increasingly arithmetical fashion but also to grapple with fundamental questions of set existence. His triumphs across a new mathematical context would be like a brilliant light to entice others into the study of the infinite, but his inability to establish CH would also cast a long shadow. Set theory had its beginnings not as some abstract foundation for mathematics but rather as a setting for the articulation and solution of the *Continuum Problem*: to determine whether there are more than two powers embedded in the continuum.

In his magisterial *Grundlagen* [1883] Cantor developed the *transfinite numbers* [Anzahlen] and the key concept of *well-ordering*. A *well-ordering* of a set is a linear ordering of it according to which every non-empty subset has a least element. No longer was the infinitary indexing of his trigonometric series investigations mere contrivance. The "symbols of infinity" became autonomous and extended as the transfinite numbers, the emergence signified by the notational switch from the ∞ of potentiality to the ω of completion as the last letter of the Greek alphabet. With this the progression of transfinite numbers could be depicted:

$$0, 1, 2, \ldots \omega, \omega + 1, \omega + 2, \ldots, \omega + \omega (= \omega \cdot 2), \ldots, \omega^2, \ldots, \omega^\omega, \ldots, \omega^{\omega^\omega}, \ldots$$

A corresponding transition from subsets of $\mathbb{R}^n$ to a broader concept of set was signaled by the shift in terminology from "point-manifold [Punktmannigfaltigkeit]" to "set [Menge]". In this new setting well-orderings conveyed the sense of sequential counting and transfinite numbers served as standards for gauging well-orderings.

---

[12] Cantor developed a bijective correspondence between $\mathbb{R}^2$ and $\mathbb{R}$ by essentially interweaving the decimal expansions of a pair of reals to define the associated real, taking care of the countably many exceptional points like $.100\ldots = .099\ldots$ by an *ad hoc* shuffling procedure. Such an argument now seems straightforward, but to have bijectively identified the plane with the line was a stunning accomplishment at the time. In a letter to Dedekind of 29 June 1877 Cantor (Ewald [1996: 860]) wrote, in French in the text, "I see it, but I don't believe it."

Cantor's work inspired a push to establish the "invariance of dimension", that there can be no *continuous* bijection of any $\mathbb{R}^n$ onto $\mathbb{R}^m$ for $m < n$, with Cantor [1879] himself providing an argument. As topology developed, the stress brought on by the lack of firm ground led Brouwer [1911] to definitively establish the invariance of dimension in a seminal paper for algebraic topology.

As Cantor pointed out, every linear ordering of a finite set is already a well-ordering and all such orderings are isomorphic, so that the general sense is only brought out by infinite sets, for which there are non-isomorphic well-orderings. Cantor called the set of natural numbers the first number class (I) and the set of numbers whose predecessors are countable the second number class (II). Cantor conceived of (II) as being bounded above according to a limitation principle and showed that (II) itself is not countable. Proceeding upward, Cantor called the set of numbers whose predecessors are in bijective correspondence with (II) the third number class (III), and so forth. Cantor took a set to be of a *higher power* than another if they are not of the same power yet the latter is of the same power as a subset of the former. Cantor thus conceived of ever higher powers as represented by number classes and moreover took every power to be so represented. With this "free creation" of numbers, Cantor [1883: 550] propounded a basic principle that was to drive the analysis of sets:

> "It is always possible to bring any *well-defined* set into the form of a *well-ordered* set."

He regarded this as a "an especially remarkable law of thought which through its general validity is fundamental and rich in consequences." Sets are to be well-ordered, and thus they and their powers are to be gauged via the transfinite numbers of his structured conception of the infinite.

The well-ordering principle was consistent with Cantor's basic view in the *Grundlagen* that the finite and the transfinite are all of a piece and uniformly comprehendable in mathematics,[13] a view bolstered by his systematic development of the arithmetic of transfinite numbers seamlessly encompassing the finite numbers. Cantor also devoted several sections of the *Grundlagen* to a justificatory philosophy of the infinite, and while this metaphysics can be separated from the mathematical development, one concept was to suggest ultimate delimitations for set theory: Beyond the transfinite was the "Absolute", which Cantor eventually associated mathematically with the collection of all ordinal numbers and metaphysically with the transcendence of God.[14]

The Continuum Problem was never far from this development and could in fact be seen as an underlying motivation. The transfinite numbers were to provide the framework for Cantor's two approaches to the problem, the approach through power and the more direct approach through definable sets of reals, these each to initiate vast research programs.

As for the approach through power, Cantor in the *Grundlagen* established that the second number class (II) is uncountable, yet *any infinite subset of* (II) *is either countable or has the same power as* (II). Hence, (II) has exactly the property that Cantor sought for the reals, and he had reduced CH to the positive assertion that the reals and (II) have the same power. The following in brief is Cantor's argument that (II) is uncountable:

Suppose that s is a (countable) sequence of members of (II), say with initial element a. Let a' be a member of s, if any, such that $a < a'$; let a'' be a member of s, if any, such that $a' < a''$; and so forth. Then however long this process continues, the supremum of these numbers, or its successor, is not a member of s.

---

[13]This is emphasized by Hallett [1984] as Cantor's "finitism".

[14]The "absolute infinite" is a varying but recurring explanatory concept in Cantor's work; see Jané [1995].

This argument was reminiscent of his [1874] argument that the reals are uncountable and suggested a correlation of the reals through their fundamental sequence representation with the members of (II) through associated cofinal sequences.[15] However, despite several announcements Cantor could never develop a workable correlation, an emerging problem in retrospect being that he could not *define* a well-ordering of the reals.

As for the approach through definable sets of reals, this evolved directly from Cantor's work on trigonometric series, the "symbols of infinity" used in the analysis of the P' operation transmuting to the transfinite numbers of the second number class (II).[16] In the *Grundlagen* Cantor studied P' for uncountable P and defined the key concept of a *perfect* set of reals (non-empty, closed, and containing no isolated points). Incorporating an observation of Ivar Bendixson [1883], Cantor showed in the succeeding [1884] that *any uncountable closed set of reals is the union of a perfect set and a countable set.* For a set A of reals, A has the *perfect set property iff* A is countable or else has a perfect subset. Cantor had shown in particular that *closed sets have the perfect set property*.

Since Cantor [1884; 1884a] had been able to show that any perfect set has the power of the continuum, he had established that "CH holds for closed sets": every closed set either is countable or has the power of the continuum. Or from his new vantage point, he had reduced the Continuum Problem to determining whether there is a closed set of reals of the power of the second number class. He was unable to do so, but he had initiated a program for attacking the Continuum Problem that was to be vigorously pursued (cf. 2.3 and 2.5).

1.3 Diagonalization and Cardinal Numbers

In the ensuing years, unable to resolve the Continuum Problem through direct correlations with transfinite numbers Cantor approached size and order from a broader perspective that *would* incorporate the continuum. He identified power with *cardinal number*, an autonomous concept beyond being *une façon de parler* about bijective correspondence, and he went beyond well-orderings to the study of linear *order types*. Cantor embraced a structured view of sets, when "well-defined", as being given together with a linear ordering of their members. Order types and cardinal numbers resulted from successive abstraction, from a set M to its order type $\overline{M}$ and then to its cardinality $\overline{\overline{M}}$.

Almost two decades after his [1874] result that the reals are uncountable, Cantor in a short note [1891] subsumed it via his celebrated diagonal argument. With it, he estab-

---

[15] After describing the similarity between ω and $\sqrt{2}$ as limits of sequences, Cantor [1887: 99] interestingly correlated the creation of the transfinite numbers to the creation of the irrational numbers, beyond merely breaking new ground in different number contexts: "The transfinite numbers are in a certain sense *new irrationalities*, and in my opinion the best method of defining the *finite* irrational numbers [via Cantor's fundamental sequences] is wholly similar to, and I might even say in principle the same as, my method of introducing transfinite numbers. One can say unconditionally: the transfinite numbers *stand or fall* with the finite irrational numbers: they are like each other in their innermost being [Wesen]; for the former like the latter are definite delimited forms or modifications of the actual infinite."

[16] Ferreirós [1995] suggests how the formulation of the second number class as a completed totality with a succeeding transfinite number emerged directly from Cantor's work on the operation P', drawing Cantor's transfinite numbers even closer to his earlier work on trigonometric series.

lished: *For any set L the collection of functions from L into a fixed two-element set has a higher cardinality than that of L.* This result indeed generalized the [1874] result, since the collection of functions from the natural numbers into a fixed two-element set has the same cardinality as the reals. Here is how Cantor gave the argument in general form:[17]

Let M be the totality of all functions from L taking only the values 0 and 1. First, L is in bijective correspondence with a subset of M, through the assignment to each $x_0 \in L$ of the function on L that assigns 1 to x_0 and 0 to all other $x \in L$. However, there cannot be a bijective correspondence between M itself and L. Otherwise, there would be a function $\phi(x, z)$ of two variables such that for every member f of M there would be a $z \in L$ such that $\phi(x, z) = f(x)$ for every $x \in L$. But then, the "diagonalizing" function $g(x) = 1 - \phi(x, x)$ cannot be a member of M since for $z_0 \in L$, $g(z_0) \neq \phi(z_0, z_0)$!

In retrospect the diagonal argument can be drawn out from the [1874] proof.[18] Cantor had been shifting his notion of set to a level of abstraction beyond sets of reals and the like, and the casualness of his [1891] may reflect an underlying cohesion with his [1874]. Whether the new proof is really "different" from the earlier one, through this abstraction Cantor could now dispense with the recursively defined nested sets and limit construction, and he could apply his argument to any set. He had proved for the first time that there is a power higher than that of the continuum and moreover affirmed "the general theorem, that the powers of well-defined sets have no maximum."[19] The diagonal argument, even to its notation, would become method, flowing later into descriptive set theory, the Gödel Incompleteness Theorem, and recursion theory.

Today it goes without saying that a function from L into a two-element set corresponds to a subset of L, so that Cantor's Theorem is usually stated as: *For any set L its power set $\mathcal{P}(L) = \{X \mid X \subseteq L\}$ has a higher cardinality than L.* However, it would be an exaggeration to assert that Cantor was working on power sets; rather, he had expanded the 19th-Century concept of *function* by ushering in arbitrary functions.[20] In any case, Cantor would now

---

[17] Actually, Cantor took L to be the unit interval of reals presumably to invoke a standard context, but he was clearly aware of the generality.

[18] Moreover, diagonalization as such had already occurred in Paul du Bois-Reymond's theory of growth as early as in his [1869]. An argument is manifest in his [1875: 365ff] for showing that for any sequence of real functions $f_0, f_1, f_2, \ldots$ there is a real function g such that for each n, $f_n(x) < g(x)$ for all sufficiently large reals x.

Diagonalization can be drawn out from Cantor's [1874] as follows: Starting with a sequence s of reals and a half-open interval I_0, instead of successively choosing delimiting *pairs* of reals in the sequence, avoid the members of s *one* at a time: Let I_1 be the left or right half-open subinterval of I_0 demarcated by its midpoint, whichever does not contain the first element of s. Then let I_2 be the left or right half-open subinterval of I_1 demarcated by its midpoint, whichever does not contain the second element of s; and so forth. Again, the nested intersection contains a real not in the sequence s. Abstracting the process in terms of reals in binary expansion, one is just generating the binary digits of the diagonalizing real.

In that letter of Cantor's to Dedekind of 7 December 1873 (Ewald [1996: 845ff]) first establishing the uncountability of the reals, there already appears, quite remarkably, a doubly indexed array of real numbers and a procedure for traversing the array downward and to the right, as in a now common picturing of the diagonal argument.

[19] Remarkably, Cantor had already conjectured in the *Grundlagen* [1883: 590] that the collection of continuous real functions has the same power as the second number class (II), and that the collection of all real functions has the same power as the third number class (III). These are consequences of the later Generalized Continuum Hypothesis and are indicative of the sweep of Cantor's conception.

[20] The "power" in "power set" is from "Potenz" in the German for cardinal exponentiation, while Cantor's

have had to confront, in his function context, a general difficulty starkly abstracted from the Continuum Problem: *From a well-ordering of a set, a well-ordering of its power set is not necessarily definable.* The diagonal argument called into question Cantor's very notion of set: On the one hand, the argument, simple and elegant, should be part of set theory and lead to new sets of ever higher cardinality; on the other hand, these sets do not conform to Cantor's principle that every set comes with a (definable) well-ordering.[21]

Cantor's *Beiträge*, published in two parts [1895] and [1897], presented his mature theory of the transfinite. In the first part he described his post-*Grundlagen* work on cardinal number and the continuum. He quickly posed *Cardinal Comparability*, whether

for cardinal numbers $\mathfrak{a}$ and $\mathfrak{b}$, $\mathfrak{a} = \mathfrak{b}$, $\mathfrak{a} < \mathfrak{b}$, or $\mathfrak{b} < \mathfrak{a}$,

as a property "by no means self-evident" and which will be established later "when we shall have gained a survey over the ascending sequence of transfinite cardinal numbers and an insight into their connection." He went on to define the addition, multiplication, and exponentiation of cardinal numbers primordially in terms of set-theoretic operations and functions. If $\mathfrak{a}$ is the cardinal number of M and $\mathfrak{b}$ is the cardinal number of N, then $\mathfrak{a}^{\mathfrak{b}}$ is the cardinal number of the collection of all functions $: N \to M$, i.e. having domain N and taking values in M. The audacity of considering arbitrary functions from a set N into a set M was encased in a terminology that reflected both its novelty as well as the old view of function as given by an explicit rule.[22] As befits the introduction of new numbers Cantor then introduced a new notation, one using the Hebrew letter aleph, $\aleph$. With $\aleph_0$ the cardinal number of the set of natural numbers Cantor observed that $\aleph_0 \cdot \aleph_0 = \aleph_0$ and that $2^{\aleph_0}$ is the cardinal number of continuum. With this he observed that the [1878] labor of associating the continuum with the plane and so forth could be reduced to a "few strokes

---

"power" is from "Mächtigkeit".

[21] This is emphasized in Lavine [1994: IV.2]. Cantor did consider power sets in a letter of 20 September 1898 to Hilbert. In it Cantor entertained a notion of "completed set", one of the guidelines being that "the collection of *all subsets* of a completed set M is a completed set." Also, in a letter of 10 October 1898 to Hilbert, Cantor pointed out, in an argument focused on the continuum, that the power set $P(S)$ is in bijective correspondence with the collection of functions from S into {0, 1}. But in a letter of 9 May 1899 to Hilbert, writing now "set" for "completed set", Cantor wrote: "…it is our common conviction that the 'arithmetic continuum' is a 'set' in this sense; the question is whether this truth is provable or whether it is an axiom. I now incline more to the latter alternative, although I would gladly be convinced by you of the former." For the first and third letters in context see Moore [2002: 45] and for the second, Ferreirós [2007: epilogue]; the letters are in Meschkowski-Nilson [1991].

[22] Cantor wrote [1895: 486]: "…by a '*covering* [Belegung] of N with M,' we understand a law by which with every element n of N a definite element of M is bound up, where one and the same element of M can come repeatedly into application. The element of M bound up with n is, in a way, a one-valued function of n, and may be denoted by $f(n)$; it is called a 'covering function [Belegungsfunktion] of n.' The corresponding covering of N will be called $f(N)$." A convoluted description! Arbitrary functions on arbitrary domains are now of course commonplace in mathematics, but several authors at the time referred specifically to Cantor's concept of covering, most notably Zermelo [1904]. Jourdain in his introduction to his English translation of the *Beiträge* wrote (Cantor [1915: 82]): "The introduction of the concept of 'covering' is the most striking advance in the principles of the theory of transfinite numbers from 1885 to 1895 …."

With Cantor initially focusing on bijective correspondence [Beziehung] and these not quite construed as functions, Dedekind was the first to entertain an arbitrary function on an arbitrary domain. He [1888: §§21,36] formulated $\phi: S \to Z$, "a mapping [Abbildung] of a system S in Z", in less convoluted terms, but did not consider the totality of such. He quickly moved to the case $Z = S$ for his theory of chains; see footnote 36.

of the pen" in his new arithmetic:

$$(2^{\aleph_0})^{\aleph_0} = 2^{\aleph_0 \cdot \aleph_0} = 2^{\aleph_0}.$$

Cantor only mentioned

$$\aleph_0, \aleph_1, \aleph_2, \ldots, \aleph_\alpha, \ldots,$$

these to be the cardinal numbers of the successive number classes from the *Grundlagen* and thus to exhaust all the infinite cardinal numbers.

Cantor went on to present his theory of *order types*, abstractions of linear orderings. He defined an arithmetic of order types and characterized the order type η of the rationals as the countable dense linear order without endpoints, introducing the "forth" part of the now familiar back-and-forth argument of model theory.[23] He also characterized the order type θ of the reals as the perfect linear order with a countable dense set; whether a realist or not, Cantor the mathematician was able to provide a characterization of the continuum.

The second *Beiträge* developed the *Grundlagen* ideas by focusing on well-orderings and construing their order types as the *ordinal numbers*. Here at last was the general proof via order comparison of well-ordered sets that ordinal numbers are comparable. Cantor went on to describe ordinal arithmetic as a special case of the arithmetic of order types and after giving the basic properties of the second number class defined $\aleph_1$ as its cardinal number. The last sections were given over to a later preoccupation, the study of ordinal exponentiation in the second number class. The operation was defined via a transfinite recursion and used to establish a normal form, and the pivotal ϵ-numbers satisfying $\epsilon = \omega^\epsilon$ were analyzed.

The two parts of the *Beiträge* are not only distinct by subject matter, cardinal number and the continuum vs. ordinal number and well-ordering, but between them there developed a wide, irreconcilable breach. In the first part nowhere is the [1891] result $\mathfrak{a} < 2^{\mathfrak{a}}$ stated even in a special case; rather, it is made clear [1895: 495] that the procession of transfinite cardinal numbers is to be secured through their construal as the alephs. However, the second *Beiträge* does not mention any aleph beyond $\aleph_1$, nor does it mention CH, which could now have been stated as

$$2^{\aleph_0} = \aleph_1.$$

(Cantor did state this in an 1895 letter.[24]) Ordinal comparability was secured, but cardinal comparability was not reduced to it. Every well-ordered set has an aleph as its cardinal number, but where is $2^{\aleph_0}$ in the aleph sequence?

Cantor's initial [1874] proof led to the Continuum Problem. That problem was embedded in the very interstices of the early development of set theory, and in fact the structures that Cantor built, while now of intrinsic interest, emerged in significant part out of efforts to articulate and solve the problem. Cantor's [1891] diagonal argument, arguably a transmutation of his initial [1874] proof, exacerbated a growing tension between having well-orderings and admitting sets of arbitrary functions (or power sets). David Hilbert, when

---

[23] See Plotkin [1993] for an analysis of the emergence of the back-and-forth argument.
[24] See Moore [1989: 99].

he presented his famous list of problems at the 1900 International Congress of Mathematicians at Paris, made the Continuum Problem the very first problem and intimated Cantor's difficulty by suggesting the desirability of "actually giving" a well-ordering of the reals.

The next, 1904 International Congress of Mathematicians at Heidelberg was to be a generational turning point for the development of set theory. Julius Kőnig delivered a lecture in which he provided a detailed argument that purportedly established that $2^{\aleph_0}$ is not an aleph, i.e. that the continuum is not well-orderable. The argument combined the now familiar inequality $\aleph_\alpha < \aleph_\alpha^{\aleph_0}$ for α of cofinality ω with a result from Felix Bernstein's Göttingen dissertation [1901: 49] which alas does not universally hold.[25] Cantor was understandably upset with the prospect that the continuum would simply escape the number context that he had devised for its analysis.

Accounts differ on how the issue was resolved. Although one has Zermelo finding an error within a day of the lecture, the weight of evidence is for Hausdorff having found the error.[26] Whatever the resolution, the torch had passed from Cantor to the next generation. Zermelo would go on to formulate his Well-Ordering Theorem and axiomatize set theory, and Hausdorff, to develop the higher transfinite in his study of order types and cofinalities.[27]

2 MATHEMATIZATION

2.1 Axiom of Choice and Axiomatization

Ernst Zermelo[28] (1871–1953), born when Cantor was establishing his trigonometric series results, had begun to investigate Cantorian set theory at Göttingen under the influence of Hilbert. In just over a month after the Heidelberg congress, Zermelo [1904] formulated what he soon called the *Axiom of Choice* (AC) and with it, established his Well-Ordering Theorem:

> Every set can be well-ordered.

Zermelo thereby shifted the notion of set away from the implicit assumption of Cantor's principle that every well-defined set is well-ordered and replaced that principle by an

---

[25] The *cofinality* of an ordinal number α is the least ordinal number β such that there is a set of form $\{\gamma_\xi \mid \xi < \beta\}$ unbounded in α, i.e. for any $\eta < \alpha$ there is an $\xi < \beta$ such that $\eta \leq \gamma_\xi < \alpha$. α is *regular* if its cofinality is itself, and otherwise α is *singular*. There concepts were not clarified until the work of Hausdorff, brought together in his [1908], discussed in 2.6.

Kőnig applied Bernstein's equality $\aleph_\alpha^{\aleph_0} = \aleph_\alpha \cdot 2^{\aleph_0}$ as follows: If $2^{\aleph_0}$ were an aleph, say $\aleph_\beta$, then by Bernstein's equality $\aleph_{\beta+\omega}^{\aleph_0} = \aleph_{\beta+\omega} \cdot 2^{\aleph_0} = \aleph_{\beta+\omega}$, contradicting Kőnig's inequality. However, Bernstein's equality fails when α has cofinality ω and $2^{\aleph_0} < \aleph_\alpha$. Kőnig's published account [1905] acknowledged the gap.

[26] See Grattan-Guinness [2000, 334] and Purkert [2002].

[27] And as with many incorrect proofs, there would be positive residues: Zermelo soon generalized Kőnig's inequality to the fundamental Zermelo-Kőnig inequality for cardinal exponentiation, which implies that the cofinality of $2^{\aleph_\alpha}$ is larger than α, and Hausdorff [1904: 571] published his recursion formula $\aleph_{\beta+1}^{\aleph_\alpha} = \aleph_{\beta+1} \cdot \aleph_\beta^{\aleph_\alpha}$, in form like Bernstein's result.

[28] Ebbinghaus [2007] is a substantive biography of Zermelo. See Kanamori [1997; 2004] for Zermelo's work in set theory.

explicit axiom about a wider notion of set, incipiently unstructured but soon to be given form by axioms.

In retrospect, Zermelo's *argument* for his Well-Ordering Theorem can be viewed as pivotal for the development of set theory. To summarize the argument, suppose that x is a set to be well-ordered, and through Zermelo's Axiom-of-Choice hypothesis assume that the power set $\mathcal{P}(x) = \{y \mid y \subseteq x\}$ has a choice function, i.e. a function γ such that for every non-empty member y of $\mathcal{P}(x)$, $\gamma(y) \in y$. Call a subset y of x a γ-*set* if there is a well-ordering R of y such that for each $a \in y$,

$$\gamma(\{z \mid z \notin y \text{ or } z R a \text{ fails}\}) = a.$$

That is, each member of y is what γ "chooses" from what does not already precede that member according to R. The main observation is that γ-sets cohere in the following sense: If y is a γ-set with well-ordering R and z is a γ-set with well-ordering S, then $y \subseteq z$ and S is a prolongation of R, or vice versa. With this, let w be the union of all the γ-sets, i.e. all the γ-sets put together. Then w too is a γ-set, and by its maximality it must be all of x and hence x is well-ordered.

The converse to this result is immediate in that if x is well-ordered, then the power set $\mathcal{P}(x)$ has a choice function.[29] Not only did Zermelo's argument analyze the connection between having well-orderings and having choice functions on power sets, it anticipated in its defining of approximations and taking of a union the proof procedure for von Neumann's Transfinite Recursion Theorem (cf. 3.1).[30]

Zermelo [1904: 516] noted without much ado that his result implies that every infinite cardinal number is an aleph and satisfies $\mathfrak{m}^2 = \mathfrak{m}$, and that it secured Cardinal Comparability — so that the main issues raised by Cantor's *Beiträge* are at once resolved. Zermelo maintained that the Axiom of Choice, to the effect that *every* set has a choice function, is a "logical principle" which "is applied without hesitation everywhere in mathematical deduction", and this is reflected in the Well-Ordering Theorem being regarded as a theorem. The axiom is consistent with Cantor's view of the finite and transfinite as unitary, in that it posits for infinite sets an unproblematic feature of finite sets. On the other hand, the Well-Ordering Theorem shifted the weight from Cantor's well-orderings with their residually temporal aspect of numbering through *successive* choices to the use of a function for making *simultaneous* choices.[31] Cantor's work had served to exacerbate a growing discord among mathematicians with respect to two related issues: whether infinite collections can be mathematically investigated at all, and how far the function concept is to be extended. The positive use of an arbitrary function operating on arbitrary subsets of a set having been made explicit, there was open controversy after the appearance of Zermelo's proof. This can be viewed as a turning point for mathematics, with the subsequent tilt-

---

[29] Namely, with $<$ a well-ordering of x, for each non-empty member y of $\mathcal{P}(x)$, let $\gamma(y)$ be the the $<$-least member of y.

[30] See Kanamori [1997] for more on the significance of Zermelo's argument, in particular as a fixed point argument.

[31] Zermelo himself stressed the importance of simultaneous choices over successive choices in criticism of an argument of Cantor's for the Well-Ordering Theorem in 1899 correspondence with Dedekind, discussed in 2.2. See Cantor [1932: 451] or van Heijenoort [1967: 117].

ing toward the acceptance of the Axiom of Choice symptomatic of a conceptual shift in mathematics.

In response to his critics Zermelo published a second proof [1908] of his Well-Ordering Theorem, and with axiomatization assuming a general methodological role in mathematics he also published the first full-fledged axiomatization [1908a] of set theory. But as with Cantor's work this was no idle structure building but a response to pressure for a new mathematical context. In this case it was not for the formulation and solution of a *problem* like the Continuum Problem, but rather to clarify a specific *proof*. In addition to codifying generative set-theoretic principles, a substantial motive for Zermelo's axiomatizing set theory was to buttress his Well-Ordering Theorem by making explicit its underlying set existence assumptions.[32] Initiating the first major transmutation of the notion of set after Cantor, Zermelo thereby ushered in a new abstract, prescriptive view of sets as structured solely by membership and governed and generated by axioms, a view that would soon come to dominate. Thus, proof played a crucial role by stimulating an axiomatization of a field of study and a corresponding transmutation of its underlying notions.

The objections raised against Zermelo's first proof [1904] mainly played on the ambiguities of a γ-set's well-ordering being only implicit, as for Cantor's sets, and on the definition of the well-ordering being impredicative — defined as a γ-set and so drawn from a collection of which it is already a member. Largely to preclude these objections Zermelo in his second [1908] proof resorted to a rendition of orderings in terms of segments and inclusion first used by Gerhard Hessenberg [1906: 674ff] and a closure approach with roots in Dedekind [1888]. Instead of extending initial segments toward the desired well-ordering, Zermelo got at the collection of its final segments by taking an *intersection* in a larger setting.[33]

With his [1908a] axiomatization, Zermelo "started from set theory as it is historically given" to seek out principles sufficiently restrictive "to exclude all contradictions" and sufficiently wide "to retain all that is valuable". However, he would transform set theory by making explicit *new* existence principles and promoting a generative point of view. Zermelo had begun working out an axiomatization as early as 1905, addressing issues raised by his [1904] proof.[34] The mature presentation is a precipitation of seven axioms, and these do not just reflect "set theory as it is historically given", but explicitly buttress his proof(s) of the Well-Ordering Theorem.

Zermelo's seven set axioms, now formalized, constitute the familiar theory Z, Zermelo set theory: Extensionality, Elementary Sets ($\emptyset$, $\{a\}$, $\{a, b\}$), Separation, Power Set, Union, Choice, and Infinity. His setting allowed for urelements, objects without members yet distinct from each other. But Zermelo focused on sets, and his Axiom of Extensional-

---

[32] Moore [1982: 155ff] supports this contention using items from Zermelo's *Nachlass*.

[33] To well-order a set M using a choice function φ on $\mathcal{P}(M)$, Zermelo defined a Θ-*chain* to be a collection Θ of subsets of M such that: (a) $M \in \Theta$; (b) if $A \in \Theta$, then $A - \{\varphi(A)\} \in \Theta$; and (c) if $Z \subseteq \Theta$, then $\bigcap Z \in \Theta$. He then took the intersection I of all Θ-chains, and observed that I is again a Θ-chain. Finally, he showed that I provides a well-ordering of M given by: $a < b$ *iff* there is an $A \in I$ such that $a \notin A$ and $b \in A$. I thus consists of the final segments of the same well-ordering as provided by the [1904] proof. Note that this second proof is less parsimonious than the [1904] proof, as it uses the power set of the power set of M.

[34] This is documented by Moore [1982: 155ff] with items from Zermelo's *Nachlass*.

ity announced the espousal of an extensional viewpoint. In line with this AC, a "logical principle" in [1904] expressed in terms of an informal choice function, was framed less instrumentally: It posited for a set consisting of non-empty, pairwise disjoint sets the existence of a *set* that meets each one in a unique element.[35] However, Separation retained an intensional aspect with its "separating out" of a new set from a given set using a *definite* property, where a property is *"definite [definit]* if the fundamental relations of the domain, by means of the axioms and the universally valid laws of logic, determine without arbitrariness whether it holds or not." But with no underlying logic formalized, the ambiguity of definite property would become a major issue. With Infinity and Power Set Zermelo provided for sufficiently rich settings for set-theoretic constructions. Tempering the logicians' extravagant and problematic "all" the Power Set axiom provided the provenance for "all" for subsets of a given set, just as Separation served to capture "all" for elements of a given set satisfying a property. Finally, Union and Choice completed the encasing of Zermelo's proof(s) of his Well-Ordering Theorem in the necessary set existence principles. Notably, Zermelo's recursive [1904] argumentation also brought him in proximity of the Transfinite Recursion Theorem and thus of Replacement, the next axiom to be adjoined in the subsequent development of set theory (cf. 3.1).

Fully two decades earlier Dedekind [1888] had provided an incisive analysis of the natural numbers and their arithmetic in terms of sets [Systeme], and several overlapping aspects can serve as points of departure for Zermelo's axiomatization.[36] The most immediate is how Dedekind's argumentation extends to Zermelo's [1908] proof of the Well-Ordering Theorem, which in the transfinite setting brings out the role of AC. Both Dedekind and Zermelo set down rules for sets in large part to articulate arguments involving simple set operations like "set of", union, and intersection. In particular, both had to argue for the equality of sets resulting after involved manipulations, and extensionality became operationally necessary. However vague the initial descriptions of sets, sets are to be determined solely by their elements, and the membership question is to be determinate.[37] The looseness of Dedekind's description of sets allowed him [1888: §66] the latitude to "prove" the existence of infinite sets, but Zermelo just stated the Axiom of Infinity as a set existence principle.

The main point of departure has to do with the larger issue of the role of proof for articulating sets. By Dedekind's time proof had become basic for mathematics, and indeed

---

[35] Russell [1906] had previously arrived at this form, his Multiplicative Axiom. The elimination of the "pairwise disjoint" by going to a choice function formulation can be established with the Union Axiom, and this is the only use of that axiom in the second, [1908] proof of the Well-Ordering Theorem.

[36] In current terminology, Dedekind [1888] considered arbitrary sets S and mappings $\phi: S \to S$ and defined a *chain* [Kette] to be a $K \subseteq S$ such that $\phi``K \subseteq K$. For $A \subseteq S$, the *chain of* $A$ is the intersection of all chains $K \supseteq A$. A set $N$ is *simply infinite iff* there is an injective $\phi: N \to N$ such that $N - \phi``N \neq \emptyset$. Letting 1 be a distinguished element of $N - \phi``N \neq \emptyset$ Dedekind considered the chain of $\{1\}$, the chain of $\{\phi(1)\}$, and so forth. Having stated an inherent induction principle, he proceeded to show that these sets have all the ordering and arithmetical properties of the natural numbers (that are established nowadays in texts for the (von Neumann) finite ordinals).

[37] Dedekind [1888: §2] begins a footnote to his statement about extensional determination with: "In what manner this determination is brought about, and whether we know a way of deciding upon it, is a matter of indifference for all that follows; the general laws to be developed in no way depend upon it; they hold under all circumstances."

his work did a great deal to enshrine proof as the vehicle for algebraic abstraction and generalization.[38] Like algebraic constructs, sets were new to mathematics and would be incorporated by setting down the rules for their proofs. Just as calculations are part of the sense of numbers, so proofs would be part of the sense of sets, as their "calculations". Just as Euclid's axioms for geometry had set out the permissible geometric constructions, the axioms of set theory would set out the specific rules for set generation and manipulation. But unlike the emergence of mathematics from marketplace arithmetic and Greek geometry, sets and transfinite numbers were neither laden nor buttressed with substantial antecedents. Like strangers in a strange land stalwarts developed a familiarity with them guided hand in hand by their axiomatic framework. For Dedekind [1888] it had sufficed to work with sets by merely giving a few definitions and properties, those foreshadowing Extensionality, Union, and Infinity. Zermelo [1908a] provided more rules: Separation, Power Set, and Choice.

Zermelo [1908], with its rendition of orderings in terms of segments and inclusion, and Zermelo [1908a], which at the end cast Cantor's theory of cardinality in terms of functions cast as set constructs, brought out Zermelo's *set-theoretic reductionism*. Zermelo pioneered the reduction of mathematical concepts and arguments to set-theoretic concepts and arguments from axioms, based on sets doing the work of mathematical objects. Zermelo's analyses moreover served to draw out what would come to be generally regarded as set-theoretic out of the presumptively logical. This would be particularly salient for Infinity and Power Set and was strategically advanced by the relegation of property considerations to Separation.

Zermelo's axiomatization also shifted the focus away from the transfinite numbers to an abstract view of sets structured solely by $\in$ and simple operations. For Cantor the transfinite numbers had become central to his investigation of definable sets of reals and the Continuum Problem, and sets had emerged not only equipped with orderings but only as the developing context dictated, with the "set of" operation never iterated more than three or four times. For Zermelo his second, [1908] proof of the Well-Ordering Theorem served to eliminate any residual role that the transfinite numbers may have played in the first proof and highlighted the set-theoretic operations. This approach to (linear) ordering was to preoccupy his followers for some time, and through this period the elimination of the use of transfinite numbers where possible, like ideal numbers, was regarded as salutary.[39] Hence, Zermelo rather than Cantor should be regarded as the creator of abstract set theory.

---

[38] Cf. the first sentence of the preface to Dedekind [1888]: "In science nothing capable of proof ought to be accepted without proof."

[39] Some notable examples: Lindelöf [1905] proved the Cantor-Bendixson result, that every uncountable closed set is the union of a perfect set and a countable set, without using transfinite numbers. Suslin's [1917], discussed in 2.5, had the unassuming title, "On a definition of the Borel sets without transfinite numbers", hardly indicative of its results, so fundamental for descriptive set theory. And Kuratowski [1922] showed, pursuing the approach of Zermelo [1908], that inclusion chains defined via transfinite recursion with intersections taken at limits can also be defined without transfinite numbers. Kuratowski [1922] essentially formulated Zorn's Lemma, and this was the main success of the push away from explicit well-orderings. Especially after the appearance of Zorn [1935] this recasting of AC came to dominate in algebra and topology.

Outgrowing Zermelo's pragmatic purposes axiomatic set theory could not long forestall the Cantorian initiative, as even $2^{\aleph_0} = \aleph_1$ could not be asserted directly, and in the 1920s John von Neumann was to fully incorporate the transfinite using Replacement (cf. 3.1).[40] On the other hand, Zermelo's axioms had the advantages of schematic simplicity and open-endedness. The generative set formation axioms, especially Power Set and Union, were to lead to Zermelo's [1930] cumulative hierarchy picture of sets, and the vagueness of the *definit* property in the Separation Axiom was to invite Thoralf Skolem's [1923] proposal to base it on first-order logic, enforcing extensionalization (cf. 3.2).

2.2 Logic and Paradox

At this point, the incursions of a looming tradition can no longer be ignored. Gottlob Frege is regarded as the greatest philosopher of logic since Aristotle for developing quantificational logic in his *Begriffsschrift* [1879], establishing a logical foundation for arithmetic in his *Grundlagen* [1884], and generally stimulating the *analytic tradition* in philosophy. The architect of that tradition was Bertrand Russell who in his earlier years, influenced by Frege and Giuseppe Peano, wanted to found all of mathematics on the certainty of logic. But from a logical point of view Russell [1903] became exercised with paradox. He had arrived at Russell's Paradox in late 1901 by analyzing Cantor's diagonal argument applied to the class of all classes,[41] a version of which is now known as Cantor's Paradox of the largest cardinal number. Russell [1903: §301] also refocused the Burali-Forti Paradox of the largest ordinal number, after reading Cesare Burali-Forti's [1897].[42] Russell's Paradox famously led to the tottering of Frege's mature formal system, the *Grundgesetze* [1893, 1903].[43]

Russell's own reaction was to build a complex logical structure, one used later to develop mathematics in Whitehead and Russell's 1910-3 *Principia Mathematica*. Russell's *ramified theory of types* is a scheme of logical definitions based on *orders* and *types* indexed by the natural numbers. Russell proceeded "intensionally"; he conceived this scheme as a classification of propositions based on the notion of *propositional function*, a notion not reducible to membership (extensionality). Proceeding in modern fashion, we may say that the universe of the *Principia* consists of *objects* stratified into disjoint types T_n, where T_0 consists of the *individuals*, $T_{n+1} \subseteq \{Y \mid Y \subseteq T_n\}$, and the types T_n for $n > 0$ are further ramified into orders O_n^i with $T_n = \bigcup_i O_n^i$. An object in O_n^i is to be defined either in terms of individuals or of objects in some fixed O_m^j for some $j < i$ and $m \leq n$, the definitions allowing for quantification only over O_m^j. This precludes Russell's Paradox and other "vicious circles", as objects consist only of previous objects and are built up

---

[40] Textbooks usually establish the Well-Ordering Theorem by first introducing Replacement, formalizing transfinite recursion, and only then defining the well-ordering using (von Neumann) ordinals; this amounts to another historical misrepresentation, but one that resonates with how acceptance of Zermelo's proof broke the ground for formal transfinite recursion.

[41] Grattan-Guinness [1974], Coffa [1979], Moore [1988], and Garciadiego [1992] describe the evolution of Russell's Paradox.

[42] Moore-Garciadiego [1981] and Garciadiego [1992] describe the evolution of the Burali-Forti Paradox.

[43] See the exchange of letters between Russell and Frege in van Heijenoort [1967: 124ff]. Russell's Paradox showed that Frege's Basic Law V is inconsistent.

through definitions referring only to previous stages. However, in this system it is impossible to quantify over all objects in a type T_n, and this makes the formulation of numerous mathematical propositions at best cumbersome and at worst impossible. Russell was led to introduce his *Axiom of Reducibility*, which asserts that *for each object there is a predicative object consisting of exactly the same objects*, where an object is *predicative* if its order is the least greater than that of its constituents. This axiom reduced consideration to individuals, predicative objects consisting of individuals, predicative objects consisting of predicative objects consisting of individuals, and so on—the *simple theory of types*. In traumatic reaction to his paradox Russell had built a complex system of orders and types only to collapse it with his Axiom of Reducibility, a fearful symmetry imposed by an artful dodger.

The mathematicians did not imbue the paradoxes with such potency. Unlike Russell who wanted to get at everything but found that he could not, they started with what could be got at and peered beyond. And as with the invention of the irrational numbers, the outward push eventually led to the positive subsumption of the paradoxes.

Cantor in 1899 correspondence with Dedekind considered the collection Ω of all ordinal numbers as in the Burali-Forti Paradox, but he used it *positively* to give mathematical expression to his Absolute.[44] First, he distinguished between two kinds of multiplicities (Vielheiten): There are multiplicities such that when taken as a unity (Einheit) lead to a contradiction; such multiplicities he called "*absolutely infinite or inconsistent multiplicities*" and noted that the "totality of everything thinkable" is such a multiplicity. A multiplicity that can be thought of without contradiction as "being together" he called a "*consistent multiplicity* or a 'set [Menge]'". Cantor then used the Burali-Forti Paradox argument to point out that the class Ω of all ordinal numbers is an inconsistent multiplicity. He proceeded to argue that every *set* can be well-ordered through a presumably recursive procedure whereby a well-ordering is defined through successive choices. The set must get well-ordered, else all of Ω would be injectible into it, so that the set would have been an inconsistent multiplicity instead.[45]

Zermelo found Russell's Paradox independently and probably in 1902,[46] but like Cantor, he did not regard the emergence of the paradoxes so much as a crisis as an overall delimitation for sets. In the Zermelian generative view [1908: 118], "... if in set theory we confine ourselves to a number of established principles such as those that constitute the basis of our proof — principles that enable us to form initial sets and to derive new sets from given ones – then all such contradictions can be avoided." For the first theorem of his axiomatic theory Zermelo [1908a] subsumed Russell's Paradox, putting it to use as is done now to establish that for any set x there is a $y \subseteq x$ such that $y \notin x$, and hence that there is no universal set.[47]

---

[44] See footnote 3 for more about the 1899 correspondence. Purkert [1989: 57ff] argues that Cantor had already arrived at the Burali-Forti Paradox around the time of the *Grundlagen* [1883]. On the interpretations supported in the text *all* of the logical paradoxes grew out of Cantor's work — with Russell shifting the weight to paradox.

[45] G.H. Hardy [1903] and Philip Jourdain [1904, 1905] also gave arguments involving the injection of Ω, but such an approach would only get codified at a later stage in the development of set theory in the work of von Neumann [1925] (cf. 3.1).

[46] See Kanamori [2004: §1].

[47] In 2.6 Hartogs's Theorem is construed as a positive subsumption of that other, the Burali-Forti Paradox.

The differing concerns of Frege-Russell logic and the emerging set theory are further brought out by the analysis of the function concept as discussed below in 2.4, and those issues are here rehearsed with respect to the existence of the null class, or empty set.[48] Frege in his *Grundlagen* [1884] eschewed the terms "set [Menge]" and "class [Klasse]", but in any case the extension of the concept "not identical with itself" was key to his definition of zero as a logical object. Ernst Schröder, in the first volume [1890] of his major work on the algebra of logic, held a traditional view that a class is merely a collection of objects, without the { } so to speak. In his review [1895] of Schröder's [1890], Frege argued that Schröder cannot both maintain this view of classes and assert that there is a null class, since the null class contains no objects. For Frege, logic enters in giving unity to a class as the extension of a *concept* and thus makes the null class viable.

It is among the set theorists that the null class, *qua* empty set, emerged to the fore as an elementary concept and a basic building block. Cantor himself did not dwell on the empty set. At one point he did write [1880: 355] that "the identity of two pointsets P and Q will be expressed by the formula $P \equiv Q$"; defined disjoint sets as "*lacking intersection*"; and then wrote [1880: 356] "for the absence of points ... we choose the letter O; $P \equiv O$ indicates that the set P contains *no single point*." (So, "$\equiv O$" is arguably more like a predication for being empty at this stage.)

Dedekind [1888: §2] deliberately excluded the empty set [Nullsystem] "for certain reasons", though he saw its possible usefulness in other contexts. Zermelo [1908a] wrote in his Axiom II: "There exists a (improper [uneigentliche]) set, the *null set [Nullmenge]* 0, that contains no element at all." Something of intension remained in the "(improper [uneigentliche])", though he did point out that because of his Axiom I, the Axiom of Extensionality, there is a single empty set. Finally, Hausdorff [1914] unequivocally opted for the empty set [Nullmenge]. However, a hint of predication remained when he wrote [1914: 3]: "... the equation $A = 0$ means that the set A has no element, vanishes [verschwindet], is empty." The use to which Hausdorff put "0" is much as "$\emptyset$" is used in modern mathematics, particularly to indicate the extension of the conjunction of mutually exclusive properties.

The set theorists, unencumbered by philosophical motivations or traditions, attributed little significance to the empty set beyond its usefulness. Although embracing both extensionality and the null class may engender philosophical difficulties for the logic of classes, the empty set became commonplace in mathematics simply through use, like its intimate, zero.

2.3 *Measure, Category, and Borel Hierarchy*

During this period Cantor's two main legacies, the investigation of definable sets of reals and the extension of number into the transfinite, were further incorporated into mathematics in direct initiatives. The axiomatic tradition would be complemented by another, one that would draw its life more directly from mathematics.

The French analysts Emile Borel, René Baire, and Henri Lebesgue took on the investigation of definable sets of reals in what was to be a paradigmatically constructive

---

[48] For more on the empty set, see Kanamori [2003a].

approach. Cantor [1884] had established the perfect set property for closed sets and formulated the concept of *content* for a set of reals, but he did not pursue these matters. With these as antecedents the French work would lay the basis for measure theory as well as *descriptive set theory,* the definability theory of the continuum.[49]

Soon after completing his thesis Borel [1898: 46ff] considered for his theory of measure those sets of reals obtainable by starting with the intervals and closing off under complementation and countable union. The formulation was axiomatic and in effect impredicative, and seen in this light, bold and imaginative; the sets are now known as the *Borel sets* and quite well-understood.

Baire in his thesis [1899] took on a dictum of Lejeune Dirichlet's that a real function is any arbitrary assignment of reals, and diverging from the 19th-Century preoccupation with pathological examples, sought a constructive approach via pointwise limits. His *Baire class 0* consists of the continuous real functions, and for countable ordinal numbers $\alpha > 0$, *Baire class α* consists of those functions f not in any previous class yet obtainable as pointwise limits of sequences $f_0, f_1, f_2, \ldots$ of functions in previous classes, i.e. $f(x) = \lim_{n \to \infty} f_n(x)$ for every real x. The functions in these classes are now known as the *Baire* functions, and this was the first stratification into a transfinite hierarchy after Cantor.[50]

Baire's thesis also introduced the now basic concept of *category*. A set of reals is *nowhere dense iff* its closure under limits includes no open set, and a set of reals is *meager* (or *of first category*) *iff* it is a countable union of nowhere dense sets — otherwise, it is *of second category*. Baire established the Baire Category Theorem: *Every non-empty open set of reals is of second category.* His work also suggested a basic property: A set of reals has the *Baire property iff* it has a meager symmetric difference with some open set. Straightforward arguments show that every Borel set has the Baire property.

Lebesgue's thesis [1902] is fundamental for modern integration theory as the source of his concept of measurability. Inspired in part by Borel's ideas but notably containing non-constructive aspects, Lebesgue's concept of measurable set through its closure under countable unions subsumed the Borel sets, and his analytic definition of measurable function through its closure under pointwise limits subsumed the Baire functions. Category and measure are quite different; there is a co-meager (complement of a meager) set of reals that has Lebesgue measure zero.[51] Lebesgue's first major work in a distinctive direction would be the seminal paper in descriptive set theory:

In the memoir [1905] Lebesgue investigated the Baire functions, stressing that they are exactly the functions definable via analytic expressions (in a sense made precise). He first established a correlation with the Borel sets by showing that they are exactly the pre-images of open intervals via Baire functions. With this he introduced the first hierarchy for the Borel sets, his *open sets of class α* not being in any previous class yet being pre-images of some open interval via some Baire class α function. After verifying various

---

[49] See Kanamori [1995] for more on the emergence of descriptive set theory. See Moschovakis [1980] or Kanamori [2003] for the mathematical development.

[50] Baire mainly studied the finite levels, particularly the classes 1 and 2. He [1898] pointed out that Dirichlet's function that assigns 1 to rationals and 0 to irrationals is in class 2 and also observed with a non-constructive appeal to Cantor's cardinality argument that there are real functions that are not Baire.

[51] See footnote 11. See Hawkins [1975] for more on the development of Lebesgue measurability. See Oxtoby [1971] for an account of category and measure in juxtaposition.

closure properties and providing characterizations for these classes Lebesgue established two main results. The first demonstrated the necessity of exhausting the countable ordinal numbers: *The Baire hierarchy is proper, i.e. for every countable α there is a Baire function of class α; correspondingly the hierarchy for the Borel sets is analogously proper.* The second established transcendence beyond countable closure for his concept of measurability: *There is a Lebesgue measurable function which is not in any Baire class; correspondingly there is a Lebesgue measurable set which is not a Borel set.*

The first result was the first of all hierarchy results, and a precursor of fundamental work in mathematical logic in that it applied Cantor's enumeration and diagonalization argument to achieve a transcendence to a next level. Lebesgue's second result was also remarkable in that he actually provided an explicitly defined set, one that was later seen to be the first example of a non-Borel analytic set (cf. 2.5). For this purpose, the reals were for the first time regarded as encoding something else, namely countable well-orderings, and this not only further embedded the transfinite into the investigation of sets of reals, but foreshadowed the later coding results of mathematical logic.

Lebesgue's results, along with the later work in descriptive set theory, can be viewed as pushing the mathematical frontier of the actual infinite past $\aleph_0$, which arguably had achieved a mathematical domesticity through increasing use in the late 19th-Century, through Cantor's second number class to $\aleph_1$. It is somewhat ironic but also revealing, then, that this grew out of work by analysts with a definite constructive bent. Baire [1899: 36] viewed the infinite ordinal numbers and hence his function hierarchy as merely *une façon de parler*, and continued to view infinite concepts only in potentiality. Borel [1898] took a pragmatic approach and seemed to accept the countable ordinal numbers. Lebesgue was more equivocal but still accepting; recalling Cantor's early attitude Lebesgue regarded the ordinal numbers as an indexing system, "symbols" for classes, but nonetheless he worked out their basic properties, even providing a formulation [1905: 149] of proof by transfinite induction. All three analysts expressed misgivings about AC and its use in Zermelo's proof.[52]

As descriptive set theory was to develop, a major concern became the extent of the *regularity properties*, those properties indicative of well-behaved sets of reals of which Lebesgue measurability, the Baire property, and the perfect set property are the prominent examples. These properties seemed to get at basic features of the extensional construal of the continuum, yet resisted inductive approaches. Early explicit uses of AC through its role in providing a well-ordering of the reals showed how it allowed for new constructions: Giuseppe Vitali [1905] established that there is a non-Lebesgue measurable set of reals, and Felix Bernstein [1908], that there is a set of reals without the perfect set property. Soon it was seen that neither of these examples have the Baire property. Thus, that the reals are well-orderable, an early contention of Cantor's, permitted constructions that precluded the universality of the regularity properties, in particular his own approach to the Continuum Problem through the perfect set property.

---

[52] See Moore [1982: 2.3].

2.4 Hausdorff and Functions

Felix Hausdorff was the first developer of the transfinite after Cantor, the one whose work first suggested the rich possibilities for a mathematical investigation of the higher transfinite. A mathematician *par excellence*, Hausdorff took that sort of mathematical approach to set theory and extensional, set-theoretic approach to mathematics that would dominate in the years to come. While the web of 19th-Century intension in Cantor's work, especially his approach toward functions, now seems rather remote, Hausdorff's work seems familiar as part of the modern language of mathematics.

In [1908] Hausdorff brought together his extensive work on *uncountable* order types.[53] Deploring all the fuss being made over foundations by his contemporaries (p.436) and with Cantor having taken the Continuum Problem as far as seemed possible, Hausdorff proceeded to venture beyond the second number class with vigor. He provided an elegant analysis of scattered linear order types (those having no dense subtype) in a transfinite hierarchy, and constructed the η_α sets, prototypes for saturated model theory. He first stated the *Generalized Continuum Hypothesis* (GCH), that $2^{\aleph_\alpha} = \aleph_{\alpha+1}$ for every α, clarified the significance of cofinality, and first considered (p.443) the possibility of an uncountable regular limit cardinal, the first *large cardinal*.

Large cardinal hypotheses posit cardinals with properties that entail their transcendence over smaller cardinals, and as it has turned out, provide a superstructure of hypotheses for the analysis of strong propositions in terms of consistency. Hausdorff observed that uncountable regular limit cardinals, also known now as *weakly inaccessible cardinals*, are a natural closure point for cardinal limit processes. In penetrating work of only a few years later Paul Mahlo [1911; 1912; 1913] investigated hierarchies of such cardinals based on higher fixed-point phenomena, the *Mahlo cardinals*. The theory of large cardinals was to become a mainstream of set theory.[54]

Hausdorff's classic text, *Grundzüge der Mengenlehre* [1914] dedicated to Cantor, broke the ground for a generation of mathematicians in both set theory and topology. A compendium of a wealth of results, it emphasized mathematical approaches and procedures that would eventually take firm root.[55] After giving a clear account of Zermelo's first, [1904] proof of the Well-Ordering Theorem, Hausdorff (p.140ff) emphasized its maximality aspect by giving synoptic versions of Zorn's Lemma two decades before Zorn [1935], one of them now known as Hausdorff's Maximality Principle.[56] Also, Hausdorff (p.304) provided the now standard account of the Borel hierarchy of sets, with the still persistent F_σ and G_δ notation. Of particular interest, Hausdorff (p.469ff, and also in [1914a]) used AC to provide what is now known as Hausdorff's Paradox, an implausible decomposition of the sphere and the source of the better known Banach-Tarski Paradox

---

[53] See Plotkin [2005] for translations and careful analyses of Hausdorff's work on ordered sets.

[54] See Kanamori [2003] for more on large cardinals.

[55] Hausdorff's mathematical attitude is reflected in a remark following his explanation of cardinal number in a revised edition [1937:§5] of [1914]: "This formal explanation says what the cardinal numbers are supposed to do, not what they are. More precise definitions have been attempted, but they are unsatisfactory and unnecessary. Relations between cardinal numbers are merely a more convenient way of expressing relations between sets; we must leave the determination of the 'essence' of the cardinal number to philosophy."

[56] Hausdorff's Maximality Principle states that if A is a partially ordered set and B is a linearly ordered subset, then there is a $\subseteq$-maximal linearly ordered subset of A including B.

from Stefan Banach and Alfred Tarski's [1924].[57] Hausdorff's Paradox was the first, and a dramatic, synthesis of classical mathematics and the Zermelian abstract view.

Hausdorff's reduction of functions through a defined ordered pair highlights the differing concerns of the earlier Frege-Russell logic and the emerging set theory.[58] Frege [1891] had two fundamental categories, *function* and *object*, with a function being "unsaturated" and supplemented by objects as arguments. A *concept* is a function with two possible values, the True and the False, and a *relation* is a concept that takes two arguments. The extension of a concept is its graph or course-of-values [Werthverlauf], which is an object, and Frege [1893: §36] devised an iterated or double course-of-values [Doppelwerthverlauf] for the extension of a relation. In these involved ways Frege assimilated relations to functions. As for the ordered pair, Frege in his *Grundgesetze* [1893: §144] provided the extravagant definition that the ordered pair of x and y is that class to which all and only the extensions of relations to which x stands to y belong.[59]

On the other hand, Peirce [1883], Schröder [1895], and Peano [1897] essentially regarded a relation from the outset as just a collection of ordered pairs. Whereas Frege was attempting an analysis of thought, Peano was mainly concerned with recasting ongoing mathematics in economical and flexible symbolism and made many reductions, e.g. construing a *sequence* in analysis as a *function* on the natural numbers. Peano from his earliest logical writings had used "(x, y)" to indicate the ordered pair in formula and function substitutions and extensions. In [1897] he explicitly formulated the ordered pair using "$(x; y)$" and moreover raised the two main points about the ordered pair: First, equation 18 of his Definitions stated the instrumental property which is all that is required of the ordered pair:

(∗) $\qquad\qquad \langle x, y \rangle = \langle a, b \rangle \text{ iff } x = a \text{ and } y = b$.

Second, he broached the possibility of reducibility, writing: "The idea of a pair is fundamental, i.e. we do not know how to express it using the preceding symbols."

In Whitehead and Russell's *Principia Mathematica* [1910-3], relations distinguished in intension and in extension were derived from "propositional" functions taken as fundamental and other "descriptive" functions derived from relations. They [1910: ∗55] like Frege defined an ordered pair derivatively, in their case in terms of classes and relations, and also for a specific purpose.[60] Previously Russell [1903: §27] had criticized Peirce and

---

[57] Hausdorff's Paradox states that a sphere can be decomposed into four pieces Q, A, B, C with Q countable and A, B, C, and $B \cup C$ all pairwise congruent. Even more implausibly, the Banach-Tarski Paradox states that a ball can be decomposed into finitely many pieces that can be rearranged by rigid motions to form two balls of the same size as the original ball. Raphael Robinson [1947] later showed that there is such a decomposition into just five pieces with one of them containing a single point, and moreover that five is the minimal number. See Wagon [1985] for more on these and similar results; they stimulated interesting developments in measure theory that, rather than casting doubt on AC, embedded it further into mathematical practice (cf. 2.6).

[58] For more on the ordered pair, see Kanamori [2003a].

[59] This definition, which recalls the Whitehead–Russell definition of the cardinal number 2, depended on Frege's famously inconsistent Basic Law V. See Heck [1995] for more on Frege's definition and use of his ordered pair.

[60] Whitehead and Russell had first defined a cartesian product by other means, and only then defined their ordered pair $x \downarrow y$ as $\{x\} \times \{y\}$, a remarkable inversion from the current point of view. They [1910: ∗56] used their ordered pair initially to define the ordinal number 2.

Schröder for regarding a relation "essentially as a class of couples," although he did not mention this shortcoming in Peano.[61] Commenting obliviously on *Principia* Peano [1911; 1913] simply reaffirmed an ordered pair as basic, defined a relation as a class of ordered pairs, and a function extensionally as a kind of relation, referring to the final version of his *Formulario Mathematico* [1905-8: 73ff.] as the source.

Capping this to and fro Norbert Wiener [1914] provided a definition of the ordered pair in terms of unordered pairs of classes only, thereby reducing relations to classes. Working in Russell's theory of types, Wiener defined the ordered pair $\langle x, y \rangle$ as

$$\{\{\{x\}, \Lambda\}, \{\{y\}\}\}$$

when x and y are of the same type and Λ is the null class (of the next type), and pointed out that this definition satisfies the instrumental property (∗) above. Wiener used this to eliminate from the system of *Principia* the Axiom of Reducibility for propositional functions of two variables; he had written a doctoral thesis comparing the logics of Schröder and Russell.[62] Although Russell praised Sheffer's stroke, the logical connective not-both, he was not impressed by Wiener's reduction. Indeed, Russell would not have been able to accept it as a genuine analysis. Unlike Russell, Willard V.O. Quine in a major philosophical work *Word and Object* [1960: §53] regarded the reduction of the ordered pair as a paradigm for philosophical analysis.

Making no intensional distinctions Hausdorff [1914: 32ff,70ff] defined an ordered pair in terms of unordered pairs, formulated functions in terms of ordered pairs, and the ordering relations as collections of ordered pairs.[63] Hausdorff thus made both the Peano [1911; 1913] and Wiener [1914] moves *in* mathematical practice, completing the reduction of functions to sets.[64] This may have been congenial to Peano, but not to Frege nor Russell, they having emphasized the primacy of functions. Following the pioneering work of Dedekind and Cantor Hausdorff was at the crest of a major shift in mathematics of which the transition from an intensional, rule-governed conception of function to an extensional, arbitrary one was a large part, and of which the eventual acceptance of the Power Set Axiom and the Axiom of Choice was symptomatic.

In his informal setting Hausdorff took the ordered pair of x and y to be

$$\{\{x, 1\}, \{y, 2\}\}$$

---

[61] In a letter accepting Russell's [1901] on the logic of relations for publication in his journal *Rivista*, Peano had pointedly written "The classes of couples correspond to relations" (see Kennedy [1975: 214]) so that relations are extensionally assimilated to classes. Russell [1903: §98] argued that the ordered pair cannot be basic and would itself have to be given sense, which would be a circular or an inadequate exercise, and "It seems therefore more correct to take an intensional view of relations ...".

[62] See Grattan-Guinness [1975] for more on Wiener's work and his interaction with Russell.

[63] He did not so define arbitrary relations, for which there was then no mathematical use, but he was the first to consider general *partial* orderings, as in his maximality principle. Before Hausdorff and going beyond Cantor, Dedekind was first to consider non-linear orderings, e.g. in his remarkably early, axiomatic study [1900] of lattices.

[64] As to historical priority, Wiener's note was communicated to the Cambridge Philosophical Society, presented on 23 February 1914, while the preface to Hausdorff's book is dated 15 March 1914. Given the pace of book publication then, it is arguable that Hausdorff came up with his reduction first.

where 1 and 2 were intended to be distinct objects alien to the situation.[65] In any case, the now-standard definition is the more intrinsic

$$\{\{x\}, \{x, y\}\}$$

due to Kazimierz Kuratowski [1921: 171]. Notably, Kuratowski's definition is a by-product of his analysis of Zermelo's [1908] proof of the Well-Ordering Theorem.[66]

2.5 Analytic and Projective Sets

A decade after Lebesgue's seminal paper [1905], descriptive set theory emerged as a distinct discipline through the efforts of the Russian mathematician Nikolai Luzin. He had become acquainted with the work of the French analysts while in Paris as a student and had addressed Baire's functions with a intriguing use of CH. What is now known as a *Luzin set* is an uncountable set of reals whose intersection with any meager set is countable, and Luzin established: CH *implies that there is a Luzin set*.[67] This would become a paradigmatic use of CH, in that a recursive construction was carried out in $\aleph_1$ steps where at each state only countable many conditions have to be attended to, in this case by applying the Baire Category Theorem. Luzin showed that the characteristic function of his set escaped Baire's function classification, and Luzin sets have since become pivotal examples of "special sets" of reals.

In Moscow Luzin began an important seminar, and from the beginning a major topic was the "descriptive theory of functions". The young Pole Wacław Sierpiński was an early participant while he was interned in Moscow in 1915, and undoubtedly this not only kindled a decade-long collaboration between Luzin and Sierpiński but also encouraged the latter's involvement in the development of a Polish school of mathematics and its interest in descriptive set theory.

Of the three regularity properties, Lebesgue measurability, the Baire property, and the perfect set property (cf. 2.3), the first two were immediate for the Borel sets. However, nothing had been known about the perfect set property beyond Cantor's own result that the closed sets have it and Bernstein's that with a well-ordering of the reals there is a set not having the property. Luzin's student Pavel Aleksandrov [1916] established the

---

[65] It should be pointed out that the definition works even when x or y is 1 or 2 to maintain the instrumental property (∗) of ordered pairs.

[66] The general adoption of the Kuratowski pair proceeded through the major developments of mathematical logic: Von Neumann initially took the ordered pair as primitive but later noted (von Neumann [1925: VI]; [1928: 338]; [1929: 227]) the reduction via the Kuratowski definition. Gödel in his incompleteness paper [1931: 176] also pointed out the reduction. In his footnote 18, Gödel blandly remarked: "Every proposition about relations that is provable in [*Principia Mathematica*] is provable also when treated in this manner, as is readily seen." This stands in stark contrast to Russell's labors in *Principia* and his antipathy to Wiener's reduction of the ordered pair. Tarski [1931: n.3] pointed out the reduction and acknowledged his compatriot Kuratowski. In his recasting of von Neumann's system, Bernays [1937: 68] also acknowledged Kuratowski [1921] and began with its definition for the ordered pair. It is remarkable that Nicolas Bourbaki in his treatise [1954] on set theory still took the ordered pair as primitive, only later providing the Kuratowski reduction in the [1970] edition.

[67] Mahlo [1913a] also established this result.

groundbreaking result that *the Borel sets have the perfect set property*, so that "CH holds for the Borel sets".[68]

In the work that really began descriptive set theory another student of Luzin's, Mikhail Suslin, investigated the *analytic sets* following a mistake he had found in Lebesgue's paper.[69] Suslin [1917] formulated these sets in terms of an explicit operation $\mathcal{A}$[70] and announced two fundamental results: *a set B of reals is Borel iff both B and* $\mathbb{R} - B$ *are analytic*; and *there is an analytic set which is not Borel*.[71] This was to be his sole publication, for he succumbed to typhus in a Moscow epidemic in 1919 at the age of 25. In an accompanying note Luzin [1917] announced the regularity properties: *Every analytic set is Lebesgue measurable, has the Baire property, and has the perfect set property*, the last result attributed to Suslin.

Luzin and Sierpiński in their [1918] and [1923] provided proofs, and the latter paper was instrumental in shifting the emphasis toward the *co-analytic sets*, i.e. sets of reals X such that $\mathbb{R} - X$ is analytic. They used well-founded relations to provide a basic *tree representation* of co-analytic sets, one from which the main results of the period flowed, and it is here that well-founded relations entered mathematical practice.[72]

After the first wave in descriptive set theory brought about by Suslin [1917] and Luzin [1917] had crested, Luzin [1925a] and Sierpiński [1925] extended the domain of study to the *projective sets*. For $Y \subseteq \mathbb{R}^{k+1}$ and with ordered k-tuples defined from the ordered pair, the *projection of Y* is

$$pY = \{\langle x_1, ..., x_k\rangle \mid \exists y(\langle x_1, ..., x_k, y\rangle \in Y)\}.$$

Suslin [1917] had essentially noted that *a set of reals is analytic iff it is the projection of a Borel subset of* $\mathbb{R}^2$.[73] Luzin and Sierpiński took the geometric operation of projection to

---

[68] After getting a partial result [1914: 465ff], Hausdorff [1916] also showed, in essence, that the Borel sets have the perfect set property.

[69] Sierpiński [1950: 28ff] describes Suslin's discovery of the mistake.

[70] A *defining system* is a family $\{X_s\}_s$ of sets indexed by finite sequences s of natural numbers. The result of the Operation $\mathcal{A}$ on such a system is that set $\mathcal{A}(\{X_s\}_s)$ defined by:

$$x \in \mathcal{A}(\{X_s\}_s) \text{ iff } (\exists f : \omega \to \omega)(\forall n \in \omega)(x \in X_{f|n})$$

where $f|n$ denotes that sequence determined by the first n values of f. For a set X of reals, X is *analytic iff* $X = \mathcal{A}(\{X_s\}_s)$ for some defining system $\{X_s\}_s$ consisting of closed sets of reals.

[71] Luzin [1925] traced the term "analytic" back to Lebesgue [1905] and pointed out how the original example of a non-Borel Lebesgue measurable set there was in fact the first example of a non-Borel analytic set.

[72] Building on the penultimate footnote, suppose that Y is a co-analytic set of reals, i.e. $Y = \mathbb{R} - X$ with $X = \mathcal{A}(\{X_s\}_s)$ for some closed sets X_s, so that for reals x,

$$x \in Y \text{ iff } x \notin X \text{ iff } (\forall f : \omega \to \omega)(\exists n \in \omega)(x \notin X_{f|n}).$$

For finite sequences s_1 and s_2 define: $s_1 \prec s_2$ iff s_2 is a proper initial segment of s_1. For a real x define: $T_x = \{s \mid x \in X_t \text{ for every initial segment } t \text{ of } s\}$. Then:

$$x \in Y \text{ iff } \prec \text{ on } T_x \text{ is a well-founded relation},$$

i.e. there is no infinite descending sequence $... \prec s_2 \prec s_1 \prec s_0$. T_x is a tree (cf. 3.5). Well-founded relations were explicitly defined much later in Zermelo [1935]. Constructions recognizable as via recursion along a well-founded relation had already occurred in the proofs that the Borel have the perfect set property in Aleksandrov [1916] and Hausdorff [1916].

[73] Borel subsets of $\mathbb{R}^k$ are defined analogously to those of $\mathbb{R}$.

be basic and defined the projective sets as those sets obtainable from the Borel sets by the iterated applications of projection and complementation. The corresponding hierarchy of projective subsets of $\mathbb{R}^k$ is defined, in modern notation, as follows: For $A \subseteq \mathbb{R}^k$,

$$A \text{ is } \Sigma_1^1 \text{ iff } A = pY \text{ for some Borel set } Y \subseteq \mathbb{R}^{k+1},$$

i.e. A is analytic[74] and for integers $n > 0$,

$$A \text{ is } \Pi_n^1 \text{ iff } \mathbb{R}^k - A \text{ is } \Sigma_n^1,$$
$$A \text{ is } \Sigma_{n+1}^1 \text{ iff } A = pY \text{ for some } \Pi_n^1 \text{ set } Y \subseteq \mathbb{R}^{k+1}, \text{ and}$$
$$A \text{ is } \Delta_n^1 \text{ iff } A \text{ is both } \Sigma_n^1 \text{ and } \Pi_n^1.$$

Luzin [1925a] and Sierpiński [1925] recast Lebesgue's use of the Cantor diagonal argument to show that the projective hierarchy is proper, and soon its basic properties were established. However, this investigation encountered basic obstacles from the beginning. Luzin [1925a] emphasized that whether the Π_1^1 sets, the co-analytic sets at the bottom of the hierarchy, have the perfect set property was a major question. In a confident and remarkably prophetic passage he declared that his efforts towards its resolution led him to a conclusion "totally unexpected", that "one does not know *and one will never know*" of the family of projective sets, although it has cardinality $2^{\aleph_0}$ and consists of "effective sets", whether every member has cardinality $2^{\aleph_0}$ if uncountable, has the Baire property, or is even Lebesgue measurable. Luzin [1925b] pointed out the specific problem of establishing whether the Σ_2^1 sets are Lebesgue measurable. Both these difficulties were also pointed out by Sierpiński [1925]. This basic impasse in descriptive set theory was to remain for over a decade, to be surprisingly resolved by penetrating work of Gödel involving metamathematical methods (cf. 3.4).

2.6 Equivalences and Consequences

In this period AC and CH began to be explored no longer as underlying axiom and primordial hypothesis but as part of mathematics. Consequences were drawn and even equivalences established, and this mathematization, like the development of non-Euclidean geometry, led eventually to a deflating of metaphysical attitudes and attendant concerns about truth and existence.

Friedrich Hartogs [1915] established an equivalence result for AC, and this was the first substantial use of Zermelo's axiomatization after its appearance. The axiomatization had initially drawn ambivalent response among commentators,[75] especially those exercised by the paradoxes, and its assimilation by structuring sets and clarifying arguments began with such uses.

As noted in 1.3, Cardinal Comparability had become a concern for Cantor by the time of his *Beiträge* [1895]; Hartogs showed in Zermelo's system *sans* AC that *Cardinal Comparability implies that every set can be well-ordered*. Thus, an evident consequence of

---

[74] Analytic subsets of $\mathbb{R}^k$ are defined as for the case $k = 1$ in terms of a defining system consisting of closed subsets of $\mathbb{R}^k$.
[75] See Moore [1982: 3.3].

every set being well-orderable also implied that well-ordering principle, and this first "reverse mathematics" result established the equivalence of the well-ordering principle, Cardinal Comparability, and AC over the base theory.

Hartogs actually established without AC what is now called *Hartogs's Theorem*: *For any set M, there is a well-orderable set E not injectible into M*. Cardinal Comparability would then imply that M is injectible into E and hence is well-orderable. For the proof Hartogs first worked out a theory of ordering relations in Zermelo's system in terms of inclusion chains as in Zermelo's [1908] proof.[76] He then used Power Set and Separation to get the set M_W of well-orderable subsets of M and the set E of equivalence classes partitioning M_W according to order-isomorphism. Finally, he showed that E itself has an inherited well-ordering and is not injectible into M.[77] Reminiscent of Zermelo's subsumption of Russell's Paradox in the denial of a universal set, Hartogs's Theorem can be viewed as a subsumption of the Burali-Forti Paradox into the Zermelian setting.

The first explicit uses of AC mostly amounted to appeals to a well-ordering of the reals, Cantor's preoccupation. Those of Vitali [1905] and Bernstein [1908] were mentioned in 2.3, and Hausdorff's Paradox [1914; 1914a], in 2.4. Georg Hamel [1905] constructed by transfinite recursion a basis for the reals as a vector space over the rationals; cited by Zermelo [1908, 114], this provided a useful basis for later work in analysis and algebra. These various results, jarring at first, broached how a well-ordering allows for a new kind of *arithmetical* approach to the continuum.

The full exercise of AC in ongoing mathematics first occurred in the pioneering work of Ernst Steinitz [1910] on abstract fields. This was the first instance of an emerging phenomenon in algebra and topology: the study of axiomatically given structures with the range of possibilities implicitly including the transfinite. Steinitz studied algebraic closures of fields and even had an explicit transfinite parameter in the transcendence degree, the number of indeterminates necessary for closure. Typical of the generality in the years to come was Hausdorff's [1932] result using well-orderings that *every vector space has a basis*. As algebra and topology developed however, such results as these came to be based on the maximal principles that Hausdorff had first broached (cf. 2.4) and began to dominate after the appearance of Zorn's Lemma [1935]. Explicit well-orderings seemed out of place at this level of organization, and Zorn's Lemma had the remarkable feature that its hypothesis was easily checked in most applications.

Poland since its reunification in 1918 featured an active school of mathematics establishing foundational results in mathematical logic, topology, and analysis, and at Warsaw Tarski and Kuratowski together with Sierpiński were making crucial contributions to set theory and the elucidation of its role in mathematics. The Polish school of mathematics carried out a penetrating investigation of the role of AC in set theory and analysis. Sierpiński's earliest publications, culminating in his survey [1918], not only dealt with specific constructions but showed how deeply embedded AC was in the informal development of cardinality, measure, and the Borel hierarchy (cf. 2.3), supporting Zermelo's

---

[76] This is better done in Kuratowski [1921]. The Hausdorff [1914] approach with an ordered pair could have been taken, but that only became standard later when more general relations were considered.

[77] As with Zermelo's Well-Ordering Theorem, textbooks usually establish Hartogs's Theorem after first introducing Replacement and (von Neumann) ordinals, and this amounts to a historical misrepresentation.

contention [1904:516] that the axiom is applied "everywhere in mathematical deduction". Tarski [1924], explicitly building his work on Zermelo's system, provided several propositions of cardinal arithmetic equivalent to AC, most notably that $\mathfrak{m}^2 = \mathfrak{m}$ for every infinite cardinal $\mathfrak{m}$. Adolf Lindenbaum and Tarski in their [1926] gave further cardinal equivalents, some related to the Hartogs [1915] result, and announced that GCH, in the form that $\mathfrak{m} < \mathfrak{n} < 2^\mathfrak{m}$ holds for no infinite cardinals $\mathfrak{m}$ and $\mathfrak{n}$, implies AC. This study of consequences led to other choice principles, further implications and sometimes converses in a continuing cottage industry.[78]

The early mathematical study of AC extended to the issue of its independence. Abraham Fraenkel's first investigations [1922] directly addressed Zermelo's axioms, pointing out the need for the Replacement Axiom and attempting an axiomatization of the *definit* property for the Separation Axiom (cf. 3.1). The latter was motivated in part by the need to better articulate independence proofs for the various axioms. Fraenkel [1922a] came to the fecund idea of starting with urelements and some initial sets closing off under set-theoretic operations to get a model. For the independence of AC he started with urelements $a_n, \overline{a_n}$ for $n \in \omega$ and the set $A = \{\{a_n, \overline{a_n}\} \mid n \in \omega\}$ of unordered pairs and argued that for any set M in the resulting model there is a co-finite $A_M \subseteq A$ such that M is invariant if members of any $\{a_n, \overline{a_n}\} \in A_M$ are permuted. This immediately implies that there is no choice function for A in the model. Finally, Fraenkel argued that the model satisfies the other Zermelo axioms, except Extensionality because of the urelements.

Fraenkel's early model building emphasized the Zermelian generative framework, anticipated well-founded recursion, and foreshadowed the later play with models of set theory. That Extensionality was not to be had precluded settling the matter, but just as for the early models of non-Euclidean or finite geometries Fraenkel's achievement lay in stimulating interest in mathematical constructions despite relaxing some basic tenet. Fraenkel tried to develop his approach from time to time, but it needed the articulation that would come with the full espousal of the satisfaction relation. In the latter 1930s Lindenbaum and Andrzej Mostowski so cast and extended Fraenkel's work. Mostowski [1939] forged a method according to post-Gödelian sensibilities, bringing out the importance of groups of permutations leaving various urelements fixed, and the resulting models as well as later versions are now known as the *Fraenkel-Mostowski models*.

Even more than AC, Sierpiński investigated CH, and summed up his researches in a monograph [1934]. He provided several notable equivalences to CH, e.g. (p.11) the plane $\mathbb{R}^2$ is the union of countably many curves, where a *curve* is a set of form $\{\langle x, y \rangle \mid y = f(x)\}$ or $\{\langle x, y \rangle \mid x = f(y)\}$ with f a real function.

Moreover, Sierpiński presented numerous consequences of CH from the literature, one in particular implying a host of others: Mahlo [1913a] and Luzin [1914] had shown that CH *implies that there is a Luzin set*, an uncountable set of reals whose intersection with any meager set is countable (cf. 2.5). To state one consequence, say that a set X of reals has *strong measure zero iff* for any sequence $\epsilon_0, \epsilon_1, \epsilon_2, \ldots$ of positive reals there is a sequence of intervals $I_0, I_1, I_2, \ldots$ such that the length of I_n is less than ϵ_n for each n and $X \subseteq \bigcup_n I_n$. Borel [1919] conjectured that such sets are countable. However, Sierpiński [1928] showed that *a Luzin set has strong measure zero*. Analogous to a Luzin set, a

---
[78] See Moore [1982], especially its 5.1, for other choice principles.

Sierpiński set is an uncountable set of reals whose intersection with any Lebesgue measure zero set is countable. Sierpiński [1924] showed that CH *implies that there is a Sierpiński set*, and emphasized [1934] an emerging duality between measure and category.

The subsequent work of Fritz Rothberger would have formative implications for the Continuum Problem. He [1938] observed that if both Luzin and Sierpiński sets exist, then they have cardinality $\aleph_1$, so that the joint existence of such sets of the cardinality of the continuum implies CH. Then in penetrating analyses of the work of Sierpinski and Hausdorff on gaps (cf. 2.1) Rothberger [1939; 1948] considered other sets and implications between cardinal properties of the continuum *independent* of whether CH holds. It became newly clarified that absent CH one can still isolate uncountable cardinals $\leq 2^{\aleph_0}$ that gauge and delimit various recursive constructions, and this approach was to blossom half a century later in the study of *cardinal characteristics (or invariants) of the continuum*.[79]

These results cast CH in a new light, as a construction principle. Conclusions had been drawn from having a well-ordering of the reals, but one given by CH allowed for recursive constructions where at any stage only countably many conditions corresponding to as many reals had to be handled. The construction of a Luzin set was a simple recursive application of the Baire Category Theorem, and later constructions took advantage of the possibility of diagonalization at each stage. However, whereas the new constructions using AC, though jarring at first, were eventually subsumed as concomitant with the acceptance of the axiom and as expressions of the richness of possibility, constructions from CH clashed with that very sense of richness for the continuum. It was the *mathematical* investigation of CH that increasingly raised doubts about its truth and certainly its provability (cf. end of 3.4).

3 CONSOLIDATION

3.1 *Ordinals and Replacement*

In the 1920s fresh initiatives structured the loose Zermelian framework with new features and corresponding developments in axiomatics: von Neumann's work with ordinals and Replacement; the focusing on well-founded sets and the cumulative hierarchy; and extensionalization in first-order logic. Von Neumann effected a counter-reformation of sorts: The transfinite numbers had been central for Cantor but peripheral to Zermelo; von Neumann reconstrued them as *bona fide* sets, now called simply the *ordinals*, and established their efficacy by formalizing transfinite recursion.

Von Neumann [1923; 1928], and before him Dimitry Mirimanoff [1917; 1917a] and Zermelo in unpublished 1915 work,[80] isolated the now familiar concept of ordinal, with the basic idea of taking precedence in a well-ordering simply to be membership. Appealing to forms of Replacement Mirimanoff and Von Neumann then established the key

---

[79] See Miller [1984] for more on special sets of reals and van Douwen [1984] as a trend setting paper for cardinal characteristics of the continuum. See Blass [2008] and Bartoszyński [2008] for recent work on cardinal characteristics.

[80] See Hallett [1984: 8.1].

instrumental property of Cantor's ordinal numbers for ordinals: *Every well-ordered set is order-isomorphic to exactly one ordinal with membership.* Von Neumann in his own axiomatic presentation took the further step of ascribing to the ordinals the role of Cantor's ordinal numbers. Thus, like Kepler's laws by Newton's, Cantor's principles of generation for his ordinal numbers would be subsumed by the Zermelian framework. For this reconstrual of ordinal numbers and already to define the arithmetic of ordinals von Neumann saw the need to establish the Transfinite Recursion Theorem, the theorem that validates definitions by transfinite recursion. The proof was anticipated by the Zermelo 1904 proof, but Replacement was necessary even for the very formulation, let alone the proof, of the theorem. With the ordinals in place von Neumann completed the restoration of the Cantorian transfinite by defining the *cardinals* as the *initial ordinals*, those ordinals not in bijective correspondence with any of its predecessors. Now, the infinite initial ordinals are denoted

$$\omega = \omega_0, \omega_1, \omega_2, \ldots, \omega_\alpha, \ldots,$$

so that ω is to be the set of natural numbers in the ordinal construal, and the identification of different intensions is signaled by

$$\omega_\alpha = \aleph_\alpha$$

with the left being a von Neumann ordinal and the right being the Cantorian cardinal number.

Replacement has been latterly regarded as somehow less necessary or crucial than the other axioms, the purported effect of the axiom being only on large-cardinality sets. Initially, Abraham Fraenkel [1921; 1922] and Thoralf Skolem [1923] had independently proposed adjoining Replacement to ensure that $E(a) = \{a, \mathcal{P}(a), \mathcal{P}(\mathcal{P}(a)), \ldots\}$ be a set when a is the particular infinite set $Z_0 = \{\emptyset, \{\emptyset\}, \{\{\emptyset\}\}, \ldots\}$ posited by Zermelo's Axiom of Infinity, since, as they pointed out, Zermelo's axioms cannot establish this. However, even $E(\emptyset)$ cannot be proved to be a set from Zermelo's axioms,[81] and if his axiom of Infinity were reformulated to accommodate $E(\emptyset)$, there would still be many finite sets a such that $E(a)$ cannot be proved to be a set.[82] Replacement serves to rectify the situation by admitting new infinite sets defined by "replacing" members of the one infinite set given by the Axiom of Infinity. In any case, the full exercise of Replacement is part and parcel of transfinite recursion, which is now used everywhere in modern set theory, and it was von Neumann's formal incorporation of this method into set theory, as necessitated by his proofs, that brought in Replacement.

That Replacement became central for von Neumann was intertwined with his taking of function, in its full extensional sense, instead of set as primitive and his establishing of a context for handling *classes*, collections not necessarily sets. He [1925; 1928a] formalized the idea that a class is *proper*, i.e. not a set, exactly when it is in bijective correspondence with the entire universe, and this exactly when it is not an element of any class. This thus brought in another move from Cantor's 1899 correspondence with

---

[81] The union of $E(Z_0)$, with membership restricted to it, models Zermelo's axioms yet does not have $E(\emptyset)$ as a member.

[82] See Mathias [2001].

Dedekind (cf. 2.2). However, von Neumann's axiomatization [1925; 1928] of function was complicated, and reverting to sets as primitive Paul Bernays (cf. his [1976]) recast and simplified von Neumann's system. Still, the formal incorporation of proper classes introduced a superstructure of objects and results distant from mathematical practice. What was to be inherited was a predisposition to entertain proper classes in the mathematical development of set theory, a willingness that would have crucial ramifications (cf. 3.6).

3.2 Well-Foundedness and the Cumulative Hierarchy

With ordinals and Replacement, set theory continued its shift away from pretensions of a general foundation toward a theory of a more definite subject matter, a process fueled by the incorporation of well-foundedness. Mirimanoff [1917] was the first to study the well-founded sets, and the later hierarchical analysis is distinctly anticipated in his work. But interestingly enough well-founded relations next occurred in the direct definability tradition from Cantor, descriptive set theory (cf. 2.5).

In the axiomatic tradition Fraenkel [1922], Skolem [1923] and von Neumann [1925] considered the salutary effects of restricting the universe of sets to the well-founded sets. Von Neumann [1929: 231,236ff] formulated in his functional terms the Axiom of Foundation, that every set is well-founded,[83] and defined the resulting hierarchy of sets in his system via transfinite recursion: In modern notation, the axiom, as is well-known, entails that the universe V of sets is stratified into cumulative ranks V_α, where

$$V_0 = \emptyset; \quad V_{\alpha+1} = \mathcal{P}(V_\alpha); \quad V_\delta = \bigcup_{\alpha<\delta} V_\alpha \text{ for limit ordinals } \delta \, ;$$

and

$$V = \bigcup_\alpha V_\alpha \, .$$

Von Neumann used this, *the cumulative hierarchy*, to establish the first relative consistency result in set theory via "inner models"; his argumentation in particular established the consistency of Foundation relative to Zermelo's axioms plus Replacement.

During this period mathematical logic gained new currency, and a tussle based on the different approaches of first- and second-order logic to set theory would lead to a substantial axiomatic development.[84] The prescient Skolem [1923] made the proposal of using for Zermelo's definite properties for the Separation Axiom those properties expressible in first-order logic with $\in$ as a binary relation symbol. After Leopold Löwenheim [1915] had broken the ground for model theory with his result about the satisfiability of a first-order sentence, Skolem [1920; 1923] had located the result solidly in first-order logic and generalized it to the Löwenheim-Skolem Theorem: *If a countable collection of first-order sentences is satisfiable, then it is satisfiable in a countable domain.* That Skolem intended for set theory to be a first-order system without a privileged interpretation for

---

[83]$\forall x(x \neq \emptyset \longrightarrow \exists y \in x(x \cap y = \emptyset))$. This is von Neumann's Axiom VI4 in terms of sets. The term "Foundation [Fundierung]" itself comes from Zermelo [1930].

[84]First-order logic is the logic of formal languages consisting of formulas built up from specified function and relation symbols using logical connectives and first-order quantifiers $\forall$ and $\exists$, quantifiers to be interpreted as ranging over the *elements* of a domain of discourse. Second-order logic has quantifiers to be interpreted as ranging over arbitrary subsets of a domain.

∈ becomes evident in the initial application of the Löwenheim-Skolem Theorem to get *Skolem's Paradox*: In first-order logic Zermelo's axioms are countable, Separation having become a *schema*, a schematic collection of axioms, one for each first-order formula; the theorem then implies the existence of countable models of the axioms although they entail the existence of uncountable sets. Skolem intended by this means to deflate the possibility of set theory becoming a foundation for mathematics. Exercised by this relativism and by the recent work of Fraenkel and von Neumann, Zermelo [1929] in his first publication in set theory in two decades proposed an axiomatization of his *definit* property in second-order terms. In direct response Skolem [1930] pointed out possible difficulties with this approach and reaffirmed his first-order formulation, completing the backdrop for a new axiomatic synthesis.

Zermelo in his remarkable [1930] offered his final axiomatization of set theory as well as a striking view of a procession of natural models that would have a modern resonance. While ostensibly a response to Skolem [1930], the dramatically new picture of sets in Zermelo [1930] reflects gained experience and the germination of ideas over a prolonged period. The main axiomatization incorporated Replacement but also the Axiom of Foundation. In contrast to Zermelo [1908a], while urelements continued to be allowed, Infinity was eschewed and Choice was regarded as part of the underlying logic. Concerning Separation and Replacement it becomes evident from how Zermelo proceeded that he regarded their applicability in a fully second-order context.

As described in above, Foundation in modern set theory ranks the universe of sets into the cumulative hierarchy $V = \bigcup_\alpha V_\alpha$. Zermelo substantially advanced this schematic generative picture with his inclusion of Foundation in an axiomatization. Replacement and Foundation focused the notion of set, with the first making possible the means of transfinite recursion and induction, and the second making possible the application of those means to get results about *all* sets. It is now almost banal that Foundation is the one axiom unnecessary for the recasting of mathematics in set-theoretic terms, but the axiom ascribes to membership the salient feature that distinguishes investigations specific to set theory as an autonomous field of mathematics. Indeed, it can be fairly said that modern set theory is at base a study couched in well-foundedness, the Cantorian well-ordering doctrines adapted to the Zermelian generative conception of sets.

In [1930] Zermelo described a range of models for set theory, each an initial segment of a cumulative hierarchy built on an initial set of urelements. Zermelo then established a categoricity of sorts for his axioms, one made possible by his second-order context. He showed that his models are characterized up to isomorphism by two cardinals, the number of their urelements and the height of their ordinals. Moreover, he established that if two models have the same number of urelements yet different heights, then one is isomorphic to an initial segment of the other's cumulative hierarchy. Grappling with Power Set and Replacement he characterized the heights of his models ("Grenzzahlen") as $\aleph_0$ or the *(strongly) inaccessible cardinals*, those uncountable regular cardinals κ that are strong limit, i.e. if $\lambda < \kappa$, then $2^\lambda < \kappa$.

Zermelo posited an endless procession of models, each a set in a next, advocating a dynamic view of sets that was a marked departure from Cantor's (and later, Gödel's) realist presumption of a fixed universe of sets. In synthesizing the sense of progression inherent in the new cumulative hierarchy picture and the sense of completion in the limit numbers, the inaccessible cardinals, he promoted the crucial idea of internal models of set theory. The open-endedness of Zermelo's original [1908a] axiomatization had been structured by Replacement and Foundation, but he advanced a new open-endedness with an eternal return of models approaching Cantor's Absolute.

In the process, inaccessible cardinals became structurally relevant. Sierpiński-Tarski [1930] had formulated these cardinals arithmetically as those uncountable cardinals that are not the product of fewer cardinals each of smaller power and observed that they are weakly inaccessible — the first large cardinal concept, from Hausdorff [1908: 443] (cf. 2.4). Be that as it may, in the early model-theoretic investigations of set theory the inaccessible cardinals provided the natural models as envisioned by Zermelo. Moreover, strong large cardinal hypotheses emerging in the 1960s were to be formulated in terms of these initial segments of the cumulative hierarchy.[85]

The journal volume containing Zermelo's paper also contained Stanisław Ulam's seminal paper [1930] on *measurable cardinals*, the most important of all large cardinals. For a set S, U is a (non-principal) *ultrafilter over S iff* U is a collection of subsets of S containing no singletons, closed under the taking of supersets and finite intersections, and such that for any $X \subseteq S$, either $X \in U$ or $S - X \in U$. For a cardinal λ, an ultrafilter U is λ-*complete iff* for any $D \subseteq U$ of cardinality less than λ, $\bigcap D \in U$. Finally, an uncountable cardinal κ is *measurable iff* there is a κ-complete ultrafilter over κ. Thus, a measurable cardinal is a cardinal whose power set is structured with a two-valued measure having strong closure property.

Measurability embodied the first large cardinal confluence of Cantor's two legacies, the investigation of definable sets of reals and the extension of number into the transfinite: Distilled from measure-theoretic considerations related to Lebesgue measure, the concept *also* entailed inaccessibility in the transfinite. Moreover, the initial airing generated a problem that was to keep the spark of large cardinals alive for the next three decades: *Can the least inaccessible cardinal be measurable?* In the 1960s consequences of, and a structural characterization of, measurability were established that became fundamental in the setting structured by the new Zermelian emphasis on well-foundedness (cf. 3.6).

3.3 *First-Order Logic and Extensionalization*

The final structuring of set theory before it was to sail forth on its independent course as a distinctive field of mathematics was its full extensionalization in first-order logic.[86] However influential Zermelo's [1930] and despite his subsequent advocacy [1931; 1935] of infinitary logic, his efforts to forestall Skolem were not to succeed, as stronger currents were at work in the direction of first-order formalization.

Hilbert effected a basic shift in the development of mathematical logic when he took Whitehead and Russell's *Principia Mathematica*, viewed it as an uninterpreted formalism, and made it an object of mathematical inquiry. The book [1928][87] by Hilbert and Wilhelm

---

[85] See Kanamori [2003: chap.5].
[86] See Goldfarb [1979] and Moore [1988a] for more on the emergence of first-order logic.
[87] The historical development is clarified by the fact that while this book was published in light of the devel-

Ackermann reads remarkably like a recent text. In marked contrast to the formidable works of Frege and Russell with their forbidding notation and all-inclusive approach, it proceeded pragmatically and upward to probe the extent of structure, making those moves emphasizing forms and axiomatics typical of modern mathematics. After a complete analysis of sentential logic it distinguished and focused on first-order logic ("functional calculus", and later "(restricted) predicate calculus") as already the source of significant problems. Thus, while Frege and Russell never separated out first-order logic, Hilbert through his mathematical investigations established it as a subject in its own right.

Hilbert in the 1920s developed proof theory, i.e. metamathematics, and proposed his program of establishing the consistency of classical mathematics. The issues here gained currency because of Hilbert's preeminence, just as mathematics in the large had been expanded in the earlier years of the century by his reliance on non-constructive proofs and transcendental methods and his advocacy of new contexts. Through this expansion the full exercise of AC had become a mathematical necessity (cf. 2.6) and arbitrary functions, and so Power Set, had become implicitly accepted in the extensive investigation of higher function spaces.

Hilbert-Ackermann [1928: 65ff,72ff] raised two crucial questions directed at the further possibilities for first-order logic: the completeness of its axioms and the Decision Problem [Entscheidungsproblem]. These as well as Hilbert's program for securing consistency were to be decisively informed by penetrating work that for set theory eventually led to its first sophisticated metamathematical result, the relative consistency of AC and GCH.

Kurt Gödel (1906–1978), born when Zermelo was devising his proofs of the Well-Ordering Theorem, virtually completed the mathematization of logic by submerging metamathematical methods into mathematics. The main vehicle was of course the direct coding, "the arithmetization of syntax", in his celebrated Incompleteness Theorem [1931]. Establishing a fundamental distinction between what is *true* about the natural numbers and what is *provable*, this theorem transformed Hilbert's consistency program and led to the undecidability of the Decision Problem from Hilbert–Ackermann [1928] and the development of recursion theory. Gödel's work showed in particular that for a (schematically definable) collection of axioms A, its *consistency*, that from A one cannot prove a contradiction, has a formal counterpart in an arithmetical formula $Con(A)$ about natural numbers. Gödel's "second" theorem asserts that if A is consistent and subsumes some elementary arithmetic of the natural numbers, then $Con(A)$ cannot be proved from A. But starting an undercurrent, the earlier Completeness Theorem [1930] from his thesis answered affirmatively a Hilbert–Ackermann [1928] question about semantic completeness, clarified the distinction between the formal syntax and semantics of first-order logic, and secured its key instrumental property with the Compactness Theorem.

Tarski [1933; 1935] then completed the mathematization of logic by providing his definition of truth, exercising philosophers to a surprising extent ever since. Through Hilbert-Ackermann [1928] and Gödel [1930] the satisfaction relation had been informal, and in that sense completeness could be said to have remained inadequately articulated. Tarski simply extensionalized truth in formal languages and provided a formal, *recur-*

---

opments of the 1920s, it has a large overlap with unpublished lecture notes for a 1917-8 course given by Hilbert at Göttingen.

sive definition of the satisfaction relation in set-theoretic terms. This new response to a growing need for a mathematical framework became the basis for model theory, but thus cast into mathematics truth would leave behind any semantics in the real meaning of the word. Tarski's [1933] was written around the same time as his [1931], a seminal paper that highlights the thrust of his initiative. In [1931] Tarski gave a precise mathematical (that is, set-theoretic) formulation of the informal concept of a (first-order) definable set of reals, thus infusing the intuitive notion of definability into ongoing mathematics. This mathematization of intuitive or logical notions was accentuated by Kuratowski-Tarski [1931], where second-order quantification over the reals was correlated with the geometric operation of projection, beginning the process of explicitly wedding descriptive set theory to mathematical logic. The eventual effect of Tarski's [1933] mathematical formulation of (so-called) semantics would be not only to make mathematics out of the informal notion of satisfiability, but also to enrich ongoing mathematics with a systematic method for forming mathematical analogues of several intuitive semantic notions.[88]

In this process of extensionalization first-order logic came to be accepted as the canonical language because of its mathematical possibilities as epitomized by the Compactness Theorem, and higher-order logics became downgraded as the workings of the power set operation in disguise. Skolem's early suggestion for set theory was thus taken up generally, and again the ways of paradox were positively subsumed, as the negative intent of Skolem's Paradox gave way to the extensive, internal use of Skolem functions from the Löwenheim-Skolem Theorem in set-theoretic constructions.

3.4 Relative Consistency

Set theory was launched on an independent course as a distinctive field of mathematics by Gödel's construction of L [1938; 1939] leading to the relative consistency of the Axiom of Choice and the Generalized Continuum Hypothesis. Synthesizing all that came before, Gödel built on the von Neumann ordinals as sustained by Replacement to formulate a relative Zermelian universe of sets based on logical definability, a universe imbued with a Cantorian sense of enumerative order.

Gödel's advances in set theory can be seen as part of a steady intellectual development. In a lecture [1933] on the foundations of mathematics Gödel propounded the axiomatic set theory "as presented by Zermelo, Fraenkel and von Neumann" as "a natural generalization of [Russell's simple] theory of types, or rather, what becomes of the theory of types if certain superfluous restrictions are removed." First, the types can be taken to be cumulative, and second, the process can be continued into the transfinite. As for how far this cumulative hierarchy of sets is to continue, "the first two or three [infinite] types already suffice to define very large [Cantorian ordinal numbers]" which can then serve to index the process, and so on. Implicitly referring to his incompleteness result Gödel noted that for a formal system S based on the theory of types a number-theoretic proposition can be constructed which is unprovable in S but becomes provable if to S is adjoined "the

---

[88] Incidentally, Tarski [1931] stated a result whose proof led to Tarski's well-known theorem [1951] that the elementary theory of real closed fields is decidable via the elimination of quantifiers.

next higher type and the axioms concerning it."[89] Thus, although he never mentioned Zermelo [1930], Gödel was entertaining its cumulative hierarchies but as motivated by the theory of types.

It is to this initiative, separately fueled by Zermelo and Gödel, that one can date how the formation of sets out of sets iterated into the transfinite as embodied by the cumulative hierarchy can be regarded as a motivation for the subject matter of set theory. In a notable inversion, what has come to be regarded as the *iterative conception* became a heuristic for motivating the axioms of set theory generally.[90] The iterative conception of sets, like Tarski's definition of truth, has exercised philosophers to a surprising extent with respect to extrinsic justifications. This has opened the door to a metaphysical appropriation in the following sense: It is as if there is some notion of set that is "there", in terms of which the axioms must find some further justification. But set theory has no particular obligations to mirror some prior notion of set arrived at *a posteriori*. Replacement and Choice for example do not quite "fit" the iterative conception,[91] but if need be, Replacement can be "justified" in terms of achieving algebraic closure of the axioms, a strong motivation in the work of Fraenkel and the later Zermelo, and choice can be "justified" in terms of Cantorian well-ordering doctrines or as a logical principle as Zermelo did.

In his first announcement [1938] about L Gödel described it as a hierarchy "which can be obtained by Russell's ramified hierarchy of types, if extended to include transfinite orders." Indeed, with L Gödel had refined the cumulative hierarchy of sets to a cumulative hierarchy of *definable* sets which is analogous to the orders of Russell's *ramified* theory. Gödel's further innovation was to continue the indexing of the hierarchy through *all* the ordinals. Von Neumann's canonical well-orderings would be the spine for a thin hierarchy of sets, and this would be the key to both the AC and CH results.

In a brief account [1939] Gödel informally presented L essentially as is done today: For any set x let $\mathrm{def}(x)$ denote the collection of subsets of x first-order definable over $\langle x, \in \rangle$.[92] Then define:

$$L_0 = \emptyset; \quad L_{\alpha+1} = \mathrm{def}(L_\alpha), \quad L_\delta = \bigcup \{L_\alpha \mid \alpha < \delta\} \text{ for limit ordinals } \delta;$$

and the *constructible universe*

$$L = \bigcup \{L_\alpha \mid \alpha \text{ is an ordinal}\}.$$

---

[89] Gödel was evidently referring to propositions like $\mathrm{Con}(S)$. In a prescient footnote, 48a, to his incompleteness paper [1931] Gödel had already written: "...the true reason for the incompleteness inherent in all formal systems of mathematics is that the formation of ever higher types can be continued into the transfinite...while in any formal system at most denumerably many of them are available. For it can be shown that the undecidable propositions constructed here become decided whenever appropriate higher types are added (for example, the type ω to the system P [Peano Arithmetic]). An analogous situation prevails for the axiom system of set theory."

[90] Shoenfield [1967: 238ff][1977], Wang [1974a], Boolos [1971], and Scott [1974] motivate the axioms of set theory in terms of an iterative concept of set based on stages of construction. Parsons [1977] raises issues about this approach.

[91] See Boolos [1971] for Replacement and Scott [1974: 214] for Choice.

[92] For a first-order formula $\varphi(v_1, \ldots, v_n)$ in $\in$, $\varphi^x(x_1, \ldots, x_n)$ is the restriction of the formula to x, i.e. each $\forall y$ is replaced by $\forall y \in x$ and each $\exists y$ is replaced by $\exists y \in x$ (with these abbreviations having the expected formal articulation). A set $y \subseteq x$ is *first-order definable over* $\langle x, \in \rangle$ if there is a first-order formula $\varphi(v_0, v_1, \ldots, v_n)$ and $a_1, \ldots, a_n$ all in x such that $y = \{z \in x \mid \varphi^x(z, a_1, \ldots, a_n)\}$.

Gödel pointed out that L "can be defined and its theory developed in the formal systems of set theory themselves." This is a remarkable understatement of arguably the central feature of the construction of L. L is a class definable in set theory via a transfinite recursion that could be based on the formalizability of def(x), the definability of definability, which was later reaffirmed by Tarski's systematic definition of the satisfaction relation in set-theoretic terms. With this, one can formalize the *Axiom of Constructibility* $V = L$, i.e. $\forall x(x \in L)$. In modern parlance, an *inner model* is a transitive class[93] containing all the ordinals such that, with membership and quantification restricted to it, the class satisfies each axiom of ZF. In summary terms, what Gödel did was to show in ZF that L is an inner model, and moreover that L satisfies AC and CH. He thus established the relative consistency Con(ZF) implies Con(ZFC + GCH).

In the approach via def(x) it is necessary to show that def(x) remains unaltered when applied in L with quantifiers restricted to L. Gödel himself would never establish this *absoluteness of first-order definability* explicitly, preferring in his one rigorous published exposition of L to take an approach that avoids def(x) altogether.

In his monograph [1940], based on 1938 lectures, Gödel provided a specific, formal presentation of L in a class-set theory emanating from that of Paul Bernays (cf. [1976]), a theory based in turn on a theory of von Neumann [1925]. Using eight binary operations producing new classes from old, Gödel generated L set by set via transfinite recursion. This veritable "Gödel numbering" with ordinals bypassed def(x) and made evident certain aspects of L. Since there is a direct, definable well-ordering of L, choice functions abound in L, and AC holds there. Of the other axioms the crux is where first-order logic impinges, in Separation and Replacement. For this, "algebraic" closure under Gödel's eight operations ensured "logical" Separation for bounded formulas,[94] and then the full exercise of Replacement (in V) secured all of the ZF axioms in L.

Gödel's proof that L satisfies GCH consisted of two separate parts. He established the implication $V = L \to$ GCH, and, in order to apply this implication within L, that L as defined within L with quantifiers restricted to L is again L itself. This latter follows from the aforementioned absoluteness of def(x), and in [1940] Gödel gave an alternate proof based on the absoluteness of his eight binary operations.

Gödel's argument for $V = L \to$ GCH rests, as he himself wrote in [1939], on "a generalization of Skolem's method for constructing enumerable models." This was the first significant use of Skolem functions since Skolem's own to establish the Löwenheim-Skolem theorem, and with it, Skolem's Paradox. Ironically, though Skolem sought through his paradox to discredit set theory based on first-order logic as a foundation for mathematics, Gödel turned paradox into method, one promoting first-order logic. Gödel [1939] specifically established:

(∗) For infinite α, every constructible subset of L_α
     belongs to some L_β for a β of the same cardinality as α.

It is straightforward to show that for infinite α, L_α has the same cardinality as that of α.

---

[93] A class C is *transitive* if members of members of C are themselves members of C, so that C is "closed under membership".

[94] That is, those first-order formulas in which all the quantifiers can be rendered as $\forall x \in y$ and $\exists x \in y$.

It follows from (∗) that in the sense of L, the power set of $L_{\aleph_\alpha}$ is included in $L_{\aleph_{\alpha+1}}$, and so GCH follows in L. To establish (∗), Gödel actually iterated the Skolem closure procedure, and made the first use of the now familiar Mostowski collapse (cf. 3.6). In an incisive 1939 lecture Gödel announced the version of (∗) for countable α as the crux of the consistency proof of CH and asserted that "this fundamental theorem constitutes the corrected core of the so-called Russellian axiom of reducibility."[95] Thus, Gödel established another connection between L and Russell's ramified theory of types. But while Russell had to *postulate* his ill-fated Axiom of Reducibility for his finite orders, Gödel was able to *derive*, with an important use of Replacement, an analogous form for his transfinite hierarchy that asserts that the types are delimited in the hierarchy of orders.

The synthesis at L extended to the resolution of difficulties in descriptive set theory (cf. end of 2.5). Gödel [1938] announced, in modern terms: *If $V = L$, then (a) there is a Δ^1_2 set of reals that is not Lebesgue measurable, and (b) there is a a Π^1_1 set of reals without the perfect set property.* Thus, the descriptive set theorists were confronting an obstacle insurmountable in ZFC! Gödel [1938] listed each of these impossibility results on an equal footing with his AC and GCH results. Unexpected, they were the first instances of metamathematical methods resolving outstanding mathematical problems that exhibited no prior connection to such methods. When eventually confirmed and refined, the results were seen to turn on a Σ^1_2 well-ordering of the reals in L defined via reals coding well-founded structures and thus connected to the well-founded tree representation of a Π^1_1 set (cf. 2.5).[96]

Set theory had progressed to the point of establishing, in addition to a consistent resolution of CH, a consistent possibility for a definable well-ordering of the reals as Cantor had wanted, one that synthesizes the two historical sources of well-foundedness. Put into a broader historical context, formal definability was brought into descriptive set theory by Tarski [1931], and by Kuratowski-Tarski [1931] and Kuratowski [1931] which pursued the basic connection between existential number quantifiers and countable unions and between existential real quantifiers and projection and used these "logical symbols" to aid in the classification of sets in the Borel and projective hierarchies. Gödel's results (a) and (b) constitute the first real synthesis of abstract and descriptive set theory, in that the axiomatic framework is brought to bear on the investigation of definable sets of reals.

Gödel brought into set theory a method of construction and argument which affirmed several features of its axiomatic presentation. Most prominently, Gödel showed how first-order definability can be formalized and used in a transfinite recursive construction to

---

[95] See Gödel [1939a: 141].

[96] When every real is in L, this Σ^1_2 well-ordering is also Δ^1_2 and does not satisfy Fubini's Theorem for Lebesgue measurable subsets of the plane, and this is one way to confirm (a). What may have been Gödel's original argument for (b) is given in Kanamori [2003: 170].

Texts establish (b) indirectly via the Kondô Π^1_1 Uniformization Theorem, and this leads to a historical point about Gödel the working mathematician. As 1938 correspondence with von Neumann makes evident, Gödel was working on one-to-one continuous images of Π^1_1 sets, and his [1938] actually states the results (a) and (b) in these terms. In a 1939 letter, von Neumann informed Gödel of Kondô [1939], the paper containing the uniformization result, from which it is immediate that the Σ^1_2 sets are exactly the one-to-one continuous images of Π^1_1 sets. In a replying letter to von Neumann of 20 March 1939 Gödel wrote: "The result of Kondô is of great interest to me and will definitely allow an important simplification in the consistency proof of [(a)] and [(b)] of the attached offprint." See Gödel [2003] for the Gödel–von Neumann correspondence.

establish striking new mathematical results. This significantly contributed to a lasting ascendancy for first-order logic which beyond its *sufficiency* as a logical framework for mathematics was seen to have considerable *operational efficacy*. Moreover, Gödel's construction buttressed the incorporation of Replacement and Foundation into set theory. Replacement was immanent in the arbitrary extent of the ordinals for the indexing of L and in its formal definition via transfinite recursion. As for Foundation, underlying the construction was the well-foundedness of sets. Gödel in a footnote to his 1939 note wrote: "In order to give A [the axiom $V = L$] an intuitive meaning, one has to understand by 'sets' all objects obtained by building up the simplified hierarchy of types on an empty set of individuals (including types of arbitrary transfinite orders)."

How Gödel transformed set theory can be broadly cast as follows: On the larger stage, from the time of Cantor, sets began making their way into topology, algebra, and analysis so that by the time of Gödel, they were fairly entrenched in the structure and language of mathematics. But how were sets viewed among set *theorists*, those investigating sets as such? Before Gödel, the main concerns were what sets *are* and how sets and their axioms can serve as a reductive basis for mathematics. Even today, those preoccupied with ontology, questions of mathematical existence, focus mostly upon the set theory of the early period. After Gödel, the main concerns became what sets *do* and how set theory is to advance as an autonomous field of mathematics. The cumulative hierarchy picture was in place as subject matter, and the metamathematical methods of first-order logic mediated the subject. There was a decided shift toward epistemological questions, e.g. what can be proved about sets and on what basis.

As a pivotal figure, what was Gödel's own stance? What he *said* would align him more with his predecessors, but what he *did* would lead to the development of methods and models. In a 1944 article on Russell's mathematical logic, in a 1947 article on Cantor's continuum problem (and in a 1964 revision), and in subsequent lectures and correspondence, Gödel articulated his philosophy of "conceptual realism" about mathematics. He espoused a staunchly objective "concept of set" according to which the axioms of set theory are true and are descriptive of an objective reality schematized by the cumulative hierarchy. Be that as it may, his actual mathematical work laid the groundwork for the development of a range of models and axioms for set theory. Already in the early 1940s Gödel worked out for himself a possible model for the negation of AC, and in a 1946 address he described a new inner model, the class of ordinal definable sets.

In later years Gödel speculated about the possibility of deciding propositions like CH with large cardinal hypotheses based on the heuristics of *reflection* and later, *generalization*. Already in that 1946 address he suggested[97] the consideration of "stronger and stronger axioms of infinity," and reflection as follows: "Any proof of a set-theoretic theorem in the next higher system above set theory (i.e. any proof involving the concept of truth ...) is replaceable by a proof from such an axiom of infinity." This ties in with the class of all ordinal numbers cast as Cantor's Absolute: A largeness property ascribable to the class might be used to derive some set-theoretic proposition; but any such property confronts the antithetical contention that the class is mathematically incomprehendable, fostering the synthetic move to a large cardinal posited with the property.

---

[97] See Gödel [1990: 151].

In the expository article [1947] on the Continuum Problem Gödel presumed that CH would be shown independent from ZF and speculated more concretely about possibilities with large cardinals. He argued that the axioms of set theory do not "form a system closed in itself" and so the "very concept of set on which they are based suggests their extension by new axioms that assert the existence of still further iterations of the operation of 'set of' ", citing Zermelo [1930] and echoing its theme. In an unpublished footnote 20 toward a 1966 revision of [1947] Gödel was to acknowledge[98] "extremely strong axioms of infinity of an entirely new kind", generalizations of properties of ω "supported by strong arguments from analogy". This heuristic of generalization ties in with Cantor's view of the finite and transfinite as unitary, with properties like inaccessibility and measurability technically satisfied by ω being too accidental were they not also ascribable to higher cardinals through the uniformity of the set-theoretic universe.[99]

Gödel [1947] at the end actually argued against CH by drawing on the work of Sierpinski and others (cf. 2.6) to exhibit six "paradoxical" consequences. One of them is the existence of a Luzin set of cardinality of the continuum, and three others actually follows from the existence of such a set. This brought to the fore Gödel's stance about what is true in set theory. Whether CH is proved consistent or independent of ZFC, he believed in a "truth of the matter" both from the point of view of intuitions about the continuum and from his philosophical standpoint. That CH is implausible because it led to various implausible conclusions became a prominent attitude, one that would stay with set theory through its subsequent development.

3.5 Combinatorics

Gödel's construction of L was both a culmination in all major respects of the early period in set theory and a source for much that was to follow. But for quite some time it was to remain an isolated monument in the axiomatic tradition. No doubt the intervening years of war were a prominent factor, but there was a continuing difficulty in handling definability within set theory and a stultifying lack of means for constructing models of set theory to settle issues of consistency and independence. It would take a new generation versed in emerging model-theoretic methods to set the stage for the next major methodological advances.

In the meantime, the direct investigation of the transfinite as extension of number was advanced, gingerly at first, by a new initiative. The seminal results of *infinite combinatorics* were established beginning in the 1930s. As for algebra and topology, it was natural to extend concepts over the transfinite, and significantly, the combinatorics that would have the most bearing there had their roots in the mathematization of logic.

Frank Ramsey [1930] established a special case of the Decision Problem of Hilbert-Ackermann [1928], the decidability of validity for the $\exists\forall$ formulas with identity. For this purpose he established a basic generalization of the pigeonhole principle. In a move that transcended purpose and context he also established an infinite version implicitly

---

[98] See Gödel [1990: 260ff].

[99] See Wang [1974: §§1,4] for more on Gödel's view on heuristics as well as the criteria of *intrinsic necessity* and *pragmatic success* for accepting new axioms.

applying the now familiar Kőnig's Lemma for trees. Stated more generally for graphs by Dénes Kőnig [1927: 121] the lemma had also figured implicitly in Löwenheim [1915]. In what follows we affirm the general terminology for the formulation of Ramsey's results and then Kőnig's Lemma, anticipating extensions into the transfinite.

For ordinals α, β, and δ and $n \in \omega$ the *partition property*

$$\beta \longrightarrow (\alpha)^n_\delta$$

is the assertion that for any partition $f: [\beta]^n \to \delta$ of the n-element subsets of β into δ cells there is an $H \subseteq \beta$ of order type α *homogeneous* for the partition, i.e. all the n-element subsets of H lie in the same cell. Ramsey showed that for any k, n, and r all in ω, there is a $m \in \omega$ such that $m \longrightarrow (k)^n_r$. Skolem [1933] sharpened Ramsey's argument and thereby lowered the possibilities for the m's, but to this day the least such m's, the *Ramsey numbers*, have not been determined except in the simplest cases. Ramsey's infinite version is: $\omega \longrightarrow (\omega)^n_r$ for every $n, r \in \omega$. This partition property and its variants have been adapted to a variety of situations, and today *Ramsey theory* is a thriving field of combinatorics.[100]

A *tree* is a partially ordered set T such that the predecessors of any element are well-ordered. The αth *level* of T consists of those elements whose predecessors have order type α, and the *height* of T is the least α such that the αth level of T is empty. A *chain* of T is a linearly ordered subset, and an *antichain* is a subset consisting of pairwise incomparable elements. A *branch* of T is a maximal chain, and a *cofinal branch* of T is a branch with elements at every non-empty level of T. Finally, for a cardinal κ, a κ-*tree* is a tree of height κ each of whose levels has cardinality less than κ, and

κ has the *tree property iff* every κ-tree has a cofinal branch.

Finite trees of course are quite basic to current graph theory and computer science. With infinite trees the concerns are rather different, typically involving cofinal branches. Kőnig's Lemma asserts that ω *has the tree property*.

The first systematic study of transfinite trees was carried out in Đuro Kurepa's thesis [1935], and several properties emerging from his investigations, particularly for ω_1-trees, would later become focal in the combinatorial study of the transfinite.

An *Aronszajn tree* is an ω_1-tree without a cofinal branch,

i.e. a counterexample to the tree property for ω_1. Kurepa [1935: §9,thm 6] gave Nachman Aronszajn's result that *there is an Aronszajn tree*.

A *Suslin tree* is an ω_1-tree with no uncountable chains or antichains.

Kurepa [1935: appendix] reduced a hypothesis growing out of a problem of Suslin [1920] about the characterizability of the order type of the reals to a combinatorial property of ω_1 as follows: *Suslin's Hypothesis holds iff there are no Suslin trees.*

A *Kurepa tree* is an ω_1-tree with at least ω_2 cofinal branches,

---

[100] See the text Graham-Rothschild-Spencer [1990] and the compendium Nešetřil-Rödl [1990] for the recent work on Ramsey Theory.

and *Kurepa's Hypothesis* deriving from Kurepa [1942: 143], is the assertion that such trees exist. Much of this would be rediscovered, and both Suslin's Hypothesis and Kurepa's Hypothesis would be resolved three decades later with the advent of forcing, several of the resolutions in terms of large cardinal hypotheses.[101] Kurepa's work also anticipated another development from a different quarter:

Paul Erdős, although an itinerant mathematician for most of his life, was the prominent figure of a strong Hungarian tradition in combinatorics, and through some seminal results he introduced major initiatives into the detailed combinatorial study of the transfinite. Erdős and his collaborators simply viewed the transfinite numbers as a combinatorially rich source of intrinsically interesting problems, the concrete questions about graphs and mappings having a natural appeal through their immediacy. One of the earliest advances was Erdős-Tarski [1943] which concluded enticingly with an intriguing list of six combinatorial problems, the positive solution to any, as it was to turn out, amounting to the existence of a large cardinal. In a footnote various implications were noted, one of them being essentially that *for inaccessible κ, the tree property for κ implies $\kappa \longrightarrow (\kappa)_2^2$*, generalizing Ramsey's $\omega \longrightarrow (\omega)_2^2$ and making explicit the Kőnig Lemma property needed. While Kurepa was investigating distinctive properties of uncountable trees, Erdős-Tarski [1943] was evidently motivated by strong properties of ω to formulate direct combinatorial generalizations to inaccessible cardinals by analogy.[102] The situation would become considerably clarified, but only two decades later.[103]

The detailed investigation of partition properties began in earnest in the 1950s, with Erdős and Richard Rado's [1956] being representative. For a cardinal κ, let κ^+ denote its successor cardinal and set $\exp_0(\kappa) = \kappa$ and $\exp_{n+1}(\kappa) = 2^{\exp_n(\kappa)}$. What became known as *the* Erdős-Rado Theorem asserts: *For any infinite cardinal κ and $n \in \omega$,*

$$\exp_n(\kappa)^+ \longrightarrow (\kappa^+)_\kappa^{n+1}.$$

This was established using the basic tree argument underlying Ramsey's results, whereby a homogeneous set is not constructed recursively, but a tree is constructed such that its branches provide homogeneous sets, and a counting argument ensures that there must be a homogeneous set of sufficient cardinality. Kurepa [1937; 1939] in effect had actually established the case $n = 1$ and shown that $\exp_1(\kappa)^+$ was the least possible. The $\exp_n(\kappa)^+$ was also shown to be the least possible in the general case, and so unlike for the Ramsey numbers in the finite case an exact analysis was quickly achieved in the transfinite. This was to be a recurring phenomenon, that the gross features of transfinite cardinality make its combinatorics actually easier than in the analogous finite situation. And notably, iterated cardinal exponentiation figured prominently, so that shedding deeper concerns the power set operation became further domesticated in the arithmetic of combinatorics. In fact, assuming GCH simplified results and formulations, and this was often done, as in Erdős, András Hajnal, and Rado's [1965], representative of the 1960s. Increasingly, a

---

[101] See Todorčević [1984] for a wide-ranging account of transfinite trees.

[102] On the other hand, Kurepa [1935: §10.3] did ask whether every inaccessible cardinal has the tree property, a question only resolved by work of Hanf (cf. 3.6).

[103] The details of implications asserted at the end of Erdős-Tarski [1943] were worked out in an influential seminar conducted by Tarski and Mostowski at Berkeley in 1958-9, and appeared in Erdős-Tarski [1961].

myriad of versions have been investigated in the larger terrain without GCH.[104]

Still among the Hungarians, Géza Fodor [1956] established the now familiar *regressive function lemma* for stationary sets: *If λ regular and uncountable, S is stationary in λ,*[105] *and $f: S \to \lambda$ is regressive (i.e. $f(\xi) < \xi$ for $\xi \in S$), then there is an $\alpha < \lambda$ such that $\{\xi \in S \mid f(\xi) = \alpha\}$ is stationary in λ.* It is a basic fact and a simple exercise now, but then it was the culmination of a progression of results beginning with a special case established in Aleksandrov-Urysohn [1929] and getting to the right largeness notion of stationarity. The contrast with how the lemma's earlier precursors were considered difficult and even paradoxical is striking, indicative of both the novelty of uncountable cofinality and the great leap forward that set theory has made.

3.6 Model-Theoretic Methods

Model theory began in earnest with the method of diagrams of Abraham Robinson's thesis [1951] and the related method of constants from Leon Henkin's thesis which gave a new proof [1949] of the Gödel Completeness Theorem. Tarski had set the stage with his definition of truth and more generally his casting of formal languages and structures in set-theoretic terms, and with him established at the University of California at Berkeley a large part of the development in the 1950s and 1960s would take place there. The construction of models freely used transfinite methods and soon led to new questions in set theory, but also set theory was to be decisively advanced by the infusion of model-theoretic methods.

The first relevant result was a generalization accreditable to Mostowski [1949] of the Mirimanoff–von Neumann result that every well-ordered set is order-isomorphic to exactly one ordinal with membership. A binary relation R is *extensional on X iff* for any $x \neq y$ both in X there is a $z \in X$ such that zRx iff $\neg zRy$. Recall that x is *transitive iff* members of members of x are themselves members of x, so that x is "closed under membership". *If R is a well-founded relation on a set X and extensional on X, there is a unique isomorphism of $\langle X, R \rangle$ onto $\langle T, \in \rangle$ where T is transitive, i.e. a bijection $\pi: X \to T$ such that for any $x, y \in X$, xRy iff $\pi(x) \in \pi(y)$.* T is the *transitive collapse* of X, and π the *collapsing isomorphism*. Thus, the linearity of well-orderings has been relaxed to well-foundedness and an analogue of the Axiom of Extensionality, and the transitive sets become canonical representatives as ordinals are for well-orderings. Gödel [1939: 222] had made the first substantial use of the transitive collapse; Mostowski [1949: 147] established the general result much later; and John Shepherdson [1951: 171] in a structured setting that brought out a further necessary hypothesis for classes X: R is *set-like*, i.e. for any $x \in X$, $\{y \mid yRx\}$ is a set. The initial applications in Mostowski [1949] and Shepherdson [1953] were to establish the independence of the assertion that there is a transitive set M which with $\in$ restricted to it is a model of set theory. While the Mirimanoff–von

---

[104] The results of Erdős-Hajnal-Rado [1965] were extended in Byzantine detail to the general situation without GCH by the book Erdős-Hajnal-Máté-Rado [1984]. See Hajnal–Larson [2008] for recent work on partition relations.

[105] A set $C \subseteq \lambda$ is *closed unbounded in λ iff* C contains its limit points, i.e. those $0 < \alpha < \lambda$ such that $C \cap \alpha = \alpha$, and is cofinal, i.e. $\bigcup C = \lambda$. A set $S \subseteq \lambda$ is *stationary in λ iff* for any C closed unbounded in λ, $S \cap C$ is not empty.

Neumann result was basic to the analysis of number in the transfinite, the transitive collapse result grew in significance from specific applications and came to epitomize how well-foundedness made possible a coherent theory of models of set theory.

The relationship between ZFC and Bernays-Gödel (BG), the class-set theory brought into prominence by its use in Gödel [1940], was clarified during this period. As analyzed in Hao Wang [1949], BG can be construed as an extension of ZFC via the introduction of class variables intended to range over subcollections of V and correlative axioms, together with a comprehension principle asserting for each formula φ with just one set variable v free and no class variables quantified that there is a corresponding class $\{v \mid \varphi\}$. Ilse Novak [1950] and Barkley Rosser and Wang in their [1950] established that *if ZFC is consistent, then so is BG* by providing model-theoretic interpretations of BG relative to ZFC. Then Mostowski [1950] showed that BG is a conservative extension of ZF, i.e. any sentence σ without class variables provable in BG is already provable in ZFC. Subsequently, Joseph Shoenfield [1954] showed how to convert directly a proof of such a σ in BG into a proof in ZFC. These results reinforced the impression that, as far as the axiomatic tradition from Zermelo through Gödel is concerned, there is essentially one set theory, and one might as well work in the parsimonious ZFC.

Shepherdson [1951; 1952; 1953] studied "inner" models of set theory, with [1952] giving a rigorous first-order account of the results of Zermelo [1930]. The term is now reserved for the case mentioned in 3.4: A transitive class containing all the ordinals such that, with membership and quantification restricted to it, the class satisfies each axiom of ZF. The archetypal inner model is Gödel's L, and $L \subseteq M$ for any inner model M since the construction of L carried out in M is again L. Because of this Shepherdson [1953] observed that the relative consistency of hypotheses like the negation of CH cannot be established via inner models.

Hajnal [1956; 1961] and Azriel Levy [1957; 1960] developed generalizations of L that were to become basic in a richer setting. For a set A, Hajnal formulated the *constructible closure $L(A)$ of A*, i.e. the smallest inner model M such that $A \in M$, and Levy formulated the *class $L[A]$ of sets constructible relative to A*, i.e. the smallest inner model M such that for every $x \in M$, $A \cap x \in M$.[106] $L(A)$ realizes the algebraic idea of building up a model starting from a set of generators, and $L[A]$ the idea of building up a model using A construed as a predicate. $L(A)$ may not satisfy AC since e.g. it may not have a well-ordering of A, yet $L[A]$ always satisfies that axiom. This distinction was only to surface later, as both Hajnal and Levy took A to be a set of ordinals, when $L(A) = L[A]$. Hajnal and Levy (and also Shoenfield [1959], who formulated a special version of Levy's construction) used these models to establish conditional independence results of the sort: if the failure of CH is consistent, then so is that failure together with $2^\lambda = \lambda^+$ for sufficiently large cardinals λ.

After Richard Montague [1956; 1961] applied reflection phenomena to investigate fi-

---

[106] To formulate $L(A)$, define: $L_0(A) =$ the smallest transitive set $\supseteq \{A\}$ (to ensure that the resulting class is transitive); $L_{\alpha+1} = \text{def}(L_\alpha(A))$ (where def is as in 3.4); $L_\delta = \bigcup_{\alpha<\delta} L_\alpha(A)$ for limit $\delta > 0$; and finally $L(A) = \bigcup_\alpha L_\alpha(A)$. To formulate $L[A]$, first let $\text{def}^A(x)$ denote the collection of subsets of x first-order definable over $\langle x, \in, A \cap x \rangle$, i.e. $A \cap x$ is now allowed as a predicate in the definitions. Then define: $L_0[A] = \emptyset$; $L_{\alpha+1}[A] = \text{def}^A(L_\alpha[A])$; $L_\delta[A] = \bigcup_{\alpha<\delta} L_\alpha[A]$ for limit $\delta > 0$; and finally $L[A] = \bigcup_\alpha L_\alpha[A]$.

nite axiomatizability for set theory, Levy [1960a; 1960b] also formulated reflection principles and established their broader significance. The *Reflection Principle* for ZF, drawn from Montague [1961: 99] and from Levy [1960a: 234], asserts: *For any (first-order) formula* $\varphi(v_1, \ldots, v_n)$ *and any ordinal* β, *there is a limit ordinal* $\alpha > \beta$ *such that for any* $x_1, \ldots, x_n \in V_\alpha$,

$$\varphi[x_1, \ldots, x_n] \text{ iff } \varphi^{V_\alpha}[x_1, \ldots, x_n],$$

i.e. the formula holds exactly when it holds with all the quantifiers restricted to V_α. Levy showed that this schema is equivalent to the conjunction of the Replacement schema together with Infinity in the presence of the other axioms of ZF. Moreover, he formulated reflection principles in local form that characterized cardinals in the Mahlo hierarchy (2.4), conceptually the least large cardinals after the inaccessible cardinals. Then William Hanf and Dana Scott in their [1961] posited analogous reflection principles for higher-order formulas, leading to what are now called the *indescribable cardinals*, and eventually Levy [1971] carried out a systematic study of the sizes of these cardinals.[107] The model-theoretic reflection idea thus provided a coherent scheme for viewing the bottom of an emerging hierarchy of large cardinals as a generalization of Replacement and Infinity, one that resonates with the procession of models in Zermelo [1930]. The heuristic of reflection had been broached in 1946 remarks by Gödel (cf. 3.4), and another point of contact is the formulation of the concept of *ordinal definable set* in those remarks. With the class of ordinal definable sets formalized by $OD = \bigcup_\alpha \text{def}(V_\alpha)$, the adequacy of this definition is based on some form of the Reflection Principle for ZF. With $\text{tc}(y)$ denoting the smallest transitive set $\supseteq y$, let $HOD = \{x \mid \text{tc}(\{x\}) \subseteq OD\}$. the class of *hereditarily ordinal definable sets*. As adumbrated by Gödel, HOD is an inner model in which AC, though not necessarily CH, holds. The basic results about this inner model were to be rediscovered several times.[108] In these several ways reflection phenomena both as heuristic and as principle became incorporated into set theory, bringing to the forefront what was to become a basic feature of the study of well-foundedness.

The set-theoretic generalization of first-order logic allowing transfinitely indexed logical operations was to lead to the solution of the problem of whether the least inaccessible cardinal can be measurable (cf. 3.2). Extending familiarity by abstracting to a new domain Tarski [1962] defined the *strongly compact* and *weakly compact* cardinals by ascribing natural generalizations of the key compactness property of first-order logic to the corresponding infinitary languages. These cardinals had figured in Erdős-Tarski [1943] (cf. 3.5) in combinatorial formulations that was later seen to imply that *a strongly compact cardinal is measurable*, and *a measurable cardinal is weakly compact*. Tarski [1962] pointed out that his student William Hanf (cf. [1964]) established, using the satisfaction relation for infinitary languages, that *there are many inaccessible cardinals (and Mahlo cardinals) below a weakly compact cardinal. A fortiori, the least inaccessible cardinal is not measurable*. This breakthrough was the first result about the size of measurable cardinals since Ulam's original paper [1930] and was greeted as a spectacular success for metamathematical methods. Hanf's work radically altered size intuitions about prob-

---

[107] See Kanamori [2003: §6].
[108] See Myhill-Scott [1971], especially p. 278.

lems coming to be understood in terms of large cardinals and ushered in model-theoretic methods into the study of large cardinals beyond the Mahlo cardinals.[109]

Weak compactness was soon seen to have a variety of characterizations; most notably in terms of 3.5, κ is weakly compact *iff $\kappa \to (\kappa)^2_2$ iff $\kappa \to (\kappa)^n_\lambda$ for every $n \in \omega$ and $\lambda < \kappa$ iff κ is inaccessible and has the tree property*. Erdős and Hajnal [1962] noted that the study of stronger partition properties had progressed to the point where a combinatorial proof that the least inaccessible cardinal is not measurable could have been given before Hanf came to his argument. However, model-theoretic methods quickly led to far stronger conclusions, particularly through the connection that had been made in Ehrenfeucht-Mostowski [1956] between partition properties and *sets of indiscernibles*.[110]

The concurrent emergence of the *ultraproduct construction* in model theory set the stage for the development of the modern theory of large cardinals in set theory. With a precursor in Skolem's [1933a; 1934] construction of a non-standard model of arithmetic the ultraproduct construction was brought to the forefront by Tarski and his students after Jerzy Łoś's [1955] adumbration of its fundamental theorem. This new method of constructing concrete models brought set theory and model theory even closer together in a surge of results and a lasting interest in ultrafilters. Measurable cardinals had been formulated (cf. 3.2) in terms of ultrafilters construed as two-valued measures; Jerome Keisler [1962] struck on the idea of taking the ultrapower of a measurable cardinal κ by a κ-complete ultrafilter over κ to give a new proof of Hanf's result, seeing the crucial point that the completeness property led to a well-founded, and so in his case well-ordered, structure.

Then Scott [1961] made the further, crucial move of taking the ultrapower of the universe V itself by such an ultrafilter. The full exercise of the transitive collapse as a generalization of the correlation of ordinals to well-ordered sets now led to an inner model $M \neq V$ and an elementary embedding $j: V \to M$.[111] With this Scott established: *If there is a measurable cardinal, then $V \neq L$.* Large cardinal hypotheses thus assumed a new significance as a means for maximizing possibilities away from Gödel's delimitative construction. Also, the Cantor-Gödel realist view of a fixed set-theoretic universe notwithstanding, Scott's construction fostered the manipulative use of inner models in set theory. The construction provided one direction and Keisler [1962a] the other of a new characterization that established a central structural role for measurable cardinals: *There is an elementary embedding $j: V \to M$ for some inner model $M \neq V$ iff there is a measurable cardinal.* This result is not formalizable in ZFC because of the use of the satisfaction relation and the existential assertion of a proper class, but technical versions are. Despite the lack of formalizability such existential assertions have been widely entertained since, and with this set theory in practice could be said to have overleaped the bounds of ZFC. On the other hand, that the existence of a class elementary embedding is equivalent to the existence of a certain set, the witnessing ultrafilter for a measurable cardinal, can be

---

[109] See Kanamori [2003: §4] for these results about strongly and weakly compact cardinals.

[110] See Kanamori [2003: §§7, 8, 9] for more on partition relations and sets of indiscernibles, particularly their role in the formulation the set of natural numbers $0^\#$ and its role of transcendence over L.

[111] That is, for any formula $\varphi(v_1, \ldots, v_n)$ and sets $x_1, \ldots, x_n$, $\varphi(x_1, \ldots, x_n) \longleftrightarrow \varphi^M(j(x_1), \ldots, j(x_n))$, i.e. the formula holds of the x_is exactly when it holds of the $j(x_i)$s with the quantifiers restricted to M. Thus elementary embeddings are just the extension of algebraic monomorphisms to the preservation of logical properties.

considered a means of formalization in ZFC, one that would be paradigmatic for such reductions.

Work of Petr Vopěnka, who started the active Prague seminar in set theory in the spring of 1963, would be closely connected to that of Scott. Aware of the limitations of inner models for establishing independence results Vopěnka (cf. [1965]) embarked on a systematic study of (mostly ill-founded) class models of Bernays-Gödel set theory using ultrapower and direct limit constructions. Vopěnka not only established [1962] Scott's result on the incompatibility of measurability and constructibility via different means, but he and his student Karel Hrbáček in their [1966] soon established a global generalization for inner models $L(A)$: *If there is a strongly compact cardinal, then $V \neq L(A)$ for any set A.*

Through model-theoretic methods set theory was brought to the point of entertaining elementary embeddings into well-founded models,[112] soon to be transfigured by a new method for getting well-founded *extensions* of well-founded models.

4 INDEPENDENCE

4.1 *Forcing*

Paul Cohen (1934–2007), born just before Gödel established his relative consistency results, established the independence of AC from ZF and the independence of CH from ZFC [1963; 1964]. That is, Cohen established that Con(ZF) implies Con(ZF + ¬AC) and Con(ZFC) implies Con(ZFC + ¬CH). These results delimited ZF and ZFC in terms of the two fundamental issues at the beginnings of set theory. But beyond that, Cohen's proofs were soon to flow into method, becoming the inaugural examples of *forcing*, a remarkably general and flexible method for extending models of set theory. Forcing has strong intuitive underpinnings and reinforces the notion of set as given by the first-order ZF axioms with conspicuous uses of Replacement and Foundation. If Gödel's construction of L had launched set theory as a distinctive field of mathematics, then Cohen's method of forcing began its transformation into a modern, sophisticated one.[113]

Cohen's approach was to start with a model M of ZF and adjoin a set G, one that would exhibit some desired new property. He realized that this had to be done in a minimal fashion in order that the resulting structure also model ZF, and so imposed restrictive conditions on both M and G. He took M to be a countable standard model, i.e. a countable transitive set that together with the membership relation restricted to it is a model of ZF.[114] The ordinals of M would then coincide with the predecessors of some ordinal ρ, and M would be the cumulative hierarchy $M = \bigcup_{\alpha < \rho} V_\alpha \cap M$. Cohen then established a system of terms to denote members of the new model, finding it convenient to use a ramified

---

[112]See Keisler-Tarski [1964] for a comprehensive account of the theory of large cardinals through the use of ultrapowers in the early 1960s.

[113]According to Scott (Bell [1985: ix]): "Set theory could never be the same after Cohen, and there is simply no comparison whatsoever in the sophistication of our knowledge about models of set theory today as contrasted to the pre-Cohen era."

[114]The existence of such a model is an avoidable assumption in formal relative consistency proofs via forcing.

language: For each $x \in M$ let $\dot{x}$ be a corresponding constant; let $\dot{G}$ be a new constant; and for each $\alpha < \rho$ introduce quantifiers $\forall_\alpha$ and $\exists_\alpha$. Then develop a hierarchy of terms as follows: $\dot{M}_0 = \{\dot{G}\}$, and for limit ordinals $\delta < \rho$, $\dot{M}_\delta = \bigcup_{\alpha<\delta} \dot{M}_\alpha$. At the successor stage, let $\dot{M}_{\alpha+1}$ be the collection of terms $\dot{x}$ for $x \in V_\alpha \cap M$ and "abstraction" terms corresponding to formulas allowing parameters from $\dot{M}_\alpha$ and quantifiers $\forall_\alpha$ and $\exists_\alpha$. It is crucial that this ramified language with abstraction terms be entirely formalizable in M, through a systematic coding of symbols. Once a set G is provided from the outside, a model $M[G] = \bigcup_{\alpha<\rho} M_\alpha[G]$ would be determined by the terms, where each $\dot{x}$ is to be interpreted by x for $x \in M$ and $\dot{G}$ is to be interpreted by G, so that: $M_0[G] = \{G\}$; for limit ordinals $\delta < \rho$, $M_\delta[G] = \bigcup_{\alpha<\delta} M_\alpha[G]$; and $M_{\alpha+1}[G]$ consists of the sets in $V_\alpha \cap M$ together with sets interpreting the abstraction terms as the corresponding definable subsets of $M_\alpha[G]$ with $\forall_\alpha$ and $\exists_\alpha$ ranging over this domain.

But what properties can be imposed on G to ensure that $M[G]$ be a model of ZF? Cohen's key idea was to tie G closely to M through a system of sets in M called *conditions* that would approximate G. While G may not be a member of M, G is to be a subset of some $Y \in M$ (with $Y = \omega$ a basic case), and these conditions would "force" some assertions about the eventual $M[G]$ e.g. by deciding some of the membership questions, whether $x \in G$ or not, for $x \in Y$. The assertions are to be just those expressible in the ramified language, and Cohen developed a corresponding *forcing relation* $p \Vdash \varphi$, "p forces φ", between conditions p and formulas φ, a relation with properties reflecting his approximation idea. For example, if $p \Vdash \varphi$ and $p \Vdash \psi$, then $p \Vdash \varphi \& \psi$. The conditions are ordered according to the constraints they impose on the eventual G, so that if $p \Vdash \varphi$, and q is a stronger condition, then $q \Vdash \varphi$. Scott actually provided the now common forcing symbol $\Vdash$, and with Cohen having worked with prenex formulas, Scott showed how to proceed generally by separating out negation with: $p \Vdash \neg\varphi$ iff for no stronger condition q does $q \Vdash \varphi$. It was crucial to Cohen's approach that the forcing relation, like the ramified language, be definable in M.

The final ingredient is that the whole scaffolding is given life by incorporating a certain kind of set G. Stepping out of M and making the only use of its countability, Cohen enumerated the formulas of the ramified language in a countable sequence and required that G be completely determined by a countable sequence of stronger and stronger conditions $p_0, p_1, p_2, \ldots$ such that for every formula φ of the ramified language exactly one of φ or $\neg\varphi$ is forced by some p_n. Such a G is called a *generic* set. Cohen was able to show that the resulting $M[G]$ does indeed satisfy the axioms of ZF: Every assertion about $M[G]$ is already forced by some condition; the forcing relation is definable in M; and so the ZF axioms, holding in M, most crucially Power Set and Replacement, can be applied to derive corresponding forcing assertions about ZF axioms holding in $M[G]$.

The foregoing outline in its main features reflects how forcing was viewed by July 1963 and presented by Cohen himself in a course in Spring 1965.[115] He first described the case when $G \subseteq \omega$ and the conditions p are functions from some finite subset of ω into $\{0, 1\}$ and $p \Vdash \dot{n} \in \dot{G}$ if $p(n) = 1$ and $p \Vdash \dot{n} \notin \dot{G}$ if $p(n) = 0$. Today, a G so adjoined to M is called a *Cohen real over M*. Cohen established the independence of CH by adjoining a set which can be construed as a sequence of many Cohen reals. He established the

---

[115] See Cohen [1966].

independence of AC by a version of the above scheme where in addition to $\dot{G}$ there are also new constants $\dot{G}_i$ for $i \in \omega$, with $\dot{G}$ to be interpreted by a set X of Cohen reals, each an interpretation of some $\dot{G}_i$. The point is that X is not well-orderable in the extension.

The appeal to a countable model in Cohen's approach is a notable positive subsumption of Skolem's Paradox (cf. 3.2) into a new method. Remarkably, Skolem [1923: 229] had entertained the possibility of adjoining a new subset of the natural numbers to a countable model of Zermelo's system and getting a new model, adding in a footnote that "it is quite probable" that the Continuum Hypothesis is not decided by Zermelo's axioms. Just as starting with a countable standard model is not formally necessary for relative consistency results, other features of Cohen's argument would soon be reformulated, reorganized, and generalized, but the main thrust of his constructive approach through definability and genericity would remain. Cohen's particular achievement lies in devising a concrete procedure for extending well-founded models of set theory in a minimal fashion to well-founded models of set theory with new properties but without altering the ordinals.[116] Set theory had undergone a sea-change, and beyond how the subject was enriched, it is difficult to convey the strangeness of it.

The creation of forcing is a singular phenomenon in the development of set theory not only since it raised the level of the subject dramatically but also since it could well have occurred decades earlier. But however epochal Cohen's advance there was a line of development for which it did provide at least a semblance of continuity: Interest in independence results for weak versions of AC had been on the rise from the mid-1950's, with more and more sophisticated Fraenkel-Mostowski models being constructed.[117] Solomon Feferman, who was one of the first to whom Cohen explained his ideas for the independence proofs in the process of their development, was the first after him to establish results by forcing; Levy soon followed; and among their first results were new independences from ZF for weak versions of AC (Feferman-Levy [1963], Feferman [1965]). Cohen [1965: 40] moreover acknowledged the similarities between his AC independence result and the previous Fraenkel-Mostowski models. In fact, consistencies first established via Fraenkel-Mostowski models were soon "transferred" to consequence of ZF via forcing by correlating urelements with generic sets.[118]

After an initial result by Feferman [1963], Levy [1963; 1965; 1970] also probed the limits of ZFC definability, establishing consistency results about definable sets of reals and well-orderings and in descriptive set theory. Intriguingly, inaccessible cardinals were brought in to overcome a technical hurdle in this study; Levy [1963: IV] applied the defining properties of such a cardinal to devise its "collapse" to $\aleph_1$ by making every smaller ordinal countable, and this forcing is now known as the *Levy collapse*.

Forcing was quickly generalized and applied to achieve wide-ranging results, particularly by Robert Solovay. He above all epitomized this period of great expansion in set

---

[116]Scott continued (Bell [1985: ix]): "I knew almost all the set-theoreticians of the day, and I think I can say that no one could have guessed that the proof would have gone in just this way. Model-theoretic methods had shown us how many *non-standard* models there were; but Cohen, starting from very primitive first principles, found the way to keep the models *standard* (that is, with a well-ordered collection of ordinals)."

[117]See Moore [1982: 5.1].

[118]See Felgner [1971] and Jech [1973] for more on the independence of weak versions of AC and transfers, and Pincus [1972] for a strong transfer theorem.

theory with his mathematical sophistication and fundamental results about and with forcing, and in the areas of large cardinals and descriptive set theory. Just weeks after Cohen's breakthrough Solovay [1963; 1965] elaborated the independence of CH by characterizing the possibilities for the size of 2^κ for regular κ and made the first exploration of a spectrum of cardinals. Then William Easton [1964; 1970] established the definitive result for powers of regular cardinals: *Suppose that GCH holds and F is a class function from the class of regular cardinals to cardinals such that for $\kappa \leq \lambda$, $F(\kappa) \leq F(\lambda)$ and the cofinality of $F(\kappa)$ is greater than κ. Then there is a forcing extension preserving cofinalities in which $2^\kappa = F(\kappa)$ for every regular κ.* Thus, as Solovay had seen locally, the *only* restriction beyond monotonicity on the power function for regular cardinals is that given by the Zermelo–König inequality.[119] Easton's result vitally infused forcing not only with the introduction of proper classes of forcing conditions but the now basic idea of a product analysis and the now familiar concept of *Easton support*. Through its reduction Easton's result focused interest on the possibilities for powers of *singular* cardinals, and this *Singular Cardinals Problem* together with the *Singular Cardinals Hypothesis* would stimulate the further development of set theory much as the Continuum Problem and the Continuum Hypothesis had stimulated its early development.[120]

In the Spring of 1964 Solovay [1965b; 1970] established a result remarkable for its mathematical depth and revelatory of what standard of argument was possible with forcing: *If there is an inaccessible cardinal, then in an inner model of a forcing extension every set of reals is Lebesgue measurable, has the Baire property, and has the perfect set property.* Like Cohen's results, this contextually decided issues dating back to the turn of the century and before; as described in 2.3 the regularity properties of sets of reals was a major concern of the early descriptive set theorists. Classical counterexamples show that Solovay's inner model cannot have a well-ordering of the reals, and so AC fails there. However, he established that the model satisfies the Principle of Dependent Choices, a principle sufficient for the formal rendition of the traditional theory of measure and category. Thus, Solovay's work vindicated the early descriptive set theorists in the sense that the regularity properties can consistently hold for all sets of reals in a *bona fide* model for mathematical analysis. For his result Solovay applied the Levy collapse of an inaccessible cardinal and built on its definability properties as first exploited by Levy [1963: IV]; for the Lebesgue measurability he introduced a new kind of forcing beyond Cohen's direct ways of adjoining new sets of ordinals or collapsing cardinals, that of adding a *random real*. Solovay's work not only opened the door to a wealth of different forcing arguments, but to this day his original definability arguments remain vital to descriptive set theory.

The perfect set property, central to Cantor's direct approach to the Continuum Problem through definability (1.2,2.3,2.5), led to the first acknowledged instance of a new phenomenon in set theory: the derivation of *equiconsistency* results with large cardinal hypotheses based on the complementary methods of forcing and inner models. A large cardinal hypothesis is typically transformed into a proposition about sets of reals by forcing that "collapses" that cardinal to $\aleph_1$ or "enlarges" the power of the continuum to that

---

[119] See 1.3, especially footnote 27.

[120] The Singular Cardinal Hypothesis asserts that 2^λ for singular λ is the least possible with respect to the powers 2^μ for $\mu < \lambda$, as given by monotonicity and the Zermelo-König inequality.

cardinal. Conversely, the proposition entails the same large cardinal hypothesis in the clarity of an inner model. Solovay's result provided the forcing direction from an inaccessible cardinal to the proposition that every set of reals has the perfect set property (and $\aleph_1$ is regular). But Ernst Specker [1957:210] had in effect established that if this obtains, then $\aleph_1$ (of V) is inaccessible in L. Thus, Solovay's use of an inaccessible cardinal was actually necessary, and its collapse to $\aleph_1$ complemented Specker's observation. Other propositions figuring in the initial applications of inaccessibility to forcing turned out to require inaccessibility, further integrating it into the interstices of set theory.[121] The emergence of such equiconsistency results is a subtle transformation of earlier hopes of Gödel (cf. 3.4): Propositions can be positively subsumed if there are enough ordinals, how many being specified by positing a large cardinal.[122] On the other hand, forcing quickly led to the conclusion that there could be no direct implication for CH: Levy and Solovay (Levy [1964], Solovay [1965a], Levy-Solovay [1967]) established that measurable cardinals neither imply nor refute CH, with an argument generalizable to most inaccessible large cardinals. Rather, the subsumption for many other propositions would be in terms of consistency, the methods of forcing and inner models being the operative modes of argument. In a new synthesis of the two Cantorian legacies, hypotheses of length concerning the extent of the transfinite are correlated with hypotheses of width concerning sets of reals.

It was the incisive work of Scott and Solovay through this early period that turned Cohen's breakthrough into a general method of wide applicability. Scott simplified Cohen's original formulation as noted above; Solovay made the important move to general partial orders and generic filters; and they together developed, with vicissitudes, the formulation in terms of Boolean-valued models.[123] These models forcibly showed how to avoid Cohen's ramified language as well as his dependence on a countable model. With their elegant algebraic trappings and seemingly more complete information they held the promise of being the right approach to independence results. But Shoenfield [1971] showed that forcing with partial orders can get at the gist of the Boolean approach in a straightforward manner. Moreover, Boolean-valued models were soon found to be too abstract and unintuitive for establishing *new* consistency results, so that within a few years set theorists were generally working with partial orders. It is a testament to Cohen's concrete

---

[121] The original application of the Levy collapse in Levy [1963: IV] also turned out to require an inaccessible cardinal (Levy [1970: 131ff]) – a remarkable turn of events for an apparently technical artifact at the beginning of forcing.

Many years later, Saharon Shelah [1980; 1984] was able to establish the necessity of Solovay's inaccessible for the proposition that every set of reals is Lebesgue measurable; on the other hand, Shelah also showed that the inaccessible is not necessary for the proposition that every set of reals has the Baire property.

[122] There is a telling antecedent in the result of Gerhard Gentzen [1936; 1943] that the consistency strength of arithmetic can be exactly gauged by an ordinal ε_0, i.e. transfinite induction up to that ordinal in a formal system of notations. Although Hilbert's program of establishing consistency by finitary means could not be realized, Gentzen provided an exact analysis in terms of ordinal length. Proof theory blossomed in the 1960s with the analysis of other theories in terms of such lengths, the proof theoretic ordinals.

[123] Vopěnka had developed a similar concept in a reworking [1964] of the independence of CH. The concept was generalized and simplified in a series of papers on the so-called ∇-models from the active Prague seminar founded by Vopěnka (see Hájek [1971: 78]), culminating in the exposition Vopěnka [1967]. However, the earlier papers did not have much impact, partly because of an involved formalism in which formulas were valued in a complete lattice rather than Boolean algebra.

approach that in this return from abstraction even the use of ramified languages has played an essential role in careful forcing arguments at the interface of recursion theory and set theory.

4.2 Envoi

Building on his Lebesgue measurability result Solovay soon reactivated the classical descriptive set theory program (cf. 2.5) of investigating the extent of the regularity properties by providing characterizations for the Σ_2^1 sets, the level at which Gödel established from $V = L$ the failure of the properties (cf. 3.4), and showed in particular that the regularity properties for these sets follow from the existence of a measurable cardinal. Thus, although measurable cardinals do not decide CH, they do establish the perfect set property for Σ_2^1 sets (Solovay [1969]) so that "CH holds for the Σ_2^1 sets" – a vindication of Gödel's hopes for large cardinals through a direct implication. Donald Martin and Solovay in their [1969] then applied large cardinal hypotheses weaker than measurability to push forward the old tree representation ideas of the classical descriptive set theorists, with the hypotheses cast in the new role of securing well-foundedness in this context.

The method of forcing as part of the axiomatic tradition together with the transmutations of Cantor's two legacies, large cardinals furthering the extension of number into the transfinite and descriptive set theory investigating definable sets of reals, established set theory as a sophisticated field of mathematics, a study of well-foundedness expanded into one of consistency strength. With the further development of forcing through increasingly sophisticated iteration techniques questions raised in combinatorics and over a broad landscape would be resolved in terms of consistency, sometimes with equiconsistencies in terms of large cardinals. The theory of large cardinals would itself be much advanced with the heuristics of reflection and generalization and sustained through increasing use in the study of consistency strength. In the most distinctive and intriguing development of contemporary set theory, the investigation of the determinacy of games, large cardinals would be further integrated into descriptive set theory. They were not only used to literally incorporate well-foundedness of inner models into the study of tree representations, historically first context involving well-foundedness, but also to provide the exact hypotheses, with Woodin cardinals, for gauging consistency strength.[124]

Stepping back to gaze at modern set theory, the thrust of mathematical research should deflate various possible metaphysical appropriations with an onrush of new models, hypotheses, and results. Shedding much of its foundational burden, set theory has become an intriguing field of mathematics where formalized versions of truth and consistency have become matters for manipulation as in algebra. As a study couched in well-foundedness ZFC together with the spectrum of large cardinals serves as a court of adjudication, in terms of relative consistency, for mathematical statements that can be informatively contextualized in set theory by letting their variables range over the set-theoretic universe. Thus, set theory is more of an open-ended framework for mathematics rather than an elucidating foundation. It is as a field *of* mathematics that both proceeds with its own internal

---

[124] See Kanamori [2003] for these recent developments.

questions and is capable of contextualizing over a broad range which makes of set theory an intriguing and highly distinctive subject.

ACKNOWLEDGEMENTS

This is a revision with significant changes of the author's The Mathematical Development of Set Theory from Cantor to Cohen, *The Bulletin of Symbolic Logic*, volume 2, 1996, pages 1–71, and appears here with the permission of the *Bulletin*.

BIBLIOGRAPHY

[Addison *et al.*, 1965] J. W. Addison Jr., L. Henkin, and A. Tarski, eds. *The Theory of Models*, Proceedings of the 1963 International Symposium at Berkeley, North-Holland, Amsterdam 1965.

[Aleksandrov, 1916] P. S. Aleksandrov. Sur la puissance des ensembles mesurables B, *Comptes Rendus Hebdomadaires des Séances de l'Académie des Sciences, Paris* 162 (1916), 323–325.

[Aleksandrov and Urysohn, 1929] P. S. Aleksandrov and P. S. Urysohn. Mémoire sur les espaces topologiques compacts, *Verhandelingen der Koninklijke Nederlandse Akademie van Wetenschappen Tweed Reeks. Afdeling Natuurkunde* 14 (1929), 1-96.

[Baire, 1898] R. Baire. Sur les fonctions discontinues qui se rattachent aux fonctions continues, *Comptes Rendus Hebdomadaires des Séances de l'Académie des Sciences, Paris* 129 (1898), 1621–1623.

[Baire, 1899] R. Baire. Sur les fonctions de variables réelles, *Annali di Matematica Pura ed Applicata* (3)3 (1899), 1-123.

[Banach, 1967] S. Banach. In S. Hartman and E. Marczewski, eds., *Oeuvres*, vol. 1, Państwowe Wydawnictwo Naukowe, Warsaw 1967.

[Banach and Tarski, 1924] S. Banach and A. Tarski. Sur la décomposition des ensembles de points en parties respectivement congruentes, *Fundamenta Mathematicae* 6 (1924), 244–277. Reprinted in Banach [1967], 118–148. Reprinted in Tarski [1986] vol. 1, 121–154.

[Bar-Hillel *et al.*, 1961] Y. Bar-Hillel, E. I. J. Poznanski, M. O. Rabin, and A. Robinson, eds. *Essays on the Foundations of Mathematics*, Magnes Press, Jerusalem 1961.

[Bartoszyński, 2008] T. Bartoszyński. Invariants of Meaure and Category, in: Foreman–Kanamori [2008].

[Bell, 1985] J. L. Bell. *Boolean-Valued Models and Independence Proofs in Set Theory*, second edition, Oxford Logic Guides #12, Oxford University Press, Oxford 1985.

[Benacerraf and Putnam, 1964] P. Benacerraf and H. Putnam, eds. *Philosophy of Mathematics. Selected Readings*. Prentice Hall, Englewood Cliffs, N.J. 1964.

[Benacerraf and Putnam, 1983] P. Benacerraf and H. Putnam. Second edition of [1964]. Cambridge University Press, Cambridge 1983.

[Bendixson, 1883] I. Bendixson. Quelques théorèmes de la théorie des ensembles de points, *Acta Mathematica* 2 (1883), 415–429.

[Bernays, 1937] P. Bernays. A system of axiomatic set theory, Part I, *The Journal of Symbolic Logic* 2 (1937), 65–77.

[Bernays, 1976] P. Bernays. A system of axiomatic set theory, in Gert H. Müller (editor), *Sets and Classes. On the Work of Paul Bernays*, North-Holland, Amsterdam 1976, 1–119. Individually appeared in various parts in *The Journal of Symbolic Logic* 2 (1937), 65–77; 6 (1941), 1–17; 7 (1942), 65–89 and 133–145; 8 (1943), 89–106; 3 (1948), 65–79; and 19 (1954), 81–96.

[Bernstein, 1908] F. Bernstein. Zur Theorie der trigonometrischen Reihen, *Berichte über die Verhandlungen der Königlich Sächsischen Gesellschaft der Wissenschaften zu Leipzig, Mathematische-Physische Klasse* 60 (1908), 325–338.

[Blass, 1984] A. R. Blass. Existence of bases implies the Axiom of Choice, in Baumgartner, James E., Donald A. Martin, and Saharon Shelah (editors), *Axiomatic Set Theory*, Contemporary Mathematics vol. 31, American Mathematical Society, Providence 1984, 31–33.

[Blass, 2008] A. R. Blass. Combinatorial Cardinal Charateristics of the Continuum, in: Foreman–Kanamori [2008].

[Boolos, 1971] G. S. Boolos. The iterative concept of set, *Journal of Philosophy* 68 (1971), 215–231. Reprinted in Benacerraf-Putnam [1983], 486–502, and in [1998] below, 13–29.

[Boolos, 1998] G. S. Boolos. *Logic, Logic, and Logic*, Harvard University Press, Cambridge 1998.
[Borel, 1898] E. Borel. *Leçons sur la théorie des fonctions*, Gauthier-Villars, Paris 1898.
[Borel, 1919] E. Borel. Sur la classification des ensembles de mesure nulle, *Bulletin de la Société Mathématique de France* 47 (1919), 97–125.
[Bourbaki, 1954] N. Bourbaki. *Eléments de Mathématique. I. Théorie des Ensembles*, Hermann, Paris 1954.
[Bourbaki, 1970] N. Bourbaki. *Eléments de Mathematique. I. Théorie des Ensembles*, combined edition, Hermann, Paris 1970.
[Brouwer, 1911] L. E. J. Brouwer. Beweis der Invarianz der Dimensionenzahl, *Mathematische Annalen* 70 (1911), 161–165. Reprinted in [1976] below, 430–434.
[Brouwer, 1976] L. E. J. Brouwer. Freudenthal, Hans (editor), *Collected Works*, vol. 2, North-Holland, Amsterdam 1976.
[Burali-Forti, 1897] C. Burali-Forti. Una questione sui numeri transfini, *Rendiconti del Circolo Matematico di Palermo* 11 (1897), 154–164. Translated in van Heijenoort [1967], 104–111.
[Cantor, 1872] G. Cantor. Über die Ausdehnung eines Satzes aus der Theorie der trignometrischen Reihen, *Mathematische Annalen* 5 (1872), 123–132. Reprinted in [1932] below, 92–102.
[Cantor, 1874] G. Cantor. Über eine Eigenschaft des Inbegriffes aller reellen algebraischen Zahlen, *Journal für die reine und angewandte Mathematik* 77 (1874), 258–262. Reprinted in [1932] below, 115–118. Translated with commentary in Ewald [1996], 839–843.
[Cantor, 1878] G. Cantor. Ein Beitrag zur Mannigfaltigkeitslehre, *Journal für die reine und angewandte Mathematik* 84 (1878), 242–258. Reprinted in [1932] below, 119–133.
[Cantor, 1879] G. Cantor. Über einen Satz aus der Theorie der stetigen Mannigfaltigkeiten, *Nachrichten von der Königlichen Gesellschaft der Wissenschaften und der Georg-Augusts-Universität zu Göttingen* (1879), 127–135. Reprinted in [1932] below, 134–138.
[Cantor, 1880] G. Cantor. Über unendliche, lineare Punktmannigfaltigkeiten. II, *Mathematische Annalen* 17 (1880), 355–358. Reprinted in [1932] below, 145–148, but excluding the referenced remark.
[Cantor, 1883] G. Cantor. Über unendliche, lineare Punktmannigfaltigkeiten. V. *Mathematische Annalen* 21 (1883), 545–591. Published separately as *Grundlagen einer allgemeinen Mannigfaltigkeitslehre. Ein mathematisch-philosophischer Versuch in der Lehre des Unendlichen*, B.G. Teubner, Leipzig 1883. Reprinted in [1932] below, 165–209. Translated with commentary in Ewald [1996], 878–920.
[Cantor, 1884] G. Cantor. Über unendliche, lineare Punktmannigfaltigkeiten. VI, *Mathematische Annalen* 23 (1884), 453–488. Reprinted in [1932] below, 210–246.
[Cantor, 1884a] G. Cantor. De la puissance des ensembles parfaits de points, *Acta Mathematica* 4 (1884), 381–392. Reprinted in [1932] below, 252–260.
[Cantor, 1887] G. Cantor. Mitteilungen zur Lehre vom Transfiniten, *Zeitschrift für Philosophie und philosophische Kritik* 91 (1887), 81–125 and 252–270. Reprinted in [1932] below, 378–439.
[Cantor, 1891] G. Cantor. Über eine elementare Frage der Mannigfaltigkeitslehre, *Jahresbericht der Deutschen Mathematiker-Vereinigung* I (1891), 75–78. Reprinted in [1932] below, 278–280. Translated with commentary in Ewald [1996], 920–922.
[Cantor, 1895] G. Cantor. Beiträge zur Begründung der transfiniten Mengenlehre. I, *Mathematische Annalen* 46 (1895), 481–512. Translated in [1915] below. Reprinted in [1932] below, 282–311.
[Cantor, 1897] G. Cantor. Beiträge zur Begründung der transfiniten Mengenlehre. II, *Mathematische Annalen* 49 (1897), 207–246. Translated in [1915] below. Reprinted in [1932] below, 312–351.
[Cantor, 1915] G. Cantor. *Contributions to the Founding of the Theory of Transfinite Numbers*. Translation of [1895] and [1897] above with introduction and notes, by Philip E.B. Jourdain, Open Court, Chicago 1915. Reprinted Dover, New York 1965.
[Cantor, 1932] G. Cantor. E. Zermelo, ed., *Gesammelte Abhandlungen mathematicschen und philosophischen Inhalts*, Julius Springer, Berlin 1932. Reprinted in Springer, Berlin 1980.
[Cavaillès, 1962] J. Cavaillès. *Philosophie Mathématique*. Paris, Hermann 1962. Includes French translation of Noether-Cavaillès [1937].
[Coffa, 1979] J. A. Coffa. The humble origins of Russell's paradox, *Russell* 33–34 (1979), 31–37.
[Cohen, 1963] P. J. Cohen. The independence of the Continuum Hypothesis. I, *Proceedings of the National Academy of Sciences U.S.A.* 50 (1963), 1143–1148.
[Cohen, 1964] P. J. Cohen. The independence of the Continuum Hypothesis. II, *Proceedings of the National Academy of Sciences U.S.A.* 51 (1964), 105–110.
[Cohen, 1965] P. J. Cohen. Independence results in set theory, in Addison-Henkin-Tarski [1965], 39-54.
[Dauben, 1979] J. W. Dauben. *Georg Cantor. His Mathematics and Philosophy of the Infinite*, Harvard University Press, Cambridge 1979. Paperback edition 1990.

[Dedekind, 1871] R. Dedekind. Über die Komposition der binären quadratischen Formen, supplement X to Dirichlet's *Vorlesungen über Zahlentheorie*, second edition, F. Vieweg, Braunschweig 1871. Reprinted in [1932] below, vol. 3, 399–407. Excerpt translated with commentary in Ewald [1996], 762-765.
[Dedekind, 1872] R. Dedekind. *Stetigkeit und irrationale Zahlen*, F. Vieweg, Braunschweig 1872. Fifth, 1927 edition reprinted in [1932] below, vol. 3, 315–334. Translated in [1963] below, 1-27. Translated with commentary in Ewald [1996], 765–779.
[Dedekind, 1888] R. Dedekind. *Was sind und was sollen die Zahlen?*, F. Vieweg, Braunschweig 1888. Sixth, 1930 edition reprinted in [1932] below, vol. 3, 335–390. Second, 1893 edition translated in [1963] below, 29-115. Third, 1911 edition translated with commentary in Ewald [1996], 787–833.
[Dedekind, 1900] R. Dedekind. Über die von drei Moduln erzeugte Dualgruppe, *Mathematische Annalen* 53 (1900), 371–403. Reprinted in [1932] below, vol. 2, 236–271.
[Dedekind, 1932] R. Dedekind. Fricke, Robert, Emmy Noether, and Öystein Ore (editors), *Gesammelte mathematische Werke*, F. Vieweg, Braunschweig 1932. Reprinted by Chelsea, New York 1969.
[Dedekind, 1963] R. Dedekind. *Essays on the Theory of Numbers*. Translations by Wooster W. Beman, Dover, New York 1963. (Reprint of original edition, Open Court, Chicago 1901).
[du Bois-Reymond, 1869] P. du Bois-Reymond. Bemerkungen über die verschidenen Werthe, welche eine Function zweier reellen Variabeln erhält, wenn man diese entweder nach einander oder gewissen Beziehungen gemäss gleichzeitig verschwinden lässt, *Journal für die reine und angewandte Mathematik* 70 (1869), 10–45.
[du Bois-Reymond, 1875] P. du Bois-Reymond. Über asymptotische Werthe, infinitäre Approximationen und infinitäre Auflösung von Gleichungen, *Mathematische Annalen* 8 (1875), 363–414.
[Dugac, 1976] P. Dugac. *Richard Dedekind et les fondements des mathématiques*, Collection des travaux de l'Académie Internationale d'Histoire des Sciences #24, Vrin, Paris 1976.
[Easton, 1964] W. B. Easton. Powers of regular cardinals. Ph.D. thesis, Princeton University 1964. Abstracted as: Proper classes of generic sets, *Notices of the American Mathematical Society* 11 (1964), 205. Published in abridged form as [1970] below.
[Easton, 1970] W. B. Easton. Powers of regular cardinals, *Annals of Mathematical Logic* 1 (1970), 139–178.
[Ebbinghaus, 2007] H.-D. Ebbinghaus. *Ernst Zermelo. An Approach to his Life and Work*, Springer, Berlin 2007.
[Ehrenfeucht and Mostowski, 1956] A. Ehrenfeucht and A. M. Mostowski. Models of axiomatic theories admitting automorphisms, *Fundamenta Mathematicae* 43 (1956), 50–68. Reprinted in Mostowski [1979], 494–512.
[Erdős and Hajnal, 1962] P. Erdős and A. Hajnal. Some remarks concerning our paper "On the structure of set mappings", *Acta Mathematica Academiae Scientiarum Hungaricae* 13 (1962), 223–226.
[Erdős et al., 1984] P. Erdős, A. Hajnal, A. Máté, and R. Rado. *Combinatorial Set theory: Partition Relations for Cardinals*, North-Holland, Amsterdam 1984.
[Erdős et al., 1965] P. Erdős, A. Hajnal, and R. Rado. Partition relations for cardinal numbers, *Acta Mathematica Academiae Scientiarum Hungaricae* 16 (1965), 93–196.
[Erdős and Rado, 1956] P. Erdős and R. Rado. A partition calculus in set theory, *Bulletin of the American Mathematical Society* 62 (1956), 427–489. Reprinted in Gessel-Rota [1987], 179–241.
[Erdős and Tarski, 1943] P. Erdős and A. Tarski On families of mutually exclusive sets, *Annals of Mathematics* 44 (1943), 315–329. Reprinted in Tarski [1986] vol. 2, 591–605.
[Erdős and Tarski, 1961] P. Erdős and A. Tarski. On some problems involving inaccessible cardinals, in: Bar-Hillel–Poznanski–Rabin–Robinson [1961], 50–82. Reprinted in Tarski [1986] vol. 4, 79–111.
[Ewald, 1996] W. Ewald, ed. *From Kant to Hilbert: A Source Book in the Foundations of Mathematics*, two volumes, Clarendon Press, Oxford 1996.
[Feferman, 1963] S. Feferman. Some applications of the notions of forcing and generic sets (summary), in Addison–Henkin–Tarski [1965], 89–95.
[Feferman, 1965] S. Feferman. Some applications of the notions of forcing and generic sets, *Fundamenta Mathematicae* 56 (1965), 325-345.
[Feferman and Levy, 1963] S. Feferman and A. Levy. Independence results in set theory by Cohen's method. II (abstract), *Notices of the American Mathematical Society* 10 (1963), 593.
[Felgner, 1971] U. Felgner. *Models of ZF-Set Theory*, Lecture Notes in Mathematics #223, Springer, Berlin 1971.
[Ferreirós, 1995] J. Ferreirós. "What fermented in me for years": Cantor's discovery of transfinite numbers, *Historia Mathematica* 22 (1995), 33–42.
[Ferreirós, 2007] J. Ferreirós. *Labyrinth of Thought: A History of Set Theory and its Role in Modern Mathematics*, second edition, Birkhauser, Basel 2007.

[Fodor, 1956] G. Fodor. Eine Bemerking zur Theorie der regressiven Funktionen, *Acta Scientiarum Mathematicarum (Szeged)* 17 (1956), 139–142.
[Foreman and Kanamori, 2008] M. Foreman and A. Kanamori. *The Handbook of Set Theory*, Springer, Berlin 2008.
[Fraenkel, 1921] A. A. Fraenkel. Über die Zermelosche Begründung der Mengenlehre (abstract), *Jahresbericht der Deutschen Mathematiker-Vereinigung* 30II (1921), 97–98.
[Fraenkel, 1922] A. A. Fraenkel. Zu den Grundlagen der Cantor-Zermeloschen Mengenlehre, *Mathematische Annalen* 86 (1922), 230–237.
[Fraenkel, 1922a] A. A. Fraenkel. Über den Begriff 'definit' und die Unabhängigkeit des Auswahlaxioms, *Sitzungsberichte der Preussischen Akademie der Wissenschaften, Physikalisch-mathematische Klasse* (1922), 253–257. Translated in van Heijenoort [1967], 284–289.
[Fraenkel, 1930] A. A. Fraenkel. Georg Cantor, *Jahresbericht der Deutschen Mathematiker-Vereinigung* 39 (1930), 189–266. Published separately as *Georg Cantor*, B.G. Teubner, Leipzig 1930. Published in abridged form in Cantor [1932], 452–483.
[Fraenkel, 1953] A. A. Fraenkel. *Abstract Set Theory*, North-Holland, Amsterdam 1953.
[Frege, 1879] G. Frege. *Begriffsschrift, eine der arithmetischen nachgebildete Formelsprache des reinen Denkens*, Nebert, Halle 1879. Reprinted Hildesheim, Olms 1964. Translated in van Heijenoort [1967], 1–82.
[Frege, 1884] G. Frege. *Die Grundlagen der Arithmetik, eine logisch-mathematische Untersuchung über den Begriff der Zahl*, Wilhelm Köbner, Breslau 1884. Translated with German text by John L. Austin, as *The Foundations of Arithmetic, A logico-mathematical enquiry into the concept of number*; Blackwell, Oxford 1950. Later editions without German text, Harper, New York.
[Frege, 1891] G. Frege. Function und Begriff, Hermann Pohle, Jena 1891. Translated in [1952] below, 21–41.
[Frege, 1893] G. Frege. *Grundgesetze der Arithmetik, Begriffsschriftlich abgeleitet.* Vol. 1, Hermann Pohle, Jena 1893. Reprinted Hildesheim, Olms 1962. Partially translated in [1964] below.
[Frege, 1895] G. Frege. Kritische Beleuchtung einiger Punkte in E. Schröders *Vorlesungen über die Algebra der Logik, Archiv für systematische Philosophie* 1 (1895), 433–456. Translated in [1952] below, 86–106.
[Frege, 1903] G. Frege. Vol. 2 of [1893]. Reprinted Hildesheim, Olms 1962.
[Frege, 1952] G. Frege. P. Geach and M. Black, trans. and eds., *Translations from the Philosophical Writings of Gottlob Frege*, Blackwell, Oxford 1952. Second, revised edition 1960. Latest edition, Rowland & Littlewood, Totowa 1980.
[Frege, 1964] G. Frege. M. Furth, trans. and ed., *The Basic Laws of Arithmetic. Exposition of the System*, University of California Press, Berkeley 1964.
[Garciadiego, 1992] A. Garciadiego. *Bertrand Russell and the Origins of the Set-theoretic "Paradoxes"*, Birkhäuser, Boston 1992.
[Gentzen, 1936] G. Gentzen. Die Widerspruchsfreiheit der reinen Zahlentheorie, *Mathematicshe Annalen* 112 (1936), 493–565. Translated in [1969] below, 132–213.
[Gentzen, 1943] G. Gentzen. Beweisbarkeit und Unbeweisbarkeit von Anfangsfällen der transfiniten Induktion in der reinen Zahlentheorie, *Mathematicshe Annalen* 119 (1943), 140–161. Translated in [1969] below, 287–308.
[Gentzen, 1969] G. Gentzen. M. E. Szabo, ed., *The Collected Papers of Gerhard Gentzen*, North-Holland, Amsterdam 1969.
[Gessel and Rota, 1987] I. Gessel and G.-C. Rota, eds. *Classic Papers in Combinatorics*, Birkhäuser, Boston 1987.
[Gödel, 1930] K. F. Gödel. Die Vollständigkeit der Axiome des logischen Funktionenkalküls, *Monatshefte für Mathematik und Physik* 37 (1930), 349–360. Reprinted and translated in [1986] below, 102–123.
[Gödel, 1931] K. F. Gödel. Über formal unentscheidbare Sätze der *Principia Mathematica* und verwandter Systeme I, *Monatshefte für Mathematik und Physik* 38 (1931), 173–198. Reprinted and translated with minor emendations by the author in [1986] below, 144–195.
[Gödel, 1933] K. F. Gödel. The present situation in the foundations of mathematics, in [1995] below, 45–53, and the page references are to these.
[Gödel, 1938] K. F. Gödel. The consistency of the Axiom of Choice and of the Generalized Continuum Hypothesis, *Proceedings of the National Academy of Sciences U.S.A.* 24 (1938), 556–557. Reprinted in [1990] below, 26–27.
[Gödel, 1939] K. F. Gödel. Consistency-proof for the Generalized Continuum-Hypothesis, *Proceedings of the National Academy of Sciences U.S.A.* 25 (1939), 220–224. Reprinted in [1990] below, 28–32.
[Gödel, 1939a] K. F. Gödel. Vortrag Göttingen, text and translation in [1995] below, 126–155, and the page references are to these.

[Gödel, 1940] K. F. Gödel. *The Consistency of the Axiom of Choice and of the Generalized Continuum Hypothesis with the Axioms of Set Theory*, Annals of Mathematics Studies #3, Princeton University Press, Princeton 1940. Reprinted in [1990] below, 33–101.
[Gödel, 1947] K. F. Gödel. What is Cantor's Continuum Problem?, *American Mathematical Monthly* 54 (1947), 515–525. Errata 55 (1948), 151. Reprinted in [1990] below, 176–187. Revised and expanded version in Benacerraf-Putnam [1964], 258–273. This version reprinted with minor emendations by the author in [1990] below, 254–270.
[Gödel, 1986] K. F. Gödel. S. Feferman, et al., eds., *Kurt Gödel, Collected Works*, vol. I, Oxford University Press, Oxford 1986.
[Gödel, 1990] K. F. Gödel. S. Feferman, et al., eds., *Kurt Gödel, Collected Works*, vol. II, Oxford University Press, New York 1990.
[Gödel, 1995] K. F. Gödel. S. Feferman, et al., eds., *Kurt Gödel, Collected Works*, vol. III, Oxford University Press, New York 1995.
[Gödel, 2003] K. F. Gödel. S. Feferman and J. W. Dawson, eds., *Kurt Gödel, Collected Works*, vol. V, Clarendon Press, Oxford 2003.
[Goldfarb, 1979] W. D. Goldfarb. Logic in the Twenties: the nature of the quantifier, *The Journal of Symbolic Logic* 44 (1979), 351–368.
[Graham et al., 1990] R. L. Graham, B. L. Rothschild, and J. H. Spencer. *Ramsey Theory*, second edition, Wiley & Sons, New York 1990.
[Grattan-Guinness, 1978] I. Grattan-Guinness. Wiener on the logics of Russell and Schröder. An account of his doctoral thesis, and of his subsequent discussion of it with Russell, *Annals of Science* 32 (1978), 103–132.
[Grattan-Guinness, 1978a] I. Grattan-Guinness. How Bertrand Russell discovered his paradox, *Historia Mathematica* 5 (1978), 127–137.
[Grattan-Guinness, 2000] I. Grattan-Guinness. *The Search for Mathematical Roots 1870–1940: Logics, Set Theories and the Foundations of Mathematics from Cantor through Russell and Gödel*, Princeton University Press, Princeton 2000.
[Gray, 1994] R. Gray. Georg Cantor and Transcendental Numbers, *American Mathematical Monthly* 101 (1994), 819–832.
[Hájek, 1971] P. Hájek. Sets, semisets, models, in: Scott [1971], 67–81.
[Hajnal, 1956] A. Hajnal. On a consistency theorem connected with the generalized continuum problem, *Zeitschrift für Mathematische Logik und Grundlagen der Mathematik* 2 (1956), 131–136.
[Hajnal, 1961] A. Hajnal. On a consistency theorem connected with the generalized continuum problem, *Acta Mathematica Academiae Scientiarum Hungaricae* 12 (1961), 321–376.
[Hajnal and Larson, 2008] A. Hajnal and J. A. Larson. Partition relations, in: Foreman–Kanamori [2008].
[Hallett, 1984] M. Hallett. *Cantorian Set Theory and Limitation of Size*, Oxford Logic Guides #10, Clarendon Press, Oxford 1984.
[Hamel, 1905] G. Hamel. Eine Basis aller Zahlen und die unstetigen Lösungen der Funktional-gleichung: $f(x+y) = f(x) + f(y)$, *Mathematische Annalen* 60 (1905), 459–462.
[Hanf, 1964] W. P. Hanf. Incompactness in languages with infinitely long expressions, *Fundamenta Mathematicae* 53 (1964), 309–324.
[Hartogs, 1915] F. Hartogs. Über das Problem der Wohlordnung, *Mathematische Annalen* 76 (1915), 436–443.
[Hardy, 1903] G. H. Hardy. A theorem concerning the infinite cardinal numbers, *The Quarterly Journal of Pure and Applied Mathematics* 35 (1903), 87–94; reprinted in [1979] below, vol. 7, 427–434.
[Hardy, 1979] G. H. Hardy. I. W. Busbridge and R. A. Rankin, eds., *Collected Papers of G.H. Hardy*, Clarendon, Oxford 1979.
[Hausdorff, 1908] F. Hausdorff. Grundzüge einer Theorie der geordneten Mengen, *Mathematische Annalen* 76, 436–443, 1908.
[Hausdorff, 1914] F. Hausdorff. *Grundzüge der Mengenlehre*, de Gruyter, Leipzig 1914. Reprinted by Chelsea, New York 1965.
[Hausdorff, 1914a] F. Hausdorff. Bemerkung über den Inhalt von Punktmengen, *Mathematische Annalen* 75 (1914), 428–433.
[Hausdorff, 1916] F. Hausdorff. Die Mächtigkeit der Borelschen Mengen, *Mathematische Annalen* 77 (1916), 430–437, 1916.
[Hausdorff, 1932] F. Hausdorff. Zur Theorie der linearen metrischen Räume, *Journal für die reine und angewandte Mathematik* 167 (1932), 294–311.
[Hausdorff, 1937] F. Hausdorff. *Mengenlehre*, third, revised edition of [1914]. Translated by John R. Auman as *Set Theory*, Chelsea, New York 1962.
[Hawkins, 1975] T. W. Hawkins. *Lebesgue's Theory of Integration. Its Origins and Development*, second edition, Chelsea, New York 1975.

[Heck, 1995] R. G. Heck, Jr. Definition by induction in Frege's *Grundgesetze der Arithmetik*, in: William Demopoulos (editor) *Frege's Philosophy of Mathematics*, Harvard University Press, Cambridge 1995, 295–233.
[Henkin, 1949] L. Henkin. The completeness of the first-order functional calculus, *The Journal of Symbolic Logic* 14 (1949), 159-166.
[Hessenberg, 1906] G. Hessenberg. *Grundbegriffe der Mengenlehre,* Vandenhoeck & Ruprecht, Göttingen 1906. Reprinted in *Abhandlungen der Fries'schen Schule, Neue Folge* 1 (1906), 479–706.
[Hilbert, 1890] D. Hilbert. Über die Theorie der algebraischen Formen, *Mathematische Annalen* 36 (1890), 473-534. Translated in [1978] below, 143–224.
[Hilbert, 1900] D. Hilbert. Mathematische Probleme. Vortrag, gehalten auf dem internationalem Mathematiker-Kongress zu Paris. 1900. *Nachrichten von der Königlichen Gesellschaft der Wissenschaften zu Göttingen* (1900), 253–297. Translated in the *Bulletin of the American Mathematical Society* 8 (1902), 437–479.
[Hilbert, 1926] D. Hilbert. Über das Unendliche, *Mathematische Annalen* 95 (1926), 161–190. Translated in van Heijenoort [1967], 367–392. Partially translated in Benacerraf-Putnam [1983], 183–201.
[Hilbert, 1978] D. Hilbert. M. Ackerman, trans., *Hilbert's Invariant Theory Papers*, Math Sci Press, Brookline.
[Hilbert and Ackermann, 1928] D. Hilbert and W. Ackermann. *Grundzüge der theoretischen Logik*, Julius Springer, Berlin 1928. Second edition, 1938; third edition, 1949. Second edition was translated by Hammond, Lewis M., George G. Leckie, and F. Steinhardt as *Principles of Mathematical Logic*, Chelsea, New York 1950.
[Howard and Rubin, 1998] P. Howard and J. E. Rubin. *Consequences of the Axiom of Choice*, Mathematical Surveys and Monographs, vol. 59, American Mathematical Society, Providence, 1998.
[Jané, 1995] I. Jané. The role of the absolute infinite in Cantor's conception of set, *Erkenntnis* 42 (1995), 375–402.
[Jech, 1973] T. J. Jech. *The Axiom of Choice*, North-Holland, Amsterdam 1973.
[Jech, 2003] T. J. Jech. *Set Theory*, the third millennium edition, revised and expanded, Springer, Berlin 2003.
[Jourdain, 1904] P. E. B. Jourdain. On the transfinite cardinal numbers of well-ordered aggregates, *Philosophical Magazine* 7 (1904), 61–75.
[Jourdain, 1905] P. E. B. Jourdain. On a proof that every aggregate can be well-ordered, *Mathematische Annalen* 60, 465–470.
[Kac and Ulam, 1968] M. Kac and S. M. Ulam. *Mathematics and Logic*, Praeger, New York 1968.
[Kanamori, 1995] A. Kanamori. The emergence of descriptive set theory, in: Hintikka, Jaakko (editor) *Essays on the Development of the Foundations of Mathematics*, Kluwer, Dordrecht 1995, 241–266.
[Kanamori, 1997] A. Kanamori. The mathematical import of Zermelo's well-ordering theorem, *The Bulletin of Symbolic Logic* 3 (1997), 281–311.
[Kanamori, 2003] A. Kanamori. *The Higher Infinite. Large Cardinals in Set Theory from their Beginnings*, second edition, Springer, Berlin 2003.
[Kanamori, 2003a] A. Kanamori. The empty set, the singleton, and the ordered pair, *The Bulletin of Symbolic Logic* 9 (2003), 273–298.
[Kanamori, 2004] A. Kanamori. Zermelo and set theory, *The Bulletin of Symbolic Logic* 10 (2004), 487–553.
[Kanamori, 2007] A. Kanamori. Gödel and set theory, *The Bulletin of Symbolic Logic* 13 (2007), 153–188.
[Kechris and Louveau, 1987] A. S. Kechris and A. Louveau. *Descriptive Set Theory and the Structure of Sets of Uniqueness*, London Mathematical Society Lecture Note Series #128, Cambridge University Press, Cambridge 1987.
[Keisler, 1962] H. J. Keisler. Some applications of the theory of models to set theory, in: Nagel, Ernest, Patrick Suppes, and Alfred Tarski (editors), *Logic, Methodology and Philosophy of Science*, Proceedings of the 1960 [and first] International Congress (Stanford, California), Stanford University Press, Stanford 1962, 80–86.
[Keisler, 1962a] H. J. Keisler. The equivalence of certain problems in set theory with problems in the theory of models (abstract), *Notices of the American Mathematical Society* 9 (1962), 339-340.
[Keisler and Tarski, 1964] H. J. Keisler and A. Tarski. From accessible to inaccessible cardinals, *Fundamenta Mathematicae* 53 (1964), 225–308. Corrections 57 (1965), 119. Reprinted in Tarski [1986] vol. 4, 129–213.
[Kennedy, 1975] H. C. Kennedy. Nine letters from Giuseppe Peano to Bertrand Russell, *Journal of the History of Philosophy* 13 (1975), 205–220.
[Kőnig, 1927] D. Kőnig. Über eine Schlussweise aus dem Endlichen ins Unendliche: Punktmengen. Kartenfärben. Verwandtschaftsbeziehungen. Schachspiel. *Acta Litterarum ac Scientiarum Regiae Universitatis Hungaricae Francisco-Josephinae, sectio scientiarum mathematicarum* 3 (1927), 121–130.
[Kőnig, 1905] J. (G.) Kőnig. Zum Kontinuum-Problem, in: Krazer, A. (editor) *Verhandlungen des Dritten Internationalen Mathematiker-Kongresses in Heidelberg vom 8. bis 13. August 1904*. B.G. Teubner, Leipzig 1905, 144–147. Reprinted in *Mathematische Annalen* 60 (1905), 177–180 and Berichtigung, 462.

[Kunen and Vaughan, 1984] K. Kunen and J. E. Vaughan, eds. *Handbook of Set-Theoretic Topology*, North-Holland, Amsterdam 1984.

[Kuratowski, 1921] K. Kuratowski. Sur la notion de l'ordre dans la théorie des ensembles, *Fundamenta Mathematicae* 2 (1921), 161–171. Reprinted in [1988] below, 1–11.

[Kuratowski, 1922] K. Kuratowski. Une méthode d'élimination des nombres transfinis des raisonnements mathématiques, *Fundamenta Mathematicae* 3 (1922), 76–108.

[Kuratowski, 1931] K. Kuratowski. Evaluation de la class Borélienne ou projective d'un ensemble de points à l'aide des symboles logiques, *Fundamenta Mathematicae* 17 (1931), 249–272.

[Kuratowski, 1988] K. Kuratowski. K. Borsuk et al., eds., *Selected Papers*, Państwowe Wydawnictwo Naukowe, Warsaw 1988.

[Kuratowski and Tarski, 1931] K. Kuratowski and A. Tarski. Les opérations logiques et les ensembles projectifs, *Fundamenta Mathematicae* 17 (1931), 240–248. Reprinted in Tarski [1986] vol. 1, 551–559, and in Kuratowski [1988], 367–375. Translated in Tarski [1983], 143–151.

[Kurepa, 1935] Đ. R. Kurepa. *Ensembles ordonnés et ramifiés*, Thèse, Paris. Published as: *Publications mathématiques de l'Université de Belgrade* 4 (1935), 1-138.

[Kurepa, 1942] Đ. R. Kurepa. propos d'une généralisation de la notion d'ensembles bien ordonnés, *Acta Mathematica* 75 (1942), 139–150.

[Kurepa, 1959] Đ. R. Kurepa. On the cardinal number of ordered sets and of symmetrical structures in dependence on the cardinal numbers of its chains and antichains, *Glasnik Matematičko-fizički i astronomski, Periodicum mathematico-physicum et astronomicum* 14 (1959), 183–203.

[Lavine, 1994] S. M. Lavine. *Understanding the Infinite*, Harvard University Press, Cambridge 1994.

[Lebesgue, 1902] H. Lebesgue. Intégrale, longueur, aire, *Annali di Matematica Pura ed Applicata* (3)7 (1902), 231–359. Reprinted in [1972] below, vol. 1, 203–331.

[Lebesgue, 1905] H. Lebesgue. Sur les fonctions représentables analytiquement, *Journal de mathématiques pures et appliquées* (6)1 (1905), 139-216. Reprinted in [1972] below, vol. 3, 103–180.

[Lebesgue, 1972] H. Lebesgue. *Oeuvres Scientifiques*, Kundig, Geneva 1972.

[Levy, 1957] A. Levy. Indépendance conditionelle de $V = L$ et d'axiomes qui se rattachent au système de M. Gödel, *Comptes Rendus Hebdomadaires des Séances de l'Académie des Sciences, Paris* 245 (1957), 1582–1583.

[Levy, 1960] A. Levy. Axiom schemata of strong infinity in axiomatic set theory, *Pacific Journal of Mathematics* 10 (1960), 223–238.

[Levy, 1960a] A. Levy. Principles of reflection in axiomatic set theory, *Fundamenta Mathematicae* 49 (1960), 1–10.

[Levy, 1963] A. Levy. Independence results in set theory by Cohen's method. I, III, IV (abstracts), *Notices of the American Mathematical Society* 10 (1963), 592–593.

[Levy, 1964] A. Levy. Measurable cardinals and the continuum hypothesis (abstract), *Notices of the American Mathematical Society* 11 (1964), 769-770.

[Levy, 1965] A. Levy. Definability in axiomatic set theory I, in: Bar-Hillel, Yehoshua (editor), *Logic, Methodology and Philosophy of Science*, Proceedings of the 1964 International Congress, Jerusalem. North-Holland, Amsterdam 1965, 127–151.

[Levy, 1970] A. Levy. Definability in axiomatic set theory II, in: Bar-Hillel, Yehoshua (editor), *Mathematical Logic and Foundations of Set Theory*, North-Holland, Amsterdam 1970, 129–145.

[Levy and Solovay, 1967] A. Levy and R. M. Solovay. Measurable cardinals and the Continuum Hypothesis, *Israel Journal of Mathematics* 5 (1967), 234–248.

[Lindelöf, 1905] E. Lindelöf. Remarques sur un théorème fondamental de la théorie des ensembles, *Acta Mathematica* 29 (1905), 183-190.

[Lindenbaum and Tarski, 1926] A. Lindenbaum and A. Tarski. Communication sur les recherches de la théorie des ensembles, *Sprawozdana z Posiedzeń Towarzystwa Naukowego Warszawskiego, Wydział III, Nauk Matematyczno-fizycznych (Comptes Rendus des Séances de la Société des Sciences et des Lettres de Varsovie, Classe III, Sciences Mathématiques et Physiques)* 19 (1926), 299–330. Reprinted in Tarski [1986] vol. 1, 171–204.

[Liouville, 1844] J. Liouville. Des remarques relatives 1° à des classes très-étendues de quantités dont la valeur n'est ni rationelle ni même réductible à des irrationnelles algébriques; 2° à un passage du livre des *Principes* où Newton calcule l'action exercée par une sphère sur un point extérieur, *Comptes Rendus Hebdomadaires des Séances de l'Académie des Sciences, Paris* 18 (1844), 883–885.

[Liouville, 1851] J. Liouville. Sur des classes très-étendues de quantités dont la valeur n'est ni algébrique ni même réductible à des irrationnelles algébriques, *Journal de mathématiques pures et appliquées* 16 (1851), 133–142.

[Łoś, 1955] J. Łoś. Quelques remarques, théorèmes et problèmes sur les classes définissables d'algèbres, in: Skolem, Thoralf, *et al.* (editors), *Mathematical Interpretation of Formal Systems*, North-Holland, Amsterdam 1955, 98–113.

[Löwenheim, 1915] L. Löwenheim. Über Möglichkeiten im Relativkalkul, *Mathematische Annalen* 76 (1915), 447–470. Translated in van Heijenoort [1967], 228–251.

[Luzin, 1914] N. N. Luzin. Sur un problème de M. Baire, *Comptes Rendus Hebdomadaires des Séances de l'Académie des Sciences, Paris* 158 (1914), 1258–1261.

[Luzin, 1917] N. N. Luzin. Sur la classification de M. Baire, *Comptes Rendus Hebdomadaires des Séances de l'Académie des Sciences, Paris* 164 (1917), 91–94.

[Luzin, 1925] N. N. Luzin. Sur un problème de M. Emile Borel et les ensembles projectifs de M. Henri Lebesgue; les ensembles analytiques, *Comptes Rendus Hebdomadaires des Séances de l'Académie des Sciences, Paris* 180 (1925), 1318–1320.

[Luzin, 1925a] N. N. Luzin. Sur les ensembles projectifs de M. Henri Lebesgue, *Comptes Rendus Hebdomadaires des Séances de l'Académie des Sciences, Paris* 180 (1925), 1572–1574.

[Luzin, 1925b] N. N. Luzin. Les propriétés des ensembles projectifs, *Comptes Rendus Hebdomadaires des Séances de l'Académie des Sciences, Paris* 180 (1925), 1817–1819.

[Luzin and Sierpiński, 1918] N. N. Luzin and W. Sierpiński. Sur quelques propriétés des ensembles (A), *Bulletin de l'Académie des Sciences de Cracovie, Classe des Sciences Mathématiques et Naturelles, Série A* (1918), 35–48. Reprinted in Sierpiński [1975], 192–204.

[Luzin and Sierpiński, 1923] N. N. Luzin and W. Sierpiński. Sur un ensemble non mesurable B, *Journal de Mathématiques Pures et Appliquées* (9)2 (1923), 53-72. Reprinted in Sierpiński [1975], 504-519.

[Mahlo, 1911] P. Mahlo. Über lineare transfinite Mengen, *Berichte über die Verhandlungen der Königlich Sächsischen Gesellschaft der Wissenschaften zu Leipzig, Mathematische-Physische Klasse* 63 (1911), 187–225.

[Mahlo, 1912] P. Mahlo. Zur Theorie und Anwendung der ρ_0-Zahlen, *Berichte über die Verhandlungen der Königlich Sächsischen Gesellschaft der Wissenschaften zu Leipzig, Mathematische-Physische Klasse* 64 (1912), 108–112.

[Mahlo, 1913] P. Mahlo. Zur Theorie und Anwendung der ρ_0-Zahlen II, *Berichte über die Verhandlungen der Königlich Sächsischen Gesellschaft der Wissenschaften zu Leipzig, Mathematische-Physische Klasse* 65 (1913), 268–282.

[Mahlo, 1913a] P. Mahlo. Über Teilmengen des Kontinuums von dessen Mächtigkeit, *Berichte über die Verhandlungen der Königlich Sächsischen Gesellschaft der Wissenschaften zu Leipzig, Mathematische-Physische Klasse* 65 (1913), 283–315.

[Martin and Solovay, 1969] D. A. Martin and R. M. Solovay. A basis theorem for Σ_3^1 sets of reals, *Annals of Mathematics* 89 (1969), 138-159.

[Mathias, 2001] A. R. D. Mathias. Slim models of Zermelo set theory, *The Journal of Symbolic Logic* 66 (2001), 487–496.

[Meschkowski, 1983] H. Meschkowski. *Georg Cantor. Leben. Werk und Wirkung*, Bibliographisches Institut, Mannheim 1983.

[Meschkowski and Nilson, 1991] H. Meschkowski and W. Nilson. *Briefe / Georg Cantor*, Springer, Berlin 1991.

[Miller, 1984] A. W. Miller. Special subsets of the real line, in: Kunen-Vaughan [1984], 201–233.

[Mirimanoff, 1917] D. Mirimanoff. Les antinomies de Russell et de Burali-Forti et le problème fondamental de la théorie des ensembles, *L'Enseignment Mathématique* 19 (1917), 37–52.

[Mirimanoff, 1917a] D. Mirimanoff. Remarques sur la théorie des ensembles et les antinomies Cantoriennes. I, *L'Enseignment Mathématique* 19 (1917), 209–217.

[Montague, 1956] R. M. Montague. Zermelo-Fraenkel set theory is not a finite extension of Zermelo set theory (abstract), *Bulletin of the American Mathematical Society* 62 (1956), 260.

[Montague, 1961] R. M. Montague. Fraenkel's addition to the axioms of Zermelo, in: Bar-Hillel–Poznanski–Rabin–Robinson [1961], 91–114.

[Moore, 1982] G. H. Moore. *Zermelo's Axiom of Choice. Its Origins, Development and Influence*, Springer, New York 1982.

[Moore, 1988] G. H. Moore. The roots of Russell's paradox, *Russell* 8 (new series) (1988), 46–56.

[Moore, 1988a] G. H. Moore. The emergence of first-order logic, in: Aspray, William, and Philip Kitcher (editors), *History and Philosophy of Modern Mathematics*, Minnesota Studies in the Philosophy of Science, vol. 11, University of Minnesota Press, Minneapolis 1988, 95–135.

[Moore, 1989] G. H. Moore. Towards a history of Cantor's Continuum Problem, in: Rowe, David E., and John McCleary (editors), *The History of Modern Mathematics*, vol. 1: *Ideas and their Reception*, Academic Press, Boston 1989, 79–121.

[Moore, 2002] G. H. Moore. Hilbert on the infinite: The role of set theory in the evolution of Hilbert's thought, *Historia Mathematica* 29 (2002), 40–64.
[Moore and Garciadiego, 1981] G. H. Moore and A. R. Garciadiego. Burali-Forti's paradox: A reappraisal of its origins, *Historia Mathematica* 8 (1981), 319–350.
[Moschovakis, 1980] Y. Moschovakis. *Descriptive Set Theory*, North-Holland, Amsterdam 1980.
[Mostowski, 1939] A. M. Mostowski. Über die Unabhängigkeit des Wohlordnungssatzes vom Ordnungsprinzip, *Fundamenta Mathematicae* 32 (1939), 201–252. Translated in [1979] below, vol. 1, 290–338.
[Mostowski, 1949] A. M. Mostowski. An undecidable arithmetical statement, *Fundamenta Mathematicae* 36 (1949), 143–164. Reprinted in [1979] below, vol. 1, 531–552.
[Mostowski, 1950] A. M. Mostowski. Some impredicative definitions in the axiomatic set theory, *Fundamenta Mathematicae* 37 (1950), 111–124; correction in 38 (1952), 238.
[Mostowski, 1979] A. M. Mostowski. *Foundational Studies. Selected Works*, North-Holland, Amsterdam 1979.
[Myhill and Scott, 1971] J. R. Myhill and D. S. Scott. Ordinal definability, in: Scott [1971], 271–278.
[Nešetřil and Rödl, 1990] J. Nešetřil and V. Rödl. *Mathematics of Ramsey Theory*, Springer, Berlin 1990.
[Noether and Cavaillès, 1937] E. Noether and J. Cavaillès, eds. *Briefwechsel Cantor-Dedekind*, Hermann, Paris 1937. Translated into French in Cavaillès [1962].
[Novak, 1950] I. L. Novak. A construction for models of consistent systems, *Fundamenta Mathematicae* 37 (1950), 87–110.
[Oxtoby, 1971] J. C. Oxtoby. *Measure and Category. A Survey of the Analogies between Topological and Measure Spaces*, Springer, New York 1971.
[Parsons, 1977] C. Parsons. What is the iterative concept of set? in: Butts, Robert E., and Jaakko Hintikka (editors), *Logic, Foundations of Mathematics and Computability Theory*. Proceedings of the Fifth International Congress of Logic, Methodology, and the Philosophy of Science (London, Ontario 1975), The University of Western Ontario Series in Philosophy and Science, vol. 9; D. Reidel, Dordrecht 1977, 335–367. Reprinted in Benacerraf-Putnam [1983], 503–529.
[Peano, 1897] G. Peano. Studii di logica matematica, *Atti della Accademia delle scienze di Torino, Classe di scienze fisiche, matematiche e naturali* 32 (1897), 565–583. Reprinted in [1958] below, 201–217.
[Peano, 1905-8] G. Peano. *Formulario Mathematico*, Bocca, Torino 1905-8. Reprinted Edizioni Cremonese, Rome 1960.
[Peano, 1911] G. Peano. Sulla definizione di funzione. *Atti della Accademia nazionale dei Lincei, Rendiconti, Classe di scienze fisiche, matematiche e naturali* 20-I (1911), 3–5.
[Peano, 1913] G. Peano. Review of: A.N. Whitehead and B. Russell, *Principia Mathematica*, vols. I,II, *Bollettino di bibliografia e storia delle scienze matematiche Loria* 15 (1913), 47–53, 75–81. Reprinted in [1958] below, 389–401.
[Peano, 1958] G. Peano. *Opere Scelte*, vol. 2, Edizioni Cremonese, Rome 1958.
[Peckhaus, 1990] V. Peckhaus. "Ich habe mich wohl gehütet alle Patronen auf einmal zu verschiessen." Ernst Zermelo in Göttingen, *History and Philosophy of Logic* 11 (1990), 19–58.
[Peirce, 1883] C. S. Peirce. A theory of probable inference. Note B. The logic of relatives, *Studies in Logic by Members of the John Hopkins University*. Boston 1883, 187–203. Reprinted in: Hartshorne, Charles, and Paul Weiss (editors), *Collected Papers of Charles Sanders Peirce*, vol. 3; Harvard University Press, Cambridge 195–209.
[Pincus, 1972] D. Pincus. Zermelo-Fraenkel consistency results by Fraenkel-Mostowski methods, *The Journal of Symbolic Logic* 37 (1972), 721–743.
[Plotkin, 1993] J. M. Plotkin. Who put the "Back" in "Back-and-Forth?" in: Crossley, Remmel, Shore and Sweedler (editors), *Logical Methods, In Honor of Anil Nerode's Sixtieth Birthday*, Birkhäuser, Boston 1993.
[Plotkin, 2005] J. M. Plotkin. *Hausdorff on Ordered Sets*, American Mathematical Society, Providence 2005.
[Purket, 1989] W. Purkert. Cantor's views on the foundations of mathematics, in: Rowe, David E., and John McCleary (editors), *The History of Modern Mathematics*, vol. 1: *Ideas and their Reception*, Academic Press, Boston 1989, 49–65.
[Purket, 2002] W. Purket. Grundzüge der Mengenlehre – Historische Einführung, in: Felix Hausdorff, *Gesammelte Werke*, vol. 2, 1–89.
[Purket and Ilgauds, 1987] W. Purkert and H. J. Ilgauds. *Georg Cantor: 1845–1918*, Birkhäuser, Basel 1987.
[Quine, 1960] W. V. O. Quine. *Word and Object*, MIT Press, Cambridge 1960.
[Ramsey, 1930] F. P. Ramsey. On a problem of formal logic, *Proceedings of the London Mathematical Society* (2)30 (1930), 264–286. Reprinted in Gessel-Rota [1987], 2–24.
[Robinson, 1947] R. M. Robinson. On the decomposition of spheres, *Fundamenta Mathematicae* 34 (1947), 246–270.
[Robinson, 1951] A. Robinson. *On the Metamathematics of Algebra*, North-Holland, Amsterdam 1951.

[Rosser and Wang, 1950] J. B. Rosser and H. Wang. Non-standard models for formal logics, *The Journal of Symbolic Logic* 15 (1950), 113–129.
[Rothberger, 1938] F. Rothberger. Eine Äquivalenz zwischen der Kontinuumhypothese und der Existenz der Lusinschen und Sierpińskischen Mengen, *Fundamenta Mathematicae* 30 (1938), 215–217.
[Rothberger, 1939] F. Rothberger. Sur un ensemble toujours de première categorie qui est depourvu de la properté λ, *Fundamenta Mathematicae* 32 (1939), 294–300.
[Rothberger, 1948] F. Rothberger. On some problems of Hausdorff and Sierpiński, *Fundamenta Mathematicae* 35 (1948), 29–46.
[Rubin and Rubin, 1985] H. Rubin and J. E. Rubin. *Equivalents of the Axiom of Choice, II*. Amsterdam, North-Holland 1985. Revised, expanded version of their *Equivalents of the Axiom of Choice*, North-Holland, Amsterdam 1963.
[Russell, 1901] B. Russell. Sur la logique des relations avec des applications à la théorie des séries, *Revue de mathématiques (Rivista di matematica)* 7, 115–148; partly reprinted in Gregory H. Moore (editor), *The Collected Papers of Bertrand Russell*, vol. 3, Routledge, London 1993, 613–627; translated in same, 310–349.
[Russell, 1903] B. Russell. *The Principles of Mathematics*, Cambridge University Press, Cambridge 1903. Later editions, George Allen & Unwin, London.
[Russell, 1906] B. Russell. On some difficulties in the theory of transfinite numbers and order types, *Proceedings of the London Mathematical Society* (2)4 (1906), 29–53. Reprinted in [1973] below, 135–164.
[Russell, 1959] B. Russell. *My Philosophical Development*, George Allen & Unwin, London 1959.
[Russell, 1973] B. Russell. D. Lackey, ed., *Essays in Analysis*, George Braziller, New York 1973.
[Schröder, 1890] E. Schröder. *Vorlesungen über die Algebra der Logik (exakte Logik)*. Vol. 1. Leipzig, B.G. Teubner 1890. Reprinted in [1966] below.
[Schröder, 1895] E. Schröder. *Vorlesungen über die Algebra der Logik (exakte Logik)*. Vol. 3: *Algebra und Logik der Relative*. Leipzig, B.G. Teubner 1895. Reprinted in [1966] below.
[Schröder, 1966] E. Schröder. *Vorlesungen über die Algebra der Logik* (three volumes), Chelsea, New York 1966.
[Scott, 1961] D. S. Scott. Measurable cardinals and constructible sets, *Bulletin de l'Académie Polonaise des Sciences, Série des Sciences Mathématiques, Astronomiques et Physiques* 9 (1961), 521–524.
[Scott, 1971] D. S. Scott. *Axiomatic Set Theory*, Proceedings of Symposia in Pure Mathematics vol. 13, part 1, American Mathematical Society, Providence 1971.
[Scott, 1974] D. S. Scott. Axiomatizing set theory, in: Jech, Thomas J. (editor) *Axiomatic Set Theory*. Proceedings of Symposia in Pure Mathematics vol. 13, part 2, American Mathematical Society, Providence 1974, 207–214.
[Shelah, 1980] S. Shelah. Going to Canossa (abstract). *Abstracts of papers presented to the American Mathematical Society* 1 (1980), 630.
[Shelah, 1984] S. Shelah. Can you take Solovay's inaccessible away? *Israel Journal of Mathematics* 48 (1984), 1–47.
[Shepherdson, 1951] J. C. Shepherdson. Inner models for set theory — Part I, *The Journal of Symbolic Logic* 16 (1951), 161–190.
[Shepherdson, 1952] J. C. Shepherdson. Inner models for set theory — Part II, *The Journal of Symbolic Logic* 17 (1952), 225–237.
[Shepherdson, 1953] J. C. Shepherdson. Inner models for set theory — Part III, *The Journal of Symbolic Logic* 18 (1953), 145–167.
[Shoenfield, 1954] J. R. Shoenfield. A relative consistency proof, *The Journal of Symbolic Logic* 19 (1954), 21–28.
[Shoenfield, 1959] J. R. Shoenfield. On the independence of the axiom of constructibility, *American Journal of Mathematics* 81 (1959), 537–540.
[Shoenfield, 1967] J. R. Shoenfield. *Mathematical Logic*, Addison-Wesley, Reading 1967.
[Shoenfield, 1971] J. R. Shoenfield. Unramified forcing, in: Scott [1971], 357–381.
[Shoenfield, 1977] J. R. Shoenfield. Axioms of set theory, in: Barwise, K. Jon (editor), *Handbook of Mathematical Logic*, North-Holland, Amsterdam 1977, 321–344.
[Sierpiński, 1918] W. Sierpiński. L'axiome de M. Zermelo et son rôle dans la Théorie des Ensembles et l'Analyse, *Bulletin de l'Académie des Sciences de Cracovie, Classe des Sciences Mathématiques et Naturelles, Série A* (1918), 97–152. Reprinted in [1975] below, 208–255.
[Sierpiński, 1924] W. Sierpiński. Sur l'hypothèse du continu ($2^{\aleph_0} = \aleph_1$), *Fundamenta Mathematicae* 5 (1924), 177–187. Reprinted in [1975] below, 527–536.
[Sierpiński, 1925] W. Sierpiński. Sur une class d'ensembles, *Fundamenta Mathematicae* 7 (1925), 237–243. Reprinted in [1975] below, 571–576.

[Sierpiński, 1928] W. Sierpiński. Sur un ensemble non dénombrable, dont toute image continue est de mesure nulle, *Fundamenta Mathematicae* 11 (1928), 302–304. Reprinted in [1975] below, 702–704.

[Sierpiński, 1934] W. Sierpiński. *Hypothèse du Continu*, Monografie Matematyczne vol. 4, Warsaw 1934. Second, revised edtion, Chelsea, New York 1956.

[Sierpiński, 1950] W. Sierpiński. *Les ensembles projectifs et analytiques*, Mémorial des Sciences Mathématiques #112, Gauthier-Villars, Paris 1950.

[Sierpiński, 1975] W. Sierpiński. S. Hartman et al., eds., *Oeuvres Choisies*. vol. 2, Warsaw, Państwowe Wydawnictwo Naukowe 1975.

[Sierpiński, 1976] W. Sierpiński. S. Hartman et al., eds., *Ouevres Choisies*. vol. 3, Państwowe Wydawnictwo Naukowe, Warsaw 1976.

[Sierpiński and Tarski, 1930] W. Sierpiński and A. Tarski. Sur une propriété caractéristique des nombres inaccessibles, *Fundamenta Mathematicae* 15 (1930), 292–300. Reprinted in Sierpiński [1976], 29-35, and in Tarski [1986] vol. 1, 289–297.

[Skolem, 1920] T. Skolem. Logisch-kombinatorische Untersuchungen über die Erfüllbarkeit oder Beweisbarkeit mathematischer Sätze nebst einem Theoreme über dichte Mengen, *Videnskaps-selskapets Skrifter, I. Matematisk-Naturvidenskabelig Klass* (1920, #4), 1-36. Reprinted in [1970] below, 103–136. Partially translated in van Heijenoort [1967], 252–263.

[Skolem, 1923] T. Skolem. Einige Bemerkungen zur axiomatischen Begründung der Mengenlehre, in: *Matematikerkongressen i Helsingfors den 4–7 Juli 1922, Den femte skandinaviska matematikerkongressen, Redogörelse*, Akademiska-Bokhandeln, Helsinki 1923, 217-232. Reprinted in [1970] below, 137–152. Translated in van Heijenoort [1967], 290–301.

[Skolem, 1930] T. Skolem. Einige Bemerkungen zu der Abhandlung von E. Zermelo: "Über die Definitheit in der Axiomatik", *Fundamenta Mathematicae* 15 (1930), 337–341. Reprinted in [1970] below, 275–279.

[Skolem, 1933] T. Skolem. Ein kombinatorischer Satz mit Anwendung auf ein logisches Entscheidungsproblem, *Fundamenta Mathematicae* 20 (1933), 254–261. Reprinted in [1970] below, 337–344.

[Skolem, 1933a] T. Skolem. Über die Unmöglichkeit einer vollständigen Charakterisierung der Zahlenreihe mittels eines endlichen Axiomensystems, *Norsk Matematisk Forenings Skrifter* 2(#10) (1933), 73–82. Reprinted in [1970] below, 345–354.

[Skolem, 1934] T. Skolem. Über die Nicht-charakterisierbarkeit der Zahlenreihe mittels endlich oder abzählbar unendlich vieler Assagen mit ausschliesslich Zahlenvariablen, *Fundamenta Mathematicae* 23 (1934), 150–161. Reprinted in [1970] below, 355–366.

[Skolem, 1970] T. Skolem. J. E. Fenstad, ed., *Selected Works in Logic*, Univesitetsforlaget, Oslo 1970.

[Solovay, 1963] R. M. Solovay. Independence results in the theory of cardinals. I, II (abstracts), *Notices of the American Mathematical Society* 10 (1963), 595.

[Solovay, 1965] R. M. Solovay. $2^{\aleph_0}$ can be anything it ought to be (abstract), in: Addison-Henkin-Tarski [1965], 435.

[Solovay, 1965a] R. M. Solovay. Measurable cardinals and the continuum hypothesis (abstract), *Notices of the American Mathematical Society* 12 (1965), 132.

[Solovay, 1965b] R. M. Solovay. The measure problem (abstract), *Notices of the American Mathematical Society* 12 (1965), 217.

[Solovay, 1969] R. M. Solovay. The cardinality of Σ_2^1 sets of reals, in: Bulloff, Jack J., Thomas C. Holyoke, and Samuel W. Hahn (editors), *Foundations of Mathematics*. Symposium papers commemorating the sixtieth birthday of Kurt Gödel. Springer, Berlin 1969, 58–73.

[Solovay, 1970] R. M. Solovay. A model of set theory in which every set of reals is Lebesgue measurable, *Annals of Mathematics* 92 (1970), 1–56.

[Specker, 1957] E. Specker. Zur Axiomatik der Mengenlehre (Fundierungs- und Auswahlaxiom), *Zeitschrift für mathematische Logik und Grundlagen der Mathematik* 3 (1957), 173–210.

[Steinitz, 1910] E. Steinitz. Algebraische Theorie der Körper, *Journal für die reine und angewandte Mathematik* 137 (1910), 167–309.

[Suslin, 1917] M. Y. Suslin. Sur une définition des ensembles mesurables B sans nombres transfinis, *Comptes Rendus Hebdomadaires des Séances de l'Académie des Sciences, Paris* 164 (1917), 88–91.

[Suslin, 1920] M. Y. Suslin. Problème 3, *Fundamenta Mathematicae* 1 (1920), 223.

[Tarski, 1924] A. Tarski. Sur quelques théorèmes qui équivalent à l'axiome du choix, *Fundamenta Mathematicae* 5 (1924), 147–154. Reprinted in [1986] below, vol. 1, 41–48.

[Tarski, 1931] A. Tarski. Sur les ensembles définissables de nombres réels, *Fundamenta Mathematicae* 17 (1931), 210–239. Translated in Tarski [1983], 110–142.

[Tarski, 1933] A. Tarski. Pojęcie prawdy w językach nauk dedukcyjnych (The concept of truth in the languages of deductive sciences), *Prace Towarzystwa Naukowego Warszawskiego, Wydział III, Nauk Matematyczno-fizycznych (Travaux de la Société des Sciences et des Lettres de Varsovie, Classe III, Sciences Mathématiques et Physiques* #34 (1933). See also [1935] below.
[Tarski, 1935] A. Tarski. Der Wahrheitsbegriff in den formalisierten Sprachen (German translation of [1933] with a postscript), *Studia Philosophica* 1 (1935), 261–405. Reprinted in [1986] below, vol. 2, 51–198. Translated in [1983] below, 152–278.
[Tarski, 1951] A. Tarski. *A Decision Method for Elementary Algebra and Geometry* (prepared by J.C.C. McKinsey), University of California Press, Berkeley 1951. Second revised edition.
[Tarski, 1962] A. Tarski. Some problems and results relevant to the foundations of set theory, in: Nagel, Ernest, Patrick Suppes, and Alfred Tarski (editors). *Logic, Methodology and Philosophy of Science*. Proceedings of the 1960 [and first] International Congress (Stanford, California), Stanford University Press, Stanford 1962, 125–135. Reprinted in [1986] below, vol. 4, 115–125.
[Tarski, 1983] A. Tarski. *Logic, Semantics, Metamathematics. Papers from 1923 to 1938*. Translations by J.H. Woodger, second edition, Hackett, Indianapolis 1983.
[Tarski, 1986] A. Tarski. S. R. Givant and R. N. McKenzie, eds., *Collected Papers*, Basel, Birkhäuser 1986.
[Todorčević, 1984] S. Todorčević. Trees and linearly ordered sets, in: Kunen–Vaughan [1984], 235–293.
[Ulam, 1930] S. M. Ulam. Zur Masstheorie in der allgemeinen Mengenlehre, *Fundamenta Mathematicae* 16 (1930), 140–150. Reprinted in [1974] below, 9–19.
[Ulam, 1974] S. M. Ulam. W. A. Beyer, J. Mycielski, and G.-C. Rota, eds., *Sets, Numbers, and Universes. Selected Works*, MIT Press, Cambridge 1974.
[van Douwen, 1984] E. K. van Douwen. The integers and topology, in: Kunen–Vaughan [1984], 111-168.
[van Heijenoort, 1967] J. van Heijenoort, ed. *From Frege to Gödel: A Source Book in Mathematical Logic, 1879-1931*. Harvard University Press, Cambridge 1967.
[Vitali, 1905] G. Vitali. Sul problema della misura dei gruppi di punti di una retta, Tip. Gamberini e Parmeggiani, Bologna 1905.
[von Neumann, 1923] J. von Neumann. Zur Einführung der transfiniten Zahlen, *Acta Litterarum ac Scientiarum Regiae Universitatis Hungaricae Francisco-Josephinae, sectio scientiarum mathematicarum* 1 (1923), 199–208. Reprinted in [1961] below, 24–33. Translated in van Heijenoort [1967], 346–354.
[von Neumann, 1925] J. von Neumann. Eine Axiomatisierung der Mengenlehre, *Journal für die reine und angewandte Mathematik* 154 (1925), 219–240. Berichtigung 155, 128. Reprinted in [1961] below, 34–56. Translated in van Heijenoort [1967], 393–413.
[von Neumann, 1928] J. von Neumann. Über die Definition durch transfinite Induktion und verwandte Fragen der allgemeinen Mengenlehre, *Mathematische Annalen* 99 (1928), 373–391. Reprinted in [1961] below, 320–338.
[von Neumann, 1928a] J. von Neumann. Die Axiomatisierung der Mengenlehre, *Mathematische Zeitschrift* 27 (1928), 669–752. Reprinted in [1961] below, 339–422.
[von Neumann, 1929] J. von Neumann. Über eine Widerspruchsfreiheitsfrage in der axiomaticschen Mengenlehre, *Journal für die reine und angewandte Mathematik* 160 (1929), 227–241. Reprinted in [1961] below, 494–508.
[von Neumann, 1961] J. von Neumann. A. H. Taub, ed., *John von Neumann, Collected Works*, vol. 1, Pergamon Press, New York 1961.
[Vopěnka, 1962] P. Vopěnka. Construction of models of set theory by the method of ultraproducts (in Russian), *Zeitschrift für mathematische Logik und Grundlagen der Mathematik* 8 (1962), 293–304.
[Vopěnka, 1964] P. Vopěnka. The independence of the Continuum Hypothesis (in Russian), *Commentationes Mathematicae Universitatis Carolinae* 5 Supplement I (1964), 1-48. Translated in *American Mathematical Society Translations* 57 (1966), 85-112.
[Vopěnka, 1965] P. Vopěnka. Construction of a model for Gödel-Bernays set theory for which the class of natural numbers is a set of the model and a proper class in the theory, in: Addison-Henkin-Tarski [1965], 436–437.
[Vopěnka, 1967] P. Vopěnka. The general theory of ∇-models, *Commentationes Mathematicae Universitatis Carolinae* 8 (1967), 145-170.
[Wagon, 1985] S. Wagon. *The Banach-Tarski Paradox*, Encyclopedia of Mathematics and Its Applications, vol. 24, Cambridge University Press, Cambridge 1985. Paperback edition 1993.
[Wang, 1974] H. Wang. *From Mathematics to Philosophy*, Humanities Press, New York 1974.
[Wang, 1974a] H. Wang. The concept of set, in Wang [1974], 181–223. Reprinted in Benacerraf-Putnam [1983], 530–570.
[Whitehead and Russell, 1910] A. N. Whitehead and B. Russell. *Principia Mathematica*, vol. 1, Cambridge University Press, Cambridge 1910.

[Whitehead and Russell, 1912] A. N. Whitehead and B. Russell. *Principia Mathematica*, vol. 2, Cambridge University Press, Cambridge 1912.
[Whitehead and Russell, 1913] A. N. Whitehead and B. Russell. *Principia Mathematica*, vol. 3, Cambridge University Press, Cambridge 1913.
[Wiener, 1914] N. Wiener. A simplification of the logic of relations, *Proceedings of the Cambridge Philosophical Society* 17 (1914), 387–390. Reprinted in van Heijenoort [1967], 224–227.
[Wittgenstein, 1956] L. Wittgenstein. G. H. von Wright, R. Rhees, and G. E. M. Anscombe, eds., *Remarks on the Foundations of Mathematics,* Basil Blackwell, Oxford 1956. Second printing 1967.
[Zermelo, 1904] E. Zermelo. Beweis, dass jede Menge wohlgeordnet werden kann (Aus einem an Herrn Hilbert gerichteten Briefe), *Mathematische Annalen* 59 (1904), 514–516. Translated in van Heijenoort [1967], 139–141.
[Zermelo, 1908] E. Zermelo. Neuer Beweis für die Möglichkeit einer Wohlordnung, *Mathematische Annalen* 65 (1908), 107–128. Translated in van Heijenoort [1967], 183–198.
[Zermelo, 1913] E. Zermelo. Über eine Anwendung der Mengenlehre auf die Theorie des Schachspiels, in Hobson, Ernest W., and A.E.H. Love (editors), *Proceedings of the Fifth International Congress of Mathematicians,* Cambridge 1912, vol. 2, Cambridge University Press, Cambridge, 1913, 501–504.
[Zermelo, 1929] E. Zermelo. Über den Begriff der Definitheit in der Axiomatik, *Fundamenta Mathematicae* 14 (1929), 339–344.
[Zermelo, 1930] E. Zermelo. Über Grenzzahlen und Mengenbereiche: Neue Untersuchungen über die Grundlagen der Mengenlehre, *Fundamenta Mathematicae* 16 (1930), 29–47. Translated with commentary by Michael Hallett in Ewald [1996], 1208–1233.
[Zermelo, 1931] E. Zermelo. Über Stufen der Quantifikation und die Logik des Unendlichen, *Jahresbericht der Deutschen Mathematiker-Vereinigung* 41 (1931), 85–88.
[Zermelo, 1935] E. Zermelo. Grundlagen einer allgemeinen Theorie der mathematicschen Satzsysteme, *Fundamenta Mathematicae* 25 (1935), 136–146.
[Zorn, 1935] M. Zorn. A remark on method in transfinite algebra, *Bulletin of the American Mathematical Society* 41 (1935), 667–670.

ALTERNATIVE SET THEORIES

Peter Apostoli, Roland Hinnion, Akira Kanda and Thierry Libert

INTRODUCTION

Alternatives to what is nowadays understood as Set Theory remain objects of study in mathematical logic. This chapter is not intended to cover all the aspects of the subject. The aim was merely to give the reader an idea of some lines of research, those familiar to the authors of this essay. And the motivation for writing such an essay was precisely the existence of unforeseen relationships between works by different authors, with different perspectives and motivations. Alternative set theories are not as peculiar as they might seem to be.

This chapter is made of three parts that can be read independently. The first was primarily written by Th. Libert as an introduction to the subject, in connection with what is said in the other parts. The second, which is R. Hinnion's work, will survey a variety of set-theoretic systems mostly related to "Positive Set Theory"; and the third part written by P. Apostoli and A. Kanda will present in details their own work on "Rough Set Theory".

PART I

TOPOLOGICAL SOLUTIONS TO THE FREGEAN PROBLEM

1 THE NAÏVE NOTION OF SET

Set theory was created by Georg Cantor, so we start with the 'definition' of the *naïve* notion of set, as given in the final presentation of his lifework:

> «*A set is a collection into a whole of definite distinct objects of our intuition or of our thought. The objects are called the elements (members) of the set.*» [Translated from German.]

By 'into a whole' is meant the consideration of a set as an *entity*, an *abstract object*, which in turn can be collected to define other sets, etc. This *abstraction* step marks the birth of set theory as a mathematical discipline.

The *logical* formulation of the naïve notion of set, however, was first explicitly presented at the end of 19th century by one of the founders of modern symbolic logic, Gottlob Frege, in his attempt to derive number theory from logic. As widely known, the resulting formal system was proved to be inconsistent by Russell in 1902.

We shall commence by reviewing some basic features of Frege's theory in order to frame and motivate our investigations.

2 THE ABSTRACTION PROCESS

First of all, Frege's original predicate calculus is *second-order*. To simplify matters, let us say here that there are two types of variables ranging over mutually exclusive domains of discourse, one for *objects* $(u, v, \ldots)$, another for *concepts* $(P, Q, \ldots)$, where a *concept* P is defined to be any unary predicate $P(x)$ whose argument x ranges over *objects*.

Frege's system is characterized by a *type-lowering* correlation: with each concept P is associated an abstract object, the *extension* of the concept, which is now familiarly denoted by $\{x \mid P\}$, and is meant to be the collection of all objects x that *fall under* the concept P. This correspondence between concepts and objects is governed by the following principle, known as

Basic Law V:
$$\forall P \forall Q \,(\, \{x \mid P\} = \{x \mid Q\} \;\longleftrightarrow\; \forall u(P(u) \equiv Q(u))\,).$$

The *equality* symbol $=$ on the left-hand side is the identity between objects, which Frege takes as primitive. The right side is the *material equivalence* of concepts, where $\equiv$ is an abbreviation for 'having the same truth value', which is – unless otherwise mentioned – taken to be the material biconditional $\leftrightarrow$.

We shall call this objectification of concepts *abstraction*. It should be stressed that Frege internalizes this process in the language by explicitly making use of an *abstractor* $\{\cdot \mid -\}$ to *name* the extension of a concept.

3 SETS AND MEMBERSHIP

Those objects that are extensions of concepts are called *sets*. Frege then defines what it is for an object to be a *member* of a set: u is a member of v, now denoted by $u \in v$, if and only if u falls under some concept of which v is the extension, i.e., $\exists P(v = \{x \mid P\} \wedge P(u))$. Note incidentally that both second-order and the use of the abstractor are required for that definition, or for the one of the concept 'being a set', that is $Set(v) :\equiv \exists P(v = \{x \mid P\})$.

Given the definition of membership, an immediate consequence of Basic Law V is the

Law of Extensions:
$$\forall P\; \forall u(u \in \{x \mid P\} \equiv P(u))$$

from which by Existential Introduction follows the well-known

Principle of Naïve Comprehension:
$$\forall P \, \exists v \forall u (u \in v \equiv P(u)).$$

According to the Law of Extensions, '$\in$' may just be regarded as an allegory for *predication*, this latter being now a proper object of the language.

Another significant rule derivable from Basic Law V is the

Principle of Extensionality:
$$\forall v \forall w (Set(v) \wedge Set(w) \longrightarrow (\forall u (u \in v \equiv u \in w) \to v = w)).$$

Sets, thought of as collections, are thus completely determined by their members. By combining the Law of Extensions and the Principle of Extensionality, it is shown that any set v is at least the extension of the concept $P(x) :\equiv x \in v$, i.e., $\forall v (Set(v) \to v = \{x \mid x \in v\})$. Note that there is no presumption that all objects are sets. As our aim is merely to study pure and abstract set-theoretic systems, we shall however assume this from now on, that is to say, $\forall v \, Set(v)$.

4 FIRST-ORDER VERSIONS

Second-order logic and the use of an abstractor are by no means necessary to render an account of naïve set theory. First-order versions of Frege's calculus are obtained by taking $\in$ as primitive notion in the language, retaining the Principle of Extensionality, and restricting either the Law of Extensions or the Principle of Naïve Comprehension to concepts definable by first-order formulas (possibly with parameters).

In choosing the Law of Extensions the language is still assumed to be equipped with an abstractor, which yields what we call the

Abstraction Scheme:

For each formula $\varphi(x)$ of the language *with* abstractor,
$$\forall u (u \in \{x \mid \varphi\} \equiv \varphi(u)).$$

By the choice of the Principle of Naïve Comprehension, it is understood that the language is no longer equipped with an abstractor, which gives the

Comprehension Scheme:

For any formula $\varphi(x)$ of the language *without* abstractor,
$$\exists v \forall u (u \in v \equiv \varphi(u)).$$

First-order comprehension with extensionality is often presented as the *ideal* formalization of set theory. However that may be, it is inconsistent. Note that yet it was not pointless to insist here on the distinction between abstraction and comprehension as Part II will describe a consistent context where these clearly appear as two different ways of axiomatizing set theory.

5 RUSSELL'S PARADOX

Set Theory originated in Cantor's result showing that some infinities are definitely bigger than others. Paradoxically enough, it is precisely this rather positive result that resulted in the inconsistency of Frege's system, and so in the incoherence of naïve set theory.

In modern terms, Cantor proved that the domain $\mathscr{P}(U)$ of all 'subsets' of any given domain of discourse U cannot be put into one-to-one correspondence to U. But this clearly contradicted what the left-to-right direction of Basic Law V was asserting, at least in its original *second-order* formulation, identifying each concept with the 'subset' of all objects that fall under it.

Inspired by Cantor's diagonal argument, Russell finally presented an elementary proof of the incoherence of naïve set theory by pointing out that the mere existence of $\{x \mid x \notin x\}$ is simply and irrevocably devastating. Still more dramatically, thinking of membership as predication, as hinted above, one could reformulate the theory of concepts and extensions without even explicitly referring to the *mathematical* concept of set as collection. That Russell's paradox could be so formulated in terms of most basic *logical* concepts came as a shock.

6 SOLUTION ROUTES

If one believes in the soundness of logic as used in mathematics throughout the ages, then one must admit that some collections are not '*objectifiable*'. The decision as to which concepts to disqualify or disregard is as difficult as it is counter-intuitive. This is attested by the diversity of diagnoses and systems advocated. Roughly, the various proposals may be divided into two categories according to whether $\{x \mid x \in x\}$ is accepted as a set or not. This distinction is, of course, more emblematic than well-established.

The second category encompasses the so-called *type-theoretic* approaches, those involving *syntactical* criteria to select admissible concepts by prohibiting circularity in definitions. One famous system associated, namely Quine's New Foundations, is discussed in details in [Forster, 1995]. In this chapter we will rather be concerned with *type-free* approaches, and mainly with ones that belong to the first category. Within those systems admitting $\{x \mid x \in x\}$ as a set there is no alternative but to tamper with the use of $\neg$ or with the definition of $\equiv$. It is the former alternative that is explored herein and particularly in Part II where non-classical interpretations of $\neg$ are even considered. For a solution route in which it is the definition of $\equiv$ that is altered while classical negation is maintained, the reader is referred to [Aczel and Feferman, 1980].

We are not going to elaborate on the axiomatic aspect of the systems tackled in this part, but rather insist on their semantic characterization as unifying framework. As usual, the underlying set theory required for such considerations is tacitly assumed to be the Zermelo-Fraenkel set theory ZF (with choice and some large cardinal assumptions if necessary). In other words, in what follows, whenever we use the terms set, subsets, etc., it is in reference to their common use in mathematics. When we want to talk about sets as objects of study within some set-theoretic system (including ZF), we will rather use the term *abstract sets*.

7 FREGE STRUCTURE

According to Basic Law V, a set-theoretic universe U for Frege's *naïve* set theory appears as a solution to $U \simeq \mathscr{P}(U)$, where $\simeq$ is an abbreviation for 'there exists a bijection'. By Cantor's theorem, such a solution cannot exist.

What we call a *Frege structure* is a solution $\mathcal{U}$ to an equation $U \simeq \mathscr{P}_*(U)$, where $\mathscr{P}_*(U)$ is any given set of *distinguished* subsets of U. Note that by a solution $\mathcal{U}$ to such an equation we really mean a set U *together with* a bijection $f : U \longrightarrow \mathscr{P}_*(U)$. Naturally, with any Frege structure $\mathcal{U} \equiv \langle U; f \rangle$ is associated an abstract set-theoretic universe whose membership relation $\in_\mathcal{U}$ is defined by $u \in_\mathcal{U} v$ if and only if $u \in f(v)$, for any $u, v \in U$. Accordingly, we shall call $f`v = \{u \in U \mid u \in_\mathcal{U} v\}$ the *extension*[1] of v in $\mathcal{U}$, and say that a subset $A \subseteq U$ is *collectable* if it lies in the range of f, that is if $A \in \mathscr{P}_*(U)$. Notice that, as f is injective, the abstract set-theoretic structure thus defined is obviously *extensional*, being understood that the interpretation of $=$ in $\mathcal{U}$ is the identity on U.

Finding pertinent – from one set-theoretic point of view or another – solutions to reflexive equations $U \simeq \mathscr{P}_*(U)$ is what we call the *Fregean problem*. We can relate the existence of such pertinent solutions for some $\mathscr{P}_*(U) \subsetneq \mathscr{P}(U)$ to the emergence of various abstract set-theoretic systems, which can then be characterized by the nature of $\mathscr{P}_*(U)$ precisely. Let us start with a well-known example.

8 THE LIMITATION OF SIZE DOCTRINE

There are solutions to the equation $U \simeq \mathscr{P}_{<\omega}(U)$, where $\mathscr{P}_{<\omega}(U)$ is the set of *finite* subsets of U, and it is well known that such solutions yield typical models of ZF *without* infinity. The best example is provided by V_ω, the set of so-called hereditarily finite sets, which actually satisfies $V_\omega = \mathscr{P}_{<\omega}(V_\omega)$. Now, if one wants a model of infinity as well, this is still possible by invoking the existence of a strongly inaccessible cardinal κ, so that V_κ, the set of hereditarily κ-finite sets – i.e., of cardinality strictly less that κ –, which satisfies $V_\kappa = \mathscr{P}_{<\kappa}(V_\kappa)$, is now itself a model of ZF. Notice that the axioms of ZF are just formulated *ad hoc* to make possible the iterative construction of the V_α's, and thanks to the *axiom of foundation* the universe coincides with $\bigcup \{V_\alpha \mid \alpha \text{ ordinal}\}$, the so-called *cumulative hierarchy*.[2] Furthermore, the existence of these *canonical* models satisfying $U \simeq \mathscr{P}_{<\kappa}(U)$

---

[1] It is worth stressing the difference between Frege's definition of the *extension of a concept*, which is the corresponding abstract set as object, and the *extension of an abstract set* in a set-theoretic structure as defined here.

[2] If need be, we would remind the reader of the definition of the V_α's, α an ordinal: $V_{\beta+1} := \mathscr{P}(V_\beta)$, for any β, and $V_\lambda := \bigcup \{V_\gamma \mid \gamma < \lambda\}$, if λ is a limit ordinal.

clearly shows that ZF is just the theory of *hereditarily small* and *iterative* sets; it is the reason why the guiding principle of ZF for avoidance of the paradoxes is often referred to as the so-called *limitation of size doctrine*. Note that the iterative conception can be dropped: variants of ZF in which the axiom of foundation fails have been used for proving independence results – e.g., permutation models – and for modelling circular phenomena – e.g., anti-foundation axioms, as in [Aczel, 1988] & [Barwise and Moss, 1996]. On the other hand, it was shown in [Church, 1974] that there are also some extensions of ZF admitting a universal set, and so transgressing the principle of limitation of size. As we shall see, there are not only alternative proposals violating this latter but a variety of them based upon a very different principle. The underlying idea is the following.

9 ADDING STRUCTURE

A natural way of specifying a class of subsets of a given set, that is $\mathscr{P}_*(U)$, consists in adding some structure on it and then looking at particular subsets defined in terms of the underlying structure. For reasons that are going to be motivated, the structure we are interested in here is a *topology* and $\mathscr{P}_*(U)$ will be taken to be $\mathscr{P}_{op}(U)$, the set of *open* subsets, or $\mathscr{P}_{cl}(U)$, the set of *closed* ones. It is then fairly easy to concoct solutions to $U \simeq \mathscr{P}_*(U)$, indeed. In fact, one can even solve this equation when we further require the bijection to be an *homeomorphism*, which we indicate in the text by replacing $\simeq$ by $\cong$ – being understood that $\mathscr{P}_*(U)$ has then been equipped with some suitable topology derived from the one of U – and which is a natural requirement as we are now dealing with *structured* objects.

Interestingly, the existence of such topological solutions has shown to be intimately related with the consistency problem of various set theories, particularly those based on so-called *positive* abstraction/comprehension principles, i.e., special cases of the abstraction/comprehension scheme corresponding to certain *negation-free* formulas; these are precisely discussed in Part II. Of course, the absence of negation in formulas defining sets is attested in the models by the fact that the complement of an open (resp. closed) set is not open (resp. closed) in general. But there is exactly one situation in which this holds, namely when the topology is generated by a single equivalence relation, and this is treated in Part III.

For a more detailed and general introduction to *topological set theory* we refer the reader to [Libert and Esser, 2005], where many references on the subject can be found. We shall content ourselves here with explaining what might be the philosophical principle – if any – supporting this line of research. To do that, a somewhat heuristic presentation of what a topological space is will be helpful.

10 TOPOLOGY AND INDISCERNIBILITY

Formally, a topological space is a set U equipped with a *topology*, which can be defined in many ways, and which is actually meant to materialize some notion of *indiscernibility* on U. The indiscernibility comes into play precisely whenever one is looking at a point $x \in U$. Then, all that one is actually able to see is a 'spot', that is some subset N of U to which x belongs. This is commonly referred to as a *neighbourhood* of x. Particularly, the topology is *discrete* when one is able to perfectly see each point, i.e., $\{x\}$ is a neighbourhood of x, for any x; there is no indiscernibility in that case. But in general, in a topological space, points appear as spots, spots are *local observations*, and these can possibly be refined.

With this in mind, most of the basic topological notions – if not all – are easily and convincingly explainable. To illustrate this, we shall only focus here on the concept of open/closed subset. Let A be a subset of a topological space U, and let $x \in U$.

- We shall say that x is *necessarily* in A, and write '$x \in_\Box A$', if one can actually see x in A, i.e., if there is some neighbourhood N of x such that $N \subseteq A$. This could be rephrased by saying that '$x \in A$' is *observable*, or *affirmative*. The *interior* of A is then the collection of its *observable* members, that is $A^\Box := \{x \in U \mid x \in_\Box A\}$, and A is said to be *open* when $A^\Box = A$, i.e., $\forall x(x \in A \Leftrightarrow x \in A^\Box \Leftrightarrow x \in_\Box A)$ – in words, when the 'real' membership correspond to the $\Box$-membership.

- Dually, we shall say that x is *possibly* in A, and write '$x \in_\Diamond A$', if x is not necessarily in the complement of A, i.e., if for any neighbourhood N of x, $N \cap A \neq \emptyset$. This could be rephrased, for instance, by saying that '$x \in A$' is *not refutable*. The *closure* of A is the collection of its *possible* members, that is $A^\Diamond := \{x \in U \mid x \in_\Diamond A\}$, and A is said to be *closed* when $A^\Diamond = A$, i.e., $\forall x(x \in A \Leftrightarrow x \in A^\Diamond \Leftrightarrow x \in_\Diamond A)$ – so when the $\Diamond$-membership correspond to the 'real' membership.

Note that in view of this, if we informally think of $A \subseteq U$ as the *extension* of some property $\phi(x)$ regarding the elements of U, i.e., $A = \{x \in U \mid \phi(x)\}$, then open subsets would actually correspond to *observable* properties, or say *affirmative* assertions, which are those properties/assertions that are true precisely in the circumstances when they can be observed/affirmed; whereas closed subsets would correspond to *refutative* ones, those that are false precisely in the circumstances when they can be refuted. Naturally, an assertion is refutative if and only if its negation is affirmative.

11 INDISCERNIBILITY AS A LIGHTNING DISCHARGER (?)

Now, given there exist pertinent solutions to the Fregian problem when $\mathscr{P}_*(U)$ is taken to be $\mathscr{P}_{op}(U)$ or $\mathscr{P}_{cl}(U)$ for some *necessarily* non-discrete topology on U, it is tempting to argue that some form of indiscernibility was inherent in the naïve conception of set. As a matter of fact, in the set-theoretic structure corresponding to such a solution, it is really the indiscernibility associated with the topology that governs the collecting process: taking all its *observable* members in consideration, or respectively all its *possible* members, each subset of U is indeed collectable! And then the objectification of affirmative concepts, or respectively refutative ones, is guaranteed in such a set-theoretic structure.

But topologies can be very different, and so can be affirmative assertions or refutative ones. It has resulted in a diversity of 'topological' set-theoretic systems which have a corresponding variety of merits and defects. As mentioned, some of them will be presented or further explored in Part II and III. We would then let the reader judge the relevance of the different proposals therein.

Also, topologies are often related to modal considerations, as suggested by the notations and the terminology we adopted in the previous section. Accordingly, some of the set-theoretic systems considered might be revisited from a modal perspective. One example of such a move is given in Part III; and another one can be found in [Baltag, 1999], which is mainly a modal formulation of previous techniques and results related to 'hyperuniverses' — see Part II.

Part II

Partial, Paradoxical and Double Sets

12 INTRODUCTION

Many solutions to the well-known paradoxes of naïve set theory have been proposed. For mathematicians, the most convenient is some variant of the Zermelo-Fraenkel system (notation: ZF), in a rather pragmatic line: the axioms state the existence of the set of all natural numbers and further guarantee the possibility of those constructions precisely needed in mathematics! Should one find a "philosophical" principle behind this, it would be the limitation of size doctrine: the sets are those collections that are 'not too large'. This at once excludes from the field of studied objects as simply definable collections as the universe $V := \{x \mid x = x\}$, the filters of type $\{x \mid a \in x\}$, and, of course, the Russell set $\{x \mid \neg x \in x\}$, etc. Alternative set theories try to reincorporate these apparently dangerous objects, and for one as the Russell set this requires to modify the underlying logic or the concepts of extension/co-extension. We will only focus on the second option in this part, and treat theories concerning *partial* sets, *paradoxical* sets and *double* sets; we have also included *positive* sets as these, however classical w.r.t. the extension/co-extension concepts, are strongly linked to partial and paradoxical sets as we shall see. Note also that the borderline between the two above mentioned options is porous, since partial sets and paradoxical sets in classical logic may also be seen as naïve sets in respectively paracomplete and paraconsistent logics (see [Hinnion, 1994; Libert, 2004; Libert, 2005] for more references on the subject).

Actually — at least in our mind — alternative set theories are *not* intended to replace the usual ZF-like ones, but rather to extend them, so that in addition to the consistency problems, the possibility of 'containing ZF' is a main point (see [Hinnion, 2003]). We will try to clarify the main ideas, motivations and results, and invite the interested reader to find further information in the references. We treat the subjects in the following order: partial sets, positive sets, paradoxical sets, double sets. In all cases we work in classical logic with equality.

13 PARTIAL SETS

Linked to the idea of 'partial information', this line of research finds its source in Gilmore's pioneer work [Gilmore, 1974].

In classical logic, any set partitions the universe V into two parts: its *extension*, the collection of its members, and its *co-extension*, the collection of its non-members. A *partial set* will rather cut V into possibly three parts: we will only assume here that the extension and co-extension are disjoint. The remaining part will correspond to those objects for which the membership w.r.t. the set is (still) 'undetermined'. Also is there the idea that the information, being incomplete, is supposed to increase with time in such a way that both the extension and co-extension grow. That explains why the properties used to define partial sets will have to be 'positive', as those properties precisely stay true when information increases. A partial set, say $x = \{t \mid P(t)\}$, can then be seen as a 'double list': the first list contains those objects t for which we got the information that P is true, while the second list contains those t for which we got the information that P is false, which will be written $\overline{P}(t)$. Note that this is *not* the classical negation $\neg P(t)$; all we have is $\overline{P}(t) \rightarrow \neg P(t)$. Basically, this 'bar' operator will act as a non-classical negation, but will stay very close to the classical one, namely in its behavior w.r.t. the connectives $\vee, \wedge$, the quantifiers $\exists, \forall$, and the symmetry between extension and co-extension. To make all this more precise, we now discuss one variant of Gilmore's partial set theory that is representative and historically gave the impulse for further research on positive and paradoxical sets.

The language has as extra-logical symbols the binary relational ones $\in, \notin, =, \neq$, and also an abstractor $\{\cdot \mid -\}$. We insist on the fact that $\notin$ and $\neq$ are primitive symbols not corresponding to the classical negation of $\in$ and $=$; also is $=$ ruled classically. We will further use the letters $x, y, z, t, \ldots$ for variables; $\tau, \sigma, \ldots$ for terms; and $\varphi, \psi, \ldots$ for formulas. Positive formulas and terms are build up by the following rules:

(1) any variable is a positive term;

(2) if τ and σ are positive terms, then $\tau \in \sigma$, $\tau \notin \sigma$, $\tau = \sigma$, $\tau \neq \sigma$ are positive formulas;

(3) if φ and ψ are positive formulas, then so are $\varphi \vee \psi$, $\varphi \wedge \psi$, $\forall x \varphi$, $\exists x \varphi$;

(4) $\bot$ and $\top$ are positive formulas;[3]

---

[3] We conveniently add these false and true constant symbols in the language, with their obvious interpretation.

(5) if φ is a positive formula, then $\{x \mid \varphi\}$ is a positive term.

Only positive formulas will be used to construct partial sets, but naturally will we accept general (i.e., not necessarily positive) formulas in our language, and those are constructed via the extra-rule:

(6) if φ is a formula, then so is $\neg \varphi$.

The 'bar' operator for positive formulas is inductively defined as follows:

- $\overline{\tau \in \sigma}$ is $\tau \notin \sigma$,
- $\overline{\tau = \sigma}$ is $\tau \neq \sigma$,
- $\overline{\varphi \vee \psi}$ is $\overline{\varphi} \wedge \overline{\psi}$,
- $\overline{\exists x \varphi}$ is $\forall x \overline{\varphi}$,
- $\overline{\overline{\varphi}}$ is φ,
- $\overline{\bot}$ is $\top$.

Obviously we get also immediately: $\overline{\tau \notin \sigma}$ is $\tau \in \sigma$, $\overline{\tau \neq \sigma}$ is $\tau = \sigma$, $\overline{\forall x \varphi}$ is $\exists x \overline{\varphi}$, $\overline{\top}$ is $\bot$.

Finally, the axioms of our partial set theory are:

(i) The 'partial case' axioms:
$$\begin{cases} \neg \, (x \in y \wedge x \notin y) \\ \neg \, (x = y \wedge x \neq y) \end{cases}$$

Notice that these axioms imply $\neg(\varphi \wedge \overline{\varphi})$, for any positive formula φ.

(ii) The abstraction axioms:
$$\forall \vec{y} \, \forall z [\, (z \in \{x \mid \varphi(x,\vec{y})\} \leftrightarrow \varphi(z,\vec{y})) \wedge (z \notin \{x \mid \varphi(x,\vec{y})\} \leftrightarrow \overline{\varphi}(z,\vec{y})) \,]$$

for each *positive* formula φ with $x, \vec{y}$ as free variables.[4]

This expresses that the elements of the partial set $\{x \mid \varphi(x,\vec{y})\}$ are those objects x satisfying $\varphi(x,\vec{y})$, while the co-elements satisfy $\overline{\varphi}(x,\vec{y})$.

Gilmore showed that this theory has a pure term model (i.e., a model whose universe is made of all positive terms without free variables). But surprisingly this theory disproves the natural axiom of extensionality (ref. [Gilmore, 1974],[Hinnion, 1994]):
$$\forall t \, [(t \in x \leftrightarrow t \in y) \wedge (t \notin x \leftrightarrow t \notin y)] \to x = y,$$

which expresses that sets having the same extension and co-extension should be equal.

---
[4] Naturally, $\vec{y}$ stands for a possible list of parameters $y_1, y_2, \ldots, y_k$.

The lack of extensionality is a great weakness of the system. Gilmore himself, after some further attempts to improve the system, followed another path based on functions as primitive objects [Gilmore, 2001; Gilmore, 2005] and developed convincing arguments in favour of *intensionality* instead of extensionality, which one can surely understand from the 'partial information' point of view. Indeed, extensionality would allow to identify two partial sets on the basis of their respectively coinciding extensions and co-extensions, but this coincidence could just be incidental, and cease in the future! So intensional criteria for identification seem much more reasonable: these would identify terms $\{x \mid \varphi(x)\}$ and $\{x \mid \psi(x)\}$ only if they 'have the same meaning', i.e., if the formulas φ and ψ are 'sufficiently equivalent' (this, of course, has to be made precise; one can also imagine several degrees of equivalence). It should be noticed that this way of thinking supposes that the sets have a name indicating their meaning, i.e., that the sets are terms, and so that one imperatively expects pure term models. This path seems promising, and is at present a subject of research (see [Hinnion, 2007]). But let us now come back to the usual 'idealistic' set theoretical point of view.

It appears that the problem with extensionality has its source in the too rich language that is used. This language indeed allows to express positively many negative properties! For example, do we get easily from our theory that $\neg(\tau \in \tau)$ and $\neg(\tau \notin \tau)$ if τ is the Russell set, so that for any given positive formula $\varphi(x)$, the positive formula $\{x \mid \varphi(x)\} = \{x \mid \tau \in \tau\}$ is actually equivalent to $\neg(\exists x\, \varphi(x) \vee \exists x\, \overline{\varphi}(x))$, a rather negative one!

A possible solution could be to renounce to the abstractor, i.e., to look at this theory, but at the pure first-order level. So the language is like before, but without rule (5); the 'partial case' axioms are kept; and the abstraction axioms are re-formulated as *comprehension* axioms:

$$\forall \vec{y}\, \exists t\, \forall x\, [(x \in t \leftrightarrow \varphi(x, \vec{y})) \wedge (x \notin t \leftrightarrow \overline{\varphi}(x, \vec{y}))].$$

Since the eighties it was conjectured that this first-order partial set theory is consistent with extensionality, but rather surprisingly that is still an open problem. Even worse, the techniques that could be applied subsequently for positive sets and paradoxical sets simply do not work at all here, and *a fortiori* is the possibility of 'containing ZF' a complete mystery [Hinnion, 1994; Hinnion, 2003].

All this led to the exploration of classical first-order positive set theory as a simplification of the partial analogue. Before we treat that case, let us mention that some authors opted for another modification of the language, namely keeping the abstractor but suppressing the symbols $=$ and $\neq$ (as in [Brady, 1971], for instance). On that path, extensionality is to be formulated by:

$$x \doteqdot y \rightarrow x \doteq y,$$

where '$x \doteqdot y$' stands for $\forall t\, [(t \in x \leftrightarrow t \in y) \wedge (t \notin x \leftrightarrow t \notin y)]$ ('downwards indiscernibility'), and '$x \doteq y$' for $\forall z\, [(x \in z \leftrightarrow y \in z) \wedge (x \notin z \leftrightarrow y \notin z)]$ ('upwards indiscernibility'). Thanks to the existence of the filters $\{x \mid a \in x\}$, this extensionality principle is actually equivalent to $x \doteqdot y \leftrightarrow x \doteq y$, so that $\doteq$ plays perfectly the role of an equality (as equivalence with substitution). It was shown in [Brady, 1971] that the corresponding theory, with that extensionality principle, has a pure term model; and the same holds for the corresponding versions for positive sets (see [Hinnion and Libert, 2003]) and for paradoxical sets (see [Brady and Routley, 1989]).

14 POSITIVE SETS

Initially seen as a simplification of partial set theories, positive set theory quickly appeared as an interesting subject on its own. The consistency problems (with extensionality) stayed surprisingly unsolved until E. Weydert discovered somewhat incidentally an unpublished Ph.D. thesis by R.J. Malitz, where the problem was not completely solved but where the adequate new ideas appeared, namely, the use of topological ingredients. In that work [Malitz, 1976] the motivations were of philosophical order and completely different from Gilmore's ones.

To allow further discussion, let us present the simplest form of positive set theory. As usual, we adopt the classical first-order language of set theory with $\in$ and $=$ as sole non-logical symbols, including $\bot$ and $\top$ as logical constants for the false and the true. The so-called positive formulas are built up from $\bot, \top$, atomic formulas of type $x \in y, x = y$, connectives $\vee, \wedge$, and quantifiers $\exists, \forall$. Here, the axioms we consider are the following:

- extensionality:
  $\forall t(t \in x \leftrightarrow t \in y) \to x = y$

- positive comprehension:
  $\forall \vec{y} \exists z \forall x(x \in z \leftrightarrow \varphi(x, \vec{y}))$, for each positive formula $\varphi(x, \vec{y})$.

Obviously, this is a simplification of the partial case in the sense that one just forgets the abstractor, the relations $\notin$ and $\neq$, and the 'bar' operator. On the other hand, it can also be seen as locating the cause of the paradoxes in the presence of the negation $\neg$ in formulas defining sets. Thus, for the Russell set $\{x \mid \neg x \in x\}$, we impute the problem to $\neg$, and not, for instance, to the non-stratification of the formula $x \in x$ as Quine would do in his 'New Foundations'. As said, the consistency of this theory was only solved after Weydert's revelation of Malitz's work, but then led to an intensive exploration of the field, with several surprising results.

The constructed models (see [Weydert, 1989; Forti and Hinnion, 1989]) are in fact typical examples of topological set-theoretic structures discussed in Part I. They appear as compact uniform spaces homeomorphic to the set of their closed subsets. These structures, subsequently called '*hyperuniverses*', have been deeply investigated by M. Forti and F. Honsell (see [Forti and Honsell, 1996] for instance). Actually, they all model much more than the simple positive theory described above: *modulo* a large cardinal assumption in the metatheory ZFC, one can indeed produce extensional models for so-called *generalized* positive comprehension that also satisfy a relevant infinity axiom, so that the class of all hereditarily well-founded sets in these models can in turn interpret ZFC! O. Esser described and studied that first-order generalization of the simple positive theory given above (see [Esser, 1999; Esser, 2004]). This was called GPK_∞^+ for historical reasons, and its axioms are the following:

- Extensionality: as before.

- Comprehension for 'bounded positive formulas', where these are build up as positive formulas, but we may also use 'bounded quantification' of type $\forall x \in y$.

- The 'closure principle', stating that any class (i.e. definable collection) is included in a least set (naturally called the closure of that class), which can be expressed by the following first-order axiom scheme:

For any formula φ (so not necessarily bounded positive!),

$$\forall \vec{y}\, \exists x\, [\forall z(\varphi(z,\vec{y}) \to z \in x) \wedge \forall t((\forall z(\varphi(z,\vec{y}) \to z \in t)) \to x \subset t)].$$

In this, x is the so-called *closure* of the class $\{z \mid \varphi(z,\vec{y})\}$. (Note that the symbol '+' in GPK_∞^+ precisely refers to this closure principle.)

- The following axiom of infinity:

$$\exists x(x \neq \emptyset \wedge WF(x) \wedge \forall y \in x\, \{y\} \in x),$$

where $WF(x)$ expresses that x is a well-founded set, i.e.,

$$\forall y \ni x\, \exists y' \in y \quad y \cap y' = \emptyset.$$

Thus this axiom says that there exists an infinite well-founded set.

Furthermore, O. Esser proved (*inter alia*) that:

- GPK_∞^+ disproves the axiom of choice (this shows a rather unexpected similarity with Quine's New Foundations),

- in the theory GPK_∞^+ (so not just in the known models) the class of all hereditarily well-founded sets interprets ZF,

- GPK_∞^+ and a very natural extension of Kelley-Morse interpret each other; so that the interpretative power of GPK_∞^+ is exactly evaluated.

All this shows that GPK_∞^+ is an outstanding alternative set theory, as it satisfies all the expectations usually attached to that kind of theory [Hinnion, 2003].

To give some intuition to the reader, and without going in too much details and technical developments, we now describe a 'small' model for GPK^+ (so without the axiom of infinity).

Define inductively N_k, for any natural number k, as follows:
$$\begin{cases} N_0 = \{\emptyset\} \\ N_{k+1} = \mathcal{P} N_k \end{cases} \text{ where } \mathcal{P}x \text{ is the powerset of } x.$$

And then define, from the unique surjection $S_1 : N_1 \longrightarrow N_0$, the surjections $S_{k+1} : N_{k+1} \longrightarrow N_k$ by the following rule:
$$S_{k+1}(x) := \{S_k(y) \mid y \in x\}.$$

This yields a projective system:
$$N_0 \xleftarrow{S_1} N_1 \xleftarrow{S_2} N_2 \xleftarrow{S_3} N_3 \ldots$$

which has a limit:
$$N_\omega := \left\{ x \in \prod_{k \in \omega} N_k \;\middle|\; \forall k \in \omega \; S_{k+1}(x_{k+1}) = x_k \right\}$$

where ω is the set of all natural numbers and $\prod_{k \in \omega} N_k$ is the usual cartesian product of all N_k's. In other words, N_ω just selects those sequences $(x_0, x_1, x_2, \ldots)$ that satisfy $S_{k+1}(x_{k+1}) = x_k$.

Now equip N_ω with the binary relation $\in_\omega$ defined by:
$$x \in_\omega y \quad \text{iff} \quad \forall k \in \omega \; x_k \in y_{k+1}.$$

One can show that $(N_\omega, \in_\omega)$ is a model of GPK^+ [Hinnion, 1990].

We shall just give two examples of extraordinary sets that exist in N_ω. Consider the sequence $z := (\emptyset, \{\emptyset\}, \{\{\emptyset\}\}, \ldots)$. One can easily check that $z \in N_\omega$ and that $\forall x \in N_\omega \; x \in_\omega z \leftrightarrow x = z$, so that z is nothing but an auto-singleton in $(N_\omega, \in_\omega)$. Now consider $v := (N_0, N_1, N_2, \ldots)$. Then $v \in N_\omega$ and $\forall x \in N_\omega \; x \in_\omega v$, so that v is the universal set in $(N_\omega, \in_\omega)$.

This small model for GPK^+ will allow us to explain easier the problems for constructing the corresponding models for paradoxical sets, and so provides a good transition to the next section. But before leaving the positive sets, let us mention that versions with abstractor have also been studied and that the problems with extensionality are analogue there to those already met in the partial case (see [Hinnion and Libert, 2003; Hinnion, 2006]). For instance it is easy to see that, assuming extensionality, the term $\tau := \{x \mid \{t \mid x \in x\} = \{t \mid \bot\}\}$, though it is positive, is just a substitute for the Russell set. It should also be said that there exist natural topological models for positive abstraction too (see [Libert, 2008]).

15 PARADOXICAL SETS

It was soon noticed that the *paradoxical* set theory, with abstractor but without extensionality, obtained as the dual of the *partial* one described in Section 13 —

that is, just by keeping the abstraction axioms (ii), but replacing the 'partial case' axioms (i) by the dual 'paradoxical case' axioms (i)': $x \in y \vee x \notin y$ & $x = y \vee x \neq y$ — is equally consistent (see [Crabbé, 1992]).

Several variants have also been studied, among which 'Hyper Frege' appeared as the most powerful one. First only vaguely suggested in [Hinnion, 2003], it got a precise definition thanks to Th. Libert in [Libert, 2003], and could finally be modelled in its form with an axiom of infinity in [Esser, 2003]. A topological model for that theory, but without that axiom of infinity, was originally presented in [Hinnion, 1994] (see also [Libert, 2005] for another approach). Basically, Hyper Frege is the natural paraconsistent counterpart of the system GPK^+ described in Section 14. The language is first-order, with primitive symbols $\in$, $\notin$, $=$, and the axioms are the following.

(1) The 'paradoxical case' axioms:

$$x \in y \vee x \notin y \quad \& \quad x = y \vee x \neq y.$$

Note that if one wants a 'natural' $\neq$, it suffices to *define* it by

$$x \neq y \quad \text{iff} \quad \exists t(t \in x \wedge t \notin y) \vee \exists t(t \in y \wedge t \notin x),$$

and this will spontaneously satisfy $x = y \vee x \neq y$. But the axioms can perfectly be stated without worrying at all about a reasonable $\neq$.

(2) Extensionality: as for the partial case.

(3) Comprehension axioms for 'bounded positive formulas'. More precisely: For every pair φ, ψ of bounded positive formulas (i.e., build up from atomic formulas of type $\bot$, $\top$, $x \in y$, $x \notin y$, $x = y$, by means of $\vee$, $\wedge$, $\exists$, $\forall$, and bounded quantifications $\forall x \in y$, $\forall x \notin y$), one takes the axiom:

$$\forall x(\varphi \vee \psi) \rightarrow \exists y \, \forall x[(x \in y \leftrightarrow \varphi) \wedge (x \notin y \leftrightarrow \psi)].$$

Notice that this version is stronger than the more natural one that would only consider pairs φ, ψ, where ψ is $\overline{\varphi}$.

(4) The following 'closure principle' (in words): for every pair φ, ψ of formulas such that $\forall x(\varphi \vee \psi)$, there is a 'least paradoxical' set y such that $\forall x(\varphi \rightarrow x \in y)$ and $\forall x(\psi \rightarrow x \notin y)$; where '$y$ is less paradoxical than z' is defined by $\forall t(t \in y \rightarrow t \in z) \wedge \forall t(t \notin y \rightarrow t \notin z)$.

This system of axioms is denoted HF (for Hyper Frege). If one adds to this an adequate axiom of infinity — which we are not going to detail here but only mention that it asserts that there exists an infinite, classical, well-founded set — then one gets a stronger theory HF_∞ in which the class of all hereditarily *classical* well-founded sets interprets ZF, indeed! The original construction that allowed to model HF [Hinnion, 1994] is a projective limit very similar to the one briefly described in Section 14 for the 'small' model of GPK^+. This, however, presents several problems when worked out beyond ω. In fact, a different approach was necessary to overcome these problems and get a model of HF_∞ (see [Esser, 2003][5]).

16 DOUBLE SETS

The theories considered so far are all closely related to positive comprehension or abstraction. This is no longer the case for the 'double extension' set theories of this last section, which, however, surely justify their presence in this part because of the modification of the concept of extension itself, and also because of their surprising strength: the strongest and original versions were by far too strong as they are inconsistent, but the weakest versions that one can reasonably think (at present) to be consistent are still strong enough to interpret ZF.

Created by A. Kisielewicz [Kisielewicz, 1989], the *double extension set theory* got several variants and the most recent ones presented rather welcome simplifications (*inter alia* to be first-order). The situation stayed mysterious — w.r.t. the consistency problems and the interpretation of ZF — until R. Holmes found the highly non-trivial argument showing the inconsistency of the strongest forms [Holmes, 2004], as well as the relative interpretation of ZF in some of the weakest forms [Holmes, 2005]. We now briefly describe one of these theories. The language is first-order with equality, but presents the particularity of having two primitive membership relations: $\in$ and $\in'$. From a philosophical point of view, this suggests that any set would have two aspects, or (more concretely) two extensions. For usual sets, these should coincide, but for dangerous ones like Russell's, these extensions must be distinct!

Technically, this idea of a double extension allows to avoid the immediate paradox $R \in R \leftrightarrow \neg R \in R$, where R is the Russell set, by replacing one of these symbols $\in$ by the other $\in'$, so that one only gets $R \in R \leftrightarrow \neg R \in' R$. Those sets having a classical behavior w.r.t. the extension — i.e., those for which both extensions coincide — are called *regular*. Formally,

$$x \text{ is regular iff } \forall t(t \in x \leftrightarrow t \in' x).$$

The axiom of extensionality considered here is very particular one, as it mixes both extensions:

$$\forall z(z \in x \leftrightarrow z \in' y) \to x = y.$$

Another specific important notion we need is the following: we say that x is *partially contained* in y iff $\forall z(z \in x \to z \in y) \lor \forall z(z \in' x \to z \in' y)$ — so this corresponds to the usual inclusion, but for at least one of the two epsilons.

At last, what we call the *dual* φ^* of a formula φ is obtained by replacing any occurrence of $\in$ in φ by $\in'$ as well as any occurrence of $\in'$ by $\in$. And a formula

---

[5]The reader will find much more details in the references, but should be careful about the notations: some authors use $\in^+$ and $\in^-$ instead of $\in$ and $\notin$ respectively, etc.

is called *uniform* if it contains no instance of $\in'$. Now, the comprehension-scheme of the double set theory considered here can be expressed as follows:
For any *uniform* formula $\varphi(x, \vec{z})$, if each z_i is partially contained in some regular set, then
$$\exists y\, \forall x[x \in' y \leftrightarrow \varphi(x,\vec{z})) \wedge (x \in y \leftrightarrow \varphi^\star(x,\vec{z}))].$$
Naturally, we refer to this set y as $\{x \mid \varphi(x,\vec{z})\}$. Notice the condition on the parameters $\vec{z} = z_1, z_2, \ldots, z_k$.

To get some familiarity with this system, let us just have a look at the Russell set in this context. Consider $R = \{x \mid \neg x \in x\}$. Then comprehension just yields: $R \in' R \leftrightarrow \neg R \in R$ and $R \in R \leftrightarrow \neg R \in' R$, so that R belongs to R in one sense but not in the other, and this is not a contradiction.

In this theory, one can prove a lot of very surprising results, as the following ones:

- with a carefully adapted notion of $\in$-ordinal and $\in'$-ordinal, one can create two classes of ordinals and prove that exactly one of these two classes has only hereditarily regular elements; we will not detail here what this means exactly, but roughly speaking it guarantees that such ordinals have the usual expected behavior of von Neumann ordinals;

- precisely this allows then to prove the existence of an infinite ordinal of that type, so that one gets a relevant axiom of infinity. Note that this is very extraordinary, as such an axiom has usually to be explicitly added because it is not deductible from the others (e.g., for ZF, GPK^+,...). Furthermore, the theory is purely *syntactic* — the axioms contain no mathematical essences — and so the situation is somewhat analogue to the one of Quine's New Foundations, which is also a purely syntactic theory proving an axiom of infinity [Specker, 1953];

- one can then also construct the von Neumann hierarchy based on the 'good' ordinals, and finally get a suitable class of hereditarily well-founded regular sets, which is shown to interpret ZF ([Holmes, 2005]).

So double extension set theory really appears as a fascinating axiomatic theory. Naturally, the main open problem still remains its consistency, like for Quine's New Foundation...

PART III

PROXIMITY SPACES OF EXACT SETS

17 INTRODUCTION

Alternatives to first-order set theory may depart from ZFC in base logic, identity theory or extra-logical principles. This chapter contains a survey of a variety of alternatives to standard set theory, all of which are united in maintaining classical, first-order logic.

Theories that modify the base logic of set theory might be considered the most "radical" departures from standard set theory. Examples include natural-deduction-based set theories (i.e. those based upon the sequent calculus presentation of partial first-order logic), as well as linear and affine set theory (which are based upon substructural logics).

Less radical departures — e.g. topological approaches such as those of parts I and II of this chapter — uphold classical logic while rejecting certain identity-theoretic principles of ZFC, such as the identity of indiscernibles. Other less radical departures include the partial, positive, paradoxical and double set theories presented in section II. These theories modify the set existence principles of ZFC and raise the prospect of enriching the universe of standard set theory with "new" sets otherwise banned under the doctrine of the limitation of size.

The development of positive set theory has lead recently to a re-evaluation of Gilmore's classical theory of partial sets. The fact that partial sets can be studied classically is an important plank in our (conservative) proposal to delimit our survey so as to exclude set theories based upon partial logic. However, substructural set theory and natural-deduction-based set theory are prima facie viable alternative foundations for mathematics precisely since they have simple cut elimination consistency proofs. Alternative set theories which seek to uphold classical logic need to match these results to be serious contenders for the foundations of mathematics. Accordingly, the topological approaches to set theory presented in this part admit of transparent semantic consistency proofs in the form of a concretely presented canonical model.

18 TOWARDS MODAL SET THEORY

Kripkean semantics for modal logic [Kripke, 1963] extends point set theory with modal operators induced by a binary "accessibility" relation on a universe of points. Abstract set theory [Cantor, 1962] — which studies sets of *sets*, more generally than set of points or families of sets of points — also extends the theory of point sets, with a type-lowering correspondence between a universe and its power set under which concepts (subsets of the universe) are comprehended as sets (elements of the universe). Since the rapid development of modal logic [Chellas, 1980] in the 1960's, philosophers have sought a unification of the concepts of modal logic with those of abstract set theory. Typically, e.g., [Fine, 1981; Parsons, 1977; Parsons, 1981], this is attempted by basing axiomatic set theory upon modal quantifier logic instead of standard first order logic. These approaches regard axiomatic set theory to be an unproblematic starting point for the investigation of modal set theory and the extension of the language of set theory by modal operators as analogous to the extension of quantifier logic by modal operators.

However, one limitation of this approach stems from the thorny fact that the consistency of axiomatic set theory is still an open mathematical question. What if modal notions *underlie* set theoretic comprehension? In that case, the difficulty in finding a model for Zermelo and Fraenkel's axioms [Fraenkel, 1921; Fraenkel, 1922; Fraenkel and Bar-Hillel, 1958] is naturally to be expected. [Apostoli and Kanda, forthcoming] explored this question and proposed an alternative marriage of modal logic and abstract set theory based upon Rough Set Theory [Orlowska, 1985; Pawlak, 1982], an extension of the theory of point sets obtained by defining interior and closure operators over subsets of a universe U of points, typically those of the partition topology associated with an equivalence relation on U.

By placing an approximation space $(U, \equiv)$ in a type-lowering retraction with its power set 2^U, [Apostoli and Kanda, forthcoming] showed that a concept forms a set just in case it is $\equiv$-exact. Set-theoretic comprehension in $(U, \equiv)$ is thus governed by the method of upper and lower approximations of RST. Modal concepts indeed underlie abstract set theory, raising serious questions regarding the philosophical motivation for the standard approaches to "modal set theory". The naïve extention of the language of axiomatic set theory to modal quantifier logic ignores the conceptual priority of modality in abstract set theory.

This paper is organized as follows. Section one introduces the notion of a proximity (or tolerance) space and its associated ortho-lattice of parts, providing some motivating examples from, e.g., mathematics and physics. Then, generalizing the developments of [Apostoli and Kanda, 2000], section two introduces axiomatically the general notion of a Proximal Frege Structure and its associated modal ortho-latice of exact sets. Model constructions [Apostoli and Kanda, forthcoming] ensuring the consistency of these notions are then summarized. Some key properties of these models which are independent of the basic axioms of PFS are discussed and an open question regarding the tolerance relation of "matching" is raised. The paper concludes by airing the task of axiomatizing abstract set theory as formalizations of the general notion of a PFS.

19 PROXIMITY STRUCTURES

Let $U \neq \emptyset$ and $\sim \subseteq U \times U$ be a tolerance (reflexive, symmetric) relation on U. The pair $(U, \sim)$ is called an *proximity structure*. When in addition $\sim$ is an equivalence relation, $(U, \sim)$ is called an *approximation structure*.[6] For each point $u \in U$, let $[u]_\sim$ denote the class of successors of u under $\sim$, i.e.,

$$[u]_\sim =_{df} \{x \in U \mid u \sim x\}.$$

$\sim$-classes $[u]_\sim$ are called ($\sim$-) *granules*, or *elementary* subsets, of U. Let $A \subseteq U$ and

$$Int_\sim(A) =_{df} \bigcup\{[u]_\sim \mid [u]_\sim \subseteq A\},$$
$$Cl_\sim(A) =_{df} \bigcup\{[u]_\sim \mid [u]_\sim \cap A \neq \emptyset\}.$$

---

[6] As indicated in the above Introduction, the symbol "$\equiv$" is often used to denote tolerance relations which are also equivalence relations.

Then $Int_\sim(A)$ and $Cl_\sim(A)$ are called the *lower* and *upper approximations* of A, respectively (in contexts where $\sim$ is given, the subscripted "$\sim$" is usually suppressed). A is called $\sim$-*exact* iff it is the union of a family of $\sim$-granules, i.e., iff

$$A = \bigcup_{u \in X} [u]_\sim$$

for some $X \subseteq U$. Note that if $\sim$ is an equivalence relation, then A is $\sim$-exact iff $Cl(A) = A = Int(A)$. It is natural to regard $\sim$-exact subsets of U as the *parts* of U and elementary subsets as the *atomic parts* of U. $\mathcal{C}(\sim)$ denotes the family of $\sim$-exact subsets of U. Then $(U, \mathcal{C}(\sim))$ is called a *proximity space*.[7] When $\sim$ is an equivalence relation, $(U, \mathcal{C}(\sim))$ is called an *approximation space*. The reason for using the term "proximity", here is, as we shall see, it is helpful to think of $x \sim y$ as meaning "x is near y".

Let $S = (U, \mathcal{C}(\sim))$ be a proximity space and $A, B \subseteq U$. Following [Bell, 1986], define

$$A \bigvee_S B =_{df} A \cup B,$$
$$A \bigwedge_S B =_{df} Int(A \cap B),$$
$$A^c =_{df} Cl(U - A).$$

I.e., the *join* of A and B is their set theoretic union, their *meet* is the *interior* of their intersection and the *complement* A^c of A is the exterior of $U - A$. Then

(1) $(\mathcal{C}(\sim), \bigvee_S, \bigwedge_S, {}^c, \emptyset, U)$

is a complete ortholattice [Bell, 1983; Bell, 1986; Birkhoff, 1960] of exact subsets. That is, for any $A, B \in \mathcal{C}(\sim)$,

1. $(A^c)^c = A$,

2. $A \bigvee_S A^c = U$,

3. $A \bigwedge_S A^c = \emptyset$,

4. $A \subseteq B \Rightarrow B^c \subseteq A^c$.

Any discrete space is a proximity space in which $\sim$ is the identity relation. More generally, a proximity space S is a topological space if and only if its proximity relation is transitive, and in that case S is almost (quasi) discrete in the sense that its lattice of parts is isomorphic to the lattice of parts of a discrete space.

Proximity spaces admit of several interpretations which serve to reveal their significance. Quoting directly from [Bell, 1986]:

---

[7]Proximity structures and spaces, also known as *tolerance approximation spaces*, *generalized approximation spaces* or *parameterized approximation spaces*, are studied in [Skowron and Stepaniuk, 1996; Skowron and Stepaniuk, 1994].

(a) S may be viewed as a space or field of perception, its points as locations in it, the relation $\sim$ as representing *the indiscernibility of locations*, the quantum at a given location as the *minimum perceptibilium* at that location, and the parts of S as the perceptibly specifiable subregions of S. This idea is best illustrated by assigning the set U a metric δ, choosing a fixed $\varepsilon > 0$ and then defining $x \sim y \Leftrightarrow \delta(x,y) \leq \varepsilon$.

(b) S may be thought of as the set of *outcomes of an experiment* and $\sim$ as the relation of equality *up to the limits of experimental error*. The quantum at an outcome is then "the outcome within a specified margin of error" of experimental practice.

(c) S may be taken to be the set of states of a quantum system and $s \sim t$ as the relation: "a measurement of the system in a state s has a non-zero probability of leaving the system in state t, or *vice versa*." More precisely, we take a Hilbert space H, put $S = H - \{0\}$, and define the proximity relation $\sim$ on S by $s \sim t \Leftrightarrow \langle s,t \rangle \neq 0$ (s is not orthogonal to t). It is then readily shown that the lattice of parts of S is isomorphic to the ortholattice of closed subspaces of H. Consequently, *[complemented] lattices of parts of proximity spaces include the [complemented] lattices of closed subspaces of Hilbert spaces* — the lattices associated with Birkhoff and von Neumann's "quantum logic".

(d) S may be taken to be the set of *hyperreal numbers* in a model of Robinson's nonstandard analysis (see, e.g., Bell and Machover [Bell and Machover, 1977]) and $\sim$ is the relation of infinitesimal nearness. In this case $\sim$ is *transitive*.

(e) S may be taken to be the *affine line* in a model of synthetic differential geometry (see Kock [Kock, 1981]). In this case there exist many square zero infinitesimals in S, i.e., elements $\varepsilon \neq 0$ such that $\varepsilon^2 = 0$, and we take $x \sim y$ to mean that the difference $x - y$ is such an infinitesimal, i.e., $(x-y)^2 = 0$. Unlike the situation in (d), the relation $\sim$ here is *not* generally transitive.

20 PROXIMAL FREGE STRUCTURES

According to the principle of comprehension in set theory, every "admissible" concept forms an element of the universe called a "set". Frege [Frege, 1884; Frege, 1903] represented this principle by postulating the existence of an "extension function" assigning objects to concepts. Models of set theory which establish a type-lowering correspondence between a universe and its power set are thus called "Frege structures" [Aczel, 1980; Bell, 2000]. [Apostoli and Kanda, 2000; Apostoli and Kanda, forthcoming] considered the idea of basing a Frege structure upon an approximation space so that the admissible concepts are precisely the exact subsets of the universe. This section generalizes the development of the resulting "Proximal Frege Structure" to arbitrary tolerance relations. Most of the results of [Apostoli and Kanda, 2000] hold in this more general setting and so are not given special mention.

Let $(U, \sim)$ be an proximity structure and $\ulcorner \cdot \urcorner : 2^U \to U$, $\llcorner \cdot \lrcorner : U \to 2^U$ be functions, called *down* and *up* (for type-lowering and type-raising), respectively. Assume further that:

1. $(\ulcorner \cdot \urcorner, \llcorner \cdot \lrcorner)$ is a retraction pair, i.e., $\ulcorner \llcorner u \lrcorner \urcorner = u$ (i.e., $\ulcorner \cdot \urcorner \circ \llcorner \cdot \lrcorner = 1_U$); thus $\ulcorner \cdot \urcorner$ is a retraction and $\llcorner \cdot \lrcorner$ is the adjoining section.

2. The operator $\llcorner \cdot \lrcorner \circ \ulcorner \cdot \urcorner$ is the operator $Cl_\sim$ over 2^U. This is that for every $X \subseteq U$, $\llcorner \ulcorner X \urcorner \lrcorner$ is $\sim$-exact and
$$\llcorner \ulcorner X \urcorner \lrcorner = Cl(X).$$

3. The $\sim$-exact subsets of U are precisely the $X \subseteq U$ for which $\llcorner \ulcorner X \urcorner \lrcorner = X$. They are fixed-point of the operator $\llcorner \cdot \lrcorner \circ \ulcorner \cdot \urcorner$.

Then $\mathfrak{F} = (U, \sim, \ulcorner \cdot \urcorner, \llcorner \cdot \lrcorner)$ is called a *(generalized) PFS*. Elements of U are called $\mathfrak{F}$-sets.

The family $\mathcal{C}(\sim)$ of $\sim$-exact subsets of U is precisely the image of U under $\llcorner \cdot \lrcorner$. In algebraic terms $\mathcal{C}(\sim)$ is the kernel of the retraction mapping. Further we have the isomorphism $\mathcal{C}(\sim) \approx U$ given by:

$$i : \mathcal{C}(\sim) \to U : X \mapsto \ulcorner X \urcorner, \quad j : U \to \mathcal{C}(\sim) : u \mapsto \llcorner u \lrcorner.$$

In summary: $\mathcal{C}(\sim) \approx U \triangleleft 2^U$, where $U \triangleleft 2^U$ asserts the existence of a retraction pair holding between 2^U and U.

As a simple example of a PFS, we offer the following two point structure
$$(U, \sim, \ulcorner \cdot \urcorner, \llcorner \cdot \lrcorner),$$
where $U = \{0, 1\}$, $\sim = U \times U$, $\ulcorner \emptyset \urcorner = 0$, $\ulcorner X \urcorner = 1$ ($X \subseteq U, X \neq \emptyset$), $\llcorner 0 \lrcorner = \emptyset$ and $\llcorner 1 \lrcorner = U$. A less trivial example [Apostoli and Kanda, forthcoming] of a PFS based upon an equivalence relation, $\mathfrak{G}$, is described in the sequel.

Let $\mathfrak{F} = (U, \sim, \ulcorner \cdot \urcorner, \llcorner \cdot \lrcorner)$ be a generalized PFS. Writing "$u_1 \in_\mathfrak{F} u_2$" for "$u_1 \in \llcorner u_2 \lrcorner$", U is thus interpreted [Apostoli and Kanda, 2000] as a universe of $\mathfrak{F}$-sets; $\llcorner \cdot \lrcorner$ supports the relation of set membership holding between $\mathfrak{F}$-sets (elements of U). Writing "$\{u : X(u)\}$" to denote $\ulcorner X \urcorner$, $\mathfrak{F}$ thus validates the Principle of Naïve Comprehension

(2) $\quad (\forall u)(u \in_\mathfrak{F} \{u : X(u)\} \leftrightarrow X(u))$

for $\sim$-exact subsets X of U. Note that, while "$\{u \in U \mid X(u)\}$" denotes a *subset* of U, the expression "$\{u : X(u)\}$" denotes an *element* of U. We thus distinguish the $\equiv$-class $[u]_\equiv$ of an $\mathfrak{F}$-set u from the $\mathfrak{F}$-set

$$\ulcorner [u]_\equiv \urcorner = \{x : u \equiv x\}$$

that represents $[u]_\equiv$; the latter is denoted $\overline{\{u\}}$, and is called the "$\equiv$-set of u".
Further, let $x, y \in U$; then,

$$(\forall u)(u \in_\mathfrak{F} x \leftrightarrow u \in_\mathfrak{F} y) \leftrightarrow x = y,$$

i.e., the principle of extensionality holds for $\mathfrak{F}$-sets.

Let $x, y \in U$. Define x to be set-theoretically *indiscernible* from y, symbolically, $x \equiv_\mathfrak{F} y$, iff x and y are elements of precisely the same $\mathfrak{F}$-sets:

$$x \equiv_\mathfrak{F} y \Leftrightarrow_{df} (\forall u)(x \in_\mathfrak{F} u \leftrightarrow y \in_\mathfrak{F} u).$$

Set-theoretic indiscernibility is thus an equivalence relation on U and a congruence for the $\sim$-exact subsets of U. Further, define

$$x \equiv_\sim y \Leftrightarrow_{df} [x]_\sim = [y]_\sim.$$

Note that since $\sim$ is a tolerance relation on U, all $\sim$-exact subsets of U are relationally closed under $\equiv_\sim$. Indeed, $x \equiv_\mathfrak{F} y$ iff $x \equiv_\sim y$, i.e., $\equiv_\sim$ is just set-theoretic indiscernibility. Also,

$$x \equiv_\mathfrak{F} y \Rightarrow x \sim y \quad (x, y \in U)$$

holds generally but the converse principle

$$x \sim y \Rightarrow x \equiv_\mathfrak{F} y \quad (x, y \in U)$$

holds just in case $\sim$ is an equivalence relation. Thus, when $\sim$ is an equivalence relation, it may always be interpreted as set-theoretic indiscernibility.

21 THE ORTHOLATTICE OF EXACT SETS

Let $\mathfrak{F} = (U, \sim, \ulcorner \cdot \urcorner, \llcorner \cdot \lrcorner)$ be a PFS based upon a tolerance relation $\sim$. Since elements of U represent exact subsets of U, the complete ortholattice given (defined) by 1 is isomorphic to

(3) $(U, \ulcorner \vee \urcorner, \ulcorner \wedge \urcorner, \ulcorner c \urcorner, \ulcorner \emptyset \urcorner, \ulcorner U \urcorner)$

under the restriction $\ulcorner \cdot \urcorner \upharpoonright C(\sim)$ of the type-lowering retraction to $\sim$-exact subsets of U. Here, $\ulcorner \vee \urcorner, \ulcorner \wedge \urcorner, \ulcorner c \urcorner$, denote the definitions of join and meet natural to $\mathfrak{F}$-sets, e.g.,

$$u_1 \ulcorner \vee \urcorner u_2 =_{df} \ulcorner \llcorner u_1 \lrcorner \vee_S \llcorner u_1 \lrcorner \urcorner = \ulcorner \llcorner u_1 \lrcorner \cup \llcorner u_1 \lrcorner \urcorner$$
$$u_1 \ulcorner \wedge \urcorner u_2 =_{df} \ulcorner \llcorner u_1 \lrcorner \wedge_S \llcorner u_1 \lrcorner \urcorner$$
$$u \ulcorner c \urcorner =_{df} \ulcorner \llcorner u \lrcorner^c \urcorner.$$

We define "$u_1 \ulcorner \subseteq \urcorner u_2$" to be "$\llcorner u_1 \lrcorner \subseteq \llcorner u_2 \lrcorner$", i.e., *inclusion* is the partial ordering naturally associated with the ortholattice of $\mathfrak{F}$-sets given in 3. Usually, the corner quotes are suppressed in naming these operations.

Let $a \in U$. Since unions of $\sim$-exact subsets are $\sim$-exact,

$$\{x \in U \mid (\exists y \in U)(a \sim y \wedge x \in_{\mathfrak{F}} y)\}$$

is an exact subset of U. Thus we define the *outer penumbra* of a, symbolically, $\Diamond a$, to be the $\mathfrak{F}$-set $\bigvee[a]_{\sim}$. Similarly, since closures of intersections of $\sim$-exact subsets are $\sim$-exact,

$$Cl(\{x \in U \mid (\forall y \in U)(a \sim y \to x \in_{\mathfrak{F}} y)\})$$

is an exact subset of U. Define the *inner penumbra*, $\Box a$, to be the $\mathfrak{F}$-set $\bigwedge[a]_{\sim}$. These operations, called the *penumbral modalities*, were interpreted in [Apostoli and Kanda, 2000; Apostoli and Kanda, forthcoming] using David Lewis' counterpart semantics for modal logic [Lewis, 1968]. Given $\mathfrak{F}$-sets a and b, we call b a *counterpart* of a whenever $a \sim b$. Then $\Box a$ ($\Diamond a$) represents the set of $\mathfrak{F}$-sets that belong to all (some) counterparts of a. In this sense, we can say that an $\mathfrak{F}$-set x *necessarily* (*possibly*) belongs to a just in case x belongs to $\Box a$ ($\Diamond a$). An $\mathfrak{F}$-set u is said to be (*penumbrally*) *open* (*closed*) iff $u = \Box u$ ($u = \Diamond u$), respectively. For example, the empty $\mathfrak{F}$-set is open and the universe is closed.

When augmented by the penumbral modal operators, the complete ortholattice of $\mathfrak{F}$-sets given by 3 forms an extensive, idempotent modal ortholattice

(4) $(U, \ulcorner \vee \urcorner, \ulcorner \wedge \urcorner, \ulcorner c \urcorner, \ulcorner \emptyset \urcorner, \ulcorner U \urcorner, \Diamond, \Box)$,

which fails, however, to satisfy the principle of monotonicity characteristic of Kripkean modal logic. Curiously, in addition,

$$\Box \Diamond u \subseteq \Box u \quad (u \in U).$$

When $\sim$ is an equivalence relation, the lattice given by 4 is a modal Boolean algebra (called the "penumbral" modal algebra in [Apostoli and Kanda, 2000; Apostoli and Kanda, forthcoming]), an example of an "abstract" approximation space in the sense of [Cattaneo, 1998] and a "generalized" approximation space in the sense of [Yao, 1998].

22 MODELS OF PFS

An example of a PFS

$$\mathfrak{G} = (M_{max}, \equiv, \ulcorner \cdot \urcorner, \llcorner \cdot \lrcorner)$$

based upon the equivalence relation $\equiv$ of set theoretic indiscernibility was constructed in [Apostoli and Kanda, forthcoming] with the theory of Sequences of Finite Projections (SFP) objects, a branch of Domain Theory [Scott, 1976] which studies the asymptotic behaviour of ω-sequences of monotone (order preserving) projections between finite partial orders.[8] First, a complete partial order (cpo) D_∞ satisfying

$$D_\infty \approx_{CSFP} [D_\infty \to T]_C$$

---

[8]See also [P. Apostoli, 2004] for the details of this construction.

is constructed [Scott, 1976] as the inverse limit of a recursively defined sequence of projections of finite partial orders, where $\approx_{\mathcal{CSFP}}$ is continuous (limit preserving) order isomorphism of cpo's in the category $\mathcal{CSFP}$ of SFP objects and continuous functions, $[D_\infty \to T]_C$ is the cpo of all continuous (limit preserving) functions from D_∞ to T under the information order associated with the nesting of partial characteristic functions and T is the domain of three-valued truth

$$\begin{array}{ccc} \textit{true} & & \textit{false} \\ \diagdown & & \diagup \\ & \bot & \end{array}$$

under the information ordering $\leq_k$ (the bottom value $\bot$ represents a truth-value gap as in partial logic [Blamey, 1986; Feferman, 1984; Gilmore, 1986]).

Then [Apostoli and Kanda, forthcoming], since D_∞ is an SFP object, each monotone function $f : D_\infty \to T$ is maximally approximated by a unique continuous function c_f in $[D_\infty \to T]_C$, whence c_f in D_∞ under representation. Then, the complete partial order M of monotone functions from D_∞ to T is constructed as a solution for the reflexive equation

$$M \approx_\mathcal{M} \prec M \to T \succ$$

where $\approx_\mathcal{M}$ is order isomorphism of cpo's in the category $\mathcal{M}$ of cpo's and monotone functions, and $\prec M \to X \succ$ is the set of all "hyper-continuous" functions from M to T. A monotone function $f : M \to T$ is said to be *hyper-continuous* iff for every $m \in M, f(m) = f(c_m)$. In words, *hyper-continuous functions are those monotone functions which can not distinguish m from c_m*. Note that a monotone function $f : M \to T$ is hyper-continuous just in case

$$c_x = c_y \Rightarrow f(x) = f(y) \quad (x, y \in M).$$

I.e., over M, the equivalence relation of sharing a common maximal continuous approximation is a congruence for all hyper-continuous functions.

Writing "$x \in y$" for $y(x) = true$ and "$x \notin y$" for $y(x) = false$, M may be interpreted as a universe of partial sets-in-extension. Finally, let M_{max} be the set of maximal elements of M. Then [Apostoli and Kanda, forthcoming] we have

$$(\forall x, y \in M_{max})[x \in y \vee x \notin y].$$

M_{max} is thus a classical (bivalent) subuniverse of M. Let $\equiv$ be the relation of set-theoretic indiscernibility, defined for $x, y \in M_{max}$ by

$$x \equiv y \Leftrightarrow_{df} (\forall z \in M_{max})[x \in z \leftrightarrow y \in z].$$

Then we have the fundamental result [Apostoli and Kanda, forthcoming] that set-theoretic indiscernibility over M_{max} is the relation of sharing a common maximal continuous approximation.

A natural example of a PFS based upon a non-transitive tolerance relation on M_{max} can now be given. Let $x, y \in M_{max}$. x *matches* y iff there is a $m \in M$ such that $c_x, c_y \leq m$. Matching is thus a tolerance relation over M_{max} which expresses the compatibility of the maximal continuous approximations of $\mathfrak{G}$-sets: two elements of M_{max} match iff their respective maximal continuous approximations yield, for any given argument, $\leq_k$-comparable truth values, i.e., they agree on the *classical* (non-$\bot$) truth values they take for a given argument. Since matching is "hyper-continuous" (a congruence for $\equiv$) in both x and y, all subsets of M_{max} which are exact with respect to matching are $\equiv$-exact, whence they may be comprehended as $\mathfrak{G}$-sets. Thus M_{max} forms a generalized PFS under the tolerance relation of matching.

23 ON THE DISCERNIBILITY OF THE DISJOINT

The above axioms for PFS's based upon an equivalence relation fall short of articulating all of the important structure of $\mathfrak{G}$. For example, distinct disjoint $\mathfrak{G}$-sets are discernible; in particular, the empty $\mathfrak{G}$-set is a "singularity" in having no counterparts other than itself [Apostoli and Kanda, forthcoming]. Further, since the complements of indiscernible $\mathfrak{G}$-sets are indiscernible, it follows that the universal $\mathfrak{G}$-set is also a singularity in this sense. These properties are logically independent of the basic axioms and may be falsified on the two-point PFS presented above. For example, the "discernibility of the disjoint" asserts the existence of infinitely many pairwise distinct granules of $\mathfrak{F}$-sets and its adoption entails Peano's axioms[9] for second order arithmetic.

Let $\mathfrak{F} = (U, \equiv, \ulcorner \cdot \urcorner, \llcorner \cdot \lrcorner)$ be a PFS based upon an equivalence relation $\equiv$. Then [Apostoli and Kanda, 2000], $\mathfrak{F}$ is said to validate the *Principle of the Discernibility of the Disjoint* iff

(5) $\quad x \cap y = \ulcorner \emptyset \urcorner \Rightarrow \neg x \equiv y \quad (x, y \in U, x \neq \ulcorner \emptyset \urcorner)$.

Suppose $\mathfrak{F}$ satisfies 5. Then distinct $\equiv$-sets are discernible, i.e.,

$$\overline{\{x\}} \equiv \overline{\{y\}} \Rightarrow \overline{\{x\}} = \overline{\{y\}} \quad (x, y \in U)$$

---

[9]Attributing his postulates to Dedekind, Peano [Peano, 1889] axiomatized the arithmetic of the positive natural numbers in terms of three primitive notions, the predicate N ("is a natural number"), 1 ("one") and $'$ ("successor"), as well as logical notions, including identity, predication and quantification over "properties" (concepts). Starting from 0 rather than 1, Peano's postulates for the natural numbers may be formulated in second order logic as follows:
(A1) $N(0)$ ("0 is a natural number").
(A2) $N(x) \rightarrow N(x')$ ("the successor of any natural number is a natural number").
(A3) $(\forall x \in N)(x' \neq 0)$ ("0 is not the successor of any natural number").
(A4) $(\forall x, y \in N)(x' = y' \rightarrow x = y)$ ("No two natural numbers have the same successor").
(A5) $(\forall P)(P(0) \land (\forall x \in N)(P(x) \rightarrow P(x')). \rightarrow .(\forall y \in N)P(y))$ ("Any property which belongs to 0 and also to the successor of any natural number to which it belongs, belongs to all natural numbers").
The second order theory comprised of axioms A1 - A5 is called (second order) "Peano Arithmetic".

and these penumbrally open $\mathfrak{F}$-sets comprise a "reduct" of U in the sense that they may be discerned with respect to their elementhood in $\equiv$-sets. It follows that

$$\overline{\{x\}} \equiv \overline{\{y\}} \Rightarrow x \equiv y \quad (x, y \in U)$$

whence the operation of forming $\equiv$-sets provides a quasi-discrete generalization of Zermelo's [Zermelo, 1908] representation of the successor function of Peano Arithmetic as the operation of forming singleton sets.

Let $L = \{\in\}$ be the first-order language of axiomatic set theory. Note that $\equiv$ may be defined in L as set-theoretic indiscernibility ($\equiv_{\mathfrak{F}}$). Interpreting the identity sign " $=$ " of Peano Arithmetic as indiscernibility $\equiv$, first-order definitions of Peano's primitives N, 0 and ' are given in L as follows: 0 is represented by the empty $\mathfrak{F}$-set; ' is the operation of forming $\equiv$-sets; finally, following the Frege-Dedekind definition of the set of natural numbers, N will be defined as "the least inductive exact set":

(6)
$$\begin{array}{rcl} 0 & =_{df} & \{x : \neg x \equiv x\} \quad (\text{i.e., } \ulcorner\emptyset\urcorner) \\ x' & =_{df} & \{v : x \equiv v\} \quad (\text{i.e., } \overline{\{x\}}) \\ IND(x) & \Leftrightarrow_{df} & 0 \in x \wedge (\forall z)(z \in x \rightarrow z' \in x) \\ N & =_{df} & \{x : (\forall z)(IND(z) \rightarrow x \in z)\}, \end{array}$$

where as usual "inductive" means closed under '. The admissibility of N relies upon the fact that the L formula

$$(\forall z)(IND(z) \rightarrow x \in z)$$

defines an exact subset of U, the intersection of all inductive exact subsets. Note that these are first-order definitions of Peano's second order notions. Finally, note that the admissibility of the indiscernibility relation $\equiv$ as an interpretation of "identity" in Peano Arithmetic resides precisely in the fact that $\equiv$ is an equivalence relation which satisfies the principle of the substitutivity of identicals for all formulas of Peano Arithmetic. Substitutivity is ensured by the fact that the L-formulas interpreting Peano Arithmetic in $\mathfrak{F}$ are molecular combinations of atomic identity formulas of the form "$t \equiv s$", for some terms s and t of Peano Arithmetic, and thus define exact subsets of N.

Peano's axiom for the arithmetic of the natural numbers may now be symbolized in L as follows:

$$\begin{array}{ll} A1^* & N(\emptyset) \\ A2^* & (\forall z)(N(z) \rightarrow N(z')) \\ A3^* & (\forall x \in N)(\neg x' \equiv \emptyset) \\ A4^* & (\forall x \in N)(\forall y \in N)(x' \equiv y' \rightarrow x \equiv y) \\ A5^* & (\forall x)(IND(x) \rightarrow (\forall y \in N)(y \in x)). \end{array}$$

THEOREM 1 [Apostoli and Kanda, forthcoming]. *Suppose $\mathfrak{F}$ validates the Principle (5) of the Discernibility of the Disjoint. Then, $\mathfrak{F}$ is a model of $A1^*$ - $A5^*$.*

The "truth-in-$\mathfrak{F}$" of Peano's axioms follows from the general proof-theoretic result [Apostoli and Kanda, forthcoming] that A1* - A5* may be derived in first-order logic from an effective first-order schema symbolizing the Principle (2) of Naïve Comprehension for $\equiv$-exact concepts, together with the Principle (5) of the Discernibility of the Disjoint expressed as a sentence of L.

24 PLENITUDE

Another property of $\mathfrak{G}$ established in [Apostoli and Kanda, forthcoming] is the following principle of Plenitude. Let $\mathfrak{F} = (U, \equiv, \ulcorner \cdot \urcorner, \llcorner \cdot \lrcorner)$ be a PFS based upon an equivalence relation $\equiv$. In [Apostoli and Kanda, 2000], $\mathfrak{F}$ was said to be a *plenum* iff the following two conditions hold for all $a, b \in U$: (A) $\Box a \equiv \Diamond a$ and (B) $a \subseteq b$ and (C) $a \equiv b$ entails for all $c \in U$,

$$a \subseteq c \subseteq b \Rightarrow c \equiv b.$$

[Apostoli and Kanda, forthcoming] showed that $\mathfrak{G}$ is a plenum and, further, if $\mathfrak{F}$ is a plenum, then

$$([a]_\equiv, \ulcorner \vee \urcorner, \ulcorner \wedge \urcorner, \ulcorner c \urcorner, \Box a, \Diamond a)$$

is a complete Boolean algebra with the least (greatest) element $\Box a$ ($\Diamond a$). Thus, the universe of a plenum factors into a family of granules $[a]_\equiv$, each of which is a complete Boolean algebra.[10] We conclude by asking a question: does M_{max} satisfy conditions (A) and (B) – thus forming a "generalized plenum" whose granules are complete ortho-lattices – under the non-transitive tolerance relation of matching?

25 CONCLUSION

Our development of the notion of a generalized PFS has been axiomatic and informal. The model construction of [Apostoli and Kanda, forthcoming] ensures the consistency of these informal axioms. It further provides a natural example of a PFS based upon the non-transitive tolerance relation of "matching". The task of presenting various axiomatic set theories as consistent "formalizations" of generalized PFS's is a task aired here for future research. E.g., the Principle of Naïve Comprehension for exact concepts (2) given in Section 20 may be symbolized by both effective and noneffective axiom schemes in L. Characterizing the proof theoretic strength of theories which adjoin various comprehension schemes for exact concepts to the first-order theory of a tolerance (or equivalence) relation remains an open problem in the foundations of mathematics.

---

[10]E.g., though M_{max} has hyper-continuum many elements, it factors into continuum many such granules.

BIBLIOGRAPHY

[Aczel and Feferman, 1980] P. Aczel and S. Feferman. Consistency of the unrestricted abstraction principle using an intensional equivalence operator. In J. P. Seldin and J. R. Hindley, editors, *To H. B. Curry: Essays on combinatory logic, lambda calculus and formalism*, pages 67–98. Acedemic Press, New York, 1980.

[Aczel, 1980] P. Aczel. Frege structure and the notions of proposition, truth and set. In K. Kunen J. Barwise, H. Keisler, editor, *The Kleene Symposium*, pages 31–59. North-Holland, 1980.

[Aczel, 1988] P. Aczel. *Non-well-founded sets*. Number 14 in CSLI Lecture Notes. Stanford, 1988.

[Apostoli and Kanda, forthcoming] P. Apostoli and A. Kanda. Parts of the continuum: towards a modern ontology of science. forthcoming in The Poznan Studies in the Philosophy of Science and the Humanities, ed. L. Nowak.

[Apostoli and Kanda, 2000] P. Apostoli and A. Kanda. Approximation spaces of type-free sets. In Y.Y. Yao W. Ziarko, editor, *Proc. of Rough Sets and Current Trends in Computing 2000*, volume 2005 of *Lecture Notes in Artificial Intelligence*, pages 98–105. Springer-Verlag, Berlin Heidelberg New York, 2000.

[Baltag, 1999] A. Baltag. STS: A structural theory of sets. *Logic Journal of the IGPL*, 7:481–515, 1999.

[Barwise and Moss, 1996] J. Barwise and L. Moss. *Vicious Circles*. Number 60 in CSLI Lecture Notes. Stanford, 1996.

[Bell and Machover, 1977] J.L. Bell and M. Machover. *A Course in Mathematical Logic*. North Holland, 1977.

[Bell, 1983] J.L. Bell. Orthologic, forcing and the manifestation of attributes. In *Proc. of the Southeast Asian Conference on Logic*, volume III of *Studies in Logic*. North Holland, 1983.

[Bell, 1986] J.L. Bell. A new approach to quantum logic. *Brit. J. Phil. Sci.*, 37:83–99, 1986.

[Bell, 2000] J.L. Bell. Set and classes as many. *J. Philos. Logic*, 29:595–681, 2000.

[Birkhoff, 1960] G. Birkhoff. *Lattice Theory*, volume XXV of *Amer. Math. Colloq. Publs.* 3rd edition, 1960.

[Blamey, 1986] S. Blamey. Partial logic. In F. Guenthner D. Gabbay, editor, *Handbook of Philosophical Logic*, volume III, pages 1–70. D. Reidel Publishing Company, 1986.

[Brady and Routley, 1989] R. T. Brady and R. Routley. The non-triviality of extensional dialectical set theory. In G. Priest, R. Routley, and J. Norman, editors, *Paraconsistent Logic*, pages 415–436. Philosophia Verlag, Munich, 1989.

[Brady, 1971] R. T. Brady. The consistency of the axioms of abstraction and extensionality in a three-valued logic. *Notre Dame J. Formal Logic*, 12:447–453, 1971.

[Cantor, 1962] G. Cantor. *Gesammelte Abhandlungen Mathematischen und Philosophischen Inhalts*. Springer, Berlin. Reprinted by Olms, Hildesheim (1962).

[Cattaneo, 1998] G. Cattaneo. Abstract approximation spaces for rough theories. In J. Kacprzyk L. Polkowski, A. Skowron, editor, *Rough Sets in Knowledge Discovery: Methodology and Applications*, volume 18 of *Studies in Fuzziness and Soft Computing*. Springer-Verlag, Berlin Heidelberg New York, 1998.

[Chellas, 1980] B.F. Chellas. *An Introduction to Modal Logic*. Cambridge University Press, Cambridge, 1980.

[Church, 1974] A. Church. Set theory with a universal set. In L. Henkin, editor, *Proceedings of the Tarski symposium*, volume XXV of *Proceedings of symposia in Pure Mathematics*, pages 297–308. American Mathematical Society, 1974.

[Crabbé, 1992] M. Crabbé. Soyons positifs: la complÃľtude de la thÃľorie naÃŕve des ensembles. *Cahiers du Centre de Logique (UniversitÃľ catholique de Louvain)*, 7:51–68, 1992.

[Esser, 1999] O. Esser. On the consistency of a positive theory. *Math. Logic Quart.*, 45:105–116, 1999.

[Esser, 2003] O. Esser. A strong model of paraconsistent logic. *Notre Dame J. Formal Logic*, 44:149ï£¡–156, 2003.

[Esser, 2004] O. Esser. *Une théorie positive des ensembles*. Number 13 in Cahiers du centre de logique. Centre national de recherches en logique, Academia-Bruylant, Louvain-La-Neuve, 2004.

[Feferman, 1984] S. Feferman. Towards useful type-free theories. I. *J. Symbolic Logic*, 49:75–111, 1984.

[Fine, 1981] K. Fine. First-order modal theories, i - sets. *Nous*, 15:177–205, 1981.

[Forster, 1995] T. E. Forster. *Set Theory with a Universal Set - Exploring an Untyped Universe*. Clarendon Press - Oxford, second edition edition, 1995.

[Forti and Hinnion, 1989] M. Forti and R. Hinnion. The consistency problem for positive comprehension principles. *J. Symbolic Logic*, 54:1401–1418, 1989.

[Forti and Honsell, 1996] M. Forti and F. Honsell. A general construction of hyperuniverses. *Theoretical Computer Science*, 156:203–215, 1996.

[Fraenkel and Bar-Hillel, 1958] A.A. Fraenkel and Y. Bar-Hillel. *Foundations of Set Theory*. Amsterdam, 1958.

[Fraenkel, 1921] A.A. Fraenkel. Über die zermelosche begründung der mengenlehre. volume 30 of *Jahresbericht der Deutschen Mathematiker-Vereiningung*, pages 97–98. 1921.

[Fraenkel, 1922] A.A. Fraenkel. Zu den grundlagen der cantor-zermeloschen mengenlehre. *Mathematische Annalen*, 86:230–237, 1922.

[Frege, 1903] G. Frege. *Grundgesetze der Arithmetik*, volume 1,2. Verlag, Hermann Pohle, Jena. Reprinted at Hildesheim (1893, 1903).

[Frege, 1884] G. Frege. *Die Grundlagen der Arithmetik. Eine logisch mathematische Untersachung uber den Begridd der Zahl*. William Koebner, Breslau., 1884. English translation by Austin, J.L.: The Foundations of Arihmetic. Basil Blackwell, Oxford (1950).

[Gilmore, 2005] P. Gilmore. *Logicism renewed: logical foundations for mathematics and computer science*. Association for Symbolic Logic, Lecture Notes in Logic, volume 23, 2005.

[Gilmore, 1974] P. Gilmore. The consistency of partial set theory without extensionality. In *Axiomatic Set Theory*, volume 13, Part II of *Proceedings of Symposia in Pure Mathematics*, pages 147–153. Amer. Math. Soc., Providence, R.I., 1974.

[Gilmore, 1986] P.C. Gilmore. Natural deduction based set theories: A new resolution of the old paradoxes. *J. Symbolic Logic*, 51:394–411, 1986.

[Gilmore, 2001] P. Gilmore. An intensional type theory: motivation and cut-elimination. *J. Symbolic Logic*, 66:383–400, 2001.

[Hinnion, 2007] R. Hinnion. Intensional solutions to the identity problem for partial sets. *Reports on Mathematical Logic*, 42:47–69, 2007.

[Hinnion and Libert, 2003] R. Hinnion and Th. Libert. Positive abstraction and extensionality. *J. Symbolic Logic*, 68:828–836, 2003.

[Hinnion, 1990] R. Hinnion. Stratified and positive comprehension seen as superclass rules over ordinary set theory. *Z. Math. Logik Grundlagen Math.*, 36:519–534, 1990.

[Hinnion, 1994] R. Hinnion. Naive set theory with extensionality in partial logic and in paradoxical logic. *Notre Dame J. Formal Logic*, 35:15–40, 1994.

[Hinnion, 2003] R. Hinnion. About the coexistence of classical sets with non-classical ones. *Logic Log. Philos.*, 11:79–90, 2003.

[Hinnion, 2006] R. Hinnion. Intensional positive set theory. *Reports on Mathematical Logic*, 40:107–125, 2006.

[Holmes, 2005] M.R. Holmes. The structure of the ordinals and the interpretation of ZF in double extension set theory. *Studia Logica*, 79: 357–372, 2005.

[Holmes, 2004] M.R. Holmes. Paradoxes in double extension set theories. *Studia Logica*, 77:41–57, 2004.

[Kisielewicz, 1989] A. Kisielewicz. Double extension set theory. *Reports on Math. Logic*, 23:81–89, 1989.

[Kock, 1981] A. Kock. Synthetic differential geometry. volume 51 of *London Math. Soc. Lecture Notes*. Cambridge University Press, 1981.

[Kripke, 1963] S. Kripke. Semantical analysis of modal logic i. normal modal propositional calculi. *Z. Math. Logik Grundlagen Math.*, 9, 1963.

[Lewis, 1968] D. Lewis. Counterpart theory and quantified modal logic. *J. of Philosophy*, 65:113–126, 1968. Reprinted in Loux, M.J. (ed.): The Possible and the Actual. Cornell University Press, Ithica, New York (1979).

[Libert, 2008] Th. Libert. Positive abstraction and extensionality revisited. *Logique et Analyse*, 51(202), 2008.

[Libert and Esser, 2005] Th. Libert and O. Esser. On topological set theory. *Math. Logic Quart.*, 51:263–273, 2005.

[Libert, 2003] Th. Libert. ZF and the axiom of choice in some paraconsistent set theories. *Logic Log. Philos.*, 11:91–114, 2003.

[Libert, 2004] Th. Libert. Semantics for naive set theory in many-valued logics: Technique and historical account. In J. van Benthem and G. Heinzmann, editors, *The Age of Alternative Logics*. Kluwer Academics, 2004. to appear.

[Libert, 2005] Th. Libert. Models for a paraconsistent set theory. *J. Appl. Log.*, 3:15–41, 2005.

[Malitz, 1976] R. J. Malitz. *Set Theory in which the Axiom of Foundation Fails*. PhD thesis, University of California, Los Angeles, 1976.

[Orlowska, 1985] E. Orlowska. Semantics of vague concepts. In NewYork Plenum Press, editor, *Foundations of Logic and Linguistics, Problems and Their Solutions*. G. Dorn and P. Weingartner, 1985.

[P. Apostoli, 2004] L. Polkowski P. Apostoli, A. Kanda. First steps towards computably infinite information systems. In M. Inuiguchi L. Polkowski D. Dubois, J. Grzymala-Busse, editor, *Rough Sets and Fuzzy Sets. Transactions in Rough Sets. Vol. 2*, Lecture Notes in Computer Science, pages 161–198. Springer-Verlag, Berlin Heidelberg New York, 2004.

[Parsons, 1977] C. Parsons. What is the iterative conception of set? In R.E. Butts and J. Hintikka, editors, *Logic, Foundation of Mathematics, and Computability Theory*, pages 335–367. D. Reidel, 1977. Reprinted in Parsons, C.: Mathematics in Philosophy: Selected Essays. Cornell University Press, Ithica, New York (1983).

[Parsons, 1981] C. Parsons. Modal set theories. *J. Symbolic Logic*, 46:683–684, 1981.

[Pawlak, 1982] Z. Pawlak. Rough sets, algebraic and topological approaches. *International Journal of Computer and Information Sciences*, 11:341–356, 1982.

[Peano, 1889] G. Peano. *Arithmetices Principia Nova Methodo Exposita*. Rome, 1889.

[Scott, 1976] D. Scott. Data types as lattices. *SIAM Journal on Computing*, 5:522–587, 1976.

[Skowron and Stepaniuk, 1994] A. Skowron and J. Stepaniuk. Generalized approximation spaces. In *Proc. 3rd Int. Workshop on Rough Sets and Soft Computing. San Jose USA (Nov. 10-12)*, pages 156–163, 1994.

[Skowron and Stepaniuk, 1996] A. Skowron and J. Stepaniuk. Tolerance approximation spaces. *Fundementa Informaticae*, 27:245–253, 1996.

[Specker, 1953] E.P. Specker. The axiom of choice in quine's new foundations for mathematical logic. *Proc. Nat. Acad. Sci. U.S.A.*, 39:972–975, 1953.

[van Heijenoort, 1967] J. van Heijenoort. *From Frege to Godel: A Source Book in Mathematical Logic, 1879 - 1931*, Harvard University Press, 1967.

[Weydert, 1989] E. Weydert. *How to Approximate the Naive Comprehension Scheme Inside of Classical Logic*. PhD thesis, Bonner Mathematische Schriften Nr.194, Bonn 1989.

[Yao, 1998] Y.Y. Yao. On generalizing pawlak approximation operators. In L. Polkowski and A. Skowron, editors, *Rough Sets and Current Trends in Computing 1998*, volume 1414 of *Lecture Notes in Artificial Intelligence*, pages 289–307. Springer-Verlag, Berlin Heidelberg New York, 1998.

[Zermelo, 1908] E. Zermelo. Untersuchungen über die grundlagen der mengenlehre i. *Math. Ann.*, 65:261–281, 1908. Translated in [van Heijenoort, 1967] as: Investigations in the Foundations of Set Theory I.

PHILOSOPHIES OF PROBABILITY

Jon Williamson

1 INTRODUCTION

The concept of probability motivates two key questions.

First, how is probability to be defined? Probability was axiomatised in the first half of the 20th century ([Kolmogorov, 1933]); this axiomatisation has by now become well entrenched, and in fact the main leeway these days is with regard to the type of domain on which probability functions are defined. Part I introduces three types of domain: variables (§2), events (§3), and sentences (§4).

Second, how is probability to be applied? In order to know how probability can be applied we need to know what probability means: how probabilities can be measured and how probabilistic predictions say something about the world. Part II discusses the predominant interpretations of probability: the frequency (§6), propensity (§7), chance (§§8, 10), and Bayesian interpretations (§9).

In Part III, we shall focus on one interpretation of probability, objective Bayesianism, and look more closely at some of the challenges that this interpretation faces. Finally, Part IV draws some lessons for the philosophy of mathematics in general.

Part I
Frameworks for Probability

2 VARIABLES

The most basic framework for probability involves defining a probability function relative to a finite set V of variables, each of which takes finitely many possible values. I shall write $v@V$ to indicate that v is an assignment of values to V.

A *probability function* on V is a function P that maps each assignment $v@V$ to a non-negative real number and which satisfies *additivity*:

$$\sum_{v@V} P(v) = 1.$$

This restriction forces each probability $P(v)$ to lie in the unit interval $[0, 1]$.

Handbook of the Philosophy of Science. Philosophy of Mathematics
Volume editor: Andrew D. Irvine. General editors: Dov M. Gabbay, Paul Thagard and John Woods.
© 2009 Elsevier B.V. All rights reserved.

The *marginal probability function* on $U \subseteq V$ induced by probability function P on V is a probability function Q on U which satisfies

$$Q(u) = \sum_{v@V, v \sim u} P(v)$$

for each $u@U$, and where $v \sim u$ means that v is consistent with u, i.e., u and v assign the same values to $U \cap V = U$. The marginal probability function Q on U is uniquely determined by P. Marginal probability functions are usually thought of as extensions of P and denoted by the same letter P. Thus P can be construed as a function that maps each $u@U \subseteq V$ to a non-negative real number. P can be further extended to assign numbers to conjunctions tu of assignments where $t@T \subseteq V, u@U \subseteq V$: if $t \sim u$ then tu is an assignment to $T \cup U$ and $P(tu)$ is the marginal probability awarded to $tu@(T \cup U)$; if $t \not\sim u$ then $P(tu)$ is taken to be 0.

A *conditional probability function* induced by P is a function R from pairs of assignments of subsets of V to non-negative real numbers which satisfies (for each $t@T \subseteq V, u@U \subseteq V$):

$$R(t|u)P(u) = P(tu),$$

$$\sum_{t@T} R(t|u) = 1.$$

Note that $R(t|u)$ is not uniquely determined by P when $P(u) = 0$. If $P(u) \neq 0$ and the first condition holds, then the second condition, $\sum_{t@T} R(t|u) = 1$, also holds. Again, R is often thought of as an extension of P and is usually denoted by the same letter P.

Consider an example. Take a set of variables $V = \{A, B\}$, where A signifies *age of vehicle* taking possible values *less than 3 years*, *3-10 years* and *greater than 10 years*, and B signifies *breakdown in the last year* taking possible values *yes* and *no*. An assignment $b@B$ is of the form $B = yes$ or $B = no$. The assignments $a@A$ are most naturally written $A < 3, 3 \leq A \leq 10$ and $A > 10$. According to the above definition a probability function P on V assigns a non-negative real number to each assignment of the form ab where $a@A$ and $b@B$, and these numbers must sum to 1. For instance,

$$P(A < 3 \cdot B = yes) = 0.05$$

$$P(A < 3 \cdot B = no) = 0.1$$

$$P(3 \leq A \leq 10 \cdot B = yes) = 0.2$$

$$P(3 \leq A \leq 10 \cdot B = no) = 0.2$$

$$P(A > 10 \cdot B = yes) = 0.35$$

$$P(A > 10 \cdot B = no) = 0.1.$$

This function P can be extended to assignments of subsets of V, yielding $P(A > 10) = P(A > 10 \cdot B = yes) + P(A > 10 \cdot B = no) = 0.35 + 0.1 = 0.45$ for example, and to conjunctions of assignments in which case inconsistent assignments are awarded probability 0, e.g., $P(B = yes \cdot B = no) = 0$. The function P can then be extended to yield conditional probabilities and, in this example, the probability of a breakdown conditional on age greater than 10 years, $P(B = yes|A > 10)$, is $P(A > 10 \cdot B = yes)/P(A > 10) = 0.35/0.45 \approx 0.78$.

3 EVENTS

While the definition of probability over assignments to variables is straightforward, simplicity is gained at the expense of generality. By moving from variables to abstract events we can capture generality. The main definition proceeds as follows.[1]

Abstract *events* are construed as subsets of an *outcome space* Ω, which represents the possible outcomes of an experiment or observation. For example, if the age of a vehicle were observed, the outcome space might be $\Omega = \{0, 1, 2, \ldots\}$, and $\{0, 1, 2\} \subseteq \Omega$ represents the event that the vehicle's age is less than three years.

An *event space* $\mathcal{F}$ is a set of subsets of Ω. $\mathcal{F}$ is a *field* if it contains Ω and is closed under the formation of complements and finite unions; it is a *σ-field* if it is also closed under the formation of countable unions.

A *probability function* is a function P from a field $\mathcal{F}$ to the non-negative real numbers that satisfies *countable additivity*:

- if $E_1, E_2, \ldots \in \mathcal{F}$ partition Ω (i.e., $E_i \cap E_j = \emptyset$ for $i \neq j$ and $\bigcup_{i=1}^{\infty} E_i = \Omega$) then $\sum_{i=1}^{\infty} P(E_i) = 1$.

In particular, $P(\Omega) = 1$. The triple $(\Omega, \mathcal{F}, P)$ is called a *probability space*.

The variable framework is captured by letting Ω contain all assignments to V and taking $\mathcal{F}$ to be the set of all subsets of Ω, which corresponds to the set of disjunctions of assignments to V. Given variable $A \in V$, the function that maps $v@V$ to the value that v assigns to A is called a *simple random variable* in the event framework.

4 SENTENCES

Logicians tend to define probability over logical languages (see, e.g., [Paris, 1994]). The simplest such framework is based around the propositional calculus, as follows.

A *propositional variable* is a variable which takes two possible values, *true* or *false*. A set $\mathcal{L}$ of propositional variables constitutes a *propositional language*. The sentences $S\mathcal{L}$ of $\mathcal{L}$ include the propositional variables, together with the *negation* $\neg \theta$ of each sentence $\theta \in S\mathcal{L}$ (which is true iff θ is false) and each *implication* of the form $\theta \to \varphi$ for $\theta, \varphi \in S\mathcal{L}$ (which is true iff θ is false or both θ and φ are true). The *conjunction* $\theta \wedge \varphi$ is defined to be $\neg(\theta \to \neg\varphi)$ and is true iff both θ and φ are

---

[1] [Billingsley, 1979] provides a good introduction to the theory behind this approach.

true; the *disjunction* $\theta \vee \varphi$ is defined to be $\neg\theta \to \varphi$ and is true iff either θ or φ are true. An assignment l of values to $\mathcal{L}$ *models* sentence θ, written $l \models \theta$, if θ is true under l. A sentence θ is a *tautology*, written $\models \theta$, if it is true whatever the values of the propositional variables in θ, i.e., if each assignment to $\mathcal{L}$ models θ.

A *probability function* is then a function P from a set $S\mathcal{L}$ of sentences to the non-negative real numbers that satisfies *additivity*:

- if $\theta_1, \ldots, \theta_n \in S\mathcal{L}$ satisfy $\models \neg(\theta_i \wedge \theta_j)$ for $i \neq j$ and $\models \theta_1 \vee \cdots \vee \theta_n$ then $\sum_{i=1}^{n} P(\theta_i) = 1$.

If the language $\mathcal{L}$ is finite then the sentence framework can be mapped to the variable framework. $V = \mathcal{L}$ is a finite set of variables each of which takes finitely many values. A sentence $\theta \in SV$ can be identified with the set of assignments v of values to V which model θ. P thus maps sets of assignments and, in particular, individual assignments, to real numbers. P is additive because of additivity on sentences. Hence P induces a probability function over assignments to V.

The sentence framework can also be mapped to the event framework. Let Ω contain all assignments to L, and let $\mathcal{F}$ be the field of sets of the form $\{l : l \models \theta\}$ for $\theta \in S\mathcal{L}$.[2] By defining $P(\{l : l \models \theta\}) = P(\theta)$ we get a probability function.[3]

Part II
Interpretations of Probability

5 INTERPRETATIONS AND DISTINCTIONS

The definitions of probability given in Part I are purely formal. In order to apply the formal concept of probability we need to know how probability is to be interpreted. The standard interpretations of probability will be presented in the next few sections.[4] These interpretations can be categorised according to the stances they take on three key distinctions:

Single-Case / Repeatable: A variable is *single-case* (or *token-level*) if it can only be assigned a value once. It is *repeatable* (or *repeatably instantiatable* or *type-level*) if it can be assigned values more than once. For example, variable A standing for *age of car with registration AB01 CDE on January 1st 2010* is single-case because it can only ever take one value (assuming the car in question exists). If, however, A stands for *age of vehicles selected at*

---
[2] These sets are called *cylinder sets* when $\mathcal{L}$ is infinite — see [Billingsley, 1979, p. 27].
[3] This depends on the fact that every probability function on the field of cylinders which is *finitely additive* (i.e., which satisfies $\sum_{i=1}^{n} P(E_i) = 1$ for partition $E_1, \ldots, E_n$ of Ω) is also countably additive. See [Billingsley, 1979, Theorem 2.3].
[4] For a more detailed exposition of the interpretations see [Gillies, 2000].

random in London in 2010 then A is repeatable: it gets reassigned a value each time a new vehicle is selected.[5]

Mental / Physical: Probabilities are *mental* — or *epistemological* ([Gillies, 2000]) or *personalist* — if they are interpreted as features of an agent's mental state, otherwise they are *physical* — or *aleatory* ([Hacking, 1975]).

Subjective / Objective: Probabilities are *subjective* (or *agent-relative*) if two agents with the same evidence can disagree as to a probability value and yet neither of them be wrong. Otherwise they are *objective*.[6]

There are four main interpretations of probability: the frequency theory (discussed in §6), the propensity theory (§7), chance (§8) and Bayesianism (§9).

6 FREQUENCY

The *Frequency* interpretation of probability was propounded by [Venn, 1866] and [Reichenbach, 1935] and developed in detail in [von Mises, 1928] and [von Mises, 1964]. Von Mises' theory can be formulated in our framework as follows. Given a set V of repeatable variables one can repeatedly determine the values of the variables in V and write down the observations as assignments to V. For example, one could repeatedly select cars and determine their age and whether they broke down in the last year, writing down $A < 3 \cdot B = no, A < 3 \cdot B = yes, A > 10 \cdot B = yes$, and so on. Under the assumption that this process of measurement can be repeated *ad infinitum*, we generate an infinite sequence of assignments $\mathcal{V} = (v_1, v_2, v_3, \ldots)$ called a *collective*.

Let $|v|_\mathcal{V}^n$ be the number of times assignment v occurs in the first n places of $\mathcal{V}$, and let $Freq_\mathcal{V}^n(v)$ be the frequency of v in the first n places of $\mathcal{V}$, i.e.,

$$Freq_\mathcal{V}^n(v) = \frac{|v|_\mathcal{V}^n}{n}.$$

Von Mises noted two things. First, these frequencies tend to stabilise as the number n of observations increases. Von Mises hypothesised that

Axiom of Convergence: $Freq_\mathcal{V}^n(v)$ tends to a fixed limit as $n \longrightarrow \infty$, denoted by $Freq_\mathcal{V}(v)$.

Second, gambling systems tend to be ineffective. A gambling system can be thought of as function for selecting places in the sequence of observations on which

---

[5]'Single-case variable' is clearly an oxymoron because the value of a single-case variable does not vary. The value of a single-case variable may not be known, however, and one can still think of the variable as taking a range of possible values.

[6]Warning: some authors, such as [Popper, 1983, §3.3] and [Gillies, 2000, p. 20], use the term 'objective' for what I call 'physical'. However, their terminology has the awkward consequence that the interpretation of probability commonly known as 'objective Bayesianism' (described in Part III) does not get classed as 'objective'.

to bet, on the basis of past observations. Thus a *place selection* is a function $f(v_1,\ldots,v_n) \in {0,1}$, such that if $f(v_1,\ldots,v_n) = 0$ then no bet is to be placed on the $n+1$-st observation and if $f(v_1,\ldots,v_n) = 1$ then a bet is to be placed on the $n+1$-st observation. So betting according to a place selection gives rise to a sub-collective $\mathcal{V}_f$ of $\mathcal{V}$ consisting of the places of $\mathcal{V}$ on which bets are placed. In practice we can only use a place selection function if it is simple enough for us to compute its values: if we cannot decide whether $f(v_1,\ldots,v_n)$ is 0 or 1 then it is of no use as a gambling system. According to Church's thesis a function is computable if it belongs to the class of functions known as *recursive* functions ([Church, 1936]). Accordingly we define a *gambling system* to be a recursive place selection. A gambling system is said to be effective if we are able to make money in the long run when we place bets according to the gambling system. Assuming that stakes are set according to frequencies of $\mathcal{V}$, a gambling system f can only be effective if the frequencies of $\mathcal{V}_f$ differ to those of $\mathcal{V}$: if $Freq_{\mathcal{V}_f}(v) > Freq_{\mathcal{V}}(v)$ then betting *on* v will be profitable in the long run; if $Freq_{\mathcal{V}_f}(v) < Freq_{\mathcal{V}}(v)$ then betting *against* v will be profitable. We can then explicate von Mises' second observation as follows:

Axiom of Randomness: Gambling systems are ineffective: if $\mathcal{V}_f$ is determined by a recursive place selection f, then for each v, $Freq_{\mathcal{V}_f}(v) = Freq_{\mathcal{V}}(v)$.

Given a collective $\mathcal{V}$ we can then define — following von Mises — the probability of v to be the frequency of v in $\mathcal{V}$:

$$P(v) \stackrel{df}{=} Freq_{\mathcal{V}}(v).$$

Clearly $Freq_{\mathcal{V}}(v) \geq 0$. Moreover $\sum_{v@V} |v|_{\mathcal{V}}^n = n$ so $\sum_{v@V} Freq_{\mathcal{V}}^n(v) = 1$ and, taking limits, $\sum_{v@V} Freq_{\mathcal{V}}(v) = 1$. Thus P is indeed a well-defined probability function.

Suppose we have a statement involving probability function P on V. If we also have a collective $\mathcal{V}$ on V then we can interpret the statement to be saying something about the frequencies of $\mathcal{V}$, and as being true or false according to whether the corresponding statement about frequencies is true or false respectively. This is the frequency interpretation of probability. The variables in question are repeatable, not single-case, and the interpretation is physical, relative to a collective of potential observations, not to the mental state of an agent. The interpretation is objective, not subjective, in the sense that once the collective is fixed then so too are the probabilities: if two agents disagree as to what the probabilities are, then at most one of the agents is right.

7 PROPENSITY

Karl Popper initially adopted a version of von Mises' frequency interpretation ([Popper, 1934, Chapter VIII]), but later, with the ultimate goal of formulating an interpretation of probability applicable to single-case variables, developed what is called the *propensity* interpretation of probability ([Popper, 1959]; [Popper, 1983, Part II]). The propensity theory can be thought of as the frequency theory together with the following law:[7]

Axiom of Independence: If collectives V_1 and V_2 on V are generated by the same repeatable experiment (or repeatable conditions) then for all assignments v to V, $Freq_{V_1}(v) = Freq_{V_2}(v)$.

In other words frequency, and hence probability, attaches to a repeatable experiment rather than a collective, in the sense that frequencies do not vary with collectives generated by the same repeatable experiment. The repeatable experiment is said to have a propensity for generating the corresponding frequency distribution.

In fact, despite Popper's intentions, the propensity theory interprets probability defined over repeatable variables, not single-case variables. If, for example, V consists of repeatable variables A and B, where A stands for *age of vehicles selected at random in London in 2010* and B stands for *breakdown in the last year of vehicles selected at random in London in 2010*, then V determines a repeatable experiment, namely the selection of vehicles at random in London in 2010, and thus there is a natural propensity interpretation. Suppose, on the other hand, that V contains single-case variables A and B, standing for *age of car with registration AB01 CDE on January 1st 2010* and *breakdown in last year of car with registration AB01 CDE on January 1st 2010*. Then V defines an experiment, namely the selection of car AB01 CDE on January 1st 2010, but this experiment is not repeatable and does not generate a collective — it is a single case. The car in question might be selected by several different repeatable experiments, but these repeatable experiments need not yield the same frequency for an assignment v, and thus the probability of v is not determined by V. (This is known as the *reference class problem*: we do not know from the specification of the single case how to uniquely determine a repeatable experiment which will fix probabilities.) In sum, the propensity theory is, like the frequency theory, an objective, physical interpretation of probability over repeatable variables.

8 CHANCE

The question remains as to whether one can develop a viable objective interpretation of probability over single-case variables — such a concept of probability is often called *chance*.[8] We saw that frequencies are defined relative to a collective and propensities are defined relative to a repeatable experiment; however, a single-case variable does not determine a unique collective or repeatable experiment and

---

[7][Popper, 1983, pp. 290 and 355]. It is important to stress that the axioms of this section and the last had a different status for Popper than they did for von Mises. Von Mises used the frequency axioms as part of an operationalist definition of probability, but Popper was not an operationalist. See [Gillies, 2000, Chapter 7] on this point. Gillies also argues in favour of a propensity interpretation.

[8]Note that some authors use 'propensity' to cover a physical chance interpretation as well as the propensity interpretation discussed above.

so neither approach allows us to attach probabilities directly to single-case variables. What then does fix the chances of a single-case variable? The view finally adopted by Popper was that the 'whole physical situation' determines probabilities ([Popper, 1990, p. 17]). The physical situation might be thought of as 'the complete situation of the universe (or the light-cone) at the time' ([Miller, 1994, p. 186]), the complete history of the world up till the time in question ([Lewis, 1980, p. 99]),[9] or 'a complete set of (nomically and/or causally) relevant conditions ... which happens to be instantiated in that world at that time' ([Fetzer, 1982, p. 195]). Thus the chance, on January 1st 2010, of car with registration AB01 CDE breaking down in the subsequent year, is fixed by the state of the universe at that date, or its entire history up till that date, or all the relevant conditions instantiated at that date. However the chance-fixing 'complete situation' is delineated, these three approaches associate a unique chance-fixer with a given single-case variable. (In contrast, the frequency / propensity theories do not associate a unique collective / repeatable experiment with a given single-case variable.) Hence we can interpret the probability of an assignment to the single-case variable as the chance of the assignment holding, as determined by its chance-fixer.

Further explanation is required as to how one can measure probabilities under the chance interpretation. Popper's line is this: if the chance-fixer is a set of relevant conditions and these conditions are repeatable, then the conditions determine a propensity and that can be used to measure the chance ([Popper, 1990, p. 17]). Thus if the set of conditions relevant to car AB01 CDE breaking down that hold on January 1st 2010 also hold for other cars at other times, then the chance of AB01 CDE breaking down in the next year can be equated with the frequency with which cars satisfying the same set of conditions break down in the subsequent year. The difficulty with this view is that it is hard to determine all the chance-fixing relevant conditions, and there is no guarantee that enough individuals will satisfy this set of conditions for the corresponding frequency to be estimable.

9 BAYESIANISM

The *Bayesian* interpretation of probability also deals with probability functions defined over single-case variables. But in this case the interpretation is mental rather than physical: probabilities are interpreted as an agent's rational degrees of belief.[10] Thus for an agent, $P(B = yes) = q$ if and only if the agent believes that $B = yes$ to degree q and this ascription of degree of belief is rational in the sense outlined below. An agent's degrees of belief are construed as a guide to her actions: she believes $B = yes$ to degree q if and only if she is prepared to place a bet of qS on $B = yes$, with return S if $B = yes$ turns out to be true. Here S is an unknown stake, which may be positive or negative, and q is called a *betting*

---

[9]See §§10, 20.

[10]This interpretation was developed in [Ramsey, 1926] and [de Finetti, 1937]. See [Howson and Urbach, 1989] and [Earman, 1992] for recent expositions.

quotient. An agent's *belief function* is the function that maps an assignment to the agent's degree of belief in that assignment.

An agent's betting quotients are called *coherent* if one cannot choose stakes for her bets that force her to lose money whatever happens. (Such a set of stakes is called a *Dutch book*.) It is not hard to see that a coherent belief function is a probability function. First $q \geq 0$, for otherwise one can set S to be negative and the agent will lose whatever happens: she will lose $qS > 0$ if the assignment on which she is betting turns out to be false and will lose $(q-1)S > 0$ if it turns out to be true. Moreover $\sum_{v@V} q_v = 1$, where q_v is the betting quotient on assignment v, for otherwise if $\sum_v q_v > 1$ we can set each $S_v = S > 0$ and the agent will lose $(\sum_v q_v - 1)S > 0$ (since exactly one of the v will turn out true), and if $\sum_v q_v < 1$ we can set each $S_v = S < 0$ to ensure positive loss.

Coherence is taken to be a necessary condition for rationality. For an agent's degrees of belief to be rational they must be coherent, and hence they must be probabilities. *Subjective Bayesianism* is the view that coherence is also sufficient for rationality, so that an agent's belief function is rational if and only if it is a probability function. This interpretation of probability is subjective because it depends on the agent as to whether $P(v) = q$. Different agents can choose different probabilities for v and their belief functions will be equally rational. *Objective Bayesianism*, discussed in detail in Part III, imposes further rationality constraints on degrees of belief — not just coherence. Very often objective Bayesianism constrains degree of belief in such a way that only one value for $P(v)$ is deemed rational on the basis of an agent's evidence. Thus, objective Bayesian probability varies as evidence varies but two agents with the same evidence often adopt the same probabilities as their rational degrees of belief.[11]

Many subjective Bayesians claim that an agent should update her degrees of belief by *Bayesian conditionalisation*: her new degrees of belief should be her old degrees of belief conditional on new evidence, $P_{t+1}(v) = P_t(v|u)$ where u represents the evidence that the agent has learned between time t and time $t+1$. In cases where $P_t(v|u)$ is harder to quantify than $P_t(u|v)$ and $P_t(v)$ this conditional probability may be calculated using *Bayes' theorem*: $P(v|u) = P(u|v)P(v)/P(u)$, which holds for any probability function P. Note that Bayesian conditionalisation is more appropriate as a constraint on subjective Bayesian updating than on objective Bayesian updating, because it disagrees with the usual principles of objective Bayesianism ([Williamson, 2008b]). 'Bayesianism' is variously used to refer to the Bayesian interpretation of probability, the endorsement of Bayesian conditionalisation or the use of Bayes' theorem.

---

[11]Objective Bayesian degrees of belief are uniquely determined on a finite set of variables; on infinite domains subjectivity can creep in (§19).

10 CHANCE AS ULTIMATE BELIEF

The question still remains as to whether one can develop a viable notion of chance, i.e., an objective single-case interpretation of probability. While the Bayesian interpretations are single-case, they either define probability relative to the whimsy of an agent (subjective Bayesianism) or relative to an agent's evidence (objective Bayesianism). Is there a probability of my car breaking down in the next year, where this probability does not depend on me or my evidence?

Bayesians typically have two ways of tackling this question.

Subjective Bayesians tend to argue that although degrees of belief may initially vary widely from agent to agent, if agents update their degrees of belief by Bayesian conditionalisation then their degrees of belief will converge in the long run: chances are these long run degrees of belief. Bruno de Finetti developed such an argument to explain the apparent existence of physical probabilities ([de Finetti, 1937]; [Gillies, 2000, pp. 69–83]). He showed that prior degrees of beliefs converge to frequencies under the assumption of *exchangeability*: given an infinite sequence of single-case variables $A_1, A_2, \ldots$ which take the same possible values, an agent's degrees of belief are *exchangeable* if the degree of belief $P(v)$ she gives to assignment v to a finite subset of variables depends only on the values in v and not the variables in v — for example $P(a_1^1 a_2^0 a_3^1) = P(a_3^0 a_4^1 a_5^1)$ since both assignments assign two 1s and one 0. Suppose the actual observed assignments are $a_1, a_2, \ldots$ and let $\mathcal{V}$ be the collective of such values (which can be thought of as arising from a single repeatable variable A). De Finetti showed that $P(a_n | a_1 \cdots a_{n-1}) \longrightarrow Freq_{\mathcal{V}}(a)$ as $n \longrightarrow \infty$, where a is the assignment to A of the value that occurs in a_n. The chance of a_n is then identified with $Freq_{\mathcal{V}}(a)$. The trouble with de Finetti's account is that since degrees of belief are subjective there is no reason to suppose exchangeability holds. Moreover, a single-case variable A_n can occur in several sequences of variables, each with a different frequency distribution (the reference class problem again), in which case the chance distribution of A_n is ill-defined. Haim Gaifman and Marc Snir took a slightly different approach, showing that as long as agents give probability 0 to the same assignments and the evidence that they observe is unrestricted, then their degrees of belief must converge ([Gaifman and Snir, 1982, §2]). Again, the problem here is that there is no reason to suppose that agents will give probability 0 to the same assignments. One might try to provide such a guarantee by bolstering subjective Bayesianism with a rationality constraint that says that agents must be *undogmatic*, i.e., they must only give probability 0 to logically impossible assignments. But this is not a feasible strategy in general, since this constraint is inconsistent with the constraint that degrees of belief be probabilities: in the more general event or sentence frameworks the laws of probability force some logical possibilities to be given probability 0.[12]

Objective Bayesians have another recourse open to them: objective Bayesian probability is fixed by an agent's evidence, and one can argue that chances are those degrees of belief fixed by some suitable all-encompassing evidence. Thus

---

[12]See [Gaifman and Snir, 1982, Theorem 3.7], for example.

the problem of producing a well-defined notion of chance is reducible to that of developing an objective Bayesian interpretation of probability. I shall call this the *ultimate belief* notion of chance to distinguish it from physical notions such as Popper's (§8), and discuss this approach in §20.

11 APPLYING PROBABILITY

In sum, there are four key interpretations of probability: frequency and propensity interpret probability over repeatable variables while chance and the Bayesian interpretation deal with single-case variables; frequency and propensity are physical interpretations while Bayesianism is mental and chance can be either mental or physical; all the interpretations are objective apart from Bayesianism which can be subjective or objective.

Having chosen an interpretation of probability, one can use the probability calculus to draw conclusions about the world. Typically, having made an observation $u@U \subseteq V$, one determines the conditional probability $P(t|u)$ to tell us something about $t@T \subseteq (V \backslash U)$: a frequency, propensity, chance or appropriate degree of belief.

Part III
Objective Bayesianism

12 SUBJECTIVE AND OBJECTIVE BAYESIANISM

In Part 4 we saw that probabilities can either be interpreted physically — as frequencies, propensities or physical chances — or they can be interpreted mentally, with Bayesians arguing that an agent's degrees of belief ought to satisfy the axioms of probability. Some Bayesians are strict subjectivists, holding that there are no rational constraints on degrees of belief other than the requirement that they be probabilities ([de Finetti, 1937]). Thus subjective Bayesians maintain that one may give probability 0 — or indeed any value between 0 and 1 — to a coin toss yielding heads, even if one knows that the coin is symmetrical and has yielded heads in roughly half of all its previous tosses. The chief criticism of strict subjectivism is that practical applications of probability tend to demand more objectivity; in science some beliefs are considered more rational than others on the basis of available evidence. This motivates an alternative position, objective Bayesianism, which posits further constraints on degrees of belief, and which would only deem the agent to be rational in this case if she gave a probability of a half to the toss yielding heads ([Jaynes, 1988]).

Objective Bayesianism holds that the probability of u is the degree to which an agent ought to believe u and that this degree is more or less objectively determined by the agent's evidence. Versions of this view were put forward by [Bernoulli, 1713];

[Laplace, 1814] and [Keynes, 1921]. More recently Jaynes claimed that an agent's probabilities ought to satisfy constraints imposed by evidence but otherwise ought to be as non-committal as possible. Moreover, Jaynes argued, this principle could be explicated using Shannon's information theory ([Shannon, 1948]): the agent's probability function should be that probability function, from all those that satisfy constraints imposed by evidence, that maximises entropy ([Jaynes, 1957]). This has become known as *the Maximum Entropy Principle* and has been taken to be the foundation of the objective Bayesian interpretation of probability by its proponents ([Rosenkrantz, 1977; Jaynes, 2003]).

In the next section, I shall sketch my own version of objective Bayesianism. This version is discussed in detail in chapter 4 of [Williamson, 2005a]. In subsequent sections we shall examine a range of important challenges that face the objective Bayesian interpretation of probability.

13 OBJECTIVE BAYESIANISM OUTLINED

While Bayesianism requires that degrees of belief respect the axioms of probability, objective Bayesianism imposes two further norms. An empirical norm requires that an agent's degrees of belief be calibrated with her evidence, while a logical norm holds that where degrees of belief are underdetermined by evidence, they should be as equivocal as possible:

Empirical: An agent's empirical evidence should constrain her degrees of belief. Thus if one knows that a coin is symmetrical and has yielded heads roughly half the time, then one's degree of belief that it will yield heads on the next throw should be roughly $\frac{1}{2}$.

Logical: An agent's degrees of belief should also be fixed by her lack of evidence. If the agent knows nothing about an experiment except that it has two possible outcomes, then she should award degree of belief $\frac{1}{2}$ to each outcome.

Jakob Bernoulli pointed out that where they conflict, the empirical norm should override the logical norm:

> three ships set sail from port; after some time it is announced that one of them suffered shipwreck; which one is guessed to be the one that was destroyed? If I considered merely the number of ships, I would conclude that the misfortune could have happened to each of them with equal chance; but because I remember that one of them had been eaten away by rot and old age more than the others, had been badly equipped with masts and sails, and had been commanded by a new and inexperienced captain, I consider that this ship, more probably than the others, was the one to perish. ([Bernoulli, 1713, §IV.II])

One can prioritise the empirical norm over the logical norm by insisting that

Empirical: An agent's degrees of belief, represented by probability function $P_\mathcal{E}$, should satisfy any constraints imposed by her evidence $\mathcal{E}$.

Logical: The agent's belief function $P_\mathcal{E}$ should otherwise be as non-committal as possible.

The empirical norm can be explicated as follows. Evidence $\mathcal{E}$ might contain a number of considerations that bear on a degree of belief: the symmetry of a penny might incline one to degree of belief $\frac{1}{2}$ in heads, past performance (say 47 heads in a hundred past tosses) may incline one to degree of belief 0.47, the mint may report an estimate of the frequency of heads on its pennies to be 0.45, and so on. These considerations may be thought of as conflicting reports as to the probability of heads. Intuitively, any individual report, say 0.47, is compatible with the evidence, and indeed intermediary degrees of belief such as 0.48 seem reasonable. On the other hand, a degree of belief that falls outside the range of reports, say 0.9, does not seem warranted by the evidence. Thus evidence constrains degree of belief to lie in the smallest closed interval that contains all the reports.

As mentioned in §12, the logical norm is explicated using the Maximum Entropy Principle: entropy is a measure of the lack of commitment of a probability function, so $P_\mathcal{E}$ should be the probability function, out of all those that satisfy constraints imposed by $\mathcal{E}$, that has maximum entropy. Justifications of the Maximum Entropy Principle are well known — see [Jaynes, 2003], [Paris, 1994] or [Paris and Vencovská, 2001] for example.

We can thus put the two norms on a more formal footing. Given a domain V of finitely many variables, each of which takes finitely many values, an agent with evidence $\mathcal{E}$ should adopt as her belief function the probability function $P_\mathcal{E}$ on V determined as follows:

Empirical: $P_\mathcal{E}$ should satisfy any constraints imposed by her evidence $\mathcal{E}$: $P_\mathcal{E}$ should lie in the smallest closed convex set $\mathbb{E}$ of probability functions containing those probability functions that are compatible with the reports in $\mathcal{E}$.[13]

Logical: $P_\mathcal{E}$ should otherwise be as non-committal as possible: $P_\mathcal{E}$ should be a member of $\mathbb{E}$ that maximises entropy $H(P) = -\sum_{v@V} P(v) \log P(v)$.

It turns out that there is a unique entropy maximiser on a closed convex set of probability functions: the degrees of belief $P_\mathcal{E}$ that an agent should adopt are uniquely determined by her evidence $\mathcal{E}$. Thus on a finite domain there is no room for subjective choice of degrees of belief.

---

[13] See [Williamson, 2005a, §5.3] for more detailed discussion of this norm. There it is argued that $\mathbb{E}$ is constrained not only by quantitative evidence of physical probability but also evidence of qualitative relations between variables such as causal relations. See §18 on this point.

14 CHALLENGES

While objective Bayesianism is popular amongst practitioners — e.g., in statistics, artificial intelligence, physics and engineering — it has not been widely accepted by philosophers, however, largely because there are a number of perceived problems with the interpretation. Several of these problems have in fact already been resolved, but other challenges remain. In the remainder of this part of the paper we shall explore the key challenges and assess the prospects of objective Bayesianism.

In §15 we shall see that one challenge is to motivate the adoption of a logical norm. Objective Bayesianism has also been criticised for being language dependent (§16) and for being impractical from a computational point of view (§17). Handling qualitative evidence poses a significant challenge (§18), as does extending objective Bayesianism to infinite event or sentence frameworks (§19). The question of whether objective Bayesianism can be used to provide an interpretation of objective chance is explored in §20, while §21 considers the application of objective Bayesianism to providing semantics for probability logic.

Jaynes points out that the Maximum Entropy Principle is a powerful tool but warns

> Of course, it is as true in probability theory as in carpentry that introduction of more powerful tools brings with it the obligation to exercise a higher level of understanding and judgement in using them. If you give a carpenter a fancy new power tool, he *may* use it to turn out more precise work in greater quantity; or he may just cut off his thumb with it. It depends on the carpenter ([Jaynes, 1979, pp. 40–41 of the original 1978 lecture]).

15 MOTIVATION

The first key question concerns the motivation behind objective Bayesianism. Recall that in §12 objective Bayesianism was motivated by the need for objective probabilities in science. Many Bayesians accept this desideratum and indeed accept the empirical norm (so that degrees of belief are constrained by evidence of frequencies, symmetries, etc.) but do not go as far as admitting a logical norm. The ensuing position, according to which degrees of belief reflect evidence but need not be maximally non-committal, is sometimes called *empirically-based subjective probability*. It yields degrees of belief that are more objective (i.e., more highly constrained) than those of strictly subjective Bayesianism, yet not as objective as those of objective Bayesianism — there is generally still some room for subjective choice of degrees of belief. The key question is thus: what grounds are there for going beyond empirically-based subjective probability and adopting objective Bayesianism?

Current justifications of the logical norm fail to address this question. Jaynes' original justification of the Maximum Entropy Principle ran like this: *given that degrees of belief ought to be maximally non-committal*, Shannon's information theory shows us that they are entropy-maximising probabilities ([Jaynes, 1957]). This type of justification assumes from the outset that some kind of logical norm is desired. On the other hand, axiomatic derivations of the Maximum Entropy Principle take the following form: *given that we need a procedure for objectively determining degrees of belief from evidence*, and given various desiderata that such a procedure should satisfy, that procedure must be entropy maximisation ([Paris and Vencovská, 1990; Paris, 1994; Paris and Vencovská, 2001]). This type of justification takes objectivity of rational degrees of belief for granted. Thus the challenge is to augment current justifications, perhaps by motivating non-committal degrees of belief or by motivating the strong objectivity of objective Bayesianism as opposed to the partial objectivity yielded by empirically-based subjective probability.

One possible approach is to argue that empirically-based subjective probability is *not objective enough* for many applications of probability. Many applications of probability follow a Bayesian statistical methodology: produce a *prior* probability function P_t, collect some evidence u, and draw predictions using the *posterior* probability function $P_{t+1}(v) = P_t(v|u)$. Now the prior function is determined before empirical evidence is available; this is matter of subjective choice for empirically-based subjectivists. However, the ensuing conclusions and predictions may be sensitive to this initial choice, rendering them subjective too. Yet such relativism is anathema in science: a disagreement between agents about a hypothesis should be arbitrated by evidence; it should be a fact of the matter, not mere whim, as to whether the evidence confirms the hypothesis.

That argument is rather inconclusive however. The proponent of empirically-based subjective probability can counter that scientists have simply over-estimated the extent of objectivity in science, and that subjectivity needs to be made explicit. Even if one grants a need for objectivity, one could argue that it is a pragmatic need: it just makes science simpler. The objective Bayesian must accept that it cannot be empirical warrant that motivates the selection of a particular belief function from all those compatible with evidence, since all such belief functions are equally warranted by available empirical evidence. In the absence of any non-empirical justification for choosing a particular belief function, such a function can only be considered objective in a *conventional* sense. One can drive on the right or the left side of the road; but we must all do the same thing; by convention in the UK we choose the left. That does not mean that the left is objectively correct or most warranted — either side will do.

A second line of argument offers explicitly pragmatic reasons for selecting a particular belief function. If probabilities are subjective then measuring probabilities must involve elicitation of degrees of belief from agents. As developers of expert systems in AI have found, elicitation and the associated consistency-checking are prohibitively time-consuming tasks (the inability of elicitation to keep pace with the demand for expert systems is known as *Feigenbaum's bottleneck*). If a subjective approach is to be routinely applied throughout science it is clear that a similar bottleneck will be reached. On the other hand, if degrees of belief are objectively determined by evidence then elicitation is not required — degrees of belief are calculated by maximising entropy. Objective Bayesianism is thus to be preferred for reasons of efficiency.

Indeed many Bayesian statisticians now (often tacitly) appeal to non-committal objective priors rather than embark on a laborious process of introspection, elicitation or analysis of sensitivity of posterior to choice of prior.

A third motivating argument appeals to caution. In many applications of probability the risks attached to bold predictions that turn out wrong are high. For instance, a patient's symptoms may narrow her condition down to meningitis or 'flu, but there may be no empirical evidence — such as information about relative prevalence — to decide between the two. In this case, the risks associated with meningitis are so much higher than those associated with 'flu, that a non-committal belief function seems more appropriate as a basis for action than a belief function that gives the probability of meningitis to be zero, even though both are compatible with available information. (With a non-committal belief function one will not dismiss the possibility of meningitis, but if one gives meningitis probability zero one will disregard it.) High-risk applications thus favour cautious conclusions, non-committal degrees of belief and an objective Bayesian approach.

I argue in [Williamson, 2007b] that the appeal to caution is the most decisive motivation for objective Bayesianism, although pragmatic considerations play a part too.

16 LANGUAGE DEPENDENCE

The Maximum Entropy Principle has been criticised for being language or representation dependent: it has been argued that the principle awards the same event different probabilities depending on the way in which the problem domain is formulated.

John Maynard Keynes surveyed several purported examples of language dependence in his discussion of Laplace's Principle of Indifference ([Keynes, 1921]). This latter principle advocates assigning the same probability to each of a number of possible outcomes in the absence of any evidence which favours one outcome over the others. Keynes added the condition that the possible outcomes must be indivisible ([Keynes, 1921, §4.21]). The Maximum Entropy Principle makes the same recommendation in the absence of evidence and so inherits any language dependence of the Principle of Indifference.

A typical example of language dependence proceeds as follows ([Halpern and Koller, 1995, §1]). Suppose an agent's language can be represented by the propositional language $\mathcal{L} = \{C\}$ with just one propositional variable C which asserts that a particular book is colourful. The agent has no evidence and so by the Principle of Indifference (or equally by the Maximum Entropy Principle) assigns $P(C) = P(\neg C) = 1/2$. But now consider a second language $\mathcal{L}' = \{R, B, G\}$ where R signifies that the book is red, B that it is blue and G that it is green. An agent with no evidence will give $P(\pm R \wedge \pm B \wedge \pm G) = 1/8$. Now $\neg C$ is equivalent to $\neg R \wedge \neg B \wedge \neg G$, yet the former is given probability $\frac{1}{2}$ while the latter is given probability $\frac{1}{8}$. Thus the probability assignments of the Principle of Indifference and the Maximum Entropy Principle depend on choice of language.

[Paris and Vencovská, 1997] offer the following resolution. They argue that the Maximum Entropy Principle has been misapplied in this type of example: if an agent refines the propositional variable C into $R \vee B \vee G$ one should consider not $\mathcal{L}'$ but $\mathcal{L}'' = \{C, R, B, G\}$ and make the agent's evidence, namely $C \leftrightarrow R \vee B \vee G$, explicit. If we do that then the probability function on $\mathcal{L}''$ with maximum entropy, out of all those that satisfy the evidence (i.e., which assign $P(C \leftrightarrow R \vee B \vee G) = 1$), will yield a value $P(\neg C) = 1/2$. This is just the same value as that given by the Maximum Entropy Principle on $\mathcal{L}$ with no evidence. Thus there is no inconsistency.

This resolution is all well and good if we are concerned with a single agent who refines her language. But the original problem may be construed rather differently. If *two* agents have languages $\mathcal{L}$ and $\mathcal{L}'$ respectively, and no evidence, then they assign two different probabilities to what we know (but they don't know) is the same proposition. There is no getting round it: probabilities generated by the Maximum Entropy Principle depend on language as well as evidence.

Interestingly, language dependence in this latter multilateral sense is not confined to the Maximum Entropy Principle. As [Halpern and Koller, 1995] and [Paris and Vencovská, 1997] point out, there is no non-trivial principle for selecting rational degrees of belief which is language-independent in the multilateral sense. More precisely, suppose we want a principle that selects a set $\mathbb{O}_\mathcal{E}$ of probability functions that are optimally rational on the basis of an agent's evidence $\mathcal{E}$. If $\mathbb{O}_\mathcal{E} \subseteq \mathbb{E}$, i.e., if every optimally rational probability function must satisfy constraints imposed by $\mathcal{E}$, and if $\mathbb{O}_\mathcal{E}$ ignores irrelevant information inasmuch as $\mathbb{O}_{\mathcal{E} \cup \mathcal{E}'}(\theta) = \mathbb{O}_\mathcal{E}(\theta)$ whenever $\mathcal{E}'$ involves no propositional variables in sentence θ, then the only candidate for $\mathbb{O}_\mathcal{E}$ that is multilaterally language independent is $\mathbb{O}_\mathcal{E} = \mathbb{E}$ ([Halpern and Koller, 1995, Theorem 3.10]). Only empirically-based subjective probability is multilaterally language independent.

So much the better for empirically-based subjective probability and so much the worse for objective Bayesianism, one might think. But such an inference is too quick. It takes the desirability of multilateral language independence for granted. I argue in [Williamson, 2005a, Chapter 12] that an agent's language constitutes empirical evidence:[14] evidence of natural kinds, evidence concerning which variables are relevant to which, and perhaps even evidence of which partitions are amenable to the Principle of Indifference. For example, having dozens of words for snow in one's language says something about the environment in which one lives. Granted that language itself is a kind of evidence, and granted that an agent's degrees of belief should depend on her evidence, language independence becomes a rather dubious desideratum.

Note that while [Howson, 2001, p. 139] criticises the Principle of Indifference on

---

[14][Halpern and Koller, 1995, §4] also suggest this tack, although they do not give their reasons. Interestingly, though, they do show in §5 that relaxing the notion of language independence leads naturally to an entropy-based approach.

account of its language dependence, the example he cites can be used to support the case *against* language independence as a desideratum. Howson considers two first-order languages with equality: $\mathcal{L}_1$ has just a unary predicate U while $\mathcal{L}_2$ has unary U together with two constants t_1 and t_2. The explicit evidence $\mathcal{E}$ is just 'there are exactly 2 individuals', while sentence θ is 'something has the property U'. $\mathcal{L}_1$ has three models of $\mathcal{E}$, which contain 0, 1 and 2 instances of U respectively, so $P(\theta) = 2/3$. In $\mathcal{L}_2$ individuals can be distinguished by constants and thus there are eight models of $\mathcal{E}$ (if constants can name the same individual), six of which satisfy θ so $P(\theta) = 3/4 \neq 2/3$. While this is a good example of language dependence, the question remains whether language dependence is a problem here. As Howson himself hints, $\mathcal{L}_1$ might be an appropriate language for talking about bosons, which are indistinguishable, while $\mathcal{L}_2$ is more suited to talk about classical particles, which are distinguishable and thus able to be named by constants. Hence choice of language $\mathcal{L}_2$ over $\mathcal{L}_1$ indicates distinguishability, while conversely choice of $\mathcal{L}_1$ over $\mathcal{L}_2$ indicates indistinguishability. In this example, then, language betokens implicit evidence. Of course all but the the most ardent subjectivists agree that an agent's degrees of belief ought to be influenced by her evidence. Therefore language independence becomes an inappropriate desideratum.

In sum, while the Principle of Indifference and the Maximum Entropy Principle have both been dismissed on the grounds of language dependence, it seems clear that some dependence on language is to be expected if degrees of belief are to adequately reflect implicit as well as explicit evidence. So much the better for objective Bayesianism, and so much the worse for empirically-based subjective probability which is language-invariant.

17 COMPUTATION

There are important concerns regarding the application of objective Bayesianism. One would like to apply objective Bayesianism in artificial intelligence: when designing an artificial agent it would be very useful to have normative rules which prescribe how the agent's beliefs should change as it gathers information about its world. However, there has seemed to be little prospect of fulfilling this hope, for the following reason. Maximising entropy involves finding the parameters $P(v)$ that maximise the entropy expression, but the number of such parameters is exponential in the number of variables in the domain, thus the size of the entropy maximisation problem quickly gets out of hand as the size of the domain increases. Indeed [Pearl, 1988, p. 468] has influentially criticised maximum entropy methods on account of their computational difficulties.

The computational problem poses a serious challenge for objective Bayesianism. However, recent techniques for more efficient entropy maximisation have largely addressed this issue. While no technique offers efficient entropy maximisation in all circumstances (entropy maximisation is an NP-complete problem), techniques exist that offer efficiency in a wide range of natural circumstances. I shall sketch the theory of *objective Bayesian nets* here — this is developed in detail in [Williamson,

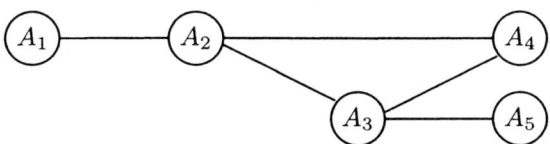

Figure 1. A constraint graph.

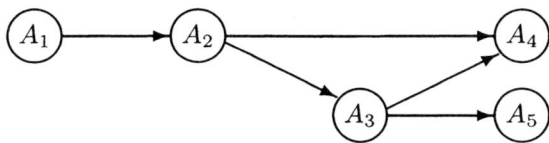

Figure 2. A directed constraint graph.

2005a, §§5.5–5.7] and [Williamson, 2005b].[15]

Given a set V of variables and some evidence $\mathcal{E}$ involving V which consists of a set of constraints on the agent's belief function P, one wants to find the probability function P, out of all those that satisfy the constraints in $\mathcal{E}$, that maximises entropy. This can be achieved via the following procedure. First form an undirected graph on vertices V by linking pairs of variables that occur in the same constraint with an edge. For example, if $V = \{A_1, A_2, A_3, A_4, A_5\}$ and $\mathcal{E}$ contains a constraint involving A_1 and A_2 (e.g., $P(a_2^1|a_1^0) = 0.9$), a constraint involving A_2, A_3 and A_4, a constraint involving A_3 and A_5 and a constraint involving just A_4, then the corresponding undirected constraint graph appears in Fig. 1. The undirected constraint graph has the following crucial property: if a set Z of variables separates $X \subseteq V$ from $Y \subseteq V$ in the graph then the maximum entropy function P will render X and Y probabilistically independent conditional on Z.

Next transform the undirected constraint graph into a directed constraint graph, Fig. 2 in the case of our example.[16] The independence property ensures that the directed constraint graph can be used as a graph in a *Bayesian net* representation of the maximum entropy function P. A Bayesian net offers the opportunity of a more efficient representation of a probability function P: in order to determine P, one only needs to determine the parameters $P(a_i|par_i)$, i.e., the probability distribution of each variable conditional on its parents, rather than the parameters $P(v)$, i.e., the joint probability distribution over all the variables. Depending on the structure of the directed graph, there may be far fewer parameters in the Bayesian net representation. In the case of our example, if we suppose that each variable has two possible values then the Bayesian net representation requires 11

---

[15] Maximum entropy methods have recently been applied to natural language processing, and other techniques for entropy maximisation have been tailored to that context — see [Della Pietra et al., 1997] for example.

[16] The algorithm for this transformation is given in [Williamson, 2005a, §5.7].

parameters rather than the 32 parameters $P(v)$ for each assignment v of values to V. For problems involving more variables the potential savings are very significant.

Roughly speaking, efficiency savings are greatest when each variable has few parents in the directed constraint graph, and this occurs when each constraint in $\mathcal{E}$ involves relatively few variables. Note that when dealing with large sets of variables it tends to be the case that while one might make a large number of observations, each observation involves relatively few variables. For example, one might use hospital data as empirical observations pertaining to a large number of health-related variables, each department of the hospital contributing some statistics; while there might be a large number of such statistics, each statistic is likely to involve relatively few variables, namely those variables that are relevant to the department in question; such observations would yield a sparse constraint graph and an efficient Bayesian net representation. Hence this method for reducing the complexity of entropy maximisation offers efficiency savings that are achievable in a wide range of natural situations.

A Bayesian net that represents the probability function produced by the Maximum Entropy Principle is called an *objective Bayesian net*. See [Nagl et al., 2008] for an application for the objective Bayesian net approach to cancer prognosis and systems biology.

18 QUALITATIVE KNOWLEDGE

The Maximum Entropy Principle has been criticised for yielding the wrong results when the agent's evidence contains qualitative causal information ([Pearl, 1988, p. 468]; [Hunter, 1989]). Daniel Hunter gives the following example:

> The puzzle is this: Suppose that you are told that three individuals, Albert, Bill and Clyde, have been invited to a party. You know nothing about the propensity of any of these individuals to go to the party nor about any possible correlations among their actions. Using the obvious abbreviations, consider the eight-point space consisting of the events $ABC, AB\bar{C}, A\bar{B}C$, etc. (conjunction of events is indicated by concatenation). With no constraints whatsoever on this space, MAXENT yields equal probabilities for the elements of this space. Thus $Prob(A) = Prob(B) = 0.5$ and $Prob(AB) = 0.25$, so A and B are independent. It is reasonable that A and B turn out to be independent, since there is no information that would cause one to revise one's probability for A upon learning what B does. However, suppose that the following information is presented: Clyde will call the host before the party to find out whether Al or Bill or both have accepted the invitation, and his decision to go to the party will be based on what he learns. Al and Bill, however, will have no information about whether or not Clyde will go to the party. Suppose, further, that we are told the probability that Clyde will go conditional

> on each combination of Al and Bill's going or not going. For the sake of specificity, suppose that these conditional probabilities are ... $[P(C|AB) = 0.1, P(C|A\bar{B}) = 0.5, P(C|\bar{A}B) = 0.5, P(C|\bar{A}\bar{B}) = 0.8]$.
>
> When MAXENT is given these constraints ... A and B are no longer independent! But this seems wrong: the information about Clyde should not make A's and B's actions dependent ([Hunter, 1989, p. 91])

But this counter-intuitive conclusion is attributable to a misapplication of the Maximum Entropy Principle. The conditional probabilities are allowed to constrain the entropy maximisation process but the knowledge that Al's and Bill's decisions are causes of Clyde's decision is simply ignored. This failure to consider the qualitative causal evidence leads to the counter-intuitive conclusion.

Keynes himself had stressed the importance of taking qualitative knowledge into account and the difficulties that ensue if qualitative information is ignored:

> Bernoulli's second axiom, that in reckoning a probability we must take everything into account, is easily forgotten in these cases of statistical probabilities. The statistical result is so attractive in its definiteness that it leads us to forget the more vague though more important considerations which may be, in a given particular case, within our knowledge ([Keynes, 1921, p. 322]).

Indeed, in the party example, the temptation is to consider only the definite probabilities and to ignore the important causal evidence.

The party example and Keynes' advice highlight an important challenge for objective Bayesianism. In order that objective Bayesianism can be applied, all evidence — *qualitative* as well as quantitative — must be taken into account. However, objective Bayesianism as outlined in §13 depends on evidence taking *quantitative* form: evidence must be explicated as a set of quantitative constraints on degrees of belief in order to narrow down a set of probability functions that satisfy those constraints. Thus the general challenge for objective Bayesianism is to show how qualitative evidence can be converted into precise quantitative constraints on degrees of belief.

To some extent this challenge has already been met. In the case where qualitative evidence takes the form of causal constraints, as in Hunter's party example above, I advocate a solution which exploits the following asymmetry of causality. Learning of the existence of a common cause of two events may warrant a change in the degrees of belief awarded to them: one may reason that if one event occurs, then this may well be because the common cause has occurred, in which case the other event is more likely — the two events become more dependent than previously thought. On the other hand, learning of the existence of a common effect would not warrant a change in degrees of belief: while the occurrence of one event may make the common effect more likely, this has no bearing on the other cause. This asymmetry motivates what I call the *Causal Irrelevance Principle*: if the agent's language contains a variable A that is known not to be a cause of any of the other variables, then her degrees of belief concerning these other variables should be the same as the degrees of belief she should adopt were she not to have A in her language (as long as any quantitative evidence involving A is compatible with those degrees of belief). The Causal Irrelevance Principle allows one to transfer qualitative causal evidence into quantitative constraints on degrees of belief — if domain $V = U \cup \{A\}$ then we have constraints of the form $P^V_{|U} = P^U$, i.e., the agent's belief function defined on V, when restricted to U, should be the same as the belief function defined just on U. By applying the Causal Irrelevance Principle, qualitative causal evidence as well as quantitative information can be used to constrain the entropy maximisation process. It is not hard to see that use of the principle avoids counter-intuitive conclusions like those in Hunter's example: knowledge that Clyde's decision is a common effect of Al's and Bill's decision ensures that Al's and Bill's actions are probabilistically independent, as seems intuitively plausible. See [Williamson, 2005a, §5.8] for a more detailed analysis of this proposal.

Thus the challenge of handling qualitative evidence has been met in the case of causal evidence. Moreover, by treating logical influence analogously to causal influence one can handle qualitative logical evidence using the same strategy ([Williamson, 2005a, Chapter 11]). But the challenge has not yet been met in other cases of qualitative evidence. In particular, I claimed in §16 that choice of language implies evidence concerning the domain. Clearly work remains to be done to render such evidence explicit and quantitative, so that it can play a role in the entropy maximisation process.

There is another scenario in which the challenge is only beginning to be met. Some critics of the Maximum Entropy Principle argue that objective Bayesianism renders learning from experience impossible, as follows. The Maximum Entropy Principle will, in the absence of evidence linking them, render outcomes probabilistically independent. Thus observing outcomes will not change degrees of belief in unobserved outcomes if there is no evidence linking them: observing a million ravens, all black, will not shift the probability of the next raven being black from $\frac{1}{2}$ (which is the most non-committal value given only that there are two outcomes, black or not black). So, the argument concludes, there is no learning from experience. The problem with this argument is that we do have evidence that connects the outcomes — the qualitative evidence that we are repeatedly sampling ravens to check whether they are black — but this evidence is mistakenly being ignored in the application of the Maximum Entropy Principle. Qualitative evidence should be taken into account so that learning from experience becomes possible — but how? [Carnap, 1952] and [Carnap, 1971] addressed the problem, as have [Paris and Vencovská, 2003]; [Williamson, 2007a] and [Williamson, 2008c] more recently. Broadly speaking, the idea behind this line of work is to take the maximally non-committal probability function to be one which permits learning from experience, as opposed to the maximum entropy probability function which does not. The difficulty with this approach is that it does genuinely seem to be the maximum entropy function that is most non-committal. An altogether different approach, developed in [Williamson, 2008b, §5], is to argue that learning from experience should result from the empirical norm rather than the logical norm: observing a million ravens, all black, does not merely impose the constraint that the agent should fully believe that *those* ravens are black — it also imposes the constraint that the agent should strongly if not fully believe that *other* (unobserved) ravens are also black. Then the agent's belief function should as usual be a function, from all those that satisfy these constraints, that has maximum entropy. This alternative approach places the problem of learning from experience firmly in the province of statistics rather than inductive logic.

19 INFINITE DOMAINS

The Maximum Entropy Principle is most naturally defined on a finite domain — for example, a space of finitely many variables each of which takes finitely many values, as in §2. The question thus arises as to whether one can extend the applicability of objective Bayesianism to infinite domains. In the variable framework, one might be interested in domains with infinitely many variables, or domains of variables with an infinite range. Alternatively, one might want to apply objective Bayesianism to full generality of the mathematical framework of §3, or to infinite logical languages (§4). This challenge has been confronted, but at the expense of some objectivity, as we shall now see.

There are two lines of work here, one of which proceeds as follows. [Paris and Vencovská, 2003] treat problems involving countable logical languages as limiting cases of finite problems. Consider a countably infinite domain $V = \{A_1, A_2, \ldots\}$ of variables taking finitely many values, and schematic evidence $\mathcal{E}$ which may pertain to infinitely many variables. If $V_n = \{A_1, \ldots, A_n\}$ and $\mathcal{E}_n$ is that part of $\mathcal{E}$ that involves only variables in V_n, then $P_{\mathcal{E}_n}^{V_n}(u)$ can be found by maximising entropy as usual (here $u@U \subseteq V_n$). Interestingly — see [Paris and Vencovská, 2003] — the limit $\lim_{n \to \infty} P_{\mathcal{E}_n}^{V_n}(u)$ exists, so one can define $P_{\mathcal{E}}^{V}(u)$ to be this limit. [Paris and Vencovská, 2003] show that this approach can be applied to very simple predicate languages and conjecture that it is applicable more generally to predicate logic.

In the transition from the finite to the infinite, the question arises as to whether countable additivity (introduced in §3) holds. [Paris and Vencovská, 2003] make no demand that this axiom hold. Indeed, it seems that the type of schematic evidence that they consider cannot be used to express the evidence that an infinite set of outcomes forms a partition. Thus the question of countable additivity cannot be formulated in their framework. In fact, even if one were to extend the framework to formulate the question, the strategy of taking limits would be unlikely to yield probabilities satisfying countable additivity. If the only evidence is that $E_1, \ldots, E_n$ partition the outcome space, maximising entropy will give each event the same probability $1/n$. Taking limits will assign members of an infinite partition probability $\lim_{n \to \infty} 1/n = 0$. But then $\sum_{i=1}^{\infty} P(E_i) = 0 \neq 1$, contradicting countable additivity.

However, not only is countable additivity important from the point of view of mathematical convenience, but according to the standard betting foundations for Bayesian interpretations of probability introduced in §9, countable additivity *must* hold: an agent whose betting quotients are not countably additive can be Dutch booked ([Williamson, 1999]). Once we accept countable additivity, we are forced either to concede that the strategy of taking limits has only limited applicability, or to reject the method altogether in favour of some alternative, as yet unformulated, strategy. Moreover, as argued in [Williamson, 1999], we are forced to accept a certain amount of subjectivity: a countably additive distribution of probabilities over a countably infinite partition must award some member of the partition more probability than some other member; but if evidence does not favour any member over any other then it is just a matter of subjective choice as to how one skews the distribution.

The other line of work deals with uncountably infinite domains. [Jaynes, 1968, §6] presents essentially the following procedure. First find a non-negative real function $P_=(x)$, which we may call the *equivocator* or *invariance* function, that represents the invariances of the problem in question: if $\mathcal{E}$ offers nothing to favour x over y then $P_=(x) = P_=(y)$. Next, find a probability function P satisfying $\mathcal{E}$ that is closest to the invariance function $P_=$, in the sense that it minimises *cross-entropy* distance $d(P, P_=) = \int P(x) \log P(x)/P_=(x) dx$. It is this function that one ought to take as one's belief function $P_\mathcal{E}$.[17]

This approach generalises entropy maximisation on discrete domains. In the case of finite domains $P_=$ can be taken to be the probability function found by maximising entropy subject to no constraints; the probability function $P_\mathcal{E} \in \mathbb{E}$ that is closest to it is just the probability function in $\mathbb{E}$ that has maximum entropy. If the set of variables admits n possible assignments of values, the equivocator $P_=$ can be taken as the function that gives value $1/n$ to each possible assignment v; this is a probability function so $P_\mathcal{E} = P_=$ if there is no evidence whatsoever. In the case of countably infinite domains $P_=$ may not be a probability function: as discussed above $P_=$ must award the same value, k say, to each member of a countable partition; however, such a function cannot be a probability function since countable additivity fails; therefore one must choose a probability function closest to $P_=$. Here we might try to minimise $d(P, P_=) = \sum P(v) \log P(v)/P_=(v) = \sum P(v) \log P(v) - \log k \sum P(v) = \sum P(v) \log P(v) - \log k$; this is minimised just when the entropy $-\sum P(v) \log P(v)$ is maximised. Of course entropy may well be infinite on an infinite partition, so this approach will not work in general; nevertheless a refinement of this kind of approach can yield a procedure for selecting $P_\mathcal{E} \in \mathbb{E}$ that is decisive in many cases ([Williamson, 2008a]).

By drawing this parallel with the discrete case we can see where problems for objectivity arise in the infinite case: even if the set $\mathbb{E}$ of probability functions compatible with evidence is closed and convex, there may be no probability function

---

[17]Objective Bayesian statisticians have developed a whole host of techniques for obtaining invariance functions and uninformative probability functions — see, e.g., [Kass and Wasserman, 1996]. [Berger and Pericchi, 2001] discuss the use of such priors in statistics.

in $\mathbb{E}$ closest to $P_=$ or there may be more than one probability function closest to $P_=$. This latter case, non-uniqueness, means subjectivity: the agent can exercise arbitrary choice as which distribution of degrees of belief to select. Subjectivity can also enter at the first stage, choice of $P_=$, since there may be cases in which several different functions represent the invariances of a problem.[18]

But does such subjectivity really matter? Perhaps not. Although objective Bayesianism often yields objectivity, it can hardly be blamed where little is to be found. If there is nothing to decide between two belief functions, then subjectivity simply does not matter. Under such a view, all the Bayesian positions — strict subjectivism, empirically-based subjective probability and objective Bayesianism — accept the fact that selection of degrees of belief can be a matter of arbitrary choice, they just draw the line in different places as to the extent of subjectivity. Strict subjectivists allow most choice, drawing the line at infringements of the axioms of probability.[19] Proponents of empirically-based subjective probability occupy a half-way house, allowing extensive choice but insisting that evidence of physical probabilities as well as the axioms of probability constrain degrees of belief. Objective Bayesians go furthest by also using logical constraints to narrow down the class of acceptable degrees of belief.

Moreover, arguably the infinite is just a tool to help us reason about the large but finite and discrete universe in which we live ([Hilbert, 1925]). Just as we create infinite continuous geometries to reason about finite discrete space, we create continuous probability spaces to reason about discrete situations. In which case if subjectivity infects the infinite then we can only conclude that the infinite may not be as effective a tool as we would like for probabilistic reasoning. Such relativity merely urges caution when idealising to the infinite; it does not tell against objective Bayesianism.

20 FULLY OBJECTIVE PROBABILITY

We see then that objectivity is a matter of degree and that while subjectivity may infect some problems, objective Bayesianism yields a high degree of objectivity. We have been focussing on what we might call *epistemic objectivity*, the extent to which an agent's degrees of belief are determined by her evidence. In applications of probability a high degree of epistemic objectivity is an important desideratum: disagreements as to probabilities can be attributed to differences in evidence; by agreeing on evidence consensus can be reached on probabilities.

While epistemic objectivity requires uniqueness relative to evidence, there are stronger grades of objectivity. In particular, the strongest grade of objectivity, *full objectivity*, i.e., uniqueness simpliciter, arouses philosophical interest. Are prob-

---

[18] See [Gillies, 2000, pp. 37–49]; [Jaynes, 1968, §§6–8] and [Jaynes, 1973]. The determination of invariant measures has become an important topic in statistics — see [Berger and Pericchi, 2001].

[19] Subjectivists usually slip in a few further constraints: e.g., known truths must be given probability 1, and degrees of belief should be updated by Bayesian conditionalisation.

abilities uniquely determined, independently of evidence? If two agents disagree as to probabilities must at least one of them be wrong, even if they disagree as to evidence? Intuitively many probabilities are fully objective: there seems to be a fact of the matter as to the probability that an atom of cobalt-60 will decay in 5 years, and there seems to be a fact of the matter as to the chance that a particular roulette wheel will yield a black on the next spin. (A qualification is needed. Chances cannot be quite fully objective inasmuch as they depend on time. There might now be a probability just under 0.5 of cobalt-60 atom decaying in the next five years; after the event, if it has decayed its chance of decaying in that time-frame is 1. Thus chances need to be indexed by time.)

As indicated in §10, objective Bayesianism has the wherewithal to meet the challenge of accounting for intuitions about full objectivity. By considering some ultimate evidence $\hat{\mathcal{E}}$ one can define fully objective probability $\hat{P} = P_{\hat{\mathcal{E}}}$ in terms of the degrees of belief one ought to adopt if one were to have this ultimate evidence. This is the *ultimate belief* notion of chance.

What should be included in $\hat{\mathcal{E}}$? Clearly it should include all information relevant to the domain at time t. To be on the safe side we can take $\hat{\mathcal{E}}$ to include all facts about the universe that are determined by time t — the entire history of the universe up to and including time t. (Remember: this challenge is of philosophical rather than practical interest.)

While the ultimate belief notion of chance is relatively straightforward to state, much needs to be done to show that this type of approach is viable. One needs to show that this notion can capture our intuitions about chance. Moreover, one needs to show that that account is coherent — in particular, one might have concerns about circularity: if probabilistic beliefs are beliefs about probability, yet probability is defined in terms of probabilistic beliefs, then probability appears to be defined in terms of itself.

However, this apparent circularity dissolves when we examine the premisses of this circularity argument more closely. Indeed, at most one premiss can be true. In our framework, 'probability is defined in terms of probabilistic beliefs' is true if we substitute 'fully objective single-case probability' or 'chance' for 'probability' and 'degrees of belief' for 'probabilistic beliefs': chance is defined in terms of degrees of belief. But then the first premiss is false. Degrees of belief are not beliefs about chance, they are partial beliefs about elements of a domain — variables, events or sentences. According to this reading 'probabilistic' modifies 'belief', isolating a type of belief; it does not specify the object of belief. On the other hand, if the first premiss is to be true and 'probabilistic beliefs' are construed as beliefs about probability, then the second premiss is false since chance is not here defined in terms of beliefs about probability. Thus neither reading permits the conclusion that probability is defined in terms of itself.

Note that Bayesian statisticians often consider probability distributions over probability parameters. These *can* be interpreted as degrees of belief about chances, where chances are special degrees of belief. But there is no circularity here either. This is because the degrees of belief about chances are of a higher order than the chances themselves. Consider, for instance, a degree of belief that a particular coin toss will yield heads. The present chance of the coin toss yielding heads can be defined using such degrees of belief. One can then go on to formulate the higher-order degree of belief that the chance of heads is 0.5. But this degree of belief is not used in the (lower order) definition of the chance itself, so there is no circularity. (One can go on to define higher and higher order chances and degrees of belief — regress, rather than circularity, is the obvious problem.)

One can make a stronger case for circularity though. One can read the empirical norm of §13 as saying that degrees of belief ought to be set to chances where they are known (see [Williamson, 2005a, §5.3]). Under such a reading the concept of rational degree of belief appeals to the notion of chance, yet in this section chances are being construed as special degrees of belief; circularity again. Here circularity is not an artifice of ambiguity of terms like 'probabilistic beliefs'. However, as before, circularity does disappear under closer investigation. One way out is to claim that there are two notions of chance in play: a physical notion which is used in the empirical norm, and an ultimate belief notion which is defined in terms of degrees of belief. But this strategy would not appeal to those who find a physical notion of chance metaphysically or epistemologically dubious. An alternative strategy is to argue that any notion of chance in the formulation of an empirical norm is simply *eliminable*. One can substitute references to chance with references to the *indicators* of chance instead. Intuitively, symmetry considerations, physical laws and observed frequencies all provide some evidence as to chances; one can simply say that an agent's degrees of belief should be appropriately constrained by her evidence of symmetries, laws and frequencies. While this may lead to a rather more complicated formulation of the empirical norm, it is truer to the epistemological route to degrees of belief — the agent has direct evidence of the indicators of chances rather than the chances themselves. Further, it shows how these indicators of chances can actually provide evidence for chances: evidence of frequencies constrains degrees of belief, and chances are just special degrees of belief. Finally, this strategy eliminates circularity, since it shows how degrees of belief can be defined independently of chances. It does, however, pose the challenge of explicating exactly how frequencies, symmetries and so on constrain degrees of belief — a challenge that (as we saw in §18) is not easy to meet.

The ultimate belief notion of chance is not quite fully objective: it is indexed by time. Moreover, if we want a notion of chance defined over infinite domains then, as the arguments of §19 show, subjectivity can creep in, for example in cases — if such cases ever arise — in which the entire history of the universe fails to differentiate between the members of an infinite partition. This mental, ultimate belief notion of chance is arguably more objective than the influential physical notion of chance put forward by David Lewis however ([Lewis, 1980; Lewis, 1994]). Lewis accepts a version of the empirical norm which he calls the *Principal Principle*: evidence of chances ought to constrain degrees of belief. However Lewis does not go on to advocate the ultimate belief notion of chance presented here: 'chance is [not] the credence warranted by our total available evidence ... if our total evidence came from misleadingly unrepresentative samples, that wouldn't affect chance in any way' ([Lewis, 1994, p. 475]). (Unrepresentative samples do not seem to me to be a real problem for the ultimate belief approach, because the entire history of the universe up to the time in question is likely to contain more information pertinent to an event than simply a small sample frequency — plenty of large samples of relevant events, and plenty of relevant qualitative information, for instance.) Lewis instead takes chances to be products of the best system of laws, the best way of systematising the universe. The problem is that the criteria for comparing systems of laws — a balance between simplicity and strength — seem to be subjective. What counts as simple for a rocket scientist may be complicated for a robot and vice versa.[20] This is not a problem that besets the ultimate belief account: as Lewis accepts, there does seem to be a fact of the matter as to how evidence should inform degrees of belief. Thus an ultimate belief notion of chance, despite being a mental rather than physical notion, suffers less from subjectivity than Lewis' theory.

Note that Lewis' approach also suffers from a type of circularity known as *undermining*. Because chances for Lewis are analysed in terms of laws, they depend not only on the past and present state of the universe, but also on the future of the universe: 'present chances are given by probabilistic laws, plus present conditions to which those laws are applicable, and ... those laws obtain in virtue of the fit of candidate systems to the whole of history' ([Lewis, 1994, p. 482]). Of course, non-actual futures (i.e., series of events which differ from the way in which the universe will actually turn out) must have positive chance now, for otherwise the notion of chance would be redundant. Thus there is now a positive chance of events turning out in the future in such a way that present chances turn out differently. But this yields a paradox: present chances cannot turn out differently to what they actually are. [Lewis, 1994] has to modify the Principal Principle to avoid a formal contradiction, but this move does not resolve the intuitive paradox. In contrast, under the ultimate belief account present chances depend on just the past and the present state of the universe, not the future, so present chances cannot undermine themselves.

21 PROBABILITY LOGIC

There are increasing demands from researchers in artificial intelligence for formalisms for normative reasoning that combine probability and logic. Purely probabilistic techniques work quite well in many areas but fail to exploit logical relationships that obtain in particular problems. Thus, for example, probabilistic techniques are applied widely in natural language processing ([Manning and Schütze, 1999]), with some success, yet largely without exploiting logical sentence structure. On the other hand, purely logical techniques take problem structure into account

---

[20]In response [Lewis, 1994, p. 479] just plays the optimism card: 'if nature is kind to us, the problem needn't arise.'

without being able to handle the many uncertainties inherent in practical problem solving. Thus automated proof systems for mathematical reasoning ([Quaife, 1992; Schumann, 2001]) depend heavily on implementing logics but often fail to prioritise searches that are most likely to be successful. It is natural to suppose that systems which combine probability and logic will yield improved results. Formalisms that combine probability and logic would also be applicable to many new problems in bioinformatics ([Durbin et al., 1999]), from inducing protein folding from noisy relational data to forecasting toxicity from uncertain evidence of deterministic chemical reactions in cell metabolism.

In a *probability logic*, or *progic* for short, probability is combined with logic in one or more of the following two ways:

External: probabilities are *attached to* sentences of a logical language,

Internal: sentences *incorporate* statements about probabilities.

In an *external* progic, entailment relationships take the form

$$\varphi_1^{X_1}, \ldots, \varphi_n^{X_n} \models \psi^Y.$$

Here $\varphi_1, \ldots, \varphi_n, \psi \in S\mathcal{L}$ are sentences of a logical language $\mathcal{L}$ which does not contain probabilities and $X_1, \ldots, X_n, Y \subseteq [0,1]$ are sets of probabilities. For example if $\mathcal{L} = \{A_1, A_2, A_3, A_4, A_5\}$ is a propositional language on propositional variables $A_1, \ldots, A_5$, we might be interested in what set Y of probabilities to attach to the conclusion in

$$A_1 \wedge \neg A_2^{.9}, (\neg A_4 \vee A_3) \to A_2^{.2}, A_5 \vee A_3^{.3}, A_4^{.7} \models A_5 \to A_1^Y.$$

In an *internal* progic, entailment relationships take the form

$$\varphi_1, \ldots, \varphi_n \models \psi$$

where $\varphi_1, \ldots, \varphi_n, \psi \in S\mathcal{L}_P$ are sentences of a logical language $\mathcal{L}_P$ which contains probabilities. $\mathcal{L}_P$ might be a first-order language with equality containing a (probability) function P, predicates U_1, U_2, U_3 and constants sorted into individuals t_i, events e_i and real numbers $x_i \in [0,1]$, and we might want to know whether

$$P(e_1) = x_1 \vee U_1(t_3), \neg P(e_2) = x_1 \to W(t_5) \models U_1(t_5).$$

Note that an internal progic might have several probability functions, each with a different interpretation.

In a *mixed* progic, the probabilities may appear both internally and externally. An entailment relationship takes the form

$$\varphi_1^{X_1}, \ldots, \varphi_n^{X_n} \models \psi^Y$$

where $\varphi_1, \ldots, \varphi_n, \psi \in S\mathcal{L}_P$ are sentences of a logical language $\mathcal{L}_P$ which contains probabilities.

There are two main questions to be dealt with when providing semantics for a progic: how are the probabilities to be interpreted? what is the meaning of the entailment relation symbol $\approx$?

The *standard probabilistic semantics* remains neutral about the interpretation of the probabilities and deals with entailment thus:

External: $\varphi_1^{X_1}, \ldots, \varphi_n^{X_n} \approx \psi^Y$ holds if and only if every probability function P that satisfies the left-hand side (i.e., $P(\varphi_1) \in X_1, \ldots, P(\varphi_n) \in X_n$) also satisfies the right-hand side (i.e., $P(\psi) \in Y$).

Internal: $\varphi_1, \ldots, \varphi_n \approx \psi$ if and only if every $\mathcal{L}_P$-model of the left-hand side in which P is interpreted as a probability function is also a model of the right-hand side.

The difficulty with the standard semantics for an external progic is that of *underdetermination*. Given some premiss sentences $\varphi_1, \ldots, \varphi_n$ and their probabilities $X_1, \ldots, X_n$ we often want to know what *single* probability y to give to a conclusion sentence ψ of interest. However, the standard semantics may give no answer to this question: often $\varphi_1^{X_1}, \ldots, \varphi_n^{X_n} \approx \psi^Y$ for a nonsingleton $Y \subseteq [0, 1]$, because probability functions that satisfy the left-hand side disagree as to the probability they award to ψ on the right-hand side. The premisses underdetermine the conclusion. Consequently an alternative semantics is often preferred.

According to the *objective Bayesian semantics* for an external progic on a finite propositional language $L = \{A_1, \ldots, A_N\}$, $\varphi_1^{X_1}, \ldots, \varphi_n^{X_n} \approx \psi^Y$ if and only if an agent whose evidence is summed up by the constraints on the left-hand side (so who ought to believe φ_1 to degree in $X_1, \ldots, \varphi_n$ to degree in X_n) ought to believe ψ to degree in Y. As long as the constraints $\varphi_1^{X_1}, \ldots, \varphi_n^{X_n}$ are consistent, there will be a unique function P that maximises entropy and a unique $y \in [0, 1]$ such that $P(\psi) = y$, so there is no problem of underdetermination.

I shall briefly sketch just three of the principal proposals in this area.[21]

Colin Howson put forward his account of the relationship between probability and logic in [Howson, 2001]; [Howson, 2003] and [Howson, 2008]. Howson interprets probability as follows: 'the agent's probability is the odds, or the betting quotient, they currently believe fair, with the sense of 'fair' that there is no calculable advantage to either side of a bet at those odds' ([Howson, 2001, 143]). The connection with logic is forged by introducing the concept of consistency of betting quotients: a set of betting quotients is consistent if it can be extended to a single-valued function on all the propositions of a given logical language $\mathcal{L}$ which satisfies certain regularity properties. Howson then shows that an assignment of betting quotients is consistent if and only if it is satisfiable by a probability function ([Howson, 2001, Theorem 1]). Having developed a notion of consistency, Howson shows that this leads naturally to an external progic with the standard semantics: consequence is defined in terms of satisfiability by probability functions, as outlined above ([Howson, 2001, 150]).

---

[21][Williamson, 2002] presents a more comprehensive survey.

In [Halpern, 2003], Joseph Halpern studies the standard semantics for internal progics. In the propositional case, $\mathcal{L}$ is a propositional language extended by permitting linear combinations of probabilities $\sum_{i=1}^{n} a_i P_i(\psi_i) > b$, where $a_1, \ldots, a_n, b \in \mathbb{R}$ and $P_1, \ldots, P_n$ are probability functions each of which represents the degrees of belief of an agent and which are defined over sentences ψ of $\mathcal{L}$ ([Halpern, 2003, §7.3]). This language allows nesting of probabilities: for example $P_1(\neg(P_2(\varphi) > 1/3)) > 1/2$ represents 'with degree more than a half, agent 1 believes that agent 2's degree of belief in φ is less than or equal to $\frac{1}{3}$.' Note, though, that the language cannot represent probabilistic independencies, which are expressed using multiplication rather than linear combination of probabilities, such as $P_1(\varphi \wedge \psi) = P_1(\varphi)P_1(\psi)$. Halpern provides a possible-worlds semantics for the resulting logic: given a space of possible worlds, a probability measure $\mu_{w,i}$ over this space for each possible world and agent, and a valuation function π_w for each possible world, $P_1(\psi) > 1/2$ is true at a world w if the measure $\mu_{w,1}$ of the set of possible words at which ψ is true is greater than half, $\mu_{w,1}(\{w' : \pi_{w'}(\psi) = 1\}) > 1/2$. Consequence is defined straightforwardly in terms of satisfiability by worlds.

Halpern later extends the above propositional language to a first-order language and introduces frequency terms $\|\psi\|_X$, interpreted as 'the frequency with which ψ holds when variables in X are repeatedly selected at random' ([Halpern, 2003, §10.3]). Linear combinations of frequencies are permitted, as well as linear combinations of degrees of belief. When providing the semantics for this language, one must provide an interpretation for frequency terms, a probability measure over the domain of the language.

In [Paris, 1994], Jeff Paris discusses external progics in detail, in conjunction with the objective Bayesian semantics. In the propositional case, Paris proposes a number of common sense desiderata which ought to be satisfied by any method for picking out a most rational belief function for the objective Bayesian semantics, and goes on to show that the Maximum Entropy Principle is the *only* method that satisfies these desiderata ([Paris, 1994, Theorem 7.9]; [Paris and Vencovská, 2001]). Later Paris shows how an external progic can be defined over the sentences of a first order logic — such a function is determined by its values over quantifier-free sentences ([Paris, 1994, Chapter 11]; [Gaifman, 1964]). Paris then introduces the problem of learning from experience: what value should an agent give to $P(U(t_{n+1})|\pm U(t_1) \wedge \cdots \wedge \pm U(t_n))$, that is, to what extent should she believe a new instance of U, given n observed instances ([Paris, 1994, Chapter 12])? As mentioned in §§18, 19, [Paris and Vencovská, 2003] and [Williamson, 2008a] suggest that the Maximum Entropy Principle may be extended to the first-order case to address this problem, though by appealing to rather different strategies.

In the case of the standard semantics one might look for a traditional proof theory to accompany the semantics:

External: Given $\varphi_1, \ldots \varphi_n \in SL, x_1, \ldots, x_n \in [0,1]$, find a mechanism for generating all ψ^Y such that $\varphi_1^{X_1}, \ldots, \varphi_n^{X_n} \approx \psi^Y$.

Internal: Given $\varphi_1, \ldots \varphi_n \in S\mathcal{L}_P$, find a mechanism for generating all $\psi \in S\mathcal{L}_P$

such that $\varphi_1,\ldots,\varphi_n \approx \psi$.

In a sense this is straightforward: the premises imply the conclusion just if the conclusion follows from the premises and the axioms of probability by deductive logic. [Fagin et al., 1990] produced a traditional proof theory for the standard probabilistic semantics, for an internal propositional progic. As with propositional logic, deciding satisfiability is NP-complete. [Halpern, 1990] discusses a progic which allows reasoning about both degrees of belief and frequencies. In general, no complete axiomatisation is possible, though axiom systems are provided in special cases where complete axiomatisation is possible. [Abadi and Halpern, 1994] consider first-order degree of belief and frequency logics separately, and show that they are highly undecidable. [Halpern, 2003] presents a general overview of this line of work.

[Paris and Vencovská, 1990] made a start at a traditional proof theory for a type of objective Bayesian progic, but express some scepticism as to whether the goal of a traditional proof system can be achieved.

A traditional proof theory, though interesting, is often not what is required in applications of an external progic. To reiterate, given some premiss sentences $\varphi_1,\ldots,\varphi_n$ and sets of probabilities $X_1,\ldots,X_n$ we often want to know what set of probabilities Y to give to a conclusion sentence ψ of interest — not to churn out all ψ^Y that follow from the premises. Objective Bayesianism provides semantics for this problem, and it is an important question as to whether there is a calculus that accompanies this semantics:

Obprogic: Given $\varphi_1,\ldots,\varphi_n, X_1,\ldots,X_n, \psi$, find an appropriate Y such that $\varphi_1^{X_1},\ldots,\varphi_n^{X_n} \approx \psi^Y$.

By 'appropriate Y' here we mean the narrowest such Y: the entailment trivially holds for $Y = [0,1]$; a maximally specific Y will be of more interest.

It is known that even finding an approximate solution to this problem is NP-complete ([Paris, 1994, Theorem 10.6]). Hence the best one can do is to find an algorithm that is scalable in a range of natural problems, rather than tractable in every case. The approach of [Williamson, 2005a] deals with the propositional case but does not take the form of a traditional logical proof theory, involving axioms and rules of inference. Instead, the proposal is to apply the computational methods of §17 to find an objective Bayesian net — a Bayesian net representation of the P that satisfies constraints $P(\varphi_1) \in X_1, \ldots, P(\varphi_n) \in X_n$ and maximises entropy — and then to use this net to calculate $P(\psi)$. The advantage of using Bayesian nets is that, if sufficiently sparse, they allow the efficient representation of a probability function and efficient methods for calculating marginal probabilities of that function. In this context, the net is sparse and the method scalable in cases where each sentence involves few propositional variables in comparison with the size of the language.

Consider an example. Suppose we have a propositional language $\mathcal{L} = \{A_1, A_2, A_3, A_4, A_5\}$ and we want to find Y such that

$$A_1 \wedge \neg A_2,^{.9} (\neg A_4 \vee A_3) \to A_2,^{.2} A_5 \vee A_3,^{.3} A_4^{.7} \approx A_5 \to A_1.^Y$$

According to our semantics we must find P that maximises

$$H = -\sum P(\pm A_1 \wedge \pm A_2 \wedge \pm A_3 \wedge \pm A_4 \wedge \pm A_5) \log P(\pm A_1 \wedge \pm A_2 \wedge \pm A_3 \wedge \pm A_4 \wedge \pm A_5)$$

subject to the constraints,

$$P(A_1 \wedge \neg A_2) = .9, P((\neg A_4 \vee A_3) \to A_2) = .2, P(A_5 \vee A_3) = .3, P(A_4) = .7.$$

One could find P by directly using numerical optimisation techniques or Lagrange multiplier methods. However, this approach would not be feasible on large languages — already we would need to optimise with respect to 2^5 parameters $P(\pm A_1 \wedge \pm A_2 \wedge \pm A_3 \wedge \pm A_4 \wedge \pm A_5)$.

Instead take the approach of §17:

Step 1: Construct an undirected *constraint graph*, Fig. 1, by linking variables that occur in the same constraint.

As mentioned, the constraint graph satisfies a key property, namely, separation in the constraint graph implies conditional independence for the entropy maximising probability function P. Thus A_2 separates A_5 from A_1, so $A_1 \perp\!\!\!\perp A_5 \mid A_2$, ($P$ renders A_1 probabilistically independent of A_5 conditional on A_2).

Step 2: Transform this into a *directed constraint graph*, Fig. 2.

Now *D*-separation, a directed version of separation ([Pearl, 1988, §3.3]), implies conditional independence for P. Having found a directed acyclic graph which satisfies this property we can construct a Bayesian net by augmenting the graph with conditional probability distributions:

Step 3: Form a Bayesian network by determining parameters $P(A_i | par_i)$ that maximise entropy.

Here the par_i are the states of the parents of A_i. Thus we need to determine $P(A_1), P(A_2|\pm A_1), P(A_3|\pm A_2), P(A_4|\pm A_3 \wedge \pm A_2), P(A_5|\pm A_3)$. This can be done by reparameterising the entropy equation in terms of these conditional probabilities and then using Lagrange multiplier methods or numerical optimisation techniques. This representation of P will be efficient if the graph is sparse, that is, if each constraint sentence φ_i involves few propositional variables in comparison with the size of the language.

Step 4: Simplify ψ into a disjunction of mutually exclusive conjunctions $\bigvee \sigma_j$ (e.g., full disjunctive normal form) and calculate $P(\psi) = \sum P(\sigma_j)$ by using standard Bayesian net algorithms to determine the marginals $P(\sigma_j)$.

In our example,

$$\begin{aligned}
P(A_5 \to A_1) &= P(\neg A_5 \vee A_1) \\
&= P(\neg A_5 \wedge A_1) + P(A_5 \wedge A_1) + P(\neg A_5 \wedge \neg A_1) \\
&= P(\neg A_5|A_1)P(A_1) + P(A_5|A_1)P(A_1) + P(\neg A_5|\neg A_1)P(\neg A_1) \\
&= P(A_1) + P(\neg A_5|\neg A_1)(1 - P(A_1)).
\end{aligned}$$

We thus require only two Bayesian net calculations to determine $P(A_1)$ and $P(\neg A_5|\neg A_1)$. These calculations can be performed efficiently if the graph is sparse and ψ involves few propositional variables relative to the size of the domain.

A major challenge for the objective Bayesian approach is to see whether potentially efficient procedures can be developed for first-order predicate logic.[Williamson, 2008a] takes a step in this direction by showing that objective Bayesian nets, and a generalisation, *objective credal nets*, can in principle be applied to first-order predicate languages.

Part IV
Implications for the Philosophy of Mathematics

Probability theory is a part of mathematics; it should be uncontroversial then that the philosophy of probability is relevant to the philosophy of mathematics. Unfortunately, though, philosophers of mathematics tend to pass over the philosophy of probability, viewing it as a branch of the philosophy of science rather than the philosophy of mathematics. Here I shall attempt to redress the balance by suggesting ways in which the philosophy of probability can suggest new directions to the philosophy of mathematics in general.

22 THE ROLE OF INTERPRETATION

One potential interaction concerns the existence of mathematical entities. Philosophers of probability tackle the question of the existence of probabilities within the context of an interpretation. Questions like 'what are probabilities?' and 'where are they?' receive different answers according to the interpretation of probability under consideration. There is little dispute that axioms of probability admit of more than one interpretation: Bayesians argue convincingly that rational degrees of belief satisfy the axioms of probability; frequentists argue convincingly that limiting relative frequencies satisfy the axioms (except the axiom of countable additivity). The debate is not so much about finding *the* interpretation of probability, but about which interpretation is *best* for particular applications of probability — applications as diverse as those in statistics, number theory, machine learning, epistemology and the philosophy of science. Now according to the Bayesian interpretation probabilities are mental entities, according to frequency theories they are features of collections of physical outcomes, and according to propensity theories they are features of physical experimental set-ups or of single-case events. So we see that an interpretation is required before one can answer questions about existence. The uninterpreted mathematics of probability is treated in an *if-then*-ist way: if the axioms hold then Bayes' theorem holds; degrees of rational belief satisfy the axioms; therefore degrees of rational belief satisfy Bayes' theorem.

The question thus arises as to whether it may *in general* be most productive to ask what mathematical entities are *within the context of an interpretation*. It may make more sense to ask 'what kind of thing is a Hilbert space in the epistemic interpretation of quantum mechanics?' than 'what kind of thing is a Hilbert space?' In mathematics it is crucial to ask questions at the right level of generality; so too in the philosophy of mathematics.

Such a shift in focus from abstraction towards interpretation introduces important challenges. For example, the act of interpretation is rarely a straightforward matter — it typically requires some sort of idealisation. While elegance plays a leading role in the selection of mathematics, the world is rather more messy, and any mapping between the two needs a certain leeway. Thus rational degrees of belief are idealised as real numbers, even though an agent would be irrational to worry about the $10^{10^{10}}$-th decimal place of her degree of belief; frequencies are construed as limits of finite relative frequencies, even though that limit is never actually reached. When assessing an interpretation, the suitability of its associated idealisations are of paramount importance. If it makes a *substantial* difference what the $10^{10^{10}}$-th decimal place of a degree of belief is, then so much the worse for the Bayesian interpretation of probability. Similarly when interpreting arithmetic or set theory: if it matters that a large collection of objects is not in fact denumerable then one should not treat it as the domain of an interpretation of Peano arithmetic; if it matters that the collection is not in fact an object distinct from its members then one should not treat it as a set. A first challenge, then, is to elucidate the role of idealisation in interpretations.

A second challenge is to demarcate the interpretations that imbue existence on mathematical entities from those that don't. While some interpretations construe mathematical entities as worldly things, some construe mathematical entities in terms of other uninterpreted mathematical entities. To take a simple example, one may appeal to affine transformations to interpret the axioms of group theory. In order to construe this group as existing, one must go on to say something about the existence of the transformations: one needs a chain of interpretations that is grounded in worldly things. In the absence of such grounding, the interpretation fails to impart existence. These interpretations within mathematics are rather different from the interpretations that are grounded in our messy world, in that they tend not to involve idealisation: the transformations really do form a group. But of course the line between world and mathematics can be rather blurry, especially in disciplines like theoretical physics: are quantum fields part of the world, or do they require further interpretation?[22]

---

[22][Corfield, 2003, Part IV] discusses interpretations within mathematics.

This shift in focus from abstraction to interpretation is ontological, but not epistemological. That mathematical entities must be interpreted to exist does not mean that uninterpreted mathematics does not qualify as knowledge. Taking an *if-then*-ist view of uninterpreted mathematics, knowledge is accrued if one knows that the consequent does indeed follow from the antecedent, and the role of proof is of course crucial here.[23]

23 THE EPISTEMIC VIEW OF MATHEMATICS

But there is undoubtedly more to mathematics than a collection of *if-then* statements and a further analogy with Bayesianism suggests a more sophisticated philosophy. Under the Bayesian view probabilities are rational degrees of belief, a feature of an agent's epistemic state; they do not exist independently of agents. According to objective Bayesianism probabilities are also objective, in the sense that two agents with the same background information have little or no room for disagreement as to the probabilities. This objectivity is a result of the fact that an agent's degrees of belief are heavily constrained by the extent and limitations of her empirical evidence.

Perhaps mathematics is also purely epistemic, yet objective. Just as Bayesianism considers probabilistic beliefs to be a type of belief — point-valued degrees of belief — rather than beliefs about agent-independent probabilities, mathematical beliefs may also be a type of belief, rather than beliefs about uninterpreted mathematical entities. Just as probabilistic beliefs are heavily constrained, so too mathematical beliefs are heavily constrained. Perhaps so heavily constrained that mathematics turns out to be fully objective, or nearly fully objective (there may be room for subjective disagreement about some principles, such as the continuum hypothesis).[24]

The constraints on mathematical beliefs are the bread and butter of mathematics. Foremost, of course, mathematical beliefs need to be useful. They need to generate good predictions and explanations, both when applied to the real world, i.e., to interpreted mathematical entities, and when applied within mathematics itself. The word 'good' itself encapsulates several constraints: predictions and explanations must achieve a balance of being accurate, interesting, powerful, simple and fruitful, and must be justifiable using two modes of reasoning: proof and interpretation. Finally sociological constraints may have some bearing (e.g. mathematical beliefs need to further mathematicians in their careers and power struggles; the development of mathematics is no doubt constrained by the fact that the most popular conferences are in beach locations) — the question is how big a role such

---

[23] See [Awodey, 2004] for a defence of a type of *if-then*-ism.

[24] [Paseau, 2005] emphasises the interpretation of mathematics. In his terminology, I would be suggesting a *re*interpretation of mathematics in terms of rational beliefs. This notion of *re*interpretation requires there to be some natural or default interpretation that is to be superseded. But as [Paseau, 2005, pp. 379–380] himself notes, it is by no means clear that there is such a default interpretation.

constraints play.

The objective Bayesian analogy then leads to an *epistemic* view of mathematics characterised by the following hypotheses:[25]

Convenience: Mathematical beliefs are convenient, because they admit good explanations and predictions within mathematics itself and also within its grounding interpretations.

Explanation: We have mathematical beliefs because of this convenience, not because uninterpreted mathematical entities correspond to physical things that we experience, nor because such entities correspond to platonic things that we somehow intuit.

Objectivity: The strength of the constraints on mathematical beliefs renders mathematics an objective, or nearly objective, activity.

Under the epistemic view, then, mathematics is like an axe. It is a tool whose design is largely determined by constraints placed on it.[26] Just as the design of an axe is roughly determined by its use (chopping wood) and demands on its strength and longevity, so too mathematics is roughly determined by its use (prediction and explanation) and high standard of certainty as to its conclusions. No wonder that mathematicians working independently end up designing similar tools.

24 CONCLUSION

If probability is to be applied it must be interpreted. Typically we are interested in single-case probabilities — e.g., the probability that I will live to the age of 80, the probability that my car will break down today, the probability that quantum mechanics is true. The Bayesian interpretation tells us what such probabilities are: they are rational degrees of belief.

Subjective Bayesianism has the advantage that it is easy to justify — the Dutch book argument is all that is needed. But subjective Bayesianism does not successfully capture our intuition that many probabilities are objective.

If we move to objective Bayesianism what we gain in terms of objectivity, we pay for in terms of hard graft to address the challenges outlined in Part III. (For this reason, many Bayesians are subjectivist in principle but tacitly objectivist in practice.) These are just challenges though; none seem to present insurmountable problems. They map out an interesting and important research programme rather than reasons to abandon any hope of objectivity.

---

[25] An analogous epistemic view of causality is developed in [Williamson, 2005a, Chapter 9].

[26] [Marquis, 1997, p. 252] discusses the claim that mathematics contains tools or instruments as well as an independent reality of uninterpreted mathematical entities. The epistemic position, however, is purely instrumentalist: there are tools but no independent reality. As Marquis notes, the former view has to somehow demarcate between mathematical objects and tools — by no means an easy task.

The two principal ideas of this chapter — that of interpretation and that of objectively-determined belief — are key if we are to understand probability. I have suggested that they might also offer some insight into mathematics in general.

Acknowledgements

I am very grateful to Oxford University Press for permission to reprint material from [Williamson, 2005a] in Part I and Part II of this chapter, and for permission to reprint material from [Williamson, 2006] in Part IV. I am also grateful to the Leverhulme Trust for a research fellowship supporting this research.

BIBLIOGRAPHY

[Abadi and Halpern, 1994] Abadi, M. and Halpern, J. Y. (1994). Decidability and expressiveness for first-order logics of probability. *Information and Computation*, 112(1):1–36.
[Awodey, 2004] Awodey, S. (2004). An answer to Hellman's question: 'Does category theory provide a framework for mathematical structuralism?'. *Philosophia Mathematica (3)*, 12:54–64.
[Berger and Pericchi, 2001] Berger, J. O. and Pericchi, L. R. (2001). Objective Bayesian methods for model selection: introduction and comparison. In Lahiri, P., editor, *Model Selection*, volume 38 of *Monograph Series*, pages 135–207. Beachwood, Ohio. Institute of Mathematical Statistics Lecture Notes.
[Bernoulli, 1713] Bernoulli, J. (1713). *Ars Conjectandi*. The Johns Hopkins University Press, Baltimore, 2006 edition. Trans. Edith Dudley Sylla.
[Billingsley, 1979] Billingsley, P. (1979). *Probability and measure*. John Wiley and Sons, New York, third (1995) edition.
[Carnap, 1952] Carnap, R. (1952). *The continuum of inductive methods*. University of Chicago Press, Chicago IL.
[Carnap, 1971] Carnap, R. (1971). A basic system of inductive logic part 1. In Carnap, R. and Jeffrey, R. C., editors, *Studies in inductive logic and probability*, volume 1, pages 33–165. University of California Press, Berkeley CA.
[Church, 1936] Church, A. (1936). An unsolvable problem of elementary number theory. *American Journal of Mathematics*, 58:345–363.
[Corfield, 2003] Corfield, D. (2003). *Towards a philosophy of real mathematics*. Cambridge University Press, Cambridge.
[de Finetti, 1937] de Finetti, B. (1937). Foresight. its logical laws, its subjective sources. In Kyburg, H. E. and Smokler, H. E., editors, *Studies in subjective probability*, pages 53–118. Robert E. Krieger Publishing Company, Huntington, New York, second (1980) edition.
[Della Pietra et al., 1997] Della Pietra, S., Della Pietra, V. J., and Lafferty, J. D. (1997). Inducing features of random fields. *IEEE Transactions on Pattern Analysis and Machine Intelligence*, 19(4):380–393.
[Durbin et al., 1999] Durbin, R., Eddy, S., Krogh, A., and Mitchison, G. (1999). *Biological sequence analysis: probabilistic models of proteins and nucleic acids*. Cambridge University Press, Cambridge.
[Earman, 1992] Earman, J. (1992). *Bayes or bust?* MIT Press, Cambridge MA.
[Fagin et al., 1990] Fagin, R., Halpern, J. Y., and Megiddo, N. (1990). A logic for reasoning about probabilities. *Information and Computation*, 87(1-2):277–291.
[Fetzer, 1982] Fetzer, J. H. (1982). Probabilistic explanations. *Philosophy of Science Association*, 2:194–207.
[Gaifman, 1964] Gaifman, H. (1964). Concerning measures in first order calculi. *Israel Journal of Mathematics*, 2:1–18.
[Gaifman and Snir, 1982] Gaifman, H. and Snir, M. (1982). Probabilities over rich languages. *Journal of Symbolic Logic*, 47(3):495–548.

[Gillies, 2000] Gillies, D. (2000). *Philosophical theories of probability*. Routledge, London and New York.
[Hacking, 1975] Hacking, I. (1975). *The emergence of probability*. Cambridge University Press, Cambridge.
[Halpern, 1990] Halpern, J. Y. (1990). An analysis of first-order logics of probability. *Artificial Intelligence*, 46:311–350.
[Halpern, 2003] Halpern, J. Y. (2003). *Reasoning about uncertainty*. MIT Press, Cambridge MA.
[Halpern and Koller, 1995] Halpern, J. Y. and Koller, D. (1995). Representation dependence in probabilistic inference. In Mellish, C. S., editor, *Proceedings of the 14th International Joint Conference on Artificial Intelligence (IJCAI 95)*, pages 1853–1860. Morgan Kaufmann, San Francisco CA.
[Hilbert, 1925] Hilbert, D. (1925). On the infinite. In Benacerraf, P. and Putnam, H., editors, *Philosophy of mathematics: selected readings*. Cambridge University Press (1983), Cambridge, second edition.
[Howson, 2001] Howson, C. (2001). The logic of Bayesian probability. In Corfield, D. and Williamson, J., editors, *Foundations of Bayesianism*, pages 137–159. Kluwer, Dordrecht.
[Howson, 2003] Howson, C. (2003). Probability and logic. *Journal of Applied Logic*, 1(3-4):151–165.
[Howson, 2008] Howson, C. (2008). Can logic be combined with probability? probably. *Journal of Applied Logic*, doi:10.1016/j.jal.2007.11.003.
[Howson and Urbach, 1989] Howson, C. and Urbach, P. (1989). *Scientific reasoning: the Bayesian approach*. Open Court, Chicago IL, second (1993) edition.
[Hunter, 1989] Hunter, D. (1989). Causality and maximum entropy updating. *International Journal in Approximate Reasoning*, 3:87–114.
[Jaynes, 1957] Jaynes, E. T. (1957). Information theory and statistical mechanics. *The Physical Review*, 106(4):620–630.
[Jaynes, 1968] Jaynes, E. T. (1968). Prior probabilities. *IEEE Transactions Systems Science and Cybernetics*, SSC-4(3):227.
[Jaynes, 1973] Jaynes, E. T. (1973). The well-posed problem. *Foundations of Physics*, 3:477–492.
[Jaynes, 1979] Jaynes, E. T. (1979). Where do we stand on maximum entropy? In Levine, R. and Tribus, M., editors, *The maximum entropy formalism*, page 15. MIT Press, Cambridge MA.
[Jaynes, 1988] Jaynes, E. T. (1988). The relation of Bayesian and maximum entropy methods. In Erickson, G. J. and Smith, C. R., editors, *Maximum-entropy and Bayesian methods in science and engineering*, volume 1, pages 25–29. Kluwer, Dordrecht.
[Jaynes, 2003] Jaynes, E. T. (2003). *Probability theory: the logic of science*. Cambridge University Press, Cambridge.
[Kass and Wasserman, 1996] Kass, R. E. and Wasserman, L. (1996). The selection of prior distributions by formal rules. *Journal of the American Statistical Association*, 91:1343–1370.
[Keynes, 1921] Keynes, J. M. (1921). *A treatise on probability*. Macmillan (1948), London.
[Kolmogorov, 1933] Kolmogorov, A. N. (1933). *The foundations of the theory of probability*. Chelsea Publishing Company (1950), New York.
[Laplace, 1814] Laplace (1814). *A philosophical essay on probabilities*. Dover (1951), New York. Pierre Simon, marquis de Laplace.
[Lewis, 1980] Lewis, D. K. (1980). A subjectivist's guide to objective chance. In *Philosophical papers*, volume 2, pages 83–132. Oxford University Press (1986), Oxford.
[Lewis, 1994] Lewis, D. K. (1994). Humean supervenience debugged. *Mind*, 412:471–490.
[Manning and Schütze, 1999] Manning, C. D. and Schütze, H. (1999). *Foundations of statistical natural language processing*. MIT Press, Cambridge MA.
[Marquis, 1997] Marquis, J.-P. (1997). Abstract mathematical tools and machines for mathematics. *Philosophia Mathematica (3)*, 5:250–272.
[Miller, 1994] Miller, D. (1994). *Critical rationalism: a restatement and defence*. Open Court, Chicago IL.
[Nagl et al., 2008] Nagl, S., Williams, M., and Williamson, J. (2008). Objective Bayesian nets for systems modelling and prognosis in breast cancer. In Holmes, D. and Jain, L., editors, *Innovations in Bayesian networks: theory and applications*. Springer.

[Paris, 1994] Paris, J. B. (1994). *The uncertain reasoner's companion.* Cambridge University Press, Cambridge.
[Paris and Vencovská, 1990] Paris, J. B. and Vencovská, A. (1990). A note on the inevitability of maximum entropy. *International Journal of Approximate Reasoning,* 4:181–223.
[Paris and Vencovská, 1997] Paris, J. B. and Vencovská, A. (1997). In defence of the maximum entropy inference process. *International Journal of Approximate Reasoning,* 17:77–103.
[Paris and Vencovská, 2001] Paris, J. B. and Vencovská, A. (2001). Common sense and stochastic independence. In Corfield, D. and Williamson, J., editors, *Foundations of Bayesianism,* pages 203–240. Kluwer, Dordrecht.
[Paris and Vencovská, 2003] Paris, J. B. and Vencovská, A. (2003). The emergence of reasons conjecture. *Journal of Applied Logic,* 1(3–4):167–195.
[Paseau, 2005] Paseau, A. (2005). Naturalism in mathematics and the authority of philosophy. *British Journal for the Philosophy of Science,* 56:377–396.
[Pearl, 1988] Pearl, J. (1988). *Probabilistic reasoning in intelligent systems: networks of plausible inference.* Morgan Kaufmann, San Mateo CA.
[Popper, 1934] Popper, K. R. (1934). *The Logic of Scientific Discovery.* Routledge (1999), London. With new appendices of 1959.
[Popper, 1959] Popper, K. R. (1959). The propensity interpretation of probability. *British Journal for the Philosophy of Science,* 10:25–42.
[Popper, 1983] Popper, K. R. (1983). *Realism and the aim of science.* Hutchinson, London.
[Popper, 1990] Popper, K. R. (1990). *A world of propensities.* Thoemmes, Bristol.
[Quaife, 1992] Quaife, A. (1992). *Automated development of fundamental mathematical theories.* Kluwer, Dordrecht.
[Ramsey, 1926] Ramsey, F. P. (1926). Truth and probability. In Kyburg, H. E. and Smokler, H. E., editors, *Studies in subjective probability,* pages 23–52. Robert E. Krieger Publishing Company, Huntington, New York, second (1980) edition.
[Reichenbach, 1935] Reichenbach, H. (1935). *The theory of probability: an inquiry into the logical and mathematical foundations of the calculus of probability.* University of California Press (1949), Berkeley and Los Angeles. Trans. Ernest H. Hutten and Maria Reichenbach.
[Rosenkrantz, 1977] Rosenkrantz, R. D. (1977). *Inference, method and decision: towards a Bayesian philosophy of science.* Reidel, Dordrecht.
[Schumann, 2001] Schumann, J. M. (2001). *Automated theorem proving in software engineering.* Springer-Verlag.
[Shannon, 1948] Shannon, C. (1948). A mathematical theory of communication. *The Bell System Technical Journal,* 27:379–423 and 623–656.
[Venn, 1866] Venn, J. (1866). *Logic of chance: an essay on the foundations and province of the theory of probability.* Macmillan, London.
[von Mises, 1928] von Mises, R. (1928). *Probability, statistics and truth.* Allen and Unwin, London, second (1957) edition.
[von Mises, 1964] von Mises, R. (1964). *Mathematical theory of probability and statistics.* Academic Press, New York.
[Williamson, 1999] Williamson, J. (1999). Countable additivity and subjective probability. *British Journal for the Philosophy of Science,* 50(3):401–416.
[Williamson, 2002] Williamson, J. (2002). Probability logic. In Gabbay, D., Johnson, R., Ohlbach, H. J., and Woods, J., editors, *Handbook of the logic of argument and inference: the turn toward the practical,* pages 397–424. Elsevier, Amsterdam.
[Williamson, 2005a] Williamson, J. (2005a). *Bayesian nets and causality: philosophical and computational foundations.* Oxford University Press, Oxford.
[Williamson, 2005b] Williamson, J. (2005b). Objective Bayesian nets. In Artemov, S., Barringer, H., d'Avila Garcez, A. S., Lamb, L. C., and Woods, J., editors, *We Will Show Them! Essays in Honour of Dov Gabbay,* volume 2, pages 713–730. College Publications, London.
[Williamson, 2006] Williamson, J. (2006). From Bayesianism to the epistemic view of mathematics. *Philosophia Mathematica (III),* 14(3):365–369.
[Williamson, 2007a] Williamson, J. (2007a). Inductive influence. *British Journal for the Philosophy of Science,* 58(4):689–708.
[Williamson, 2007b] Williamson, J. (2007b). Motivating objective Bayesianism: from empirical constraints to objective probabilities. In Harper, W. L. and Wheeler, G. R., editors, *Probability and Inference: Essays in Honour of Henry E. Kyburg Jr.,* pages 151–179. College Publications, London.

[Williamson, 2008a] Williamson, J. (2008a). Objective Bayesian probabilistic logic. *Journal of Algorithms in Cognition, Informatics and Logic*, in press.

[Williamson, 2008b] Williamson, J. (2008b). Objective Bayesianism, Bayesian conditionalisation and voluntarism. *Synthese*, in press.

[Williamson, 2008c] Williamson, J. (2008c). Objective Bayesianism with predicate languages. *Synthese*, 163(3):341–356.

ON COMPUTABILITY

Wilfried Sieg

1 INTRODUCTION

Computability is perhaps the most significant and distinctive notion modern logic has introduced; in the guise of decidability and effective calculability it has a venerable history within philosophy and mathematics. Now it is also the basic theoretical concept for computer science, artificial intelligence and cognitive science. This essay discusses, at its heart, methodological issues that are central to any mathematical theory that is to reflect parts of our physical or intellectual experience. The discussion is grounded in historical developments that are deeply intertwined with meta-mathematical work in the foundations of mathematics. How is that possible, the reader might ask, when the essay is concerned *solely* with computability? This introduction begins to give an answer by first describing the context of foundational investigations in logic and mathematics and then sketching the main lines of the systematic presentation.

1.1 Foundational contexts

In the second half of the 19th century the issues of decidability and effective calculability rose to the fore in discussions concerning the nature of mathematics. The divisive character of these discussions is reflected in the tensions between Dedekind and Kronecker, each holding broad methodological views that affected deeply their scientific practice. Dedekind contributed perhaps most to the radical transformation that led to modern mathematics: he introduced abstract axiomatizations in parts of the subject (e.g., algebraic number theory) and in the foundations for arithmetic and analysis. Kronecker is well known for opposing that high level of structuralist abstraction and insisting, instead, on the decidability of notions and the effective construction of mathematical objects from the natural numbers. Kronecker's concerns were of a traditional sort and were recognized as perfectly legitimate by Hilbert and others, as long as they were positively directed towards the effective solution of mathematical problems and not negatively used to restrict the free creations of the mathematical mind.

At the turn of the 20th century, these structuralist tendencies found an important expression in Hilbert's book *Grundlagen der Geometrie* and in his essay *Über den Zahlbegriff*. Hilbert was concerned, as Dedekind had been, with the consistency of the abstract notions and tried to address the issue also within a broad set theoretic/logicist framework. The framework could have already been sharpened at that point by adopting the contemporaneous development of Frege's *Begriffsschrift*, but that was not done until the late 1910s, when Russell and Whitehead's work had been absorbed in the Hilbert School. This rather circuitous development is apparent from Hilbert and Bernays' lectures [1917/18] and the many foundational lectures Hilbert gave between 1900 and the summer semester of 1917. Apart from using a version of *Principia Mathematica* as the frame for formalizing mathematics in a direct way, Hilbert and Bernays pursued a dramatically different approach with a sharp focus on meta-mathematical questions like the semantic completeness of logical calculi and the syntactic consistency of mathematical theories.

In his *Habilitationsschrift* of 1918, Bernays established the semantic completeness for the sentential logic of *Principia Mathematica* and presented a system of provably independent axioms. The completeness result turned the truth-table test for validity (or logical truth) into an effective criterion for provability in the logical calculus. This latter problem has a long and distinguished history in philosophy and logic, and its pre-history reaches back at least to Leibniz. I am alluding of course to the decision problem ("Entscheidungsproblem"). Its classical formulation for first-order logic is found in Hilbert and Ackermann's book *Grundzüge der theoretischen Logik*. This problem was viewed as *the* main problem of mathematical logic and begged for a rigorous definition of mechanical procedure or finite decision procedure.

How intricately the "Entscheidungsproblem" is connected with broad perspectives on the nature of mathematics is brought out by an amusingly illogical argument in von Neumann's essay *Zur Hilbertschen Beweistheorie* from 1927:

> ... it appears that there is no way of finding the general criterion for deciding whether or not a well-formed formula a is provable. (We cannot at the moment establish this. Indeed, we have no clue as to how such a proof of undecidability would go.) ... the undecidability is even a *conditio sine qua non* for the contemporary practice of mathematics, using as it does heuristic methods, to make any sense. The very day on which the undecidability does not obtain any more, mathematics as we now understand it would cease to exist; it would be replaced by an absolutely mechanical prescription (eine absolut mechanische Vorschrift) by means of which anyone could decide the provability or unprovability of any given sentence.
>
> Thus we have to take the position: it is generally undecidable, whether a given well-formed formula is provable or not.

If the underlying conceptual problem had been attacked directly, then something like Post's unpublished investigations from the 1920s would have been carried out in Göttingen. A different and indirect approach evolved instead, whose origins can be traced back to the use of calculable number theoretic functions in finitist consistency proofs for parts of arithmetic. Here we find the most concrete beginning of the history of modern computability with close ties to earlier mathematical and later logical developments.

There is a second sense in which "foundational context" can be taken, not as referring to work in the foundations of mathematics, but directly in modern logic and cognitive science. Without a deeper understanding of the nature of calculation and underlying processes, neither the scope of undecidability and incompleteness results nor the significance of computational models in cognitive science can be explored in their proper generality. The claim for logic is almost trivial and implies the claim for cognitive science. After all, the relevant logical notions have been used when striving to create artificial intelligence or to model mental processes in humans. These foundational problems come strikingly to the fore in arguments for Church's or Turing's Thesis, asserting that an informal notion of effective calculability is captured fully by a particular precise mathematical concept. Church's Thesis, for example, claims in its original form that the effectively calculable number theoretic functions are exactly those functions whose values are computable in Gödel's equational calculus, i.e., the general recursive functions.

There is general agreement that Turing gave the most convincing analysis of effective calculability in his 1936 paper *On computable numbers — with an application to the Entscheidungsproblem*. It is Turing's distinctive philosophical contribution that he brought the computing agent into the center of the analysis and that was for Turing a human being, proceeding mechanically.[1] Turing's student Gandy followed in his [1980] the outline of Turing's work in his analysis of machine computability. Their work is not only closely examined in this essay, but also thoroughly recast. In the end, the detailed conceptual analysis presented below yields rigorous characterizations that dispense with theses, reveal human and machine computability as axiomatically given mathematical concepts and allow their systematic reduction to Turing computability.

1.2 Overview

The core of section 2 is devoted to decidability and calculability. Dedekind introduced in his essay *Was sind und was sollen die Zahlen?* the general concept of a "(primitive) recursive" function and proved that these functions can be made explicit in his logicist framework. Beginning in 1921, these obviously calculable functions were used prominently in Hilbert's work on the foundations of mathematics, i.e., in the particular way he conceived of finitist mathematics and its role in consistency proofs. Hilbert's student Ackermann discovered already before 1925 a non-primitive recursive function that was nevertheless calculable. In 1931, Herbrand, working on Hilbert's consistency problem, gave a very general and open-ended characterization of "finitistically calculable number-theoretic functions" that included also the Ackermann function. This section emphasizes the

---
[1]The Shorter Oxford English Dictionary makes perfectly clear that *mechanical*, when applied to a person or action, means "performing or performed without thought; lacking spontaneity or originality; machine-like; automatic, routine."

broader intellectual context and points to the rather informal and epistemologically motivated demand that, in the development of logic and mathematics, certain notions (for example, proof) *should* be decidable by humans and others *should not* (for example, theorem). The crucial point is that the core concepts were deeply intertwined with mathematical practice and logical tradition before they came together in Hilbert's consistency program or, more generally, in meta-mathematics.

In section 3, entitled *Recursiveness and Church's Thesis*, we see that Herbrand's broad characterization was used in Gödel's 1933 paper reducing classical to intuitionist arithmetic. It also inspired Gödel to give a definition of "general recursive functions" in his 1934 Princeton Lectures. Gödel was motivated by the need for a rigorous and adequate notion of "formal theory" so that a general formulation of his incompleteness theorems could be given. Church, Kleene and Rosser investigated Gödel's notion that served subsequently as the rigorous concept in Church's first published formulation of his thesis in [Church, 1935]. Various arguments in support of the thesis, given by Church, Gödel and others, are considered in detail and judged to be inadequate. They all run up against the same stumbling block of having to characterize elementary calculation steps rigorously and without circles. That difficulty is brought out in a conceptually and methodologically clarifying way by the analysis of "reckonable function" ("regelrecht auswertbare Funktion") given in Hilbert and Bernays' 1939 book.

Section 4 takes up matters where they were left off in the third section, but proceeds in a quite different direction: it returns to the original task of characterizing mechanical procedures and focuses on computations and combinatory processes. It starts out with a look at Post's brief 1936 paper, in which a human worker operates in a "symbol space" and carries out very simple operations. Post hypothesized that the operations of such a worker can effect all mechanical or, in his terminology, combinatory processes. This hypothesis is viewed as being in need of continual verification. It is remarkable that Turing's model of computation, developed independently in the same year, is "identical". However, the contrast in methodological approach is equally, if not more, remarkable. Turing took the calculations of human computers or "computors" as a starting-point of a detailed analysis and reduced them, appealing crucially to the agents' sensory limitations, to processes that can be carried out by Turing machines. The restrictive features can be formulated as *boundedness* and *locality* conditions. Following Turing's approach, Gandy investigated the computations of machines or, to indicate the scope of that notion more precisely, of "discrete mechanical devices" that can compute in parallel. In spite of the great generality of his notion, Gandy was able to show that any machine computable function is also Turing computable.

Both Turing and Gandy rely on a restricted *central* thesis, when connecting an informal concept of calculability with a rigorous mathematical one. I sharpen Gandy's work and characterize "Turing Computors" and "Gandy Machines" as discrete dynamical systems satisfying appropriate axiomatic conditions. Any Turing computor or Gandy machine turns out to be computationally reducible to a Turing machine. These considerations constitute the core of section 5 and lead to the conclusion that computability, when relativized to a particular kind of computing device, has a standard methodological status: no thesis is needed, but rather the recognition that the axiomatic conditions are correct for the intended device. The proofs that the characterized notions are equivalent to Turing computability establish then important mathematical facts.

In section 6, I give an "Outlook on Machines and Mind". The question, whether there are concepts of effectiveness broader than the ones characterized by the axioms for Gandy machines and Turing computors, has of course been asked for both physical and mental processes. I discuss the seemingly sharp conflict between Gödel and Turing expressed by Gödel, when asserting: i) Turing tried (and failed) in his [1936] to reduce all mental processes to mechanical ones, and ii) the human mind infinitely surpasses any finite machine. This conflict can be clarified and resolved by realizing that their deeper disagreement concerns the nature of machines. The section ends with some brief remarks about supra-mechanical devices: if there are such, then they cannot satisfy the physical restrictions expressed through the boundedness and locality conditions for Gandy machines. Such systems must violate either the upper bound on signal propagation or the lower bound on the size of distinguishable atomic components; such is the application of the axiomatic method.

1.3 Connections

Returning to the beginning, we see that Turing's notion of human computability is exactly right for both a convincing negative solution of the "Entscheidungsproblem" and a precise characterization of formal systems that is needed for the general formulation of the incompleteness theorems. One disclaimer and one claim should be made at this point. For many philosophers computability is of special importance because of its central role in "computational models of the human mind". This role is touched upon only indirectly through the reflections on the nature and content of Church's and Turing's theses. The disclaimer is complemented by the claim that the conceptual analysis naturally culminates in the formulation of axioms that characterize different computability notions. Thus, arguments in support of the various theses should be dismissed in favor of considerations for the adequacy of axiomatic characterizations of computations that do not correspond to deep mental procedures, but rather to strictly mechanical processes.

Wittgenstein's terse remark about Turing machines, "These machines are *humans* who calculate,"[2] captures the very feature of Turing's analysis of calculability that makes it epistemologically relevant. Focusing on the epistemology of mathematics, I will contrast this feature with two striking aspects of mathematical experience implicit in repeated remarks of Gödel's. The first "conceptional" aspect is connected to the notion of effective calculability through his assertion that

---

[2]From [1980, § 1096]. I first read this remark in [Shanker, 1987], where it is described as a "mystifying reference to Turing machines." In his later book [Shanker, 1998] that characterization is still maintained.

"with this concept one has for the first time succeeded in giving an absolute definition of an interesting epistemological notion". The second "quasi-constructive" aspect is related to axiomatic set theory through his claim that its axioms "can be supplemented without arbitrariness by new axioms which are only the natural continuation of the series of those set up so far". Gödel speculated how the second aspect might give rise to a humanly effective procedure that cannot be mechanically calculated. Gödel's remarks point to data that underlie the two aspects and challenge, in the words of Parsons[3], "any theory of meaning and evidence in mathematics". Not that I present a theory accounting for these data. Rather, I clarify the first datum by reflecting on the question that is at the root of Turing's analysis. In its sober mathematical form the question asks, *"What is an effectively calculable function?"*

2 DECIDABILITY AND CALCULABILITY

This section is mainly devoted to the *decidability* of relations between finite syntactic objects and the *calculability* of number theoretic functions. The former notion is seen by Gödel in 1930 to be derivative of the latter, since such relations are considered to be decidable just in case the characteristic functions of their arithmetic analogues are calculable. Calculable functions rose to prominence in the 1920s through Hilbert's work on the foundations of mathematics. Hilbert conceived of finitist mathematics as an extension of the Kroneckerian part of constructive mathematics and insisted programmatically on carrying out consistency proofs by finitist means only. Herbrand, who worked on Hilbert's consistency problem, gave a general and open-ended characterization of "finitistically calculable functions" in his last paper [Herbrand, 1931a]. This characterization was communicated to Gödel in a letter of 7 April 1931 and inspired the notion of general recursive function that was presented three years later in Gödel's Princeton Lectures and is the central concept to be discussed in Section 3.

Though this specific meta-mathematical background is very important, it is crucial to see that it is embedded in a broader intellectual context, which is philosophical as well as mathematical. There is, first, the normative requirement that some central features of the formalization of logic and mathematics should be decidable on a radically inter-subjective basis; this holds, in particular, for the proof relation. It is reflected, second, in the quest for showing the decidability of problems in pure mathematics and is connected, third, to the issue of predictability in physics and other sciences. Returning to the meta-mathematical background, Hilbert's Program builds on the formalization of mathematics and thus incorporates aspects of the normative requirement. Gödel expressed the idea for realizing this demand in his [1933a]:

> The first part of the problem [see fn. 4 for the formulation of "the problem"] has been solved in a perfectly satisfactory way, the solu-

---

[3] In [Parsons, 1995].

tion consisting in the so-called "formalization" of mathematics, which means that a perfectly precise language has been invented, by which it is possible to express any mathematical proposition by a formula. Some of these formulas are taken as axioms, and then certain rules of inference are laid down which allow one to pass from the axioms to new formulas and thus to deduce more and more propositions, the outstanding feature of the rules of inference being that they are purely formal, i.e., refer only to the outward structure of the formulas, not to their meaning, so that they could be applied by someone who knew nothing about mathematics, or by a machine.[4]

Let's start with a bit of history and see how the broad issue of decidability led to the question, "What is the precise extension of the class of calculable number theoretic functions?"

2.1 Decidability

Any historically and methodologically informed account of calculability will at least point to Leibniz and the goals he sought to achieve with his project of a *characteristica universalis* and an associated *calculus ratiocinator*. Similar projects for the development of artificial languages were common in 17th century intellectual circles. They were pursued for their expected benefits in promoting religious and political understanding, as well as commercial exchange. Leibniz's project stands out for its emphasis on mechanical reasoning: a universal character is to come with algorithms for making and checking inferences. The motivation for this requirement emerges from his complaint about Descartes's *Rules for the direction of the mind*. Leibniz views them as a collection of vague precepts, requiring intellectual effort as well as ingenuity from the agents following the rules. A reasoning method, such as the universal character should provide, comes by contrast with rules that completely determine the actions of the agents. Neither insight nor intellectual effort is needed, as a mechanical thread of reasoning guides everyone who can perceive and manipulate concrete configurations of symbols.

> Thus I assert that all truths can be demonstrated about things expressible in this language with the addition of new concepts not yet expressed in it — all such truths, I say, can be demonstrated *solo calculo*, or solely by the manipulation of characters according to a certain form, without any labor of the imagination or effort of the mind, just

---

[4] Cf. p. 45 of [Gödel 1933a]. To present the context of the remark, I quote the preceding paragraph of Gödel's essay: "The problem of giving a foundation of mathematics (and by mathematics I mean here the totality of the methods of proof actually used by mathematicians) can be considered as falling into two different parts. At first these methods of proof have to be reduced to a minimum number of axioms and primitive rules of inference, which have to be stated as precisely as possible, and then secondly a justification in some sense or other has to be sought for these axioms, i.e., a theoretical foundation of the fact that they lead to results agreeing with each other and with empirical facts."

as occurs in arithmetic and algebra. (Quoted in [Mates, 1986, fn. 65, 185])

Leibniz's expectations for the growth of our capacity to resolve disputes were correspondingly high. He thought we might just sit down at a table, formulate the issues precisely, take our pens and say *Calculemus*! After finitely many calculation steps the answer would be at hand, or rather visibly on the table. The thought of having machines carry out the requisite mechanical operations had already occurred to Lullus. It was pursued further in the 19th century by Jevons and was pushed along by Babbage in a theoretically and practically most ambitious way.

The idea of an epistemologically unproblematic method, turning the task of testing the conclusiveness of inference chains (or even of creating them) into a purely mechanical operation, provides a direct link to Frege's *Begriffsschrift* and to the later reflections of Peano, Russell, Hilbert, Gödel and others. Frege, in particular, saw himself in this Leibnizian tradition as he emphasized in the introduction to his 1879 booklet. That idea is used in the 20th century as a normative requirement on the fully explicit presentation of mathematical proofs in order to insure inter-subjectivity. In investigations concerning the foundations of mathematics that demand led from axiomatic, yet informal presentations to fully formal developments. As an example, consider the development of elementary arithmetic in [Dedekind 1888] and [Hilbert 1923]. It can't be overemphasized that the step from *axiomatic systems* to *formal theory* is a radical one, and I will come back to it in the next subsection.[5]

There is a second Leibnizian tradition in the development of mathematical logic that leads from Boole and de Morgan through Peirce to Schröder, Löwenheim and others. This tradition of the algebra of logic had a deep impact on the classical formulation of modern mathematical logic in Hilbert and Ackermann's book. Particularly important was the work on the decision problem, which had a longstanding tradition in algebraic logic and had been brought to a highpoint in Löwenheim's paper from 1915, *Über Möglichkeiten im Relativkalkül*. Löwenheim established, in modern terminology, the decidability of monadic first-order logic and the reducibility of the decision problem for first-order logic to its binary fragment. The importance of that mathematical insight was clear to Löwenheim, who wrote about his reduction theorem:

> We can gauge the significance of our theorem by reflecting upon the fact that every theorem of mathematics, or of any calculus that can be invented, can be written as a relative equation; the mathematical

---

[5]The nature of this step is clearly discussed in the Introduction to Frege's *Grundgesetze der Arithmetik*, where he criticizes Dedekind for not having made explicit all the methods of inference: "In a much smaller compass it [i.e., Dedekind's *Was sind und was sollen die Zahlen?*] follows the laws of arithmetic much farther than I do here. This brevity is only arrived at, to be sure, because much of it is not really proved at all. ... nowhere is there a statement of the logical laws or other laws on which he builds, and, even if there were, we could not possibly find out whether really no others were used — for to make that possible the proof must be not merely indicated but completely carried out." [Geach and Black, 119]

theorem then stands or falls according as the equation is satisfied or not. This transformation of arbitrary mathematical theorems into relative equations can be carried out, I believe, by anyone who knows the work of Whitehead and Russell. Since, now, according to our theorem the whole relative calculus can be reduced to the binary relative calculus, it follows that we can decide whether an arbitrary mathematical proposition is true provided we can decide whether a binary relative equation is identically satisfied or not. (p. 246)

Many of Hilbert's students and collaborators worked on the decision problem, among them Ackermann, Behmann, Bernays, Schönfinkel, but also Herbrand and Gödel. Hilbert and Ackermann made the connection of mathematical logic to the algebra of logic explicit. They think that the former provides more than a precise language for the following reason: "Once the logical formalism is fixed, it can be expected that a systematic, so-to-speak calculatory treatment of logical formulas is possible; that treatment would roughly correspond to the theory of equations in algebra." (p. 72) Subsequently, they call sentential logic "a developed algebra of logic". The decision problem, solved of course for the case of sentential logic, is viewed as one of the most important logical problems; when it is extended to full first-order logic it must be considered "as the main problem of mathematical logic". (p. 77) Why the decision problem should be considered as *the* main problem of mathematical logic is stated clearly in a remark that may remind the reader of Löwenheim's and von Neumann's earlier observations:

The solution of this general decision problem would allow us to decide, at least in principle, the provability or unprovability of an arbitrary mathematical statement. (p. 86)

Taking for granted the finite axiomatizability of set theory or some other fundamental theory in first-order logic, the general decision problem is solved when that for first-order logic has been solved. And what is required for its solution?

The decision problem is solved, in case a procedure is known that permits — for a given logical expression — to decide the validity, respectively satisfiability, by finitely many operations. (p. 73)

Herbrand, for reasons similar to those of Hilbert and Ackermann, considered the general decision problem in a brief note from 1929 "as the most important of those, which exist at present in mathematics" (p. 42). The note was entitled *On the fundamental problem of mathematics.*

In his paper *On the fundamental problem of mathematical logic* Herbrand presented a little later refined versions of Löwenheim's reduction theorem and gave positive solutions of the decision problem for particular parts of first-order logic. The fact that the theorems are refinements is of interest, but not the crucial reason for Herbrand to establish them. Rather, Herbrand emphasizes again and again that Löwenheim's considerations are "insufficient" (p. 39) and that his proof "is totally inadequate for our purposes" (p. 166). The fullest reason for these judgments is given in section 7.2 of his thesis, *Investigations in proof theory*, when discussing two central theorems, namely, if the formula P is provable (in first-order logic), then its negation is not true in any infinite domain (Theorem 1) and if P is not provable, then we can construct an infinite domain in which its negation is true (Theorem 2).

> Similar results have already been stated by Löwenheim, but his proofs, it seems to us, are totally insufficient for our purposes. First, he gives an intuitive meaning to the notion 'true in an infinite domain', hence his proof of Theorem 2 does not attain the rigor that we deem desirable .... Then — and this is the gravest reproach — because of the intuitive meaning that he gives to this notion, he seems to regard Theorem 1 as obvious. This is absolutely impermissible; such an attitude would lead us, for example, to regard the consistency of arithmetic as obvious. On the contrary, it is precisely the proof of this theorem ... that presented us with the greatest difficulty.
>
> We could say that Löwenheim's proof was sufficient in mathematics. But, in the present work, we had to make it 'metamathematical' (see Introduction) so that it would be of some use to us. (pp. 175–176)

The above theorems provide Herbrand with a method for investigating the decision problem, whose solution would answer also the consistency problem for finitely axiomatized theories. As consistency has to be established by using restricted meta-mathematical methods, Herbrand emphasizes that the decision problem has to be attacked exclusively with such methods. These meta-mathematical methods are what Hilbert called finitist. So we reflect briefly on the origins of finitist mathematics and, in particular, on the views of its special defender and practitioner, Leopold Kronecker.

2.2 Finitist mathematics

In a talk to the Hamburg Philosophical Society given in December 1930, Hilbert reminisced about his finitist standpoint and its relation to Kronecker; he pointed out:

> At about the same time [around 1888], thus already more than a generation ago, Kronecker expressed clearly a view and illustrated it by several examples, which today coincides essentially with our finitist standpoint. [Hilbert, 1931, 487]

He added that Kronecker made only the mistake "of declaring transfinite inferences as inadmissible". Indeed, Kronecker disallowed the classical logical inference from the negation of a universal to an existential statement, because a proof of an existential statement should provide a witness. Kronecker insisted also on the decidability of mathematical notions, which implied among other things the rejection of the general concept of irrational number. In his 1891 lectures *Über den Zahlbegriff in der Mathematik* he formulated matters clearly and forcefully:

> The standpoint that separates me from many other mathematicians culminates in the principle, that the definitions of the experiential sciences (Erfahrungswissenschaften), — i.e., of mathematics and the natural sciences, ... — must not only be consistent in themselves, but must be taken from experience. It is even more important that they must contain a criterion by means of which one can decide for any special case, whether or not the given concept is subsumed under the definition. A definition, which does not provide that, may be praised by philosophers or logicians, but for us mathematicians it is a mere verbal definition and without any value. (p. 240)

Dedekind had a quite different view. In the first section of *Was sind und was sollen die Zahlen?* he asserts that "things", any objects of our thought, can frequently "be considered from a common point of view" and thus "be associated in the mind" to form a system. Such systems S are also objects of our thought and are "completely determined when it is determined for every thing whether it is an element of S or not". Attached to this remark is a footnote differentiating his position from Kronecker's:

> How this determination is brought about, and whether we know a way of deciding upon it, is a matter of indifference for all that follows; the general laws to be developed in no way depend upon it; they hold under all circumstances. I mention this expressly because Kronecker not long ago (*Crelle's Journal*, Vol. 99, pp. 334–336) has endeavored to impose certain limitations upon the free formation of concepts in mathematics, which I do not believe to be justified; but there seems to be no call to enter upon this matter with more detail until the distinguished mathematician shall have published his reasons for the necessity or merely the expediency of these limitations. (p. 797)

In Kronecker's essay *Über den Zahlbegriff* and his lectures *Über den Zahlbegriff in der Mathematik* one finds general reflections on the foundations of mathematics that at least partially address Dedekind's request for clarification.

Kronecker views arithmetic in his [1887] as a very broad subject, encompassing all mathematical disciplines with the exception of geometry and mechanics. He thinks that one will succeed in "grounding them [all the mathematical disciplines] solely on the number-concept in its narrowest sense, and thus in casting off the modifications and extensions of this concept which were mostly occasioned by the applications to geometry and mechanics". In a footnote Kronecker makes clear that he has in mind the addition of "irrational as well as continuous quantities". The principled philosophical distinction between geometry and mechanics on the one hand and arithmetic (in the broad sense) on the other hand is based on Gauss' remarks about the theory of space and the pure theory of quantity: only the latter has "the complete conviction of necessity (and also of absolute truth)," whereas the former has also outside of our mind a reality "to which we cannot *a priori* completely prescribe its laws".

These programmatic remarks are refined in the 1891 lectures. The lecture of 3 June 1891 summarizes Kronecker's perspective on mathematics in four theses. The first asserts that mathematics does not tolerate "Systematik," as mathematical research is a matter of inspiration and creative imagination. The second thesis asserts that mathematics is to be treated as a natural science "for its objects are as real as those of its sister sciences (Schwesterwissenschaften)". Kronecker explains:

> That this is so is sensed by anyone who speaks of mathematical 'discoveries'. Since we can discover only something that already really exists; but what the human mind generates out of itself that is called 'invention'. The mathematician 'discovers', consequently, by methods, which he 'invented' for this very purpose. (pp. 232–3)

The next two theses are more restricted in scope, but have important methodological content. When investigating the fundamental concepts of mathematics and when developing a particular area, the third thesis insists, one has to keep separate the individual mathematical disciplines. This is particularly important, because the fourth thesis demands that, for any given discipline, i) its characteristic methods are to be used for determining and elucidating its fundamental concepts and ii) its rich content is to be consulted for the explication of its fundamental concepts.[6] In the end, the only real mathematical objects are the natural numbers: "True mathematics needs from arithmetic only the [positive] integers." (p. 272)

In his Paris Lecture of 1900, Hilbert formulated as an axiom that any mathematical problem can be solved, either by answering the question posed by the problem or by showing the impossibility of an answer. Hilbert asked, "What is a legitimate condition that solutions of mathematical problems have to satisfy?" Here is the formulation of the central condition:

> I have in mind in particular [the requirement] that we succeed in establishing the correctness of the answer by means of a finite number of inferences based on a finite number of assumptions, which are inherent in the problem and which have to be formulated precisely in each case. This requirement of logical deduction by means of a finite

---

[6] Kronecker explains the need for ii) in a most fascinating way as follows: "Clearly, when a reasonable master builder has to put down a foundation, he is first going to learn carefully about the building for which the foundation is to serve as the basis. Furthermore, it is foolish to deny that the richer development of a science may lead to the necessity of changing its basic notions and principles. In this regard, there is no difference between mathematics and the natural sciences: new phenomena overthrow the old hypotheses and replace them by others." (p. 233)

> number of inferences is nothing but the requirement of rigor in argumentation. Indeed, the requirement of rigor ... corresponds [on the one hand] to a general philosophical need of our understanding and, on the other hand, it is solely by satisfying this requirement that the thought content and the fruitfulness of the problem in the end gain their full significance. (p. 48)

Then he tries to refute the view that only arithmetic notions can be treated rigorously. He considers that opinion as thoroughly mistaken, though it has been "occasionally advocated by eminent men". That is directed against Kronecker as the next remark makes clear.

> Such a one-sided interpretation of the requirement of rigor soon leads to ignoring all concepts that arise in geometry, mechanics, and physics, to cutting off the flow of new material from the outer world, and finally, as a last consequence, to the rejection of the concepts of the continuum and the irrational number. (p. 49)

Positively and in contrast, Hilbert thinks that mathematical concepts, whether emerging in epistemology, geometry or the natural sciences, are to be investigated in mathematics. The principles for them have to be given by "a simple and complete system of axioms" in such a way that "the rigor of the new concepts, and their applicability in deductions, is in no way inferior to the old arithmetic notions". This is a central part of Hilbert's much-acclaimed axiomatic method, and Hilbert uses it to shift the Kroneckerian effectiveness requirements from the mathematical to the "systematic" meta-mathematical level.[7] That leads, naturally, to a distinction between "solvability in principle" by the axiomatic method and "solvability by algorithmic means". Hilbert's famous 10th Problem concerning the solvability of Diophantine equations is a case in which an algorithmic solution is sought; the

---

[7]That perspective, indicated here in a very rudimentary form, is of course central for the meta-mathematical work in the 1920s and is formulated in the sharpest possible way in many of Hilbert's later publications. Its epistemological import is emphasized, for example in the first chapter of *Grundlagen der Mathematik I*, p. 2: "Also formal axiomatics definitely requires for its deductions as well as for consistency proofs certain evidences, but with one essential difference: this kind of evidence is not based on a special cognitive relation to the particular subject, but is one and the same for all axiomatic [formal] systems, namely, that primitive form of cognition, which is the prerequisite for any exact theoretical research whatsoever." In his Hamburg talk of 1928 Hilbert stated the remarkable philosophical significance he sees in the proper formulation of the rules for the meta-mathematical "formula game": "For this formula game is carried out according to certain definite rules, in which the *technique of our thinking* is expressed. These rules form a closed system that can be discovered and definitively stated. The fundamental idea of my proof theory is none other than to describe the activity of our understanding, to make a protocol of the rules according to which our thinking actually proceeds." He adds, against Kronecker and Brouwer's intuitionism, "If any totality of observations and phenomena deserves to be made the object of a serious and thorough investigation, it is this one. Since, after all, it is part of the task of science to liberate us from arbitrariness, sentiment, and habit and to protect us from the subjectivism that already made itself felt in Kronecker's views and, it seems to me, finds its culmination in intuitionism." [van Heijenoort, 1967, 475]

impossibility of such a solution was found only in the 1970s after extensive work by Robinson, Davis and Matijasevic, work that is closely related to the developments of computability theory described here; cf. [Davis, 1973].

At this point in 1900 there is no firm ground for Hilbert to claim that Kroneckerian rigor for axiomatic developments has been achieved. After all, it is only the radical step from axiomatic to formal theories that guarantees the rigor of solutions to mathematical problems in the above sense, and that step was taken by Hilbert only much later. Frege had articulated appropriate mechanical features and had realized them for the arguments given in his concept notation. His booklet *Begriffsschrift* offered a rich language with relations and quantifiers, whereas its logical calculus required that all assumptions be listed and that each step in a proof be taken in accord with one of the antecedently specified rules. Frege considered this last requirement as a sharpening of the axiomatic method he traced back to Euclid's *Elements*. With this sharpening he sought to recognize the "epistemological nature" of theorems. In the introduction to *Grundgesetze der Arithmetik* he wrote:

> Since there are no gaps in the chains of inferences, each axiom, assumption, hypothesis, or whatever you like to call it, upon which a proof is founded, is brought to light; and so we gain a basis for deciding the epistemological nature of the law that is proved. (p. 118)

But a true basis for such a judgment can be obtained only, Frege realized, if inferences do not require contentual knowledge: their application has to be recognizable as correct on account of the form of the sentences occurring in them. Frege claimed that in his logical system "inference is conducted like a calculation" and observed:

> I do not mean this in a narrow sense, as if it were subject to an algorithm the same as ... ordinary addition and multiplication, but only in the sense that there is an algorithm at all, i.e., a totality of rules which governs the transition from one sentence or from two sentences to a new one in such a way that nothing happens except in conformity with these rules.[8] [Frege, 1984, 237]

Hilbert took the radical step to fully formal axiomatics, prepared through the work of Frege, Peano, Whitehead and Russell, only in the lectures he gave in the winter-term of 1917/18 with the assistance of Bernays. The effective presentation of formal theories allowed Hilbert to formulate in 1922 the finitist consistency program, i.e., describe formal theories in Kronecker-inspired finitist mathematics and formulate consistency in a finitistically meaningful way. In line with the Paris

---

[8] Frege was careful to emphasize (in other writings) that all of thinking "can never be carried out by a machine or be replaced by a purely mechanical activity" [Frege 1969, 39]. He went on to claim: "It is clear that the syllogism can be brought into the form of a calculation, which however cannot be carried out without thinking; it [the calculation] just provides a great deal of assurance on account of the few rigorous and intuitive forms in which it proceeds."

remarks, he viewed this in [1921/22] as a dramatic expansion of Kronecker's purely arithmetic finitist mathematics:

> We have to extend the domain of objects to be considered, i.e., we have to apply our intuitive considerations also to figures that are not number signs. Thus we have good reason to distance ourselves from the earlier dominant principle according to which each theorem of mathematics is in the end a statement concerning integers. This principle was viewed as expressing a fundamental methodological insight, but it has to be given up as a prejudice. (p. 4a)

As to the extended domain of objects, it is clear that formulas and proofs of formal theories are to be included and that, by contrast, geometric figures are definitely excluded. Here are the reasons for holding that such figures are "not suitable objects" for finitist considerations:

> ... the figures we take as objects must be completely surveyable and only discrete determinations are to be considered for them. It is only under these conditions that our claims and considerations have the same reliability and evidence as in intuitive number theory. (p. 5a)

If we take this expansion of the domain of objects seriously (as we should, I think), we are dealing not just with numbers and associated principles, but more generally with elements of inductively generated classes and associated principles of proof by induction and definition by recursion. That is beautifully described in the Introduction to Herbrand's thesis and was strongly emphasized by von Neumann in his Königsberg talk of 1930. For our systematic work concerning computability we have to face then two main questions, i) "How do we move from decidability issues concerning finite syntactic configurations to calculability of number theoretic functions?" and ii) "Which number theoretic functions can be viewed as being calculable?"

2.3 (Primitive) Recursion

Herbrand articulated in the Appendix to his [1931] (the paper itself had been written already in 1929) informed doubts concerning the positive solvability of the decision problem: "Note finally that, although at present it seems unlikely that the decision problem can be solved, it has not yet been proved that it is impossible to do so." (p. 259) These doubts are based on the second incompleteness theorem, which is formulated by Herbrand as asserting, "it is impossible to prove the consistency of a theory through arguments formalizable in the theory."

> ... if we could solve the decision problem in the restricted sense [i.e., for first-order logic], it would follow that every theory which has only a finite number of hypotheses and in which this solution is formalizable

would be inconsistent (since the question of the consistency of a theory having only a finite number of hypotheses can be reduced to this problem). (p. 258)

A historical fact has to be mentioned here: Herbrand spent the academic year 1930/31 in Germany and worked during the fall of 1930 with von Neumann in Berlin. Already in November of 1930 he learned through von Neumann about Gödel's first incompleteness theorem and by early spring of 1931 he had received through Bernays the galleys of [Gödel 1931].

Von Neumann, in turn, had learned from Gödel himself about a version of the first incompleteness theorem at the *Second Conference for Epistemology of the Exact Sciences* held from 5 to 7 September 1930 in Königsberg. On the last day of that conference a roundtable discussion on the foundations of mathematics took place to which Gödel had been invited. Hans Hahn chaired the discussion and its participants included Carnap, Heyting and von Neumann. Toward the end of the discussion Gödel made brief remarks about the first incompleteness theorem; the transcript of his remarks was published in *Erkenntnis* and as [1931a] in the first volume of his *Collected Works*. This is the background for the personal encounter with von Neumann in Königsberg; Wang reports Gödel's recollections in his [1981]:

> Von Neumann was very enthusiastic about the result and had a private discussion with Gödel. In this discussion, von Neumann asked whether number-theoretical undecidable propositions could also be constructed in view of the fact that the combinatorial objects can be mapped onto the integers and expressed the belief that it could be done. In reply, Gödel said, "Of course undecidable propositions about integers could be so constructed, but they would contain concepts quite different from those occurring in number theory like addition and multiplication." Shortly afterward Gödel, to his own astonishment, succeeded in turning the undecidable proposition into a polynomial form preceded by quantifiers (over natural numbers). At the same time but independently of this result, Gödel also discovered his second theorem to the effect that no consistency proof of a reasonably rich system can be formalized in the system itself. (pp. 654–5)

This passage makes clear that Gödel had not yet established the second incompleteness theorem at the time of the Königsberg meeting. On 23 October 1930 Hahn presented to the Vienna Academy of Sciences an abstract containing the theorem's classical formulation. The full text of Gödel's 1931-paper was submitted to the editors of *Monatshefte* on 17 November 1930.[9] The above passage makes

---

[9] As to the interaction between von Neumann and Gödel after Königsberg and von Neumann's independent discovery of the second incompleteness theorem, cf. their correspondence published in volume V of Gödel's *Collected Works*. — In the preliminary reflections of his [1931] Gödel simply remarks on p. 146 about the "arithmetization": "For meta-mathematical considerations it is of course irrelevant, which objects are taken as basic signs, and we decide to use natural numbers as such [basic signs]."

also clear something surprising, namely, that the arithmetization of syntax used so prominently in the 1931 paper was seemingly developed only after the Königsberg meeting. (There is also no hint of this technique in [Gödel, 1931a].) Given an effective coding of syntax the part of finitist mathematics needed for the description of formal theories is consequently contained in finitist number theory, and finitist decision procedures can then presumably be captured by finitistically calculable number theoretic functions. This answers the first question formulated at the end of section 2.2. Let us take now a step towards answering the second question, "Which number theoretic functions can be viewed as being calculable?"

It was Kronecker who insisted on decidability of mathematical notions and calculability of functions, but it was Dedekind who formulated in *Was sind und was sollen die Zahlen?* the general concept of a "(primitive) recursive" function. These functions are obviously calculable and Dedekind proved, what is not so important from our computational perspective, namely, that they can be made explicit in his logicist framework.[10] Dedekind considers a *simply infinite system* $(N, \varphi, 1)$ that is characterized by axiomatic conditions, now familiar as the Dedekind-Peano axioms:

$$1 \in N,$$
$$(\forall n \in N) \, \varphi(n) \in N,$$
$$(\forall n, m \in N)(\varphi(n) = \varphi(m) \to n = m),$$
$$(\forall n \in N) \, \varphi(n) \neq 1 \text{ and}$$
$$(1 \in \Sigma \, \& \, (\forall n \in N)(n \in \Sigma \to \varphi(n) \in \Sigma)) \to (\forall n \in N) \, n \in \Sigma.$$

(Σ is any subset of N.) For this and other simply infinite systems Dedekind isolates a crucial feature in theorem 126, *Satz der Definition durch Induktion*: let $(N, \varphi, 1)$ be a simply infinite system, let θ be an arbitrary mapping from a system Ω to itself, and let ω be an element of Ω; then there is exactly one mapping ψ from N to Ω satisfying the recursion equations:

$$\psi(1) = \omega,$$
$$\psi(\varphi(n)) = \theta(\psi(n)).$$

The proof requires subtle meta-mathematical considerations; i.e., an inductive argument for the existence of approximations to the intended mapping on initial segments of N. The basic idea was later used in axiomatic set theory and extended to functions defined by transfinite recursion. It is worth emphasizing that Dedekind's is a very abstract idea: show the existence of a unique solution for a functional equation! Viewing functions as given by calculation procedures, Dedekind's general point recurs in [Hilbert, 1921/22], [Skolem, 1923], [Herbrand, 1931a], and [Gödel, 1934], when the existence of a solution is guaranteed by the existence of a calculation procedure.

In the context of his overall investigation concerning the nature and meaning of number, Dedekind draws two important conclusions with the help of theorem

---

[10]However, in his [193?] Gödel points out on p. 21, that it is Dedekind's method that is used to show that recursive definitions can be defined explicitly in terms of addition and multiplication.

126: on the one hand, all simply infinite systems are similar (theorem 132), and on the other hand, any system that is similar to a simply infinite one is itself simply infinite (theorem 133). The first conclusion asserts, in modern terminology, that the Dedekind-Peano axioms are categorical. Dedekind infers in his remark 134 from this fact, again in modern terminology, that all simply infinite systems are elementarily equivalent — claiming to justify in this way his abstractive conception of natural numbers.

Dedekind's considerations served a special foundational purpose. However, the recursively defined number theoretic functions have an important place in mathematical practice and can be viewed as part of constructive (Kroneckerian) mathematics quite independent of their logicist foundation. As always, Dedekind himself is very much concerned with the impact of conceptual innovations on the development of actual mathematics. So he uses the recursion schema to define the arithmetic operations of addition, multiplication and exponentiation. For addition, to consider just one example, take Ω to be N, let ω be m and define

$$m + 1 = \varphi(m)$$
$$m + \varphi(n) = \varphi(m + n).$$

Then Dedekind establishes systematically the fundamental properties of these operations (e.g., for addition and multiplication, commutativity, associativity, and distributivity, but also their compatibility with the ordering of N). It is an absolutely elementary and rigorously detailed development that uses nothing but the schema of primitive recursion to define functions and the principle of proof by induction (only for equations) to establish general statements. In a sense it is a more principled and focused presentation of this elementary part of finitist mathematics than that given by either Kronecker, Hilbert and Bernays in their 1921/22 lectures, or Skolem in his 1923 paper, where the foundations of elementary arithmetic are established on the basis "of the recursive mode of thought, without the use of apparent variables ranging over infinite domains".

In their Lecture Notes [1921/22], Hilbert and Bernays treat elementary arithmetic from their new finitist standpoint; here, in elementary arithmetic, they say, we have "that complete certainty of our considerations. We get along without axioms, and the inferences have the character of the concretely-certain." They continue:

> It is first of all important to see clearly that this part of mathematics can indeed be developed in a definitive way and in a way that is completely satisfactory for knowledge. The standpoint we are gaining in this pursuit is of fundamental importance also for our later considerations. (p. 51)

Their standpoint allows them to develop elementary arithmetic as "an intuitive theory of certain simple figures ..., which we are going to call number signs (Zahlzeichen)". The latter are generated as $1, 1 + 1$, etc. The arithmetic operations are introduced as concrete operations on number signs. For example, $a + b$

refers to the number sign "which is obtained by first placing + after the number sign a and then the number sign b". (p. 54) Basic arithmetic theorems like associativity of addition, are obtained by intuitive considerations including also the "ordinary counting of signs". They define less-than as a relation between number signs: a is less than b, just in case a coincides with a proper part of b. Then they use the *method of descent* to prove general statements, for example, the commutativity of addition. Having defined also divisibility and primality in this concrete manner, they establish Euclid's theorem concerning the infinity of primes. They assert

> Now we can proceed in this manner further and further; we can introduce the concepts of the greatest common divisor and the least common multiple, furthermore the number congruences. (p. 62)

That remark is followed immediately by the broader methodological claim that the definition of number theoretic functions by means of recursion formulas is admissible from the standpoint of their intuitive considerations. However, "For every single such definition by recursion it has to be determined that the application of the recursion formula indeed yields a number sign as function value — for each set of arguments."[11] They consider then as an example the standard definition of exponentiation. The mathematical development is concluded with the claim

> Fermat's little theorem, furthermore the theorems concerning quadratic residues can be established by the usual methods as intuitive theorems concerning the number signs. In fact all of elementary number theory can be developed as a theory of number signs by means of concrete intuitive considerations. (p. 63)

This development is obviously carried farther than Dedekind's and proceeds in a quite different, constructive foundational framework. For our considerations concerning computability it is important that we find here in a rough form Herbrand's way of characterizing finitistically calculable functions; that will be discussed in the next subsection.

Skolem's work was carried out in 1919, but published only in 1923; there is an acknowledged Kroneckerian influence, but the work is actually carried out in a fragment of *Principia Mathematica*. Skolem takes as basic the notions "natural number", "the number $n+1$ following the number n", as well as the "recursive mode of thought". By the latter, I suppose, Skolem understands the systematic use of "recursive definitions" and "recursive proof", i.e., definition by primitive recursion and proof by induction. Whereas the latter is indeed taken as a principle, the former is not really: for each operation or relation (via its characteristic function) an appropriate descriptive function in the sense of *Principia Mathematica*

---

[11] Here is the German formulation of this crucial condition: "Es muss nur bei jeder solchen Definition durch Rekursion eigens festgestellt werden, dass tatsächlich für jede Wertbestimmung der Argumente die Anwendung der Rekursionsformel ein Zahlzeichen als Funktionswert liefert."

has to be shown to have an unambiguous meaning, i.e., to be properly defined.[12] The actual mathematical development leads very carefully, and in much greater detail than in Hilbert and Bernays' lectures, to Euclid's theorem in the last section of the paper; the paper ends with reflections on cardinality. It is Skolem's explicit goal to avoid unrestricted quantification, as that would lead to "an infinite task — that means one that cannot be completed ..." (p. 310). In the Concluding Remark that was added to the paper at the time of its publication, Skolem makes a general point that is quite in the spirit of Hilbert: "The justification for introducing apparent variables ranging over infinite domains therefore seems very problematic; that is, one can doubt that there is any justification for the actual infinite or the transfinite." (p. 332) Skolem also announces the publication of another paper, he actually never published, in which the "formal cumbrousness" due to his reliance on *Principia Mathematica* would be avoided. "But that work, too," Skolem asserts, "is a consistently finitist one; it is built upon Kronecker's principle that a mathematical definition is a genuine definition if and only if it leads to the goal by means of a *finite* number of trials." (p. 333)

Implicit in these discussions is the specification of a class PR of functions that is obtained from the successor function by explicit definitions and the schema of (primitive) recursion. The definition of the class PR emerged in the 1920s; in Hilbert's *On the Infinite* (pp. 387–8) one finds it in almost the contemporary form: it is given inductively by specifying initial functions and closing under two definitional schemas, namely, what Hilbert calls *substitution* and *(elementary) recursion*. This can be done more precisely as follows: PR contains as its initial functions the zero-function Z, the successor function S, and the projection functions P_i^n for each n and each i with $1 \leq i \leq n$. These functions satisfy the equations $Z(x) = 0, S(x) = x'$, and $P_i^n(x_1, \ldots, x_n) = x_i$, for all $x, x_1, \ldots, x_n; x'$ is the successor of x. The class is closed under the *schema of composition*: Given an m-place function ψ in PR and n-place functions $\varphi_1, \ldots, \varphi_m$ in PR, the function ϕ defined by

$$\phi(x_1, \ldots, x_n) = \psi(\varphi_1(x_1, \ldots, x_n), \ldots, \varphi_m(x_1, \ldots, x_n))$$

is also in PR; ϕ is said to be obtained by *composition* from ψ and $\varphi_1, \ldots, \varphi_m$. PR is also closed under the *schema of primitive recursion*: Given an n-place function ψ in PR, and an $n + 24$-place function φ in PR, the function ϕ defined by

$$\phi(x_1, \ldots, x_n, 0) = \psi(x_1, \ldots, x_n)$$
$$\phi(x_1, \ldots, x_n, y') = \varphi(x_1, \ldots, x_n, y, \phi(x_1, \ldots, x_n, y))$$

is a function in PR; ϕ is said to be obtained by primitive recursion from ψ and φ. Thus, a function is primitive recursive if and only if it can be obtained from some initial functions by finitely many applications of the composition and recursion schemas. This definition was essentially given in Gödel's 1931 paper together with

---

[12]That is done for addition on p. 305, for the less-than relation on p. 307, and for subtraction on p. 314.

arguments that this class contains the particular functions that are needed for the arithmetic description of *Principia Mathematica* and related systems.

By an inductive argument on the definition of PR one can see that the values of primitive recursive functions can be determined, for any particular set of arguments, by a standardized calculation procedure; thus, all primitive recursive functions are in this sense calculable. Yet there are calculable functions, which are not primitive recursive. An early example is due to Hilbert's student Ackermann; it was published in 1928, but discussed already in [Hilbert, 1925]. Here is the definition of the *Ackermann function*:

$$\phi_0(x, y) = S(y)$$
$$\phi_{n'}(x, 0) = \begin{cases} x & \text{if } n = 0 \\ 0 & \text{if } n = 1 \\ 1 & \text{if } n > 1 \end{cases}$$
$$\phi_{n'}(x, y') = \phi_n(x, \phi_{n'}(x, y)).$$

Notice that ϕ_1 is addition, ϕ_2 is multiplication, ϕ_3 is exponentiation, etc; i.e., the next function is always obtained by iterating the previous one. For each n, the function $\phi_n(x, x)$ is primitive recursive, but $\phi(x, x, x)$ is not: Ackermann showed that it grows faster than any primitive recursive function. Herbrand viewed the Ackermann function in his [1931a] as finitistically calculable.

2.4 *Formalizability and calculability*

In lectures and publications from 1921 and 1922, Hilbert and Bernays established the consistency of an elementary part of arithmetic from their new finitist perspective. The work is described together with an *Ansatz* for its extension in [Hilbert, 1923]. They restrict the attention to the quantifier-free part of arithmetic that contains all primitive recursive functions and an induction rule; that part is now called *primitive recursive arithmetic* (*PRA*) and is indeed the system **F*** of Herbrand's discussed below, when the class F of finitist functions consists of exactly the primitive recursive ones.[13]

PRA has a direct finitist justification, and thus there was no programmatic need to establish its consistency. However, the proof was viewed as a stepping-stone towards a consistency proof for full arithmetic and analysis. It is indeed the first sophisticated proof-theoretic argument, transforming arbitrary derivations into configurations of variable-free formulas. The truth-values of these formulas can be effectively determined, because Hilbert and Bernays insist on the calculability of functions and the decidability of relations. Ackermann attempted in his dissertation, published as [Ackermann, 1924], to extend this very argument to analysis. Real difficulties emerged even before the article appeared and the

---

[13]Tait argues in his [1981] for the identification of finitist arithmetic with *PRA*. This is a conceptually coherent position, but I no longer think that it reflects the historical record of considerations and work surrounding Hilbert's Program; cf. also Tait's [2002], the papers by Zach referred to, Ravaglia's Carnegie Mellon Ph.D. thesis, as well as our joint paper [2005].

validity of the result had to be restricted to a part of elementary number theory. The result is obtained also in von Neumann's [1927]. The problem of extending the restricted result was thought then to be a straightforward mathematical one. That position was clearly taken by Hilbert in his Bologna address of 1928, when he claims that the results of Ackermann and von Neumann cover full arithmetic and then asserts that there is an *Ansatz* of a consistency proof for analysis: "This [Ansatz] has been pursued by Ackermann already to such an extent that the remaining task amounts only to proving a purely arithmetic elementary finiteness theorem." (p. 4)

These difficulties were revealed, however, by the incompleteness theorems as "conceptual" philosophical ones. The straightforwardly mathematical consequence of the second incompleteness theorem can be formulated as follows: Under general conditions[14] on a theory T, T proves the conditional $(con_T \to G)$; con_T is the statement expressing the consistency of T, and G is the Gödel sentence. G states its own unprovability and is, by the first incompleteness theorem, not provable in T. Consequently, G would be provable in T, as soon as a finitist consistency proof for T could be formalized in T. That's why the issue of the formalizability of finitist considerations plays such an important role in the emerging discussion between von Neumann, Herbrand and Gödel. At issue was the extent of finitist methods and thus the reach of Hilbert's consistency program. That raises in particular the question, what are the finitistically calculable functions; it is clear that the primitive recursively defined functions are to be included. (Recall the rather general way in which recursive definitions were dicussed in Hilbert's lectures [1921/22].)

Herbrand's own [1931a] is an attempt to harmonize his proof theoretic investigations with Gödel's results. Gödel insisted in his paper that the second incompleteness theorem does not contradict Hilbert's "formalist viewpoint":

> For this viewpoint presupposes only the existence of a consistency proof in which nothing but finitary means of proof is used, and it is conceivable that there exist finitary proofs that *cannot* be expressed in the formalism of P (or of M and A).[15]

Having received the galleys of Gödel's paper, von Neumann writes in a letter of 12 January 1931:

> I absolutely disagree with your view on the formalizability of intuitionism. Certainly, for every formal system there is, as you proved, another formal one that is (already in arithmetic and the lower functional calculus) stronger. But that does not affect intuitionism at all.

---

[14]The general conditions on T include, of course, the representability conditions for the first theorem and the Hilbert-Bernays derivability conditions for the second theorem.

[15]*Collected Works I*, p. 195. P is a version of the system of *Principia Mathematica*, M the system of set theory introduced by von Neumann, and A classical analysis.

(Note that Herbrand and von Neumann, but also others at the time, use intuitionist as synonymous with finitist; even Gödel did as should be clear from his [1931a].) Denoting first-order number theory by A, analysis by M, and set theory by Z, von Neumann continues:

> Clearly, I cannot prove that every intuitionistically correct *construction* of *arithmetic* is formalizable in A or M or even in Z — for intuitionism is undefined and undefinable. But is it not a fact, that not a single construction of the kind mentioned is known that cannot be formalized in A, and that no living logician is in the position of naming such [[a construction]]? Or am I wrong, and you know an effective intuitionistic arithmetic construction whose formalization in A creates difficulties? If that, to my utmost surprise, should be the case, then the formalization should work in M or Z!

This line of argument was sharpened, when Herbrand wrote to Gödel on 7 April 1931. By then he had discussed the incompleteness phenomena extensively with von Neumann, and he had also read the galleys of [Gödel, 1931]. Herbrand's letter has to be understood, and Gödel in his response quite clearly did, as giving a sustained argument against Gödel's assertion that the second incompleteness theorem does not contradict Hilbert's formalist viewpoint.

Herbrand introduces a number of systems for arithmetic, all containing the axioms for predicate logic with identity and the Dedekind-Peano axioms for zero and successor. The systems are distinguished by the strength of the induction principle and by the class F of finitist functions for which recursion equations are available. The system with induction for all formulas and recursion equations for the functions in F is denoted here by **F**; if induction is restricted to quantifier-free formulas, I denote the resulting system by **F***. The axioms for the elements $f_1, f_2, f_3, \ldots$ in F must satisfy according to Herbrand's letter the following conditions:

1. The defining axioms for f_n contain, besides f_n, only functions of lesser index.

2. These axioms contain only constants and free variables.

3. We must be able to show, by means of intuitionistic proofs, that with these axioms it is possible to compute the value of the functions univocally for each specified system of values of their arguments.

As examples for classes F Herbrand considers the set E_1 of addition and multiplication, as well as the set E_2 of all primitive recursive functions. He asserts that many other functions are definable by his "general schema", in particular, the non-primitive recursive Ackermann function. He also argues that one can construct by diagonalization a finitist function that is not in E, if **E** contains axioms such that "one can always determine, whether or not certain defining axioms [for the elements of E] are among these axioms". It is here that the "double" use of finitist functions — straightforwardly as part of finitist mathematical practice and as a tool to describe formal theories — comes together to allow the definition of additional finitist functions; that is pointed out in Herbrand's letter to Gödel. Indeed, it is quite explicit also in Herbrand's almost immediate reaction to the incompleteness phenomena in his letter to Chevalley from 3 December 1930. (See [Sieg, 1994, 103–4].)

This fact of the open-endedness of any finitist presentation of the concept "finitist function" is crucial for Herbrand's conjecture that one cannot prove that all finitist methods are formalizable in *Principia Mathematica*. But he claims that, as a matter of fact, every finitist proof can be formalized in a system $\mathbf{F}^*$, based on a suitable class F that depends on the given proof, thus in *Principia Mathematica*. Conversely, he insists that every proof in the quantifier-free part of $\mathbf{F}^*$ is finitist. He summarizes his reflections by saying in the letter and with almost identical words in [1931a]:

> It reinforces my conviction that it is impossible to prove that every intuitionistic proof is formalizable in Russell's system, but that a counterexample will never be found. There we shall perhaps be compelled to adopt a kind of logical postulate.

Herbrand's conjectures and claims are completely in line with those von Neumann communicated to Gödel in his letters of November 1930 and January 1931. In the former letter von Neumann wrote

> I believe that every intuitionistic consideration can be formally copied, because the "arbitrarily nested" recursions of Bernays-Hilbert are equivalent to ordinary transfinite recursions up to appropriate ordinals of the second number class. This is a process that can be formally captured, unless there is an intuitionistically definable ordinal of the second number class that could not be defined formally — which is in my view unthinkable. Intuitionism clearly has no finite axiom system, but that does not prevent its being a part of classical mathematics that does have one. (*Collected Works V*, p. 339)

We know of Gödel's response to von Neumann's dicta not through letters from Gödel, but rather through the minutes of a meeting of the Schlick or Vienna Circle that took place on 15 January 1931. According to these minutes Gödel viewed as questionable the claim that the totality of all intuitionistically correct proofs is contained in *one* formal system. That, he emphasized, is the weak spot in von Neumann's argumentation. (Gödel did respond to von Neumann, but his letters seem to have been lost. The minutes are found in the Carnap Archives of the University of Pittsburgh.)

When answering Herbrand's letter, Gödel makes more explicit his reasons for questioning the formalizability of finitist considerations in a single formal system like *Principia Mathematica*. He agrees with Herbrand on the indefinability of the concept "finitist proof". However, even if one accepts Herbrand's very schematic presentation of finitist methods and the claim that every finitist proof can be formalized in a system of the form **F***, the question remains "whether the intuitionistic proofs that are required in each case to justify the unicity of the recursion axioms are all formalizable in *Principia Mathematica*". Gödel continues:

> Clearly, I do not claim either that it is certain that some finitist proofs are not formalizable in *Principia Mathematica*, even though intuitively I tend toward this assumption. In any case, a finitist proof not formalizable in *Principia Mathematica* would have to be quite extraordinarily complicated, and on this purely practical ground there is very little prospect of finding one; but that, in my opinion, does not alter anything about the possibility in principle.

At this point there is a stalemate between Herbrand's "logical postulate" that no finitist proof outside of *Principia Mathematica* will be found and Gödel's "possibility in principle" that one might find such a proof.

By late December 1933 when he gave an invited lecture to the Mathematical Association of America in Cambridge (Massachusetts), Gödel had changed his views significantly. In the text for his lecture, [Gödel, 1933], he sharply distinguishes intuitionist from finitist arguments, the latter constituting the most restrictive form of constructive mathematics. He insists that the known finitist arguments given by "Hilbert and his disciples" can all be carried out in a certain system **A**. Proofs in **A**, he asserts, "can be easily expressed in the system of classical analysis and even in the system of classical arithmetic, and there are reasons for believing that this will hold for any proof which one will ever be able to construct". This observation and the second incompleteness theorem imply, as sketched above, that classical arithmetic cannot be shown to be consistent by finitist means. The system **A** is similar to the quantifier-free part of Herbrand's system **F***, except that the provable totality for functions in F is not mentioned and that **A** is also concerned with other inductively defined classes.[16] Gödel's reasons for conjecturing that **A** contains all finitist arguments are not made explicit.

Gödel discusses then a theorem of Herbrand's, which he considers to be the most far-reaching among interesting partial results in the pursuit of Hilbert's consistency program. He does so, as if to answer the question "How do current consistency proofs fare?" and formulates the theorem in this lucid and elegant way: "If we

---

[16] The restrictive characteristics of the system **A** are formulated on pp. 23 and 24 of [1933] and include the requirement that notions have to be decidable and functions must be calculable. Gödel claims that "such notions and functions can always be defined by complete induction". Definition by complete induction is to be understood as definition by recursion, which — for the integers — is not restricted to primitive recursion. The latter claim is supported by the context of the lecture and also by Gödel's remark at the very beginning of section 9 in his Princeton Lectures, where he explains that a version of the Ackermann function is "defined inductively". The actual definition is considered "as an example of a definition by induction with respect to two variables simultaneously". That is followed by the remark, "The consideration of various sorts of functions defined by induction leads to the question what one would mean by 'every recursive function'."

take a theory which is constructive in the sense that each existence assertion made in the axioms is covered by a construction, and if we add to this theory the non-constructive notion of existence and all the logical rules concerning it, e.g., the law of excluded middle, we shall never get into any contradiction." (This implies directly the extension of Hilbert's first consistency result from 1921/22 to the theory obtained from it by adding full classical first order logic, but leaving the induction principle quantifier-free.) Gödel conjectures that Herbrand's method might be generalized, but he emphasizes that "for larger systems containing the whole of arithmetic or analysis the situation is hopeless if you insist upon giving your proof for freedom from contradiction by means of the system **A**". As the system **A** is essentially the quantifier-free part of **F***, it is clear that Gödel now takes Herbrand's position concerning the impact of his second incompleteness theorem on Hilbert's Program.

Nowhere in the correspondence does the issue of *general* computability arise. Herbrand's discussion, in particular, is solely trying to explore the limits that are imposed on consistency proofs by the second theorem. Gödel's response focuses also on that very topic. It seems that he subsequently developed a more critical perspective on the character and generality of his theorems. This perspective allowed him to see a crucial open question and to consider Herbrand's notion of a finitist function as a first step towards an answer. A second step was taken in 1934 when Gödel lectured on his incompleteness theorems at Princeton. There one finds not only an even more concise definition of the class of primitive recursive functions, but also a crucial and revealing remark as to the pragmatic reason for the choice of this class of functions.

The very title of the lectures, *On undecidable propositions of formal mathematical systems*, indicates that Gödel wanted to establish his theorems in greater generality, not just for *Principia Mathematica* and related systems. In the introduction he attempts to characterize "formal mathematical system" by requiring that the rules of inference, and the definitions of meaningful [i.e., syntactically well-formed] formulas and axioms, be "constructive"; Gödel elucidates the latter concept as follows:

> ...for each rule of inference there shall be a finite procedure for determining whether a given formula B is an immediate consequence (by that rule) of given formulas $A_1, \ldots A_n$, and there shall be a finite procedure for determining whether a given formula A is a meaningful formula or an axiom. (p. 346)

That is of course informal and imprecise, mathematically speaking. The issue is addressed in section 7, where Gödel discusses conditions a formal system must satisfy so that the arguments for the incompleteness theorems apply to it. The first of five conditions is this:

> Supposing the symbols and formulas to be numbered in a manner similar to that used for the particular system considered above, then the

class of axioms and the relation of immediate consequence shall be recursive (i.e., in these lectures, primitive recursive).

This is a precise condition which in practice suffices as a substitute for the unprecise requirement of §1 that the class of axioms and the relation of immediate consequence be constructive. (p. 361)[17]

A principled precise condition for characterizing formal systems in general is needed. Gödel defines in §9 the class of "general recursive functions"; that is Gödel's second step alluded to above and the focus of the next section.

3 RECURSIVENESS AND CHURCH'S THESIS

In Section 2 I described the emergence of a broad concept of calculable function. It arose out of a mathematical practice that was concerned with effectiveness of solutions, procedures and notions; it was also tied in important ways to foundational discussions that took place already in the second half of the 19th century with even older historical roots. I pointed to the sharply differing perspectives of Dedekind and Kronecker. It was the former who formulated in his [1888] the schema of primitive recursion in perfect generality. That all the functions defined in this way are calculable was of course clear, but not the major issue for Dedekind: he established that primitive recursive definitions determine unique functions in his logicist framework. From a constructive perspective, however, these functions have an autonomous significance and were used in the early work of Hilbert and Bernays, but also of Skolem, for developing elementary arithmetic in a deeply Kroneckerian spirit. Hilbert and Bernays viewed this as a part of finitist mathematics, their framework for meta-mathematical studies in general and for consistency proofs in particular.

An inductive specification of the class of primitive recursive functions is found in the *Zwischenbetrachtung* of section 3 in Gödel's [1931] and, even more standardly, in the second section of his [1934]. That section is entitled "Recursive functions and relations." In a later footnote Gödel pointed out that "recursive" in these lectures corresponds to "primitive recursive" as used now. It was a familiar fact by then that there are calculable functions, which are not in the class of primitive recursive functions, with Ackermann's and Sudan's functions being the best-known examples. Ackermann's results were published only in 1928, but they had been discussed extensively already earlier, e.g., in Hilbert's *On the infinite*. Herbrand's schema from 1931 defines a broad class of finitistically calculable functions including the Ackermann function; it turned out to be the starting-point of significant further developments.

Herbrand's schema is a natural generalization of the definition schemata for calculable functions that were known to him and built on the practice of the

---

[17] In the Postscriptum to [Gödel, 1934] Gödel asserts that exactly this condition can be removed on account of Turing's work.

Hilbert School. It could also be treated easily by the methods for proving the consistency of weak systems of arithmetic Herbrand had developed in his thesis. In a letter to Bernays of 7 April 1931, the very day on which he also wrote to Gödel, Herbrand contrasts his consistency proof with Ackermann's, which he mistakenly attributes to Bernays:

> In my arithmetic the axiom of complete induction is restricted, but one may use a variety of other functions than those that are defined by simple recursion: in this direction, it seems to me, that my theorem goes a little farther than yours [i.e., than Ackermann's].

The point that is implicit in my earlier discussion should be made explicit here and be contrasted with discussions surrounding Herbrand's schema by Gödel and van Heijenoort as to the programmatic direction of the schema[18]: the above is hardly a description of a class of functions that is deemed to be of fundamental significance for the question of "general computability". Rather, Herbrand's remark emphasizes that his schema captures a broader class of finitist functions and should be incorporated into the formal theory to be shown consistent.

Gödel considered the schema, initially and in perfect alignment with Herbrand's view, as a way of partially capturing the constructive aspect of mathematical practice. It is after all the classical theory of arithmetic with Herbrand's schema that is reduced to its intuitionistic version by Gödel in his [1933]; this reductive result showed that intuitionism provides a broader constructive framework than finitism. I will detail the modifications Gödel made to Herbrand's schema when introducing in [1934] the general recursive functions. The latter are the primary topic of this section, and the main issues for our discussion center around Church's Thesis.

3.1 Relative consistency

Herbrand proved in his [1931a], as I detailed above and at the end of section 2.4, the consistency of a system for classical arithmetic that included defining equations for all the finitistically calculable functions identified by his schema, but made the induction principle available only for quantifier-free formulas. In a certain sense that restriction is lifted in Gödel's [1933], where an elementary translation of full classical arithmetic into intuitionistic arithmetic is given. A system for intuitionistic arithmetic had been formulated in [Heyting, 1930a]. Gödel's central claim in the paper is this: *If a formula A is provable in Herbrand's system for classical arithmetic, then its translation A^* is provable in Heyting arithmetic. A^** is obtained from A by transforming the latter into a classically equivalent formula not containing $\vee, \rightarrow, (\exists)$. The crucial auxiliary lemmata are the following:

---

[18]Van Heijenoort analyzed the differences between Herbrand's published proposals and the suggestion that had been made, according to [Gödel, 1934], by Herbrand in his letter to Gödel. References to this discussion in light of the actual letter are found in my paper [1994]; see in particular section 3.2 and the Appendix.

(i) For all formulas A^*, Heyting arithmetic proves $\neg\neg A^* \to A^*$; and

(ii) For all formulas A^* and B^*, Heyting arithmetic proves that $A^* \to B^*$ is equivalent to $\neg(A^* \& \neg B^*)$

The theorem establishes obviously the consistency of classical arithmetic relative to Heyting arithmetic. If the statement 0=1 were provable in classical arithmetic, then it would be provable in Heyting arithmetic, as (0=1)* is identical to 0=1. From an intuitionistic point of view, however, the principles of Heyting arithmetic can't lead to a contradiction. Gödel concludes his paper by saying (p. 294 in *Collected Works I*):

> The above considerations provide of course an intuitionistic consistency proof for classical arithmetic and number theory. However, the proof is not "finitist" in the sense Herbrand gave to the term, following Hilbert.

This implies a clear differentiation of intuitionistic from finitist mathematics, and the significance of this result cannot be overestimated. Ironically, it provided a basis and a positive direction for modifying Hilbert's Program: exploit in consistency proofs the constructive means of intuitionistic mathematics that go beyond finitist ones. Gödel's result is for that very reason important and was obtained, with a slightly different argument, also by Gentzen. The historical point is made forcefully by Bernays in his contribution on Hilbert to the *Encyclopedia of Philosophy*; the systematic point, and its relation to the further development of proof theory, has been made often in the context of a generalized reductive program in the tradition of the Hilbert school; see, for example, [Sieg and Parsons, 1995] or my [2002].

I discuss Gödel's result for two additional reasons, namely, to connect the specific developments concerning computability with the broader foundational considerations of the time and to make it clear that Gödel was thoroughly familiar with Herbrand's formulation when he gave the definition of "general recursive functions" in his 1934 Princeton Lectures. Herbrand's schema is viewed, in the reductive context, from the standpoint of constructive mathematical practice as opposed to its meta-mathematical use in the description of "formal theories". That is made clear by Gödel's remark, "The definition of number-theoretic functions by recursion is unobjectionable for intuitionism as well (see H_2, 10.03, 10.04). Thus all functions f_i (Axiom Group C) occur also in intuitionistic mathematics, and we consider the formulas defining them to have been adjoined to Heyting's axioms;..."[19] The meta-mathematical, descriptive use will become the focus of our investigation, as the general characterization of "formal systems" takes center stage and is pursued via an explication of "constructive" or "effective" procedures. We will then take on the problem of identifying an appropriate mathematical concept for this informal notion, i.e., issues surrounding Church's or Turing's Thesis. To get a

---

[19] *Collected Works I*, p. 290. The paper H_2 is [Heyting, 1930a], and the numbers 10.03 and 10.04 refer to not more and not less than the recursion equations for addition.

concrete perspective on the significance of the broad issues, let me mention claims formulated by Church and Gödel with respect to Turing's work, but also point to tensions and questions that are only too apparent.

Church reviewed Turing's *On computable numbers* for the *Journal of Symbolic Logic* just a few months after its publication. He contrasted Turing's notion for effective calculability (via idealized machines) with his own (via λ-definability) and with Gödel's (via the equational calculus). "Of these [notions]," Church remarked, "the first has the advantage of making the identification with effectiveness in the ordinary (not explicitly defined) sense evident immediately..." Neither in this review nor anywhere else did Church give reasons, why the identification *is* immediately evident for Turing's notion, and why it is *not* for the others. In contrast, Gödel seemed to capture essential aspects of Turing's considerations when making a brief and enigmatic remark in the 1964 postscript to the Princeton Lectures he had delivered thirty years earlier: "Turing's work gives an *analysis* of the concept of 'mechanical procedure'.... This concept is *shown* to be equivalent with that of a 'Turing machine'."[20] But neither in this postscript nor in other writings did Gödel indicate the nature of Turing's analysis and prove that the analyzed concept is indeed equivalent to that of a Turing machine.

Gödel underlined the significance of Turing's analysis, repeatedly and emphatically. He claimed, also in [1964], that only Turing's work provided "a precise and unquestionably adequate definition of the general concept of formal system". As a formal system is for Gödel just a mechanical procedure for producing theorems, the adequacy of this definition rests squarely on the correctness of Turing's analysis of mechanical procedures. The latter lays the ground for the most general mathematical formulation and the broadest philosophical interpretation of the incompleteness theorems. Gödel himself had tried to arrive at an adequate concept in a different way, namely, by directly characterizing calculable number theoretic functions more general than primitive recursive ones. As a step towards such a characterization, Gödel introduced in his Princeton Lectures "general recursive functions" via his equational calculus "using" Herbrand's schema. I will now discuss the crucial features of Gödel's definition and contrast it with Herbrand's as discussed in Section 2.4.

3.2 Uniform calculations

In his Princeton Lectures, Gödel strove to make the incompleteness results less dependent on particular formalisms. Primitive recursive definability of axioms and inference rules was viewed as a "precise condition, which in practice suffices as a substitute for the unprecise requirement of §1 that the class of axioms and the relation of immediate consequence be constructive". A notion that would suffice *in principle* was needed, however, and Gödel attempted to arrive at a more general

---

[20] Gödel's *Collected Works I*, pp. 369–70. The emphases are mine. In the context of this paper and reflecting the discussion of Church and Gödel, I consider effective and mechanical procedures as synonymous.

notion. Gödel considers the fact that the value of a primitive recursive function can be computed by a finite procedure for each set of arguments as an "important property" and adds in footnote 3:

> The converse seems to be true if, besides recursions according to the scheme (2) [i.e., primitive recursion as given above], recursions of other forms (e.g., with respect to two variables simultaneously) are admitted. This cannot be proved, since the notion of finite computation is not defined, but it can serve as a heuristic principle.

What other recursions might be admitted is discussed in the last section of the Notes under the heading "general recursive functions".

The general recursive functions are taken by Gödel to be those number theoretic functions whose values can be calculated via elementary substitution rules from an extended set of basic recursion equations. This is an extremely natural approach and properly generalizes primitive recursiveness: the new class of functions includes of course all primitive recursive functions and also those of the Ackermann type, defined by nested recursion. Assume, Gödel suggests, you are given a finite sequence $\psi_1, \ldots, \psi_k$ of "known" functions and a symbol ϕ for an "unknown" one. Then substitute these symbols "in one another in the most general fashions" and equate certain pairs of the resulting expressions. If the selected set of functional equations has exactly one solution, consider ϕ as denoting a "recursive" function.[21] Gödel attributes this broad proposal to define "recursive" functions mistakenly to Herbrand and proceeds then to formulate two restrictive conditions for his definition of "general recursive" functions:

(1) the l.h.s. of equations is of the form $\phi(\psi_{i_1}(x_1, \ldots, x_n), \ldots, \psi_{i_l}(x_1, \ldots, x_n))$, and

(2) for every l-tuple of natural numbers the value of ϕ is "computable in a calculus".

The first condition just stipulates a standard form of certain terms, whereas the important second condition demands that for every l-tuple $k_1, \ldots, k_l$ there is exactly one m such that $\phi(k_1, \ldots, k_l) = m$ is a "derived equation". The set of derived equations is specified inductively via elementary substitution rules; the basic clauses are:

(A.1) All numerical instances of a given equation are derived equations;

(A.2) All true equalities $\psi_{i_j}(x_1, \ldots, x_n) = m$ are derived equations.

The rules allowing steps from already obtained equations to additional ones are formulated as follows:

---

[21] Kalmar proved in his [1955] that these "recursive" functions, just satisfying recursion equations, form a strictly larger class than the general recursive ones.

(**R.1**) Replace occurrences of $\psi_{i_j}(x_1,\ldots,x_n)$ by m, if $\psi_{i_j}(x_1,\ldots,x_n) = m$ is a derived equation;

(**R.2**) Replace occurrences of $\phi(x_1,\ldots,x_l)$ on the right-hand side of a derived equation by m, if $\phi(x_1,\ldots,x_l) = m$ is a derived equation.

In addition to restriction (**1**) on the syntactic form of equations, we should recognize with Gödel two novel features in this definition when comparing it to Herbrand's: first, the precise specification of *mechanical* rules for deriving equations, i.e., for carrying out numerical computations; second, the formulation of the *regularity condition* requiring computable functions to be total, but without insisting on a finitist proof. These features were also emphasized by Kleene who wrote with respect to Gödel's definition that "it consists in specifying the form of the equations and the nature of the steps admissible in the computation of the values, and in requiring that for each given set of arguments the computation yield a unique number as value" [Kleene, 1936, 727]. Gödel re-emphasized these points in later remarks, when responding to van Heijenoort's inquiry concerning the precise character of Herbrand's suggestion.

In a letter to van Heijenoort of 14 August 1964 Gödel asserts "it was exactly by specifying the rules of computation that a mathematically workable and fruitful concept was obtained". When making this claim Gödel took for granted that Herbrand's suggestion had been "formulated *exactly* as on page 26 of my lecture notes, i.e., without reference to computability". At that point Gödel had to rely on his recollection, which, he said, "is very distinct and was still very fresh in 1934". On the evidence of Herbrand's letter, it is clear that Gödel misremembered. This is not to suggest that Gödel was wrong in viewing the specification of computation rules as extremely important, but rather to point to the absolutely crucial step he had taken, namely, to disassociate general recursive functions from the epistemologically restricted notion of proof that is involved in Herbrand's formulation.

Gödel dropped later the regularity condition altogether and emphasized, "that the precise notion of mechanical procedures is brought out clearly by Turing machines producing partial rather than general recursive functions." At the earlier juncture in 1934 the introduction of the equational calculus with particular computation rules was important for the mathematical development of recursion theory as well as for the underlying conceptual motivation. It brought out clearly, what Herbrand — according to Gödel in his letter to van Heijenoort — had failed to see, namely "that the computation (for all computable functions) proceeds by exactly the same rules". Gödel was right, for stronger reasons than he put forward, when he cautioned in the same letter that Herbrand had *foreshadowed*, but not *introduced*, the notion of a general recursive function. Cf. the discussion in and of [Gödel, 193?] presented in Section 6.1.

Kleene analyzed the class of general recursive functions in his [1936] using Gödel's arithmetization technique to describe provability in the equational calculus. The uniform and effective generation of derived equations allowed Kleene to establish an important theorem that is most appropriately called "Kleene's normal form theorem": *for every recursive function φ there are primitive recursive functions ψ and ρ such that $\varphi(x_1, \ldots, x_n)$ equals $\psi(\varepsilon y.\rho(x_1, \ldots, x_n, y) = 0)$, where for every n-tuple $x_1, \ldots, x_n$ there is a y such that $\rho(x_1, \ldots, x_n, y) = 0$.* The latter equation expresses that y is (the code of) a computation from the equations that define φ for the arguments $x_1, \ldots, x_n$. The term $\varepsilon y.\rho(x_1, \ldots, x_n, y) = 0$ provides the smallest y, such that $\rho(x_1, \ldots, x_n, y) = 0$, if there is a y for the given arguments, and it yields 0 otherwise. Finally, the function ψ considers the last equation in the selected computation and determines the numerical value of the term on the r.h.s of that equation — which is a numeral and represents the value of φ for given arguments $x_1, \ldots, x_n$. This theorem (or rather its proof) is quite remarkable: the ease with which "it" allows to establish equivalences of different computability formulations makes it plausible that some stable notion has been isolated. What is needed for the proof is only that the inference or computation steps are all primitive recursive. Davis observes in his [1982, 11] quite correctly, "The theorem has made equivalence proofs for formalisms in recursive function theory rather routine, ..." The informal understanding of the theorem is even more apparent from Kleene's later formulation involving his T-predicate and result-extracting function U; see for example his *Introduction to Metamathematics*, p. 288 ff.

Hilbert and Bernays had introduced in the first volume of their *Grundlagen der Mathematik* a μ-operator that functioned in just the way the ε-operator did for Kleene. The μ-notation was adopted later also by Kleene and is still being used in computability theory. Indeed, the μ-operator is at the heart of the definition of the class of the so-called "μ-recursive functions". They are specified inductively in the same way as the primitive recursive functions, except that a third closure condition is formulated: if $\rho(x_1, \ldots, x_n, y)$ is μ-recursive and for every n-tuple $x_1, \ldots, x_n$ there is a y such that $\rho(x_1, \ldots, x_n, y) = 0$, then the function $\theta(x_1, \ldots, x_n)$ given by $\mu y.\rho(x_1, \ldots, x_n, y) = 0$ is also μ-recursive. The normal form theorem is the crucial stepping stone in proving that this class of functions is co-extensional with that of Gödel's general recursive ones.

This result was actually preceded by the thorough investigation of λ-definability by Church, Kleene and Rosser.[22] Kleene emphasized in his [1987, 491], that the approach to effective calculability through λ-definability had "quite independent roots (motivations)" and would have led Church to his main results "even if Gödel's paper [1931] had not already appeared". Perhaps Kleene is right, but I doubt it. The flurry of activity surrounding Church's *A set of postulates for the foundation of logic* (published in 1932 and 1933) is hardly imaginable without knowledge of Gödel's work, in particular not without the central notion of representability and, as Kleene points out, the arithmetization of meta-mathematics. The Princeton group knew of Gödel's theorems since the fall of 1931 through a lecture of von Neumann's. Kleene reports in [1987, 491], that through this lecture "Church and the rest of us first learned of Gödel's results". The centrality of representability

---

[22] For analyses of the quite important developments in Princeton from 1933 to 1937 see [Davis, 1982] and my [1997], but of course also the accounts given by Kleene and Rosser. [Crossley, 1975a] contains additional information from Kleene about this time.

for Church's considerations comes out clearly in his lecture on Richard's paradox given in December 1933 and published as [Church, 1934]. According to [Kleene, 1981, 59] Church had formulated his thesis for λ-definability already in the fall of 1933; so it is not difficult to read the following statement as an extremely cautious statement of the thesis:

> ... it appears to be possible that there should be a system of symbolic logic containing a formula to stand for every definable function of positive integers, and I fully believe that such systems exist.[23]

One has only to realize from the context that (i) 'definable' means 'constructively definable', so that the value of the functions can be calculated, and (ii) 'to stand for' means 'to represent'.

A wide class of calculable functions had been characterized by the concept introduced by Gödel, a class that contained all known effectively calculable functions. Footnote 3 of the Princeton Lectures I quoted earlier seems to express a form of Church's Thesis. In a letter to Martin Davis of 15 February 1965, Gödel emphasized that no formulation of Church's Thesis is implicit in that footnote. He wrote:

> ... The conjecture stated there only refers to the equivalence of "finite (computation) procedure" and "recursive procedure". However, I was, at the time of these lectures, not at all convinced that my concept of recursion comprises all possible recursions; and in fact the equivalence between my definition and Kleene's ... is not quite trivial.

At that time in early 1934, Gödel was equally unconvinced by Church's proposal to identify effective calculability with λ-definability; he called the proposal "thoroughly unsatisfactory". That was reported by Church in a letter to Kleene dated 29 November 1935 (and quoted in [Davis, 1982, 9]).

Almost a year later, Church comes back to his proposal in a letter to Bernays dated 23 January 1935; he conjectures that the λ-calculus may be a system that allows the representability of all constructively defined functions:

> The most important results of Kleene's thesis concern the problem of finding a formula to represent a given intuitively defined function of positive integers (it is required that the formula shall contain no other symbols than λ, variables, and parentheses). The results of Kleene are so general and the possibilities of extending them apparently so unlimited that one is led to conjecture that a formula can be found to represent any particular constructively defined function of positive integers whatever. It is difficult to prove this conjecture, however, or even to state it accurately, because of the difficulty in saying precisely what is meant by "constructively defined". A vague description can be

---

[23][Church, 1934, 358]. Church assumed, clearly, the converse of this claim.

given by saying that a function is constructively defined if a method can be given by which its values could be actually calculated for any particular positive integer whatever.

When Church wrote this letter, it was known in his group that all general recursive functions are λ-definable; Church established in collaboration with Kleene the converse by March 1935. (Cf. [Sieg, 1997, 156].) This mathematical equivalence result and the quasi-empirical adequacy through Kleene's and Rosser's work provided the background for the public articulation of Church's Thesis in the 1935 abstract to be discussed in the next subsection. The elementary character of the steps in computations made the normal form theorem and the equivalence argument possible. In the more general setting of his 1936 paper, Church actually tried to show that every informally calculable number theoretic function is indeed general recursive.

3.3 Elementary steps

Church, Kleene and Rosser had thoroughly investigated Gödel's notion and its connection with λ-definability by the end of March 1935; Church announced his thesis in a talk contributed to the meeting of the American Mathematical Society in New York City on 19 April 1935. I quote the abstract of the talk in full.

> Following a suggestion of Herbrand, but modifying it in an important respect, Gödel has proposed (in a set of lectures at Princeton, N.J., 1934) a definition of the term *recursive function*, in a very general sense. In this paper a definition of *recursive function of positive integers* which is essentially Gödel's is adopted. And it is maintained that the notion of an effectively calculable function of positive integers should be identified with that of a recursive function, since other plausible definitions of effective calculability turn out to yield notions that are either equivalent to or weaker than recursiveness. There are many problems of elementary number theory in which it is required to find an effectively calculable function of positive integers satisfying certain conditions, as well as a large number of problems in other fields which are known to be reducible to problems in number theory of this type. A problem of this class is the problem to find a complete set of invariants of formulas under the operation of conversion (see abstract 41.5.204). It is proved that this problem is unsolvable, in the sense that there is no complete set of effectively calculable invariants.

General recursiveness served, perhaps surprisingly, as the rigorous concept in this first published formulation of Church's Thesis. The surprise vanishes, however, when Rosser's remark in his [1984] about this period is seriously taken into account: "Church, Kleene, and I each thought that general recursivity seemed to embody the idea of effective calculability, and so each wished to show it equivalent to λ-definability" (p. 345). Additionally, when presenting his [1936a] to the American

Mathematical Society on 1 January 1936, Kleene made these introductory remarks (on p. 544): "The notion of a recursive function, which is familiar in the special cases associated with primitive recursions, Ackermann-Péter multiple recursions, and others, has received a general formulation from Herbrand and Gödel. The resulting notion is of especial interest, since the intuitive notion of a 'constructive' or 'effectively calculable' function of natural numbers can be identified with it very satisfactorily." λ-definability was not even mentioned.

In his famous 1936 paper *An unsolvable problem of elementary number theory* Church described the form of number theoretic problems to be shown unsolvable and restated his proposal for identifying the class of effectively calculable functions with a precisely defined class:

> There is a class of problems of elementary number theory which can be stated in the form that it is required to find an effectively calculable function f of n positive integers, such that $f(x_1, x_2, \ldots, x_n) = 2$ is a necessary and sufficient condition for the truth of a certain proposition of elementary number theory involving $x_1, x_2, \ldots, x_n$ as free variables.
> ...
>
> The purpose of the present paper is to propose a definition of effective calculability which is thought to correspond satisfactorily to the somewhat vague intuitive notion in terms of which problems of this class are often stated, and to show, by means of an example, that not every problem of this class is solvable. [f is the characteristic function of the proposition; that 2 is chosen to indicate 'truth' is, as Church remarked, accidental and non-essential.] (pp. 345–6)

Church's arguments in support of his proposal used again recursiveness; the fact that λ-definability was an equivalent concept added "... to the strength of the reasons adduced below for believing that they [these precise concepts] constitute as general a characterization of this notion [i.e. effective calculability] as is consistent with the usual intuitive understanding of it." (footnote 3, p. 90) Church claimed that those reasons, to be presented and examined in the next paragraph, justify the identification "so far as positive justification can ever be obtained for the selection of a formal definition to correspond to an intuitive notion". (p. 100) Why was there a satisfactory correspondence for Church? What were his reasons for believing that the most general characterization of effective calculability had been found?

To give a deeper analysis Church pointed out, in section 7 of his paper, that two methods to characterize effective calculability of number-theoretic functions suggest themselves. The first of these methods uses the notion of "algorithm", and the second employs the notion of "calculability in a logic". He argues that neither method leads to a definition that is more general than recursiveness. Since these arguments have a parallel structure, I discuss only the one pertaining to the second method. Church considers a logic **L**, that is a system of symbolic logic whose language contains the equality symbol =, a symbol { }( ) for the application of a unary function symbol to its argument, and numerals for the positive integers. For unary functions F he gives the definition:

> F is *effectively calculable* if and only if there is an expression f in the logic **L** such that: $\{f\}(\mu) = \nu$ is a theorem of **L** iff $F(m) = n$; here, μ and ν are expressions that stand for the positive integers m and n.

Church claims that F is recursive, assuming that **L** satisfies certain conditions which amount to requiring the theorem predicate of **L** to be recursively enumerable. Clearly, for us the claim then follows immediately by an unbounded search.

To argue for the recursive enumerability of **L**'s theorem predicate, Church starts out by formulating conditions *any* system of logic has to satisfy if it is "to serve at all the purposes for which a system of symbolic logic is usually intended". These conditions, Church notes in footnote 21, are "substantially" those from Gödel's Princeton Lectures for a formal mathematical system, I mentioned at the end of section 2.4. They state that (**i**) each rule must be an effectively calculable operation, (**ii**) the set of rules and axioms (if infinite) must be effectively enumerable, and (**iii**) the relation between a positive integer and the expression which stands for it must be effectively determinable. Church supposes that these conditions can be "interpreted" to mean that, via a suitable Gödel numbering for the expressions of the logic, (**i′**) each rule must be a recursive operation, (**ii′**) the set of rules and axioms (if infinite) must be recursively enumerable, and (**iii′**) the relation between a positive integer and the expression which stands for it must be recursive. The theorem predicate is then indeed recursively enumerable; but the crucial interpretative step is not argued for at all and thus seems to depend on the very claim that is to be established.

Church's argument in support of the thesis may appear to be viciously circular; but that would be too harsh a judgment. After all, the general concept of calculability is explicated by that of derivability in a logic, and Church uses (**i′**) to (**iii′**) to sharpen the idea that in a logical formalism one operates with an effective notion of immediate consequence.[24] The thesis is consequently appealed to only in a more special case. Nevertheless, it is precisely here that we encounter the major stumbling block for Church's analysis, and that stumbling block was quite clearly seen by Church. To substantiate the latter observation, let me modify a remark Church made with respect to the first method of characterizing effectively calculable functions: *If this interpretation* [what I called the "crucial interpretative step" in the above argument] *or some similar one is not allowed, it is difficult to see how the notion of a system of symbolic logic can be given any exact meaning at all.*[25] Given the crucial role this remark plays, it is appropriate to view and to

---

[24]Compare footnote 20 on p. 101 of [Church, 1936] where Church remarks: "In any case where the relation of immediate consequence is recursive it is possible to find a set of rules of procedure, equivalent to the original ones, such that each rule is a (one-valued) recursive operation, and the complete set of rules is recursively enumerable."

[25]The remark is obtained from footnote 19 of [Church, 1936, 101] by replacing "an algorithm" by "a system of symbolic logic".

formulate it as a normative requirement:

Church's central thesis. The steps of any effective procedure (governing derivations of a symbolic logic) must be recursive.

If this central thesis is accepted and a function is defined to be effectively calculable if, and only if, it is calculable in a logic, then what Robin Gandy called Church's "step-by-step argument" proves that all effectively calculable functions are recursive. These considerations can be easily adapted to Church's first method of characterizing effectively calculable functions via algorithms and provide another perspective for the "selection of a formal definition to correspond to an intuitive notion". The detailed reconstruction of Church's argument pinpoints the crucial difficulty and shows, first of all, that Church's methodological attitude is quite sophisticated and, secondly, that at this point in 1936 there is no major difference from Gödel's position. (A rather stark contrast is painted in [Davis, 1982] as well as in [Shapiro, 1991] and is quite commonly assumed.) These last points are supported by the directness with which Church recognized, in writing and early in 1937, the importance of Turing's work as making the identification of effectiveness and (Turing) computability "immediately evident".

3.4 Absoluteness

How can Church's Thesis be supported? — Let me first recall that Gödel defined the class of general recursive functions after discussion with Church and in response to Church's "thoroughly unsatisfactory" proposal to identify the effectively calculable functions with the λ-definable ones. Church published the thesis, as we saw, only after having done more mathematical work, in particular, after having established with Kleene the equivalence of general recursiveness and λ-definability. Church gives then two reasons for the thesis, namely, (i) the quasi-empirical observation that all known calculable functions can be shown to be general recursive, the *argument from coverage* and (ii) the mathematical fact of the equivalence of two differently motivated notions, the *argument from confluence*. A third reason comes directly from the 1936 paper and was discussed in the last subsection, (iii) the step-by-step *argument from a core conception*.

Remark. There are additional arguments of a more mathematical character in the literature. For example, in the Postscriptum to [1934] Gödel asserts that the question raised in footnote 3 of the Princeton Lectures, whether his concept of recursion comprises all possible recursions, can be "answered affirmatively" for recursiveness as given in section 10 "which is equivalent with general recursiveness as defined today". As to the contemporary definition he seems to point to μ-recursiveness. How could that *definition* convince Gödel that all possible recursions are captured? How could the *normal form theorem*, as Davis suggests in his [1982, 11], go "a considerable distance towards convincing Gödel" that all possible recursions are comprised by his concept of recursion? It seems to me that arguments answering these questions require crucially an appeal to Church's central thesis and are essentially reformulations of his semi-circular argument. That holds also for the appeal to the recursion theorem[26] in *Introduction to Metamathematics*, p. 352, when Kleene argues "Our methods ... are now developed to the point where they seem adequate for handling any effective definition of a function which might be proposed." After all, in the earlier discussion on p. 351 Kleene asserts: "We now have a general kind of 'recursion', in which the value $\varphi(x_1, \ldots, x_n)$ can be expressed as depending on other values of the same function in a quite arbitrary manner, provided only that the rule of dependence is describable by previously treated effective methods." Thus, to obtain a mathematical result, the "previously treated effective methods" must be identified via Church's central thesis with recursive ones. (End of Remark.)

All these arguments are in the end unsatisfactory. The quasi-empirical observation could be refuted tomorrow, as we might discover a function that is calculable, but not general recursive. The mathematical fact by itself is not convincing, as the ease with which the considerations underlying the proof of the normal form theorem allow one to prove equivalences shows a deep family resemblance of the different notions. The question, whether one or any of the rigorous notions corresponds to the informal concept of effective calculability, has to be answered independently. Finally, as to the particular explication via the core concept "calculability in a logic", Church's argument appeals semi-circularly to a restricted version of the thesis. A conceptual reduction has been achieved, but a mathematically convincing result only with the help of the central thesis. Before discussing Post's and Turing's reflections concerning calculability in the next section, I will look at important considerations due to Gödel and Hilbert and Bernays, respectively.

The concept used in Church's argument is extremely natural for number theoretic functions and is directly related to "Entscheidungsdefinitheit" for relations and classes introduced by Gödel in his [1931] as well as to the representability of functions used in his Princeton Lectures. The rules of the equational calculus allow the mechanical computation of the values of calculable functions; they must be contained in *any* system S that is adequate for number theory. Gödel made an important observation in the addendum to his brief 1936 note *On the length of proofs*. Using the general notion "f is computable in a formal system S" he considers a hierarchy of systems S_i (of order $i, 1 \leq i$) and observes that this notion of computability is independent of i in the following sense: If a function is computable in any of the systems S_i, possibly of transfinite order, then it is already computable in S_1. "Thus", Gödel concludes, "the notion 'computable' is in a certain sense 'absolute,' while almost all meta-mathematical notions otherwise known (for example, provable, definable, and so on) quite essentially depend upon the system adopted." For someone who stressed the type-relativity of provability as strongly as Gödel, this was a very surprising insight.

At the Princeton Bicentennial Conference in 1946 Gödel stressed the special

---

[26] In [Crossley, 1975a, 7], Kleene asserts that he had proved this theorem before June of 1935.

importance of general recursiveness or Turing computability and emphasized (*Collected Works II*, p. 150):

> It seems to me that this importance is largely due to the fact that with this concept one has for the first time succeeded in giving an *absolute* definition of an interesting epistemological notion, i.e., one not depending on the formalism chosen.

In the footnote added to this remark in 1965, Gödel formulated the mathematical fact underlying his claim that an absolute definition had been obtained, namely, "To be more precise: a function of integers is computable in any formal system containing arithmetic if and only if it is computable in arithmetic, where a function f is called computable in S if there is in S a computable term representing f." Thus not just higher-type extensions are considered now, but any theory that contains arithmetic, for example set theory. Tarski's remarks at this conference, only recently published in [Sinaceur, 2000], make dramatically vivid, how important the issue of the "intuitive adequacy" of general recursiveness was taken to be. The significance of his 1935 discovery was described by Gödel in a letter to Kreisel of 1 May 1968: "That my [incompleteness] results were valid for all possible formal systems began to be plausible for me (that is since 1935) only because of the *Remark* printed on p. 83 of 'The Undecidable' ... But I was completely convinced only by Turing's paper."[27]

If Gödel had been completely convinced of the adequacy of this notion at that time, he could have established the unsolvability of the decision problem for first-order logic. Given that mechanical procedures are exactly those that can be computed in the system S_1 or any other system to which Gödel's Incompleteness Theorem applies, the unsolvability follows from Theorem IX of [Gödel, 1931]. The theorem states that there are formally undecidable problems of predicate logic; it rests on the observation made by Theorem X that every sentence of the form $(\forall x)F(x)$, with F primitive recursive, can be shown in S_1 to be equivalent to the question of satisfiability for a formula of predicate logic. (This last observation has to be suitably extended to general recursiveness.)

Coming back to the conclusion Gödel drew from the absoluteness, he is right that the details of the formalisms extending arithmetic do not matter, but it is crucial that we are dealing with formalisms at all; in other words, a precise aspect of the unexplicated *formal* character of the extending theories has to come into play, when arguing for the absoluteness of the concept computability. Gödel did not prove that computability is an absolute concept, neither in [1946] nor in the earlier note. I conjecture that he must have used considerations similar to those for the proof of Kleene's normal form theorem in order to convince himself of the claim. The absoluteness was achieved then only relative to some effective

---

[27] In [Odifreddi, 1990, 65]. The content of Gödel's note was presented in a talk on June 19, 1935. See [Davis, 1982, 15, footnote 17] and [Dawson, 1986, 39]. "Remark printed on p. 83" refers to the remark concerning absoluteness that Gödel added in proof (to the original German publication) and is found in [Davis, 1965, 83].

description of the "formal" systems S and the stumbling block shows up exactly here. If my conjecture is correct, then Gödel's argument is completely parallel to Church's contemporaneous step-by-step argument for the co-extensiveness of effective calculability and general recursiveness. Church required, when explicating effective calculability as calculability in logical calculi, the inferential steps in such calculi not only to be effective, but to be general recursive. Some such condition is also needed for completing Gödel's argument.

3.5 Reckonable functions

Church's and Gödel's arguments contain a hidden and semi-circular condition on "steps", a condition that allows their parallel arguments to go through. This step-condition was subsequently moved into the foreground by Hilbert and Bernays's marvelous analysis of "calculations in deductive formalisms". However, before discussing that work in some detail, I want to expose some broad considerations by Church in a letter from 8 June 1937 to the Polish logician Josef Pepis. These considerations (also related in a letter to Post on the same day) are closely connected to Church's explication in his [1936]; they defend the central thesis in an indirect way and show how close his general conceptual perspective was to Gödel's.

In an earlier letter to Church, Pepis had described his project of constructing a number theoretic function that is effectively calculable, but not general recursive. Church explained in his response why he is "extremely skeptical". There is, he asserts, a minimal condition for a function f to be effectively calculable and "if we are not agreed on this then our ideas of effective calculability are so different as to leave no common ground for discussion". This minimal condition is formulated as follows: for every positive integer a there must exist a positive integer b such that the proposition $f(a) = b$ has a "valid proof" in mathematics. Indeed, Church argues, all existing mathematics is formalizable in *Principia Mathematica* or in one of its known extensions; consequently there must be a formal proof of a suitably chosen formal proposition. If f is not general recursive the considerations of [Church, 1936] ensure that for every definition of f within the language of *Principia Mathematica* there exists a positive integer a such that for no b the formal proposition corresponding to $f(a) = b$ is provable in *Principia Mathematica*. Church claims that this holds not only for all known extensions, but for "any system of symbolic logic whatsoever which to my knowledge has ever been proposed". To respect this quasi-empirical fact and satisfy the above minimal condition, one would have to find "an utterly new principle of logic, not only never before formulated, but also never before actually used in a mathematical proof".

Moreover, and here is the indirect appeal to the recursivity of steps, the new principle "must be of so strange, and presumably complicated, a kind that its metamathematical expression as a rule of inference was not general recursive", and one would have to scrutinize the "alleged effective applicability of the principle with considerable care". The dispute concerning a proposed effectively calculable, but non-recursive function would thus center for Church around the required new principle and its effective applicability as a rule of inference, i.e., what I called Church's central thesis. If the latter is taken for granted (implicitly, for example, in Gödel's absoluteness considerations), then the above minimal understanding of effective calculability and the quasi-empirical fact of formalizability block the construction of such a function. This is not a completely convincing argument, as Church admits, but does justify his extreme skepticism of Pepis's project. Church states "this [skeptical] attitude is of course subject to the reservation that I may be induced to change my opinion after seeing your work". So, in a real sense Church joins Gödel in asserting that in any "formal theory" (extending *Principia Mathematica*) only general recursive functions can be computed.

Hilbert and Bernays provide in the second supplement[28] to *Grundlagen der Mathematik II* mathematical underpinnings for Gödel's absoluteness claim and Church's arguments *relative* to their recursiveness conditions ("Rekursivitätsbedingungen"). They give a marvelous conceptual analysis and establish independence from particular features of formalisms in an even stronger sense than Gödel. The core notion of *calculability in a logic* is made directly explicit and a number-theoretic function is said to be reckonable ("regelrecht auswertbar") just in case it is computable (in the above sense) in *some* deductive formalism. Deductive formalisms must satisfy, however, three recursiveness conditions. The crucial one is an analogue of Church's central thesis and requires that the theorems of the formalism can be enumerated by a primitive recursive function or, equivalently, that the proof-predicate is primitive recursive. Then it is shown that a special number theoretic formalism (included in Gödel's S_1) suffices to compute the reckonable functions, and that the functions computable in this particular formalism are exactly the general recursive ones. Hilbert and Bernays's analysis is a natural capping of the development from *Entscheidungsdefinitheit* to an absolute notion of computability, because it captures the informal notion of rule-governed evaluation of number theoretic functions and explicitly isolates appropriate restrictive conditions. But this analysis does *not* overcome the major stumbling block, it puts it rather in plain view.

The conceptual work of Gödel, Church, Kleene and Hilbert and Bernays had intimate historical connections and is still of genuine and deep interest. It explicated calculability of functions by *one core notion*, namely, computability of their values in a deductive formalism via restricted elementary rules. But no one gave convincing reasons for the proposed restrictions on the steps permitted in computations. This issue was not resolved along Gödelian lines by generalizing recursions, but by a quite different approach due to Alan Turing and, to some extent, Emil Post. I reported in subsection 3.1 on Gödel's assessment of Turing's work in the Postscriptum to the 1934 Princeton Lectures. That Postscriptum was written on 3 June 1964; a few months earlier, on 28 August 1963, Gödel had formulated a brief note for the publication of the translation of his [1931] in [van Heijenoort, 1967]. That note is reprinted in *Collected Works I* (p. 195):

---

[28]The supplement is entitled, "Eine Präzisierung des Begriffs der berechenbaren Funktion und der Satz von Church über das Entscheidungsproblem."

In consequence of later advances, in particular of the fact that due to A. M. Turing's work a precise and unquestionably adequate definition of the general notion of formal system can now be given, a completely general version of Theorems VI and XI is now possible. That is, it can be proved rigorously that in *every* consistent formal system that contains a certain amount of finitary number theory there exist undecidable arithmetic propositions and that, moreover, the consistency of any such system cannot be proved in the system.

To the first occurrence of "formal system" in this note Gödel attached a most informative footnote and suggested in it that the term "formal system" should never be used for anything but this notion. For example, the transfinite iterations of formal systems he had proposed in his contribution to the Princeton Bicentennial are viewed as "something radically different from formal systems in the proper sense of the term". The properly formal systems have the characteristic property "that reasoning in them, in principle, can be completely replaced by mechanical devices". That connects back to the remark he had made in [1933a] concerning the formalization of mathematics. The question is, what is it about Turing's notion that makes it an "unquestionably adequate definition of the general notion of formal system"? My contention is that a dramatic shift of perspective overcame the stumbling block for a fundamental conceptual analysis. Let us see what that amounts to: Turing's work is *the* central topic of the next section.

4 COMPUTATIONS AND COMBINATORY PROCESSES

We saw in the previous section that the work of Gödel, Church, Kleene and Hilbert and Bernays explicated calculability of number-theoretic functions as computability of their values in some deductive formalism via elementary steps. Church's direct argument for his thesis appeals to the central thesis asserting that the elementary steps in a computation (or deduction) should be recursive. There is no reason given, why that is a correct or motivated requirement. However, if the central thesis is accepted, then every effectively calculable function is indeed general recursive.

In some sense of elementary, the steps in deductive formalisms are *not* elementary at all. Consider Gödel's equational calculus contained in all of them: it allows the substitution of variables by arbitrary numerals in one step, and arbitrarily complex terms can be replaced by their numerical values, again, in one step. In general, a human calculator cannot carry out such mechanical steps without subdividing them into more basic ones. It was a dramatic shift of perspective, when Turing and Post formulated the most basic mechanical steps that underlie the effective determination of values of number-theoretic functions, respectively the execution of combinatory processes, and that can be carried out by a human computing agent. This shift of perspective made for real progress; it is contiguous with the other work, but it points the way towards overcoming, through Turing's reflections, the stumbling block for a fundamental conceptual analysis.

In the first subsection, *Machines and workers*, I present the mechanical devices or machines Turing introduced, and I'll discuss Post's human workers who operate robot-like in a "symbol space" of marked and unmarked boxes, carrying out extremely simple actions. It is perhaps surprising that Turing's model of computation, developed independently in the same year, is "identical". In contrast to Post, Turing investigated his machines systematically; that work resulted in the discovery of the universal machine, the proof of the unsolvability of the halting problem and, what is considered to be, the definitive resolution of the *Entscheidungsproblem*.

The contrast between the methodological approaches Post and Turing took is *prima facie* equally surprising, if not even more remarkable. For Post it is a "working hypothesis" that all combinatory processes can be effected by the worker's actions, and it is viewed as being in need of continual verification. Turing took the calculations of human computers as the starting-point of a detailed analysis to uncover the underlying symbolic operations, appealing crucially to the agent's sensory limitations. These operations are so basic that they cannot be further subdivided and essentially *are* the operations carried out by Turing machines. The general restrictive features can be formulated as *boundedness* and *locality* conditions. The analysis is the topic of section 4.2 entitled *Mechanical computors*.

Turing's reductive analysis will be critically examined in section 4.3 under the heading *Turing's central thesis*. Using Post's later presentation of Turing machines we can simplify and sharpen the restrictive conditions, but also return to the purely symbolic operations required for the general issues that were central before attention focused on the effective calculability of number theoretic functions. Here we are touching on the central reason why Turing's analysis is so appropriate and leads to an adequate notion. However, Turing felt that his arguments were mathematically unsatisfactory and thought, as late as 1954, that they had to remain so. Before addressing this pivotal point in Section 5, I am going to discuss in subsection 4.5 Church's "machine interpretation" of Turing's work, but also Gandy's proposal to characterize machine computability. Following Turing's broad approach, Gandy investigated in his [1980] the computations of machines or, to indicate better the scope of that notion, of "discrete mechanical devices". According to Gandy, machines can, in particular, carry out parallel computations. In spite of the great generality of his notion, Gandy was able to show that any machine computable function is also Turing computable.

This section is focused on a sustained conceptual analysis of human computability and contrasts it briefly with that of machine computability. Here lies the key to answering the question, "What distinguishes Turing's proposal so dramatically from Church's?" After all, the naïve examination of Turing machines hardly produces the conviction that Turing computability is provably equivalent to an analyzed notion of mechanical procedure (as Gödel claimed) or makes it immediately evident that Turing computability should be identified with effectiveness in the ordinary sense (as Church asserted). A tentative answer is provided; but we'll see that a genuine methodological problem remains. It is addressed in Section 5.

4.1 Machines and workers

The list of different notions in the argument from confluence includes, of course, Turing computability. Though confluence is at issue, there is usually an additional remark that Turing gave in his [1936] the most convincing analysis of effective calculability, and that his notion is truly adequate. What is the notion of computation that is being praised? — In the next few paragraphs I will describe a two-letter Turing machine, following [Davis, 1958] rather than Turing's original presentation. (The differences are discussed in Kleene's *Introduction to Metamathematics*, p. 361, where it is also stated that this treatment "is closer in some respects to [Post, 1936]".)

A Turing machine consists of a finite, but potentially infinite tape. The tape is divided into squares, and each square may carry a symbol from a finite alphabet, say, just the two-letter alphabet consisting of 0 and 1. The machine is able to scan one square at a time and perform, depending on the content of the observed square and its own internal state, one of four operations: print 0, print 1, or shift attention to one of the two immediately adjacent squares. The operation of the machine is given by a finite list of commands in the form of quadruples $q_i s_k c_l q_m$ that express the following: If the machine is in internal state q_i and finds symbol s_k on the square it is scanning, then it is to carry out operation c_l and change its state to q_m. The deterministic character of the machine's operation is guaranteed by the requirement that a program must not contain two different quadruples with the same first two components.

Gandy in his [1988] gave a lucid informal description of a Turing machine computation without using internal states or, as Turing called them, m-configurations: "The computation proceeds by discrete steps and produces a record consisting of a finite (but unbounded) number of cells, each of which is either blank or contains a symbol from a finite alphabet. At each step the action is local and is locally determined, according to a finite table of instructions" (p. 88). How Turing avoids the reference to internal states will be discussed below; why such a general formulation is appropriate will be seen in section 4.3.

For the moment, however, let me consider the Turing machines I just described. Taking for granted a representation of natural numbers in the two-letter alphabet and a straightforward definition of when to call a number-theoretic function *Turing computable*, I put the earlier remark before you as a question: Does this notion provide "an unquestionably adequate definition of the general concept of formal system"? Is it even plausible that every effectively calculable function is Turing computable? It seems to me that a naïve inspection of the restricted notion of Turing computability should lead to "No!" as a tentative answer to the second and, thus, to the first question. However, a systematic development of the theory of Turing computability convinces one quickly that it is a powerful notion.

One goes almost immediately beyond the examination of particular functions and the writing of programs for machines computing them; instead, one considers machines corresponding to operations that yield, when applied to computable functions, other functions that are again computable. Two such functional operations are crucial, namely, composition and minimization. Given these operations and the Turing computability of a few simple initial functions, the computability of all general recursive functions follows. This claim takes for granted Kleene's 1936 proof of the equivalence between general recursiveness and μ-recursiveness. Since Turing computable functions are readily shown to be among the μ-recursive ones, it seems that we are now in exactly the same position as before with respect to the evidence for Church's Thesis. This remark holds also for Post's model of computation.

Post's combinatory processes are generated by computation steps "identical" with Turing's; Post's model was published in the brief 1936 note, *Finite combinatory processes — Formulation 1*. Here we have a worker who operates in a *symbol space* consisting of

> a two way infinite sequence of spaces or boxes, i.e., ordinally similar to the series of integers .... The problem solver or worker is to move and work in this symbol space, being capable of being in, and operating in but one box at a time. And apart from the presence of the worker, a box is to admit of but two possible conditions, i.e., being empty or unmarked, and having a single mark in it, say a vertical stroke.[29]

The worker can perform a number of *primitive acts*, namely, make a vertical stroke $[V]$, erase a vertical stroke $[E]$, move to the box immediately to the right $[M_r]$ or to the left $[M_l]$ (of the box he is in), and determine whether the box he is in is marked or not $[D]$. In carrying out a particular combinatory process the worker begins in a special box (the *starting point*) and then follows directions from a finite, numbered sequence of instructions. The i-th direction, i between 1 and n, is in one of the following forms: (1) carry out act V, E, M_r, or M_l and then follow direction j_i, (2) carry out act D and then, depending on whether the answer was positive or negative, follow direction j_i' or j_i''. (Post has a special stop instruction, but that can be replaced by stopping, conventionally, in case the number of the next direction is greater than n.) Are there intrinsic reasons for choosing Formulation 1, except for its simplicity and Post's expectation that it will turn out to be equivalent to general recursiveness? An answer to this question is not clear (from Post's paper), and the claim that psychological fidelity is aimed for seems quite opaque. Post writes at the very end of his paper,

> The writer expects the present formulation to turn out to be equivalent to recursiveness in the sense of the Gödel–Church development. Its purpose, however, is not only to present a system of a certain logical potency but also, in its restricted field, of psychological fidelity. In the

---

[29][Post, 1936, 289]. Post remarks that the infinite sequence of boxes can be replaced by a potentially infinite one, expanding the finite sequence as necessary.

> latter sense wider and wider formulations are contemplated. On the other hand, our aim will be to show that all such are logically reducible to formulation 1. We offer this conclusion at the present moment as a *working hypothesis*. And to our mind such is Church's identification of effective calculability with recursiveness. (p. 291)

Investigating wider and wider formulations and reducing them to the above basic formulation would change for Post this "hypothesis not so much to a definition or to an axiom but to a *natural law*".[30]

It is methodologically remarkable that Turing proceeded in *exactly* the opposite way when trying to support the claim that all computable numbers are machine computable or, in our way of speaking, that all effectively calculable functions are Turing computable. He did not try to extend a narrow notion reducibly and obtain in this way additional quasi-empirical support; rather, he attempted to analyze the intended broad concept and reduce it to the narrow one — *once and for all*. I would like to emphasize this, as it is claimed over and over that Post provided in his 1936 paper "much the same analysis as Turing". As a matter of fact, Post hardly offers an analysis of effective calculations or combinatory processes in this paper; it may be that Post took the context of his own work, published only much later, too much for granted.[31] There is a second respect in which Post's logical work differs almost tragically from Gödel's and Turing's, and Post recognized that painfully in the letters he wrote to Gödel in 1938 and 1939: these logicians obtained decisive mathematical results that had been within reach of Post's own investigations.[32]

By examining Turing's analysis and reduction we will find the key to answering the question I raised on the difference between Church's and Turing's proposals. Very briefly put it is this: Turing deepened Church's step-by-step argument by focusing on the mechanical operations underlying the elementary steps and by formulating well-motivated constraints that guarantee their recursiveness. Before presenting in the next subsection Turing's considerations systematically, with some simplification and added structure, I discuss briefly Turing's fundamental

---

[30][L.c., 291]

[31]The earlier remark on Post's analysis is from [Kleene, 1988, 34]. In [Gandy, 1988, 98], one finds this pertinent and correct observation on Post's 1936 paper: "Post does not analyze nor justify his formulation, nor does he indicate any chain of ideas leading to it." However, that judgment is only locally correct, when focusing on this very paper. To clarify some of the interpretative difficulties and, most of all, to see the proper context of Post's work that reaches back to the early 1920s, it is crucial to consider other papers of his, in particular, the long essay [1941] that was published only in [Davis, 1965] and the part that did appear in 1943 containing the central mathematical result (canonical production systems are reducible to normal ones). In 1994 Martin Davis edited Post's *Collected Works*. Systematic presentations of Post's approach to computability theory were given by Davis [1958] and Smullyan [1961] and [1993]. Brief, but very informative introductions can be found in [Davis, 1982, 18–22], [Gandy, 1988, 92–98], and [Stillwell, 2004]. Büchi continued in most interesting ways Post's investigations; see his *Collected Works*, in particular part 7 on computability with comments by Davis.

[32]The letters are found in volume V of Gödel's *Collected Works*; a very brief description of Post's work on canonical and normal systems is given in my Introductory Note to the correspondence.

mathematical results (in Kleene's formulation) and infer the unsolvability of the *Entscheidungsproblem*.

Let ψ_M be the unary number theoretic function that is computed by machine M, and let $T(z, x, y)$ express that y is a computation of a machine with Gödelnumber z for argument x; then $\psi_M(x) = U(\mu y.T(gn(M), x, y))$; U is the result-extracting function and $gn(M)$ the Gödelnumber of M. Both T and U are easily seen to be primitive recursive, in particular, when Turing machines are presented as Post systems; see subsection 4.3. Consider the binary function $\varphi(z, x)$ defined by $U(\mu y.T(z, x, y))$; that is a partial recursive function and is computable by a machine $\mathcal{U}$ such that $\psi_{\mathcal{U}}(z, x) = \varphi(z, x)$ on their common domain of definition. $\mathcal{U}$ can compute any unary total function f that is Turing computable: $f(x) = \psi_M(x)$, when M is the machine computing f; as $\psi_M(x) = U(\mu y.T(gn(M), x, y))$, $U(\mu y.T(gn(M), x, y)) = \varphi(gn(M), x)$, and $\varphi(gn(M), x) = \psi_{\mathcal{U}}(gn(M), x)$, we have $f(x) = \psi_{\mathcal{U}}(gn(M), x)$. Thus, $\mathcal{U}$ can be considered as a "universal machine".

A modification of the diagonal argument shows that Turing machines cannot answer particular questions concerning Turing machines. The most famous question is this: Does there exist an effective procedure implemented on a Turing machine that decides for any Turing machine M and any input x, whether the computation of machine M for input x terminates or halts? This is the *Halting Problem* as formulated by Turing in 1936; it is clearly a fundamental issue concerning computations and is unsolvable. The argument is classical and begins by assuming that there is an H that solves the halting problem, i.e., for any M and x, $\psi_H(gn(M), x) = 1$ iff M halts for argument x; otherwise $\psi_H(z, x) = 0$. It is easy to construct a machine H^* from H, such that H^* halts for x iff $\psi_H(x, x) = 0$. Let h^* be $gn(H^*)$; then we have the following equivalences: H^* halts for h^* iff $\psi_H(h^*, h^*) = 0$ iff $\psi_H(gn(H^*), h^*) = 0$ iff H^* does not halt for h^*, a contradiction. Turing used the unsolvability of this problem to establish the unsolvability of related machine problems, the self-halting and the printing problem. For that purpose he implicitly used a notion of effective reducibility; a problem P, identified with a set of natural numbers, is reducible to another problem Q just in case there is a recursive function f, such that for all $x : P(x)$ if and only if $Q(f(x))$. Thus, if we want to see whether x is in P we compute $f(x)$ and test its membership in Q. In order to obtain his negative answer to the decision problem Turing reduced in a most elegant way the halting problem to the decision problem. Thus, if the latter problem were solvable, the former problem would be.

The self-halting problem K is the simplest in an infinite sequence of increasingly complex and clearly undecidable problems, the so-called *jumps*. Notice that for a machine M with code e the set K can be defined arithmetically with Kleene's T-predicate by $(\exists y)T(e, e, y)$. K is indeed *complete* for sets A that are definable by formulas obtained from recursive ones by prefixing one existential quantifier; i.e., any such A is reducible to K. These A can be given a different and very intuitive characterization: A is either the empty set or the range of a recursive function. Under this characterization the A's are naturally called "recursively enumerable", or simply r.e.. It is not difficult to show that the recursive sets are exactly those that are r.e. and have an r.e. complement. Post's way of generating these sets by production systems thus opened a distinctive approach to recursion theory.[33]

Now that we have developed a small fraction of relevant computability theory, we return to the fundamental issue, namely, why was Turing's notion of computability exactly right to obtain a convincing negative solution of the decision problem and also for achieving a precise characterization of "formal systems"? That it was exactly right, well, that still has to be argued for. The examination of mathematical results and the cool shape of a definition certainly don't provide the reason. Let us look back at Turing's paper; it opens (p. 116) with a brief description of what is ostensibly its subject, namely, "computable numbers" or "the real numbers whose expressions as a decimal are calculable by finite means". Turing is quick to point out that the fundamental problem of explicating "calculable by finite means" is the same when considering calculable functions of an integral variable, calculable predicates, and so forth. So it is sufficient to address the question, what does it mean for a real number to be calculable by finite means? Turing admits:

> This requires rather more explicit definition. No real attempt will be made to justify the definitions given until we reach §9. For the present I shall only say that the justification lies in the fact that the human memory is necessarily limited. (p. 117)

In §9 Turing claims that the operations of his machines "include all those which are used in the computation of a number". He tries to establish the claim by answering the real question at issue, "*What are the possible processes which can be carried out in computing a number?*" The question is implicitly restricted to processes that can be carried out by a human computer. Given the systematic context that reaches back to Leibniz's "Calculemus!" this is exactly the pertinent issue to raise: the general problematic *requires* an analysis of the mechanical steps a human computer can take; after all, a positive solution to the decision problem would be provided by a procedure that in principle can be carried out by us.

Gandy made a useful suggestion, namely, calling a human carrying out a computation a "computor" and referring by "computer" to some computing machine or other. In Turing's paper, "computer" is always used for a human computing agent who proceeds mechanically; his machines, our Turing machines, consistently are just machines. The Oxford English Dictionary gives this meaning of "mechanical" when applied to a person as "resembling (inanimate) machines or their operations; acting or performed without the exercise of thought or volition; ...". When I want to stress strongly the machine-like behavior of a computor, I will

---

[33] Coming back to complex sets, one obtains the *jump hierarchy* by relativizing the concept of computation to sets of natural numbers whose membership relations are revealed by "oracles". The jump K' of K, for example, is defined as the self-halting problem, when an oracle for K is available. This hierarchy can be associated to definability questions in the language of arithmetic: all jumps are definable by some arithmetical formula, and all arithmetically definable sets are reducible to some jump. A good survey of more current work can be found in [Griffor, 1999].

even speak of a *mechanical computor*. The processes such a computor can carry out are being analyzed, and that is exactly Turing's specific and extraordinary approach: the computing agent is brought into the analysis. The question is thus no longer, "Which number theoretic functions can be calculated?" but rather, "Which number theoretic functions can be calculated by a mechanical computor?" Let's address that question with Turing and see, how his analysis proceeds. Gandy emphasizes in his [1988, 83–84], absolutely correctly as we will see, that "Turing's analysis makes no reference whatsoever to calculating machines. Turing machines appear as a result, as a codification, of his analysis of calculations by humans".

4.2 Mechanical computors

Turing imagines a computor writing symbols on paper that is divided into squares "like a child's arithmetic book". Since the two-dimensional character of this computing space is taken not to be an "essential of computation" (p. 135), Turing takes a one-dimensional tape divided into squares as the basic computing space. What determines the steps of the computor? And what elementary operations can he carry out? Before addressing these questions, let me formulate one crucial and normative consideration. Turing explicitly strives to isolate operations of the computor (p. 136) that are "so elementary that it is not easy to imagine them further divided". Thus it is crucial that symbolic configurations relevant to fixing the circumstances for the computor's actions can be recognized *immediately* or *at a glance*.

Because of Turing's first reductive step to a one-dimensional tape, we have to be concerned with either individual symbols or sequences of symbols. In the first case, only finitely many distinct symbols should be written on a square; Turing argues (p. 135) for this restriction by remarking, "If we were to allow an infinity of symbols, then there would be symbols differing to an arbitrarily small extent", and the computor could not distinguish at a glance between symbols that are "sufficiently" close. In the second and related case consider, for example, Arabic numerals like 178 or 99999999 as one symbol; then it is not possible for the computor to determine at one glance whether or not 9889995496789998769 is identical with 98899954967899998769. This restriction to finitely many observed symbols or symbol sequences will be the central part of condition (**B.1**) below and also constrains via condition (**L.1**) the operations a computor can carry out.

The behavior of a computor is determined uniquely at any moment by two factors, namely, the symbols or symbol sequences he observes, and his "state of mind" or "internal state"; what is uniquely determined is the action to be performed and the next state of mind to be taken.[34] This uniqueness requirement may be called *determinacy condition* (**D**) and guarantees that computations are deterministic. Internal states are introduced so that the computor's behavior can

---

[34]Turing argues in a similar way for bounding the number of states of mind, alleging confusion, if the states of mind were too close.

depend on earlier observations, i.e., reflect his experience.[35] A computor thus satisfies two *boundedness conditions*:

(**B.1**) There is a fixed finite bound on the number of symbol sequences a computor can immediately recognize;

(**B.2**) There is a fixed finite bound on the number of states of mind that need to be taken into account.

For a computor there are thus only boundedly many different relevant combinations of symbol sequences and internal states. Since the computor's behavior, according to (**D**), is uniquely determined by such combinations and associated operations, the computor can carry out at most finitely many different operations, and his behavior is fixed by a finite list of commands. The operations of a computor are restricted by *locality conditions*:

(**L.1**) Only elements of observed symbol sequences can be changed;

(**L.2**) The distribution of observed squares can be changed, but each of the new observed squares must be within a bounded distance L of a previously observed square.

Turing emphasizes that "the new observed squares must be immediately recognizable by the computor" and that means the distributions of the observed squares arising from changes according to (**L.2**) must be among the finitely many ones of (**B.1**). Clearly, the same must hold for the symbol sequences resulting from changes according to (**L.1**). Since some of the operations involve a change of state of mind, Turing concludes that

> The most general single operation must therefore be taken to be one of the following: (A) A possible change (a) of symbol [as in (L.1)] together with a possible change of state of mind. (B) A possible change (b) of observed squares [as in (L.2)] together with a possible change of state of mind. (p. 137)

With this restrictive analysis of the computor's steps it is rather straightforward to conclude that a Turing machine can carry out his computations. Indeed, Turing first considers machines that operate on strings ("string machines") and mimic directly the work of the computor; then he asserts referring to ordinary Turing machines ("letter machines") that

> The machines just described [string machines] do not differ very essentially from computing machines as defined in § 2 [letter machines], and corresponding to any machine of this type a computing machine

---

[35]Turing relates state of mind to memory in §1 for his machines: "By altering its m-configuration the machine can effectively remember some of the symbols which it has 'seen' (scanned) previously." Kleene emphasizes this point in [1988, 22]: "A person computing is not constrained to working from just what he sees on the square he is momentarily observing. He can remember information he previously read from other squares. This memory consists in a state of mind, his mind being in a different state at a given moment of time depending on what he remembers from before."

can be constructed to compute the same sequence, that is to say the sequence computed by the computer. (p. 138)

It should be clear that the string machines, just as Gandy asserted, "appear as a result, as a codification, of his analysis of calculations by humans". Thus we seem to have, shifting back to the calculation of values of number-theoretic functions, an argument for the claim: Any number-theoretic function F calculable by a computor, who satisfies the conditions (**D**) and (**B.1**)–(**L.2**), is computable by a Turing machine.[36] Indeed, both Gandy in his [1988] and I in my [1994] state that Turing established a theorem by the above argument. I don't think anymore, as the reader will notice, that that is correct in general; it is correct, however, if one considers the calculations as being carried out on strings of symbols from the very beginning.

Because of this last remark and an additional observation, Turing's analysis can be connected in a straightforward way with Church's considerations discussed in section 3.3. The additional observation concerns the determinacy condition (**D**): it is not needed to guarantee the Turing computability of F in the above claim. More precisely, (**D**) was used in conjunction with (**B.1**) and (**B.2**) to argue that computors can carry out only finitely many operations; this claim follows also from conditions (**B.1**)–(**L.2**) *without* appealing to (**D**). Thus, the behavior of computors can still be fixed by a finite list of commands (though it may exhibit non-determinism) and can be mimicked by a Turing machine. Consider now an effectively calculable function F and a non-deterministic computor who calculates, in Church's sense, the value of F in a logic **L**. Using the (additional) observation and the fact that Turing computable functions are recursive, F is recursive.[37] *This* argument for F's recursiveness does no longer appeal to Church's Thesis, not even to the restricted central thesis; rather, such an appeal is replaced by the assumption that the calculation in the logic is done by a computor satisfying the conditions (**B.1**)–(**L.2**).

Both Church and Gödel state they were convinced by Turing's work that effective calculability should be identified with Turing computability and thus is also co-extensional with recursiveness and λ-definability. Church expressed his views in the 1937 review of Turing's paper from which I quoted in the introduction; on account of Turing's work the identification is considered as "immediately evident". We'll look at that review once more in subsection 4.4 when turning attention to machine computability, as Church emphasizes the machine character of Turing's model. As to Gödel I have not been able to find in his published papers any

---

[36] A similar analysis is presented in [Wang, 1974, 90–95]. However, Wang does not bring out at all the absolutely crucial point of grounding the boundedness and locality conditions in the limitations of the computing subject; instead he appeals to an abstract *principle of finiteness*. Post's remarks on "finite methods" on pp. 426–8 in [Davis, 1965] are also grappling with these issues.

[37] The proof is given via considerations underlying Kleene's normal form theorem. That is done in the most straightforward way if, as discussed in the next subsection, Turing machines are described as Post systems.

reference to Turing's paper before his [1946] except in the purely mathematical footnote 44 of [Gödel, 1944]; that paper was discussed in section 3.4 and does not give a distinguished role to Turing's analysis. Rather, the "great importance of the concept of general recursiveness" is pointed to and "Turing computability" is added disjunctively, indeed just parenthetically. As we saw, Gödel judged that the importance of the concept is "largely due" to its absoluteness.

There is some relevant discussion of Turing's work in unpublished material that is now available in the *Collected Works*, namely, in Gödel's [193?, 164—175] of *CW III*), the Gibbs lecture of 1951 (pp. 304–5 and p. 309 of *CW III*), and in the letter of 2 February 1957 that was addressed, but not sent, to Ernest Nagel (pp. 145–6 of *CW V*). The first written and public articulation of Gödel's views can be found in the 1963 Addendum to his [1931] (for its publication in [van Heijenoort, 1967]) and in the 1964 Postscriptum to the Princeton Lectures (for their publication in [Davis, 1965]). In the latter, more extended note, Gödel is perfectly clear about the structure of Turing's argument. "Turing's work", he writes, "*gives an analysis* [my emphasis] of the concept 'mechanical procedure' (*alias* 'algorithm' or 'computation procedure' or 'finite combinatorial procedure'). This concept is *shown* [my emphasis] to be equivalent with that of a 'Turing machine'." In a footnote attached to this observation he called "previous equivalent definitions of computability", referring to λ-definability and recursiveness, "much less suitable for our purpose". What is not elucidated by any remark of Gödel, as far as I know, is the *result* of Turing's analysis, i.e., the explicit formulation of restrictive conditions. There is consequently no discussion of the *reasons* for the correctness of these conditions or, for that matter, of the analysis; there is also no indication of a proof establishing the equivalence between the analyzed (and presumably rigorous) notion of mechanical procedure and the concept of a Turing machine. (Gödel's views are traced with many more details in my [2006].)

A comparison of Gödel's concise description with Turing's actual argument raises a number of important issues, in particular one central question I earlier put aside: Isn't the starting-point of Turing's argument too vague and open, unless we take for granted that the symbolic configurations are of a certain kind, namely, symbol strings in our case? But even if that is taken for granted and Turing's argument is viewed as perfectly convincing, there remains a methodological problem. According to Gödel the argument consists of an analysis followed by a proof; how do we carve up matters, i.e., where does the analysis stop and the proof begin? Does the analysis stop only, when a string machine has fully captured the computor's actions, and the proof is just the proof establishing the reduction of computations by string machines to those by letter machines? Or does the analysis just lead to restrictive conditions for mechanical computors and the proof establishes the rest? To get a clearer view about these matters, I will simplify the argument and examine more closely the justificatory steps.

4.3 Turing's central thesis

The first section of this essay had the explicit purpose of exposing the broad context for the investigations of Herbrand, Gödel, Church, Kleene, Post, and Turing. There is no doubt that an analysis of human effective procedures on finite (symbolic) configurations was called for, and that the intended epistemological restrictions were cast in "mechanical" terms; *vide* as particularly striking examples the remarks of Frege and Gödel quoted in section 2.1. Thus, Turing's explication of *effective calculability* as *calculability by a mechanical computer* should be accepted. What are the general restrictions on calculation processes, and how are such constraints related to the nature of mechanical computers?

The justificatory steps in Turing's argument contain crucial appeals to boundedness and locality conditions. Turing claims that their ultimate justification lies in the necessary limitation of human memory. According to Gandy, Turing arrives at the restrictions "by considering the limitations of our sensory and mental apparatus". However, in Turing's argument only limitations of our sensory apparatus are involved, unless "state of mind" is given an irreducibly mental touch. That is technically unnecessary as Post's equivalent formulation makes clear. It is systematically also not central for Turing, as he describes in section 9 (III) of his paper, p. 139, a modified computor. There he avoids introducing "state of mind" by considering instead "a more physical and definite counterpart of it". (Indeed, if we take into account the quest for insuring "radical intersubjectivity" then internal, mental states should be externalized in any event.) Thus, Turing's analysis can be taken to appeal only to sensory limitations of the type I discussed at the beginning of section 4.2.[38] Such limitations *are* obviously operative when we work as purely mechanical computors.

Turing himself views his argument for the reduction of effectively calculable functions to functions computable by his machines as basically "a direct appeal to intuition". Indeed, he claims, p. 135, more strongly, "All arguments which can be given [for this reduction] are bound to be, fundamentally, appeals to intuition, and for that reason rather unsatisfactory mathematically." If we look at his paper [Turing, 1939], the claim that such arguments are "unsatisfactory mathematically" becomes at first rather puzzling, since he observes there that intuition is inextricable from mathematical reasoning. Turing's concept of intuition is much more general than that ordinarily used in the philosophy of mathematics. It is introduced in the 1939 paper explicitly to address the issues raised by Gödel's first

---

[38] As Turing sees memory limitations as ultimately justifying the restrictive conditions, but none of the conditions seems to be directly motivated by such a limitation, we should ask, how we can understand his claim. I suggest the following: If our memory were not subject to limitations of the same character as our sensory apparatus, we could scan (with the limited sensory apparatus) a symbolic configuration that is not immediately recognizable, read in sufficiently small parts so that their representations could be assembled in a unique way to a representation of the given symbolic configuration, and finally carry out (generalized) operations on that representation in memory. Is one driven to accept Turing's assertion as to the limitation of memory? I suppose so, if one thinks that information concerning symbolic structures is physically encoded and that there is a bound on the number of available codes.

incompleteness theorem; that is done in the context of work on ordinal logics or, what was later called, progressions of theories. The discussion is found in section 11:

> Mathematical reasoning may be regarded rather schematically as the exercise of a combination of two faculties, which we may call *intuition* and *ingenuity*. The activity of the intuition consists in making spontaneous judgements which are not the result of conscious trains of reasoning. These judgements are often but by no means invariably correct (leaving aside the question of what is meant by "correct"). ...The exercise of ingenuity in mathematics consists in aiding the intuition through suitable arrangements of propositions, and perhaps geometrical figures or drawings. It is intended that when these are really well arranged the validity of the intuitive steps which are required cannot seriously be doubted. (pp. 208–210)

Are the propositions in Turing's argument arranged with sufficient ingenuity so that "the validity of the intuitive steps which are required cannot seriously be doubted"? Or, does their arrangement allow us at least to point to central restrictive conditions with clear, adjudicable content?

To advance the further discussion, I simplify the formulation of the restrictive conditions that can be extracted from Turing's discussion by first eliminating internal states by "more physical counterparts" as Turing himself proposed. Then I turn machine operations into purely symbolic ones by presenting suitable Post productions as Turing himself did for obtaining new mathematical results in his [1950a], but also for a wonderful informal exposition of solvable and unsolvable problems in [1954]. Turing extended in the former paper Post's (and Markov's) result concerning the unsolvability of the word-problem for semi-groups to semi-groups with cancellation; on the way to the unsolvability of this problem, [Post, 1947] had used a most elegant way of describing Turing machines as production systems. The configurations of a Turing machine are given by *instantaneous descriptions* of the form $\alpha q_l s_k \beta$, where α and β are possibly empty strings of symbols in the machine's alphabet; more precisely, an *id* contains exactly one state symbol and to its right there must be at least one symbol. Such *ids* express that the current tape content is $\alpha s_k \beta$, the machine is in state q_l and scans (a square with symbol) s_k. Quadruples $q_i s_k c_l q_m$ of the program are represented by rules; for example, if the operation c_l is *print* 0, the corresponding rule is

$$\alpha q_i s_k \beta => \alpha q_m 0 \beta.$$

Such formulations can be given, obviously, for all the different operations. One just has to append s_0 to $\alpha(\beta)$ in case c_l is the operation *move to the left (right)* and $\alpha(\beta)$ is the empty string; that reflects the expansion of the only potentially infinite tape by a blank square.

This formulation can be generalized so that machines operate directly on finite strings of symbols; operations can be indicated as follows:

$$\alpha\gamma q_l\delta\beta => \alpha\gamma^*q_m\delta^*\beta.$$

If in internal state q_l a *string machine* recognizes the string $\gamma\delta$ (i.e., takes in the sequence at one glance), it replaces that string by $\gamma^*\delta^*$ and changes its internal state to q_m. The rule systems describing string machines are semi-Thue systems and, as the latter, not deterministic, if their programs are just sequences of production rules. The usual non-determinism certainly can be excluded by requiring that, if the antecedents of two rules coincide, so must the consequents. But that requirement does not remove every possibility of two rules being applicable simultaneously: consider a machine whose program includes in addition to the above rule also the rule

$$\alpha\gamma^\sharp q_l\delta^\sharp\beta => \alpha\gamma^\perp q_n\delta^\perp\beta,$$

where $\delta^\sharp$ is an initial segment of δ, and $\gamma^\sharp$ is an end segment of γ; under these circumstances both rules would be applicable to $\gamma q_l\delta$. This non-determinism can be excluded in a variety of ways, e.g., by always using the applicable rule with the largest context. In sum, the Post representation joins the physical counterparts of internal states to the ordinary symbolic configurations and forms instantaneous descriptions, abbreviated as *id*. Any *id* contains exactly one such physical counterpart, and the immediately recognizable sub-configuration of an *id* must contain it. As the state symbol is part of the observed configuration, its internal shifting can be used to indicate a shift of the observed configuration. Given this compact description, the restrictive conditions are as follows:

(**B**) (Boundedness) There is a fixed finite bound on the number of symbol sequences (containing a state symbol) a computor can immediately recognize.

(**L**) (Locality) A computor can change only an *id*'s immediately recognizable sub-configuration.

These restrictions on computations are specifically and directly formulated for Post productions. Turing tried to give, as we saw, a more general argument starting with a broader class of symbolic configurations. Here is the starting-point of his considerations together with a dimension-lowering step to symbol sequences:

> Computing is normally done by writing certain symbols on paper. We may suppose this paper is divided into squares like a child's arithmetic book. In elementary arithmetic, the two-dimensional character of the paper is sometimes used. But such a use is always avoidable, and I think that it will be agreed that the two-dimensional character of paper is no essential of computation. I assume then that the computation is carried out on one-dimensional paper, *i.e.* on a tape divided into squares. (p. 135)

This last assumption, ... *the computation is carried out on one-dimensional paper* ..., is based on an appeal to intuition in Turing's sense and makes the general argument unconvincing as a rigorous proof. Turing's assertion that effective calculability can be identified with machine computability should thus be viewed as the result of asserting a central thesis and constructing a two-part argument: the central thesis asserts that the computor's calculations are carried out on symbol sequences; the first part of the argument (using the sensory limitations of the computor) yields the *claim* that every operation (and thus every calculation) can be carried out by a suitable string machine; the second part is the rigorous proof that letter machines can simulate these machines. The *claim* is trivial, as the computor's operations *are* the machine operations.

4.4 Stronger theses

The above argumentative structure leading from computor calculations to Turing machine computations is rather canonical, once the symbolic configurations are fixed as symbol sequences and given the computor's limitations. In the case of other, for example, two or three-dimensional symbolic configurations, I do not see such a canonical form of reduction, unless one assumes again that the configurations are of a very special regular or normal shape.[39] In general, an "argumentative structure" supporting a reduction will contain then a *central thesis* in a far stronger sense, namely, that the calculations of the computor can be carried out by a precisely described device operating on a particular class of symbolic configurations; indeed, the devices should be viewed as generalized Post productions. These last considerations also indicate, how to carve up matters between analysis and proof; i.e., they allow us to answer the question asked at the end of subsection 4.2.

The diagram below represents these reflections graphically and relates them to the standard formulation of Turing's Thesis. Step 1 is given by conceptual analysis, whereas step 2 indicates the application of the central thesis for a particular class of symbolic configurations or *symcons*. (The symcon machines are Post systems operating, of course, on symcons.) The equivalence proof justifies an extremely simple description of computations that is most useful for mathematical investigations, from the construction of a universal machine and the formulation of the halting problem to the proof of the undecidability of the *Entscheidungsproblem*. It should be underlined that step 2, not the equivalence proof, is for Turing the crucial one that goes beyond the conceptual analysis; for me it is the problematic one that requires further reflection. I will address it in two different ways: inductively now and axiomatically in Section 5.

In order to make Turing's central thesis, quite in Post's spirit, inductively more convincing, it seems sensible to allow larger classes of symbolic configurations and more general operations on them. Turing himself intended, as we saw, to give an analysis of mechanical procedures on two-dimensional configurations already in 1936. In 1954 he considered even three-dimensional configurations and mechanical operations on them, starting out with examples of puzzles: square piece puzzles,

---

[39]This issue is also discussed in Kleene's *Introduction to Metamathematics*, pp. 376–381, in an informed and insightful defense of Turing's Thesis. However, in Kleene's way of extending configurations and operations, much stronger normalizing conditions are in place; e.g., when considering machines corresponding to our string machines the strings must be of the same length.

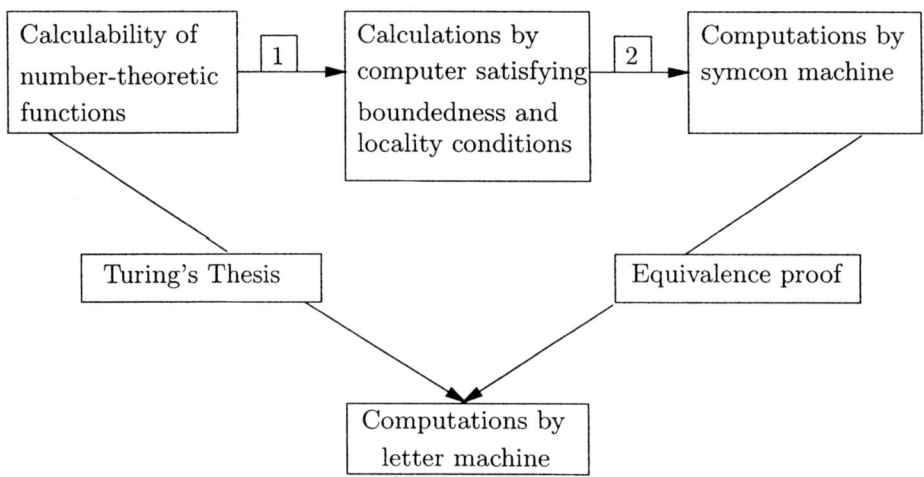

puzzles involving the separation of rigid bodies or the transformation of knots, i.e., puzzles in two and three dimensions. He viewed Post production systems as *linear* or *substitution puzzles*. As he considered them as puzzles in "normal form", he was able to formulate a suitable version of "Turing's Thesis":

> Given any puzzle we can find a corresponding substitution puzzle which is equivalent to it in the sense that given a solution of the one we can easily find a solution of the other. ... A transformation can be carried out by the rules of the original puzzle if and only if it can be carried out by substitutions ....[40]

Turing admits that this formulation is "somewhat lacking in definiteness" and claims that it will remain so; he characterizes its status as lying between a theorem and a definition: "In so far as we know *a priori* what is a puzzle and what is not, the statement is a theorem. In so far as we do not know what puzzles are, the statement is a definition which tells us something about what they are." Of course, Turing continues, one could define puzzle by a phrase beginning with "a set of definite rules", or one could reduce its definition to that of computable function or systematic procedure. A definition of any of these notions would provide one for puzzles. Neither in 1936 nor in 1954 did Turing try to characterize mathematically more general configurations and elementary operations on them. I am going to describe briefly one particular attempt of doing just that by Byrnes and me in our [1996].

Our approach was influenced by Kolmogorov and Uspensky's work on algorithms and has three distinct components: the symbolic configurations are certain finite connected and labeled graphs, we call *K(olmogorov)-graphs*; K-graphs contain a unique *distinguished element* that corresponds to the scanned square of a

---

[40][Turing, 1954, 15]

Turing machine tape; the operations *substitute neighborhoods* of the distinguished element by appropriate other neighborhoods and are given by a finite list of generalized Post production rules. Though broadening Turing's original considerations, we remain clearly within his general analytic framework and prove that letter machines can mimic K-graph machines. Turing's central thesis expresses here that K-graph machines can do the work of computors directly. As a playful indication of how K-graph machines straightforwardly can carry out human and genuinely symbolic, indeed diagrammatic algorithms, we programmed a K-graph machine to do ordinary, two-dimensional column addition. In sum then, a much more general class of symbolic configurations and operations on them is considered, and the central thesis for K-graph machines seems even more plausible than the one for string machines.

The separation of conceptual analysis and mathematical proof is essential for recognizing that the correctness of *Turing's Thesis* (taken generically) rests on two pillars, namely, on the correctness of boundedness and locality conditions for computors and on the correctness of the pertinent central thesis. The latter asserts explicitly that calculations of a computor can be mimicked by a particular kind of machine. However satisfactory one may find this line of argument, there are two weak spots: the looseness of the restrictive conditions (What are symbolic configurations? What changes can mechanical operations effect?) and the corresponding vagueness of the central thesis. We are, no matter how we turn ourselves, in a position that is methodologically not fully satisfactory.

4.5 Machine computability

Before attacking the central methodological issue in Section 5 from a different angle that is however informed by our investigations so far, let us look at the case, where reflection on limitations of computing devices leads to an important general concept of parallel computation and allows us to abstract further from particular types of configurations and operations. These considerations are based on Gandy's work in his [1980] that in its broad methodological approach parallels Turing's. At issue is machine calculability. The machines Turing associates with the basic operations of a computor can be physically realized, and we can obviously raise the question, whether these devices (our desktop computers, for example) are just doing things faster than we do, or whether they are in a principled way computationally more powerful.

It is informative first to look at Church's perspective on Turing's work in his 1937 review for the *Journal of Symbolic Logic*. Church was very much on target, though there is one fundamental misunderstanding as to the relative role of computor and machine computability in Turing's argument. For Church, computability by a machine "occupying a finite space and with working parts of finite size" is analyzed by Turing; given that the Turing machine is the outcome of the analysis, one can then observe that "in particular, a human calculator, provided with pencil and paper and explicit instructions, can be regarded as a kind of Turing machine". On account of the analysis and this observation it is for Church then "immediately clear" that (Turing-) machine computability can be identified with effectiveness. This is re-emphasized in the rather critical review of Post's 1936 paper in which Church pointed to the essential finiteness requirements in Turing's analysis: "To define effectiveness as computability by an arbitrary machine, subject to restrictions of finiteness, would seem to be an adequate representation of the ordinary notion, and if this is done the need for a working hypothesis disappears." This is right, as far as emphasis on finiteness restrictions is concerned. But Turing analyzed, as we saw, a mechanical computor, and *that* provides the basis for judging the correctness of the finiteness conditions. In addition, Church is rather quick in his judgment that "certain further restrictions" can be imposed on such arbitrary machines to obtain Turing's machines; this is viewed "as a matter of convenience" and the restrictions are for Church "of such a nature as obviously to cause no loss of generality".

Church's apparent misunderstanding is rather common; see, as a later example, Mendelson's paper [1990]. It is Turing's student, Robin Gandy who analyzes machine computability in his 1980 paper *Church's thesis and principles for mechanisms* and proposes a particular mathematical description of *discrete mechanical devices* and their computations. He follows Turing's three-step-argument of analysis, formulation of restrictive principles and proof of a "reduction theorem". Gandy shows that everything calculable by a device satisfying the restrictive principles is already computable by a Turing machine. The central and novel aspect of Gandy's analysis is the fact that it incorporates parallelism and covers cellular automata directly. This is of real interest, as cellular automata do not satisfy the locality condition (**L**); after all, the configurations affected in a single computation step are potentially unbounded.

What are discrete mechanical devices "in general"? Gandy introduces the term to make it clear that he does not deal with analogue devices, but rather with machines that are "discrete" (i.e., consist of finitely many parts) and proceed step-by-step from one state to the next. Gandy considers two fundamental physical constraints for such devices: (1) a lower bound on the size of atomic components; (2) an upper bound on the speed of signal propagation.[41] These two constraints together guarantee what the sensory limitations guarantee for computors, namely that in a given unit of time there are only a bounded number of different observable configurations (in a broad sense) and just a bounded number of possible actions on them. This justifies Gandy's contention that states of such machines "can be adequately described in finite terms", that calculations are proceeding in discrete and uniquely determined steps and, consequently, that these devices can be viewed, in a loose sense, as digital computers. If that's all, then it seems that without further ado we have established that machines in this sense are computationally not

---

[41] Cf. [Gandy, 1980, 126, but also 135–6]. For a more detailed argument see [Mundici and Sieg, section 3], where physical limitations for computing devices are discussed. In particular, there is an exploration of how space-time of computations are constrained, and how such constraints prevent us from having "arbitrarily" complex physical operations.

more powerful than computors, at least not in any principled way. However, if the concept of machine computability has to encompass "massive parallelism" then we are not done yet, and we have to incorporate that suitably into the mathematical description. And that can be done. Indeed, Gandy provided for the first time a conceptual analysis and a general description of parallel algorithms.

Gandy's characterization is given in terms of discrete dynamical systems $\langle \mathbf{S}, \mathbf{F} \rangle$, where $\mathbf{S}$ is the set of states and $\mathbf{F}$ governs the system's evolution. These dynamical systems have to satisfy four restrictive principles. The first principle pertains to the *form of description* and states that any machine $\mathbf{M}$ can be presented by such a pair $\langle \mathbf{S}, \mathbf{F} \rangle$, and that $\mathbf{M}$'s computation, starting in an initial state $\mathbf{x}$, is given by the sequence $\mathbf{x}$, $\mathbf{F}(\mathbf{x})$, $\mathbf{F}(\mathbf{F}(\mathbf{x}))$, ... Gandy formulates three groups of substantive principles, the first of which, *The Principle of Limitation of Hierarchy*, requires that the set theoretic rank of the states is bounded, i.e., the structural class $\mathbf{S}$ is contained in a fixed initial segment of the hierarchy of hereditarily finite and non-empty sets HF. Gandy argues (on p. 131) that it is natural or convenient to think of a machine in hierarchical terms, and that "for a given machine the maximum height of its hierarchical structure must be bounded". The second of the substantive principles, *The Principle of Unique Reassembly*, claims that any state can be "assembled" from "parts" of bounded size; its proper formulation requires care and a lengthy sequence of definitions. The informal idea, though, is wonderfully straightforward: any state of a concrete machine must be built up from (finitely many different types of) off-the-shelf components. Clearly, the components have a bound on their complexity. Both of these principles are concerned with the states in $\mathbf{S}$; the remaining third and central principle, *The Principle of Local Causality*, puts conditions on (the local determination of) the structural operation $\mathbf{F}$. It is formulated by Gandy in this preliminary way: "The next state, $\mathbf{Fx}$, of a machine can be reassembled from its restrictions to overlapping 'regions' s and these restrictions are locally caused." It requires that the parts from which $\mathbf{F}(\mathbf{x})$ can be assembled depend only on bounded parts of $\mathbf{x}$.

Gandy's Central Thesis is naturally the claim that any discrete mechanical device can be described as a dynamical system satisfying the above substantive principles. As to the set-up John Shepherdson remarked in his [1988, 586]: "Although Gandy's principles were obtained by a very natural analysis of Turing's argument they turned out to be rather complicated, involving many subsidiary definitions in their statement. In following Gandy's argument, however, one is led to the conclusion that that is in the nature of the situation." Nevertheless, in [Sieg and Byrnes, 1999] a greatly simplified presentation is achieved by choosing definitions appropriately, following closely the central informal ideas and using one key suggestion made by Gandy in the Appendix to his paper. This simplification does not change at all the form of presentation. However, of the four principles used by Gandy only a restricted version of the principle of local causality is explicitly retained. It is formulated in two separate parts, namely, as the principle of *Local Causation* and that of *Unique Assembly*. The separation reflects the distinction between the local determination of regions of the next state and their assembly into the next state.

Is it then correct to think that Turing's and Gandy's analyses lead to results that are in line with Gödel's general methodological expectations expressed to Church in 1934? Church reported that expectation to Kleene a year later and formulated it as follows:

> His [i.e. Gödel's] only idea at the time was that it might be possible, in terms of effective calculability as an undefined notion, to state a set of axioms which would embody the generally accepted properties of this notion, and to do something on that basis.[42]

Let's turn to that issue next.

5 AXIOMS FOR COMPUTABILITY.

The analysis offered by Turing in 1936 and re-described in 1954 was contiguous with the work of Gödel, Church, Kleene, Hilbert and Bernays, and others, but at the same time it was radically different and strikingly novel. They had explicated the calculability of number-theoretic functions in terms of their evaluation in calculi using only elementary and arithmetically meaningful steps; that put a stumbling-block into the path of a deeper analysis. Turing, in contrast, analyzed the basic processes that are carried out by computors and underlie the elementary calculation steps. The restricted machine model that resulted from his analysis almost hides the fact that Turing deals with general symbolic processes.

Turing's perspective on such general processes made it possible to restrict computations by *boundedness* and *locality* conditions. These conditions are obviously violated and don't even make sense when the values of number theoretic functions are determined by arithmetically meaningful steps. For example, in Gödel's equational calculus the replacement operations involve quite naturally arbitrarily complex terms. However, for steps of general symbolic processes the conditions are convincingly motivated by the sensory limitations of the computing agent and the normative demand of immediate recognizability of configurations; the basic steps, after all, must not be in need of further analysis. Following Turing's broad approach Gandy investigated in [1980] the *computations of machines*. Machines can in particular carry out parallel computations, and physical limitations motivate restrictive conditions for them. In spite of the generality of his notion, Gandy was able to show that any machine computable function is also Turing computable.

These analyses are taken now as a basis for further reflections along Gödelian lines. In a conversation with Church that took place in early 1934, Gödel found Church's proposal to identify effective calculability with λ-definability "thoroughly unsatisfactory", but he did make a counterproposal. He suggested "to state a set of axioms which embody the generally accepted properties of this notion (i.e.,

---

[42]Church in the letter to Kleene of November 29, 1935, quoted in [Davis, 1982, 9].

effective calculability), and to do something on that basis". Gödel did not articulate what the generally accepted properties of effective calculability might be or what might be done on the basis of an appropriate set of axioms. Sharpening Gandy's work I will give an abstract characterization of "Turing Computors" and "Gandy Machines" as discrete dynamical systems whose evolutions satisfy some well-motivated and general axiomatic conditions. Those conditions express constraints, which have to be satisfied by computing processes of these particular devices. Thus, I am taking the axiomatic method as a tool to resolve the methodological problems surrounding Church's thesis for computors and machines. The mathematical formulations that follow in section 5.1 are given in greater generality than needed for Turing computors, so that they cover also the discussion of Gandy machines. (They are also quite different from the formulation in [Gandy, 1980] or in [Sieg and Byrnes, 1999a].)

5.1 *Discrete dynamical systems*

At issue is, how we can express those "well-motivated conditions" in a sharp way, as I clearly have not given an answer to the questions: What are symbolic configurations? What changes can mechanical operations effect? Nevertheless, some aspects can be coherently formulated for computors: (i) they operate deterministically on finite configurations; (ii) they recognize in each configuration exactly one pattern (from a bounded number of different kinds of such); (iii) they operate locally on the recognized patterns; (iv) they assemble the next configuration from the original one and the result of the local operation. Discrete dynamical systems provide an elegant framework for capturing these general ideas precisely. We consider pairs $\langle \mathbf{D}, \mathbf{F} \rangle$ where $\mathbf{D}$ is a class of states (ids or syntactic configurations) and $\mathbf{F}$ an operation from $\mathbf{D}$ to $\mathbf{D}$ that transforms a given state into the next one. States are finite objects and are represented by non-empty hereditarily finite sets over an infinite set $\mathbf{U}$ of atoms. Such sets reflect states of computing devices just as other mathematical structures represent states of nature, but this reflection is done somewhat indirectly, as only the $\in$-relation is available.

In order to obtain a more adequate mathematical framework free of ties to particular representations, we consider *structural classes* $\mathbf{S}$, i.e., classes of states that are closed under $\in$-isomorphisms. After all, any $\in$-isomorphic set can replace a given one in this reflective, representational role. That raises immediately the question, what invariance properties the state transforming operations $\mathbf{F}$ should have or how the $\mathbf{F}$-images of $\in$-isomorphic states are related. Recall that any $\in$-isomorphism π between states is a unique extension of some permutation on atoms, and let $\pi(\mathbf{x})$ or $\mathbf{x}^\pi$ stand for the result of applying π to the state $\mathbf{x}$. The lawlike connections between states are given by *structural operations* $\mathbf{G}$ from $\mathbf{S}$ to $\mathbf{S}$. The requirement on $\mathbf{G}$ will fix the dependence of values on just structural features of a state, not the nature of its atoms: for all permutations π on $\mathbf{U}$ and all $\mathbf{x} \in \mathbf{S}$, $\mathbf{G}(\pi(\mathbf{x}))$ is $\in$-isomorphic to $\pi(\mathbf{G}(\mathbf{x}))$, and the isomorphism has the additional property that it is the identity on the atoms occurring in the support of $\pi(\mathbf{x})$. $\mathbf{G}(\pi(\mathbf{x}))$ and $\pi(\mathbf{G}(\mathbf{x}))$ are said to be $\in$-*isomorphic over* $\pi(\mathbf{x})$, and we write $\mathbf{G}(\pi(\mathbf{x})) \cong_{\pi(\mathbf{x})} \pi(\mathbf{G}(\mathbf{x}))$. Note that we do not require the literal identity of $\mathbf{G}(\pi(\mathbf{x}))$ and $\pi(\mathbf{G}(\mathbf{x}))$; that would be too restrictive, as the state may be expanded by new atoms and it should not matter which new atoms are chosen. On the other hand, the requirement $\mathbf{G}(\pi(\mathbf{x}))$ is $\in$-isomorphic to $\pi(\mathbf{G}(\mathbf{x}))$ would be too loose, as we want to guarantee the physical persistence of atomic components. Here is the appropriate diagram:

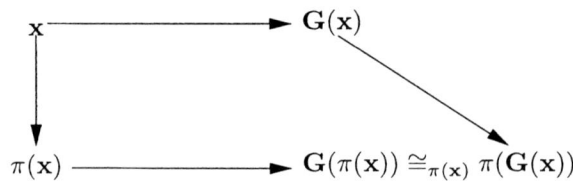

This mathematical framework addresses just point (i) in the above list of central aspects of mechanical computers. Now we turn to *patterns* and *local* operations. If $\mathbf{x}$ is a given state, regions of the next state are determined *locally* from particular *parts for* $\mathbf{x}$ on which the computor can operate.[43] *Boundedness* requires that there is only a bounded number of different kinds of parts, i.e., each part lies in one of a finite number of isomorphism types or, using Gandy's terminology, *stereotypes*. So let $\mathbf{T}$ be a fixed finite class of stereotypes. A part for $\mathbf{x}$ that is a member of a stereotype of $\mathbf{T}$ is called, naturally enough, a $\mathbf{T}$-*part for* $\mathbf{x}$. A $\mathbf{T}$-part $\mathbf{y}$ for $\mathbf{x}$ is a *causal neighborhood for* $\mathbf{x}$ given by $\mathbf{T}$, briefly $\mathbf{y} \in \mathrm{Cn}(\mathbf{x})$, if there is no $\mathbf{T}$-part $\mathbf{y}^*$ for $\mathbf{x}$ such that $\mathbf{y}$ is $\in$-embeddable into $\mathbf{y}^*$. The causal neighborhoods for $\mathbf{x}$ will also be called *patterns in* $\mathbf{x}$. Finally, the local change is effected by a structural operation $\mathbf{G}$ that works on unique causal neighborhoods. Having also given points (ii) and (iii) a proper mathematical explication, the assembly of the next state has to be determined.

The values of $\mathbf{G}$ are in general not exactly what we need in order to assemble the next state, because the configurations may have to be expanded and that involves the addition and coordination of new atoms. To address that issue we introduce *determined regions* $\mathrm{Dr}(\mathbf{z}, \mathbf{x})$ of a state $\mathbf{z}$; they are $\in$-isomorphic to $\mathbf{G}(\mathbf{y})$ for some causal neighborhood $\mathbf{y}$ for $\mathbf{x}$ and must satisfy a technical condition on the "newness" of atoms. More precisely, $\mathbf{v} \in \mathrm{Dr}(\mathbf{z}, \mathbf{x})$ if and only if $\mathbf{v} <^* \mathbf{z}$ and there is a $\mathbf{y} \in \mathrm{Cn}(\mathbf{x})$, such that $\mathbf{G}(\mathbf{y}) \cong_{\mathbf{y}} \mathbf{v}$ and $\mathrm{Sup}(\mathbf{v}) \cap \mathrm{Sup}(\mathbf{x}) \subseteq \mathrm{Sup}(\mathbf{y})$. The last condition for Dr guarantees that new atoms in $\mathbf{G}(\mathbf{y})$ correspond to new atoms in $\mathbf{v}$, and that the new atoms in $\mathbf{v}$ are new for $\mathbf{x}$. If one requires $\mathbf{G}$ to satisfy similarly $\mathrm{Sup}(\mathbf{G}(\mathbf{y})) \cap \mathrm{Sup}(\mathbf{x}) \subseteq \mathrm{Sup}(\mathbf{y})$, then the condition $\mathbf{G}(\mathbf{y}) \cong_{\mathbf{y}} \mathbf{v}$ can be strengthened

---

[43] A part $\mathbf{y}$ for $\mathbf{x}$ used to be in my earlier presentations a connected subtree $\mathbf{y}$ of the $\in$-tree for $\mathbf{x}$, briefly $\mathbf{y} <^* \mathbf{x}$, if $\mathbf{y} \neq \mathbf{x}$ and $\mathbf{y}$ has the same root as $\mathbf{x}$ and its leaves are also leaves of $\mathbf{x}$. More precisely, $\mathbf{y} \neq \mathbf{x}$ and $\mathbf{y}$ is a non-empty subset of $\{\mathbf{v} \mid (\exists \mathbf{z})(\mathbf{v} <^* \mathbf{z}\ \&\ \mathbf{z} \in \mathbf{x})\} \cup \{\mathbf{r} \mid \mathbf{r} \in \mathbf{x}\}$. Now it is just a subset, but I will continue to use the term "part" to emphasize that we are taking the whole $\in$-structure into account.

to $\mathbf{G(y)} \cong_x \mathbf{v}$. The new atoms are thus always taken from $\mathbf{U}\backslash\mathrm{Sup}(\mathbf{x})$.[44] One final definition: for given states $\mathbf{z}$ and $\mathbf{x}$ let $A(\mathbf{z},\mathbf{x})$ stand for $\mathrm{Sup}(\mathbf{z})\backslash\mathrm{Sup}(\mathbf{x})$. Note that the number of new atoms introduced by $\mathbf{G}$ is bounded, i.e., $|A(\mathbf{G(y)}, \mathrm{Sup}(\mathbf{x}))| < n$ for some natural number n (any $\mathbf{x} \in \mathbf{S}$ and any causal neighborhood $\mathbf{y}$ for $\mathbf{x}$).

So, how is the next state of a Turing computor assembled? By simple set theoretic operations, namely, difference \ and union ∪. Recalling the boundedness and locality conditions for computors we define that $\mathbf{M} = \langle \mathbf{S}; \mathbf{T}, \mathbf{G} \rangle$ is a *Turing Computor on* $\mathbf{S}$, where $\mathbf{S}$ is a structural class, $\mathbf{T}$ a finite set of stereotypes, and $\mathbf{G}$ a structural operation on ∪ $\mathbf{T}$, if and only if, for every $\mathbf{x} \in \mathbf{S}$ there is a $\mathbf{z} \in \mathbf{S}$, such that:

(**L.0**) $(\exists!\mathbf{y})\ \mathbf{y} \in \mathrm{Cn}(\mathbf{x})$,

(**L.1**) $(\exists!\mathbf{v} \in \mathrm{Dr}(\mathbf{z},\mathbf{x}))\ \mathbf{v} \cong_x \mathbf{G}(\mathbf{cn}(\mathbf{x}))$,

(**A.1**) $\mathbf{z} = (\mathbf{x}\backslash \mathrm{Cn}(\mathbf{x})) \cup \mathrm{Dr}(\mathbf{z},\mathbf{x})$.

L stands for Locality and **A** for Assembly. $(\exists!\mathbf{y})$ is the existential quantifier expressing uniqueness. $\mathbf{cn}(\mathbf{x})$ denotes the sole causal neighborhood of $\mathbf{x}$ guaranteed by **L.0**, i.e., every state is required by **L.0** to contain exactly one pattern. This pattern in state $\mathbf{x}$ yields a unique determined region of a possible next state $\mathbf{z}$; that is expressed by **L.1**. The state $\mathbf{z}$ is obtained according to the assembly condition **A.1**. It is determined up to $\in$-isomorphism over $\mathbf{x}$. A *computation by* $\mathbf{M}$ is a finite sequence of transition steps via $\mathbf{G}$ that is halted when the operation on a state $\mathbf{w}$ yields $\mathbf{w}$ as the next state. This result, for input $\mathbf{x}$, is denoted by $\mathbf{M(x)}$. A function $\mathbf{F}$ is (Turing) *computable* if and only if there is a Turing computor $\mathbf{M}$ whose computation results determine — under a suitable encoding and decoding — the values of $\mathbf{F}$ for any of its arguments. After all these definitions one can use a suitable set theoretic representation of Turing machines to establish one lemma, namely, that Turing machines are Turing computors. (See section 5.4.)

In the next subsection, we will provide a characterization of computations by machines that is as general and convincing as that of human computors. Gandy laid the groundwork in his thought-provoking paper *Church's Thesis and Principles for Mechanisms* — a rich and difficult, but unnecessarily and maddeningly complex paper. The structure of Turing's argument actually guided Gandy's analysis; however, Gandy realized through conversations with J. C. Shepherdson that the analysis "must take parallel working into account". In a comprehensive survey article published ten years after Gandy's paper, Leslie Lamport and Nancy Lynch argued that the theory of sequential computing "rests on fundamental concepts of computability that are independent of any particular computational model". They emphasized that the "fundamental formal concepts underlying distributed computing", if there were any, had not yet been developed. "Nevertheless", they wrote, "one can make some informal observations that seem to be important":

---

[44]This selection of atoms new for $\mathbf{x}$ has in a very weak sense a "global" aspect; as $\mathbf{G}$ is a structural operation, the precise choice of the atoms does not matter.

> Underlying almost all models of concurrent systems is the assumption that an execution consists of a set of discrete events, each affecting only part of the system's state. Events are grouped into processes, each process being a more or less completely sequenced set of events sharing some common locality in terms of what part of the state they affect. For a collection of autonomous processes to act as a coherent system, the processes must be synchronized. (p. 1166)

Gandy's analysis of parallel computation is conceptually convincing and provides a sharp mathematical form of the informal assumption(s) "underlying almost all models of concurrent systems". Gandy takes as the paradigmatic parallel computation, as I mentioned already, the evolution of the Game of Life or other cellular automata.

5.2 Gandy machines

Gandy uses, as Turing did, a *central thesis*: any discrete mechanical device satisfying some informal restrictive conditions can be described as a particular kind of dynamical system. Instead, I characterize a *Gandy Machine* axiomatically based on the following informal idea: the machine has to recognize the causal neighborhoods of a given state, act on them locally in parallel, and assemble the results to obtain the next state, which should be unique up to $\in$-isomorphism. In analogy to the definition of Turing computability, we call a function $\mathbf{F}$ *computable in parallel* if and only if there is a Gandy machine $\mathbf{M}$ whose computation results determine — under a suitable encoding and decoding — the values of $\mathbf{F}$ for any of its arguments. What then is the underlying notion of parallel computation?

Generalizing the above considerations for Turing computors, one notices quickly complications, when new atoms are introduced in the images of causal neighborhoods as well as in the next state: the different new atoms have to be "structurally coordinated". That can be achieved by a second local operation and a second set of stereotypes. Causal neighborhoods of type 1 are parts of neighborhoods of type 2 and the overlapping determined regions of type 1 must be parts of determined regions of type 2, so that they fit together appropriately. This generalization is absolutely crucial to allow the machine to assemble the determined regions. Here is the definition: $\mathbf{M} = \langle \mathbf{S}; \mathbf{T}_1, \mathbf{G}_1, \mathbf{T}_2, \mathbf{G}_2 \rangle$ is a *Gandy Machine on* $\mathbf{S}$, where $\mathbf{S}$ is a structural class, $\mathbf{T}_i$ a finite set of stereotypes, $\mathbf{G}_i$ a structural operation on $\cup \mathbf{T}_i$ ($i = 1$ or 2), if and only if for every $\mathbf{x} \in \mathbf{S}$ there is a $\mathbf{z} \in \mathbf{S}$ such that

(L.1) $(\forall \mathbf{y} \in \mathrm{Cn}_1(\mathbf{x}))(\exists! \mathbf{v} \in \mathrm{Dr}_1(\mathbf{z}, \mathbf{x})) \mathbf{v} \cong_\mathbf{x} \mathbf{G}_1(\mathbf{y})$;

(L.2) $(\forall \mathbf{y} \in \mathrm{Cn}_2(\mathbf{x}))(\exists \mathbf{v} \in \mathrm{Dr}_2(\mathbf{z}, \mathbf{x})) \mathbf{v} \cong_\mathbf{x} \mathbf{G}_2(\mathbf{y})$;

(A.1) $(\forall C)[C \subseteq \mathrm{Dr}_1(\mathbf{z}, \mathbf{x}) \& \cap \{ \mathrm{Sup}(\mathbf{v}) \cap A(\mathbf{z}, \mathbf{x}) | \mathbf{v} \in C \} \neq \emptyset \rightarrow$
$(\exists \mathbf{w} \in \mathrm{Dr}_2(\mathbf{z}, \mathbf{x}))(\forall \mathbf{v} \in C) \mathbf{v} <^* \mathbf{w}]$;

(A.2) $\mathbf{z} = \cup \, \mathrm{Dr}_1(\mathbf{z}, \mathbf{x})$.

The condition $\cap\{\text{Sup}(\mathbf{v}) \cap A(\mathbf{z},\mathbf{x}) | \mathbf{v} \in \mathbf{C}\} \neq \varnothing$ in (**A.1**) expresses that the determined regions **v** in **C** have new atoms in common, i.e., they *overlap*. — It might be helpful to the reader to look at section 5.4 and the description of the game of life as a Gandy machine one finds there.

The restrictions for Gandy machines, as in the case of Turing computors, amount to boundedness and locality conditions. They are justified *directly* by two physical limitations, namely, a lower bound on the size of atomic components and an upper bound on the speed of signal propagation. I have completed now all the foundational work and can describe two important mathematical facts for Gandy machines: (i) the state **z** following **x** is determined uniquely up to $\in$-isomorphism over **x**, and (ii) Turing machines can effect such transitions. The proof of the first claim contains the combinatorial heart of matters and uses crucially the assembly conditions. The proof of the second fact is rather direct. Only finitely many finite objects are involved in the transition, and all the axiomatic conditions are decidable. Thus, a search will allow us to find **z**. This can be understood as a *Representation Theorem*: any particular Gandy machine is computationally equivalent to a two-letter Turing machine, as Turing machines are also Gandy machines. The first fact for Gandy machines, **z** is determined uniquely up to $\in$-isomorphism over **x**, follows from the next theorem.[45] Before being able to formulate and prove it, we need to introduce one more concept. A collection **C** of parts for **x** is a *cover for x* just in case $\mathbf{x} \subseteq \cup \mathbf{C}$.

THEOREM. Let **M** be $\langle \mathbf{S}; \mathbf{T}_1, \mathbf{G}_1, \mathbf{T}_2, \mathbf{G}_2 \rangle$ as above and $\mathbf{x} \in \mathbf{S}$; if there are **z** and **z'** in **S** satisfying principles (**L.1-2**), (**A.1**), and if $\text{Dr}_1(\mathbf{z},\mathbf{x})$ and $\text{Dr}_1(\mathbf{z'},\mathbf{x})$ cover **z** and **z'**, then $\text{Dr}_1(\mathbf{z},\mathbf{x}) \cong_\mathbf{x} \text{Dr}_1(\mathbf{z'},\mathbf{x})$.

In the following Dr_1, Dr_1', A, and A' will abbreviate $\text{Dr}_1(\mathbf{z},\mathbf{x})$, $\text{Dr}_1(\mathbf{z'},\mathbf{x})$, $A(\mathbf{z},\mathbf{x})$, and $A(\mathbf{z'},\mathbf{x})$ respectively. Note that Dr_1 and Dr_1' are finite. Using (**L.1**) and (**L.2**) one can observe that there is a natural number m and there are sequences $\mathbf{v}_i$ and $\mathbf{v}_i'$, $i < m$, such that $\text{Dr}_1 = \{\mathbf{v}_i | i < m\}$, $\text{Dr}_1' = \{\mathbf{v}_i' | i < m\}$, and $\mathbf{v}_i'$ is the unique part of **z'** with $\mathbf{v}_i \cong_\mathbf{x} \mathbf{v}_i'$ via permutations π_i (for all $i < m$). See Figure 1, which is a picture of the situation.

To establish the Theorem, we have to find a *single* permutation π that extends to an $\in$-isomorphism over **x** for all $\mathbf{v}_i$ and $\mathbf{v}_i'$ simultaneously. Such a π must obviously satisfy for all $i < m$:

(i) $\mathbf{v}_i \cong_\mathbf{x} \mathbf{v}_i'$ via π

and, consequently,

(ii) $\pi[\text{Sup}(\mathbf{v}_i)] = \text{Sup}(\mathbf{v}_i')$.

---

[45] In [Gandy, 1980] this uniqueness up to $\in$-isomorphism over **x** is achieved in a much more complex way, mainly, because parts of a state are proper subtrees, in general non-located. Given an appropriate definition of cover, a collection **C** is called an *assembly for* **x**, if **C** is a cover for **x** and the elements of **C** are maximal. The fact that **C** is an assembly for exactly one **x**, if indeed it is, is expressed by saying that **C** *uniquely assembles to* **x**; see [Sieg and Byrnes, 1999a, 157]. In my setting, axiom (**A.2**) is equivalent to the claim that $\text{Dr}_1(\mathbf{z},\mathbf{x})$ uniquely assembles to **z**.

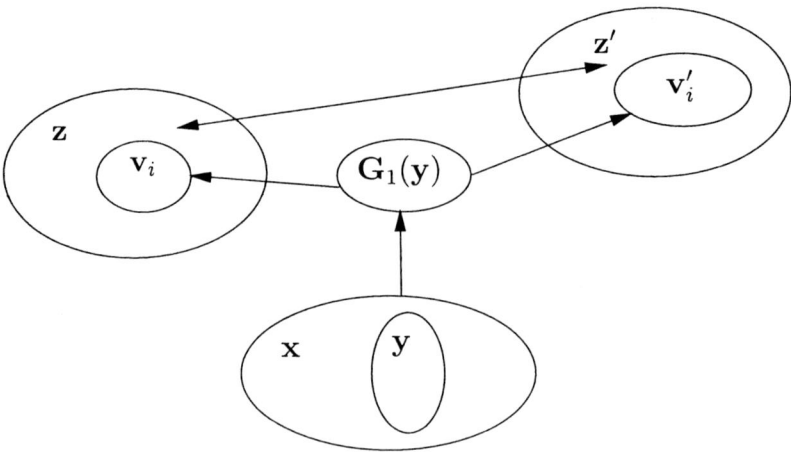

Figure 1.

As π is an $\in$-isomorphism over $\mathbf{x}$, we have:

(iii) $\pi[A] = A'$.

Condition (ii) implies for all $i < m$ and all $\mathbf{r} \in A$ the equivalence between $\mathbf{r}\in\mathrm{Sup}(\mathbf{v}_i)$ and $\mathbf{r}^\pi \in\mathrm{Sup}(\mathbf{v}'_i)$. This can also be expressed by

(ii*) $\mu(\mathbf{r})=\mu'(\mathbf{r}^\pi)$, for all $\mathbf{r}\in A$,

where $\mu(\mathbf{r}) = \{i | \mathbf{r} \in \mathrm{Sup}(\mathbf{v}_i)\}$ and $\mu'(\mathbf{r}) = \{i | \mathbf{r} \in \mathrm{Sup}(\mathbf{v}'_i)\}$; these are the *signatures* of $\mathbf{r}$ with respect to $\mathbf{z}$ and $\mathbf{z}'$.

To obtain such a permutation, the considerations are roughly as follows: (i) if the $\mathbf{v}_i$ do not overlap, then the π_i will do; (ii) if there is overlap, then an equivalence relation $\approx$ ($\approx'$) on $A(A')$ is defined by $\mathbf{r}_1 \approx \mathbf{r}_2$ iff $\mu(\mathbf{r}_1) = \mu(\mathbf{r}_2)$, and analogously for $\approx'$; (iii) then we prove that the "corresponding" equivalence classes $[\mathbf{r}]_\approx$ and $[\mathbf{s}]_{\approx'}$ (the signatures of their elements are identical) have the same cardinality. $[\mathbf{r}]_\approx$ can be characterized as $\cap\{\mathrm{Sup}(\mathbf{v}_i) \cap A | i \in \mu(\mathbf{r})\}$; similar for $[\mathbf{s}]_{\approx'}$. This characterization is clearly independent of the choice of representative by the very definition of the equivalence relation(s). With this in place, a global $\in$-isomorphism can be defined. These considerations are made precise through the proofs of the combinatorial lemma and two corollaries in the next section.

5.3 Global assembly

All considerations in this section are carried out under the assumptions of the Theorem: $\mathbf{M} = \langle \mathbf{S}; \mathbf{T}_1, \mathbf{G}_1, \mathbf{T}_2, \mathbf{G}_2 \rangle$ is an arbitrary Gandy machine and $\mathbf{x}\in\mathbf{S}$ an arbitrary state; we assume furthermore that **z** and **z′** are in **S**, the principles (**L.1-2**) and (**A.1**) are satisfied, and that Dr_1 and Dr_1' cover **z** and **z′**, because of (**A.2**). We want to show that $\mathrm{Dr}_1 \cong_\mathbf{x} \mathrm{Dr}_1'$, knowing already that there are sequences $\mathbf{v}_i$ and $\mathbf{v}_i'$ of length m, such that $\mathrm{Dr}_1 = \{\mathbf{v}_i | i<m\}$, $\mathrm{Dr}_1' = \{\mathbf{v}_i' | i<m\}$ and $\mathbf{v}_i'$ is the unique part of **z′** with $\mathbf{v}_i \cong_\mathbf{x} \mathbf{v}_i'$ via permutations π_i (for all i<m). I start out with the formulation of a key lemma concerning overlaps.

LEMMA. (**Overlap Lemma.**) Let $\mathbf{r}_0 \in A$ and $\mu(\mathbf{r}_0) \neq \varnothing$; then there is a permutation ρ on **U** with $\mathbf{v}_i \cong_\mathbf{x} \mathbf{v}_i'$ via ρ for all $i \in \mu(\mathbf{r}_0)$.

Proof. We have $\{\mathbf{v}_i | i \in \mu(\mathbf{r}_0)\} \subseteq \mathrm{Dr}_1$; as $\mathbf{r}_0$ is in A and in $\mathrm{Sup}(\mathbf{v}_i)$ for each $i \in \mu(\mathbf{r}_0) \neq \varnothing$, we have also that $\cap\{\mathrm{Sup}(\mathbf{v}_i) \cap A | i \in \mu(\mathbf{r}_0)\} \neq \emptyset$. The antecedent of (**A.1**) is satisfied, and we conclude that there is a $\mathbf{w} \in \mathrm{Dr}_2$ such that $\mathbf{v}_i <^* \mathbf{w} <^* \mathbf{z}$, for all $i \in \mu(\mathbf{r}_0)$. Using (**L.2**) we obtain a $\mathbf{w}' \in \mathrm{Dr}_2'$ with $\mathbf{w} \cong_\mathbf{x} \mathbf{w}'$. This ∈-isomorphism over **x** is induced by a permutation ρ and yields for all $i \in \mu(\mathbf{r}_0)$

$$\mathbf{v}_i^\rho <^* \mathbf{w}^\rho = \mathbf{w}' <^* \mathbf{z}'.$$

So we have, $\mathbf{v}_i \cong_\mathbf{x} \mathbf{v}_i^\rho$ and $\mathbf{v}_i^\rho <^* \mathbf{z}'$, thus — using (**L.1**) — $\mathbf{v}_i^\rho = \mathbf{v}_i'$; that holds for all $i \in \mu(\mathbf{r}_0)$. ∎

Note that the condition $\mu(\mathbf{r}) \neq \varnothing$ is satisfied in our considerations for any $\mathbf{r} \in A$, as Dr_1 is a cover of **z**; so we have for any such **r** an appropiate *overlap permutation* $\rho^\mathbf{r}$ *for* **r**. The crucial combinatorial lemma we have to establish is this:

LEMMA. (**Combinatorial Lemma.**) For $\mathbf{r}_0 \in A : |\{\mathbf{r} \in A | \mu(\mathbf{r}_0) \subseteq \mu(\mathbf{r})\}| = |\{\mathbf{s} \in A' | \mu(\mathbf{r}_0) \subseteq \mu'(\mathbf{s})\}|$.

Proof. Consider $\mathbf{r}_0 \in A$. I establish first the claim

$$\rho[\{\mathbf{r} \in A | \mu(\mathbf{r}_0) \subseteq \mu(\mathbf{r})\}] \subseteq \{\mathbf{s} \in A' | \mu(\mathbf{r}_0) \subseteq \mu'(\mathbf{s})\},$$

where ρ is an overlap permutation for $\mathbf{r}_0$. The claim follows easily from

$$\mathbf{r} \in A \& \mu(\mathbf{r}_0) \subseteq \mu(\mathbf{r}) \to \mu(\mathbf{r}_0) \subseteq \mu'(\mathbf{r}^\rho),$$

by observing that $\mathbf{r}^\rho$ is in A'. Assume, to establish this conditional indirectly, for arbitrary $\mathbf{r} \in A$ that $\mu(\mathbf{r}_0) \subseteq \mu(\mathbf{r})$ and $\neg(\mu(\mathbf{r}_0) \subseteq \mu'(\mathbf{r}^\rho))$. The first assumption implies that $\mathbf{r} \in \mathrm{Sup}(\mathbf{v}_i)$ for all $i \in \mu(\mathbf{r}_0)$, and the construction of ρ yields then:

($\heartsuit$) $\mathbf{r}^\rho \in \mathrm{Sup}(\mathbf{v}_i')$ for all $i \in \mu(\mathbf{r}_0)$.

The second assumption implies that there is a k in $\mu(\mathbf{r}_0) \setminus \mu'(\mathbf{r}^\rho)$. Obviously, $k \in \mu(\mathbf{r}_0)$ and $k \notin \mu'(\mathbf{r}^\rho)$. The first conjunct $k \in \mu(\mathbf{r}_0)$ and ($\heartsuit$) imply that $\mathbf{r}^\rho \in \mathrm{Sup}(\mathbf{v}_k')$; as the second conjunct $k \notin \mu'(\mathbf{r}^\rho)$ means that $\mathbf{r}^\rho \notin \mathrm{Sup}(\mathbf{v}_k')$, we have obtained a contradiction.

Now I'll show that $\rho[\{\mathbf{r} \in A | \mu(\mathbf{r}_0) \subseteq \mu(\mathbf{r})\}]$ cannot be a *proper* subset of $\{\mathbf{s} \in A' | \mu(\mathbf{r}_0) \subseteq \mu'(\mathbf{s})\}$. Assume, to obtain a contradiction, that it is; then there is

$\mathbf{s}^* \in A'$ that satisfies $\mu(\mathbf{r}_0) \subseteq \mu'(\mathbf{s}^*)$ and is not a member of $\rho[\{\mathbf{r} \in A | \mu(\mathbf{r}_0) \subseteq \mu(\mathbf{r})\}]$. As $\mu(\mathbf{r}_0) \subseteq \mu'(\mathbf{s}^*)$, $\mathbf{s}^*$ is in $\text{Sup}(\mathbf{v}'_i)$ for all $i \in \mu(\mathbf{r}_0)$; the analogous fact holds for all $\mathbf{r} \in A$ satisfying $\mu(\mathbf{r}_0) \subseteq \mu(\mathbf{r})$, i.e., all such $\mathbf{r}$ must be in $\text{Sup}(\mathbf{v}_i)$ for all $i \in \mu(\mathbf{r}_0)$. As $\mathbf{v}_i \cong_\mathbf{x} \mathbf{v}'_i$ via ρ for all $i \in \mu(\mathbf{r}_0)$, $\mathbf{s}^*$ must be obtained as a ρ-image of some $\mathbf{r}^*$ in $\text{Sup}(\mathbf{x})$ or in A (and, in the latter case, violating $\mu(\mathbf{r}_0) \subseteq \mu(\mathbf{r}^*)$). However, in either case we have a contradiction. The assertion of the Lemma is now immediate. ∎

Next I establish two consequences of the Combinatorial Lemma, the second of which is basic for the definition of the global isomorphism π.

COROLLARY 1. For any $I \subseteq \{0, 1, \ldots, m-1\}$ with $I \subseteq \mu(\mathbf{r}_0)$ for some $\mathbf{r}_0$ in A,

$$|\{\mathbf{r} \in A | I \subseteq \mu(\mathbf{r})\}| = |\{\mathbf{s} \in A' | I \subseteq \mu'(\mathbf{s})\}|.$$

Proof. Consider an arbitrary $I \subseteq \mu(\mathbf{r}_0)$ for some $\mathbf{r}_0$ in A. If $I = \mu(\mathbf{r}_0)$, then the claim follows directly from the Combinatorial Lemma. If $I \subset \mu(\mathbf{r}_0)$, let $\mathbf{r}^0, \ldots, \mathbf{r}^{k-1}$ be elements $\mathbf{r}$ of A with $I \subset \mu(\mathbf{r})$ and require that $\mu(\mathbf{r}^j) \neq \mu(\mathbf{r}^{j'})$, for all $j, j' < k$ and $j \neq j'$, and for every $\mathbf{r} \in A$ with $I \subset \mu(\mathbf{r})$ there is a unique $j < k$ with $\mu(\mathbf{r}) = \mu(\mathbf{r}^j)$. The Combinatorial Lemma implies, for all $j < k$,

$$|\{\mathbf{r} \in A | \mu(\mathbf{r}^j) \subseteq \mu(\mathbf{r})\}| = |\{\mathbf{s} \in A' | \mu(\mathbf{r}^j) \subseteq \mu'(\mathbf{s})\}|.$$

Now it is easy to verify the claim of Corollary 1:

$$|\{\mathbf{r} \in A | I \subseteq \mu(\mathbf{r})\}| =$$
$$|\{\mathbf{r} \in A | (\exists j < k)\mu(\mathbf{r}^j) \subseteq \mu(\mathbf{r})\}| =$$
$$|\{\mathbf{s} \in A' | (\exists j < k)\mu(\mathbf{r}^j) \subseteq \mu'(\mathbf{s})\}| =$$
$$|\{\mathbf{s} \in A' | I \subseteq \mu'(\mathbf{s})\}|.$$

This completes the proof of Corollary 1. ∎

The second important consequence of the Combinatorial Lemma can be obtained now by an inductive argument.

COROLLARY 2. For any $I \subseteq \{0, 1, \ldots, m-1\}$ with $I \subseteq \mu(\mathbf{r}_0)$ for some $\mathbf{r}_0$ in A.

$$|\{\mathbf{r} \in A | I = \mu(\mathbf{r})\}| = |\{\mathbf{s} \in A' | I = \mu'(\mathbf{s})\}|.$$

Proof. (By downward induction on $|I|$). Abbreviating $|\{\mathbf{r} \in A | I = \mu(\mathbf{r})\}|$ by ν_I and $|\{\mathbf{s} \in A' | I = \mu'(\mathbf{s})\}|$ by ν'_I, the argument is as follows:
Base case ($|I| = m$): In this case there are no proper extensions I^* of I, and we have

$$\begin{aligned} \nu_I &= |\{\mathbf{r} \in A | I = \mu(\mathbf{r})\}| \\ &= |\{\mathbf{r} \in A | I \subseteq \mu(\mathbf{r})\}|, \quad \text{as there is no proper extension of } I, \\ &= |\{\mathbf{s} \in A' | I \subseteq \mu'(\mathbf{s})\}|, \quad \text{by Corollary 1,} \\ &= |\{\mathbf{s} \in A' | I = \mu'(\mathbf{s})\}|, \quad \text{again, as there is no proper extension,} \\ &= \nu'_I \end{aligned}$$

Induction step ($|I|) < m$): Assume that the claim holds for all I^* with $n+1 \leq |I^*| \leq m$ and show that it holds for I with $|I| = n$. Using the induction hypothesis we have, summing up over all proper extensions I^* of I:

$$(\clubsuit) \ \Sigma_{I^*} \nu_{I^*} = \Sigma_{I^*} \nu'_{I^*}.$$

Now we argue as before:

$$\begin{aligned}
\nu_I &= |\{\mathbf{r} \in A | I = \mu(\mathbf{r})\}| \\
&= |\{\mathbf{r} \in A | I \subseteq \mu(\mathbf{r})\}| - \Sigma_{I^*} \nu_{I^*} \\
&= |\{\mathbf{s} \in A' | I \subseteq \mu'(\mathbf{s})\}| - \Sigma_{I^*} \nu'_{I^*}, \quad \text{by Corollary 1 and } (\clubsuit), \\
&= |\{\mathbf{s} \in A' | I = \mu'(\mathbf{s})\}| \\
&= \nu'_I
\end{aligned}$$

This completes the proof of Corollary 2. ∎

Finally, we can define an appropriate global permutation π. Given an atom $\mathbf{r} \in A$, there is an overlap permutations ρ^r, which can be restricted to

$$[\mathbf{r}]_\approx = \cap \{\text{Sup}(\mathbf{v}_i) \cap A | i \in \mu(\mathbf{r})\};$$

let ρ^* denote this restriction. Because of Corollary 2, $\rho*$ is a bijection between $[\mathbf{r}]_\approx$ and $[\rho^*(\mathbf{r})]_{\approx'}$. The desired global permutation is now defined as follows for any atom $\mathbf{r} \in \cup\{\text{Sup}(\mathbf{v}_i) | i < m\}$:

$$\pi(\mathbf{r}) = \begin{cases} \rho^*(\mathbf{r}) & \text{if } \mathbf{r} \in \cap\{\text{Sup}(\mathbf{v}_i) \cap A | i \in \mu(\mathbf{r})\} \\ \mathbf{r} & \text{otherwise} \end{cases}$$

π is a well-defined bijection with $\pi[A] = A'$ and $\mu(\mathbf{r}) = \mu'(\mathbf{r}^\pi)$. It remains to establish:

Claim: For all $i < m$, $\mathbf{v}_i \cong_\mathbf{x} \mathbf{v}'_i$ via π.

For the **Proof** consider an arbitrary $i < m$. By the basic set-up of our considerations, we have $\pi_i(\mathbf{v}_i) = \mathbf{v}'_i$. If $\mathbf{v}_i$ does not contain in its support an element of A, then π and π_i coincide; if $\mathbf{v}_i$'s support contains an element of A that is possibly even in an overlap, the argument proceeds as follows. Notice first of all that *all* elements of $[\mathbf{r}]_\approx$ are in $\text{Sup}(\mathbf{v}_i)$ as soon as one $\mathbf{r} \in A$ is in $\text{Sup}(\mathbf{v}_i)$. Taking this into account, we have by definition of π and $\mathbf{v}_i \upharpoonright [\mathbf{r}]_\approx$: $\pi(\mathbf{v}_i \upharpoonright [\mathbf{r}]_\approx) = \rho^*(\mathbf{v}_i \upharpoonright [\mathbf{r}]_\approx)$.[46] The definition of ρ^* and the fact that $\rho^r(\mathbf{v}_i) = \mathbf{v}'_i$ allow us to infer that $\rho^*(\mathbf{v}_i \upharpoonright [\mathbf{r}]_\approx) = \mathbf{v}'_i \upharpoonright [\rho^*(\mathbf{r})]_{\approx'}$. As $\mu'(\rho^*(\mathbf{r})) = \mu'(\pi_i(\mathbf{r}))[=\mu(\mathbf{r})]$ we can extend this sequence of identities by $\mathbf{v}'_i \upharpoonright [\rho^*(\mathbf{r})]_{\approx'} = \mathbf{v}'_i \upharpoonright [\pi_i(\mathbf{r})]_{\approx'}$. Consequently, as $\pi_i(\mathbf{v}_i) = \mathbf{v}'_i$, we have $\mathbf{v}'_i \upharpoonright [\pi_i(\mathbf{r})]_{\approx'} = \pi_i(\mathbf{v}_i \upharpoonright [\mathbf{r}]_\approx)$. These considerations hold for all $\mathbf{r} \in \text{Sup}(\mathbf{v}_i) \cap A$; we can conclude $\pi(\mathbf{v}_i) = \pi_i(\mathbf{v}_i)$ and, with $\pi_i(\mathbf{v}_i) = \mathbf{v}'_i$, we have $\pi(\mathbf{v}_i) = \mathbf{v}'_i$.

This concludes, finally, the argument for the Theorem.

---

[46] $\upharpoonright$ is the pruning operation; it applies to an element $\mathbf{x}$ of HF and a subset $\mathbf{Y}$ of its support: $\mathbf{x} \upharpoonright \mathbf{Y}$ is the subtree of $\mathbf{x}$ that is built up exclusively from atoms in $\mathbf{Y}$. The $\in$-recursive definition is: $(\mathbf{x} \cap \mathbf{Y}) \cup [\{\mathbf{y} \upharpoonright (\mathbf{Y} \cap Tc(\mathbf{y})) | \mathbf{y} \in \mathbf{x}\} \setminus \{\emptyset\}]$. Cf. [Sieg and Byrnes, 1999a, 155–6].

5.4 Models

There is a rich variety of models, as the game of life, other cellular automata and many artificial neural nets are Gandy machines. Let me first sketch a set theoretic presentation of a Turing machine as a Turing computor and then, even more briefly, that of the Game of Life as a Gandy machine. Consider a Turing machine with symbols $s_0, \ldots, s_k$ and internal states $q_0, \ldots, q_m$; its program is given as a finite list of quadruples of the form $q_i s_j c_k q_m$, expressing that the machine is going to perform action c_k and change into internal state q_m, when scanning symbol s_j in state q_i. The tape is identified with a set of overlapping pairs

$$\mathbf{Tp} := \{\langle b,b \rangle, \langle b,c \rangle, \ldots, \langle d,e \rangle, \langle e,e \rangle\}$$

where $b, c, \ldots, d, e$ are distinct atoms; c is the leftmost square of the tape with a possibly non-blank symbol on it, d its rightmost one. The symbols are represented by $\underline{s}_j := \{r\}^{(j+1)}, 0 \leq j \leq k$; the internal states are given by $\underline{q}_j := \{r\}^{(k+1)+(j+1)}, 0 \leq j \leq l$. The tape content is given by

$$\mathbf{Ct} := \{\langle \underline{s}_{j_0}, c \rangle, \ldots, \langle \underline{s}_{j_r}, d \rangle\}$$

and, finally, the *id* is represented as the union of $\mathbf{Tp}$, $\mathbf{Ct}$, and $\{\langle \underline{q}_i, r \rangle\}$ with r being a square of $\mathbf{Tp}$. So the structural set $\mathbf{S}$ of states is obtained as the set of all ids closed under $\in$-isomorphisms. Stereotypes (for each program line given by $q_i s_j$) consist of parts like

$$\{\langle \underline{q}_i, r \rangle, \langle \underline{s}_j, r \rangle, \langle t, r \rangle, \langle r, u \rangle\};$$

these are the causal neighborhoods on which $\mathbf{G}$ operates. Consider the program line $q_i s_j s_k q_l$ (print s_k); applied to the above causal neighborhood $\mathbf{G}$ yields

$$\{\langle \underline{q}_l, r \rangle, \langle \underline{s}_k, r \rangle, \langle t, r \rangle, \langle r, u \rangle\}.$$

For the program line $q_i s_j R q_l$ (move Right) two cases have to be distinguished. In the first case, when r is not the rightmost square, $\mathbf{G}$ yields

$$\{\langle \underline{q}_l, u \rangle, \langle \underline{s}_j, r \rangle, \langle t, r \rangle, \langle r, u \rangle\};$$

in the second case, when r is the rightmost square, $\mathbf{G}$ yields

$$\{\langle \underline{q}_l, * \rangle, \langle \underline{s}_j, r \rangle, \langle \underline{s}_0, * \rangle, \langle t, r \rangle, \langle r, * \rangle, \langle *, u \rangle\};$$

where $*$ is a new atom. The program line $q_i s_j L q_l$ (move Left) is treated similarly. It is easy to verify that a Turing machine presented in this way is a Turing Computor.

Cellular automata introduced by Ulam and von Neumann operate in parallel; a particular cellular automaton was made popular by Conway, the Game of Life. A cellular automaton is made up of many identical cells. Typically, each cell is located on a regular grid in the plane and carries one of two possible values. After each time unit its values are updated according to a simple rule that depends on its own previous value and the previous values of the neighboring cells. Cellular automata of this sort can simulate universal Turing machines, but they also yield discrete simulations of very general and complex physical processes.

Gandy considered playing Conway's Game of Life as a paradigmatic case of parallel computing. It is being played on subsets of the plane, more precisely, subsets that are constituted by finitely many connected squares. For reasons that will be obvious in a moment, the squares are also called *internal cells*; they can be in two states, *dead* or *alive*. In my presentation the internal cells are surrounded by one layer of *border cells*; the latter, in turn, by an additional layer of *virtual cells*. Border and virtual cells are dead by convention. Internal cells and border cells are jointly called *real*. The layering ensures that each real cell is surrounded by a full set of eight neighboring cells. For real cells the game is played according to the rules:

1. living cells with 0 or 1 (living) neighbor die (from isolation);

2. living cells with 4 or more (living) neighbors die (from overcrowding);

3. dead cells with exactly 3 (living) neighbors become alive.

4. In all other cases the cell's state is unchanged.

A real cell a with neighbors $a_1, \ldots, a_8$ and state $s(a)$ is given by

$$\{a, s(a), \langle a_1, \ldots, a_8 \rangle\}.$$

The neighbors are given in "canonical" order starting with the square in the leftmost top corner and proceeding clockwise; $s(a)$ is $\{a\}$ in case a is alive, otherwise $\{\{a\}\}$. The $\mathbf{T}_1$-causal neighborhoods of real cells are of the form

$$\{\{a, s(a), \langle a_1, \ldots, a_8 \rangle\}, \{a_1, s(a_1)\}, \ldots, \{a_8, s(a_8)\}\}.$$

It is obvious how to define the structural operation $\mathbf{G}_1$ on the causal neighborhoods of internal cells; the case of border cells requires attention. There is a big number of stereotypes that have to be treated, so I will discuss only one simple case that should, nevertheless, bring out the principled considerations. In the following diagram we start out with the cells that have letters assigned to them; the diagram should be thought of extending at the left and at the bottom. The v's indicate virtual cells, the b's border cells, the $\{a\}$'s internal cells that are alive, and the $*$'s new atoms that are added in the next step of the computation. Let's see how that comes about.

Consider the darkly shaded square b_3 with its neighbors, i.e., its presentation

$$\{b_3, \{\{b_3\}\}, \langle v_2, \ldots, b_2 \rangle\};$$

applying $\mathbf{G}_1$ to its causal neighborhood yields

$*_0$	$*_1$	$*_2$	$*_3$	$*_4$	$*_5$	$*_6$	$*_7$		
v_0	v_1	v_2	v_3	v_4	v_5	v_6	v_7	v_8	v_9
b_0	b_1	b_2	b_3	b_4	b_5	b_6	b_7	b_8	v_{10}
$\{a_0\}$	$\{a_1\}$	$\{a_2\}$	$\{a_3\}$	$\{a_4\}$	$\{a_5\}$	$\{a_6\}$	$\{a_7\}$	b_9	v_{11}

$$\{\{b_3, \{b_3\}, \langle v_2, \ldots, b_2 \rangle\}, \{v_3, \{\{v_3\}\}, \langle *_2, *_3, *_4, v_4, b_4, b_3, b_2, v_2 \rangle\}\},$$

where $*_2, *_3$, and $*_4$ are new atoms (and v_3 has been turned from a virtual cell into a real one, namely a border cell). Here the second set of stereotypes and the second structural operation come in to ensure that the new squares introduced by applying $\mathbf{G}_1$ to "adjacent" border cells (whose neighborhoods overlap with the neighborhood of b_3) are properly identified in the next state. Consider as the appropriate $\mathbf{T}_2$-causal neighborhood the set consisting of the $\mathbf{T}_1$-causal neighborhoods of b_2, b_3, and b_4; $\mathbf{G}_2$ applied to it yields the set with presentations of the cells v_2, v_3, and v_4.

5.5 Tieferlegung

The above considerations constitute the mathematical core of this section. They lead to the conclusion that computability, when relativized to a particular kind of computing agent or device, has a perfectly standard methodological status: no thesis is needed, but rather the recognition that the axiomatic characterization is correct for the intended computing device. The recognition that the notions do not go beyond Turing computability is then an important mathematical fact. It seems to me that we have gained in Hilbert's broad terms a deepening of the foundations via the axiomatic method, a *Tieferlegung der Fundamente*. As I mentioned earlier, Gödel advocated such an approach in a conversation with Church in early 1934 and suggested "to state a set of axioms which would embody the generally accepted properties of this notion (i.e., effective calculability), and to do something on that basis."

The sharpened version of Turing's work and a thorough-going re-interpretation of Gandy's approach allow us to fill in the blanks of Gödel's suggestion; this resolves in my view the methodological issue raised at the end of section 4. Perhaps the remarks in the 1964 Postscriptum to the Princeton Lectures of 1934 echo his earlier considerations. "Turing's work gives," according to Gödel, "an analysis of the concept of 'mechanical procedure'.... This concept is shown to be equivalent with that of a 'Turing machine'." The work, on which I reported, substantiates these remarks in the following sense: it provides an axiomatic analysis of the concept "mechanical procedure" and shows that this concept is computationally equivalent to that of a Turing machine. Indeed, it does so for two such concepts, namely, when the computing agents are computors or discrete machines; and it does so by imposing constraints on the computations these agents carry out in steps. The natural and well-motivated constraints guarantee the effectiveness of the steps in the most direct way.

The axiomatic approach captures the essential nature of computation processes in an abstract way. The difference between the two types of calculators I have been describing is reduced to the fact that Turing computors modify *one* bounded part of a state, whereas Gandy machines operate in parallel on *arbitrarily many* bounded parts. The axiomatic conditions arise from underlying analyses that lead to a particular structural view. Of course, an appeal to some informal understanding can no more be avoided in this case than in any other case of an axiomatically characterized mathematical structure intended to model broad aspects of physical or intellectual reality. The general point is this: we don't have to face anything especially mysterious for the concept of calculability; rather, we have to face the ordinary issues for the adequacy of mathematical concepts and they are, of course, non-trivial.

I have been distinguishing in other writings two aspects of mathematical experience. The first, the *quasi-constructive* aspect, has to do with the recognition of laws for accessible domains; this includes, in particular, our recognition of the correctness of the Zermelo Fraenkel axioms in set theory and their extendibility by suitable axioms of infinity. The second, the *conceptional* aspect, deals with the uncovering of abstract, axiomatically characterized notions. These abstract notions are distilled from mathematical practice for the purpose of comprehending complex connections, of making analogies precise and of obtaining a more profound understanding. Bourbaki in their [1950] expressed matters quite in Dedekind and Hilbert's spirit, when claiming that the axiomatic method teaches us

> to look for the deep-lying reasons for such a discovery [that two or several quite distinct theories lend each other "unexpected support"], to find the common ideas of these theories, ... to bring these ideas forward and to put them in their proper light. (p. 223)

Notions like group, field, topological space and differentiable manifold are abstract in this sense. Turing's analysis shows, when properly generalized, that computability exemplifies the second aspect of mathematical experience. Although Gödel used "abstract" in a more inclusive way than I do here his broad claim is pertinent also for computability, namely, "that we understand abstract terms more and more precisely as we go on using them, and that more and more abstract terms enter the sphere of our understanding." [1972, 306]

6 OUTLOOK ON MACHINES AND MIND

Turing's notion of human computability is exactly right not only for obtaining a negative solution of the Entscheidungsproblem that is conclusive, but also for achieving a precise characterization of formal systems that is needed for the general formulation of Gödel's incompleteness theorems. I argued in sections 1 and 2 that the specific intellectual context reaches back to Leibniz and requires us to focus attention on effective, indeed mechanical procedures; these procedures are to be carried out by computors without invoking higher cognitive capacities. The axioms of section 5.1 are intended for this informal concept. The question whether there are strictly broader notions of effectiveness has of course been asked for both cognitive and physical processes. I am going to address this question not in any general and comprehensive way, but rather by focusing on one central issue: the discussion might be viewed as a congenial dialogue between Gödel and Turing on aspects of mathematical reasoning that transcend mechanical procedures.

I'll start in section 6.1 by returning more fully to Gödel's view on mechanical computability as articulated in his [193?]. There he drew a dramatic conclusion from the undecidability of certain Diophantine propositions, namely, that mathematicians cannot be replaced by machines. That theme is taken up in the Gibbs Lecture of 1951 where Gödel argues in greater detail that the human mind infinitely surpasses the powers of any finite machine; an analysis of the argument is presented in section 6.2 under the heading *Beyond calculation*. Section 6.3 is entitled *Beyond discipline* and gives Turing's perspective on intelligent machinery; it is devoted to the seemingly sharp conflict between Gödel's and Turing's views on mind. Their deeper disagreement really concerns the nature of machines, and I'll end with some brief remarks on (supra-) mechanical devices in section 6.4.

6.1 Mechanical computability

In section 4.2 I alluded briefly to the unpublished and untitled draft for a lecture Gödel presumably never delivered; it was written in the late 1930s. Here one finds the earliest extensive discussion of Turing and the reason why Gödel, at the time, thought Turing had established "beyond any doubt" that "this really is the correct definition of mechanical computability". Obviously, we have to clarify what "this" refers to, but first I want to give some of the surrounding context. Already in his [1933] Gödel elucidated, as others had done before him, the mechanical feature of effective procedures by pointing to the possibility that machines carry them out. When insisting that the inference rules of precisely described proof methods have to be "purely formal" he explains:

> [The inference rules] refer only to the outward structure of the formulas, not to their meaning, so that they could be applied by someone who knew nothing about mathematics, or by a machine. This has the consequence that there can never be any doubt as to what cases the rules of inference apply to, and thus the highest possible degree of exactness is obtained. [*Collected Works III*, p. 45]

During the spring term of 1939 Gödel gave an introductory logic course at Notre Dame. The logical decision problem is informally discussed and seen in the historical context of Leibniz's "Calculemus".[47] Before arguing that results of modern logic prevent the realization of Leibniz's project, Gödel asserts that the rules of logic can be applied in a "purely mechanical" way and that it is therefore possible "to construct a machine which would do the following thing":

> The supposed machine is to have a crank and whenever you turn the crank once around the machine would write down a tautology of the calculus of predicates and it would write down every existing tautology ... if you turn the crank sufficiently often. So this machine would really replace thinking completely as far as deriving of formulas of the calculus of predicates is concerned. It would be a thinking machine in the literal sense of the word. For the calculus of propositions you can do even more. You could construct a machine in form of a typewriter such that if you type down a formula of the calculus of propositions then the machine would ring a bell [if the formula is a tautology] and if it is not it would not. You could do the same thing for the calculus of monadic predicates.

Having formulated these positive results Gödel points out that "it is impossible to construct a machine which would do the same thing for the whole calculus of predicates". Drawing on the undecidability of predicate logic established by Church and Turing, he continues with a striking claim:

> So here already one can prove that Leibnitzens [sic!] program of the "calculemus" cannot be carried through, i.e. one knows that the human mind will never be able to be replaced by a machine already for this comparatively simple question to decide whether a formula is a tautology or not.

I mention these matters to indicate the fascination Gödel had with the mechanical realization of logical procedures, but also his *penchant* for overly dramatic formulations concerning the human mind. He takes obviously for granted here that a mathematically satisfactory definition of mechanical procedures has been given.

Such a definition, Gödel insists in [193?, 166], is provided by the work of Herbrand, Church and Turing. In that manuscript he examines the relation between mechanical computability, general recursiveness and machine computability. This is of special interest, as we will see that his methodological perspective here is quite different from his later standpoint. He gives, on pp. 167–8, a perspicuous presentation of the equational calculus that is "essentially Herbrand's" and defines general recursive functions. He claims outright that it provides "the correct definition of a computable function". Then he asserts, "That this really is the correct definition of mechanical computability was established beyond any doubt by Turing." Here the referent for "this" has finally been revealed: it is the definition of general recursive functions. How did Turing establish that this is also the correct definition of *mechanical* computability? Gödel's answer is as follows:

---

[47]This is [Gödel 1939]. As to the character of these lectures, see [Dawson], p. 135.

> He [Turing] has shown that the computable functions defined in this way [via the equational calculus] are exactly those for which you can construct a machine with a finite number of parts which will do the following thing. If you write down any number $n_1, \ldots, n_r$ on a slip of paper and put the slip of paper into the machine and turn the crank, then after a finite number of turns the machine will stop and the value of the function for the argument $n_1, \ldots, n_r$ will be printed on the paper. [*Collected Works III*, p. 168]

The implicit claim is clearly that a procedure is mechanical just in case it is executable by a machine with a finite number of parts. There is no indication of the structure of such machines except for the insistence that they have only finitely many parts, whereas Turing machines are of course potentially infinite due to the expanding tape.

The literal reading of the argument for the claim "this really is the correct definition of mechanical computability was established beyond any doubt by Turing" amounts to this. The equational calculus characterizes the computations of number-theoretic functions and provides thus "the correct definition of computable function". That the class of computable functions is co-extensional with that of *mechanically* computable ones is then guaranteed by "Turing's proof" of the equivalence between general recursiveness and machine computability.[48] Consequently, the definition of general recursive functions via the equational calculus characterizes correctly the mechanically computable functions. Without any explicit reason for the first step in this argument, it can only be viewed as a direct appeal to Church's Thesis.

If we go beyond the literal reading and think through the argument in parallel to Turing's analysis in his [1936], then we can interpret matters as follows. Turing considers arithmetic calculations done by a computor. He argues that they involve only very elementary processes; these processes can be carried out by a Turing machine operating on strings of symbols. Gödel, this interpretation maintains, also considers arithmetic calculations done by a computor; these calculations can be reduced to computations in the equational calculus. This first step is taken in parallel by Gödel and Turing and is based on a conceptual analysis; cf. the next paragraph. The second step connects calculations of a computor to computations of a Turing machine. This connection is established by mathematical arguments: Turing simply states that machines operating on finite strings can be proved to be equivalent to machines operating on individual symbols, i.e., to ordinary Turing machines; Gödel appeals to "Turing's proof" of the fact that general recursiveness and machine computability are equivalent.

Notice that in Gödel's way of thinking about matters at this juncture, the *mathematical theorem* stating the equivalence of general recursiveness and machine

---

[48] In Turing's [1936] general recursive functions are not mentioned. Turing established in an Appendix to his paper the equivalence of his notion with λ-definability. As Church and Kleene had already proved the equivalence of λ-definability and general recursiveness, "Turing's Theorem" is thus established for Turing computability.

computability plays the pivotal role: It is not Turing's analysis that is appealed to by Gödel but rather "Turing's proof". The central analytic claim my interpretation attributes to Gödel is hardly argued for. On p. 13 Gödel just asserts, "... by analyzing in which manner this calculation [of the values of a general recursive function] proceeds you will find that it makes use only of the two following rules." The two rules as formulated here allow substituting numerals for variables and equals for equals. So, in some sense, Gödel seems to think that the rules of the equational calculus provide a way of "canonically" representing steps in calculations and, in addition, that his characterization of recursion is the most general one.[49] The latter is imposed by the requirement that function values have to be calculated, as pointed out in [1934, 369 top]; the former is emphasized much later in a letter to van Heijenoort of April 23, 1963, where Gödel distinguishes his definition from Herbrand's. His definition, Gödel asserts, brought out clearly what Herbrand had failed to see, namely "that the computation (for all computable functions) proceeds by exactly the same rules". [*Collected Works V*, p. 308] By contrast, Turing shifts from arithmetically meaningful steps to symbolic processes that underlie them and can be taken to satisfy restrictive boundedness as well as locality conditions. These conditions cannot be imposed directly on arithmetic steps and are certainly not satisfied by computations in the equational calculus. So, we are back precisely at the point of the discussion in section 3.

6.2 Beyond calculation

In [193?] Gödel begins the discussion by reference to Hilbert's "famous words" that "for any precisely formulated mathematical question a unique answer can be found". He takes these words to mean that for any mathematical proposition A there is a proof of either A or not-A, "where by 'proof' is meant something which starts from evident axioms and proceeds by evident inferences". He argues that the incompleteness theorems show that something is lost when one takes the step from this notion of proof to a formalized one: "... it is not possible to formalise mathematical evidence even in the domain of number theory, but the conviction about which Hilbert speaks remains entirely untouched. Another way of putting the result is this: it is not possible to mechanise mathematical reasoning; ..." Then he continues, in a way that is similar to the striking remark in the Notre Dame Lectures, "i.e., it will never be possible to replace the mathematician by a machine, even if you confine yourself to number-theoretic problems." (pp. 164–5)

The succinct argument for this conclusion is refined in the Gibbs Lecture of 1951. In the second and longer part of the lecture, Gödel gave the most sustained defense of his Platonist standpoint drawing the "philosophical implications" of the situation presented by the incompleteness theorems.[50] "Of course," he says

---

[49]This is obviously in contrast to the view he had in 1934 when defining general recursive functions; cf. section 3.2.

[50]That standpoint is formulated at the very end of the lecture as follows: p. 38 (CW III, 322/3): "Thereby [i.e., the Platonistic view] I mean the view that mathematics describes a non-

polemically, "in consequence of the undeveloped state of philosophy in our days, you must not expect these inferences to be drawn with mathematical rigor." The mathematical aspect of the situation, he claims, can be described rigorously; it is formulated as a disjunction, "Either mathematics is incompletable in this sense, that its evident axioms can never be comprised in a finite rule, that is to say, the human mind (even within the realm of pure mathematics) infinitely surpasses the powers of any finite machine, or else there exist absolutely unsolvable Diophantine problems of the type specified ..." Gödel insists that this fact is both "mathematically established" and of "great philosophical interest". He presents on pages 11–13 an argument for the disjunction and considers its conclusion as "inevitable".

The disjunction is called in footnote 15 a theorem that holds for finitists and intuitionists as an implication. Here is the appropriate implication: If the evident axioms of mathematics can be comprised in a finite rule, then there exist absolutely unsolvable Diophantine problems. Let us establish this implication by adapting Gödel's considerations for the disjunctive conclusion; the argument is brief. Assume the axioms that are evident for the human mind can be comprised in a finite rule "that is to say", for Gödel, a Turing machine can list them. Thus there exists a mechanical rule producing all the evident axioms for "subjective" mathematics, which is by definition the system of all humanly demonstrable mathematical propositions.[51] On pain of contradiction with the second incompleteness theorem, the human mind cannot prove the consistency of subjective mathematics. (This step is of course justified only if the inferential apparatus for subjective mathematics is given by a mechanical rule, and if subjective mathematics satisfies all the other conditions for the applicability of the second theorem.) Consequently, the Diophantine problem corresponding to the consistency statement cannot be proved either in subjective mathematics. That justifies Gödel's broader claim that it is undecidable "not just within some particular axiomatic system, but by *any* mathematical proof the human mind can conceive". (p. 13) In this sense the problem is *absolutely undecidable* for the human mind. So it seems that we have established the implication. However, the very first step in this argument, indicated by "that is to say", appeals to the precise concept of "finite procedure" as analyzed by Turing. Why is "that is to say" justified for Gödel? To answer this question, I examine Gödel's earlier remarks about finite procedures and finite machines.[52]

---

sensual reality, which exists independently both of the acts and [[of]] the dispositions of the human mind and is only perceived, and probably perceived very incompletely, by the human mind."

[51] This is in contrast to the case of "objective" mathematics, the system of all true mathematical propositions, for which one cannot have a "well-defined system of correct axioms" (given by a finite rule) that comprises all of it. In [Wang, 1974, 324–6], Gödel's position on these issues is (uncritically) discussed. The disjunction is presented as one of "two most interesting rigorously proved results about minds and machines" and is formulated as follows: "Either the human mind surpasses all machines (to be more precise: it can decide more number theoretic questions than any machine) or else there exist number theoretical questions undecidable for the human mind."

[52] Boolos' *Introductory Note* to the Gibbs Lecture, in particular section 3, gives a different perspective on difficulties in the argument.

Gödel stresses in the first paragraph of the Gibbs Lecture that the incompleteness theorems have taken on "a much more satisfactory form than they had had originally". The greatest improvement was made possible, he underlines, "through the precise definition of the concept of finite procedure, which plays a decisive role in these results". Though there are a number of different ways of arriving at such a definition which all lead to "exactly the same concept", the most satisfactory way is that taken by Turing when "reducing the concept of a finite procedure to that of a machine with a finite number of parts". Gödel does not indicate the character of, or an argument for, the reduction of finite procedures to procedures effected by a machine with a finite number of parts, but he states explicitly that he takes finite machine "in the precise sense" of a Turing machine. (p. 9) This reduction is pivotal for establishing the central implication rigorously, and it is thus crucial to understand and grasp its *mathematical* character. How else can we assent to the claim that the implication has been established mathematically as a theorem? In his [1964] Gödel expressed matters quite differently (and we discussed that later Gödelian perspective extensively in section 4): there he asserts that Turing in [1936] gave an analysis of mechanical procedures and showed that the analyzed concept is equivalent to that of a Turing machine. The claimed equivalence is viewed as central for obtaining "a precise and unquestionably adequate definition of the general concept of formal system" and for supporting, I would like to add in the current context, the mathematical cogency of the argument for the implication.

Gödel neither proved the mathematical conclusiveness of the reduction nor the correctness of the equivalence. So let us assume, for the sake of the argument, that the implication has been mathematically established and see what conclusions of great philosophical interest can be drawn. There is, as a first background assumption, Gödel's deeply rationalist and optimistic perspective that denies the consequent of the implication. That perspective, shared with Hilbert as we saw in section 6.1, was articulated in [193?], and it was still taken in the early 1970s. Wang reports in [1974, 324-5], that Gödel agreed with Hilbert in rejecting the possibility that there are number-theoretic problems undecidable for the human mind. Our task is then to follow the path of Gödel's reflections on the first alternative of his disjunction or the negated antecedent of our implication. That assertion states: There is no finite machine (i.e. no Turing machine) that lists all the axioms of mathematics which are evident to the human mind. Gödel argues for two related conclusions: i) the working of the human mind is not reducible to operations of the brain, and ii) the human mind infinitely surpasses the powers of any finite machine.[53]

A second background assumption is introduced to obtain the first conclusion: The brain, "to all appearances", is "a finite machine with a finite number of parts,

---

[53] This does not follow just from the fact that for every Turing machine that lists evident axioms there is another axiom evident to the human mind not included in the list. Turing had tried already in his 1939 paper, *Ordinal Logics*, to overcome the incompleteness results by strengthening theories systematically. He added consistency statements (or reflection principles) and iterated this step along constructive ordinals; Feferman perfected that line of investigation, cf. his [1988]. Such a procedure was also envisioned in [Gödel, 1946, 1-2].

namely, the neurons and their connections". (p. 15) As finite machines are taken to be Turing machines, brains are consequently also considered as Turing machines. That is reiterated in [Wang, 1974, 326], where Gödel views it as very likely that "The brain functions basically like a digital computer." Together with the above assertion this allows Gödel to conclude in the Gibbs Lecture, "the working of the human mind cannot be reduced to the working of the brain".[54] In [Wang] it is taken to be in conflict with the commonly accepted view, "There is no mind separate from matter." That view is for Gödel a "prejudice of our time, which will be disproved scientifically (perhaps by the fact that there aren't enough nerve cells to perform the observable operations of the mind)". Gödel uses the notion of a finite machine in an extremely general way when considering the brain as a finite machine with a finite number of parts. It is here that the identification of finite machines with Turing machines becomes evidently problematic: Is it at all plausible to think that the brain has a similarly fixed structure and fixed program as a particular Turing machine? The argumentation is problematic also on different grounds; namely, Gödel takes "human mind" in a more general way than just the mind of any one individual human being. Why should it be then that mind is realized through any particular brain?

The proposition that the working of the human mind cannot be reduced to the working of the brain is thus not obtained as a "direct" consequence of the incompleteness theorems, but requires additional substantive assumptions: i) there are no Diophantine problems the human mind cannot solve, ii) brains are finite machines with finitely many parts, and iii) finite machines with finitely many parts are Turing machines. None of these assumptions is uncontroversial; what seems not to be controversial, however, is Gödel's more open formulation in [193?] that it is not possible to mechanize mathematical reasoning. That raises immediately the question, what aspects of mathematical reasoning or experience defy formalization? In his note [1974] that was published in [Wang, 325–6], Gödel points to two "vaguely defined" processes that may lead to systematic and effective, but non-mechanical procedures, namely, the process of defining recursive well-orderings of integers for larger and larger ordinals of the second number class and that of formulating stronger and stronger axioms of infinity. The point was reiterated in a modified formulation [Gödel, 1972.3] that was published only later in *Collected Works II*, p. 306. The [1972.3] formulation of this note is preceded by [1972.2], where Gödel gives *Another version of the first undecidability theorem* that involves number theoretic problems of Goldbach type. This version of the theorem may be taken, Gödel states, "as an indication for the existence of mathematical yes or no questions undecidable for the human mind". (p. 305) However, he points to a *fact* that "weighs against this interpretation", namely, that "there *do* exist unexplored series of axioms which are analytic in the sense that they only explicate the concepts occurring in them". As an example he points also here to axioms of infinity, "which only explicate the content of the general concept of set". (p. 306) If the existence of such effective, non-mechanical procedures is taken as a fact or, more

---

[54]Cf. also note 13 of the Gibbs Lecture and the remark on p. 17.

cautiously, as a third background assumption, then Gödel's second conclusion is established: The human mind, indeed, infinitely surpasses the power of any finite machine.

Though Gödel calls the existence of an "unexplored series" of axioms of infinity a *fact*, he also views it as a "vaguely defined" procedure and emphasizes that it requires further mathematical experience; after all, its formulation can be given only once set theory has been developed "to a considerable extent". In the note [1972.3] Gödel suggests that the process of forming stronger and stronger axioms of infinity does not yet form a "well-defined procedure which could actually be carried out (and would yield a non-recursive number-theoretic function)": it would require "a substantial advance in our understanding of the basic concepts of mathematics". In the note [1974], Gödel offers a *prima facie* startlingly different reason for not yet having a precise definition of such a procedure: it "would require a substantial deepening of our understanding of the basic operations of the mind". (p. 325)

Gödel's *Remarks before the Princeton bicentennial conference* in 1946 throw some light on this seeming tension. Gödel discusses there not only the role axioms of infinity might play in possibly obtaining an absolute concept of demonstrability, but he also explores the possibility of an absolute mathematical "definition of definability". What is most interesting for our considerations here is the fact that he considers a restricted concept of *human definability* that would reflect a human capacity, namely, "comprehensibility by our mind". That concept should satisfy, he thinks, the "postulate of denumerability" and in particular allow us to define (in this particular sense) only countably many sets. "For it has some plausibility that all things conceivable by us are denumerable, even if you disregard the question of expressibility in some language." (p. 3) That requirement, together with the related difficulty of the definability of the least indefinable ordinal, does not make such a concept of definability "impossible, but only [means] that it would involve some extramathematical element concerning the psychology of the being who deals with mathematics." Obviously, Turing brought to bear on his definition of computability, most fruitfully, an extramathematical feature of the psychology of a human computor.[55] Gödel viewed that definition in [1946], the reader may recall, as the first "absolute definition of an interesting epistemological notion". (p. 1) His reflections on the possibility of absolute definitions of demonstrability and definability were encouraged by the success in the case of computability. Can we obtain by a detailed study of *actual* mathematical experience a deeper "understanding of the basic operations of the mind" and thus make also a "substantial advance in our understanding of the basic concepts of mathematics"?

6.3 Beyond discipline

Gödel's brief exploration in [1972.3] of the issue of defining a non-mechanical, but effective procedure is preceded by a severe critique of Turing. The critical attitude is indicated already by the descriptive and harshly judging title of the note, *A*

---

[55] Cf. Parsons' informative remarks in the *Introductory Note* to [Gödel, 1946, 148].

philosophical error in Turing's work. The discussion of Church's thesis and Turing's analysis is in general fraught with controversy and misunderstanding, and the controversy begins often with a dispute over what the intended informal concept is. When Gödel spotted a philosophical error in Turing's work, he *assumed* that Turing's argument in the 1936 paper was to show that "mental procedures cannot go beyond mechanical procedures". He considered the argument as inconclusive:

> What Turing disregards completely is the fact that *mind, in its use, is not static, but constantly developing*, i.e., that we understand abstract terms more and more precisely as we go on using them, and that more and more abstract terms enter the sphere of our understanding. [*Collected Works II*, p. 306]

Turing did not give a conclusive argument for Gödel's claim, but then it has to be added that he did not intend to argue for it. Simply carrying out a mechanical procedure does not, indeed, should not involve an expansion of our understanding. Turing viewed the restricted use of mind in computations undoubtedly as static; after all, it seems that this feature contributed to the good reasons for replacing states of mind of the human computor by "more definite physical counterparts" in section 9, part III, of his classical paper.

Even in his work of the late 1940s and early 1950s that deals explicitly with mental processes, Turing does not argue that mental procedures cannot go beyond mechanical procedures. Mechanical processes are, as a matter of fact, still made precise as Turing machine computations; machines that might exhibit intelligence have, in contrast, a more complex structure than Turing machines. Conceptual idealization and empirical adequacy are now being sought for quite different purposes, and Turing is trying to capture clearly what Gödel found missing in his analysis for a broader concept of humanly effective calculability, namely, "... that mind, in its use, is not static, but constantly developing".[56] Gödel continued the above remark in this way:

> There may exist systematic methods of actualizing this development, which could form part of the procedure. Therefore, although at each stage the number and precision of the abstract terms at our disposal

---

[56][Gödel, 1972.3] may be viewed, Gödel mentions, as a note to the word "mathematics" in the sentence, "Note that the results mentioned in this postscript do not establish any bounds of the powers of human reason, but rather for the potentialities of pure formalism in mathematics." This sentence appears in the 1964 Postscriptum to the Princeton Lectures Gödel gave in 1934; *Collected Works I*, pp. 369–371. He states in that Postscriptum also that there may be "finite non-mechanical procedures" and emphasizes, as he does in many other contexts, that such procedures would "involve the use of abstract terms on the basis of their meaning". (Note 36 on p. 370 of *Collected Works I*) Other contexts are found in volume III of the *Collected Works*, for example, the Gibbs Lecture (p. 318 and note 27 on that very page) and a related passage in "Is mathematics syntax of language?" (p. 344 and note 24) These are systematically connected to Gödel's reflections surrounding (the translation of) his Dialectica paper [1958] and [1972]. A thorough discussion of these issues cannot be given here; but as to my perspective on the basic difficulties, see the discussion in section 4 of my paper "Beyond Hilbert's Reach?".

may be *finite*, both (and, therefore, also Turing's number of *distinguishable states of mind*) may *converge toward infinity* in the course of the application of the procedure.

The particular procedure mentioned as a plausible candidate for satisfying this description is the process of forming stronger and stronger axioms of infinity. We saw that the two notes, [1972-3] and [1974], are very closely connected. However, there is one subtle and yet substantive difference. In [1974] the claim that the number of possible states of mind may converge to infinity is obtained as a consequence of the dynamic development of mind. That claim is then followed by a remark that begins, in a superficially similar way, as the first sentence in the above quotation:

> Now there may exist systematic methods of accelerating, specializing, and uniquely determining this development, e.g. by asking the right questions on the basis of a mechanical procedure.

Clearly, I don't have a full understanding of these enigmatic observations, but there are three aspects that are clear enough. First, mathematical experience has to be invoked when asking the right questions; second, aspects of that experience may be codified in a mechanical procedure and serve as the basis for the right questions; third, the answers may involve abstract terms that are incorporated into the non-mechanical mental procedure.

We should not dismiss or disregard Gödel's methodological remark that "asking the right questions on the basis of a mechanical procedure" may be part of a systematic method to push forward the development of mind. It allows us, even on the basis of a very limited understanding, to relate Gödel's reflections tenuously with Turing's proposal for investigating matters. Prima facie their perspectives are radically different, as Gödel proceeds by philosophical argument and broad, speculative appeal to mathematical experience, whereas Turing suggests attacking the problem largely by computational experimentation. That standard view of the situation is quite incomplete. In his paper *Intelligent machinery* written about ten years after [1939], Turing states what is really the central problem of cognitive psychology:

> If the untrained infant's mind is to become an intelligent one, it must acquire both discipline and initiative. So far we have been considering only discipline [via the universal machine, W.S.]. ... But discipline is certainly not enough in itself to produce intelligence. That which is required in addition we call initiative. This statement will have to serve as a definition. Our task is to discover the nature of this residue as it occurs in man, and to try and copy it in machines. (p. 21)

How can we transcend discipline? A hint is provided in Turing's 1939 paper, where he distinguishes between ingenuity and intuition. He observes that in formal logics their respective roles take on a greater definiteness. Intuition is used for "setting down formal rules for inferences which are always intuitively valid", whereas ingenuity is to "determine which steps are the more profitable for the purpose of proving a particular proposition". He notes:

> In pre-Gödel times it was thought by some that it would be possible to carry this programme to such a point that all the intuitive judgements of mathematics could be replaced by a finite number of these rules. The necessity for intuition would then be entirely eliminated. (p. 209)

The distinction between ingenuity and intuition, but also the explicit link of intuition to incompleteness, provides an entry to exploit through concrete computational work the "parallelism" of Turing's and Gödel's considerations. Copying the residue in machines is the task at hand. It is extremely difficult in the case of mathematical thinking, and Gödel would argue it is an impossible one, if machines are Turing machines. Turing would agree. Before we can start copying, we have to discover at least partially the nature of the residue, with an emphasis on "partially", through some restricted proposals for finding proofs in mathematics. Let us look briefly at the broad setting.

Proofs in a formal logic can be obtained uniformly by a patient search through an enumeration of all theorems, but additional intuitive steps remain necessary because of the incompleteness theorems. Turing suggested particular intuitive steps in his ordinal logics; his arguments are theoretical, but connect directly to the discussion of actual or projected computing devices that appears in his *Lecture to London Mathematical Society* and in *Intelligent Machinery*. In these papers he calls for intellectual searches (i.e., heuristically guided searches) and initiative (that includes, in the context of mathematics, proposing new intuitive steps). However, he emphasizes [1947, 122]:

> As regards mathematical philosophy, since the machines will be doing more and more mathematics themselves, the centre of gravity of the human interest will be driven further and further into philosophical questions of what can in principle be done etc.

Gödel and Turing, it seems, could have cooperated on the philosophical questions of what can in principle be done. They also could have agreed, so to speak terminologically, that there is a human mind whose working is not reducible to the working of any particular brain. Towards the end of *Intelligent Machinery* Turing emphasizes, "the isolated man does not develop any intellectual power", and argues:

> It is necessary for him to be immersed in an environment of other men, whose techniques he absorbs during the first twenty years of his life. He may then perhaps do a little research of his own and make a very few discoveries which are passed on to other men. From this point of view the search for new techniques must be regarded as carried out by the human community as a whole, rather than by individuals.

Turing calls this, appropriately enough, a *cultural search* and contrasts it with more limited, *intellectual searches*. Such searches, Turing says definitionally, can be carried out by individual brains. In the case of mathematics they would include searches through all proofs and would be at the center of "research into intelligence of machinery". Turing had high expectations for machines' progress in mathematics; indeed, he was unreasonably optimistic about their emerging capacities. Even now it is a real difficulty to have machines do mathematics on their own: work on Gödel's "theoretical" questions has to be complemented by sustained efforts to meet Turing's "practical" challenge. I take this to be one of the ultimate motivations for having machines find proofs in mathematics, i.e., proofs that reflect logical as well as mathematical understanding.

When focusing on proof search in mathematics it may be possible to use and expand logical work, but also draw on experience of actual mathematical practice. I distinguish two important features of the latter: i) the refined conceptual organization internal to a given part of mathematics, and ii) the introduction of new abstract concepts that cut across different areas of mathematics.[57] Logical formality per se does not facilitate the finding of arguments from given assumptions to a particular conclusion. However, strategic considerations can be formulated (for natural deduction calculi) and help to bridge the gap between assumptions and conclusion, suggesting at least a very rough structure of arguments. These logical structures depend solely on the syntactic form of assumptions and conclusion; they provide a seemingly modest, but in fact very important starting-point for strategies that promote automated proof search in mathematics.

Here is a pregnant general statement that appeals primarily to the first feature of mathematical practice mentioned above: *Proofs provide explanations of what they prove by putting their conclusion in a context that shows them to be correct.*[58] The deductive organization of parts of mathematics is the classical methodology for specifying such contexts. "Leading mathematical ideas" have to be found, proofs have to be planned: I take this to be the axiomatic method turned dynamic and local.[59] This requires undoubtedly the introduction of heuristics that reflect a deep understanding of the underlying mathematical subject matter. The broad and operationally significant claim is, that we have succeeded in isolating the leading ideas for a part of mathematics, if that part can be developed by machine — automatically, efficiently, and in a way that is furthermore easily accessible to human mathematicians.[60] This feature can undoubtedly serve as a

---

[57]That is, it seems to me, still far removed from the introduction of "abstract terms" in Gödel's discussions. They are also, if not mainly, concerned with the introduction of new mathematical objects. Cf. note 10.

[58]That is a classical observation; just recall the dual experiences of Hobbes and Newton with the Pythagorean Theorem, when reading Book 1 of Euclid's *Elements*.

[59]Saunders MacLane articulated such a perspective and pursued matters to a certain extent in his Göttingen dissertation. See his papers [1935] and [1979].

[60]To mention one example: in an abstract setting, where representability and derivability conditions, but also instances of the diagonal lemma are taken for granted as axioms, Gödel's proofs can be found fully automatically; see [Sieg and Field]. The leading ideas used to extend the basic logical strategies are very natural; they allow moving between object and meta-theoretic

springboard for the second feature I mentioned earlier, one that is so characteristic of the developments in modern mathematics, beginning in the second half of the 19th century: the introduction of abstract notions that do not have an intended interpretation, but rather are applicable in many different contexts. (Cf. section 5.5.) The above general statement concerning mathematical explanation can now be directly extended to incorporate also the second feature of actual mathematical experience. Turing might ask, whether machines can be educated to make such reflective moves on their own.

It remains a deep challenge to understand better the very nature of reasoning. A marvelous place to start is mathematics; where else do we find such a rich body of systematically and rigorously organized knowledge that is structured for intelligibility and discovery? The appropriate logical framework should undoubtedly include a *structure theory of (mathematical) proofs*. Such an extension of mathematical logic and in particular of proof theory interacts directly with a sophisticated automated search for humanly intelligible proofs. How far can this be pushed? What kind of broader leading ideas will emerge? What deeper understanding of basic operations of the mind will be gained? We'll hopefully find out and, thus, uncover with strategic ingenuity part of Turing's residue and capture also part of what Gödel considered as "humanly effective", but not mechanical — "by asking the right questions on the basis of a mechanical procedure".

6.4 (Supra-) Mechanical devices

Turing machines codify directly the most basic operations of a human computor and can be realized as physical devices, up to a point. Gödel took for granted that finite machines just are (computationally equivalent to) Turing machines. Similarly, Church claimed that Turing machines are obtained by natural restrictions from machines occupying a finite space and with working parts of finite size; he viewed the restrictions "of such a nature as obviously to cause no loss of generality". (Cf. section 4.5.) In contrast to Gödel and Church, Gandy did not take this equivalence for granted and certainly not as being supported by Turing's analysis. He characterized machines informally as discrete mechanical devices that can carry out massively parallel operations. Mathematically Gandy machines are discrete dynamical systems satisfying boundedness and locality conditions that are physically motivated; they are provably not more powerful than Turing machines. (Cf. section 5.2.) Clearly one may ask: Are there plausible broader concepts of computations for physical systems? If there are systems that carry out supra-Turing processes they cannot satisfy the physical restrictions motivating the boundedness and locality conditions for Gandy machines. I.e., such systems must violate either the upper bound on signal propagation or the lower bound on the size of distinguishable atomic components.[61]

---

considerations via provability elimination and introduction rules.

[61] For a general and informative discussion concerning "hypercomputation", see Martin Davis's paper [2004]. A specific case of "computations" beyond the Turing limit is presented through

In *Paper machines*, Mundici and I diagnosed matters concerning physical processes in the following way. Every mathematical model of physical processes comes with at least two problems, "How accurately does the model capture physical reality, and how efficiently can the model be used to make predictions?" What is distinctive about modern developments is the fact that, on the one hand, computer simulations have led to an emphasis on algorithmic aspects of scientific laws and, on the other hand, physical systems are being considered as computational devices that process information much as computers do. It seems, ironically, that the mathematical inquiry into paper machines has led to the point where (effective) mathematical descriptions of nature and (natural) computations for mathematical problems coincide.

How could we have physical processes that allow *supra-Turing computations*? If harnessed in a machine, we would have a genuinely supra-mechanical device. However, we want to be able to *effectively determine* mathematical states from other such states — that "parallel" physical states, i.e., we want to make predictions and do that in a sharply intersubjective way. If that would not be the case, why would we want to call such a physical process a computation and not just an oracle? Wouldn't that undermine the radical intersubjectivity computations were to insure? There are many fascinating open issues concerning mental and physical processes that may or may not have adequate computational models. They are empirical, broadly conceptual, mathematical and, indeed, richly interdisciplinary.

ACKNOWLEDGMENTS

The presentation given here has evolved over almost two decades, and I have drawn systematically on my earlier publications, in particular on [1994], [1996], [1997], [2002a,b] and [2007]. Indeed, parts 5.2 and 5.3 are taken from [2002b]; part 6 was published as [2007]. Pioneering papers from Dedekind, Kronecker and Hilbert through Church, Gödel, Kleene, Post and Turing to Gandy have been a source of continuing inspiration. The historical accounts by Davis, Gandy, Kleene and Rosser have been helpful in clarifying many developments, so has the correspondence between Gödel and Herbrand, as well as that between Bernays and Church. A detailed review of classical arguments for Church's and Turing's theses is found in Kleene's *Introduction to Metamathematics*, in particular, sections 62, 63 and 70; section 6.4 of [Shoenfield, 1967] contains a careful discussion of Church's Thesis. The first chapter of [Odifreddi, 1989] and Cooper's essay [1999] provide a broad perspective for the whole discussion, as does [Soare, 1999]. Much of the material was presented in talks and seminars, and I am grateful to critical responses by the many audiences; much of the work was done in collaboration, and I owe particular

---

Siegelmann's ANNs (artificial neural nets): they perform hypercomputations only if arbitrary reals are admitted as weights. Finally, there is the complex case of quantum computations; if I understand matters correctly, they allow a significant speed-up for example in Shore's algorithm, but the current versions don't go beyond the Turing limit.

debts to John Byrnes, Daniele Mundici, Mark Ravaglia and last, but undoubtedly not least, Guglielmo Tamburrini. Finally, the material was organized for four seminars I gave in November of 2004 at the University of Bologna; I am grateful to Rossella Lupacchini and Giorgio Sandri for their invitation, critical support and warm hospitality.

BIBLIOGRAPHY

[Ackermann, 1925] W. Ackermann. Begründung des "tertium non datur" mittels der Hilbertschen Theorie der Widerspruchsfreiheit. *Mathematische Annalen*, 93: 1–26, 1925.
[Ackermann, 1928] W. Ackermann. Zum Hilbertschen Aufbau der rellen Zahlen. *Mathematische Annalen*, 99: 118–133, 1928.
[Baldwin, 2004] J. Baldwin. Review of [Wolfram, 2002]. *Bulletin of Symbolic Logic*, 10(1): 112–114, 2004.
[Behmann, 1922] H. Behmann. Beiträge zur Algebra der Logik, insbesondere zum Entscheidungsproblem. *Mathematische Annalen*, 86: 163–229, 1922.
[Benacerraf and Putnam, 1983] P. Benacerraf and H. Putnam. *Philosophy of Mathematics - Selected Readings* (Second edition). Cambridge University Press, 1983.
[Bernays, 1918] P. Bernays. *Beiträge zur axiomatischen Behandlung des Logik-Kalküls*. Habilitationsschrift. Göttingen, 1918.
[Bernays, 1922] P. Bernays. Über Hilberts Gedanken zur Grundlegung der Arithmetik. *Jahresbericht der Deutschen Mathematiker Vereinigung*, 31: 10–19, 1922.
[Bernays, 1926] P. Bernays. Axiomatische Untersuchung des Aussagen-Kalkuls der "Principia mathematica". *Mathematische Zeitschrift*, 25: 305–320, 1926.
[Bernays, 1967] P. Bernays. Hilbert, David. In P. Edwards (Editor in Chief), *The Encyclopedia of Philosophy*, vol. 3, pages 496–504, 1967.
[Bernays, 1976] P. Bernays. *Abhandlungen zur Philosophie der Mathematik*. Wissenschaftliche Buchgesellschaft, Darmstadt, 1976.
[Boniface and Schappacher, 2001] J. Boniface and N. Schappacher. Sur le concept de nombre en mathématique - Cours inédit de Leopold Kronecker à Berlin (1891). *Revue d'histoire des mathématiques*, 7: 207–275, 2001.
[Büchi, 1990] J. R. Büchi. *The Collected Works of J. Richard Büchi*. S. Mac Lane and D. Siefkes (eds.), Springer Verlag, 1990.
[Byrnes and W. Sieg, 1996] J. Byrnes and W. Sieg. A graphical presentation of Gandy's parallel machines (Abstract). *Bulletin of Symbolic Logic*, 2: 452–3, 1996.
[Carnap, 1931] R. Carnap. Die logizistische Grundlegung der Mathematik. *Erkenntnis*, 2: 91–105, 1931. (Translation in [Benacerraf and Putnam]).
[Church, 1932] A. Church. A set of postulates for the foundation of logic I. *Annals of Mathematics*, 33(2): 346–366, 1932.
[Church, 1933] A. Church. A set of postulates for the foundation of logic II. *Annals of Mathematics*, 34(2): 839–864, 1933.
[Church, 1934] A. Church. The Richard Paradox. *American Mathematical Monthly*, 41: 356–361, 1934.
[Church, 1935] A. Church. An unsolvable problem of elementary number theory. Preliminary report (abstract). *Bulletin of the American Mathematical Society*, 41: 332–333, 1935.
[Church, 1936] A. Church. An unsolvable problem of elementary number theory. *American Journal of Mathematics*, 58: 345–363, 1936. (Reprinted in [Davis, 1965].)
[Church, 1936a] A. Church. A note on the Entscheidungsproblem. *Journal of Symbolic Logic*, 1(1): 40–41, 1936.
[Church, 1937] A. Church. Review of [Turing, 1936]. *Journal of Symbolic Logic*, 2(1): 42–43, 1937.
[Church, 1937a] A. Church. Review of [Post, 1936]. *Journal of Symbolic Logic*, 2(1): 43, 1937.
[Church and Kleene, 1935] A. Church and S. C. Kleene. Formal definitions in the theory of ordinal numbers. *Bulletin of the American Mathematical Society*, 41: 627, 1935.
[Church and Kleene, 1936] A. Church and S. C. Kleene. Formal definitions in the theory of ordinal numbers. *Fundamenta Mathematicae*, 28: 11–21, 1936.

[Colvin, 1997] S. Colvin. *Intelligent Machinery: Turing's Ideas and Their Relation to the Work of Newell and Simon.* M.S. Thesis, Carnegie Mellon University, Department of Philosophy, 63 pp., 1997.

[Cooper, 1999] S. B. Cooper. Clockwork or Turing Universe — Remarks on causal determinism and computability. In S. B. Cooper and J. K. Truss (eds.), *Models and Computability*, London Mathematical Society, Lecture Note Series 259, Cambridge University Press, pages 63–116, 1999.

[Copeland, 2004] J. Copeland. *The Essential Turing.* Oxford University Press, 2004.

[Crossley, 1975] J. N. Crossley (ed.). *Algebra and Logic; Papers from the 1974 Summer Research Institute of the Australasian Mathematical Society; Monash University.* Springer Lecture Notes in Mathematics 450, 1975.

[Crossley, 1975a] J. N. Crossley. Reminiscences of logicians. In [Crossley, 1975], 1–62, 1975.

[Davis, 1958] M. Davis. *Computability and Unsolvability.* McGraw-Hill, New York, 1958. (A Dover edition was published in 1973 and 1982.)

[Davis, 1965] M. Davis (ed.). *The Undecidable, Basic papers on undecidable propositions, unsolvable problems and computable functions.* Raven Press, Hewlett, New York, 1965.

[Davis, 1973] M. Davis. Hilberts tenth problem is unsolvable. *American Mathematical Monthly,* 80: 233–269, 1973. (Reprinted in the second Dover edition of [Davis, 1958].)

[Davis, 1982] M. Davis. Why Gödel didnt have Church's Thesis. *Information and Control,* 54: 3–24, 1982.

[Davis, 2004] M. Davis. The myth of hypercomputation. In C. Teuscher (ed.), *Alan Turing: Life and Legacy of a Great Thinker.* Springer, 195–211, 2004.

[Dawson, 1986] J. Dawson. A Gödel chronology. In [Gödel, 1986, 37–43].

[Dawson, 1991] J. Dawson. Prelude to recursion theory: the Gödel-Herbrand correspondence. Manuscript, 1991.

[Dawson, 1997] J. Dawson. *Logical Dilemmas.* A. K. Peters, Wellesley, Massachusetts, 1997.

[De Pisapia, 2000] N. De Pisapia. *Gandy Machines: An Abstract Model of Parallel Computation for Turing Machines, the Game of Life, and Artificial Neural Networks.* M.S. Thesis, Carnegie Mellon University, Department of Philosophy, 75 pp., 2000.

[Dedekind, 1888] R. Dedekind. *Was sind und was sollen die Zahlen?* Vieweg, Braunschweig, 1888. (Translation in [Ewald, 1996].)

[Ewald, 1996] W. Ewald (ed.). *From Kant to Hilbert: A Source Book in the Foundations of Mathematics.* Two volumes. Oxford University Press, 1996.

[Feferman, 1988] S. Feferman. Turing in the land of O(z). In [Herken, 1988, 113–147].

[Frege, 1879] G. Frege. *Begriffsschrift.* Verlag Nebert, Halle, 1879.

[Frege, 1893] G. Frege. *Grundgesetze der Arithmetik, begriffsschriftlich abgeleitet.* Jena, 1893. (Translation in [Geach and Black].)

[Frege, 1969] G. Frege. In H. Hermes, F. Kambartel, and F. Kaulbach (eds.), *Nachgelassene Schriften.* Meiner Verlag, Hamburg, 1969.

[Frege, 1984] G. Frege. In B. McGuinness (ed.), *Collected Papers on Mathematics, Logic, and Philosophy.* Oxford University Press, 1984.

[Gandy, 1980] R. Gandy. Churchs Thesis and principles for mechanisms. In J. Barwise, H. J. Keisler, and K. Kunen (eds.), *The Kleene Symposium.* North-Holland Publishing Company, Amsterdam, 123–148, 1980.

[Gandy, 1988] R. Gandy. The confluence of ideas in 1936. In [Herken, 1988, 55–111].

[Geach and Block, 1977] Geach and Block. *Translations from the Philosophical Writings of Gottlob Frege.* Blackwell, Oxford, 1977.

[Gödel, 1931] K. Gödel. Über formal unentscheidbare Sätze der Principia Mathematica und verwandter Systeme I. *Monatshefte für Mathematik und Physik,* 38: 173–198, 1931. (Translation in [Davis, 1965], [van Heijenoort, 1967] and *Collected Works I.*)

[Gödel, 1931a] K. Gödel. Diskussion zur Grundlegung der Mathematik. In *Collected Works I,* 200–204, 1931.

[Gödel, 1933] K. Gödel. Zur intuitionistischen Arithmetik und Zahlentheorie. In *Collected Works I,* 286–295, 1933.

[Gödel, 1933a] K. Gödel. The present situation in the foundations of mathematics. In *Collected Works III,* 45–53, 1933.

[Gödel, 1934] K. Gödel. On undecidable propositions of formal mathematical systems. In *Collected Works I,* 346–371, 1934.

[Gödel, 1936] K. Gödel. Über die Länge von Beweisen. In *Collected Works I,* 396–399, 1936.

[Gödel, 1946] K. Gödel. Undecidable Diophantine propositions. In *Collected Works III*, 164–175, 1946.

[Gödel, 1938] K. Gödel. Vortrag bei Zilsel. In *Collected Works III*, 86–113, 1938.

[Gödel, 1946] K. Gödel. Remarks before the Princeton bicentennial conference on problems in mathematics. In *Collected Works II*, 150–153, 1946.

[Gödel, 1951] K. Gödel. Some basic theorems on the foundations of mathematics and their implications. In *Collected Works III*, 304–323, 1951.

[Gödel, 1963] K. Gödel. Postscriptum for [1931]. In *Collected Works I*, 195, 1963.

[Gödel, 1964] K. Gödel. Postscriptum for [1934]. In *Collected Works I*, 369-371, 1964.

[Gödel, 1972] K. Gödel. Some remarks on the undecidability results. In *Collected Works II*, 305–306, 1972.

[Gödel, 1972.1] K. Gödel. The best and most general version of the unprovability of consistency in the same system. First of the three notes [1972].

[Gödel, 1972.2] K. Gödel. Another version of the first undecidability result. Second of the three notes [1972].

[Gödel, 1972.3] K. Gödel. A philosophical error in Turings work. Third of the three notes [1972].

[Gödel, 1974] K. Gödel. Note in [Wang, 1974, 325-6].

[Gödel, 1986] K. Gödel. *Collected Works I*. Oxford University Press, 1986.

[Gödel, 1990] K. Gödel. *Collected Works II*. Oxford University Press, 1990.

[Gödel, 1995] K. Gödel. *Collected Works III*. Oxford University Press, 1995.

[Gödel, 2003] K. Gödel. *Collected Works IV-V*. Oxford University Press, 2003.

[Griffor, 1999] E. R. Griffor (ed.). *Handbook of Computability Theory*. Elsevier, 1999.

[Herbrand, 1928] J. Herbrand. On proof theory. 1928. In [Herbrand, 1971, 29–34].

[Herbrand, 1929] J. Herbrand. On the fundamental problem of mathematics. 1929. In [Herbrand, 1971, 41–43].

[Herbrand, 1930] J. Herbrand. Investigations in proof theory. 1930. In [Herbrand, 1971, 44–202].

[Herbrand, 1931] J. Herbrand. On the fundamental problem of mathematical logic. 1931. In [Herbrand, 1971, 215–259].

[Herbrand, 1931a] J. Herbrand. On the consistency of arithmetic. 1931. In [Herbrand, 1971, 282–298].

[Herbrand, 1971] J. Herbrand. *Logical Writings*. W. Goldfarb (ed.). Harvard University Press, 1971.

[Herken, 1988] R. Herken (ed.). *The Universal Turing Machine — A Half-Century Survey*. Oxford University Press, 1988.

[Herron, 1995] T. Herron. *An Alternative Definition of Pushout Diagrams and their Use in Characterizing K-Graph Machines*. Carnegie Mellon University, 1995.

[Heyting, 1930] A. Heyting. Die formalen Regeln der intuitionistischen Logik. *Sitzungsberichte der Preussischen Akademie der Wissenschaften, physikalisch-mathematische Klasse*, 42–56, 1930.

[Heyting, 1930a] A. Heyting. Die formalen Regeln der intuitionistischen Mathematik. Ibid., 57–71, 158-169, 1930.

[Heyting, 1931] A. Heyting. Die intuitionistische Grundlegung der Mathematik. *Erkenntnis*, 2: 106–115, 1931. (Translation in [Benacerraf and Putnam, 1983].)

[Hilbert, 1899] D. Hilbert. *Grundlagen der Geometrie*. Teubner, Leipzig, 1899.

[Hilbert, 1900] D. Hilbert. Über den Zahlbegriff. *Jahresbericht der Deutschen Mathematiker Vereinigung*, 8: 180–194, 1900.

[Hilbert, 1901] D. Hilbert. Mathematische Probleme Vortrag, gehalten auf dem internationalen Mathematiker-Kongreß zu Paris 1900. *Archiv der Mathematik und Physik*, 1: 44-63 and 213–237, 1901. (Partial translation in [Ewald, 1996].)

[Hilbert, 1917*] D. Hilbert. Prinzipien der Mathematik. Lectures given by Hilbert and Bernays in the winter term of the academic year 1917/18. 1917*.

[Hilbert, 1921*] D. Hilbert. Grundlagen der Mathematik. Lectures given by Hilbert and Bernays in the winter term of the academic year 1921/22. 1921*.

[Hilbert, 1922] D. Hilbert. Neubegründung der Mathematik. *Abhandlungen aus dem mathematischen Seminar der Hamburgischen Universität*, 1: 157–177, 1922.

[Hilbert, 1923] D. Hilbert. Die logischen Grundlagen der Mathematik. *Mathematische Annalen*, 88: 151–165, 1923.

[Hilbert, 1926] D. Hilbert. Über das Unendliche. *Mathematische Annalen*, 95: 161–190, 1926.

[Hilbert, 1928] D. Hilbert. Die Grundlagen der Mathematik. *Abhandlungen aus dem mathematischen Seminar der Hamburgischen Universität*, 6: 65–85, 1928.
[Hilbert, 1929] D. Hilbert. Probleme der Grundlegung der Mathematik. *Mathematische Annalen*, 102: 1–9, 1929.
[Hilbert, 1931] D. Hilbert. Die Grundlegung der elementaren Zahlenlehre. *Mathematische Annalen*, 104: 485–494, 1931.
[Hilbert and Ackermann, 1928] D. Hilbert and W. Ackermann. *Grundzüge der theoretischen Logik*. Springer Verlag, Berlin, 1928.
[Hilbert and Bernays, 1934] D. Hilbert and P. Bernays. *Grundlagen der Mathematik I*. Springer Verlag, Berlin, 1934.
[Hilbert and Bernays, 1939] D. Hilbert and P. Bernays. *Grundlagen der Mathematik II*. Springer Verlag, Berlin, 1939.
[Kalmar, 1955] L. Kalmar. Über ein Problem, betreffend die Definition des Begriffes der allgemein-rekursiven Funktion. *Zeitschrift für mathematische Logik und Grundlagen der Mathematik*, 1: 93–5, 1955.
[Kleene, 1935] S. C. Kleene. General recursive functions of natural numbers. *Bulletin of the American Mathematical Society*, 41: 489, 1935.
[Kleene, 1935a] S. C. Kleene. λ-definability and recursiveness. *Bulletin of the American Mathematical Society*, 41: 490, 1935.
[Kleene, 1936] S. C. Kleene. General recursive functions of natural numbers. *Mathematische Annalen*, 112: 727–742, 1936. (Reprinted in [Davis, 1965].)
[Kleene, 1936a] S. C. Kleene. A note on recursive functions. *Bulletin of the American Mathematical Society*, 42: 544–546, 1936.
[Kleene, 1952] S. C. Kleene. *Introduction to Metamathematics*. Elsevier, Groningen, 1952.
[Kleene, 1981] S. C. Kleene. Origins of recursive function theory. *Annals of the History of Computing*, 3: 52–66, 1981.
[Kleene, 1987] S. C. Kleene. Reflections on Church's Thesis. *Notre Dame Journal of Formal Logic*, 28: 490–498, 1987.
[Kolmogorov and Uspensky, 1963] A. Kolmogorov and V. Uspensky. On the definition of an algorithm. *AMS Translations*, 21(2): 217–245, 1963.
[Kronecker, 1887] L. Kronecker. Über den Zahlbegriff. In *Philosophische Aufsätze*, Eduard Zeller zu seinem fünfzigjährigen Doctorjubiläum gewidmet. Fues, Leipzig, 271–74, 1887. (Translation in [Ewald, 1996].)
[Kronecker, 1891] L. Kronecker. Über den Zahlbegriff in der Mathematik. 1891. In [Boniface and Schappacher, 2001].
[Lamport and Lynch, 1990] L. Lamport and N. Lynch. Distributed Computing: Models and Methods. In J. van Leeuwen (ed.), *Handbook of Theoretical Computer Science*. Elsevier, Groningen, 1990.
[Löwenheim, 1915] L. Löwenheim. Über Möglichkeiten im Relativkalkül. *Mathematische Annalen*, 76: 447–470. (Translation in [van Heijenoort, 1967].)
[MacLane, 1934] S. MacLane. *Abgekürzte Beweise im Logikkalkul*. Inaugural-Dissertation, Göttingen, 1934.
[MacLane, 1935] S. MacLane. A logical analysis of mathematical structure. *The Monist*, 118–130, 1935.
[MacLane, 1979] S. MacLane. A late return to a thesis in logic. In I. Kaplansky (ed.), *Saunders MacLane Selected Papers*. Springer-Verlag, 1979.
[Mancosu, 1999] P. Mancosu. Between Russell and Hilbert: Behmann on the foundations of mathematics. *Bulletin of Symbolic Logic*, 5(3): 303–330, 1999.
[Mancosu, 2003] P. Mancosu. The Russellian influence on Hilbert and his school. *Synthese*, 137: 59–101, 2003.
[Mates, 1986] B. Mates. *The Philosophy of Leibniz*. Oxford University Press, 1986.
[Mendelson, 1990] E. Mendelson. Second thoughts about Church's Thesis and mathematical proofs. *Journal of Philosophy*, 87(5): 225–33, 1990.
[Mundici and Sieg, 1995] D. Mundici and W. Sieg. Paper machines. *Philosophia Mathematica*, 3: 5–30, 1995.
[Odifreddi, 1989] Odifreddi. *Classical Recursion Theory*. North Holland Publishing Company, Amsterdam, 1989.

[Odifreddi, 1990] Odifreddi. *About Logics and Logicians — A Palimpsest of Essays by Georg Kreisel.* Volume II: Mathematics; manuscript, 1990.

[Parsons, 1995] C. D. Parsons. Quine and Gödel on analyticity. In P. Leonardi and M. Santambrogio (eds.), *On Quine.* Cambridge University Press, 1995.

[Post, 1936] E. Post. Finite combinatory processes. Formulation I. *Journal of Symbolic Logic,* 1: 103–5, 1936.

[Post, 1941] E. Post. Absolutely unsolvable problems and relatively undecidable propositions — Account of an anticipation. 1941. (In [Davis, 1965, 340–433].)

[Post, 1943] E. Post. Formal reductions of the general combinatorial decision problem. *Amererican Journal of Mathematics,* 65(2): 197–215, 1943.

[Post, 1947] E. Post. Recursive unsolvability of a problem of Thue. *Journal of Symbolic Logic,* 12: 1–11, 1947.

[Post, 1994] E. Post. *Solvability, Provability, Definability: The Collected Works of Emil L. Post.* M. Davis (ed.). Birkhäuser, 1994.

[Ravaglia, 2003] M. Ravaglia. *Explicating the Finitist Standpoint.* Ph.D. Thesis; Department of Philosophy, Carnegie Mellon University, 2003.

[Rosser, 1935] B. Rosser. A mathematical logic without variables. *Annals of Mathematics,* 36: 127–150, 1935.

[Rosser, 1936] B. Rosser. Extensions of some theorems of Gödel and Church. *Journal of Symbolic Logic,* 1: 87–91, 1936.

[Rosser, 1984] B. Rosser. Highlights of the history of the lambda-calculus. *Annals of the History of Computing,* 6(4): 337–349, 1984.

[Shanker, 1987] S. G. Shanker. Wittgenstein versus Turing on the nature of Church's Thesis. *Notre Dame Journal of Formal Logic,* 28(4): 615–649, 1987.

[Shanker, 1998] S. G. Shanker. *Wittgenstein's Remarks on the Foundations of AI.* Routledge, London and New York, 1998.

[Shapiro, 1983] S. Shapiro. Remarks on the development of computability. *History and Philosophy of Logic,* 4: 203–220, 1983.

[Shapiro, 1994] S. Shapiro. Metamathematics and computability. In I. Grattan-Guinness (ed.), *Encyclopedia of the History and Philosophy of the Mathematical Sciences.* Routledge, London, 644–655, 1994.

[Shapiro, 2006] S. Shapiro. Computability, proof, and open-texture. In A. Olszewski, J. Wolenski, and R. Janusz (eds.), *Church's Thesis after 70 Years.* Logos Verlag, Berlin, 355–390, 2006.

[Shepherdson, 1988] J. Shepherdson. Mechanisms for computing over arbitrary structures. In [Herken, 1988, 581–601].

[Shoenfield, 1967] J. Shoenfield. *Mathematical Logic.* Addison-Wesley, Reading, Massachusetts, 1967.

[Sieg, 1994] W. Sieg. Mechanical procedures and mathematical experience. In A. George (ed.), *Mathematics and Mind.* Oxford University Press, 71–117, 1994.

[Sieg, 1996] W. Sieg. Aspects of mathematical experience. In E. Agazzi and G. Darvas (eds.), *Philosophy of mathematics today.* Kluwer, 195–217, 1996.

[Sieg, 1997] W. Sieg. Step by recursive step: Church's analysis of effective calculability. *Bulletin of Symbolic Logic,* 3: 154–80, 1997.

[Sieg, 1999] W. Sieg. Hilbert's programs: 1917–1922. *Bulletin of Symbolic Logic,* 5(1): 1–44, 1999.

[Sieg, 2002] W. Sieg. Beyond Hilberts Reach? In D. B. Malalment (ed.), *Reading Natural Philosophy.* Open Court, Chicago, 363–405, 2002.

[Sieg, 2002a] W. Sieg. Calculations by man and machine: conceptual analysis. *Lecture Notes in Logic,* 15: 390–409, 2002.

[Sieg, 2002b] W. Sieg. Calculations by man and machine: mathematical presentation. In P. Gärdenfors, J. Wolenski and K. Kijania-Placek (eds.), *In the Scope of Logic, Methodology and Philosophy of Science,* volume one of the 11th International Congress of Logic, Methodology and Philosophy of Science, Cracow, August 1999. Kluwer, Synthese Library volume 315: 247–262, 2002.

[Sieg, 2005] W. Sieg. Only two letters. *Bulletin of Symbolic Logic,* 11(2): 172–184, 2005.

[Sieg, 2006] W. Sieg. Gödel on computability. *Philosophia Mathematica,* 14: 189–207, 2006.

[Sieg, 2007] W. Sieg. On mind and Turing's machines. *Natural Computing,* 6: 187–205, 2007.

[Sieg and Byrnes, 1996] W. Sieg and J. Byrnes. K-graph machines: generalizing Turing's machines and arguments. In P. Hajek (ed.), *Gödel '96*. Lecture Notes in Logic 6, Springer Verlag, 98–119, 1996.

[Sieg and Byrnes, 1999] W. Sieg and J. Byrnes. Gödel, Turing, and K-graph machines. In A. Cantini, E. Casari, P. Minari (eds.), *Logic and Foundations of Mathematics*. Synthese Library 280, Kluwer, 57–66, 1999.

[Sieg and Byrnes, 1999a] W. Sieg and J. Byrnes. An abstract model for parallel computations: Gandy's Thesis. *The Monist*, 82(1): 150–64, 1999.

[Sieg and Field, 2005] W. Sieg and C. Field. Automated search for Gödel's proofs. *Annals of Pure and Applied Logic*, 133: 319–338, 2005.

[Sieg and Parsons, 1995] W. Sieg and C. D. Parsons. Introductory Note to [Gödel, 1938]. In *Gödel's Collected Works III*, 62–85, 1995.

[Sieg and Ravaglia, 2005] W. Sieg and M. Ravaglia. David Hilbert and Paul Bernays, Grundlagen der Mathematik. In I. Grattan-Guinness (ed.), *Landmark Writings in Western Mathematics, 1640-1940*. Elsevier, 981–999, 2005.

[Sieg and Schlimm, 2005] W. Sieg and D. Schlimm. Dedekind's analysis of number: Systems and axioms. *Synthese*, 147: 121-170, 2005.

[Sieg et al., 2002] W. Sieg, R. Sommer, and C. Talcott (eds.). *Reflections on the Foundations of Mathematics — Essays in Honor of Solomon Feferman*. Association for Symbolic Logic, Lecture Notes in Logic 15, 2002.

[Siegelmann, 1997] H. T. Siegelmann. *Neural Networks and Analog Computation — Beyond the Turing Limit*. Birkhäuser, 1997.

[Sinaceur, 2000] H. Sinaceur. Address at the Princeton University Bicentennial Conference on Problems of Mathematics (December 17-19, 1946), by Alfred Tarski. *Bulletin of Symbolic Logic*, 6(1): 1–44, 2000.

[Skolem, 1923] T. Skolem. Begründung der elementaren Arithmetik durch die rekurrierende Denkweise ohne Anwendung scheinbarer Vernderlichen mit unendlichem Ausdehnungsbereich; Skrifter utgit av Videnskapsselskapet i Kristiana, I. Matematisk-naturvidenskabelig klasse, no. 6, 1–38. (Translation in [van Heijenoort, 1967].)

[Skolem, 1923a] T. Skolem. Einige Bemerkungen zur axiomatischen Begründung der Mengenlehre. In *Matematikerkongressen I Helsingfors den 4-7 Juli 1922*. Akademiska Bokhandeln, Helsinki, 217–232, 1923. (Translation in [van Heijenoort, 1967].)

[Spruit and G. Tamburrini, 1991] L. Spruit and G. Tamburrini. Reasoning and computation in Leibniz. *History and Philosophy of Logic*, 12: 1–14, 1991.

[Smullyan, 1961] R. Smullyan. *Theory of formal systems*. Annals of Mathematics Studies 47, Princeton University Press, 1961. (A revised edition was published in 1968.)

[Smullyan, 1993] R. Smullyan. *Recursion theory for metamathematics*. Oxford University Press, 1993.

[Soare, 1996] R. Soare. Computability and recursion. *Bulletin of Symbolic Logic*, 2(3): 284–321, 1996.

[Soare, 1999] R. Soare. The history and concept of computability. In [Griffor, 1999, 3-36].

[Stillwell, 2004] J. Stillwell. Emil Post and his anticipation of Gödel and Turing. *Mathematics Magazine*, 77(1): 3–14, 2004.

[Sudan, 1927] G. Sudan. Sur le nombre transfini ω^ω. *Bulletin mathématique de la Société roumaine des sciences*, 30: 11–30, 1927.

[Tait, 1981] W. W. Tait. Finitism. *Journal of Philosophy*, 78: 524–546, 1981.

[Tait, 2002] W. W. Tait. Remarks on finitism. In [Sieg, Sommer, and Talcott, 2002, 410–419].

[Tamburrini, 1987] G. Tamburrini. *Reflections on Mechanism*. Ph.D. Thesis, Department of Philosophy, Columbia University, New York, 1987.

[Tamburrini, 1997] G. Tamburrini. Mechanistic theories in cognitive science: The import of Turing's Thesis. In M. L. Dalla Chiara, K. Doets, D. Mundici, and J. van Benthem (eds.), *Logic and Scientific Methods*. Synthese Library 259, Kluwer, 239–57, 1997.

[Turing, 1936] A. Turing. On computable numbers, with an application to the Entscheidungsproblem. *Proceedings of the London Mathematical Society*, series 2, 42: 230–265, 1936. (Reprinted in [Davis, 1965].)

[Turing, 1939] A. Turing. Systems of logic based on ordinals. *Proceedings of the London Mathematical Society*, series 2, 45: 161–228, 1939. (Reprinted in [Davis, 1965].)

[Turing, 1947] A. Turing. Lecture to the London Mathematical Society on 20 February 1947. In D. C. Ince (ed.), *Collected Works of A. M. Turing — Mechanical Intelligence*. North Holland, 87–105, 1992.

[Turing, 1948] A. Turing. Intelligent Machinery. 1948. In D. C. Ince (ed.), *Collected Works of A. M. Turing — Mechanical Intelligence*. North Holland, 107–127, 1992.

[Turing, 1950] A. Turing. Computing machinery and intelligence. *Mind*, 59: 433–460, 1950.

[Turing, 1950a] A. Turing. The word problem in semi-groups with cancellation. *Annals of Mathematics*, 52: 491–505, 1950.

[Turing, 1954] A. Turing. Solvable and unsolvable problems. *Science News*, 31: 7–23, 1954.

[Uspensky, 1992] V. A. Uspensky. Kolmogorov and mathematical logic. *Journal of Symbolic Logic*, 57: 385–412, 1992.

[Uspensky and Semenov, 1981] V. A. Uspensky and A. L. Semenov. What are the gains of the theory of algorithms: Basic developments connected with the concept of algorithm and with its application in mathematics. In A. P. Ershov and D. E. Knuth (eds.), *Algorithms in Modern Mathematics and Computer Science*. Lecture Notes in Computer Science, 122: 100–235, 1981.

[van Heijenoort, 1967] J. van Heijenoort (ed.). *From Frege to Gödel - A Source Book in Mathematical Logic, 1879-1931*. Harvard University Press, 1967.

[van Heijenoort, 1985] J. van Heijenoort. *Selected Essays*. Bibliopolis, Naples, 1985.

[van Heijenoort, 1985a] J. van Heijenoort. Jacques Herbrand's work in logic and its historical context. In [van Heijenoort, 1985, 99–122].

[von Neumann, 1927] J. von Neumann. Zur Hilbertschen Beweistheorie. *Mathematische Zeitschrift*, 26: 1–46, 1927.

[von Neumann, 1931] J. von Neumann. Die formalistische Grundlegung der Mathematik. *Erkenntnis*, 2: 116–121, 1931. (Translation in [Benacerraf and Putnam, 1983].)

[Wang, 1974] H. Wang. *From Mathematics to Philosophy*. Routledge & Kegan Paul, London, 1974.

[Wang, 1981] H. Wang. Some facts about Kurt Gödel. *Journal of Symbolic Logic*, 46: 653–659, 1981.

[Whitehead and Russell, 1910] A. N. Whitehead and B. Russell. *Principia Mathematica, vol. 1*. Cambridge University Press, 1910.

[Whitehead and Russell, 1912] A. N. Whitehead and B. Russell. *Principia Mathematica, vol. 2*. Cambridge University Press, 1912.

[Whitehead and Russell, 1913] A. N. Whitehead and B. Russell. *Principia Mathematica, vol. 3*. Cambridge University Press, 1913.

[Wittgenstein, 1980] L. Wittgenstein. *Remarks on the philosophy of psychology, vol. 1*. G. E. M. Anscombe and G. H. van Wright (eds.). Blackwell, Oxford, 1980.

[Wolfram, 2002] S. Wolfram. *A New Kind of Science*. Wolfram Media, Inc., Champaign, 2002.

[Zach, 1999] R. Zach. Completeness before Post: Bernays, Hilbert, and the development of propositional logic. *Bulletin of Symbolic Logic*, 5: 331–366, 1999.

[Zach, 2003] R. Zach. The practice of finitism: Epsilon calculus and consistency proofs in Hilbert's program. *Synthese*, 137: 211–259, 2003.

INCONSISTENT MATHEMATICS: SOME PHILOSOPHICAL IMPLICATIONS

Chris Mortensen

1 INTRODUCTION: THE PARADOXES

We begin with the paradoxes. Many puzzles that have been called paradoxes have been discovered. Some of these are trivial, such as the paradox of the Barber. Others are tricky but it is possible to discern a way through them, such as the Unexpected Examination. Others are genuinely profound in their implications. Among these, two groups were distinguished: *semantic* paradoxes such as the Liar and Grelling's; and *set-theoretic* paradoxes such as Russell's and Curry's. In the last quarter of the twentieth century, the semantic paradoxes led Routley and Priest to conclude that some contradictions are true [Priest, 1979; 1987; Priest, Routley and Norman, 1989]. This view, known as *dialetheism*, was at once highly radical and yet disarmingly simple. To describe it as radical is to allude to its reception among the body of contemporary philosophers, the large majority of whom still regard it as extreme. To describe it as simple is to allude to the appeal to simplicity in support: alternative solutions to the Liar, such as Tarski's hierarchy of languages, look unsimple by comparison. A similar observation can be made about the set-theoretic paradoxes: naïve set theory with unrestricted comprehension is simple and natural in comparison with contrived patch-ups such as Zermelo–Frankel set theory or Russell's theory of types.

The present essay is not about the semantic paradoxes, and not so much about the set-theoretic paradoxes either. Nonetheless, the example of the paradoxes hopefully softens the reader up for two points. The first point is that the idea that some contradictions might be true has considerable antiquity. Routley and Priest were in a long tradition. Some of the Ancient Greeks, notably Herakleitos and the author of the *Dissoi Logoi*, seem to have taken dialetheism seriously; and this generated a Western tradition which extends to Hegel, Marx and Engels. In the Eastern tradition there have been the *Tao-te-Ching*, Chan Buddhism in China, and Zen in Japan. The second point is that if dialetheism is true then any logic which validates the classical law Ex Contradictione Quodlibet (ECQ), *from a contradiction any conclusion may be validly deduced*, cannot be entirely correct as a description of universal principles of reasoning. This conclusion is supported by the evident artificiality of ECQ. A logic in which ECQ fails is known as *paraconsistent*, or inconsistency-tolerant.

The effect of the set-theoretic paradoxes on the nature of mathematics is conditioned by the question of *foundationalism*. If mathematics has a foundation, then set theory is a good candidate. Frege and Russell's logicist program had two pillars: that mathematics has a foundation, which is set theory, and that set theory in turn reduces to logic. If natural set theory is inconsistent then this seems to weaken the first pillar. It also seems to weaken foundationalism generally, if no better foundation can be found. In passing, it cannot be ruled out *a priori* that some other field of mathematics, such as category theory, might serve as a better foundation for mathematics than set theory. However, it seems clear that category theory employs similar strong comprehension-like principles to those of set theory, and so has similar problems with consistency (see [Hatcher, 1982]).

If consistent set theory is bought only at the cost of unsimple and artificial principles which do not look much like principles of logic or definition, then, as Russell realised, the second pillar of logicism falls too. However, what Frege and Russell did not envisage is the possibility of accepting the contradictions outright. Set-theoretic foundationalism might survive, and both pillars of logicism with it, if the alleged contradictions caused by an unrestricted comprehension principle were restricted to regions where little or no damage to mathematics ensues. The barrier is, of course, ECQ, but we have just been seeing independent reasons for rejecting that. Hence we can register a preliminary conclusion: foundationalism and logicism might be salvageable if contradictions which are true-in-mathematics are tolerated, and ECQ abandoned. Nonetheless, we will later see different reasons for rejecting both foundationalism and logicism.

The barrier that ECQ erects against liberated thinking can be described in another way. It is the idea that once a contradiction presents itself as proved in a theory, then reasoning with that theory must cease. Distinctions between different inconsistencies are impossible because any attempt to describe their structure dissolves into any other attempt. It is the doctrine that *the inconsistent has no structure*. Such a view, if true, would immediately ruin any attempt to develop a Theory of Inconsistency. This essay aims to refute that view.

2 THE ROLE OF LOGIC

It might help the reader to begin by setting aside the Platonist question of what kind of an object, mathematical or otherwise, could possibly have inconsistent properties. In its place, it is recommended to put the primacy of the proposition and the theory in which it occurs. Mathematical texts and lectures do not present abstract objects for transcendental scrutiny. They begin with assertions. Certainly, mathematical texts employ also diagrams. But mathematical texts, where they use diagrams, make assertions about them from the start. The intuitively natural epistemic method for mathematical propositions and theories is of course *proof.* Mathematical truth is, at first pass, mathematical provability. This does not restrict us to a narrow conception of provability as constructability, as the intuitionists have done: a generous inclusive methodology of proof, which can include model theory, should be our starting point. It does mean, however, that we should be open both to the possibility that the intuitionists allowed, that neither A nor not-A be provable for some A, *and* the possibility that both A and not-A be provable, for some A. It also means that we should be less inclined to ask how could an inconsistent proposition be true in mathematics. Rather we should be more inclined to wonder where that might lead. Perhaps later might come an appreciation of mathematical objects with inconsistent properties, as the truthmakers for preferred mathematical propositions, and a basis for model theory. But this metaphysical extra is certainly not necessary to make a beginning with.

Hence, our starting point is collections of propositions. More precisely, if we are to study structure, we must deal with mathematical *theories*, that is, sets of propositions closed under a deducibility relation. Deducibility relations are characteristic of logics; and it is well-known that there are many deducibility relations, since there are many logics. Hence the discussion has to be generalised to *L-theories,* that is, theories of a logic (or deducibility relation) L. An L-theory Th is *inconsistent* iff for some proposition A both $A \; \varepsilon \; Th$ and $\sim A \; \varepsilon \; Th$, where $\sim$ represents the symbol for negation (there are other symbols for special kinds of negations). Th is *incomplete* iff for some A neither $A \; \varepsilon \; Th$ nor $\sim A \; \varepsilon \; Th$. Th is *trivial* iff Th is the whole language, i.e. Th contains every proposition; otherwise Th is *nontrivial*. The members of any L-theory are also called its *theorems*, and are said to *hold* in the theory.

In the end it will be desirable to suppress the logical apparatus provided by L as much as possible. However, for the present, consideration of logic is forced upon us by the logical principle ECQ itself, which, if correct, would ensure that there is just one inconsistent theory, the trivial theory. This, in turn, would prevent any distinctions between kinds of inconsistency, between inconsistent mathematical structures. But at this point we are able to exercise some *free choice*: we can *decide* to countenance mathematical theories of logics for which ECQ fails. If there are none, then *invent* them. There are plenty of paraconsistent logics around to supply adequate logical apparatus. Thus there is a sense in which classical logic, regarded as the logic of mathematics, is *made false* by the existence of inconsistent mathematical theories. To paraphrase Marx, philosophers have hitherto attempted to understand the nature of contradiction, the point however is to change it.

Given a logic, there are two ways to construct theories of that logic: by axioms or by models. The first intentionally inconsistent arithmetical theory was Robert K. Meyer's RM3(mod 2), which was specified by a model. Its background logic was the paraconsistent 3-valued logic RM3. The theory RM3(mod 2) was inconsistent because both $0 = 2$ and $\sim (0 = 2)$ were theorems. However, Meyer constructed this theory because he wanted to study the relevant arithmetic R#, which is axiomatically constructed. The logic for R# is Anderson and Belnap's quantified relevant logic R, axiomatically presented. R# is then given by taking the classical axioms for Peano Arithmetic, replacing their classical implication connectives $\supset$ by the implication connective $\to$ of R, and closing under the deducibility relation for R. There is no suggestion that R# is inconsistent. However, by virtue of Meyer's result that R# $\subseteq$ RM3(mod 2), it follows that R# can have $0 = 2$ added as an axiom, the result being an inconsistent axiomatically-presented arithmetic which is nontrivial. Indeed, RM3(mod 2) itself has an axiomatic presentation: to R# add 0=2 together with all instances of the propositional axiom Mingle, $A \to (A \to A)$. See [Meyer 1976; Meyer and Mortensen, 1984; Mortensen, 1995].

The question can be asked: given that there are many paraconsistent logics, which is "best" for inconsistent mathematics? The answer that emerged was that it doesn't much matter which: the properties of inconsistent theories tend to be invariant over a large class of background logics. To be more exact, when theories are specified by means of models, their logical properties tend to take second place behind *mathematical calculations which are performed at the sub-atomic level* (sub-atomic relative to the atoms of logic, that is). This suggests an important idea: that mathematics is after all different from logic since logic deals with the *general* properties of propositions, predicates and identity, while mathematics deals with calculations in *particular kinds* of structures. We will be developing this theme as we proceed.

Even so, there is one paraconsistent logic which is particularly natural: closed set logic. It is well known that intuitionist logic is the logic of open sets; closed set logic is its topological dual. For many familiar logics, such as tense and modal logics, we can think of propositions as indexed by sets of points in an appropriate space, such as a set of times, or a set of possible worlds, or a phase-space. This idea can then be extended by supposing that the index set has a topological structure. If we make the stipulation that *propositions only ever hold on open sets of points*, we obtain open set logic. It is not difficult to then think of the disjunction of two propositions as holding on the union of the sets on which they hold, and conjunction as holding on the intersection. Considering negation however, it is apparent that the negation of a proposition A cannot hold on the set-theoretic complement of set of points on which A holds, since the set-theoretic complement of an open set is not in general open. It is thus customary to take for negation *the largest open set contained in the set-theoretic complement*. We can then see the familiar intuitionist property of negation emerging: *at the boundary neither A nor not-A holds*. Theories of open set logic may thus be incomplete. It is widely acknowledged that this is a natural-sounding semantics.

Applying the topological open-closed duality, we must have closed set logic. Closed set logic is the stipulation that *whatever holds, holds on closed sets of points*. The interesting case is negation. It is thus customary to take *the smallest closed set containing the set-theoretic complement*. We then have the familiar paraconsistent property of negation emerging: *at the boundary both A and not-A holds*. It is apparent that this is an equally natural semantics to that of open set logic. It is one in which ECQ fails and which supports inconsistent theories. This is as natural as the natural transformation: open $\leftrightarrow$ closed.

Unfortunately, it was soon found that inconsistency can spread for reasons other than ECQ. Curry's paradox generates triviality for naïve set theory even in the absence of negation, even in the absence of ECQ (see e.g., [Meyer, Routley and Dunn, 1979]). All that is necessary is the logical law of Contraction $(A \to (A \to B)) \to (A \to B)$ as well as Modus Ponens, Simplification and Universal Instantiation. Indeed, even weaker principles suffice, as shown by Slaney [1989] and Rogerson [2000]. Thus we must not live in a fool's paradise when constructing inconsistent theories *axiomatically*. Maybe some variant of Curry's paradox can jump up and bite us as a consequence of our axioms, even if we are sure that ECQ cannot hurt our theories.

Nevertheless, we have a guarantee from *model theory* that the spread of contradictions can be stopped short of triviality, at least for naïve set theory. This was essentially shown by Brady very early on [1971], using a model-theoretic fixed-point method derived from Gilmore [1967]. (It should be noted that Brady's result was not explicitly inconsistent, but the latter follows by a trivial manouvre, as he later realised [1989].) Similar work was done independently by Da Costa (see e.g. [1974)]. The importance of Brady's and Da Costa's result cannot be stressed enough. Brady demonstrated nontriviality in the presence of the Russell Set and the Curry set; so by brute force, *whatever* logical principles have to fail for these sets not to lead to explosion, *must* fail in Brady's construction. In a further parallel to Meyer, Brady developed the method in later papers to show that classically false ordinal equations are not provable in naïve set theory either (see [1989]). Thus, just as in arithmetic, the contradictions in naïve set theory are far away and contained, and do not interfere with serious mathematical calculations.

Hence, problems for inconsistency arising from *logic* are not insurmountable. But this is far from being an end to it. Dunn pointed out that if any classically false equation was added to *real number theory*, then every equation became provable. The proof of this is elementary algebra: from $a = b$, where a and b are distinct real numbers, we can subtract a from both sides to get $0 = (b-a)$. Each side may then be multiplied by any number we please to get $0 = r$ for any real number r. Hence by the principle of the substitutivity of identicals, every real number equals every other.

We can coin the term *mathematically trivial* for any (mathematical) theory all of whose (logically) *atomic* propositions are theorems. Now mathematical triviality implies full triviality in the presence of the rule ECQ. But in general it does not do so. Yet, it is mathematical triviality that is catastrophic for mathematics: no calculation would mean anything. And in Dunn's argument we have an example where mathematical triviality is spread by principles other than ECQ or anything else from pure logic. Conversely, if calculations are possible at all, then it is nothing short of crude classical hegemony to insist that a detour through mere *logical* principles such as ECQ ought to render the theory useless for this purpose or any other.

Correspondingly, we can define a theory to be *transparent* if it permits full substitutivity of identicals; that is, if $t_1 = t_2$ holds then Ft_1 holds iff Ft_2 holds, where F is any context. A theory is *functional* if substitutivity of identicals is restricted to *logically atomic* contexts; that is, F is any atomic context. In theories of classical logic, functionality implies transparency, but this is not so in the general case. Furthermore, Dunn's argument requires no more than functionality to work. But now we can see that it is functionality that matters more for mathematics than transparency, since functionality is what ensures that calculations can proceed. Failure of substitutivity because of logic is not such a weighty matter, while both functionality and its failure are of greater moment for mathematics.

3 PURE MATHEMATICS

It is impossible in this brief account to survey all the results of inconsistent mathematics. However, some broad outlines can be touched on. The study has tended to concentrate on techniques from model theory rather than axiomatics, and we will take that approach here. Thus we begin with a first-order language containing

(i) names for mathematical objects, such as the natural numbers, integers, real numbers, sets, topological spaces;

(ii) term-forming operations on these objects, such as $+, \times, -, \div, \prime$ (successor) ;

(iii) atomic predicates and relations, such as $=, \subseteq, \in$;

(iv) logical operations such as $\&, \vee, \sim, \supset, \rightarrow, \leftrightarrow, \forall, \exists$.

Well-formed formulae are defined in the usual way. A *model* is a triple $\langle D, L, I \rangle$, where D is a domain of mathematical objects, L is a many-valued logic, some of whose values are designated and the others undesignated, and I is an interpretation which maps names to elements of the domain, term-forming operators to (partial) operators on the domain, predicates to subsets of the domain, n-ary relations to subsets of D^n, and wffs to the values of L in accordance with the interpretations of parts of the wff to the domain or other values respectively. The *theory* associated with the model is then formed by taking the all those wffs of the model which take a designated value in the interpretation.

One device worth mentioning is the use of *extensions* and *anti-extensions* for each predicate and n-ary relation. The idea, due to Dunn and used by Priest, is that the extension and anti-extension of a predicate can overlap and in that case the predicate is counted as both true and false of those objects. However, it is not necessary to use this device, and it is less than fully general when a logic having numerous values is being used. The reader is cautioned at this stage from taking models with too much ontological seriousness. Models are to be regarded in the first instance as devices for controlling the membership of theories. Notice also in passing the implied distinction between mathematics and logic in that, with the exception of $=$ and perhaps $\in$, logic proper only enters under (iv).

To take an example, consider the language to contain names for all natural numbers (perhaps constructed in the usual way from 0 and the successor operation), arithmetical operations $+, \times, \prime$, and a single binary relation $=$. Let the domain D be the natural numbers modulo 2, and the logic L be the 3-valued paraconsistent logic RM3, with values $\{T, B, F\}$ where T and B are designated values (B is understood as "both"). Interpret names for the natural numbers as their counterparts mod 2 and term-forming operators as their corresponding operators mod 2. Atomic sentences $t_1 = t_2$ are interpreted as taking the value B if $t_1 \bmod 2 = t_2 \bmod 2$, otherwise $t_1 = t_2$ is interpreted as taking the value F. The set of sentences taking either of the designated values $\{T, B\}$ is Meyer's theory RM3(mod 2). The theory is inconsistent since the equation $0 = 2$ takes the value B while $\sim (0 = 2)$ takes T. Meyer then proved that relevant arithmetic R# $\subseteq$ RM3(mod 2), which was the basis for his finitary nontriviality proof for R# (see [Meyer, 1976]). It is obvious that Meyer's construction can be modified to produce RM3(mod n) for any number n. Since R# is contained in any of these, we can also see that no classically false equation $t_1 = t_2$ can be proved in R#. (See [Meyer and Mortensen, 1984].)

Meyer's proof that R# $\subseteq$ RM3(mod 2) was finitary in Hilbert's sense, in that it relied solely on ordinary mathematical induction over the length of formulae. Since by inspection RM3(mod 2) is nontrivial, it follows that R# can be shown to be non-trivial by finitary means. By contrast, it follows from Gödel's incompleteness theorems that there is no finitary proof of the non-triviality (equivalently, consistency) of *classical* Peano arithmetic. This was viewed with great pessimism by Hilbert, who felt that it spelt the end of his program to demonstrate the consistency of mathematics by finitary means. However, Meyer concluded that Hilbert's pessimism is unfounded, as long as we cast aside the shackles of classical logic and ECQ. A further corollary of Meyer's result was not merely that the explosive spread of contradiction in relevant arithmetic is prevented, but that *no false atomic propositions (equations) can be proved in R#*. Thus *calculation is untouched by contradiction in relevant arithmetic*. This is then a further important consequence for the philosophy of mathematics. We saw earlier that logicism might be rehabilitated from Russell's paradox by retaining naïve comprehension, as long as ECQ fails. Now we see that the Hilbert program similarly has excellent prospects for rehabilitation in logics in which ECQ fails. These include logics only slightly weaker than classical logic.

It is fairly easy to show that *extensional* part of R# (with logical operators $\&, \vee, \sim, \supset, \equiv, \exists, \forall, =$, but lacking intensional operators $\to, \leftrightarrow$) is a subset of classical Peano arithmetic PA. There was for a time the hope that they coincided exactly. This would of course imply the non-triviality of PA, and hence its consistency. That would not of course violate Gödel's second incompleteness theorem, since there is no suggestion that the proof method itself be representable in classical arithmetic. But it would be a new proof all the same, perhaps using quite different techniques from the usual. It was eventually discovered by Meyer, adapting Friedman, that R# is strictly weaker than PA, [Meyer-Friedman, 1992]. This dashed the hopes of a consistency proof for PA. Meyer himself expressed pessimism that R# was thereby shown to be less than adequate for arithmetic, since there are true extensional propositions unprovable in R#. But it seems that this makes R# all the more interesting: a genuine rival to PA in which all calculations can be performed; and in which, moreover, all primitive recursive functions are representable so that the incompleteness theorems apply. Moreover, it is hardly something that adherents to classical PA can rejoice in, since they, too, have had to live with the incompleteness theorems ever since they were proved: what is the Gödel sentence if not a true-but-unprovable statement?

The class of all mod models, for varying modulus n, has various interesting properties. Its intersection RMω has the property that its counter-theorems are recursively enumerable, but it is not known whether it is recursive or not. There are also *non-standard* mod models (see [Mortensen, 1987; 1995]). Recently, Priest [1997; 2000] has completely characterised the class of mod models, that is, he showed that all mod models take a certain form.

Of interest is the case of RM3(mod p) where p is prime, since it is known that the natural numbers mod p form a *field*; that is, division is well-defined. This raises the question of where Dunn's proof of triviality for the inconsistent real number field breaks down in mod p. The answer is that in an inconsistent mod arithmetic the equation $a = b$ holds only if the classical difference between a and b differ by an integral multiple of the modulus. Multiplying or dividing both sides by the same integral number does not disturb that, so the inconsistency does not spread everywhere.

It is well known that in the history of the calculus debate raged about whether one should take seriously the use of "very small" real numbers. By the early nineteenth century it seemed that disputes over the status of infinitesimals were resolved in favour of real numbers alone by means of the Cauchy–Weierstrass (ε, δ) technique, which quickly became the orthodoxy in mathematics departments. By 1960, however, Robinson revived infinitesimals by showing rigorously that one could develop calculus just as well with them, and that calculus based on infinitesimals is in various ways simpler to manipulate, (see [Robinson, 1966]). Now it is notorious that in working out derivatives Newton opportunistically divided by very small numbers, yet set them to zero when it was convenient to ignore them. Perhaps then one might be able to make them inconsistently both equal to zero and not equal to zero? However, the prospect that inconsistency in the real numbers spreads uncontrollably into triviality poses an obvious problem for developing inconsistent differential and integral calculus, and resorting to infinitesimals does not offer an obvious relief since the mathematical triviality proof goes over immediately to a mathematical triviality proof in the hyperreal field.

One way to avoid this is to take as one's domain something with a little less than the full structure of fields. This is accomplished by beginning with the *noninfinite hyperreal numbers*, that is, the finite real numbers together with the infinitesimals. Selecting an infinitesimal number η, define $a \approx b$ to mean that $(a - b)/\eta$ is infinitesimal or zero. One may then prove that the equivalence classes so generated form a *ring* under the induced operations. This ring serves as the domain for an inconsistent model. Taking RM3 as background logic as before, set $I(t_1 = t_2)$ to be T if $t_1 = t_2$ as real numbers, set $I(t_1 = t_2)$ to be B if t_1 and t_2 are distinct real numbers but $[t_1] = [t_2]$, and set $I(t_1 = t_2)$ to be F otherwise. Then it is easy to see that both $\eta^2 = 0$ and $\sim (\eta^2 = 0)$ hold, whereas η itself is consistently non-zero. The prospect that infinitesimals smaller that a certain level in size (i.e. infinitesimals which are even infinitesimal w.r.t. η) can be equated with zero, allows calculations in which they can be ignored, even though their "effects" remain in that division by them is retained in various contexts. Differentiation and integration can be developed, and Taylor's theorem and the fundamental theorem of the calculus can all be proved.

There is more to be said about results from pure mathematics than this. Analysis, topology and category theory have all been studied. For an extended discussion, see [Mortensen, 1995; 2000; 2002a]. However, we now proceed with our survey by turning to make some brief remarks on geometry.

4 GEOMETRY

Consider the picture below.

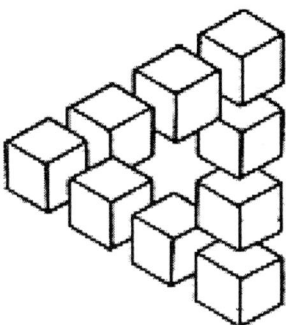

There are many others. It is notable that the beginnings of inconsistent mathematics avoided dealing with such pictorial puzzles, though now the situation is slowly being remedied. Interestingly, classical mathematics has also largely avoided dealing with them. In the classical mathematical literature there were to be found three approaches. The first, due to Thaddeus Cowan [1974], studied n-sided figures in terms of the properties of their corners, employing the theory of *braids*. Second, George Francis [1987] asked what sort of *consistent non-Euclidean* space could be inhabited by such objects. The answer, for the above figure, is $R^2 \times S^1$. Third, Roger Penrose [1991] used the theory of *cohomology groups* to obtain necessary and sufficient conditions for a picture to be of a consistent object; *a fortiori* the failure of those conditions would mean that the picture was of an inconsistent object, (see also [Penrose and Penrose, 1958]).

These were unquestionably all very perceptive approaches. However, as argued by the present writer in [1997b; 2002b; 2002c], they all shared a common deficit: *they did not explain the sense we have that we are perceiving an object with impossible properties*. This suggests a different conception of the problem, namely, to think of the brain as encoding *an inconsistent geometrical theory*. The problem would then become to write out such a theory (or rather theories, for there are many different impossible pictures with different properties). The theory in question would stand to the pictures in somewhat the way that projective geometry stands to the experience of perspective; and with somewhat the same justification, namely, that projective geometry is important to us because of the experience of having an eye.

This kind of justification of the study of inconsistency has been described as the *epistemic* or *cognitive* justification. Such justifications appeal to a human capacity, typically the capacity to reason in a logically-anomalous environment, without intellectual collapse into triviality. There is of course no suggestion that inconsistent objects exist in the physical world. Rather, it is that our perceptions construct a geometrical theory while at the same time retaining geometrical principles which are incompatible with it. It seems that in inconsistent pictures we have a clear example of the mind's ability to make constructions which are inconsistent and yet persist even when the impossibility is manifest to us. The lack of cognitive penetrability of the experiences is characteristic of the *modularity* of perceptual capacities which has been noted by various authors, e.g. [Fodor, 1983].

The details of such mathematical theories are still in an early stage of development. The interested reader is invited to consult the above references for further elucidation.

5 APPLIED MATHEMATICS

A good antidote to the error that mathematics develops in pristine logical order is to read the works of Imre Lakatos [1976]. It is particularly in applied mathematics, physics and engineering where mathematical opportunism is most apparent. Here the lack of classical rigor comes with applications built-in. Hence we can ask, as with the historical disputes over infinitesimals, whether the "logically erroneous" theory might be more accurately described as an inconsistent theory rich enough to permit useful calculations.

A good example is Dirac's Delta "function", $\delta(x)$. This had the twin properties: (i) $\delta(x) = 0$ for all $x \neq 0$, and (ii) $\int \delta(x)dx = 1$, where the integration was over the whole real line. It is apparent that there is no such function on the real numbers. Yet Dirac perceived a use for it in his version of Quantum Mechanics. In this he was followed by many of the physics community. Quantum theory developed rapidly and decisively. It was not for some forty years that Laurent Schwartz managed to put things on a consistent footing by using functionals rather than functions. There was a significant cost, however, in that the new theory was considerably more complicated. There is a fairly obvious construction for the Delta function which uses infinitesimals: draw a triangle of infinitesimal base β and infinite height $2/\beta$. This satisfies something close to the condition (i), namely $\delta(x) = 0$ for all *real* $x \neq 0$; and clearly the second condition is satisfied since the area of the triangle is 1. This was not Robinson's construction, however, since it requires second-order principles; whereas Robinson restricted himself to first-order conditions, so that his theory amounted pretty much to a copy of Schwartz'. It turns out, however, that there is an inconsistent theory which adapts the construction above of inconsistent infinitesimals, and which has the property that $\delta(x) = 0$ for all x for which $x = 0$ fails to hold. Since it is the propositions that hold that are relevant to property of functionality, we can say that the construction recovers the concept of a function, albeit an inconsistent function.

It is hardly surprising that Quantum Mechanics lends itself to inconsistent applications, since QM has long been regarded as a source of anomaly and paradox. One more application in this area is quantum measurement. In cases where an operator has a discrete spectrum, such as the energy levels of the hydrogen atom, elementary QM postulates discontinuous changes in the wave function when a measurement is made. Now discontinuity is an enemy of causality: it would be desirable to have a theory in which quantum measurement was reducible to the other familiar quantum process of unitary evolution. This is the *measurement problem*, and it is fair to say that the measurement problem remains unsolved, and is even intensified given the problem of nonlocality, Bell's theorem and Aspect's experiments. An approach using *inconsistent continuous functions* seems to allow both for continuity/causality and at the same time discrete spectra. For more details, see [Mortensen, 1997a].

The cognitive justification of paraconsistency, discussed before, is apparent in the application to information systems. Nuel Belnap [1977] famously pointed out that any control system with more than one stream of informational inputs, must allow for the possibility that its inputs may be in conflict. Furthermore, it may be impossible to shut the system down until the problem is resolved, as with an aircraft aloft. Thus there has to be a way of operating in an anomalous informational environment, which is after all what we humans manage to do. One theory taking this approach considers the problem of solving *inconsistent systems of linear equations*. Inconsistent systems of linear equations have been known about for centuries, and the standard mathematical reaction has been to throw up the hands in despair. However, it proves possible to solve some such systems of equations in an inconsistent space. Now the classical theory of *control systems* makes heavy use of systems of linear equations. This in turn suggests that if one were able to model a malfunctioning control system in terms of an inconsistent system of linear equations, there might be a way of continuing to exercise some limited control. The modelling proved not to be so difficult. According to classical control theory, when a system is functioning correctly, its internal organisation is modelled by a *transfer matrix*, which transforms a (column) vector of inputs into a vector of outputs. When the system is malfunctioning, there is a difference between the *expected* outputs and the *observed* outputs. By superimposing the observed outputs onto the expected transfer matrix, one obtains an inconsistent system of equations which can then be solved. In software modellings this has met with some limited success. A related approach has been taken by the Brazilian group around Abe [2000], who have demonstrated a paraconsistent robot, Emmy.

It should be noted that it is not being claimed here that the control system is behaving inconsistently in the real world. It is rather that the discrepancy between expected and observed creates an epistemological gap that has to be resolved. Calculations take place in a virtual space in which all the information available is used to form a composite picture with the aim of continuing to function until proper knowledge and control can be fully restored.

A final point to be noted is the shift in ontology that takes place between pure mathematics and applied mathematics. In rejecting Platonism, we were rejecting abstract truthmakers for pure mathematics. The truthmakers for applied mathematics, one would imagine, are its applications. These involve systems of physical objects and their physical quantities, the kinds of things which are causally active, changing and producing change. Physical quantities, such as 5 gram, 2 cm, 3 sec, come as a package of a number ("5") and a quantity kind or dimension ("gram"). In the present writer's view, the best account of quantities treats them as causally relevant universals. Laws of nature come out as relations between universals (see [Armstrong, 1978]). Real numbers then emerge fairly unproblematically as ratios (i.e. relations of comparison) between dimensioned quantities having the same dimension (see [Forrest and Armstrong, 1987; Bigelow, 1988; Mortensen, 1998]). It is not proposed to develop this account here, the reader is directed to these references. The point being made is that there is not necessarily an equivalence between the problem of the truthmakers for pure mathematics, and truthmakers for applied mathematics. The harder problem seems to be for pure mathematics, while applied mathematics looks rather more tractable.

6 LOGICISM AND FOUNDATIONALISM REVISITED

With this all-too-sketchy survey of what is known to date, we return to our flirtatious quarrel with logicism. The foregoing suggests that we can draw a (rough) line between logics and mathematics in the kinds of reasonings they employ. Logics deal with universally applicable principles of reasoning, centrally $(\vdash, \vDash, \&, \vee, \sim, \rightarrow, \leftrightarrow, \exists, \forall, =)$, and other constructions arising in natural language (e.g. tense, modality, adverbs). Logic applies to mathematical reasoning, certainly, but it applies to that aspect of mathematical reasoning that applies to other subject matter as well. In contrast, mathematics *distinctively* deals with concepts like those of algebra, calculus, differential equations, analysis and geometry. Somewhere in the middle between logic and mathematics lie set theory, number theory, recursion theory and parts of algebra. In the case of algebra, logicians' interests have tended to be confined to structures which can supply a plausible semantics for various sets of logical axioms, such as lattices. With only a few exceptions, logicians have not been much interested in groups, for example. This leads to the challenge to logicism: in what sense is mathematics no more than logic with definitions? *It all depends on which definitions.*

Mathematicians tend to be anti-foundationalist. The previous challenge can also be directed against foundationalism, and it explains why mathematicians have not taken logic's attempts at hegemony too seriously. Claims like "set theory is a foundation for mathematics" or "mathematics reduces to logic" look like they are saying that *all there is to mathematics* is set theory or logic. But this is precisely to suppress what is *distinctive* about mathematics. They give a false sense of what is the *nature* of mathematics.

The point can be further illustrated by considering the "reduction" of geometry to algebra. It is uncontestable that Descartes' discovery of the coordinatisation of the plane enabled an immense step forward in geometry. The methods of algebra could then be applied to the study of the plane. Space could be studied by solving equations involving real numbers and their functions. Nonetheless, it is a mistake to take this as implying that geometry is *nothing but* real number theory, as Russell seems to have thought (see e.g. [Ayer, 1972, 43]). *The two-dimensional plane is not R^2; space is not a collection of numbers.* Its parts are points, lines, curves, and planes, not sets or real numbers. We need only pay attention to our own perceptions of space to see this: we perceive areas, lines etc., we do not perceive numbers. In short, there is no *conceptual* equivalence possible between geometry and set theory. This is why a mathematician can pursue the study of space paying little or no attention to foundations: mathematics has a *conceptual autonomy* that foundations cannot supply.

From this point of view, the gap between mathematics and logic is even wider than that between mathematics and real numbers and set theory. Logicians study "and", "or", "not", "implies" and the like. Their discipline begins where mathematics leaves off in studying the behaviour of geometry, groups and the like. This makes logic look more like a small area in the corpus of mathematics, rather than a foundation for it. Furthermore, it exposes ECQ for what it is: a tool in a takeover bid to establish the hegemony of logic over mathematics.

As a piece of personal reportage I recall years ago explaining to a visiting eminent mathematician why I was inclined to reject ECQ. After listening politely, he asked: "Excuse me, but are you not denying that the null set is a subset of every set?" This confusion embodies a subtle reversal, but it is no better motivated. We may be inclined to make a limited "reduction" of set theory to logic by adopting naïve set theory and claiming that there is nothing to set theory but logic. Naïve comprehension would then be an expression of the reduction. In favour it can be said that it is certainly less *ad hoc* than rival comprehension principles. However, our eminent mathematician was reversing the order of explanation: he felt that the principles of set theory were *sui generis* and that the legitimacy of ECQ was ensured by that!

7 REVISIONISM AND DUALITY

Earlier, we referred to the topological duality between incomplete theories of open set logic, and inconsistent theories of closed set logic. There is another kind of duality, Routley-* duality. This applies between theories of logics in which the laws of Double Negation $A \leftrightarrow \sim\sim A$ and De Morgan $\sim (A \vee B) \leftrightarrow (\sim A \,\&\, \sim B)$ and $\sim (A \,\&\, B) \leftrightarrow (\sim A \vee \sim B)$ hold. Neither open set logic nor closed set logic has these laws unrestrictedly, however many of the logics in the Anderson–Belnap class of relevant logics have them. For any set of sentences S, define S^* to be $\{A : \sim A \notin S\}$. Then a simple argument shows that if Th is any theory of a logic containing Double Negation and De Morgan, then Th is inconsistent iff Th^* is incomplete. Since DN ensures that $Th^{**} = Th$, we also have that Th is incomplete iff Th^* is inconsistent.

That is, incompleteness and inconsistency as properties of theories are duals of one another in *two* senses: they are topological duals of one another, and they are Routley-* duals of one another. Duality results are of course sources of "theorems for free". As a quick illustration of free theorems, we note a dualisation of Kripke's modelling of the truth predicate in an incomplete theory. Kripke [1975] showed, using a fixed point method deriving from Gilmore [1967] and Brady [1971], that the Liar proposition L and its negation are excluded from a theory satisfying the condition for a truth predicate: $A \leftrightarrow T(A)$ where A is any proposition and $T(.)$ is the truth predicate for the name (Gödel number) of A. Kripke interpreted this as showing that the Liar proposition L ought to be regarded as neither true nor false. However, applying the Routley-* to the truth theory, we can immediately conclude that there is a theory satisfying the conditions for a truth theory to which both L and $\sim L$ belong. We might also observe that the inconsistent dual theory has certain advantages over the incomplete theory. Any theory which, like Kripke's, declares that L lacks a value, suffers from a dilemma. Either we say that the instance of the T-scheme for the Liar sentence has a value (presumably True), or it does not. If it does, then we have the oddity that none of $L, \sim L, T(L)$ and $T(\sim L)$ receive a truth value even though $L \leftrightarrow T(L)$ and $\sim L \leftrightarrow T(\sim L)$ hold in the theory. If it does not, then it odd to say that the T-scheme holds even though some of its instances fail to hold. Note that while Kripke employed a third logical value in his construction, he was clear that this was a formal device for calculation only, and that he regarded the liar sentence as lacking a value. This is perfectly reasonable as a proof device, however it seems strange that a valueless proposition could yet contribute to making a compound hold. In contrast, in the inconsistent dual, all of $L, \sim L, T(L)$ and $T(\sim L)$ take contradictory values; which is at least some reason to hold that $L \leftrightarrow T(L)$ does too.

Intuitionism and constructivism are examples of *revisionist* philosophies of mathematics, in that they declare that certain principles accepted in classical mathematics are incorrect. They aim to revise mathematics by truncating it, based on a narrower conception of what is an acceptable proof. However, revisionism leaves unanswered an important question: *why are the excluded areas yet mathe-*

matics? In their haste to offer a theory of correct proof, revisionists neglect the central question of the philosophy of mathematics: what is mathematics? This is hardly to be answered adequately by declaring those parts of mathematics that the theorists don't like, not to be mathematics at all.

The classical Hilbertian ideal of a mathematical theory is one which is complete and demonstrably consistent. Revisionist theories, by excluding aspects of classical theories, render themselves incomplete, a fact which has been long-noted in connection with intuitionism. By contrast, inconsistent mathematics is not revisionist at all. Taking a lead from the duality results, it aims to extend mathematics, not weaken it. The duals of incomplete theories are inconsistent, and they include classical consistent complete theories as subtheories, and consistent incomplete theories as sub-sub-theories. Thus inconsistent mathematics supports a principle of tolerance about what counts as mathematics, an inclusive approach not an exclusive one. Both classical mathematics and revisionist mathematics emerge as *special cases* of a more generalised conception of mathematics, which includes inconsistent mathematics as well.

8 THE ROLE OF TEXT

One further matter needs to be raised, though dealing with it fully would take much more space than we have here. If we ask what makes all of the above examples mathematics, it is apparent that the answer must have something central to do with the characteristic use of notation or symbols. That is to say, mathematics is *textually distinctive*. Importantly, this is something it shares with symbolic logic. It is apparent that the rise of symbolic logic in the twentieth century is attributable to its use of mathematical text. The question is: just how is it that this has been so efficacious? This dovetails with the broader question of just why it is that the distinctive textual features of *any* mathematics do their jobs so well? We are all familiar with examples like the advantage of the change from Roman numerals to Arabic numerals: it is clear that this is a microcosm of the general question of the distinctive nature and efficacy of mathematical text. There is something important to be explained about how mathematical meaning is carried by text.

There is another observation which is a kind of converse to this one. In his University of Adelaide PhD thesis *TheRole of Notation in Mathematics* [1988], Edwin Coleman pointed out the *varieties* of mathematical text. He drew attention the differences between a page from Euclid, a page from *Principia Mathematica*, a page from a text on business mathematics, a page from a standard calculus text, and a page from a mechanical engineering text. Consider for example the varying role of diagrams, and the presence or absence of natural language. The differences are richly textual, and yet the very stuff of mathematics. Thus, the question of the usefulness of distinctive *texts* in mathematics, is part of the question of how mathematical text generates meaning. It is the *interplay* of similarity and difference that needs to be understood.

Coleman argued that the right discipline to undertake such a study was the theory of signs, *semiotics*. Co-discovered by Peirce and Saussure, semiotics aims to study how text and other signs generate meaning. Saussure in particular had to rely somewhat more heavily on the internal *differences* within a code or system of signs, because, unlike Peirce, his account lacked a theory of extra-linguistic reference. While this is an obvious drawback in any general account of language, it can be seized on by (we) anti-Platonists as just right for any account of *mathematical* meaning, where (according to us) there are no abstract objects to be the referents. This is nothing but an application of Saussure's concept of *difference*. Ockham's Razor does the rest against Platonism. A certain amount of literature which addresses these issues in the indicated ways has grown up, including Nelson Goodman [1981], Rene Thom [1980], Brian Rotman [1987; 1990], Coleman [1988; 1990; 1992], and Mortensen and Roberts [1997].

We saw earlier that Meyer's nontriviality result serves fit to re-habilitate Hilbert's program of demonstrating that mathematics does not have false consequences. But there are problems for Hilbert's program of a different sort here. Drawing on the above, Coleman attacked Hilbert's formalism. Like Brouwer, Hilbert gave way to the despair of revisionism. In order to demonstrate mathematics to be consistent and complete, or at least without error, mathematical theories must be displayed in canonical form, as formal systems, purely symbolic and devoid of all meaning (save that generated internally). But here, as with revisionisms anywhere, we can again ask *why are the uncanonical parts yet mathematics?* Don't get me wrong. I am certainly not against reconstruction of a theory as a first order formal theory, if only because then you could automate it! But notice that in producing an "equivalent" formal theory we are suppressing a difference that is part of what has to be explained: in what sense can notationally distinct codes be equivalent, and how can textual features contribute to distinctness of code, and thus to differences of meaning?

This kind of study cuts across the inconsistency program to some extent. Nonetheless, it serves to reinforce the point that an inclusive point of view about mathematics is necessary if one is to understand what mathematics is. Revisionism inevitably reduces our view of what is possible for mathematics, and thus distorts our understanding of the phenomena.

9 CONCLUSIONS

To summarise, the following propositions have been advanced.

1. Logicism and foundationalism may well be saved if we adopt a logic lacking ECQ.

2. Similarly, *part of* Hilbert's program, to prove that mathematics has no false consequences, may well be saved if such a logic is adopted. There are many suitable logics, some of them only slightly weaker than classical logic.

3. Nonetheless, logicism and foundationalism do not explain the conceptual autonomy of mathematics from logic. In particular, geometry is conceptually separate from logic and set theory, and does not reduce to them.

4. Revisionist philosophies of mathematics, whether they be revisionist about the truths of mathematics (intuitionism) or revisionist about notation (formalism), are open to the objection that they do not account for the varieties of mathematics outside of approved canonical norms.

5. In contrast to revisionism, we must take an inclusive position, whereby inconsistent mathematics is seen as extending our conception of what is possible for mathematics rather than rejecting the corpus of existing mathematics.

6. This is just as well, since inconsistent mathematics has numerous applications beyond itself.

7. As part of comprehending the nature of mathematics, the distinctively textual aspects of mathematics, both the similarities and the differences between textual styles, have to be understood; and semiotics seems to be the best theoretical framework for this project.

Two related issues of traditional philosophy of mathematics have been placed on the backburner in this essay. One is the matter of truthmakers for pure mathematics. The other is the distinctive epistemology of mathematics, and in particular the method of *a priori* proof. Neither can be neglected in a full account. However, we might make the very limited suggestion that if the primary phenomenon to be explained for mathematics is textual, then it is not so speculative that the account ought to derive from the features of text, rather than abstract acausal objects. Certainly, the legitimacy of inconsistency ought to give pause to the Platonist. It poses the dilemma: either abandon Platonism, or admit inconsistent objects. One salient virtue in sheeting home the primary account to the theory of signs, is that it scores well on the second issue: we have a readily-understandable epistemology for signs. It can hardly be denied that getting in contact with signs, such as those on your keyboard, is a thoroughly natural activity. The same can't be said for Platonism.

BIBLIOGRAPHY

[Abe et al., 2000] Abe et al. Emmy, a Paraconsistent Robot, *Second World Congress on Paraconsistency*, Juquehy Beach, 2000.
[Armstrong, 1978] D. Armstrong. *Universals and Scientific Realism*, Cambridge, CambridgeUP, 1978
[Ayer, 1972] A. J. Ayer. *Russell*, Fontana Modern Masters, 1972.
[Belnap, 1977] N. D. Belnap. How a Computer Should Think, in G.Ryle, ed. *Contemporary Aspects of Philosophy*, Stocksfield, Oriel Press, 30-55, 1977.
[Bigelow, 1988] J. Bigelow. *The Reality of Numbers*, Oxford, The Clarendon Press, 1988.
[Brady, 1971] R. Brady. The Consistency of the Axioms of Abstraction and Extensionality in a Three-Valued Logic, *Notre Dame Journal of Formal Logic* (NDJFL) 12, 447-453, 1971

[Brady, 1989] R. Brady. The Non-triviality of Dialectical Set Theory. In [Priest et al, 1989, 437–471].

[Brady and Routley, 1989] R. Brady and R.Routley. The Non-triviality of Extensional Dialectical Set Theory. In [Priest et al., 1989, 415–436].

[Coleman, 1988] E. Coleman. *The Role of Notation in Mathematics*, PhD thesis, U. of Adelaide, 1988

[Coleman, 1990] E. Coleman. Paragraphy, *Information Design Journal*, 6, 131-146

[Coleman, 1992] E. Coleman. Presenting Mathematical Information. In R.Penman and D.Sless, eds, *Designing Information for People*, Canberra, ANU Press, 1992.

[Cowan, 1974] Th. Cowan. The Theory of Braids and the Analysis of Impossible Pictures, *Journal of Mathematical Psychology*, 11, 190-212, 1974.

[Da Costa, 1974] N. C. A. Da Costa. On the Theory of Inconsistent Formal Systems, *NDJFL*, 15, 497-510, 1974

[Dunn, 1979] J. M. Dunn. A Theorem in Three-Valued Model Theory with Connections to Number Theory, Type Theory and Relevant Logic, *Studia Logica*, 38, 149-169, 1979.

[Fodor, 1983] J. Fodor. *The Modularity of Mind*, MIT Press, 1983.

[Forrest and Armstrong, 1987] P. Forrest and D. Armstrong. The Nature of Number, *Philosophical Papers*, 16, 165-186, 1987.

[Francis, 1987] G. Francis. *A Topological Picturebook*, Springer-Verlag, 1987.

[Gilmore, 1967] P. Gilmore. The Consistency of Partial Set Theory Without Extensionality. In D.Scott (ed.) *Axiomatic Set Theory, Proceedings of Symposia in Pure Mathematics*, Los Angeles, U. of California, 1967.

[Goodman, 1981] N. Goodman. *Languages of Art*, Brighton, Harvester, 1981.

[Hatcher, 1982] W. Hatcher. *The Logical Foundations of Mathematics*, Amsterdam, North-Holland, Elsevier, 1982.

[Kripke, 1975] S. Kripke. Outline of a Theory of Truth, *The Journal of Philosophy* 72, 690-716, 1975

[Lakatos, 1976] I. Lakatos. *Proofs and Refutations*, Cambridge, Cambridge University Press, 1976.

[Meyer, 1976] R. K. Meyer. Relevant Arithmetic, *Bulletin of the Section of the Polish Academy of Science*, 5, 133-137, 1976.

[Meyer and Friedman, 1992] R. K. Meyer and H. Friedman. Whither Relevant Logic? *The Journal of Symbolic Logic*, 57, 824-831, 1992.

[Meyer and Mortensen, 1984] R. K. Meyer and C. Mortensen. Inconsistent Models for Relevant Arithmetics, *The Journal of Symbolic Logic*, 49, 917-929, 1984.

[Meyer et al., 1979] R. K. Meyer, R. Routley, and J. M. Dunn. Curry's Paradox, *Analysis*, 39, 124-128, 1979.

[Mortensen, 1987] C. Mortensen. Inconsistent Nonstandard Arithmetic, *The Journal of Symbolic Logic*, 52 , 512-18, 1987.

[Mortensen, 1988] C. Mortensen. Inconsistent Number Systems, *Notre Dame Journal of Formal Logic*, 29 (1988), 45-60, 1988.

[Mortensen, 1990] C. Mortensen. Models for Inconsistent and Incomplete Differential Calculus, *Notre Dame Journal of Formal Logic*, 31, 274-285, 1990.

[Mortensen, 1995] C. Mortensen. *Inconsistent Mathematics*, Kluwer Mathematics and Its Applications Series, 1995.

[Mortensen, 1997a] C. Mortensen. The Leibniz Continuity Condition, Inconsistency and Quantum Dynamics, *The Journal of Philosophical Logic*, 26, 377-389, 1997.

[Mortensen, 1997b] C. Mortensen. Peeking at the Impossible, *Notre Dame Journal of Formal Logic*, Vol 38 No 4 (Fall 1997), 527-534, 1977.

[Mortensen, 1998] C. Mortensen. On the Possibility of Science Without Numbers, *The Australasian Journal of Philosophy*, 182-197, 1998.

[Mortensen, 2000] C. Mortensen. Topological Separation Principles and Logical Theories, *Synthese* 125 Nos 1-2, 169-178, 2000.

[Mortensen, 2002a] C. Mortensen. Prospects for Inconsistency, in D.Batens (et.al.) eds. *Frontiers of Paraconsistent Logic*, London, Research Studies Press, 203-208, 2002.

[Mortensen, 2002b] C. Mortensen. Towards a Mathematics of Impossible Pictures. In W. Carnelli, M. Coniglio, and I. D'Ottaviano, eds, *Paraconsistency, The Logical Way to the Inconsistent*, Lecture Notes in Pure and Applied Mathematics Vol 228, New York, Marcel Dekker, 445-454, 2002.

[Mortensen, 2002c] C. Mortensen. Paradoxes Inside and Outside Language, *Language and Communication*, Vol 22, No 3, 301-311, 2002.
[Mortensen and Roberts, 1997] C. Mortensen and L. Roberts. Semiotics and the Foundations of Mathematics, *Semiotica*, 115, 1-25, 1997.
[Penrose and Penrose, 1958] L. S. Penrose and R. Penrose. Impossible Objects, a Special Kind of Illusion, *British Journal of Psychology*, 49, 1958.
[Penrose, 1991] R. Penrose. On the Cohomology of Impossible Pictures, *Structural Topology*, 17, 11-16, 1991.
[Priest, 1979] G. Priest. The Logic of Paradox *The Journal of Philosophical Logic*, 8, 219-241, 1979.
[Priest, 1987] G. Priest. *In Contradiction*, Dordrecht, Hijhoff, 1987.
[Priest, 1997] G. Priest. Inconsistent Models of Arithmetic; I, Finite Models, *The Journal of Philosophical Logic*, 26, 223-235, 1997.
[Priest, 2000] G. Priest. Inconsistent Models of Arithmetic; II, The General Case, *The Journal of Symbolic Logic*, 65, 1519-29, 2000.
[Priest et al., 1989] G. Priest, R. Routley, and J. Norman, eds. *Paraconsistent Logic, Essays on the Inconsistent*, Philosophia Verlag, 1989.
[Robinson, 1966] A. Robinson. *Nonstandard Analysis*, Amsterdam, North-Holland, 1966.
[Rogerson, 2000] S. Rogerson. Curry Paradoxes. AAL Annual Conference, U. of Sunshine Coast, Noosa, 2000.
[Slaney, 1989] J. Slaney. RWX is not Curry Paraconsistent. In [Priest et al., 1989, 472]-482].
[Rotman, 1987] B. Rotman. *Signifying Nothing: The Semiotics of Zero*, London, Macmillan, 1987.
[Rotman, 1990] B. Rotman. Towards a Semiotics of Mathematics, *Semiotica*, 72, 1-35, 1990
[Thom, 1980] R. Thom. L'espace des Signes, *Semiotica*, 29, 193-208, 1980.

MATHEMATICS AND THE WORLD

Mark Colyvan

One of the most intriguing features of mathematics is its applicability to empirical science. Every branch of science draws upon large and often diverse portions of mathematics, from the use of Hilbert spaces in quantum mechanics to the use of differential geometry in general relativity. It's not just the physical sciences that avail themselves of the services of mathematics either. Biology, for instance, makes extensive use of difference equations and statistics. The roles mathematics plays in these theories is also varied. Not only does mathematics help with empirical predictions, but it also allows elegant and economical statements of many theories. Indeed, so important is the language of mathematics to science, that it is hard to imagine how theories such as quantum mechanics and general relativity could even be stated without employing a substantial amount of mathematics.

From the rather remarkable but seemingly uncontroversial fact that mathematics is indispensable to science, some philosophers have drawn serious metaphysical conclusions. In particular, Quine [1948/1980; 1951/1980; 1981b] and Putnam [1971/1979; 1979] have argued that the indispensability of mathematics to empirical science gives us good reason to believe in the existence of mathematical entities. According to this line of argument, reference to (or quantification over) mathematical entities such as sets, numbers, functions and such is indispensable to our best scientific theories, and so we ought to be committed to the existence of these mathematical entities. To do otherwise is to be guilty of what Putnam has called "intellectual dishonesty" [Putnam, 1971/1979, p. 347]. Moreover, mathematical entities are seen to be on an epistemic par with the other theoretical entities of science, since belief in both kinds of entities is justified by the same evidence that confirms the theory as a whole. This argument is known as the Quine-Putnam indispensability argument for mathematical realism. In this chapter I will discuss this argument and some of the various attempts to defuse it.

I will also consider another topic related to mathematics and its applications: the so-called unreasonable effectiveness of mathematics. The problem here is that (pure) mathematical methods are largely *a priori* and driven by largely aesthetic considerations, and yet mathematics is in great demand in describing and even explaining the physical world. As Mark Steiner puts it "how does the mathematician — closer to the artist than the explorer — by turning away from nature, arrive at its most appropriate descriptions?" [Steiner, 1995, p. 154]. This problem and its relationship to the indispensability argument will also be examined.

Handbook of the Philosophy of Science. Philosophy of Mathematics
Volume editor: Andrew D. Irvine. General editors: Dov M. Gabbay, Paul Thagard and John Woods.
© 2009 Elsevier B.V. All rights reserved.

1 THE INDISPENSABILITY ARGUMENT

1.1 *Realism and Anti-realism in Mathematics*

There are many different ways to characterise realism and anti-realism in mathematics. Perhaps the most common way is as a thesis about the existence or non-existence of mathematical entities. Thus, according to this conception of realism, mathematical entities such as functions, numbers, and sets have mind- and language-independent existence or, as it is also commonly expressed, we *discover* rather than invent mathematical theories (which are taken to be a body of facts about the relevant mathematical objects). This is usually called *metaphysical realism*. Anti-realism, then, is the position that mathematical entities do not enjoy mind-independent existence or, alternatively, we *invent* rather than discover mathematical theories. According to this characterisation, a realist believes that Fermat's Last Theorem was true before Wiles's proof and, indeed, even before Fermat first thought of his now famous theorem. This is because, according to the realist, the integers exist independently of our knowledge of them and Fermat's theorem is a fact about them. Of course there are other characterisations of realism and anti-realism but since my interests in this chapter are largely metaphysical, I'll be content with this characterisation of realism.[1]

There are various Platonist and nominalist strategies in the philosophy of mathematics. Each of these has its own particular strengths and weaknesses. Platonist accounts of mathematics generally have the problems of providing an adequate epistemology for mathematics [Benacerraf, 1973/1983] and of explaining the apparent indeterminacy of number terms [Benacerraf, 1965/1983]. On the other hand, nominalist accounts generally have trouble providing an adequate treatment of the wide and varied applications of mathematics in the empirical sciences. There is also the challenge for nominalism to provide a uniform semantics for mathematics and other discourse [Benacerraf, 1973/1983]. Let's consider a few different strategies encountered in the literature.

An important nominalist response to these arguments is fictionalism. A fictionalist about mathematics believes that mathematical statements are, by and large, false. According to the fictionalist, mathematical statements are 'true in the story of mathematics' but this does not amount to truth simpliciter. Fictionalists take their lead from some standard semantics for literary fiction. On many accounts of literary fiction 'Sherlock Holmes is a detective' is false (because there is no such person as Sherlock Holmes), but it is 'true in the stories of Conan Doyle.' The mathematical fictionalist takes sentences such as 'seven is prime' to be false (be-

---

[1]While on matters terminological, I should also point out that, in keeping with most of the modern literature in the area, I will use the terms 'mathematical realism' and 'Platonism' interchangeably. So I take Platonism to be the view that mathematical objects exist and, what is more, that their existence is mind and language independent. I also take it that according to Platonism, mathematical statements are true or false in virtue of the properties of these mathematical objects. I do not mean to imply anything more than this. I do not, for instance, intend Platonism to imply that mathematical objects are causally inert, that they are not located in space-time, or that they exist necessarily.

cause there is no such entity as seven) but 'true in the story of mathematics.' The fictionalist thus provides a distinctive response to the challenge of providing a uniform semantics — all the usually accepted statements of mathematics are false.[2] The problem of explaining the applicability of mathematics is more involved, and I will leave a discussion of this until later (see section 4).

In recent times many Platonist strategies have responded to the epistemological challenge by placing mathematical objects firmly in the physical realm. Thus Penelope Maddy in *Realism in Mathematics* [1990a] argued that we can see sets. When we see six eggs in a carton we are seeing *the set* of six eggs. This account provides mathematics with an epistemology consistent with other areas of knowledge by giving up one of the core doctrines of traditional Platonism — that mathematical entities are abstract. In response to the apparent indeterminacy of the reduction of numbers to sets, one popular Platonist strategy is to identify a given natural number with a certain position in *any* ω-sequence. Thus, it doesn't matter that three can be represented as $\{\{\{\emptyset\}\}\}$ in Zermelo's ω-sequence and $\{\emptyset, \{\emptyset\}, \{\emptyset, \{\emptyset\}\}\}$ in von Neumann's ω-sequence. What is important, according to this account, is that the structural properties are identical. This view is usually called *structuralism* since it is the structures that are important, not the items that constitute the structures.[3]

These are not meant to be anything more than cursory sketches of some of the available positions. Some of these positions will arise again later, but for now I will be content with these sketches and move on to discuss indispensability arguments and how these arguments are supposed to deliver mathematical realism.

1.2 Indispensability Arguments

An indispensability argument, as Hartry Field points out, "is an argument that we should believe a certain claim ... because doing so is indispensable for certain purposes (which the argument then details)" [Field, 1989, p. 14]. Clearly the strength of the argument depends crucially on what the as yet unspecified purpose is. For instance, few would find the following argument persuasive: We should believe that whites are morally superior to blacks because doing so is indispensable for the purpose of justifying black slavery. Similarly, few would be convinced by the argument that we ought to believe that God exists because to do so is indispensable to the purpose of enjoying a healthy religious life. The "certain purposes" of which Field speaks must be chosen very carefully. Although the two arguments just mentioned count as indispensability arguments, they are implausible because 'enjoying a healthy religious life' and 'justifying black slavery' are *not* the right

---

[2]This is not quite right. Since fictionalists take the domain of quantification to be empty, they claim that all existentially quantified statements (and statements about what are apparently denoting terms) are false, but that all universally quantified sentences are true. So, for example, 'there is an even prime number' is taken to be false while 'every number has a successor' is taken to be true.

[3]See, for example, Hellman [1989], Resnik [1997], and Shapiro [1997].

sort of purposes to ensure the cogency of the respective arguments. This raises the very interesting question: Which purposes *are* the right sort for cogent arguments?

I know of no easy answer to this question, but fortunately an answer is not required for a defence of the class of indispensability arguments with which I am concerned. I will restrict my attention largely to arguments that address indispensability to *our best scientific theories*. I will argue that this *is* the right sort of purpose for cogent indispensability arguments. I will also be concerned primarily with indispensability arguments in which the "certain claim" of which Field speaks is an existence claim. We may thus take a scientific indispensability argument to rest upon the following major premise:

ARGUMENT 1 *Scientific Indispensability Argument.* If apparent reference to some entity (or class of entities) ξ is indispensable to our best scientific theories, then we ought to believe in the existence of ξ.

In this formulation, the purpose, if you like, is that of doing science. This is a rather ill-defined purpose, and I deliberately leave it ill defined for the moment. But to give an example of one particularly important scientific indispensability argument with a well-defined purpose, consider the argument that takes providing explanations of empirical facts as its purpose. I'll call such an argument an *explanatory* indispensability argument.

Although indispensability arguments are typically associated with realism about mathematical objects, it's important to realise that they do have a much wider usage. What is more, this wider usage is fairly uncontroversial. To see this, we need only consider an example of an explanatory indispensability argument used for non-mathematical purposes.

Most astronomers are convinced of the existence of so called "dark matter" to explain (among other things) certain facts about the rotation curves of spiral galaxies.[4] This is an indispensability argument. Anyone unconvinced of the existence of dark matter is not unconvinced of the cogency of the general form of the argument being used; it's just that they are inclined to think that there are better explanations of the facts in question.

It's not too hard to see that this form of argument is very common in both scientific and everyday usage. Indeed, in these examples, it amounts to no more than an application of inference to the best explanation. This is not to say, of course, that inference to the best explanation is completely uncontroversial. Philosophers of science such as Bas van Fraassen [1980] and Nancy Cartwright [1983] reject unrestricted usage of this style of inference. Typically, rejection of inference to the best explanation results in some form of anti-realism (agnosticism, about theoretical entities in van Fraassen's case and anti-realism about scientific laws in Cartwright's case). Such people will have little sympathy for indispensability arguments. Scientific realists, on the other hand, are generally committed to inference to the best explanation, and *they* are the main target of the indispensability

---

[4]These are graphs of radial angular speed versus mean distance from the centre of the galaxy for stars in a particular galaxy.

argument.[5] Indispensability arguments about mathematics urge scientific realists to place mathematical entities in the same ontological boat as (other) theoretical entities. That is, it invites them to embrace Platonism.[6]

The use of indispensability arguments for defending mathematical realism is usually associated with Quine and Putnam. Quine's version of the indispensability argument is to be found in many places. For instance, in 'Success and Limits of Mathematization' he says:

> Ordinary interpreted scientific discourse is as irredeemably committed to abstract objects — to nations, species, numbers, functions, sets — as it is to apples and other bodies. All these things figure as values of the variables in our overall system of the world. The numbers and functions contribute just as genuinely to physical theory as do hypothetical particles. [Quine, 1981b, pp.149–150]

Here he draws attention to the fact that abstract entities, in particular mathematical entities, are as indispensable to our scientific theories as the theoretical entities of our best physical theories.[7] Elsewhere [Quine, 1951/1980] he suggests that anyone who is a realist about theoretical entities but anti-realist about mathematical entities is guilty of holding a "double standard." For instance, Quine points out that the position that scientific claims, but not mathematical claims, are supported by empirical data is untenable:

> The semblance of a difference in this respect is largely due to overemphasis of departmental boundaries. For a self-contained theory which we can check with experience includes, in point of fact, not only its various theoretical hypotheses of so-called natural science but also such portions of logic and mathematics as it makes use of. [Quine, 1963/1983, p. 367]

He is claiming here that those portions of mathematical theories that are employed by empirical science enjoy whatever empirical support the scientific theory as a whole enjoys. (I will have more to say on this matter in section 5.2.)

Hilary Putnam also once endorsed this argument:

> [Q]uantification over mathematical entities is indispensable for science, both formal and physical; therefore we should accept such quantifica-

---

[5] Indeed, one of the most persuasive arguments *for* scientific realism is generally taken to appeal to inference to the best explanation. This argument is due to J.J.C. Smart [1963].

[6] I'm not claiming here that the indispensability argument for mathematical entities is simply an instance of inference to the best explanation; I'm just noting that inference to the best explanation is a kind of indispensability argument, so those who accept inference to the best explanation are at least sympathetic to this style argument.

[7] I often speak of certain entities being dispensable or indispensable to a given theory. Strictly speaking it's not the entities themselves that are dispensable or indispensable, but rather it's the *postulation of* or *reference to* the entities in question that may be so described. Having said this, though, for the most part I'll continue to talk about *entities* being dispensable or indispensable, eliminable or non-eliminable and occurring or not occurring. I do this for stylistic reasons.

tion; but this commits us to accepting the existence of the mathematical entities in question. This type of argument stems, of course, from Quine, who has for years stressed both the indispensability of quantification over mathematical entities and the intellectual dishonesty of denying the existence of what one daily presupposes. [Putnam, 1971/1979, p. 347]

Elsewhere he elaborates on this "intellectual dishonesty":

> It is like trying to maintain that God does not exist and angels do not exist while maintaining at the very same time that it is an objective fact that God has put an angel in charge of each star and the angels in charge of each of a pair of binary stars were always created at the same time! [Putnam, 1979, p. 74]

Both Quine and Putnam, in these passages, stress the indispensability of mathematics to science. It thus seems reasonable to take science, or at least whatever the goals of science are, as the purpose for which mathematical entities are indispensable. But, as Putnam also points out [1971/1979, p. 355], it is doubtful that there is a single unified goal of science — the goals include explanation, prediction, retrodiction, and so on. Thus, we see that we may construct a variety of indispensability arguments, all based on the various goals of science. As we've already seen, the explanatory indispensability argument is one influential argument of this style, but it is important to bear in mind that it is not the only one.

To state the Quine-Putnam indispensability argument, we need merely replace 'ξ' in argument 1 with 'mathematical entities'. For convenience of future reference I will state the argument here in an explicit form.

ARGUMENT 2 *The Quine-Putnam Indispensability Argument.*

1. We ought to have ontological commitment to all and only those entities that are indispensable to our best scientific theories;

2. Mathematical entities are indispensable to our best scientific theories.

   Therefore:

3. We ought to have ontological commitment to mathematical entities.

A number of questions about this argument need to be addressed. The first is: The conclusion has normative force and clearly this normative force originates in the first premise, but why should an argument about ontology be normative? This question is easily answered, for I take most questions about ontology to be really questions about what we *ought to believe* to exist. The Quine-Putnam indispensability argument, as I've presented it, certainly respects this view of ontology. Indeed, I take it that indispensability arguments are essentially normative. For example, if you try to turn the above Quine-Putnam argument into a descriptive argument, so that the conclusion is that mathematical entities exist, you find you must have something like 'All and only those entities that are indispensable to our best theories exist' as the crucial first premise. This premise, it seems to me, is much more controversial than the normative one. As we shall see, this normativity arises in the doctrine of naturalism, on which I will have more to say shortly.

The next question is: How are we to understand the phrase 'indispensable to our best scientific theory'? In particular, what does 'indispensable' mean in this context? Much hangs on this question, and I'll need to treat it in some detail. I'll do this in the next section. In the meantime, take it to intuitively mean 'couldn't get by without' or the like. In fact, whatever sense it is in which electrons, neutron stars, and viruses are indispensable to their respective theories will do.[8]

The final question is: Why believe the first premise? That is, why should we believe in the existence of entities indispensable to our best scientific explanations? Answering this question is not easy. Briefly, I will argue that the crucial first premise follows from the doctrines of *naturalism* and *holism*. Before I embark on this task, I should point out that the first premise, as I've stated it, is a little stronger than required. In order to gain the given conclusion, all that is really required in the first premise is the 'all,' not the 'all and only.' I include the 'all and only,' however, for the sake of completeness and also to help highlight the important role naturalism plays in questions about ontology, since it is naturalism that counsels us to look to science and *nowhere else* for answers to ontological questions.

Although I'll have more to say about naturalism and holism (in section 3), it will be useful here to outline the argument from naturalism and holism to the first premise of argument 2. Naturalism, for Quine at least, is the philosophical doctrine that there is no first philosophy and the philosophical enterprise is continuous with the scientific enterprise. What is more, science, thus construed (i.e., with philosophy as a continuous part) is taken to be the complete story of the world. This doctrine arises out of a deep respect for scientific methodology and an acknowledgment of the undeniable success of this methodology as a way of answering fundamental questions about all nature of things. As Quine suggests, its source lies in "unregenerate realism, the robust state of mind of the natural scientist who has never felt any qualms beyond the negotiable uncertainties internal to science" [Quine, 1981a, p. 72]. For the metaphysician this means looking to our best scientific theories to determine what exists, or, perhaps more accurately, what we ought to believe to exist. Naturalism, in short, rules out unscientific ways of determining what exists. For example, I take it that naturalism would rule out believing in the transmigration of souls for mystical reasons. It would not, however, rule out belief in the transmigration of souls if this were required by our best scientific theories.

Naturalism, then, gives us a reason for believing in the entities in our best scientific theories and no other entities. Depending on exactly how you conceive of

---

[8]If you think that there is *no* sense in which electrons, neutron stars, and viruses are indispensable to their respective theories, then the indispensability argument is unlikely to have any appeal.

naturalism, it may or may not tell you whether to believe in *all* the entities of your best scientific theories. I take it that naturalism does give us *some* (defeasible) reason to believe in all such entities. This is where the holism comes to the fore — in particular, confirmational holism. Confirmational holism is the view that theories are confirmed or disconfirmed as wholes. So, if a theory is *confirmed* by empirical findings, the *whole* theory is confirmed. In particular, whatever mathematics is made use of in the theory is also confirmed. Furthermore, as Putnam [1971/1979] has stressed, the same evidence that is appealed to in justifying belief in the mathematical components of the theory is appealed to in justifying the empirical portion of the theory (if indeed the empirical can be separated from the mathematical). Taking naturalism and holism together, then, we have the first premise of argument 2.

Before concluding this section, I would like to outline one other indispensability argument that appears in the literature: Michael Resnik's [1995] pragmatic indispensability argument. This argument focuses on the purpose of 'doing science' and is a response to some problems raised for the Quine-Putnam indispensability argument by Penelope Maddy and Elliott Sober. Although I won't discuss these problems here (I do so a little later on, in section 5), one point is important in understanding Resnik's motivation. Resnik wishes to avoid the Quine-Putnam argument's reliance on confirmational holism.

Resnik presents the argument in two parts. The first is an argument for the conditional claim that if we are justified in drawing conclusions from and within science, then we are justified in taking mathematics used in science to be true. He presents this part of the argument as follows:

> 1) In stating its laws and conducting its derivations science assumes the existence of many mathematical objects and the truth of much mathematics.
> 2) These assumptions are indispensable to the pursuit of science; moreover, many of the important conclusions drawn from and within science could not be drawn without taking mathematical claims to be true.
> 3) So we are justified in drawing conclusions from and within science only if we are justified in taking the mathematics used in science to be true. [Resnik, 1995, pp. 169–170]

He then combines the conclusion of this argument with the argument that we *are* justified in drawing conclusions from and within science, since this is the only way we know of doing science. And clearly we *are* justified in doing science. The conclusion, then, is that we are justified in taking whatever mathematics is used in science to be true.[9]

This argument clearly fits the mould of the scientific indispensability argument that I outlined earlier. It differs from the Quinean argument in that it doesn't

---

[9]In fact, Resnik draws the additional (stronger) conclusion that mathematics is true, arguing that this follows from the weaker conclusion, since to assent to the weaker conclusion while denying the stronger invites a kind of Moore's paradox. (Moore's paradox is the paradox of asserting 'P but I don't believe P.')

rely on confirmational holism. Resnik pinpoints the difference rather nicely in the following passage:

> This argument is similar to the confirmational argument except that instead of claiming that the evidence for science (one body of statements) is also evidence for its mathematical components (another body of statements) it claims that the justification for doing science (one act) also justifies our accepting as true such mathematics as science uses (another act). [Resnik, 1995, p. 171]

This argument has some rather attractive features. For instance, since it doesn't rely on confirmational holism, it doesn't require confirmation of any scientific theories in order for belief in mathematical objects to be justified. Indeed, even if *all* scientific theories were disconfirmed, we would (presumably) still need mathematics to do science, and since doing science is justified we would be justified in believing in mathematical objects. This is clearly a very powerful argument and one with which I have considerable sympathy. Although I won't have much more to say about this argument in what follows, it is important to see that a cogent argument in the general spirit of the Quine-Putnam argument can be maintained without recourse to confirmational holism.

2 WHAT IS IT TO BE INDISPENSABLE?

The Quine-Putnam indispensability argument may be stated as follows: We have good reason to believe our best scientific theories and there are no grounds on which to differentiate scientific entities from mathematical entities, so we have good reason to believe in mathematical entities, since they, like the relevant scientific entities, are indispensable to the theories in which they occur. Furthermore, it is exactly the same evidence that confirms the scientific theory as a whole, that confirms the mathematical portion of the theory and hence the mathematical entities contained therein. The concept of *indispensability* is doing a great deal of work in this argument and so we need to have a clear understanding of what is meant by this term.

I've already pointed out that one way an entity can be indispensable is that it can be indispensable for explanation (in which case the resulting argument is an instance of inference to the best explanation). But I think there are other ways in which an entity can be indispensable to a theory.[10] In order to come to a clear understanding of how 'indispensability' is to be understood, I will consider a case where there should be no disagreement about the dispensability of the entity in question. I shall then analyse this case to see what leads us to conclude that the entity in question is dispensable.

---

[10] Quine actually speaks of entities existentially quantified over in the canonical form of our best theories, rather than indispensability. (See [Quine, 1948/1980] for details.) Still, the debate continues in terms of *indispensability*, so we would be well served to clarify this latter term.

Consider an empirically adequate and consistent theory Γ and let 'ξ' be the name of some entity neither mentioned, predicted, nor ruled out by Γ. Clearly we can construct a new theory Γ^+ from Γ by simply adding the sentence 'ξ exists' to Γ. It is reasonable to suppose that ξ plays no role in the theory Γ^+;[11] it is merely predicted by it. I propose that there should be no disagreement here when I say that ξ is dispensable to Γ^+, but let us investigate why this is so.

On one interpretation of 'dispensable' we could argue that ξ is not dispensable since its removal from Γ^+ results in a different theory, namely, Γ.[12] This is not a very helpful interpretation though, since *all* entities are indispensable to the theories in which they occur under this reading. Another interpretation of 'dispensable' might be that ξ is dispensable to Γ^+ since there exists another theory, Γ, with the same empirical consequences as Γ^+ in which ξ does not occur.[13] This interpretation can also be seen to be inadequate since it may turn out that *no* theoretical entities are indispensable under this reading. This result follows from Craig's theorem.[14] If the vocabulary of the theory can be partitioned in the way that Craig's theorem requires (cf. footnote 14), then the theory can be reaxiomatised so that any given theoretical entity is eliminated.[15] I claim, therefore, that this interpretation of 'dispensable' is unacceptable since it fails to account for why ξ in particular is dispensable.

This leads to the following explication of 'dispensable':

An entity is dispensable to a theory iff the following two conditions hold:

(1) There exists a modification of the theory in question resulting in a second theory with exactly the same observational consequences as the first, in which the entity in question is neither mentioned nor predicted.

(2) The second theory must be preferable to the first.

In the preceding example, then, ξ is dispensable since Γ makes no mention of ξ and Γ is preferable to Γ^+ in that the former has fewer ontological commitments, all other things being equal. (Assuming, of course, that fewer ontological commitments is better.[16])

---

[11]The reason I hedge a bit here is that if Γ asserts that all entities have positive mass, for instance, then the existence of ξ helps account for some of the "missing mass" of the universe. Thus, ξ *does* play a role in Γ^+. I know of no way of ruling out such cases; hence the hedge.

[12]More correctly, we should say that we can remove all sentences asserting or implying the existence of ξ from Γ^+.

[13]Modulo my concerns in footnote 11.

[14]This theorem states that relative to a partition of the vocabulary of an axiomatisable theory T into two classes, τ and ω (theoretical and observational say), there exists an axiomatisable theory T' in the language whose only non-logical vocabulary is ω, of all and only the consequences of T that are expressible in ω alone.

[15]Naturally, the question of whether such partitioning is possible is important and somewhat controversial. If it is not possible, it will be considerably more difficult to eliminate theoretical entities from scientific theories. Let's grant for the sake of argument, at least, that such a partitioning *is* possible.

[16]One way in which you might think that fewer ontological commitments is *not* better, is if ξ

Now, it might be argued that on this reading once again every theoretical entity is dispensable, since by Craig's theorem we can eliminate any reference to any entity and the resulting theory will be better, since it doesn't have ontological commitment to the entity in question. This is mistaken though, since the reason for preferring one theory over another is a complicated question — it is not simply a matter of empirical adequacy combined with a principle of ontological parsimony. We thus need to consider some aspects of confirmation theory and its role in indispensability decisions.

Quine clearly had the hypothetico-deductive method in mind as his model of scientific theory confirmation. Philosophy of science has moved on since then; now semantic conceptions of theories and confirmation prevail. But the details of the theory of confirmation need not concern us. All that really matters for present purposes is that in order to decide whether one theory is better than another we appeal to desiderata for good theories and these (for the scientific realist, at least) typically amount to more than mere empirical adequacy.

There's no doubt that a good theory should be empirically adequate; that is, it should agree with (most) observations. Second, all other things being equal, we'd prefer our theories to be consistent, both internally and with other major theories. This is not the whole story though. As we have already seen, Γ and Γ^+ have the same degree of empirical adequacy and consistency (by construction), and yet we are inclined to prefer the former over the latter. Typically such a deadlock is settled by appeal to additional desiderata such as:

(1) **Simplicity/Parsimony:** Given two theories with the same empirical adequacy, we generally prefer that theory which is simpler both in its statement and in its ontological commitments.

(2) **Unificatory/Explanatory Power:** Philip Kitcher [1981] argues rather convincingly for scientific explanation being unification; that is, accounting for a maximum of observed phenomena with a minimum of theoretical devices. Whether or not you accept Kitcher's account, we still require that a theory not simply predict certain phenomena, but explain why such predictions are expected. Furthermore, the best theories do so with a minimum of theoretical devices.

(3) **Boldness/Fruitfulness:** We expect our best theories not to simply predict everyday phenomena, but to make bold predictions of novel entities and phenomena that lead to fruitful future research.

---

actually exists. In this case it seems that Γ^+ is the better theory since it best describes reality. This, however, is to gloss over the important question of how we come to know that ξ exists. If there is some evidence of ξ's existence, then Γ^+ will indeed be the better theory, since it will be empirically superior. If there is no such evidence for the existence of ξ, then it seems entirely reasonable to prefer Γ over Γ^+ as I suggest. It is the latter I had in mind when I set up this case. Indeed, the former case is ruled out by construction. I am not concerned with whether ξ actually exists or not — just that there be no empirical evidence for it.

(4) **Formal Elegance:** This is perhaps the hardest feature to characterise (and no doubt the most contentious). However, there is at least *some* sense in which our best theories have aesthetic appeal. For instance, it may well be on the grounds of formal elegance that we rule out *ad hoc* modifications to a failing theory.

I will not argue in detail for each of these, except to say that despite the notorious difficulties involved in explicating what we mean by terms such as 'simplicity' and 'elegance,' most scientific realists, at least, do look for such virtues in our best theories.[17] Otherwise, we could never choose between two theories such as Γ and Γ^+. I do not claim that this list of desiderata is comprehensive nor do I claim that it is minimal;[18] I merely claim that these sorts of criteria are typically appealed to in the literature to distinguish good theories and I have no objections to such appeals.

In the light of the preceding discussion then, we see that to claim that an entity is dispensable is to claim that a modification of the theory in which it is posited can be made in such a way as to eliminate the entity in question and result in a theory that is better overall (or at least not worse) in terms of simplicity, elegance, and so on. Thus, we see that the argument I presented at the end of the previous section that *any* theoretical entity is dispensable does indeed fail, as I claimed. This is because in most cases the benefit of ontological simplicity obtained by the elimination of the entity in question will be more than offset by losses in other areas.

While it seems reasonable to suppose that the elimination from the body of scientific theory of physical entities such as electrons would result in an overall reduction in the previously described virtues of that theory, it is not so clear that the elimination of mathematical entities would have the same impact. Someone might argue that mathematics is certainly a very effective language for the expression of scientific ideas, in that it simplifies the calculations and statement of much of science, but to do so at the expense of introducing into one's ontology the whole gamut of mathematical entities simply isn't a good deal.

One response to this is to deny that it is a high price at all. After all, a powerful and efficient language is the cornerstone of any good theory. If you have to introduce a few more entities into the theory to get this power and efficiency, then so be it. Although I have considerable sympathy with this line of thought, a more persuasive response is available.

Elsewhere [Colyvan, 1999b; Colyvan, 2001a; Colyvan, 2002] I have argued that mathematics plays an *active* role in many of the theories that make use of it. That is, mathematics is not just a tool that makes calculations easier or simplifies the statement of the theory; it makes important contributions to all of the desiderata of good theories I mentioned earlier. Let me give just one brief example here of

---

[17] And recall that the main target of the indispensability argument is scientific realists.
[18] For instance, it may be possible to explain formal elegance in terms of simplicity and unificatory power.

how mathematics can help provide unification.[19]

Consider a physical system described by the differential equation:

(1) $y - y'' = 0$

(where y is a real-valued function of a single real variable). Equations such as these describe physical systems exhibiting (unconstrained) growth and we can solve them with a little elementary real algebra. But now consider a strikingly similar differential equation that describes certain periodic behaviour:

(2) $y + y'' = 0$

(where, again, y is a real-valued function of a single real variable). Somewhat surprisingly, the same real algebra cannot be used to solve equations such as (2) — we are pushed to complex methods.[20]

Now since complex algebra is a generalisation of real algebra, we can employ the same (complex) method for solving both (1) and (2). Thus we see how complex methods may be said to unify, not only the mathematical theory of differential equations, but also the various physical theories that employ differential equations. But the unification doesn't stop there. The exponential function, which is a solution to (1), is very closely related to the sine and cosine functions, which are solutions to (2). This relationship is spelled out via the definitions of the complex sine and cosine functions. Without complex methods, we would be forced to consider phenomena described by (1) and (2) as completely disparate and, moreover, we would have no unified approach to solving the respective equations. I see this is a striking example of the unification brought to science by mathematics — by complex numbers, in this case. (It is by no means the only such case though; detours into complex analysis are commonplace in modern mathematics — even for what are essentially real-valued phenomenon.)

3 NATURALISM AND HOLISM

With a more precise understanding of what indispensability amounts to, let us now turn to the doctrines required to support the Quine-Putnam indispensability argument. Although a great deal of Quine's philosophy is interconnected, making the isolation of particular doctrines very difficult, I will argue that the two essential theses for our purposes — confirmational holism and naturalism — can be disentangled from the rest of the Quinean web.

---

[19] Although if you are inclined towards the view that explanation is unification that I mentioned earlier, then the following case might be thought to be one in which the mathematics is playing an explanatory role.

[20] Of course, in this simple case we can solve the equations in question by other means (such as by inspection) but the fact remains that complex methods are needed to provide a systematic and unified approach to all such differential equations. See [Boyce and DiPrima, 1986] for details.

3.1 Introducing Naturalism

Naturalism, in its most general form, is the doctrine that we ought to seek accounts of the nature of reality that are not "other-worldly" or "unscientific," but to be more precise than this is to immediately encounter trouble. For instance, David Papineau points out that "nearly everybody nowdays wants to be a 'naturalist', but the aspirants to the term nevertheless disagree widely on substantial questions of philosophical doctrine" [Papineau, 1993, p. 1]. In one way this is not at all surprising, for, after all, there is no compulsion for all naturalists to agree on other philosophical stances, distinct from naturalism, and such stances, when combined with naturalism, presumably yield different results. It all depends on what you mix your naturalism with.

There is, however, another reason for disagreement among naturalistic philosophers: Different philosophers use the word 'naturalism' to mean different things. Naturalism involves a certain respect for the scientific enterprise — that much is common ground — but exactly how this is cashed out is a matter of considerable debate. For instance, for David Armstrong naturalism is the doctrine that "nothing but Nature, the single, all-embracing spatio-temporal system exists" [Armstrong, 1978, Vol. 1, p. 138], whereas, for Quine, naturalism is the "abandonment of the goal of a first philosophy" [Quine, 1981a, p. 72].

One issue on which naturalistic philosophers disagree, and which is of fundamental importance for our purposes, is the ontological status of mathematical entities. We've already seen how the Quine-Putnam indispensability argument legitimates belief in mind-independent mathematical objects, and that this argument depends on naturalism. On the other hand, philosophers such as David Armstrong cite naturalism as grounds for rejecting belief in any such mind-independent abstract objects.

While there is no way of preventing philosophers from mixing their naturalism with other philosophical doctrines (so long as the mix is coherent), there is good reason for requiring that the various, often contrary, positions that fly under the banner of naturalism be disentangled, from one another. This is a very large task but we can at least try to identify the difference between the varieties of naturalism that may be used to undermine mathematical realism and the Quinean variety.

3.2 Quinean Naturalism

Quine's aphoristic characterisations of naturalism are well known. In 'Five Milestones of Empiricism' he tells us that naturalism is the

> abandonment of the goal of a first philosophy. It sees natural science as an inquiry into reality, fallible and corrigible but not answerable to any supra-scientific tribunal, and not in need of any justification beyond observation and the hypothetico-deductive method. [Quine, 1981a, p. 72]

And that:

> [t]he naturalistic philosopher begins his reasoning within the inherited world theory as a going concern. He tentatively believes all of it, but believes also that some unidentified portions are wrong. He tries to improve, clarify, and understand the system from within. He is the busy sailor adrift on Neurath's boat. [Quine, 1981a, p. 72]

The aphorisms are useful, but they also mask a great deal of the subtlety and complexity of Quinean naturalism. Indeed, the subtleties and complexities of naturalism are far greater than one would expect for such a widely held and intuitively plausible doctrine. We would do well to spend a little time attempting to better understand Quinean naturalism.

As I see it, there are two strands to Quinean naturalism. The first is a normative thesis concerning how philosophy ought to approach certain fundamental questions about our knowledge of the world. The advice here is clear: look to science (and nowhere else) for the answers. Science, although incomplete and fallible, is taken to be the best guide to answering all such questions. In particular, "first philosophy" is rejected. That is, Quine rejects the view that philosophy precedes science or oversees science. This thesis has implications for the way we should answer metaphysical questions: We should determine our ontological commitments by looking to see which entities our best scientific theories are committed to. Thus, I take it that naturalism tells us (1) we ought to grant real status only to the entities of our best scientific theories and (2) we ought to (provisionally) grant real status to all the entities of our best scientific theories. For future reference I'll call this first strand of Quinean naturalism the *no-first-philosophy thesis* and its application to metaphysics the *Quinean ontic thesis*.

It is worth pointing out that the Quinean ontic thesis is distinct from a thesis about how we determine the ontological commitments of *theories*. According to this latter thesis, the ontological commitments of theories are determined on the basis of the domain of quantification of the theory in question.[21] Call this thesis *the ontological commitments of theories thesis*. One could quite reasonably believe the ontological commitments of theories thesis without accepting the Quinean ontic thesis. For instance, I take it that Bas van Fraassen [1980] accepts that our current physics is committed to entities such as electrons and the like, but it does not follow that he believes that it is rational to believe in these entities in order to believe the theory. The ontological commitments of theories thesis is purely descriptive, whereas the Quinean ontic thesis is normative. From here on I shall be concerned only with the Quinean ontic thesis, but it is worth bearing in mind the difference, because I don't think that the ontological commitments of theories thesis rightfully belongs to the doctrine of naturalism. It is *an* answer to the question of how we determine the ontological commitments of theories, but it is not the only naturalistic way such questions can be answered.

The second strand of Quinean naturalism is a descriptive thesis concerning the subject matter and methodology of philosophy and science. Here naturalism

---

[21] See [Quine, 1948/1980, pp. 12–13] for details.

tells us that philosophy is continuous with science and that together they aim to investigate and explain the world around us. What is more, it is supposed that this science–philosophy coalition is up to the task. That is, all phenomena are in principle explicable by science. For future reference I'll call this strand the *continuity thesis*.

Although it is instructive to distinguish the two strands of Quinean naturalism in this way, it is also important to see how intimately intertwined they are. First, there is the intriguing interplay between the two strands. The no-first-philosophy thesis tells us that we ought to believe our best scientific theories and yet, according to the continuity thesis, philosophy is part of these theories. This raises a question about priority: In the case of a conflict between philosophy and science, which gets priority? Philosophy does not occupy a privileged position. That much is clear. But it also appears, from the fact that philosophy is seen as part of the scientific enterprise, that science (in the narrow sense — i.e., excluding philosophy) occupies no privileged position either.

The second important connection between the two strands is the way in which the continuity thesis lends support to no-first-philosophy thesis. The traditional way in which first philosophy is conceived is as an enterprise that is prior and distinct from science. Philosophical methods are seen to be a priori while those of science are *a posteriori*. But accepting the continuity thesis rules out such a view of the relationship between philosophy and empirical science. Once philosophy is located within the scientific enterprise, it is more difficult to endorse the view that philosophy oversees science. I'm not claiming that the continuity thesis entails the no-first-philosophy thesis, just that it gives it a certain plausibility.[22]

Now to the question of why one ought to embrace naturalism. I won't embark on a general defence of naturalism — that would be far too ambitious. I take it that almost everyone accepts some suitably broad sense of this doctrine.[23] But subscribing to some form or another does not entail subscribing to Quinean naturalism. Again, I won't try anything so ambitious as defend Quinean naturalism here.[24] Still it is useful to see what's at issue.

Let's start by marking out the common ground. Naturalists of all ilks agree that we should look only to science when answering questions about the nature of reality. What is more, they all agree that there is at least prima facie reason to accept all the entities of our best scientific theories. That is, they all agree that there is a metaphysical component to naturalism. So they are inclined to accept the first part of the Quinean ontic thesis (the 'only' part) and are inclined to, at least provisionally, accept the second part (the 'all' part). (Most naturalists believe that naturalism entails scientific realism but they are inclined to be a little

---

[22] Indeed, the continuity thesis cannot entail the no-first-philosophy thesis since the former is descriptive and the latter normative.

[23] Again it is worth bearing in mind that the primary targets of the indispensability argument are scientific realists disinclined to believe in mathematical entities. These scientific realists typically subscribe to some form of naturalism.

[24] See [Colyvan, 2001a, chap. 2 and 3] for a limited defence.

reluctant to embrace *all* the entities of our best scientific theories.)[25] What I take to be the distinctive feature of Quinean naturalism is the view that our best scientific theories are continuous with philosophy and are not to be overturned by first philosophy. It is this feature that blocks any first-philosophy critique of the ontological commitments of science. Consequently, it is this feature of Quinean naturalism that is of fundamental importance to the indispensability argument.

3.3 Holism

Holism comes in many forms. Even in Quine's philosophy there are at least two different holist theses. The first is what is usually called *semantic holism* (although Quine calls it *moderate holism* [1981a, p. 71]) and is usually stated, somewhat metaphorically, as the thesis that the unit of meaning is the whole of the language. As Quine puts it:

> The idea of defining a symbol in use was ... an advance over the impossible term-by-term empiricism of Locke and Hume. The statement, rather than the term, came with Bentham to be recognized as the unit accountable to an empiricist critique. But what I am now urging is that even in taking the statement as unit we have drawn our grid too finely. The unit of empirical significance is the whole of science. [Quine, 1951/1980, p. 42]

Semantic holism is closely related to Quine's denial of the analytic/synthetic distinction and his thesis of indeterminacy of translation. He argues for the former in a few places, but most notably in 'Two Dogmas of Empiricism' [Quine, 1951/1980], while the latter is presented in *Word and Object* [Quine, 1960].

The other holist thesis found in Quine's writings is *confirmational holism* (also commonly referred to as the Quine/Duhem thesis). As Fodor and Lepore point out [Fodor and Lepore, 1992, pp. 39–40], the Quine/Duhem thesis receives many different formulations by Quine and it is not clear that all these formulations are equivalent. For example, in *Pursuit of Truth* Quine writes:[26]

> [T]he falsity of the observation categorical[27] does not conclusively refute the hypothesis. What it refutes is the conjunction of sentences that was needed to imply the observation categorical. In order to retract that conjunction we do not have to retract the hypothesis in question; we could retract some other sentence of the conjunction instead. This is the important insight called *holism*. [Quine, 1992, pp. 13–14]

And in a much quoted passage from 'Two Dogmas of Empiricism', he suggests that "our statements about the external world face the tribunal of sense experience not

---

[25]For example, Keith Campbell [1994] advocates "selective realism", and Quine restricts commitment to indispensable entities.
[26]Cf. Duhem [1962, p. 187] for a similar statement of the thesis.
[27]By 'observation categorical' Quine simply means a statement of the form 'whenever P, then Q.' For example, 'where there's smoke, there's fire.'

individually but only as a corporate body" [Quine, 1951/1980, p. 41]. Elsewhere, in a similar vein, he tells us:

> As Pierre Duhem urged, it is the system as a whole that is keyed to experience. It is taught by exploitation of its heterogeneous and sporadic links with experience, and it stands or falls, is retained or modified, according as it continues to serve us well or ill in the face of continuing experience. [Quine, 1953/1976, p. 222]

In the last two of these three passages Quine emphasizes the *confirmational* aspects of holism — it's the whole body of theory that is tested, not isolated hypotheses. In the first passage he emphasizes *disconfirmational* aspects of holism — when our theory conflicts with observation, any number of alterations to the theory can be made to resolve the conflict. Despite the difference in emphasis, I take it that these theses are equivalent (or near enough). Moreover, I take it that they are all true, modulo some quibbles about how much theory is required to face the tribunal at any time.

Although Quine was inclined to argue for confirmational holism from (the more controversial) semantic holism, this is not the only way to establish the former. Both Duhem [1962] and Lakatos [1970] have argued for confirmational holism without any (obvious) recourse to semantic considerations. They emphasize the simple yet undeniable point that there is more than one way in which a theory, faced with recalcitrant data, can be modified to conform with that data. Consequently, certain core doctrines of a theory may be held onto in the face of recalcitrant data by making suitable alterations to auxiliary hypotheses. Indeed, in its most general form, confirmational holism is little more than a point about logic.

Before leaving the doctrine of holism, I wish to consider one last question: Might one accept confirmational holism as stated, but reject the claim that mathematical propositions are one with the rest of science? That is, might it not be possible to pinpoint some semantic difference between the mathematical propositions employed by science and the rest, with empirical confirmation and disconfirmation reserved for the latter? Carnap [1937], with his appeal to "truth by convention," suggested precisely this. Quine, of course, denies that this can be done [1936/1983; 1951/1980; 1963/1983], but exploring the reasons for his denial would take us deep into issues in the philosophy of language. For our purposes, it will suffice to note that there is no obvious way of disentangling the purely mathematical propositions from the main body of science. Our empirical theories have the so-called empirical parts intimately intertwined with the mathematical. A cursory glance at any physics book will confirm this, where one is likely to find mixed statements such as: 'planets travel in elliptical orbits'; 'the curvature of space-time is not zero'; 'the work done by the force on the particle is given by $W = \int_a^b \mathbf{F} \cdot d\mathbf{r}$.'

Thus, even if you reject Quine's semantic holism and you think that mathematical and logical language is different in kind from empirical language, you need not reject confirmational holism. In order to reject confirmational holism, you would need (at the very least) to separate the mathematical vocabulary from the empirical in all of our best scientific theories. Clearly this task is not trivial.[28] If you still feel some qualms about confirmational holism, though, you may rest assured — this doctrine will be called into question when we consider some of the objections to the indispensability argument.

3.4 The First Premise Revisited

Let's return to the question of how confirmational holism and Quinean naturalism combine to yield the first premise of the Quine-Putnam indispensability argument. First, you might wonder whether holism is required for the argument. After all, (Quinean) naturalism alone delivers something very close to the crucial first premise. (More specifically, the Quinean ontic thesis is very suggestive of the required premise.) As a matter of fact, I think that the argument can be made to stand without confirmational holism: It's just that it is more secure *with* holism. The problem is that naturalism is somewhat vague about ontological commitment to the entities of our best scientific theories. It quite clearly rules out entities *not* in our best scientific theories, but there seems room for dispute about commitment to some of the entities that *are* in these theories. Holism helps to block such a move since, according to holism, it is the whole theory that is granted empirical support.

So, naturalism tells us to look to our best scientific theories for our ontological commitments. We thus have provisional support for all the entities in these theories and no support for entities not in these theories. For reasons of parsimony, however, we may wish to grant real status to only those entities that are *indispensable* to these theories. However, we are unable to pare down our ontological commitments further by appealing to some distinction based on empirical support because, according to holism, all the entities in a confirmed theory receive such support. In short, holism blocks the withdrawal of the provisional support supplied by naturalism. And that gives us the first premise of the Quine-Putnam indispensability argument.

4 THE HARD ROAD TO NOMINALISM: FIELD'S PROJECT

In the last twenty five years, the indispensability argument has suffered attacks from seemingly all directions. Charles Chihara [1973] and Hartry Field [1980] raised doubts about the indispensability of mathematics to science, then Elliot Sober [1993], Penelope Maddy [1992; 1995; 1997] and others have expressed concerns about whether we really ought to be committed to the indispensable entities of our best scientific theories.

These attacks can be divided into two kinds: *hard-road strategies* and *easy-road strategies*. The hard-road strategies seek to show that mathematics, despite

---

[28] As we shall see, Hartry Field [1980] undertakes this task for reasons not unrelated to those I've aired here.

initial appearances, is in fact dispensable to science. That is, the hard road to nominalism is to attempt to demonstrate the falsity of the second premise of the indispensability argument. As we shall see in this section, there is a great deal of quite technical work associated with this enterprise — much of which is yet to be carried out. The alternative, the easy road, tackles the first premise and attempts to show that we need not be committed to *all* the indispensable entities of our best scientific theories. This latter strategy, if successful, would avoid the many difficulties associated with the hard road. I begin this section by considering Hartry Field's hard-road strategy, then in the next I consider a couple of attempts at finding an easy road to nominalism.

Field's distinctive fictionalist philosophy of mathematics has been very influential in the 25 years since the publication of *Science Without Numbers* [Field, 1980]. This influence is no accident; it's a tribute to the plausibility of the account of mathematics offered by Field and his unwillingness to dodge the issues associated with the applications of mathematics. Furthermore, unlike other nominalist philosophies of mathematics,[29] Field's nominalism is not revisionist:

> I do not propose to reinterpret any part of classical mathematics; instead, I propose to show that the mathematics needed for application to the physical world does not include anything which even *prima facie* contains references to (or quantifications over) abstract entities like numbers, functions, or sets. Towards that part of mathematics which does contain references to (or quantification over) abstract entities — and this includes virtually all of conventional mathematics — I adopt a fictional attitude: that is, I see no reason to regard this part of mathematics as *true*. [Field, 1980, pp. 1–2]

He accepts the Quinean backdrop discussed in section 3 and agrees that if mathematics were indispensable to our best scientific theories, we would have good reason to grant mathematical entities real status. Field, however, denies that mathematics is indispensable to science. In effect he accepts the burden of proof in this debate. That is, he accepts that he must show (1) how it is that mathematical discourse may be used in its various applications in physical science and (2) that it is possible to do science without reference to mathematical entities. This is indeed an ambitious project and certainly one deserving careful attention, for if it succeeds, the indispensability argument is no longer a way of motivating mathematical realism.

4.1 Science without Numbers

Before discussing the details of Field's project, it is important to understand something of its motivation. Field is driven by two things. First, there are well known

---

[29]For example, see [Chihara, 1973], where mathematical discourse is *reinterpreted* so as to be about linguistic entities rather than mathematical entities.

prima facie difficulties with Platonism — namely, the two Benacerraf problems [Benacerraf, 1965/1983; Benacerraf, 1973/1983] — which nominalism avoids [Field, 1989, p. 6].[30] Second, he is motivated by certain rather attractive principles in the philosophy of science: (1) we ought to seek *intrinsic* explanations whenever this is possible and (2) we ought to eliminate arbitrariness from theories [Field, 1980, p. ix]. In relation to (1), Field says, "one wants to be able to explain the behaviour of the physical system *in terms of the intrinsic features of that system*, without invoking extrinsic entities (whether non-mathematical or mathematical) whose properties are irrelevant to the behaviour of the system being explained" (emphasis in original) [Field, 1984/1989, p. 193]. He also points out that this concern is orthogonal to nominalism [Field, 1980, p. 44]. As for (2), this too is independent of nominalism. Coordinate-independent (tensor) methods used in most field theories are considered more attractive by Platonists and nominalists alike. These motivations are important for a full understanding of Field's project; the project is driven by more than just nominalist sympathies.

Now to the details of Field's project. There are two parts to the project. The first is to justify the use of mathematics in its various applications in empirical science. If one is to present a believable, fictional account of mathematics, one must present some account of how mathematics may be used with such effectiveness in its various applications in physical theories. To do this, Field argues that mathematical theories don't have to be true to be useful in applications; they merely need to be *conservative*. Conservativeness is, roughly, that if a mathematical theory is added to a nominalist scientific theory, no nominalist consequences follow that wouldn't follow from the scientific theory alone. I'll have more to say about this shortly. The second part of Field's project is to demonstrate that our best scientific theories can be suitably nominalised. To do this, he is content to nominalise a large fragment of Newtonian gravitational theory. Although this is a far cry from showing that *all* our *current* best scientific theories can be nominalised, it is certainly not trivial. The hope is that once one sees how the elimination of reference to mathematical entities can be achieved for a typical physical theory, it will seem plausible that the project could be completed for the rest of science.

One further point that is important to bear in mind is that Field is interested in undermining what he takes to be the only good argument for Platonism. He is thus justified in using Platonistic methods. His strategy is to show *Platonistically* that abstract entities are not needed in order to do empirical science. If his project is successful, "[P]latonism is left in an unstable position: it entails its own unjustifiability" [Field, 1980, p. 6]. I'll now discuss the first part of his project.

Field's account of how mathematical theories might be used in scientific theories, even when the mathematical theory in question is false, is crucial to his fictionalism about mathematics. Field, of course, does provide such an account, the key to which is the concept of *conservativeness*, which may be defined (roughly) as follows:

---

[30]Or, rather, nominalism trades these problems for a different set of problems — most notably, to disarm the indispensability argument.

A mathematical theory M is said to be conservative if, for any body of nominalistic assertions S and any particular nominalistic assertion C, then C is not a consequence of $M + S$ unless it is a consequence of S.

A few comments are warranted here in relation to definition 4.1. First, as it stands, the definition is not quite right; it needs refinement in order to avoid certain technical difficulties. For example, we need to exclude the possibility of the nominalistic theory containing the assertion that there are no abstract entities. Such a situation would render $M + S$ inconsistent. There are natural ways of performing the refinements required, but the details aren't important here. (See [Field, 1980, pp. 11–12] for details.)[31] Second, 'nominalistic assertion' is taken to mean an assertion in which all the variables are explicitly restricted to non-mathematical entities (for reasons I suggested earlier). Third, Field is at times a little unclear about whether he is speaking of semantic entailment or syntactic entailment (e.g., [Field, 1980, pp. 16–19]; in other places (e.g., [Field, 1980, p. 40], and [Field, 1985/1989]) he is explicit that it *is* semantic entailment he is concerned with.[32] Finally, the key concept of conservativeness is closely related to (semantic) consistency.[33] Field, however, cannot (and does not) cash out consistency in model-theoretic terms (as is usually the case), for obviously such a construal depends on models, and these are not available to a nominalist. Instead, Field appeals to a primitive sense of possibility.

Now if it could be proved that all of mathematics were conservative, then its truth or falsity would be irrelevant to its use in empirical science. More specifically, if some mathematical theory were false but conservative, it would not lead to false nominalistic assertions when conjoined with some nominalist, empirical theory, unless such false assertions were consequences of the empirical theory alone. As Field puts it, "mathematics does not need to be true to be good" [Field, 1985/1989, p. 125]. Put figuratively, conservativeness ensures that the alleged falsity of the mathematical theory does not "infect" the whole theory.

Field provides a number of reasons for thinking that mathematical theories are conservative. These reasons include several formal proofs of the conservativeness of set theory.[34] Here I just wish to demonstrate the plausibility of the conservativeness claim by showing how closely related conservativeness is to consistency. First, for pure set theory (i.e., set theory without urelements[35]) conservativeness follows from consistency alone [Field, 1980, p. 13]. In the case of impure set theory, the conservativeness claim is a little stronger than consistency. An impure set theory could be consistent but fail to be conservative because it implied con-

---

[31]There are, however, more serious worries about Field's formulation of the conservativeness claim. See [Urquhart, 1990] for details.

[32]Of course, this is irrelevant if the logic in question is first-order. But since Field was at one stage committed to second-order logic, the semantic–syntactic issue is non-trivial. See [Shapiro, 1983] and [Field, 1985/1989] for further details. See also footnote 39 of this chapter.

[33]Conservativeness entails consistency and, in fact, conservativeness can be defined in terms of consistency.

[34]See [Field, 1980, pp. 16–19] and [Field, 1992] for details.

[35]A urelement is an element of a set that is not itself a set.

clusions about concrete entities that were not logically true. Field sums up the situation (emphasis in original):

> [S]tandard mathematics *might* turn out not to be conservative ..., for it might conceivably turn out to be inconsistent, and if it is inconsistent it certainly isn't conservative. We would however regard a proof that standard mathematics was inconsistent as extremely surprising, and as showing that standard mathematics needed revision. Equally, it would be extremely surprising if it were to be discovered that standard mathematics implied that there are at least 10^6 non-mathematical objects in the universe, or that the Paris Commune was defeated; and were such a discovery to be made, all but the most unregenerate rationalists would take this as showing that standard mathematics needed revision. *Good* mathematics *is* conservative; a discovery that accepted mathematics isn't conservative would be a discovery that it isn't good. [Field, 1980, p. 13]

It is also worth noting that Field claims that there is a disanalogy between mathematical theories and theories about unobservable physical entities. The latter he suggests *do* facilitate new conclusions about observables and hence are not conservative [Field, 1980, p. 10]. The disanalogy is due to the fact that conservativeness is also closely related to necessary truth. In fact, conservativeness follows from necessary truth. Field remarks that "[c]onservativeness might loosely be thought of as 'necessary truth without the truth'" [Field, 1988/1989, p. 241].

With conservativeness established, it is permissible for a fictionalist about mathematics to use mathematics in a nominalistic scientific theory, despite the falsity of the former. It remains to show that our current best scientific theories can be purged of their references to abstract objects. Field's strategy for eliminating all references to mathematical objects from empirical science is to appeal to the representation theorems of measurement theory. Although the details of this are fairly technical, no account of Field's project is complete without at least an indication of how this is done. It is also of considerable interest in its own right. Furthermore, as Michael Resnik points out, this part of his project provides a very nice account of applied mathematics, which should be of interest to *all* philosophers of mathematics, realists and anti-realists alike [Resnik, 1983, p. 515]. In light of all this, it would be remiss of me not to at least outline this part of Field's project.

Field's project is modelled on Hilbert's axiomatisation of Euclidean geometry [Hilbert, 1899/1971]. The central idea is to replace all talk of distance and location, which require quantification over real numbers, with the comparative predicates 'between' and 'congruent,' which require only quantification over space-time points. It will be instructive to present this case in a little more detail. My treatment here follows [Field, 1980, pp. 24–29].

For present purposes, the important feature of Hilbert's theory is that it contains the following relations:

1. The three-place *between* relation (where 'y' is between 'x' and 'z' is written

'$y\,\text{Bet}\,xz$'), which is intuitively understood to mean that x is a point on the line segment with endpoints y and z.

2. The four-place *segment-congruence* relation (where 'x and y are congruent to z and w' is written '$xy\,\text{Cong}\,zw$'), which is intuitively understood to mean that the distance from point x to point y is the same as the distance from point z to point w.

The notion of (Euclidean) distance appealed to in the segment-congruence relation is *not* part of Hilbert's theory; in fact, it cannot even be defined in the theory. But this does not mean that Hilbert's theory is deficient in any sense, for he proved in a broader mathematical theory the following representation theorem:

THEOREM 3 *Hilbert's Representation Theorem. For any model of Hilbert's axiom system for space S, there exists a function $\text{d}: S \times S \to \mathbb{R}^+ \cup \{0\}$ which satisfies the following two homomorphism conditions:*

(a) *For any four points x, y, z, and w, $xy\,\text{Cong}\,zw$ iff $\text{d}(xy) = \text{d}(zw)$;*

(b) *For any three points x, y, and z, $y\,\text{Bet}\,xz$ iff $\text{d}(xy) + \text{d}(yz) = \text{d}(xz)$.*

From this it is easy to show that any Euclidean theorem about length would be true if restated as a theorem about any function d satisfying the conditions of theorem 3. In this way we can replace quantification over numbers with quantification over points. As Field puts it (emphasis in original):

> So *in the geometry itself* we can't talk about numbers, and hence we can't talk about distances ... ; but we have a metatheoretic proof which associates claims about distances ... with what we can say in the theory. Numerical claims then, are abstract counterparts of purely geometric claims, and the equivalence of the abstract counter-part with what it is an abstract counterpart of is established in the broader mathematical theory. [Field, 1980, pp. 27]

Hilbert also proved a uniqueness theorem corresponding to theorem 3. This theorem states that if there are two functions d_1 and d_2 satisfying the conditions of theorem 3, then $\text{d}_1 = k\text{d}_2$ where k is some arbitrary positive constant. This, claims Field, provides a satisfying explanation of why geometric laws formulated in terms of distance are invariant under multiplication by a positive constant (and that this is the only transformation under which they are invariant). Field claims that this is one of the advantages of this approach: The invariance is given an explanation in terms of the intrinsic facts about space [Field, 1980, pp. 27].

With the example of Hilbert's axiomatisation of Euclidean space in hand, Field then does for Newtonian space-time what Hilbert did for $\mathbb{R}^2$. This in itself is non-trivial, but Field is required to do much more, since he must dispense with *all* mention of physical quantities. He does this by appeal to relational properties, which compare space-time points with respect to the quantity in question. For example, rather than saying that some space-time point has a certain gravitational potential, Field compares space-time points with respect to the 'greater gravitational potential' relation.[36] The details of this and the more technical task of how to formulate differential equations involving scalar quantities (such as gravitational potential) in terms of the spatial and scalar relational primitives need not concern us here. (The details can be found in [Field, 1980, pp. 55–91].) The important point is that Field is able to derive an extended representation theorem:[37]

THEOREM 4 *Field's Extended Representation Theorem. For any model of a theory N with space-time S that uses comparative predicates but not numerical functors there are:*

(a) *a 1-1 spatio-temporal co-ordinate function* $\Phi : S \to \mathbb{R}^4$, *which is unique up to generalised Galilean transformation,*

(b) *a mass density function* $\rho : S \to \mathbb{R}^+ \cup \{0\}$, *which is unique up to a positive multiplicative transformation, and*

(c) *a gravitation potential function* $\Psi : S \to \mathbb{R}$, *which is unique up to positive linear transformation,*

all of which are structure preserving (in the sense that the comparative relations defined in terms of these functions coincide with the comparative relations used in N); moreover, the laws of Newtonian gravitational theory in their functorial form hold if Φ, ρ, and Ψ are taken as denotations of the relevant functors.

There are many complaints against Field's project, ranging from the complaint that it is not genuinely nominalist [Resnik, 1985a; Resnik, 1985b] since it makes use of space-time points, to technical difficulties such as the complaint that it is hard to see how Field's project can be made to work for general relativity where the space-time manifold has non-constant curvature [Urquhart, 1990, p. 151] and for theories where the represented objects are not space-time points, but mathematical objects [Malament, 1982].[38] Other complaints revolve around issues concerning the appropriate logic for the project — should it be first- or second-order? — and various problems associated with each option.[39] Finally, Field's project has been

---

[36] Of course there is the task of getting the axiomatisation of the gravitational potential relation such that the desired representation and uniqueness theorems are forthcoming. But much of Field's work has, in effect, been done for him by workers in measurement theory [Field, 1980, pp. 57–58].

[37] The statement of the theorem here is from [Field, 1985/1989, pp. 130–131].

[38] For example, in classical Hamiltonian mechanics the represented objects are possible dynamical states. Similar problems, it seems, will arise in any phase-space theory, and the prospects look even dimmer for quantum mechanics [Malament, 1982, pp. 533–534]. See also [Balaguer, 1998, chap. 6] for an indication of how the nominalisation of quantum mechanics might proceed.

[39] See, for example, [Shapiro, 1983; Urquhart, 1990; Maddy, 1990b; Maddy, 1990c] in this regard. See also [Field, 1990], where Field seemingly retreats from his earlier endorsement of second-order logic as a result of subsequent debate. The interested reader is also referred to [Burgess and Rosen, 1997], (especially pp. 118–123 and pp. 190–196) for a nice survey and discussion of criticisms of Field's project.

criticised because it seems unlikely that his nominalised science is able to properly account for progress [Baker, 2001; Burgess, 1983] and unification [Colyvan, 1999b; Colyvan, 2001a] in science.

While such debates are of considerable interest, I will not pursue them here. It would seem that the consensus of informed opinion on Field's project is that the various technical difficulties it faces leaves a serious question over its likely success.

Although I am not yet convinced that Field's project will be successful, I have no doubt about the importance of his project. Indeed, I, like Field, believe that the correct philosophical stance with regard to the realism/anti-realism debate in mathematics hangs on the outcome of his project. However, not everyone takes this view. In the next section I turn to some criticisms of the first premise of indispensability argument which are in some ways more fundamental than Field's. The authors I discuss in the next section argue that even if mathematics turns out to be indispensable to our best scientific theories, that does not mean we need to treat mathematics realistically (or as having been confirmed). If they are right about this, then Field's project is irrelevant to whether mathematical objects ought to be considered real or not.

5 THE EASY ROAD TO NOMINALISM: REJECTING HOLISM

Now I turn to some of the attacks on the first premise. There are many such attacks and I don't have space to do justice to them all here. Instead, I'll focus on just two influential ones that give the flavour of this style of critique of the indispensability argument.[40] What is common to the following critiques of the indispensability argument is that, in different ways, each rejects holism. That is they offer arguments against the Quinean thesis that we ought to be committed to all the indispensable entities of our best scientific theories.

5.1 Maddy

One-time mathematical realist Penelope Maddy has advanced some serious objections to the indispensability argument. Indeed, so serious are these objections, that she has renounced the realism she so enthusiastically argued for in [Maddy, 1990a].[41] That realism crucially depended on indispensability arguments. Although her objections to indispensability arguments are largely independent of one another, there is a common thread that runs through each of them. Maddy's arguments draw attention to problems of reconciling the naturalism and confirmational holism required for the Quine-Putnam indispensability argument. In particular, she points out how a holistic view of scientific theories has problems

---

[40] Jody Azzouni [2004], Mark Balaguer [1998, chap. 7], Colin Cheyne [2001] and Joseph Melia [2000] are others who have recently argued against the first premise of the indispensability argument.

[41] She implicitly renounces the set theoretic realism of *Realism in Mathematics* in many places, but she explicitly renounces it in [Maddy, 1997].

explaining the legitimacy of certain aspects of scientific and mathematical practices — practices that presumably *ought to be* legitimate given the high regard for scientific methodology that naturalism endorses.[42]

The first objection to the indispensability argument, and in particular to confirmational holism, is that the actual attitudes of working scientists towards the components of well-confirmed theories vary "from belief to grudging tolerance to outright rejection" [Maddy, 1992, p. 280]. In 'Taking Naturalism Seriously' [Maddy, 1994] Maddy presents a detailed and concrete example that illustrates these various attitudes. The example is the history of atomic theory from early last century, when the (modern) theory was first introduced, until early this century, when atoms were finally universally accepted as real. The puzzle for the Quinean "is to distinguish between the situation in 1860, when the atom became 'the fundamental unit of chemistry', and that in 1913, when it was accepted as real" [Maddy, 1994, p. 394]. After all, if the Quinean ontic thesis is correct, then scientists ought to have accepted atoms as real once they became indispensable to their theories (presumably around 1860), and yet renowned scientists such as Poincaré and Ostwald remained sceptical of the reality of atoms until as late as 1904.

For Maddy the moral to be drawn from this episode in the history of science is that "the scientist's attitude toward contemporary scientific practice is rarely so simple as uniform belief in some overall theory" [Maddy, 1994, p. 395]. Furthermore, she claims that "[s]ome philosophers might be tempted to discount this behavior of actual scientists on the grounds that experimental confirmation is enough, but such a move is not open to the naturalist" [Maddy, 1992, p. 281], presumably because "naturalism counsels us to second the ontological conclusions of natural science" [Maddy, 1995, p. 251]. She concludes:

> If we remain true to our naturalistic principles, we must allow a distinction to be drawn between parts of a theory that are true and parts that are merely useful. We must even allow that the merely useful parts might in fact be indispensable, in the sense that no equally good theory of the same phenomena does without them. Granting all this, the indispensability of mathematics in well-confirmed scientific theories no longer serves to establish its truth. [Maddy, 1992, p. 281]

The next problem for indispensability, Maddy suggests, follows on from the last. Once one rejects the picture of a scientific theory as a homogeneous unit, there's a need to address the question of whether the mathematical portions of theories

---

[42]I should mention that Maddy does not claim to be advancing a nominalist philosophy of mathematics; her official position is neither Platonist nor nominalist. Instead, she rejects this metaphysical approach to the philosophy of mathematics in favour of a more methodologically-based approach. This results in a position she calls *set theoretic naturalism*. See [Maddy, 1997] for details. Despite her official stance on the realism/anti-realism issue, I include her here among the "easy roaders" because she, like the others in this camp, rejects the first premise of the indispensability argument. It is because of this that she in turn rejects Platonism. This is enough to make her an easy roader, or at least a travelling companion of the easy roaders.

fall within the true elements of the confirmed theories. To answer this question, Maddy points out first that much mathematics is used in theories that make use of hypotheses that are explicitly false, such as the assumption that water is infinitely deep in the analysis of water waves or that matter is continuous in fluid dynamics. Furthermore, she argues that these hypotheses are indispensable to the relevant theory, since the theory would be unworkable without them. It would be foolish, however, to argue for the reality of the infinite simply because it appears in our best theory of water waves [Maddy, 1995, p. 254].

Next she looks at instances of mathematics appearing in theories not known to contain explicitly false simplifying assumptions and she claims that "[s]cientists seem willing to use strong mathematics whenever it is useful or convenient to do so, without regard to the addition of new *abstracta* to their ontologies, and indeed, even more surprisingly, without regard to the additional physical structure presupposed by that mathematics" [Maddy, 1995, p. 255]. In support of this claim she looks at the use of continuum mathematics in physics. It seems the real numbers are used purely for convenience. No regard is given to the addition of uncountably many extra entities (from the rationals, say) or to the seemingly important question of whether space and time (which the reals are frequently used to model) are in fact continuous or even dense. Nor is anyone interested in devising experiments to test the density or continuity of space and time. She concludes that "[t]his strongly suggests that *abstracta* and mathematically-induced structural assumptions are not, after all, on an epistemic par with physical hypotheses" [Maddy, 1995, p. 256].

Maddy begins her third line of objection by noting what she takes to be an anomaly in Quinean naturalism, namely, that it seems to respect the methodology of empirical science but not that of mathematics. It seems that, by the indispensability argument, mathematical ontology is legitimised only insofar as it is useful to empirical science. This, claims Maddy, is at odds with actual mathematical practice, where theorems of mathematics are believed because they are proved from the relevant axioms, *not* because such theorems are useful in applications [Maddy, 1992, p. 279]. Furthermore, she claims that such a "simple" indispensability argument leaves too much mathematics unaccounted for. Any mathematics that does not find applications in empirical science is apparently without ontological commitment. Quine himself suggests that we need some unapplied mathematics in order to provide a simplificatory rounding out of the mathematics that is applied, but "[m]agnitudes in excess of such demands, e.g. $\beth_\omega$ or inaccessible numbers"[43] should be looked upon as "mathematical recreation and without ontological rights"

---

[43] $\beth_\omega = \bigcup_{\alpha \in \omega} \beth_\alpha$, where $\beth_\alpha = 2^{\beth_{\alpha-1}}$, α is an ordinal and $\beth_0 = \aleph_0$. See [Enderton, 1977, pp. 214–215] for further details.

A cardinal number κ is said to be inaccessible iff the following conditions hold: (a) $\kappa > \aleph_0$ (some texts omit this condition) (b) $\forall \lambda < \kappa \; 2^\lambda < \kappa$ and (c) It is not possible to represent κ as the supremum of fewer than κ smaller ordinals (i.e., κ is *regular*). For example, $\beth_\omega$ satisfies (a) and (b) but not (c). $\aleph_0$ satisfies (b) and (c) but obviously not (a). Inaccessible numbers have to be postulated (by large cardinal axioms) in much the same way as the axiom of infinity postulates (a set of cardinality) $\aleph_0$.

[Quine, 1986, p. 400].[44]

Maddy claims that this is a mistake, as it is at odds with Quine's own naturalism. Quine is suggesting we reject some portions of accepted mathematical theory on non-mathematical grounds. Instead, she suggests the following modified indispensability argument:[45]

> [T]he successful application of mathematics gives us good reason to believe that there are mathematical things. Then, given that mathematical things exist, we ask: By what methods can we best determine precisely what mathematical things there are and what properties these things enjoy? To this, our experience to date resoundingly answers: by mathematical methods, the very methods mathematicians use; these methods have effectively produced all of mathematics, including the part so far applied in physical science. [Maddy, 1992, p. 280]

This modified indispensability argument and, in particular, the respect it pays to mathematical practice, she finds more in keeping with the spirit, if not the letter, of Quinean naturalism.

She then goes on to consider how this modified indispensability argument squares with mathematical practice. She is particularly interested in some of the independent questions of set theory such as Cantor's famous continuum hypothesis: Does $2^{\aleph_0} = \aleph_1$? and the question of the Lebesgue measurability of Σ_2^1 sets.[46] One aspect

---

[44]Later Quine refined his position on the higher reaches of set theory and other parts of mathematics, which are not, nor are ever likely to be, applicable to natural science. For instance, in his last book, he suggested:

> They are couched in the same vocabulary and grammar as applicable mathematics, so we cannot simply dismiss them as gibberish, unless by imposing an absurdly awkward gerrymandering of our grammar. Tolerating them, then, we are faced with the question of their truth or falsehood. Many of these sentences can be dealt with by the laws that hold for applicable mathematics. Cases arise, however (notably the axiom of choice and the continuum hypothesis), that are demonstrably independent of prior theory. It seems natural at this point to follow the same maxim that natural scientists habitually follow in framing new hypotheses, namely, simplicity: economy of structure and ontology. [Quine, 1995, p. 56]

A little later, after considering the possibility of declaring such sentences meaningful but neither true nor false, he suggests:

> I see nothing for it but to make our peace with this situation. We may simply concede that every statement in our language is true or false, but recognize that in these cases the choice between truth and falsity is indifferent both to our working conceptual apparatus and to nature as reflected in observation categoricals. [Quine, 1995, p. 57]

Elsewhere [Quine, 1992, pp. 94–95] he expresses similar sentiments.

[45]This suggestion was in fact made earlier by Hartry Field [1980, pp. 4–5], but of course he denies that any portion of mathematics is indispensable to science so he had no reason to develop the idea.

[46]Σ_2^1 sets are part of the projective hierarchy of sets, obtained by repeated operations of projection and complementation on open sets. The Σ_2^1 sets, in particular, are obtained from the open sets (denoted Σ_0^1) by taking complements to obtain the Π_0^1 sets, taking projections of these to obtain the Σ_1^1 sets, taking complements of these to obtain the Π_1^1 sets and finally, taking the

of mathematical realism that Maddy finds appealing is that independent questions such as these ought to have determinate answers, despite their independence from the usual ZFC axioms. The problem though, for indispensability-motivated mathematical realism, is that it is hard to make sense of what working mathematicians are doing when they try to settle such questions, or so Maddy claims.

For example, in order to settle the question of the Lebesgue measurability of the Σ_2^1 sets, new axioms have been proposed as supplements to the standard ZFC axioms. Two of these competing axiom candidates are Gödel's axiom of constructibility, $V = L$, and large cardinal axioms, such as MC (there exists a measurable cardinal). These two candidates both settle the question at hand, but with different answers. MC implies that all Σ_2^1 sets are Lebesgue measurable, whereas $V = L$ implies that there exists a non-Lebesgue measurable Σ_2^1 set. The consensus of informed opinion is that $V \neq L$ and that some large cardinal axiom or other is true,[47] but the reasons for this verdict seem to have nothing to do with applications in physical science. Indeed, much of the appeal of large cardinal axioms is that they are less restrictive than $V = L$, so to oppose such axioms would be "mathematically counterproductive" [Maddy, 1995, p. 265]. These are clearly intra-mathematical arguments that make no appeal to applications.

Furthermore, if the indispensability argument is cogent, it is not unreasonable to expect that physical theories would have some bearing on developments in set theory, since they are both part of the same overall theory. For example, Maddy claims that if space-time is not continuous, as some physicists are suggesting,[48] this could undermine much of the need for set theory (at least in contexts where it is interpreted literally) beyond cardinality $\aleph_0$. Questions about the existence of large cardinals would be harder to answer in the positive if it seemed that indispensability considerations failed to deliver cardinalities as low as $\beth_1$. Maddy thus suggests that indispensability-motivated mathematical realism advocates set theorists looking at developments in physics (e.g., theories of quantum gravity) in order to tailor set theory to best accord with such developments.[49] Given that set theorists in general do not do this, a serious revision of mathematical practice is being advocated by supporters of the indispensability argument, and this, Maddy claims, is a violation of naturalism [Maddy, 1992, p. 289]. She concludes:

> In short, legitimate choice of method in the foundations of set theory does not seem to depend on physical facts in the way indispensability theory requires. [Maddy, 1992, p. 289]

Maddy's sustained critique of the indispensability argument is a serious challenge for any defender of the indispensability argument. And I think it's fair to say

---

projections of these to obtain the Σ_2^1 sets. See [Maddy, 1990a, chap. 4] (and references contained therein) for further details and an interesting discussion of the history of the question of the Lebesgue measurability of these sets.

[47]There are, of course, some notable supporters of $V = L$, in particular, Quine [1992, p. 95] and Keith Devlin [1977].

[48]For example, Richard Feynman [1965, pp. 166–167] suggests this.

[49]Cf. [Chihara, 1990, p. 15] for similar sentiments.

that a defence of the indispensability argument in the light of Maddy's arguments will need to address issues about the role of naturalism and the precise role of mathematics in specific episodes in the history of science. Maddy quite rightly draws attention to the diverse roles mathematics plays in science and the different attitudes scientists can have towards the mathematics they use. Independently of whether Maddy's critique of the indispensability argument is deemed successful, this move to a more careful attitude towards both the history and the particular details of mathematics in applications is a welcome one.

Let me note one issue that Maddy's critique raises: the role of naturalism in debates about ontology and scientific practice. An important part of Maddy's strategy for undermining the indispensability argument is to show that confirmational holism flies in the face of naturalism. For instance, in her case study of early atomic theory, she shows how prominent scientists such as Poincaré and Ostwald did not take the indispensability of atoms to the theory in question to imply the reality of atoms. That is, Maddy takes it that working scientists do not take the holistic attitude to confirmation that Quine would like. This, claims Maddy, shows that naturalism and holism are in conflict. But what is the conception of naturalism being invoked here? At times Maddy suggests that naturalism implies that "if philosophy conflicts with [scientific] practice, it is the philosophy that must give" [Maddy, 1998a, p. 176]. And, indeed, much of Maddy's case against Quine seems to rely on such a reading. But this is certainly not Quine's conception of naturalism. There is much ground between first philosophy, which Quine rejects, and this philosophy-last style naturalism[50] that Maddy seems to endorse. For instance, there is the position that science and philosophy are continuous with one another and as such there is *no* high court of appeal. On this view, the philosopher of science has much to contribute to discussions of both scientific methodology and ontological conclusions, as does the scientific community. It may be that you're inclined to give more credence to the views of the scientific community in the eventuality of disagreement between scientists and philosophers, but even this does not imply that it is philosophy that must always give. I take it that this view of science and philosophy as continuous, without either having the role of "high court," is in fact the view that Quine intends. As it turns out, this is also the version of naturalism that Maddy subscribes to (as she points out in more careful statements of her position [Maddy, 1998a, p. 178]). Rather than 'philosophy must give' in the earlier passage, she really just means that *first* philosophy must give.

Now returning to the issue of prominent scientists not adhering to holism. If we understand naturalism as 'philosophy last', then the naturalistic philosopher must, with the scientists in question, reject holism. But if we take naturalism to be the rejection of first philosophy, then there is room to mount a naturalistic critique of the scientists in question. One needs to take care not to attract the charge of practicing first philosophy whilst mounting this critique, but there is at least

---

[50]Elsewhere [Colyvan, 2001a] I've referred to this variety of naturalism as "rubber stamp naturalism", since the only role it gives to philosophy is that of rubber stamping approval of all scientific practice.

room for a critique. Moreover, the question of whether the Poincaré and Ostwald were correct in their instrumentalism about atomic theory will not be decided by appeal to any general principle that tells us to always side with prominent scientists. Maddy is quite right to focus attention on the historical details and on the role of naturalism here. In the end, I don't think that Maddy's objections are as telling against the indispensability argument as may first appear.[51] But irrespective of what Maddy's arguments mean for the fate of the indispensability argument, the debate has certainly been shifted in very interesting and fruitful directions.

5.2 Sober

Elliott Sober's [1993] objection to the indispensability argument is framed from the viewpoint of *contrastive empiricism*, so it will be necessary to first consider some of the details of this theory in order to evaluate the force of Sober's objection. As will become apparent, though, contrastive empiricism has some difficulties that I'm inclined to think cannot be overcome. This robs Sober's objection of much — but not all — of its force. Finally, I will recast the objection without the contrastive empiricism framework and show that this version of the objection also faces significant difficulties.

Contrastive empiricism is best understood as a position between scientific realism and Bas van Fraassen's [1980] constructive empiricism. The central idea of contrastive empiricism is the appeal to the *Likelihood Principle* as a means of choosing between theories.

The Likelihood Principle Observation O favours hypothesis H_1 over hypothesis H_2 iff $P(O|H_1) > P(O|H_2)$.

It's clear from principle 5.2 that the support a hypothesis receives is a relative matter. As Sober puts it (emphasis in original):

> The Likelihood Principle entails that the degree of support a theory enjoys should be understood relatively, not absolutely. A theory competes with other theories; observations reduce our uncertainty about this competition by discriminating among alternatives. The evidence we have for the theories we accept is evidence that favours those theories *over others*. [Sober, 1993, p. 39]

According to Sober, though, evidence can never favour one theory over all possible competitors since "[o]ur evidence is far less powerful, the range of alternatives that we consider far more modest" [Sober, 1993, p. 39].

Another consequence of principle 5.2 is that some observational data may fail to discriminate between two theories. For instance, contrastive empiricism cannot discriminate between standard geological and evolutionary theory, and Gosse's theory that the earth was created about 4,000 years ago with all the fossil records

---

[51]See [Colyvan, 1998a; Colyvan, 2001a; Resnik, 1995; Resnik, 1997] for some replies to Maddy on these issues.

and so on in place. Indeed, Sober's account cannot rule out any cleverly formulated sceptical hypothesis. Furthermore, Sober is reluctant to appeal to simplicity or parsimony as non-observational signs of truth, and so such sceptical problems are taken to be scientifically insoluble. This is one important way in which contrastive empiricism departs from standard scientific realism (and, arguably, standard scientific methodology).

Although according to contrastive empiricism "science attempts to solve discrimination problems" [Sober, 1993, p. 39] and the burden of solving these problems is placed firmly on the observational data, there is no restriction to hypotheses about observables, as in van Fraassen's constructive empiricism (emphasis in original):

> Contrastive empiricism differs from constructive empiricism in that the former does not limit science to the task of assigning truth values to hypotheses that are strictly about observables. What the hypotheses are *about* is irrelevant; what matters is that the competing hypotheses make different claims about what we can observe. Put elliptically, the difference between the two empiricisms is that constructive empiricism focuses on *propositions*, whereas contrastive empiricism focuses on *problems*. The former position says that science can assign truth values only to *propositions* of a particular sort; the latter says that science can solve *problems* only when they have a particular character. [Sober, 1993, p. 41]

Much more could be said about contrastive empiricism, but we have seen enough to motivate Sober's objection to the indispensability argument.

Sober's main objection is that if mathematics is confirmed along with our best empirical hypotheses, there must be mathematics-free competitors (or at least alternative mathematical theories as competitors):

> Formulating the indispensability argument in the format specified by the Likelihood Principle shows how unrealistic that argument is. For example, do we really have alternative hypotheses to the hypotheses of arithmetic? If we could make sense of such alternatives, could they be said to confer probabilities on observations that differ from the probabilities entailed by the propositions of arithmetic themselves? I suggest that both these questions deserve negative answers. [Sober, 1993, pp. 45–46]

It is important to be clear about what Sober is claiming. He is *not* claiming that indispensability arguments are fatally flawed. He is not unfriendly to the general idea of ontological commitment to the indispensable entities of our best scientific theories. He simply denies that "a mathematical statement inherits the observational support that accrues to the empirically successful scientific theories in which it occurs" [Sober, 1993, p. 53]. This is enough, though, to place him at odds with the Quine-Putnam version of the indispensability argument.

In reply to this objection, I wish to first point out that there *are* alternatives to number theory. Frege showed us how to express most numerical statements required by empirical science without recourse to quantifying over numbers.[52] Furthermore, depending on how much analysis you think Hartry Field has successfully nominalised, there are alternatives to that also. (At the very least he has suggested that there are nominalist alternatives to differential calculus.)[53]

I take the crux of Sober's objection then to be the second of his two questions, and I agree with him here that the answer to this question deserves a negative answer. I don't think that Field's version of Newtonian mechanics and standard Newtonian mechanics would confer different probabilities on any observational data, but so much the worse for contrastive empiricism. The question of which is the better theory will be decided on the grounds of simplicity, elegance, and so on — grounds explicitly ruled out by contrastive empiricism. Supporters of the indispensability argument do not propose to settle all discrimination problems by purely empirical means, so it should come as no surprise to find that they run into trouble when forced into the straight-jacket of contrastive empiricism.

You might be inclined to think that since a mathematised theory such as Newtonian mechanics and Field's nominalist counterpart have the same empirical consequences, it can't be said that the mathematics receives empirical support. According to this view, the mathematised version is preferred on the *a priori* grounds of simplicity, elegance and so on, *not* on empirical grounds. In reply to this, I simply point out that there is nothing special about the mathematical content of theories in this respect. As I've already mentioned, we prefer standard evolutionary theory and geology over Gosse's version of creationism and we do so for the same apparently a priori reasons. It would be a very odd view, however, that denied evolutionary theory and geology received empirical support. Surely the right thing to say here is that evolutionary theory and geology receive both empirical support *and* support from a priori considerations. I'm inclined to say the same for the mathematical cases.[54]

Another objection to the whole contrastive empiricism approach to theory choice is raised by Geoffrey Hellman and considered by Sober [1993]. The objection is that often a theory is preferred over alternatives, not because it makes certain (correct) predictions that the other theories assign very low probabilities to, but rather, because it is the *only* theory to address such phenomena at all.[55] Sober

---

[52] For example, 'There are two Fs' or 'the number of the Fs is two' is written as:
$$(\exists x)(\exists y)(((Fx \wedge Fy) \wedge x \neq y) \wedge (\forall z)(Fz \supset (z = x \vee z = y))).$$

[53] This is only considering sensible alternatives. There are, presumably, many rather bad theories that do without mathematics. Perhaps most pseudosciences such as astrology and palm reading do without all but the most rudimentary mathematics.

[54] It is perhaps best to speak of the 'scientific justification of theories,' where this includes empirical support *and* support from *a priori* considerations. This is clearly the sort of support that our best scientific theories receive, so we see that Sober's concentration on purely empirical support might be thought to skew the whole debate. Thanks to Bernard Linsky for a useful discussion on this point.

[55] Hellman [1999] gives the example of relativistic physics correctly predicting the relationship between total energy and relativistic mass. In pre-relativistic physics no such relationship is even

points out that the relevance of this to the question of the indispensability of mathematics is that presumably "stronger mathematical assumptions facilitate empirical predictions that cannot be obtained from weaker mathematics" [Sober, 1993, p. 52].[56] If this objection stands, then the central thesis of contrastive empiricism is thrown into conflict with actual scientific practice. For a naturalist this almost amounts to serious trouble. Indeed, Sober admits that "[i]f this point were correct, it would provide a quite general refutation of contrastive empiricism" [Sober, 1993, p. 52]. I believe that Hellman's point is correct, but first let's consider Sober's reply.

Sober's first point is that when scientists are faced with a theory with no relevant competitors, they can contrast the theory in question with its own negation. He considers the example of Newtonian physics correctly predicting the return of Halley's comet, something on which other theories were completely silent. Sober claims, however, that "alternatives to Newtonian theory can be constructed from Newtonian laws themselves" [Sober, 1993, p. 52]. For example, Newton's law of universal gravitation:[57]

$$F = \frac{Gm_1 m_2}{r^2}$$

competes with:

$$F = \frac{Gm_1 m_2}{r^3}$$

and

$$F = \frac{Gm_1 m_2}{r^4}$$

and many others. There is no doubt that such alternatives *can be* constructed and contrasted with Newtonian theory, but surely we are not interested in what scientists *could do*; we are interested in *actual scientific practice*.

Sober takes this a step further and claims that this *is* standard scientific practice for such cases [Sober, 1993, pp. 52–53]. He offers no evidence in support of this last claim, and without a thorough investigation of the history of relevant episodes in the history of science it seems rather implausible. Were scientists really interested in debating whether it should be r^2, r^3, or r^4 in the law of universal gravitation?[58] The relevant debate would have surely been over retaining the existing theory or adopting Newtonian theory. At the very least, Sober needs to present some evidence to suggest that scientists are inclined to contrast a theory with its own negation when nothing better is on offer.

---

postulated, indeed, questions about such a relationship cannot even be posed.

[56] For example, [Hellman, 1992] argues that the weaker constructivist mathematics, such as that of the intuitionists, will not allow the empirical predictions facilitated by the stronger methods of standard analysis.

[57] Here F is the size of the gravitational force exerted on two particles of mass m_1 and m_2 separated by a distance r, and G is the gravitational constant.

[58] Not to mention $r^{2.000000001}$ or $r^{1.999999999}$. (Although it seems that cases such as these *were* considered when the problems with Mercury's perihelion came to light [Roseveare, 1983], they were considered only in order to save the essentials of Newtonian theory, which, by that stage, was already a highly confirmed theory.)

In his second point in response to Hellman's objection, Sober considers the possibility of "strong" mathematics allowing empirical predictions that cannot be replicated using weaker mathematics. Sober points out that strong mathematics also allows the formulation of theories that make false predictions, and that this is ignored by the indispensability argument (emphasis in original):

> It is a striking fact that mathematics allows us to construct theories that make *true* predictions and that we could not construct such predictively *successful* theories without mathematics. It is less often noticed that mathematics allows us to construct theories that make *false* predictions and that we could not construct such predictively *unsuccessful* theories without mathematics. If the authority of mathematics depended on its empirical track record, *both* these patterns should matter to us. The fact that we do not doubt the mathematical parts of empirically *un*successful theories is something we should not forget. Empirical testing does not allow one to ignore the bad news and listen only to the good. [Sober, 1993, p. 53]

It may be useful at this point to spell out the dialectic thus far. Hellman's point is that contrastive empiricism does not account for cases where a theory is preferred because it makes predictions that no other theory is able to address one way or another. If this is accepted, then contrastive empiricism as a representation of how theory choice is achieved seems at best only part of the story, and at worst completely misguided. Furthermore, if it is reasonable to prefer some theory because it correctly predicts new phenomena that other theories are silent on, then it is reasonable to accept strong mathematical hypotheses, since theories employing strong mathematics are able to predict just such phenomena.

I take it that Sober's reply runs like this: Contrastive empiricism can accommodate the Hellman examples of scientific theories that address new phenomena. This is done by contrasting such theories with their negations. Thus, a general undermining of contrastive empiricism is avoided. This reply, however, seems to allow that strong mathematics is confirmed, because such theories correctly predict empirical phenomena that theories employing weaker mathematics cannot address. So the cost of saving contrastive empiricism from the Hellman objection is that Sober's original point against the empirical confirmation of mathematics now fails. Here is where the second part of Sober's reply is called upon. The point here is simply that the case of strong mathematics is different from that of bold new physical theories in that strong mathematics can also facilitate false predictions that competing theories are silent on. Thus, the mathematics cannot share the credit for the successful empirical predictions, since it won't share the blame for unsuccessful empirical predictions.

There are a couple of interesting issues raised by this rejoinder. First, the rejoinder is in the context of a defence of contrastive empiricism and yet it is not an argument for that thesis. Nor is it an argument depending on contrastive empiricism. It seems like a new objection to the use of indispensability arguments to gain conclusions about mathematical entities. What is more, this objection appears to be independent of contrastive empiricism and as such is the more substantial part of his objection to the indispensability argument.

So far I've suggested that Sober is wrong about scientists contrasting bold new theories with their negations. At the very least Sober needs to give some evidence to support his claim that scientists do this.[59] Indeed, it would be interesting to investigate some candidate cases in detail to shed some light on this issue, but fortunately this is not necessary for our purposes, since even if I grant Sober his first point (that contrastive empiricism can accommodate Hellman's examples of bold new theories), the second part of Sober's reply also runs into trouble.

Sober claims, in effect, that mathematical theories cannot enjoy the confirmation received by theories that make bold new true predictions because the mathematics is not disconfirmed when it is employed by a theory that makes bold new false predictions. I've already noted that this point is stated independently of contrastive empiricism. Indeed, I take this to be a separate worry about the indispensability argument as applied to mathematical entities. Also bear in mind that it is important to Sober's case that there be a difference between mathematical hypotheses and non-mathematical hypotheses in this respect.

This last claim, though, is false. Many non-mathematical hypotheses can be employed by false theories and not be held responsible for the disconfirmation. Hypotheses about electrons (notoriously) have been employed by many false theories, and yet we are unwilling to blame electrons for the lack of empirical support for the theories in question. Astrologers refer to the orbits of the planets in grossly false theories about human behaviour, and yet we do not blame the planets for the lack of empirical support for astrology. It is surely one of the important tasks of scientists to decide which parts of a disconfirmed theory are in need of revision and which are not. Sober would have us throw out the baby with the bathwater, it seems.

Hellman [1999] points out that this partial asymmetry between confirmation and disconfirmation is a consequence of confirmational holism. When a theory is confirmed, the *whole* theory is confirmed. When it is disconfirmed, it is rarely the fault of every part of the theory, and so the guilty part is to be found and dispensed with. It's analogous to a sensitive computer program. If the program delivers the correct results, then every part of the program is believed to be correct. However, if the program is not working, it is often because of only one small error. The job of the computer programmer (in part) is to seek out the faulty part of the program and correct it. Furthermore, the programmer will resort to wholesale changes to the program only if no other solution presents itself. This is especially evident when one part of the program *is* working. In such a case the programmer seeks to make a small *local* change in the defective part of the program. Changing the programming language, for instance, is *not* such a change.

---

[59]It is worth pointing out that he must provide evidence that contrasting theories with their negations is a general phenomenon. Even if there are only one or two counterexamples, contrastive empiricism is in trouble.

Now if we return to Sober's charge that mathematics cannot enjoy the credit for confirmation of a theory if it cannot share the blame for disconfirmation, we see that blaming mathematics for the failure of some theory is never going to be a small local change, due to the simple fact that mathematics is used almost everywhere in science. What is more, much of that science is working perfectly well. Blaming the mathematics is like a programmer blaming the computer language. And, similarly, claiming that mathematics cannot share the credit is like claiming that the computer language cannot share the credit for the successful program. In some cases it may well be the fault of the mathematics or the programming language, but it is not a good strategy to start with changes to these.

Furthermore, we see that mathematics is not alone in this respect. Many clearly empirical hypotheses share this feature of apparent immunity from blame for disconfirmation. Michael Resnik points out that conservation principles seem immune from liability for much the same reasons as mathematics. He goes even further to express doubts about whether such principles could be tested at all in the contrastive empiricist framework and "yet we do not want to be forced to deny them empirical content or to hold that the general theories containing them have not been tested experimentally" [Resnik, 1995, p. 168]. Another untestable empirical hypothesis is the hypothesis that space-time is continuous rather than discrete and dense.

To sum up, then. I agree with Sober that there is a problem of reconciling contrastive empiricism with the indispensability argument, but for the most part this is because of general problems with the former. In particular, contrastive empiricism fails to give an adequate account of a theory being adopted because it correctly predicts phenomena that its competitors are unable to speak to at all. I agree with Hellman here that this looks like the kind of role mathematics plays in theory selection. Strong mathematics allows the formulation of theories that address phenomena on which other theories are completely silent. Sober's rejoinder is that mathematical hypotheses are different from other scientific hypotheses, in that mathematical hypotheses allow false predictions just as readily as true ones, and yet mathematics remains blameless for the former. This rejoinder is in effect a new argument against the indispensability argument applied to mathematical entities and, what is more, it is independent of the framework of contrastive empiricism. Nevertheless, the rejoinder faces problems of its own. First, it seems to misrepresent the type of holism at issue — the holism at issue has an asymmetry between confirmation and disconfirmation built into it. Second, it seems clear that mathematics is not alone in its apparent immunity from blame in cases of disconfirmation.

I should mention Sober's claim that the main point of his objection can be separated to some extent from the contrastive empiricist epistemology. He does not, however, seem to have the residual worry that I discussed in mind. He is concerned that you might think that contrastive empiricism can't be right because it ignores nonempirical criteria such as simplicity. He then suggests that "even proponents of such nonempirical criteria should be able to agree that *empirical* considerations must be mediated by likelihoods" [Sober, 1993, p. 55]. Sober is suggesting that at the very least we discriminate between empirical hypotheses by appeal to likelihoods and that his objection goes through granting only this.[60] But why should we accept that all discriminations between empirical hypotheses must be mediated by likelihoods? After all, we have already seen that we cannot discriminate between the hypothesis that space-time is continuous and the hypothesis that space-time is discrete and dense on empirical grounds and yet these are surely both empirical hypotheses. So Sober's objections to the indispensability argument fail because they depend crucially on accepting the Likelihood Principle as the only arbiter on empirical matters. The independent residual point I identified fails because it doesn't take account of the asymmetric character of confirmational holism.

6 THE UNREASONABLE EFFECTIVENESS OF MATHEMATICS

In this section I'll turn my attention to another important issue that arises in the context of philosophy of applied mathematics. This is the issue of how mathematics manages to be so "unreasonably" suited to the business of science. The physicist Eugene Wigner once remarked that

> [t]he miracle of the appropriateness of the language of mathematics for the formulation of the laws of physics is a wonderful gift which we neither understand nor deserve. [Wigner, 1960, p. 14]

Steven Weinberg is another physicist who finds the applicability of mathematics puzzling:

> It is very strange that mathematicians are led by their sense of mathematical beauty to develop formal structures that physicists only later find useful, even where the mathematician had no such goal in mind. [...] Physicists generally find the ability of mathematicians to anticipate the mathematics needed in the theories of physics quite uncanny. It is as if Neil Armstrong in 1969 when he first set foot on the surface of the moon had found in the lunar dust the footsteps of Jules Verne. [Weinberg, 1993, p. 125]

And it's not only physicists who have waxed lyrical on the applicability of mathematics. Charles Darwin remarked that:

> I have deeply regretted that I did not proceed far enough at least to understand something of the great leading principles of mathematics, for men thus endowed seem to have an extra sense. [Darwin, 1958]

In each case the author seems to be suggesting something mysterious — even miraculous — about the applicability of mathematics. Indeed, this puzzle, which

---

[60]Since, according to the indispensability argument, mathematics *is* empirical, and yet we cannot discriminate between mathematical and non-mathematical theories by appeal to likelihoods.

Wigner calls 'the unreasonable effectiveness of mathematics', is often remarked upon by physicists and applied mathematicians[61] but receives surprisingly little attention in the philosophical literature.[62] It is hard to say why this puzzle has not caught the imagination of the philosophical community. It is not because it's unknown in philosophical circles. On the contrary, it is very well known; it just does not get discussed. This lack of philosophical attention, I believe, is due (in part) to the fact that the way the problem is typically articulated seems to presuppose a formalist philosophy of mathematics.[63]

Given the decline of formalism as a credible philosophy of mathematics in the latter half of the twentieth century, and given the rise of anti-realist philosophies of mathematics that pay great respect to the applicability of mathematics in the physical sciences (such as Hartry Field's fictionalism [Field, 1980]), it is worth reconsidering Wigner's puzzle to see to what extent, if any, it relies on a particular philosophy of mathematics. The central task of this paper is to argue that although Wigner set the puzzle up in language that suggested an anti-realist philosophy of mathematics, it appears that the puzzle is independent of any particular philosophy of mathematics. At least, a version of the puzzle can be posed for two of the most influential, contemporary philosophies of mathematics: one realist, the other anti-realist.

6.1 What is the Puzzle?

Mark Steiner is one of the few philosophers to take interest in Wigner's puzzle [Steiner, 1989; Steiner, 1995; Steiner, 1998]. Steiner has quite rightly suggested that Wigner's "puzzle" is in fact a whole family of puzzles that are not distinguished by Wigner; it depends on what you mean by 'applicability' when talking of the applications of mathematics. Steiner claims that it is important to distinguish the different senses of 'applicability' because some of the associated puzzles are easily solved while others are not. For example, Steiner argues that the problem of the (semantic) applicability of mathematical theorems[64] was explained

---

[61] For example: Paul Davies [1992, pp. 140–60]; Freeman Dyson [1964]; Richard Feynman [1965, p. 171]; R.W. Hamming [1980]; Steven Weinberg [1986] and many others in [Mickens, 1990].

[62] Though, that may be starting to change. See [Azzouni, 2000; Wilson, 2000] for some relatively recent discussion of this topic.

[63] Saunders Mac Lane, for example, explicitly takes the puzzle to be a puzzle for formalist philosophies of mathematics [Mac Lane, 1990]. Others have taken the problem to be a problem for anti-realist philosophies of mathematics generally. See, for example, [Davies, 1992, pp. 140–60] and [Penrose, 1989, pp. 556–7]. One exception here is Philip Kitcher [Kitcher, 1984, pp. 104–5] who presents it as a problem for Platonism. I will discuss, what is in essence, Kitcher's problem in section 6.2.

[64] This is the problem of explaining the validity of mathematical reasoning in both pure and applied contexts — to explain, for instance, why the truth of (i) there are 11 Lennon-McCartney songs on the Beatles' 1966 album Revolver, (ii) there are 3 non-Lennon-McCartney songs on that same album, and (iii) $11 + 3 = 14$, implies that there are 14 songs on Revolver. (The problem is that in (i) and (ii) '11' and '3' seem to act as names of predicates and yet in (iii) '11' and '3' apparently act as names of objects. What we require is a constant interpretation of the mathematical vocabulary across such contexts.

adequately by Frege [1995]. There is, according to Steiner, however, a problem which Frege did not address. This is the problem of explaining the appropriateness of mathematical concepts for the description of the physical world. Of particular interest here are cases where the mathematics seems to be playing a crucial role in making predictions. Moreover, Steiner has argued for his own version of Wigner's thesis. According to Steiner, the puzzle is not simply the extraordinary appropriateness of mathematics for the formulation of physical theories, but concerns the role mathematics plays in the very discovery of those theories. In particular, this requires an explanation that is in keeping with the methodology of mathematics — a methodology that does not seem to be guided at every turn by the needs of physics.

The problem is epistemic: why is mathematics, which is developed primarily with aesthetic considerations in mind, so crucial in both the discovery and the statement of our best physical theories? Put this way the problem may seem like one aspect of a more general problem in the philosophy of science — the problem of justifying the appeal to aesthetic considerations such as simplicity, elegance, and so on. This is not the case though. Scientists and philosophers of science invoke aesthetic considerations to help decide between two theories that are empirically equivalent. Aesthetics play a much more puzzling role in the Wigner/Steiner problem. Here aesthetic considerations are largely responsible for the development of mathematical theories. These, in turn, (as I will illustrate shortly) play a crucial role in the discovery of our best scientific theories. In particular, novel *empirical* phenomena are discovered via mathematical analogy. In short, aesthetic considerations are not just being invoked to decide between empirically equivalent theories; they seem to be an integral part of the process of scientific discovery.

Steiner's statement of the puzzle is clearer and more compelling, so when I speak of Wigner's puzzle, I will have Steiner's version in mind. I will thus concentrate on cases where the mathematics seems to be playing an active role in the discovery of the correct theory — not just in providing the framework for the statement of the theory. I'll illustrate this puzzle by presenting one rather classic case and refer the interested reader to Steiner's article [1989] and book [1998] for further examples.[65] In the case I'll consider here, we see how Maxwell's equations predicted electromagnetic radiation.

Maxwell found that the accepted laws for electromagnetic phenomena prior to about 1864, namely Gauss's law for electricity, Gauss's law for magnetism, Faraday's law, and Ampère's law, jointly contravened the conservation of electric charge. Maxwell thus modified Ampère's law to include a *displacement current*, which was not an electric current in the usual sense (a so-called *conduction current*), but a rate of change (with respect to time) of an electric field. This modification was made on the basis of formal mathematical analogy, *not* on the basis

---

[65]Steiner distinguishes between two quite different, but equally puzzling, ways in which mathematics has facilitated the discovery of physical theories: *Pythagorean analogy* and *formalist analogy*. Although this distinction is of considerable interest, it has little bearing on the main thesis of this section, so I will set it aside. See [Steiner, 1998, pp. 2–11] for details.

of empirical evidence.[66] The analogy was with Newtonian gravitational theory's conservation of mass principle. The modified Ampère law states that the curl of a magnetic field is proportional to the sum of the conduction current and the displacement current. More specifically:

$$(3) \quad \nabla \times \mathbf{B} = \frac{4\pi}{c}\mathbf{J} + \frac{1}{c}\frac{\partial}{\partial t}\mathbf{E}.$$

Here $\mathbf{E}$ and $\mathbf{B}$ are the electric and magnetic field vectors respectively, $\mathbf{J}$ is the current density, and c is the speed of light in a vacuum.[67] When this law (known as the Maxwell-Ampère law) replaces the original Ampère law in the above set of equations, they are known as Maxwell's equations and they provide a wonderful unity to the subject of electromagnetism.

The interesting part of this story for the purposes of the present discussion, though, is that Maxwell's equations were formulated on the assumption that the charges in question moved with a constant velocity, and yet such was Maxwell's faith in the equations, he assumed that they would hold for *any* arbitrary system of electric fields, currents, and magnetic fields. In particular, he assumed they would hold for charges with accelerated motion and for systems with zero conduction current. An unexpected consequence of Maxwell's equations followed in this more general setting: a changing magnetic field would produce a changing electric field and *vice versa*. Again from the equations, Maxwell found that the result of the interactions between these changing fields on one another is a wave of electric and magnetic fields that can propagate through a vacuum. He thus predicted the phenomenon of electromagnetic radiation. Furthermore, he showed that the speed of propagation of this radiation is the speed of light. This was the first evidence that light was an electromagnetic phenomenon.[68]

It seems that these predictions (which were eventually confirmed experimentally by Heinrich Hertz in 1888) can be largely attributed to the mathematics, since the predictions were being made for circumstances beyond the assumptions of the equations' formulation. Moreover, the formulation of the crucial equation (the Maxwell-Ampère law) for these predictions was based on formal mathematical analogy. Cases such as this *do* seem puzzling, at least when presented a certain way. The question on which I wish to focus is whether the puzzlement is an artifact of the presentation (because some particular philosophy of mathematics is

---

[66] Indeed, there was very little (if any) empirical evidence at the time for the displacement current.

[67] The first term on the right of equation 3 is the conduction current and the second on the right is the displacement current.

[68] Actually the story is a little more complicated than this. Maxwell originally had a mechanical model of electromagnetism in which the displacement current was a physical effect. (For the details of the relevant history, see [Chalmers, 1973], [Hunt, 1971] and [Siegel, 1991].) This, however, does not change the fact that there was little (if any) empirical evidence for the displacement current and the reasoning that led to the prediction of electromagnetic radiation went beyond the assumptions on which either the equations or the mechanical model were based [Steiner, 1998, pp. 77–8].

explicitly or implicitly invoked), or whether these cases are puzzling *simpliciter*. I will argue that it is the latter.

6.2 Is the Puzzle Due to a Particular Philosophy of Mathematics?

Applicability has long been the Achilles' heel of anti-realist accounts of mathematics. For example, if you believe that mathematics is some kind of formal game — as Hilbert did — then you need to explain why mathematical theories are needed to such an extent in our descriptions of the world. After all, other games, like chess, do not find themselves in such demand. Or if you think that mathematics is a series of conditionals — '2+2=4' is short for 'If the Peano-Dedekind axioms hold then 2+2=4' — the same challenge stands.

In Wigner's article he seems to be taking a distinctly anti-realist point of view (my italics):

> [M]athematics is the science of skillful operations with concepts and rules *invented* just for that purpose. [Wigner, 1960, p. 2]

Others, such as Reuben Hersh, also adopt anti-realist language when stating the problem (again, my italics):[69]

> There is no way to deny the obvious fact that arithmetic was *invented* without any special regard for science, including physics; and that it turned out (unexpectedly) to be needed by every physicist. [Hersh, 1990, p. 67]

Some, such as Paul Davies [1992, pp. 140–60] and Roger Penrose [1989, pp. 556–7], have suggested that the unreasonable effectiveness of mathematics in the physical sciences is evidence for realism about mathematics. That is, there is only a puzzle here if you think we invent mathematics and then find that this invention is needed to describe the physical world. Things aren't that simple though. There are contemporary anti-realist philosophies of mathematics that pay a great deal of attention to applications, and it is not clear that these suffer the same difficulties that formalism faces. Furthermore, it is not clear that realist philosophies of mathematics are home free. In what follows I will argue that there are puzzles for both realist and anti-realist philosophies of mathematics with regard to accounting for the unreasonable effectiveness of mathematics.

I will consider two philosophies of mathematics that we've already encountered: one influential realist philosophy of mathematics — Quinean realism [Quine, 1981b] and—and one equally influential anti-realist position — Hartry Field's fictionalism [Field, 1980]. Both of these philosophical positions are motivated by, and pay careful attention to, the role mathematics plays in physical theories. It

---

[69] Also recall Weinberg's reference to Jules Verne in the passage I quoted earlier in this section and Steiner's remark (quoted at the beginning of this chapter) about the mathematician being more like an artist than an explorer.

is rather telling, then, that each suffers similar problems accounting for Wigner's puzzle.

Recall that the Quinean realist is committed to realism about mathematical entities because of the indispensable role such entities play in our best scientific theories. Now, granted this, it might be thought that the Quinean realist has a response to Wigner. The Quinean could follow the lead of scientific realists such as J.J.C. Smart who put pressure on anti-realists by exposing their inability to explain the applications of electron theory, say. It's no miracle, claim scientific realists, that electron theory is remarkably effective in describing all sorts of physical phenomena such as lightning, electromagnetism, the generation of x-rays in Roentgen tubes and so on. Why is it no miracle? Because electrons exist and are at least partially causally responsible for the phenomena in question. Furthermore, it's no surprise that electron theory is able to play an active role in novel discoveries such as superconductors. Again this is explained by the existence of electrons and their causal powers. There is, however, a puzzle here for the anti-realist. As Smart points out:

> Is it not odd that the phenomena of the world should be such as to make a purely instrumental theory true? On the other hand, if we interpret a theory in a realist way, then we have no need for such a cosmic coincidence: it is not surprising that galvanometers and cloud chambers behave in the sort of way they do, for if there really are electrons, etc., this is just what we should expect. A lot of surprising facts no longer seem surprising. [Smart, 1963, p. 39]

There is an important disanalogy, however, between the case of electrons and the case of sets. Electrons have causal powers — they can bring about changes in the world. Mathematical entities such as sets are usually taken to be causally idle — they are Platonic in the sense that they do not exist in space-time nor do they have causal powers. So how is it that the positing of such Platonic entities reduces mystery?[70] Colin Cheyne and Charles Pigden [1996] have suggested that in light of this, the Quinean is committed to causally active mathematical entities. While I dispute the cogency of Cheyne's and Pigden's argument (see [Colyvan, 1998b]), I agree that there is a puzzle here. The puzzle is this: on Quine's view, mathematics is seen to be part of a true description of the world because of the indispensable role mathematics plays in physical theories, but the Quinean account gives us no indication as to *why* mathematics is indispensable to physical science. That is, Quine does not explain why mathematics is required in the formulation of our best physical theories and, even more importantly, he does not explain why mathematics is so often required for the discovery of these theories. Indispensability is simply taken as brute fact.

It might be tempting to reply, on behalf of Quine, that mathematics is indispensable because it's true. This, however, will not do. After all, there are presumably

---

[70] A few people have pointed to this problem in Quine's position (see [Balaguer, 1998, pp. 110–1], [Field, 1998, p. 400], [Kitcher, 1984, pp. 104–5] and [Shapiro, 1997, p. 46]).

many truths that are not indispensable to our best scientific theories. What is required is an account of why mathematical truths, in particular, are indispensable to science. Moreover, we require an account of why mathematical methods which, as Steiner points out [1995, p. 154], are closer to those of the artist's than those of the explorer's, are reliable means of finding the mathematics that science requires. It is these issues, lying at the heart of the Wigner/Steiner puzzle, that Quine does not address.

The above statement of the problem for Quine can easily be extended to any realist philosophy of mathematics that takes mathematical entities to be causally inert. This suggests that one way to solve the puzzle in question is to follow Cheyne's and Pigden's suggestion and posit causally *active* mathematical entities (*a la* early Maddy [1990a] or Bigelow [1988]). Now such physicalist strategies may or may not solve Wigner's puzzle.[71] But it is not my concern here to decide which realist philosophies fall foul of Wigner's puzzle and which do not. My concern is to demonstrate that realist philosophies of mathematics do not, *in general*, escape the problem. In particular, I have shown that Quine's influential realist philosophy of mathematics, at least if taken to be about abstract objects, succumbs to Wigner's puzzle.

Now consider Field's [1980] philosophy of mathematics in light of this problem. Recall that Field responds to Quine's argument by claiming that mathematics is, in fact, dispensable to our best physical theories. He adopts a fictional account of mathematics in which all the usually accepted sentences of mathematics are literally false, but *true-in-the-story* of accepted mathematics. There is no doubt that Field's partial nominalisation of Newtonian gravitational theory sheds considerable light on the role of mathematics in that theory, and perhaps on applied mathematics more generally. But it is interesting to note that despite Field's careful attention to the applications of mathematics, he leaves himself open to Wigner's puzzle. Field explains why we can use mathematics in physical theories — because mathematics is conservative. He also explains why mathematics often finds its way into physical theories — because mathematics simplifies calculations and the statement of these theories. What he fails to provide is an account of why mathematics leads to simpler theories and simpler calculations. Moreover, Field gives us no reason to expect that mathematics will play an active role in the prediction of novel phenomena.[72]

If I'm correct that facilitating novel scientific predictions (via mathematical analogy) is at least partly why we consider mathematics indispensable to science, then Field has not fully accounted for the indispensability of mathematics until he has provided an account of the active role mathematics plays in scientific discovery. So although Field did not set out to provide a solution to *this* particular problem of applicability (i.e. the Steiner/Wigner problem), it seems that, nevertheless, he is obliged to. (Indeed, this was the basis of my criticism of Field in [Colyvan,

---

[71] It's not clear to me that they do.
[72] I discuss this matter in more detail in [Colyvan, 1999b] and in [Colyvan, 2001a, chap. 4]. John Burgess raises similar issues in [1983].

1999b].) On the other hand, if this shortcoming of his project is seen (as I'm now suggesting) as part of the more general problem of applicability — a problem that Quine too faces — Field's obligation in this regard is not so pressing. In short, it's a problem for everyone.

Now the fact that Field *does not* provide a solution to Wigner's puzzle does not mean that he *cannot* do so. But whether he can provide a solution or not, *the puzzle needs to be discussed* and that is all I am arguing for here. Still, let me put to rest one obvious response Field may be tempted with.[73] He might appeal to the structural similarities between the empirical domain under consideration and the mathematical domain used to model it, to explain the applicability of the latter. So, for example, the applicability of real analysis to flat space-time is explained by the structural similarities between $\mathbb{R}^4$ (with the Minkowski metric) and flat space-time. There is no denying that this is right, but this response does not give an account of why mathematics leads to novel predictions and facilitates simpler theories and calculations. Appealing to structural similarities between the two domains does not explain, for example, why mathematics played such a crucial role in the prediction of electromagnetic radiation. Presumably certain mathematical structures in Maxwell's theory (which predict electromagnetic radiation) are similar to the various physical systems in which electromagnetic radiation is produced (and it would seem that there are no such structural similarities with the pre-Maxwell theory). But then Wigner's puzzle is to explain the role mathematical analogy played in the development of Maxwell's theory. The fact that Maxwell's theory is structurally similar to the physical system in question is simply irrelevant to this problem.

To sum up this section then. I agree with Steiner that the applicability of mathematics presents a general problem. What I hope to have shown is that the problem exists for at least two major contemporary positions in the philosophy of mathematics. Moreover, the two positions I discuss — Field's and Quine's — I take to be the two that are the most sensitive to the applications of mathematics in the physical sciences. The fact that these two influential positions do not seem to be able to explain Wigner's puzzle, clearly does not mean that *every* philosophy of mathematics suffers the same fate. It does show, however, that Wigner's puzzle is not merely a difficulty for unfashionable formalist theories of mathematics.

While the problems I've discussed in this paper for both Quine and Field are not new, they can now be seen in a new light. Previously each problem was seen as a difficulty for the particular account in question (in the context of the realism/anti-realism debate). That is, whenever these problems were discussed (and I include myself here [Colyvan, 1999b]), they were presented as reasons to reject one account in favour of another. If what I'm suggesting now is correct, that is the wrong way of looking at it. There are striking similarities between the problem that Burgess

---

[73]Mark Balaguer seems to have something like this response in mind when he says that "I do not think it would be very difficult to solve this general problem of applicability [of mathematics]" [Balaguer, 1998, p. 144]. It should also be mentioned that if this response were successful, it would also be available to realist philosophies of mathematics.

and I have pointed out for Field and the problem that Balaguer and others have pointed out for Quine. I claim that these problems are best seen as manifestations of the unreasonable effectiveness of mathematics. Moreover, these difficulties seem to cut across the realism/anti-realism debate and thus deserve careful attention from contemporary philosophers of all stripes — realists and anti-realists alike.

7 APPLIED MATHEMATICS: THE PHILOSOPHICAL LESSONS AND FUTURE DIRECTIONS

Let me close with some general comments about the philosophy of applied mathematics. Although much of the recent work on the applications of mathematics has had a fairly narrow focus on the indispensability argument, there is much of value to immerge from this work that transcends such a focus. For a start, both Maddy's [1997] and Field's [1980; 1989] critique of the indispensability argument (and the subsequent discussion of these two) suggests that we need to pay careful attention to the details of the way mathematics is used in various physical applications; it is not sufficient to simply note that mathematics is used in science. We need to consider whether the mathematics is merely providing a convenient model of the system in question or is it doing more? For example, is the mathematics contributing to the explanatory power of the theory? Is it helping to unify the theory in question? What attitudes do scientists in the area in question take towards the mathematics they use? Indeed, what attitude do these same scientists take towards the theory itself? All in all, the applications of mathematics to physical science is a much more nuanced affair than perhaps was appreciated by some earlier writers.

Also we should not forget that mathematics finds many and varied applications in areas of science other than physics. Although most discussions of applied mathematics begin and end with physics, careful attention to other branches of science such as biology and chemistry are of considerable interest here. It is not clear that mathematics plays the same kind of role in, say, the biological sciences.[74] For instance, it may be that the biological sciences are less satisfied with unification-style explanations (if they are explanations) — which mathematics is rather well suited to. Instead, there is some reason to suggest that biology is more interested in causal explanations [Colyvan and Ginzburg, 2003]. Furthermore, in the biological sciences there is the issue of abuse of mathematics and overmathematicising.[75] One rarely encounters such issues in physics, yet mathematical models in ecology, for instance, are treated with considerable suspicion by many ecologists. One concern is that the mathematics is obscuring ecological detail or invoking simplifications that are not well supported by ecological theory. This again suggests

---

[74]See, for instance, [Ginzburg and Colyvan, 2004; May, 2004] for recent discussions of the role of mathematics in the biological sciences.

[75]Some mathematical ecologists are even charged with "physics envy". (This is the "crime" of invoking sophisticated mathematical methods, that would be appropriate in physics but allegedly inappropriate in ecology.)

the possibility of a significant difference between the use of mathematics in the biological sciences and its use in the physical sciences.

Finally, the Wigner problem of the applicability always lurks in the background. It simply won't do to pass it off as a problem for Platonism, or formalism or any other particular philosophy of mathematics. As I've argued above, it is a problem for everyone. Moreover, a solution to this problem is likely to involve both careful attention to the details of the scientific and mathematical theories in question, and also careful attention to the history of science. For instance, it might turn out that my example in the previous section of Maxwell's positing of the displacement current (and the consequent prediction of electro-magnetic radiation) rides roughshod over historical or mathematical details — details that once brought to light, help us to understand why mathematics is apparently so unreasonably effective here. I should also add that Steiner's [1998] recent work on this topic suggests that the many and varied ways that mathematics is utilised in scientific theories makes the prospects of a unified solution to the problem of applied mathematics look dim. It may be that we'll need to look at the problem case by case.[76]

This brief overview of some of the issues in the philosophy of applied mathematics should give those interested in the topic considerable joy. There are some fascinating issues for future work — issues that cut deep into other fascinating issues in theories of explanation, the nature of scientific analogies, philosophy of biology and, of course, the history of science and mathematics. And no doubt there are many other issues I haven't addressed here that lead in equally interesting directions.[77]

BIBLIOGRAPHY

[Armstrong, 1978] D.M. Armstrong, *Universals and Scientific Realism*, Cambridge University Press, Cambridge (1978).

---

[76]There is also the issue of inconsistent mathematics and its applications. Inconsistent theories, such as the early calculus, were remarkably successful in applications. This suggests that consistency is not as important as many classically-minded logicians and philosophers of mathematics would have us believe. There are substantial issues here in need of further exploration. See [Mortensen, 1995] for a nice treatment of non-trivial inconsistent mathematical theories. See also Mortensen's chapter in this volume.

[77]Some of the material in this chapter has been previously published. I gratefully acknowledge Oxford University Press for permission to reproduce material from [Colyvan, 2001a], the editors of *The Stanford Encyclopedia of Philosophy* for permission to reproduce material from my [Colyvan, 2004], *Philosophia Mathematica* for permission to reproduce sections of [Colyvan, 1998a], *Mind* for permission to reproduce a section of [Colyvan, 2002], and Kluwer Academic Publishers for permission to reproduce sections of [Colyvan, 1999a; Colyvan, 1999b; Colyvan, 2001b] in *Erkenntnis*, *Philosophical Studies*, and *Synthese* respectively. The relevant copyrights remain with the publishers in question. I'd also like to thank the Center for Philosophy of Science at the University of Pittsburgh where I held a Visiting Research Fellowship in the winter term of 2004 and where some of the work on this chapter was carried out. Thanks especially to my colleagues there John Norton and Alan Chalmers for many interesting discussions. Work on this chapter was funded by an Australian Research Council Discovery Grant (grant number DP0209896).

[Azzouni, 2000] J. Azzouni, *Applying mathematics: an attempt to design a philosophical problem*, Monist, **83** (2000), 209–227.
[Azzouni, 2004] J. Azzouni, *Deflating Existential Consequence: A Case for Nominalism*, Oxford University Press, New York (2004).
[Baker, 2001] A.R. Baker, *Mathematics indispensability and scientific progress*, Erkenntnis, 55 (2001), 85–116.
[Balaguer, 1998] M. Balaguer, *Platonism and Anti-Platonism in Mathematics*, Oxford University Press, New York (1998).
[Benacerraf, 1965/1983] P. Benacerraf, *What numbers could not be*, reprinted in Philosophy of Mathematics Selected Readings, second edition, P. Benacerraf and H. Putnam, eds., Cambridge University Press, Cambridge (1983) (first published in 1965), 272–294.
[Benacerraf, 1973/1983] P. Benacerraf, *Mathematical truth*, reprinted in Philosophy of Mathematics Selected Readings, second edition, P. Benacerraf and H. Putnam eds., Cambridge University Press, Cambridge (1983) (first published in 1973), 403–420.
[Bigelow, 1988] J. Bigelow, *The Reality of Numbers: A Physicalist's Philosophy of Mathematics*, Clarendon Press, Oxford (1988).
[Boyce and DiPrima, 1986] W.E. Boyce and R.C. DiPrima, *Elementary Differential Equations and Boundary Value Problems*, fourth edition, John Wiley, New York (1986).
[Burgess, 1983] J. Burgess, *Why I am not a nominalist*, Notre Dame Journal of Formal Logic, **24** (1) (1983), 93–105.
[Burgess and Rosen, 1997] J. Burgess and G. Rosen, *A Subject with No Object: Strategies for Nominalistic Interpretation of Mathematics*, Clarendon Press, Oxford (1997).
[Campbell, 1994] K. Campbell *Selective realism in the philosophy of physics*, The Monist, **77** (1994), 27–46.
[Carnap, 1937] R. Carnap, *The Logical Syntax of Language*, Routledge and Kegan Paul, London (1937).
[Cartwright, 1983] N. Cartwright, *How the Laws of Physics Lie*, Oxford University Press, New York (1983).
[Chalmers, 1973] A.F. Chalmers, *Maxwell's methodology and his application of it to electromagnetism*, Studies in History and Philosophy of Science, **4** (1973), 107–64.
[Cheyne and Pigden, 1996] C. Cheyne and C. Pigden, *Pythagorean powers or a challenge to Platonism*, Australasian Journal of Philosophy, **74** (4) (1996), 639–645.
[Cheyne, 2001] C. Cheyne, *Knowledge, Cause, and Abstract Objects: Causal Objections to Platonism*, Kluwer, Dordrecht (2001).
[Chihara, 1973] C.S. Chihara, *Ontology and the Vicious-Circle Principle*, Cornell University Press, Ithaca NY (1973).
[Chihara, 1990] C.S. Chihara, *Constructibility and Mathematical Existence*, Clarendon Press, Oxford (1990).
[Colyvan, 1998a] M. Colyvan, *In defence of indispensability*, Philosophia Mathematica (3), **6** (1998), 39–62.
[Colyvan, 1998b] M. Colyvan, *Is Platonism a bad bet?*, Australasian Journal of Philosophy, **76** (1998), 115–119.
[Colyvan, 1998c] M. Colyvan, *Can the eleatic principle be justified?*, The Canadian Journal of Philosophy, **28** (1998), 313–336.
[Colyvan, 1999a] M. Colyvan, *Contrastive empiricism and indispensability*, Erkenntnis, **51** (1999), 323–332.
[Colyvan, 1999b] M. Colyvan, *Confirmation theory and indispensability* Philosophical Studies, **96** (1999), 1–19.
[Colyvan, 2001a] M. Colyvan, *The Indispensability of Mathematics*, Oxford University Press, New York (2001).
[Colyvan, 2001b] M. Colyvan, *The miracle of applied mathematics*, Synthese, 127 (2001), 265–278.
[Colyvan, 2002] M. Colyvan, *Mathematics and aesthetic considerations in science*, Mind, **111** (2002), 69–74.
[Colyvan, 2004] M. Colyvan, *Indispensability arguments in the philosophy of mathematics*, Stanford Encyclopedia of Philosophy, E.N. Zalta, ed. (Fall 2004 Edition), Stanford University, Stanford, URL= <http://plato.stanford.edu/archives/fall2004/entries/mathphil-indisp/>.
[Colyvan and Ginzburg, 2003] M. Colyvan and L.R. Ginzburg, *The Gallilean turn in population ecology*, Biology & Philosophy, **18** (2003), 401–414.

[Darwin, 1958] C. Darwin, *Autobiography and Selected Letters*, Francis Darwin ed., Dover, New York (1958).
[Davies, 1992] P. Davies, *The Mind of God*, Penguin, London (1992).
[Devlin, 1977] K. Devlin, *The Axiom of Constructibility*, Lecture Notes in Mathematics, Vol. 617. Springer-Verlag, Berlin (1977).
[Duhem, 1962] P. Duhem, *The Aim and Structure of Physical Theory*, Princeton University Press, Princeton (1954) (first published in 1906).
[Dyson, 1964] F.J. Dyson, *Mathematics in the physical sciences*, Scientific American, **211** (3) (1964), 128–146.
[Enderton, 1977] H.B. Enderton, *Elements of Set Theory*, Academic Press, New York (1977).
[Feynman, 1965] R. Feynman, *The Character of Physical Law*, BBC, London (1965).
[Field, 1980] H. Field, *Science Without Numbers: A Defence of Nominalism*, Blackwell Publishers, Oxford (1980).
[Field, 1984/1989] H. Field, *Can we dispense with space-time?*, Realism, Mathematics and Modality, Blackwell Publishers, Oxford (1989) (first published in 1984), 171–226.
[Field, 1985/1989] H. Field, *On conservativeness and incompleteness*, Realism, Mathematics and Modality, Blackwell Publishers, Oxford (1989) (first published in 1985), 125–146.
[Field, 1988/1989] H. Field, *Realism mathematics and modality*, Realism, Mathematics and Modality, Blackwell Publishers, Oxford, 1989 (first published in 1988), 227–281.
[Field, 1989] H. Field, *Realism, Mathematics and Modality*. Blackwell Publishers, Oxford (1989).
[Field, 1990] H. Field, *Mathematics without truth (a reply to Maddy)*, Pacific Philosophical Quarterly, **71** (3) (1990), 206–222.
[Field, 1992] H. Field, *A nominalistic proof of the conservativeness of set theory*, Journal of Philosophical Logic, **21** (2) (1992), 111–123.
[Field, 1998] H. Field, *Mathematical objectivity and mathematical objects*, Contemporary Readings in the Foundations of Metaphysics, S. Laurence and C. Macdonald, eds., Blackwell Publishers, Oxford (1998), 387–403.
[Fodor and Lepore, 1992] J. Fodor and E. Lepore, *Holism: A Shopper's Guide*, Blackwell Publishers, Cambridge (1992).
[Ginzburg and Colyvan, 2004] L. Ginzburg and M. Colyvan, *Ecological Orbits: How Planets Move and Populations Grow*, Oxford University Press (2004).
[Hamming, 1980] R.W. Hamming, *The unreasonable effectiveness of mathematics*, American Mathematics Monthly, **87** (1980), 81–90.
[Hellman, 1989] G. Hellman, *Mathematics without Numbers: Towards a Modal-Structural Interpretation*, Clarendon Press, Oxford (1989).
[Hellman, 1992] G. Hellman, *The boxer and his fists: the constructivist in the arena of quantum physics*, Proceedings of the Aristotelian Society, Supplement, LXVI, (1992), 61–77.
[Hellman, 1999] G. Hellman, *Some ins and outs of indispensability: a modal-structural perspective*, Logic and Foundations of Mathematics, A. Cantini, E. Casari and P. Minari, eds., Kluwer, Dordrecht. 1999, 25–39.
[Hersh, 1990] R. Hersh, *Inner vision outer truth*, Mathematics and Science, in R.E. Mickens, ed., World Scientific Press, Singapore (1990), 64–72.
[Hilbert, 1899/1971] D. Hilbert, *Foundations of Geometry*, Open Court, La Salle ILL (1971) (first published in 1899).
[Hunt, 1971] B.J. Hunt, *The Maxwellians*, Cornell University Press, Ithaca NY (1991).
[Kitcher, 1981] P. Kitcher, *Explanatory unification*, Philosophy of Science, **48** (1981), 507–531.
[Kitcher, 1984] P. Kitcher, *The Nature of Mathematical Knowledge*, Oxford University Press, New York (1984).
[Lakatos, 1970] I. Lakatos, *Falsification and the methodology of scientific research programmes*, Criticism and the Growth of Knowledge, I. Lakatos and A. Musgrave, eds., Cambridge University Press, Cambridge (1970), 91–195.
[Mac Lane, 1990] S. Mac Lane, *The reasonable effectiveness of mathematics*, Mathematics and Science, R.E. Mickens, ed., World Scientific Press, Singapore (1990), 115–135.
[Maddy, 1990a] P. Maddy. *Realism in Mathematics*, Clarendon Press, Oxford (1990).
[Maddy, 1990b] P. Maddy, *Physicalistic Platonism*, Physicalism in Mathematics, A.D. Irvine, ed., Kluwer, Dordrecht (1990), 259–289.
[Maddy, 1990c] P. Maddy, *Mathematics and Oliver Twist*, Pacific Philosophical Quarterly,**71** (3) (1990), 189–205.

[Maddy, 1992] P. Maddy, *Indispensability and practice*, Journal of Philosophy, **89** (1992), 275–289.
[Maddy, 1994] P. Maddy, *Taking naturalism seriously*, Logic, Methodology and Philosophy of Science IX, D. Prawitz, B. Skyrms and D. Westerståhl, eds., Elsevier, Amsterdam (1994), 383–407.
[Maddy, 1995] P. Maddy, *Naturalism and ontology*, Philosophia Mathematica (3), **3** (3) (1995), 248–270.
[Maddy, 1997] P. Maddy, *Naturalism in Mathematics*, Clarendon Press, Oxford (1997).
[Maddy, 1998a] P. Maddy, *Naturalizing mathematical methodology*, Philosophy of Mathematics Today, M. Schirn, ed., Clarendon Press, Oxford (1998), 175–193.
[Malament, 1982] D. Malament, *Review of Field's Science Without Numbers*, Journal of Philosophy, **79** (1982), 523–534.
[May, 2004] R.M. May, *Uses and abuses of mathematics in biology*, Science, **303** (6 February 2004), 790–793.
[Melia, 2000] J. Melia, *Weaseling away the indispensability argument*, Mind, **109** (2000), 455–479.
[Mickens, 1990] R.E. Mickens, ed., *Mathematics and Science*, World Scientific Press, Singapore (1990).
[Mortensen, 1995] C. Mortensen, *Inconsistent Mathematics*, Kluwer, Dordrecht (1995).
[Papineau, 1993] D. Papineau, *Philosophical Naturalism*, Blackwell Publishers, Oxford (1993).
[Penrose, 1989] R. Penrose, *The Emperor's New Mind: Concerning Computers, Minds and the Laws of Physics*, Vintage Press, London (1990).
[Putnam, 1971/1979] H. Putnam, *Philosophy of logic*, reprinted in *Mathematics Matter and Method: Philosophical Papers Vol. 1*, second edition, Cambridge University Press, Cambridge (1979) (first published in 1971), 323–357.
[Putnam, 1979] H. Putnam, *What is mathematical truth?*, Mathematics Matter and Method: Philosophical Papers Vol. 1, second edition, Cambridge University Press, Cambridge (1979), 60–78.
[Quine, 1936/1983] W.V. Quine, *Truth by convention*, reprinted in Philosophy of Mathematics Selected Readings, second edition, P. Benacerraf and H. Putnam, eds., Cambridge University Press, Cambridge (1983) (first published in 1936), 329–354.
[Quine, 1948/1980] W.V. Quine, *On what there is*, reprinted in From a Logical Point of View, second edition, Harvard University Press, Cambridge MA (1980) (first published 1948), 1–19.
[Quine, 1951/1980] W.V. Quine, *Two dogmas of empiricism*, reprinted in From a Logical Point of View, second edition. Harvard University Press, Cambridge MA, 1980 (first published in 1951), 20–46.
[Quine, 1953/1976] W.V. Quine, *On mental entities*, reprinted in The Ways of Paradox and Other Essays, revised edition, Harvard University Press, Cambridge, MA (1976) (first published in 1953), 221–227.
[Quine, 1960] W.V. Quine, *Word and Object*, Massachusetts Institute of Technology Press and John Wiley and Sons, New York (1960).
[Quine, 1963/1983] W.V. Quine, *Carnap and logical truth*, reprinted in Philosophy of Mathematics Selected Readings, second edition, P. Benacerraf and H. Putnam, eds., Cambridge University Press, Cambridge (1983) (first published in 1963), 355–376.
[Quine, 1981a] W.V. Quine, *Five milestones of empiricism*, Theories and Things, Harvard University Press, Cambridge, MA (1981), 67–72.
[Quine, 1981b] W.V. Quine, *Success and limits of mathematization*, Theories and Things, Harvard University Press, Cambridge, MA (1981), 148–155.
[Quine, 1986] W.V. Quine, *Reply to Charles Parsons*, The Philosophy of W.V. Quine, L. Hahn and P. Schilpp, eds., Open Court, La Salle ILL (1986), 396–403.
[Quine, 1992] W.V. Quine, *Pursuit of Truth*, revised edition, Harvard University Press, Cambridge MA (1992).
[Quine, 1995] W.V. Quine, *From Stimulus to Science*, Harvard University Press, Cambridge MA (1995).
[Resnik, 1983] M.D. Resnik, *Review of Hartry Field's Science Without Numbers*, Noûs, **17** (1983), 514–519.
[Resnik, 1985a] M.D. Resnik, *How nominalist is Hartry Field's nominalism?*, Philosophical Studies, **47** (1985), 163–181.

[Resnik, 1985b] M.D. Resnik, *Ontology and logic: remarks on Hartry Field's anti-platonist philosophy of mathematics*, History and Philosophy of Logic, **6** (1985), 191–209.

[Resnik, 1995] M.D. Resnik, *Scientific vs. mathematical realism: the indispensability argument*, Philosophia Mathematica (3), **3** (2) (1995), 166–174.

[Resnik, 1997] M.D. Resnik, *Mathematics as a Science of Patterns*, Clarendon Press, Oxford (1997).

[Roseveare, 1983] N.T. Roseveare, *Mercury's Perihelion from Le Verrier to Einstein*, Clarendon Press, Oxford (1983).

[Shapiro, 1983] S. Shapiro, *Conservativeness and incompleteness*, Journal of Philosophy, **80** (9) (1983), 521–531.

[Shapiro, 1997] S. Shapiro, *Philosophy of Mathematics: Structure and Ontology*, Oxford University Press, Oxford (1997).

[Siegel, 1991] D.M. Siegel, *Innovation in Maxwell's Electromagnetic Theory*, Cambridge University Press, Cambridge (1991).

[Smart, 1963] J.J.C. Smart, *Philosophy and Scientific Realism*, Routledge and Kegan Paul, London (1963).

[Sober, 1993] E. Sober, *Mathematics and indispensability*, Philosophical Review, **102** (1) (1993), 35–57.

[Steiner, 1989] M. Steiner, *The application of mathematics to natural science*, Journal of Philosophy, **86** (9) (1989), 449–480.

[Steiner, 1995] M. Steiner, *The applicabilities of mathematics*, Philosophia Mathematica (3), **3** (2) (1995), 129–156.

[Steiner, 1998] M. Steiner, *The Applicability of Mathematics as a Philosophical Problem*, Harvard University Press, Cambridge MA (1998).

[Urquhart, 1990] A. Urquhart, *The logic of physical theory* Physicalism in Mathematics, A.D. Irvine, ed., Kluwer, Dordrecht (1990), 145–154.

[van Fraassen, 1980] B.C. van Fraassen, *The Scientific Image*, Clarendon Press, Oxford (1980).

[Weinberg, 1986] S. Weinberg, *Lecture on the applicability of mathematics*, Notices of the American Mathematical Society, **33** (1986), 725–728.

[Weinberg, 1993] S. Weinberg, *Dreams of a Final Theory*, Vintage Press, London (1993).

[Wigner, 1960] E.P. Wigner, *The unreasonable effectiveness of mathematics in the natural sciences*, Communications on Pure and Applied Mathematics, **13** (1960), 1–14.

[Wilson, 2000] M. Wilson, *The unreasonable uncooperativeness of mathematics in the physical sciences*, Monist, **83** (2000), 296–314.

INDEX

σ-field, 495

a posteriori knowledge, 3, 5, 6, 8, 157, 179–186, 199, 213–226
a priori knowledge, 3, 5–7, 9, 18, 22, 33
    saying is believing, 18
Abelard, A., 234
aboutness
    thick vs. thin, 46, 92
'absolute' rest, motion, simultaneity, 223
absoluteness, 572–574, 576, 617
abstract algebra, 112
abstract objects, 94–98, 238
abstraction axioms, 470
abstraction principle, 236
abstraction scheme, 463
Ackermann function, 555, 559
Ackermann, W., 302, 537, 543, 556
Aczel, P., 481
additivity, 493, 496
agent-relative, 497
aggregates, 256
aleatory, 497
Aleksandrov, P., 418
algebraic theories, 353–354
algorithm, 586
analysis, 214–216
analytic geometry, 241
analytic sets, 419
analytic truths, 169–170
analyticity, 4, 5, 7, 8, 18, 20, 22, 24, 25, 33
anti-foundation axioms, 466
anti-platonism
    mathematical, 76–86
anti-realism, vii, 347

mathematical, 35–98
apartness relation, 325
Apostoli, P., x, 479, 481, 482, 484–488
applicability of mathematics, 133
    to empirical science, 84–86
approximation space, 479, 480
Archimedes, 164
Aristotelian realism (or Aristotelianism), viii, 103
Aristotle, 1–2, 53, 105, 131, 135, 138, 160–168, 172, 176, 178, 192, 196, 203, 215–216, 225, 239, 356
arithmetic, 239
Armstrong's
    abstract objects, 664
    naturalism, 664
Armstrong, D. M., 53, 132, 642
Aronszajn tree, 435
Australian school, 110
axiom, 241
Axiom of Choice (AC), 314 405, 406
Axiom of Constructibility, 431, 680
Axiom of Convergence, 497
Axiom of Dependent Choice, 324
Axiom of Extensionality, 470, 476
Axiom of Foundation, 425
Axiom of Independence, 499
Axiom of Infinity, 473
Axiom of Randomness, 498
Axiom of Reducibility, 336, 411
Axiom of Replacement, 423, 431
axiomatic method, 2
axiomatic set theory, 279
axiomatization, 119, 241
axioms, 242
    for arithmetic, 175, 178–179

for geometry, 171
    Frege–Hilbert dispute, 297, 298
Ayer, A. J., 44, 213
Ayer–Hempel–Carnap, 49
Azzouni, J., 49, 85, 384, 676

Baire Category Theorem, 413
Baire property, 413
Baire, R., 412, 413
Balaguer, M., vii, 349, 367, 373–374,
    381, 676
Banach, S., 381
Banach–Tarski Paradox, 415
Bar Theorem, 326
Bar-Hillel, Y., 479
Barcan formula, 388
Basic Law V, 462, 464, 465
Bayes' theorem, 501
Bayesian, 141, 500
Bayesian conditionalisation, 501
Bayesian net, 511
Bayesianism, x
Beall, JC, 49
Behmann, H., 543
belief, 345–349
belief function, 501
Bell, J. L., 480, 481
Belnap, N., 641
Benacerraf, P., 42, 61–64, 66, 112,
    113, 157, 199, 206, 352–353,
    355, 365, 373, 377, 384–385
Bentham, J., 357–359
Berkeley models, 384, 388
Bernays, P., 302, 424, 431, 536, 543,
    552, 555, 563, 568, 575, 576
Bernoulli, 504
Bernstein, F., 421
betting quotient, 501
BHK interpretation, 329
Bigelow, J., 110, 642
Birkhoff, G., 480
Bishop, E. A., 311, 332
Blamey, S., 485
Boethius, 356

boldness
    of theories, 661
Bolyai, J., 173
Bonevac, D., ix, 352–353, 378
BonJour, L., 182, 184
Borel, E., 412, 413
Bostock, D., viii, 245
boundedness, 538, 598
    condition, 585, 590, 593, 596, 599
Bourbaki, 114, 132
Bourbakism, 308
Boyer, C. B., 198
Brady, R. T., 472, 635
bridges of Königsberg, 111
Brittan, G., 263
Brouwer, L. E. J., 39–40, 243, 320
Brown, J., 243
Burali–Forti Paradox, 410
Burgess, J., 348, 350, 357, 367, 371,
    378, 389

calculability, 535, 540, 555–577
Campbell, K., 667
Cantor's diagonal argument, 464
Cantor's paradox, 18, 410
Cantor's theorem, vii, 402, 465
Cantor, G., 14–16, 131, 198, 201, 246,
    248, 300, 317, 359, 379, 381,
    396, 461, 464, 478, 679
cardinal characteristics, 423
cardinal comparability, 403
cardinal number, 244, 401
Carnap, R., 44, 148, 263, 350, 375,
    668
Cartesian dualism, 52
Cartwright, N., 654
Cassirer, E., 231, 249, 250, 266, 267
Casullo, A., 184
categorical concept, 233
categorical theories, 210–212
Cauchy, A.-L., 114, 216, 364
causal inertness of abstract objects,
    52, 85, 93
causal irrelevance principle, 513

Čeitin, G. S., 329
ceteris paribus principles, 361–363
chance, 499
Chellas, B., 478
Cheyne, C., 135, 676
Chihara, C., 45–46, 80, 84, 159, 190, 191, 194–195, 201, 212, 346, 359, 669
choice sequences, 318
Church, A., 28, 538, 564, 568, 570–572, 575, 576, 586, 593, 611, 622
Church's Thesis, 323, 537, 561, 564, 569, 572, 573, 576, 586
Church-Turing Thesis, 314
classical logic, 469
closed set logic, 634
closure principle, 473, 475
Cohen real, 442
Cohen, P., 441
coherence, 501
Coleman, E., 645
collective, 497
Colyvan, M., xi, 75, 134, 371
combinatory process, 538, 577, 580
Compactness Theorem, 428
Completeness Theorem, 428
compositionality, 376
comprehension, 473
    axiom, 109, 471, 475
    scheme, 463
computability, 535, 537, 539, 576, 577, 594, 610, 611, 617
    theory, x
computable function, 611
computation, 538
computor, 584–587, 612, 618
conceivability, 171–173, 217–218
concept, 462
conceptualism, 158
conditional probability function, 494
confirmation holism, 55–56
conjectures, 142
conjunction, 495

conservative extensions, 209–210
conservativeness, 366–373, *672*, 671–673
constraint graph, 525
constructed objects, 381
constructible objects, 381, 383–385
constructible universe, 431
construction, 232
constructive empiricism, *see* empiricism, constructive
constructive mathematics, 381–382, 685
constructive proof, 311
constructivism, viii, 167, 200, 311
constructivist, 243
contact with abstract objects, 51–54
contingency, 371
    of mathematics, 56–58, 93–94
continuity, 215–216, 242
continuum, 131, 253, 260
    hypothesis (CH), 17, 60, 68, 77–78, 91, 249, 367, 399, 404, 679
    problem, 399, 433
contrastive empiricism, *see* empiricism, contrastive
contrivance, 359
control systems, 641
convenience, 529
conventionalism, 44, 45, 81
Copi, I. M., 200
countable additivity, 495
counterpart semantics, 484
Cowan, T., 639
Craig, W., *660*
creative activity, 347
cross-entropy, 516
cumulative hierarchy, 425
Curry's paradox, 635
Curry, H. B., 44
curvature of space, 173–175
cylinder sets, 496

Dalen, D. van, 318

dark matter, 654
Darwin, C., 689
Davis, M., 547, 566, 568, 572, 573
De Finetti, 502
decidability, 535, 540–544
decision problem, 428, 536, 542, 543, 549, 574
Dedekind, R., 9, 12, 15, 26, 42, 178, 198, 210, 215–216, 254, 364, 396, 408, 412, 535, 537, 545, 609
deductivism, 38, 45, 46, 78–80, 346–348, 351, 353, 388
definition, 175, 367 see analysisFrege–Hilbert dispute, 297, 298
    implicit, 297, 298
deflationary fictionalism, 349 350
Dehaene, S., 40
dependence of a variable, 277
Descartes, R., 168, 172, 205, 240, 246, 643
descriptions, 368
descriptive aid
    mathematics, 86
descriptive set theory, 414, 418
Dever, J., 377
Devlin, K., 680
diagram, 111, 139, 240
dialethism, 631
Dirac Delta Function, 640
directed constraint graph, 525
discernibility of the disjoint, 486, 488
discovery vs. invention in mathematics, 93
discrete, 467
disjunction, 496
    property, 316
distributed computing, 599
doctrine of the limitation of size, 478
double extension set theory, 476, 477
double set theory, 468, 476, 478
du Bois-Reymond, D. P. G., 316
dualism, 52
duality, 644

Duhem, P., 667, 667, 668
Dummett, M., 40, 167, 222, 389
Dunn, J. M., 635, 638
Dutch book, 501

Easton, W., 443
Eculid, 211
Edidin, A., 184
effective calculability, 570, 571, 575, 587
effectively calculable function, 540
Einstein, A., 115, 119, 172, 234
Eklund, M., 350
Eleatic principle, 134
elegance, 661
elementary mathematics, 110
eliminability, 367–368
empirical, 504, 505
    scrutability, 353, 365
    strategy, 359
    -based subjective probability, 506
empiricism, 104
    constructive, 682
    contrastive, 682–683
empiricists, 246
Entscheidungsproblem, 539, 582, 591, 609
epistemic, 529
    objectivity, 517
epistemological, 497
    argument against platonism, 50–61
epistemology, 136
equational calculus, 564–566, 573, 577, 611, 612
Equivocator, 516
Erdös, P., viii
Erdös, P., 435
Erdmann, B., 39
Escher, M. C., 171
Esser, O., 473, 476
Euclid, 1–2, 10–12, 19, 24, 112, 113, 138, 164, 171, 173–175, 234, 239

Euclidean geometry, x
Euler, L., 111, 144
event space, 495
evolutionary theory, 682
ex contradictione quodlibet, 631
exceptionalism, 346
exchangeable, 502
existence
    mathematical — as consistency, 298, 299
    non-spatiotemporal, 95–98
existence property, 316
existential theories, 353
experience, 159–160
experiential equivalence, 350, 351
experimental mathematics, 141
explanation, 529
    as unification, 661
    inference to the best, 654
    intrinsic, 671
explanatory power, 661
extension of a concept, 465
extension of an abstract set, 465
extensionality, 471–473, 475
    principle, 472
external, 521–523
externalism (concerning knowledge), 182–185

fabulous entities, 358–359
facts
    of the matter, 94–98
    physical, 85
    purely nominalistic, 85
    purely platonisitc, 85
Fan Theorem, 326
Feferman, S., 201, 337, 443, 485
Feigenbaum's bottleneck, 507
Fermat's last theorem, 142, 652
Fetzer, J., 125
Feynman, R., 680
fictionalism, ix, 211–213, 226, 345–389, 652
    deflationary, 349–350

    Field, *see* Field, fictionalism
    free-range, 363, 365, 373
    Hermeneutic, 375–377
    hermeneutic, 348
    instrumentalist, 358–360, 366
    mathematical, 35, 46–48, 76–81, 91–94, 98
    relative reflexive, 375–377
    representational, 360–363, 366
    revolutionary, 348
fictitious objects, 358–359
Field, 495
    conservativeness, 671, *671–673*
    consistency, 672
    critics of, 675–676
    entailment, 672
    fictionalism, 670
    indispensability, 653, 669–670, 679
    motivation for nominalism, 670–671
    nominalisation, 671, *673–675*
    Platonistic methods, 671
    representation theorem, 675
Field, H., 46–47, 56, 57, 77, 78, 84, 92, 130, 158, 196, 202, 207–213, 345, 348, 353, 359, 363, 366–373, 378, 695
figuralism, 374–377
Fine, K., 478
finitary arithmetic, 239
finitely additive, 496
finitism, 303, 311, 337, 552, 557–559, 562
finitist function, 557
finitist mathematics, 540, 544–550, 562
finitist proof, 558
finitistically calculable functions, 553
Finsler, P., 319
Fodor, J., 667
forcing, 441
formal sciences, 123
formal system, 577, 610
formal theory, 538, 542, 548
formalism, ix, x, 36, 38, 44–45, 237

game, 44
metamathematical, 44
Forrest, P., 642
Forti, M., 472
Fosen, G., 348
foundation, 426
foundationalism, 632
Fraenkel, A., 421, 424, 479
Fraenkel–Mostowski models, 421, 443
Francis, G., 639
Franklin, J., viii
free-range fictionalism, 363, 365, 373
freedom, 379
Frege structure, 465, 481
Frege, G., 8–10, 14–17, 23–25, 39–41, 44, 45, 62, 76, 82, 83, 89, 90, 109, 127, 129, 157, 161, 163, 176–178, 188–189, 194–196, 203–204, 207, 235, 254, 294, 410, 416, 462, 464, 481, 536, 548, 684
Fregean problem, 465, 467
frequency, 497
fruitfulness (of axioms), 192–193
full conception of the natural numbers (FCNN), 63–68, 71–75
full objectivity, 517
full-blooded platonism (FBP), 35, 40–41, 49, 59–61, 68–75, 91–94, 98, 373–374
function
μ recursive, 567
primitive, 537
calculable, 561
effectively calculable, 588
finitist, 561
finitistically calculable, 537, 561
general recursive, 564, 565, 569, 572
primitive recursive, 555, 561
recursive, 537, 569
Turing computable, 579
functionality, 636

Gödel's first incompleteness theorem, 249, 306
Gödel, K., viii, 26–27, 40, 41, 51, 52, 69, 157, 159, 191–193, 200, 243, 319, 564, 370–569, 572, 574, 576, 586, 608–610, 613–618, 620, 622, 680
Gabbay, D., xi
Gaifman, 502
Galileo, 361
Gambling system, 498
game formalism, 45
Gandy Machine, 538, 597, 599–601, 606
Gandy's Thesis, 596
Gandy, R., 537, 538, 572, 579, 584, 586, 593–596, 608, 622
general recursive function, 561, 611
general recursiveness, 575
generic set, 442
Gentzen, G., 307, 308
geometry, 104, 239, 367, 387, 639
Gilmore, P., 469–472, 478, 485
Goldbach's conjecture, 148
Goldman, A., 182
Goodstein, R. L., 339
Gosse, E., 682, 684
Gray, J., 265.
Greek mathematics, 164–166, 211–212
group, 132, 642
group theory, ix, 112

Hajnal, A., 438
Hale, B., 52, 55, 57, 184, 236
Hallett, M., 248
Halpern, 523
Hamel, G., 421
Hanf, W., 439
Hankel, H., 293
Hardy, G. H., viii
Hart, W. D., vii
Hausdorff's paradox, 415
Hausdorff, F., 412, 414, 416, 417
Hawthorne, N., 346, 358

Heath, T. L., 171
Heine, H. E., 293
Hellman, E., 689
Hellman, G., 42, 45, 46, 79, 346, 378, *684*
Hempel, C., 44, 346
Henkin, L., 437
Herbrand, J., 537, 538, 540, 543–544, 549, 551, 556, 559, 564, 566, 611
hereditarily finite sets, 192–196, 201
hermeneutic fictionalism, 348, 375–377
Hersh, R., 40, 693
Heyting, A., 39, 40, 316, 327
Hilbert program, 637, 646
Hilbert space, 481
Hilbert, D., 16, 25, 41, 44, 45, 177, 211, 237, 239, 337, 367, 369, 379, 381, 387, 535, 540, 544, 546–548, 551, 552, 555, 575, 576, 608, 613, 615, *673–674*
    Grundlagen der Geometrie, 296
Hinnion, R., x, 468, 470–472, 474, 475
Hintikka, J., viii
Hoare, C. A. R., 125
Hodes, H., 375
Hofstadter, D., 247
Hofweber, T., 347, 350
holism, 657, 667
    confirmational, 667
    moderate, *see* holism, semantic
    semantic, 667
Holland, R. A., 263
Holmes, M. R., 476, 477
homeomorphism, 466
homoiomerous, 109, 129
Honsell, F., 472
Horgan, T., 367
Howson, 510, 522
Hume, D., 7, 25, 114, 168–169, 233, 235
Hunter, 512

Husserl, E., 39
Hyper Frege, 475
hyper-continuous function, 485
hyperuniverse, 468

idea, 168, 233
ideal, 233
ideal elements, 239
idealisation, 118, 162–163, 189–191, 220, 360–363
idealism, 365
idealist, 243
identity, 235, 237
    of indiscernibles, 235, 478
if-thenism, *see* deductivism, 45
Ignorabimusstreit, 317
implication, 495
inaccessible numbers, 678, 678
incommensurability of the diagonal, 113
incompleteness, 556
Incompleteness Theorem, 428, 559, 610, 614
    Gödel, K., 306
inconsistency, 363–365, 374
inconsistent mathematics, 631, 698
indescribable cardinals, 439
indeterminacy, 355–356, 373
indiscernibility, 466
    of locations, 481
indispensability, 134, 197–202
    of mathematics to empirical science, 84–86
indispensability argument
    general, 653
    pragmatic, 658–659
    Quine-Putnam, 656
    scientific, 654
infant cognition, 137
infinite, 232, 241, 245
infinitesimals, 29, 31, 32
infinity, 164–168, 190–191, 196–198, 203–204, 219, 363, 374, 378, 387

axiom, 473
infintesimal nearness, 481
inner penumbra, 484
insight, 138
instantiation, 274
    rules, 275
instrumentalism, 207–209
instrumentalist fictionalism, 358–360, 366
instrumentalist strategy, 359
intended objects or structures, 67–69, 71–75, 77
intensionality, 471
internal, 521–523
internal properties of mathematical objects, 42–43
internalism (concerning knowledge), 182–185
intuition, 232, 243
    mathematical, 52–55
    (Gödel's), 192–193
intuitionism, x, 39–40, 167, 201, 311, 633
intuitionistic logic, 381–382, 389
invariance, 516
invention vs. discovery in mathematics, 93
Inwagen, P. van, 48

Jaynes, J., 148, 506
Juhl, C., 367

Kalderon, M., 347
Kamp, H., 379
Kanamori, A., x
Kanda, A., x, 479, 481, 482, 484–488
Kant, I., viii, 3–7, 135, 168–170, 176, 213, 363–364, 374
Kantianism, viii
Katz, J., 52, 56–59
Keynes, J. M., 508, 513
Kisielewicz, A., 476
Kitcher, P., 37–38, 83–84, 158, 179–191, 203, 225, 661
Kladeron, M., 345

Kleene's normal form theorem, 566, 569, 572
Kleene, S. C., 324, 538, 566, 568, 576, 582, 596
Kline, M., 198
knowledge of mathematical objects, 50–61
Kock, A., 481
Kolmogorov, A., 329, 592
Kőnig, D., 434
Kripke, S., 170, 378, 380, 381, 478
Kronecker, L., 314, 319, 535, 544–548
Kuratowski, K., 418, 421
Kurepa tree, 435
Kurepa, R., 436

Löweinheim–Skolem Theorem, 425
Löwenhcim, L., 425, 542
Lakatos, I., 640, 668
language
    definition of, 97
    relativity, 508
Laplace, 508
lattice, 642
Law of extensions, 462, 463
Lebesgue measure, 413
Lebesgue, H., 412, 413, 679, 680
Leibniz, G. W., 172, 216, 234, 541–542, 610
Lepore, E., 667
Levy collapse, 443
Levy, A., 438
Lewis, D., 38, 52, 56–59, 132, 484, 519, 520
Libert, T., x, 472, 474, 475
Liebniz, G., 114
likelihood principle, 682
limitation of size doctrine, 465, 466, 468
Lindenbaum, A., 421
Link, G., 346
Linsky, B., 41, 70, 684
Liouville, J., 398
Lobatchevsky, N. I., 173

local observations, 467
locality condition, 538, 585, 590, 593, 596, 599
Locke, J., 39
logic, ix
    alternative, 218–222
    nature of, 213, 216–225
    second order, 209
logical, 504, 505
logicism, viii, x, 40, 205, 271, 346, 347, 632
logicists, 235
Lowenheim, L., 544
lower approximation, 480
Luzin set, 418
Luzin, N., 418

Machover, M., 481
Maddy's
    indispensability, 679
    mathematical fictions, *677–678*
    mathematical practice, *678–680*
    problems with indispensability, 669, *676*
    scientific fictions, *677*
    set theoretic realism, 653
    $V = L$, 680
Maddy, P., 37, 51, 53–54, 91, 158, 179, 191–196, 203, 225, 243, 352
Mahlo cardinals, 415, 439
Mahlo, P., 415
make-believe, 345, 375–377
Malament, D., 84, 367
Malitz, R. J., 472
Mancosu, P., 198
Manfredi, P. A., 184
manifold, 254, 272
Marginal probability function, 494
Markov's Principle, 330
Markov, A. A., 329
Markovian constructivism, 311
marsupial constructions, 384–385
mathematical

anti-platonism, 76–86
anti-realism, 35–98
fictionalism, 695
intuition, 52–55
knowledge, vii, viii, x, xi, 50–61
physicalism, 36–38
physics, 233
platonism, 40–44, 50–75
realism, 35–98
triviality, 635
truth, 68
mathematics
    as descriptive aid, 86
    as invention or discovery, 93
    in biology, 697
Maximum Entropy Principle, 505–510, 512–515, 523
Maxwell, J. C., 691
Mayberry, J., 244, 257
McCarty, C., viii
measurable cardinal, 427
measurement, 114, 133, 369
mechanical
    computability, 610
    procedure, 538, 574, 608, 611, 615, 617, 619
    process, 618
Meinong, A., 37, 48–49, 81
Meinongianism, 37, 48–50, 81
Melia, J., 676
Mental, 497
Mental / Physical, 497
mereology, 115
metamathematics, 282, 304
Meyer, R. K., 633, 635, 637
Mill, J. S., viii, 37, 83, 87, 91, 158, 168–359
Minervan constructions, 383–384
minimum perceptibility, 481
Mirimanoff, D., 423
mixed, 521
modal fictionalism, 377
modal strategy, 359
modal structuralism, 346

model theory, 242
models, 496
    of PFS, 484
Montague, R., 438
Moore's paradox, 658
Moore, A., 245
Mortensen, C., x
Mostowski, A. M., 421, 437
multiple reductions objection to platonism, 61–69

naïve notion of set, 461
naïve set theory, 463, 465, 468
natural sciences, ix
naturalism, 657, *664*
    Quinean, *see* Quine, naturalism
naturalized epistemology, 352
naturalized platonism, 53–54
NBG set theory, 205
necessary truth, 6
necessity, 204, 362, 372–373
    of mathematics, 56–58, 93–94
negation, 495
neighbourhood, 467
Nerlich, G., 251
Neugebauer, O., 208
Neumann, J. von, 423
new colours, 171, 184–185
new constructivism, 311
new foundations, 472
Newton, I., 114, 120, 172, 201, 211–212, 216, 223–685
Newton-Smith, W. H., 201
Newtonian mechanics, 234
niminalization of empirical science, 77
no-class theory, 349, 350
nominalism, 106, 130, 158, 202–207, 335, 348, 350, 356, 366–373
    easy road to, 669
    hard road to, 669
nominalistic content of empirical science, 85–86
nominalistic scientific realism, 85–86

nominalization of empirical science, 46, 84
non-deductive logic, 142
non-Euclidean geometry, ix, 234
non-spatiotemporal existence, 95–98
non-uniqueness objection to platonism, 61–69
non-uniqueness platonism (NUP), 67–69, 73–75
noncognitivism, 345
notions, 296
Nozik, R., 182
null set, 194
number, 113, 238
number theory, 462
numerals, 238
numerical ordinals, 214–215
numerical quantifiers, 203–205, 209–210, 214
numerically definite comparisons, 206, 215–216

object-platonism, 41–44
objective, 497
Objective Bayesian net, 510, 512
objective Bayesian semantics, 522
objective Bayesianism, 501
objective credal nets, 526
objectivity, 529
obprogic, 524
Ockham's Razor, 87–90
Ockham, William of, 350, 357–358
ontological commitment, 346, 349–353, 357, 377–379, 387
ontological parsimony, 87–90
open models, 386–387
operations (physical vs. mathematical), 176, 187–191
operations research, 124
ordinal number, 244, 404
ortholattice, 483, 484
    of exact sets, 483
ostensible commitment, 351, 358, 377
Ostwald, J., 677, 681

outcome space, 495
outer penumbra, 484

Papineau, D., 664
paraconsistent, 632
paradox, 364, 631
paradoxical case axioms, 475
paradoxical set theory, 468, 472, 474, 478
parallel computation, 578, 599, 607
Paris, 523
parsimony, 661
    ontological, 87–90
    Quine, *see* Quine, parsimony
Parsons, C., 43, 52, 55, 184, 191, 192, 198–199, 263, 478
Parsons, T., 378
part-whole relation, 235
partial set, 468, 469, 471, 478
partition property, 434
pattern recognition, 136
patterns, 363
    mathematical, 42
Peano arithmetic, 487
Peano, G., 178, 364, 388, 410, 416
Peirce, C. S., 360, 366, 542, 646
Penrose, R., 639
penumbral modality, 484
perfect set property, 401, 418
permutable models, 385
personalist, 497
PFS, 484, 486, 488
physical, 497
physicalism
    mathematical, 36–38
physicalistic platonism, 37, 53–54
Pigden, C., 135
place selection, 498
Plato, 1, 24, 40, 61, 157, 160–161, 192, 213, 239, 356, 378
Platonic realism (or Platonism), vii
Platonism, viii, 106, 107, 127, 135, 352–355, 369, 373–374, *see* realism, mathematical

full-blooded (FBP), 35, 40–41, 49, 68–75, 91–94, 98
    mathematical, 40–44, 50–75
    naturalized, 53–54
    non-uniqueness (NUP), 67–69, 73–75
    object, 41–44
    physicalistic, 37, 53–54
    plenitudinous, 41, 49, 68–75
plenitude, 488
plenitudinous platonism, 41, 49, 68–75
plenum, 488
pleonastic propositions, 346
Poincaré, H., 41, 200, 313, 334, 359, 379–381, 677, 681
Polya, G., 142, 151
Popper, 498–500, 503
Porphyry, 356
positive comprehension, 472
positive set, 468, 472, 478
Post worker, 580
Post, E., 28, 536, 538, 576, 578–582, 589–591
Posterior, 507
potential infinite, 246
powers of relations, 205–206
pragmatism, 372
predicament, 354–355
predicate calculus, 462
predicative theories, 200–201, 212
predicativism, 311
pretense, 367, 375
Priest, G., 49, 631
Prime Number Theorem, vii
Principal Principle, 519, 520
Principia Mathematica, 312, 410, 416, 427, 645
Principle of extensionality, 463
Principle of Indifference, 508–510
Principle of naïve comprehension, 463
Prior, 507
probability function, 493, 495, 496
probability logic, 521

probability space, 495
probability theory, x
Progic, 521
progression, 112
projective sets, 419
proof, 139, 632
proof theory, 307
proofs of correctness of computer programs, 125
propensity, 499
propositional language, 495
propositional variable, 495
proximal Frege structure, 479, 481
proximity space, 480
proximity structure, 479
psychologism, 36–50, 81–86
Ptolemy, 208
Putnam
    goals of science, 656
    indispensability, *655–656*, 658
    intellectual dishonesty, *656*
Putnam, H., 35, 45, 76–78, 82, 84–86, 88, 90, 134, 158, 191, 196–199, 201, 220–222, 226, 346, 348
Pythagorean Theorem, vii

quantification, 379
quantifiers, 277
quantity, 104, 110
quantum
    field theory, 367
    logic, 219–222, 481
    mechanics, 641
        nominalization of, 84
    theory, 172
Quine(an)
    confirmational holism, *667*
    continuity thesis, 666
    –Duhem Thesis, *see* holism, confirmational
    first philosophy, 665
    indispensability, *655*
    naturalism, 657–658, 664, *664*

    new foundation, 464, 477
    ontic commitments, 665
    –Putnam indispensability argument, 84–86
    quantification, 659
    realism, 693
    semantic holism, 667
    unapplied mathematics, 678–679
    $V = L$, 680
Quine, W. V. O., 35, 37, 40, 48, 52, 55–56, 76–78, 82, 84–86, 88, 90, 134, 158, 169, 191, 195–199, 201, 202, 218, 226, 353, 355, 359, 378, 381–383, 472

Rado, R., 436
Ramsey, F., 434
ranging-over idea, 277
ratio, 112
rationalists, 246
real number system
    axioms, 299
real numbers, 197–202, 211–213, 396
realism, 158, 191–192
    mathematical, 35–98, 652
    metaphysical, 652
    selective, 667
    set theoretic, 653
realist, 243
reckonable function, 538, 575–577
recollection (Plato's theory), 159, 161
recursive, 498
    function, 559
reduction, 346, 347, 351, 357, 377, 388
reference class problem, 499, 502
Reflection Principle for ZF, 438
regularity property, 414
Reichenbach, H., 218, 219
Reichenbach, R., 250, 263
relation, 108, 296
relational structures, 243
relative reflexive fictionalism, 375–377
repeatable, 496

repeatably instantiatable, 496
replacement, 426
representational fictionalism, 360–363, 366
Resnik
    indispensability, 659
Resnik, M., 37, 39, 41–43, 52, 55–56, 58, 59, 63, 64, 84, 105, 110, 346, 367
Restall, G., 41, 70–74
revisionism, 644
revolutionary fictionalsim, 348
Riemann hypothesis, 142, 145
Riemann, B., 173
Robertson, H. P., 265
Robinson, A., 437, 638
Roscelin, 356
Rosen, G., 46, 85, 350, 356, 367, 376, 378, 389
Rosser, B., 538, 568
Rothberger, F., 421
Rotman, B., 255
rough set theory, 479
Routley, R., 49, 472, 631, 635
rules, 238
Russell set, 468, 471, 472, 476, 477
Russell's paradox, xi, 17, 24, 295, 300, 301, 410, 464
Russell, B., 19–21, 109, 119, 193, 198, 203–204, 225, 250, 300, 312, 334, 346, 349, 350, 357–358, 364, 368, 410, 416, 462, 464, 536, 632, 643

Sainsbury, R. M., 200
Salmon, N., 48
schema, 233
Schiffer, S., 346–347
scholastics, 138
Schröder, E., 412, 542
sciences of complexity, 124
Scott, D. S., 439, 440, 446, 484, 485
second-order logic, 462, 463
second-order set theory, 366–367, 387

semantics, 352–353, 375–377
sequences of finite projections, see SFP
series, 214–215
set theory, x, 243, 272, 366–367
set-theoretic indiscernibility, 483
sets, 109, 132
settled models, 383, 386
Shanin, N. A., 329
Shannon, C., 507
Shapiro, S., 41, 42, 52, 58, 59, 64, 84, 105, 110, 115, 210, 370–378
Shepherdson, J., 437, 438
Sieg, W., x, 597–610, 621–623
Sierpiński, W., 418, 421
Simons, P., ix
simple random variable, 495
simplicity, 661
simply infinite system, 258
simulation, 140
single-case / repeatable, 496
Singular Cardinals Hypothesis, 444
Singular Cardinals Problem, 444
Skolem's Paradox, 425
Skolem, T., 337, 338, 424, 425, 551, 553
Smart, J. C. C., 655, 694
Snir, M., 502
Sober, E., 367, 669, *682–689*
social challenges, 180–181, 183–187
Solovay, R. M., 443, 444, 446
space and time, 232, 235, 369
Specker, E., 444, 477
Spurr, J., xi
Stalnaker, R., 350
standard probabilistic semantics, 522
Stanley, J., 367, 375–377
Steiner, M., 42, 52, 55–56, 690–696
Steinitz, E., 421
Steps 1, 2, 3, and 4, 525
strong nets, 384
structural property, 109
structuralism, 41–44, 58–59, 64–66, 110, 114, 365, 653
structuralist models, 385–386, 388

structure, 110, 114
subitization, 137
subjective, 497
subjective / Objective, 497
subjective Bayesianism, 501
substitutional quantification, 378–379
success, 345
Summerfield, D. M., 184
sundials, x
supervenience, 346, 347, 368
Suppes, P., 367
Suslin tree, 435
Suslin, M., 419
symbolic manipulation, 133, 145
symmetry, 112, 115, 117
synthesis, 232
synthetic *a priori*, 232, 263, 264, 266
Szabo, Z., 348, 352, 378

Tarski, A., 283, 352–354, 369, 379, 421, 574
Tautology, 496
tertium non datur, 312
Thagard, P., xi
The Maximum Entropy Principle, 504
theoretic virtues, 661–662
theories, 633
theories of inconsistent mathematics, x
Third Man Argument, 135
Thomas, C. J., 293
Thomasson, A., 48
Tiles, M., viii
Token-level, 496
topological set theory, 466
topology, 112, 466, 468
transfinite arithmetic, 300
transfinite numbers, 399
transparency, 635
tree property, 435
Troelstra, A., 318
truth, 346–348, 352–353, 366, 377–378
    mathematical, 68

truth by convention, 668
Turing computer, 538, 597, 599, 601, 606
Turing machine, 313, 538, 579, 586, 606, 609, 612, 622
Turing's Thesis, 537, 564, 578, 587–592
Turing, A., 28, 538, 580, 584–586, 588, 590, 596, 608, 609, 611, 617–620, 622
type theoretic, 464
type-level, 496
type-neutrality, 205–206

Ulam, S., 427
ultimate belief, 503, 518
ultrafinitist, 133
ultraproduct construction, 440
undecidability, 610
undecidable mathematical sentences, *see* continuum hypothesis, 91–92
undefinability theorem, 283
underdetermination, 522
undermining, 520
understanding, 138
understudy properties, 385–386
undogmatic, 502
unificatory power, 661
Uniform Continuity Theorem, 326
uninstantiated universals, 106
unit-making properties, 109
universals, 105
unreasonable effectiveness of mathematics, 689–698
upper approximation, 480
Urquhart, A., 366

vagueness, 360–361
Vaihinger, H., 363–365, 373
van Fraassen, B., 654, 665, 682, 683
Velleman, D., 355
vicious circle principle, 200, 313, 335
Vitali, G., 421
Von Mises, R., 498, 499

von Neumann, J., 306, 536, 550, 556, 557, 653

Walton, K., 345
Wang, H., 550, 615, 616
warrants (for knowledge), 179–180, 182–185
weak counterexamples, 318
weak nets, 385
Weierstrass, K., 114, 216
Weinberg, S., 689
well-founded set, 425, 473
well-ordering, 399
Well-ordering Theorem, 405
Weydert, E., 472
Weyl, H., 336
Whedon, J., 380–382
Whitehead, A. N., 263, 312, 536
Wiener, N., 417
Wigner, E., 689
William of Ockham, 350, 357–358
Williamson, J., x
Wittgenstein, L., 45, 205, 237
Woods, C., xi
Woods, J., xi
Wright, C., 52, 55, 57, 236, 371

Yablo, S., 46, 85, 348, 354–356, 371, 374–378, 384
Yessenin-Volpin, A. S., 337, 339

Zalta, E., 41, 48, 70, 75
Zermelo's set theory, 407
Zermelo, E., 21–22, 405, 407, 487, 653
Zermelo–Fraenkel axiom, 479
Zermelo-Fraenkel set theory, *see* ZF
ZF, 193, 198, 199, 465, 466, 468, 471
ZFC, 478
ZFU, 366–373
Zorn's Lemma, 415